Environmental Neuroscience

Simone Kühn
Editor

Environmental Neuroscience

Springer

Editor
Simone Kühn
Center for Environmental Neuroscience
Max Planck Institute for Human Development
Berlin, Germany

Department of Psychiatry and Psychotherapy
University Medical Center Hamburg-Eppendorf
Hamburg, Germany

ISBN 978-3-031-64698-0 ISBN 978-3-031-64699-7 (eBook)
https://doi.org/10.1007/978-3-031-64699-7

Cover art by Sandra Weber (StudioDigitialAI, pixabay, Photoshop Elements 2024)

This Springer imprint is published by the registered company Springer Nature Switzerland AG
The registered company address is: Gewerbestrasse 11, 6330 Cham, Switzerland

Preface

Neuroscientific research on brain plasticity has convincingly demonstrated that our activities and experiences can shape the function as well as the structure of the adult human brain (Lövden et al., 2013). Most of this plasticity research has examined the effects that lifestyle factors such as smoking, drinking, physical exercise, media use, and other habits have on the brain. This focus on lifestyle factors in previous research has led to an oversight of the impact of the physical environment on brain plasticity. This is surprising, since in animal studies experience-dependent alterations of the brain in the form of adult neurogenesis and synaptic plasticity have mostly been investigated using so-called "enriched environments" (Diamond et al., 1964) (➔ chapter "The Fundamentals of Environmental Neuroscience" by Berman and Bratman). In 1947, Donald Hebb was one of the first to use an experimental paradigm comparing rats that could roam freely with rats held in standard housing conditions (Hebb, 1947). The question that the emerging discipline of human environmental neuroscience (➔ chapter "The Rise and Future of Environmental Neuroscience and Environmental Psychology" by McCunn) raises is: What does an enriched environment look like for humans? What are environments that facilitate brain plasticity and/or mental health in humans?

The importance of the new field of environmental neuroscience is underscored by the fact that our planet is undergoing profound changes that are already—and will increasingly—affect the habitats of its inhabitants. To name just a few, urbanization processes lead to more and more individuals living in urban compared to rural areas worldwide, and climate change brings increasing temperatures and altered precipitation patterns leading to drought, more frequent wildfires, floods, and heavier winds, all of which will affect our living environments. This lends urgency to the need for a better understanding of how these changes in physical living environments affect human beings. Plenty of epidemiological research has suggested that physical living environment has a major impact on mental health (➔ chapter "The CITY & the PSYCHE—A Chronology" by Mehran, Kluge and Adli).

Therefore, we see the need to unravel the neural pathways and mechanisms by which this effect comes about. Gaining knowledge on what the "active ingredients" of the physical environment are, which human senses are most prominently involved, and which brain networks play an important role (➔ chapter "Interplay Between the Physical Environment and the Human Brain: A Review of MRI Research" by Beyer et al.) are key goals of the emerging discipline (Kühn & Gallinat, 2023). Within the scope of this book we refer to how the environment impacts humans as the "adaptation pathway". Here we highlight attempts to quantify exposures to the physical environment under the concept of the "exposome" (➔ chapter "Advancing the Neural Exposome" by Veasey) and based on geospatial mapping (➔ chapter "Human Brain and Behavior in the Geospatial Context: Why and How" by Paus).

In contrast to the adaptation pathway, the so-called "mitigation pathway" aims to capture how humans act on the environment. In this section, we focus on how neuroscientific methodology can be utilized to understand and predict the emergence of pro-environmental behavior (➔ chapters "Leveraging the Social Neuroscience of Prosocial Behavior to Advance Our Understanding of Pro-environmental Behavior" by Lamm et al. and "Neuroscientific Approaches to Understand Barriers and Promotors of Pro-environmental Behavior" by Krebs and Sawe).

Much of the existing research published under the umbrella of environmental neuroscience is focused on the external environment, e.g., contrasting populations from rural vs urban settings, or relating geographical parameters such as green space around the home environment to brain function or brain structure. However, today's people spend only a small fraction of their daily time outdoors and an estimated 80–90% indoors (Klepeis et al., 2001), which is why the built environment and indoor design may play an even more significant role (➔ chapters "Aesthetics of Architecture: Emerging Insights from the Brain" by Vartanian, "The Built Environment and the Brain: Review of Emerging Methods to Investigate the Impact of Viewing Architectural Design" by Tawil and Kühn, and "Applications of Biometrics in Architectural and Environmental Design" by Lavdas and Sussman). Justas in the outdoors, it seems as if natural elements in indoor design may have a positive impact on human well-being (➔ chapter "The Potential of Biophilic Design and Nature to Improve Health, Creativity and Well-Being" by Sjövall and Spiers).

In the next section of the book, we focus on potential differences in special populations. Based on the assumption that the relationship between the environment and human mental and brain health is very likely not invariant across the lifespan, we focus on infancy and childhood (➔ chapter "Environmental Exposures and Child Neurodevelopment" by Peuters et al.) and older age (➔ chapter "Neighbourhood Urban Environments and Cognitive Health in Ageing Populations" by Soloveva et al.). Then we turn towards a more clinical perspective by focusing on mental health and psychiatric populations (➔ chapter "The CITY & the PSYCHE—A Chronology" by Mehran and Adli), as well as on neurological conditions (➔ chapter "Environmental Neurorehabilitation: Harnessing the Power of the Environment in Rehabilitation Following Neurological Conditions" by Norwood et al.) and the potential to use the environment for prevention or rehabilitation. In addition, we explore the extent to which there is and can be experiential equity in how the environment is perceived by and impacts humans (➔ chapter "Experiential Equity: Reconstructing Cities Through Intersecting Theories" by Mazumder).

We next highlight different methodological approaches and discuss how those might be best utilized in the emerging field of environmental neuroscience. We start with methods to simulate environments by means of extended reality, referring mostly to virtual (VR) and augmented reality (AR) (➔ chapter "Extended Reality in Environmental Neuroscience Research" by Li et al.), the use of geographic information systems (GIS) approaches to quantify the geographical properties of living environments (➔ chapter "Neuroscience and Geographic Information Systems to Investigate the Impact of Global Warming on Mood Disorders and Brain Plasticity in Urban Areas" by Vieira Ruas et al.), and how GIS can be used to enhance data acquisition in the daily life of participants through geographical ecological momentary assessment (GEMA), in order to ensure to accurately capture *where* in space individuals spend most of their time (➔ chapter "In the Right Place at the Right Time: Leveraging Geographic Ecological Momentary Assessment to Analyze Contexts and Health" Kestens and Kingsbury). Finally, we review central neuroscientific methods used to capture brain function and structure, namely magnetic resonance imaging (MRI) (➔ chapter "Mapping the Human Brain with Computational Anatomy" by Draganski, Joost and Kherif) and electroencephalography (EEG) (➔ chapter "EEG for the Study of Environmental Neuroscience" by Grassini).

Many studies within the emerging field of environmental neuroscience hope to contribute to what has been called "evidence-based design" and to inform landscape and urban planning, as well as the field of architecture, to make living environments more salutogenic.

We end with a sociological and historical perspective, elaborating on how environmental neuroscience developed and how it is embedded in the research landscape, critically discussing the field's aim to influence real-life city planning and policy making and the recommendation to not forget the moral and ethical implications of this "discipline in the making" (➔ chapter "The Cognitive Neuroscience of Nature?: Sociological and Historical Perspective on Environmental Neuroscience" by Fitzgerald).

Berlin, Germany — Simone Kühn
Hamburg, Germany

References

Diamond, M. C., Krech, D., & Rosenzweig, M. R. (1964, August). The effects of an enriched environment on the histology of the rat cerebral cortex. *The Journal of Comparative Neurology, 123*, 111–120. http://www.ncbi.nlm.nih.gov/pubmed/14199261

Hebb, D. O. (1947). The effects of early experience on problemsolving at maturity. *American Psychologist, 2*, 206–307.

Klepeis, N. E., Nelson, W. C., Ott, W. R., Robinson, J. P., Tsang, A. M., Switzer, P., Behar, J. V., Hern, S. C., & Engelmann, W. H. (2001). The National Human Activity Pattern Survey (NHAPS): A resource for assessing exposure to environmental pollutants. *Journal of Exposure Science & Environmental Epidemiology, 11*(3), 231–252. https://doi.org/10.1038/sj.jea.7500165

Kühn, S., & Gallinat, J. (2023). Environmental neuroscience unravels the pathway from the physical environment to mental health. *Nature Mental Health.* https://doi.org/10.1038/s44220-023-00137-6

Lövden, M., Wenger, E., Martensson, J., Lindenberger, U., & Backman, L. (2013, November). Structural brain plasticity in adult learning and development. *Neuroscience & Biobehavioral Reviews, 37*(9 Pt B), 2296–2310. https://doi.org/10.1016/j.neubiorev.2013.02.014

Contents

About the Editor

Simone Kühn, PhD, is the director of the Center for Environmental Neuroscience at the Max Planck Institute for Human Development in Berlin and professor for Neural Plasticity in Psychiatric Disease at the University Medical Center Hamburg-Eppendorf in Germany. Her primary research focuses on how the human brain is shaped by the physical environment in healthy individuals and in psychiatric patients. She has examined brain plasticity in response to lifestyle factors and habits using various brain imaging methods. She studied psychology at Columbia University in New York, USA and Potsdam University, Germany and received her doctorate in psychology from the University of Leipzig. As a post-doc she worked at University of Gent in Belgium, Charité University Medicine in Berlin, and at the Institute of Cognitive Neuroscience, University College London.

Part I
Theoretical Background

The Fundamentals of Environmental Neuroscience

Marc G. Berman and Gregory N. Bratman

Abstract In this chapter, we explore the fundamentals of the field of environmental neuroscience. We start by defining the field of environmental neuroscience and then proceed to describing its roots and outlining some of its distinguishing features. We define what we mean by an environment, including those factors that are considered in this work, focus on the effects of natural environments, and note why many researchers do the same. We then discuss some of the mechanisms through which natural environments may affect brain processing through the perception of different visual and acoustic features and the centrality of attentional processes. We close by discussing some of the pitfalls and challenges that environmental neuroscientists face and how those challenges may be overcome in the future.

Keywords Environmental neuroscience · Natural environments · Perception · Cognition · Visual features

What Is Environmental Neuroscience?

Environmental neuroscience is a field devoted to the scientific study of the bidirectional relationships between neural processing and the environment (Berman et al., 2019a, b). It has its roots in early neuroscience in nonhuman animals (Berman et al., 2019a). Work by Hebb (1949) showed that rodents reared normally vs. those raised in more depleted environments showed differences in their visual perception abilities. Later, Greenough and colleagues showed that enriched environments filled with toys, running wheels, and free play lead to structural and functional brain

M. G. Berman (✉)
Department of Psychology, University of Chicago, Chicago, IL, USA
e-mail: bermanm@uchicago.edu

G. N. Bratman
School of Environmental and Forest Sciences, The University of Washington, Seattle, WA, USA

S. Kühn (ed.), *Environmental Neuroscience*,
https://doi.org/10.1007/978-3-031-64699-7_1

differences and subsequent changes in behavior in rodents compared to less enriching environments (Benefiel & Greenough, 1998; West & Greenough, 1972), thus emphasizing the direct effect of an environment on neural processing. The Nobel Prize-winning work of Hubel and Wiesel (1962) demonstrated that the mammalian visual cortex has neurons that only fire for specific directions and orientations of light. Interestingly, Blakemore and Cooper (1970) followed up on that work and found that if you raised kittens in artificial environments that only had vertical bars of light, then the neurons in the visual cortex of those kittens only fired for vertical bars of light and not for other directions. More recent research has demonstrated that plasticity in rodent olfactory bulbs relates to changes in the olfactory environments that they experience, as the circuit components within the bulb shift (including the incorporation of new neurons on a daily basis) in response to the experiences of different odors from the environment that are the most behaviorally relevant to the organism (Wu et al., 2020). Taken together, these and other studies offer a demonstration of experience-dependent plasticity (Blakemore & Cooper, 1970): our neural systems are altered and tuned depending on the type of environment we are in. This plasticity was also demonstrated in humans from work by Gauthier and colleagues showing how the fusiform face area may change in activity with increases in expertise with training Gauthier et al. (1999). These are all powerful examples of environmental neuroscience, though these researchers did not necessarily characterize the studies or themselves in this way.

Unlike the studies on nonhuman animals, many environmental neuroscience studies on humans are cross-sectional for practical and ethical reasons (e.g., we cannot easily move people to different environments to study their responses). This is an important consideration and one reason why it is important to take a multidisciplinary approach to these questions, which environmental neuroscience sets out to do.

The pathways through which enriched environments affect brain structure and function in humans have been broadly categorized as including sensory, motor, social, and cognitive dimensions (Kempermann, 2019). For example, in one study, children with autism who were 3–12 years of age were randomly assigned to an "enriched" vs. "control" group in which a variety of sensorimotor exercises and exposures were administered as the treatment over the course of 6 months. The results showed that there were marked improvements in autism symptom severity and cognitive performance in those assigned to the more enriched environments (Woo & Leon, 2013). Another work, though cross-sectional, has found an association between certain structures of the brain and living near forested environments (Kühn et al., 2017)—though causal relationships between these factors could not be established due to the study design.

Environmental neuroscience is also a multiscale science spanning many spatial and temporal scales: on the spatial side from neurons to brains and from communities to cities and on the temporal side from milliseconds to seconds and from hours to years (Berman et al., 2019b). In this sense, the field aims to understand how momentary and more long-term environmental exposures affect brain processing and behavior and how long such effects last. It also aspires to find ways to model these complex and multifaceted interactions that feedback onto themselves. For example, not only are we adapting to the environment but we are also altering the environment, which is then altering our cognitive and neural processing (Doell et al., 2023).

Why Is Environmental Neuroscience Important?

It is often ignored how much of our behavior is driven by environmental factors. Humans have an incredible amount of autonomy in how they function and navigate the world. Unlike many other nonhuman animals, humans seem to be less stimulus-driven and can maintain goals that are abstract and somewhat disconnected from immediate survival. Humans may not be completely unique in these abilities, but it would not be far-fetched to claim that we humans may be less stimulus-driven than many other species. This ability may bias us into thinking that the environment does not matter that much at all, but this is a fallacy. As the famed psychologist Kurt Lewin discussed in the 1930s, human behavior is a function of the person (*P*) and the environment (*E*) or behavior = f (*P*,*E*) (Lewin, 1936). It seems that much of psychology started to focus more on the person rather than the environment. There were a few important studies and theories that have stressed the importance of situational/environmental factors such as Walter Mischel's (1968) writing on how situational factors can affect dispositional factors or the classic work by Asch (1951). Many of these studies focus on social environmental factors but not physical environment factors. However, as discussed previously, the physical environment can also have consequential effects on brain processing and behavior. Therefore, the importance of environmental neuroscience lies in the many ways that the surrounding environment can have profound impacts on the brain and behavior. In addition, the field tries to quantify these relationships that might expand upon the Lewin equation (Berman et al., 2019a, b) and create first principles mathematical models of environment–behavior relationships (Stier et al., 2021b; Stier et al., 2024).

What Do We Mean by an Environment?

Relevant conditions and settings to study using environmental neuroscience include a variety of physical and social factors (i.e., the "total environment") that can influence developmental trajectories in both good and bad ways (Tulve et al., 2016). A large body of evidence has shown how social environments and experiences can lead to adverse outcomes through complex, sometimes nonlinear pathways (Olvera-Alvarez et al., 2018; Hertzman & Boyce, 2010). These include early life and psychosocial stress, adverse childhood experiences, and socioeconomic position (Agorastos et al., 2019; Nusslock & Miller, 2016). There are complex sets of impacts from the aspects of physical environments and exposures as well (Slavich, 2020; van den Bosch & Meyer-Lindenberg, 2019).

Taken together, these exposures can lead to a variety of ill- or well-being outcomes. It is important to consider where effective interventions might be introduced into these environments in ways that can buffer against their impacts. For example, what are the ways to decrease stress with physical environment interventions, and what are the underlying mechanisms of these effects that are illuminated through environmental neuroscience? In addition to buffering against negative impacts, what are the ways that other elements of the physical environment might help support positive emotional well-being (Park et al., 2022), and how can environmental neuroscience help us understand this set of causal pathways?

These questions span a variety of considerations, including those that involve an examination of how some social or physical factors (e.g., socioeconomic position) can "get under the skin" (Sapolsky, 2015) and thereby change the ways in which stimuli are subsequently processed (e.g., increased reactivity or vulnerability) (Slavich, 2020; Bratman & Olvera-Alvarez, 2022). Here too, environmental neuroscience can play a role by elucidating processes that might help explain individual differences in responses to different sets of stimuli (e.g., increased amygdala activity related to a stressor) and how nature exposure may subsequently have differential effects. In addition to prior experiences and living conditions, personality traits may play a role in these different pathways and effect modifications as well (Kandler & Rauthmann, 2022)—understanding these mechanisms on the individual scale is critical if population-level interventions (e.g., designing urban green spaces) are to be designed effectively. In these and other ways, environmental neuroscience offers an important avenue to study aspects of the total environment, including how the duration and prevalence of exposures come together to impact an organism.

What Do We Mean by the Natural Environment, and Why Focus on It?

Much of the environmental neuroscience research has focused on the natural environment. In the context of environmental psychology, nature has been defined as "encompassing elements and phenomena of Earth's lands, waters, and biodiversity, across spatial scales and degrees of human influence, from a potted plant or a small urban creek or park to expansive, 'pristine' wilderness with its dynamics of fire, weather, geology, and other forces" (Bratman et al., 2019). A reason for this focus is the influence of psychoevolutionary theory and the huge environmental shift away from nature and toward the urban for humans. By 2050, it is estimated that two-third of the world's population will be living in urban areas (6.7 billion vs. 3.1 billion; Our World in Data (OWID) based on UN World Urbanization Prospects 2018 and others: processed by Our World in Data). This is an incredible change that really began in the late 1800s/early 1900s. Although this urbanization has come with many benefits, there of course are issues. While many researchers focus on the energy, economic, and ecological impacts of this mass migration to cities, environmental neuroscience is concerned with the psychological impacts of such migration. More specifically, environmental psychologists and environmental neuroscientists have focused on the ways in which interactions with natural environments can be salubrious for both our mental and physical health (Kaplan, 1995; Mitchell & Popham, 2008; Kardan et al., 2015; Bratman et al., 2015; Frumkin et al., 2017) (➔ chapter "The CITY & the PSYCHE—A Chronology" by Mehran, Kluge and Adli ➔ chapter "Environmental Neurorehabilitation: Harnessing the Power of the Environment in Rehabilitation Following Neurological Conditions" by Norwood et al.).

Environmental neuroscience offers insights into the different ways in which environmental stimuli (in particular, natural stimuli) are processed and how they impact our thoughts, feelings, and behavior. These influences from the environment can come as a result of a variety of different processes, ranging from initial affective and cognitive responses ("bottom up") to judgments and responses that are tied to executive function and cognitive appraisals as well as to prior experiences and associations ("top down") (Fig. 1).

As an example of these "bottom-up" processes, perceiving stimuli with more curved edges can cause people to think more about topics related to spirituality and their life journey (Schertz et al., 2018; Schertz et al., 2020; see Fig. 2). This was true even when you removed semantics from the images that looked like Jackson Pollock paintings (Schertz et al., 2020; see Fig. 3). Furthermore, different low-level visual features, such as curved edges and fractalness, predict how much people will like scenes with mixed natural and urban content (Berman et al., 2014; Kardan et al., 2015) and scenes that are completely architectural such as building facades and interiors that have no overt nature in them (Coburn et al., 2019). These insights contribute to further understanding of the current theories in environmental psychology, such as attention restoration theory (ART; Kaplan, 1995; Kaplan & Berman, 2010), the perceptual fluency account (Joye et al., 2016), and stress

Fig. 1 The image on the left has more horizontal and vertical symmetry and is rated as being more orderly than the one on the right that is rated as being more disorderly. Being exposed to a series of figures, like the one on the right, caused participants to cheat more often and by a larger magnitude than did being exposed to images like the one on the left. (Adapted from Kotabe et al. (2016))

Fig. 2 Adapted from Schertz et al., 2018. Images (**a**) and (**b**) have the same naturalness rating, but image (**a**) has more curved edges. Images (**c**) and (**d**) have the same urban rating but differ in curved edges, where image (**c**) has more curved edges. Images (**a**) and (**c**) have the same number of curved edges and images (**b**) and (**d**) have the same number of curved edges. Being exposed to images with more curved edges, caused participants to think more about thoughts related to spirituality and their life journey

Fig. 3 The above image contains all the curved edges from an original image from a park but with the semantic content obscured through a procedure to scramble the edges. When participants are exposed to images like this that have more curved edges, it causes them to think more about topics related to spirituality and their life journey. (Adapted from Schertz et al. (2020))

reduction theory (SRT; Ulrich et al., 1991) as perceiving different environmental features may have consequential effects on behavior (Schertz & Berman, 2019).

According to ART (Kaplan, 1995; Kaplan & Berman, 2010), human attention comes in two forms. One is known as directed attention (sometimes referred to as top-down or endogenous attention), and this is the kind of attention where you as the individual person decides on what you are going to attend. It is believed that this kind of attention is fatigable or depletable, that you can only direct your attention for so long, and then one becomes mentally fatigued and cannot direct one's attention any longer. This differs from involuntary attention (sometimes referred to as bottom-up or exogenous attention), where any interesting or important stimulation in the environment automatically captures our attention and we do not have control over it. Bright lights, loud noises, flying objects, etc. automatically capture our attention without much if any volitional effort. It is believed that this kind of attention is less fatigable or depletable. It is hypothesized that many natural environments have interesting stimulation to capture involuntary attention while not requiring a lot of directed attention (not needing to be so vigilant), and, thus, interacting with most natural environments can restore this precious directed attention resource. The perceptual fluency hypothesis posits that natural environments and naturally occurring stimuli are easier for human brains to process and therefore lead to positive affective responses and maybe lower stress. This theory has similarities to ART, in that both claim that natural stimulation is more fluently processed (or is as softly fascinating as in ART), but the difference is that ART claims that this more fluent processing leads to less exerted effort, which then leads to cognitive benefits, whereas the perceptual fluency hypothesis posits that this effortless processing leads to improved affect, which increases attention (Schertz & Berman, 2019). SRT (Ulrich et al., 1991) posits that interacting with natural environments can reduce stress and improve mood through an initial, positive affective response, which allows people to maintain higher levels of sustained attention that then leads to overall cognitive improvements. The relationship of mood improvements with cognitive impacts after nature interactions has been questioned since changes in mood are not always correlated with changes in cognitive performance (Stenfors et al., 2019; Schertz & Berman, 2019).

Much of the environmental neuroscience research is related to "top-down" processes that include subjective judgments and appraisals of the environment or a subsequent shift in thinking about one's own cognition or relationship with others. This holds promise for a greater understanding of the pathways related to pro-environmental decision-making (Sawe & Knutson, 2015), emotion regulation (Bratman et al., 2021a), and potentially theories of social connection and restoration (Hartig, 2021). In addition, these top-down processes can often interact in interesting ways with more low-level features. For example, Van Hedger et al. (2019) showed that for an unaltered nature and urban sounds, the more entropic the sound, the more it was preferred. However, when the sounds were altered, in such a way that you could not identify whether the sound was natural or artificial, then the more entropic the sound was, the more it was disliked. These results suggested that for sounds, there were no necessarily inherent sound features that were more preferred but that the context mattered as well, i.e., whether the sound was originating from a natural source or not. This is an example for how top-down information can affect the processing of bottom-up information, which has also been shown more generally by many researchers in different contexts (Oliva & Torralba, 2007; Chun & Jiang, 1999).

Emerging work has demonstrated the potential for the environment to decrease rumination, potentially through a shift in attentional allocation (Bratman et al., 2015; Bratman et al., 2021b), and a conceptual framework has been proposed about the ways in which the impacts of nature contact and connectedness may extend to other aspects of emotion regulation as well (Bratman et al., 2021a; Korpela et al., 2018; Richardson, 2019). This may apply to an intentional choice of visiting a location as a place for regulation and coping (Korpela et al., 2001; Korpela et al., 2020) or may relate to a real-time influence of the environment on processes related to metacognition. As an example of the latter, Tost et al. (2019) found that greater affective benefits from exposure to urban green spaces were experienced by individuals who were less able to recruit those areas of their prefrontal cortex involved in emotion regulation during separate presentations of aversive stimuli in the context of a neuroimaging protocol. In other words, for these individuals, urban green spaces may serve as an external resource to help them with their emotion regulation, when they are not able to perform that function as well themselves.

Difficulties and Pitfalls of Environmental Neuroscience

Thus far, we have not cited much evidence for the direct neural effects of the environment on humans. There are a couple of reasons for this. First, it is difficult to expose individuals to real or realistic environments while they are in a magnetic resonance imaging (MRI) scanner, for example, where participants lay down in an enclosed space that is extremely small and artificial. The MRI scanner is also extremely noisy and requires participants to remain perfectly still while brain hemodynamics are being measured. This makes it difficult to have a realistic immersive experience while being in the MRI scanner (➔ chapter "Mapping the Human Brain with Computational Anatomy" by Draganski, Joost and Kherif). Some of this may be aided by technologies such as functional near-infrared spectroscopy (fNIRS) that allows participants to move around and even be in real-world environments while brain hemodynamics are being measured. There is also the possibility of using virtual reality (➔ chapter "Extended Reality in Environmental Neuroscience Research" by Li et al.) in MRI scanners (Qian et al., 2021), but the MRI scanner is still highly constraining and noisy.

Second, as mentioned earlier in this chapter, there are many dimensions that differentiate between natural and urban environments, for example. These environments differ in terms of many characteristics such as the number of curved edges, entropy, fractalness, and differences in hues. This makes it difficult to know what might be driving brain differences between the two environments, i.e., whether there is something intrinsic to natural vs. urban environments or whether it is simply a difference in the number of curved edges or color palette. This is a problem for behavioral studies as well, i.e., whether the changes in working memory performance are due to the environment (i.e., nature vs. urban) or something more basic, such as perceiving more curved edges, which may not have to do with the environment per se (i.e., urban environments with more curved edges could also produce similar effects). This requires studies that try to isolate portions of the environment that are different while trying to preserve similarities (for example, manipulating the number of curved edges between two environments but keeping the preference between them similar).

A major confounding factor in many studies of nature vs. urban environments is the effect of preference, as nature scenes tend to be more preferred over urban scenes. This means that any difference between the two environments may not be intrinsic to nature vs. urban environments but may rather reflect a difference in preferences (i.e., maybe just viewing anything more preferred vs. less preferred could lead to similar effects such as viewing an attractive face vs. a less attractive face). These confounds present challenges to studies that attempt to isolate causal mechanisms. A recent work has demonstrated that the affective benefits that came from viewing nature imagery vs. urban imagery have been tied to preferences over image category (Meidenbauer et al., 2020) and that when preference was equated between nature and urban scenes, they each changed affect to an equal extent (Meidenbauer et al., 2020). Interestingly, there was a subset of nature imagery that was so highly preferred that it was virtually impossible to find images from other categories that were as preferred, which may have made it difficult to study such imagery in a way that would isolate the confounding factors. Another research study showed stronger approach behaviors toward nature imagery and avoidance of urban imagery, even when the picture sets were equated for aesthetic preference (Schiebel et al., 2022). This effect differs from the affect results from Meidenbauer et al. (2020) that did depend on aesthetic preference. Relatedly, in order to equate images on many dimensions, such as curved edges, fractalness, etc., many of the factors producing the effects in the first place should be controlled. This is indeed a challenge but one that environmental neuroscientists must take seriously in order to make correct causal inferences.

The third problem in environmental neuroscience can be the interpretation of the neural results obtained by the researchers and the inferences that can be made from them. For example, if a hypothetical researcher shows nature and urban images to participants and finds that viewing urban images activates the amygdala more than viewing natural images, then a researcher may be tempted to conclude that viewing urban images causes feelings of fear in individuals and that is causing the positive effects of interacting with nature. This, however, is a logical fallacy (i.e., affirming the consequent in logic). Researchers cannot conclude that if amygdala activity is observed, then the mental process is something related to fear; in the neuroimaging literature, this is known as reverse inference (Poldrack, 2011; Poldrack, 2006). There are some potential solutions to these problems, such as using large database methods (e.g., Neurosynth), to determine how often certain brain activity patterns are related to different cognitive functions (Yarkoni et al., 2011) in a probabilistic way. This still does not solve the reverse inference problem completely, but it does help as a researcher could claim that such activity patterns are related to certain mental processes with X probability. Another way to fight reverse inference is to also include behavioral measures that might be related to a researcher's cognitive processes of interest. If the same neuroimaging researchers had measures of how anxious or fearful their participants were when seeing these different images, then they could then correlate the amygdala activity with those behavioral measures of fear and anxiety. If they then found that indeed the amygdala activity did correlate with behavioral measures of fear and anxiety, then that would lead the researchers to be able to make a stronger inference that the amygdala activity was tied to fear/anxiety and not to something like arousal or excitement. In fact, it is crucial to pair changes in any neural activity measure, perfusion measure, or structural brain measure with changes in behavior that will help mitigate problems of reverse inference.

Fourth, there is often a tendency in environmental neuroscience and environmental psychology to pit urban environments against natural environments and to criticize the former and celebrate the latter. Importantly, environmental neuroscience also investigates the ways in which urban environments are related to improved or worsened psychological well-being. Some have posited that urban environments are related to poorer mental health (Lederbogen et al., 2011; van Os et al., 2010), but more recent research has shown that depression rates are lower in more populous urban centers (Stier et al., 2021b). This improved mental health was hypothesized to be driven by increases in individuals' social network sizes as cities got larger, which is consistent with other works showing that larger cities have disproportionate increases in wealth and innovation that are driven by the increased flow of information (Bettencourt, 2013). This is not always good, e.g., as COVID-19 spreads faster in more populated cities, similarly, so many things spread faster in these larger socioeconomic networks (Stier et al., 2021a). It has also been found that larger cities have lower implicit racial biases, which may also be driven by increased contact with outgroup members (Stier et al., in press). So, while urban life may seem to be related to worse mental health, a more careful examination suggests that this may be too simple of a story. From an environmental neuroscience perspective, an interesting approach may be to try and retain the effects of the good aspects of urban life on residents' mental health while mitigating some of the negatives, maybe through increasing urban green spaces. We would argue against such a simple dichotomy between urban and natural environments (Brancato et al., 2022).

Future Directions and Opportunities

There is great promise in environmental neuroscience and the field's ability to understand how the environment affects and interacts with brain processes to produce behavior. Environmental neuroscience could help lead to research that improves human psychological health at large scales while also helping humans adjust and attempt to solve the climate crisis (Doell et al., 2023). Below are just a few possible future directions for environmental neuroscience.

Substantial challenges exist for the ecological validity of environmental neuroscience research when attempting to study many psychological processes and perceptions in real-world contexts. As mentioned earlier, certain neuroimaging techniques, like functional MRI (fMRI), are incredible technologies, but they pose certain challenges when attempting to measure neural activity in real-world contexts (i.e., because the MRI environment is extremely artificial and noisy). Development of new tools and methodologies holds some promise for potentially examining neural activity and mechanisms while participants are experiencing real natural environments, including fNIRS assessments, where participants can wear the technology while they traverse real environments. In addition, progress is being made to study the effects of immersive environments using virtual reality technology in MRI scanners or with electroencephalogram (EEG) (➔ chapter "EEG for the Study of Environmental Neuroscience" by Grassini).

In addition, researchers can utilize ecological momentary assessment (EMA) surveys (e.g., mobile phone apps) (➔ chapter "In the Right Place at the Right Time: Leveraging Geographic Ecological Momentary Assessment to Analyze Contexts and Health" Kestens and Kingsbury) to study individuals as they traverse real-world natural and urban environments (Schertz et al., 2022; Schertz et al., 2023). These EMA surveys can also be coupled to mobile physiology equipment (such as fNIRS) or to other psychophysiological measurements of autonomic nervous system (ANS) activity during these assessments.

Some highly interesting works have also utilized twin studies to examine how heritable environmental preferences may be. For example, Chang et al. (2022) found that people's preferences for nature was about 40% heritable and that it is possible that this might also influence how much of a benefit individuals might attain from interacting with nature. Another work by Belsky et al. (2009) stressed the importance of gene-by-environment interactions, i.e., G x E, where certain individuals may carry "plasticity genes" that make them more or less receptive to the environment in a better or worse way. More specifically, an s/s (short allele) genotype for the *5-HTTLPR* gene was related to a lower risk of developing depression if one grew up in a low-risk and positive caregiving environment but was more likely to develop depression if one grew up in a high-risk and negative caregiving environment compared to participants who had the l/l (long allele) genotype (Belsky et al., 2009). It is then possible that the s/s individuals may be more affected by all different kinds of environments, maybe even natural environments. With a greater quantification of environmental factors and genetic factors, environmental neuroscientists may gain more insight into the ways that humans and other organisms are affected by their environment.

Environmental neuroscientists have also begun to utilize more "big datasets." For example, work from the mappiness project utilized data that included millions of ratings of scenes from thousands of participants (MacKerron & Mourato, 2013; Seresinhe et al., 2019) to map where people are the happiest. Often, these locations are more natural. Schertz et al. (2021) utilized cellphone tracing data from hundreds of thousands of participants to examine how park visits may be related to crime when controlling for sociodemographic factors and found that in neighborhoods where participants visited a park more often, there was less crime. Another work using large behavioral datasets showed that the kind of environment that people grow up in and how entropic their roads are (e.g., Prague, Czech Republic, has a much more entropic road network than does Chicago, IL, USA, which is a grid) predicts how well they will do on a navigation video game and the entropy of the video game level (Coutrot et al., 2022). In particular, if you grew up in a city like Chicago, you did better on video game levels that were more of a grid, and if you grew up in a city like Prague, you did better on video game levels that were more entropic (Coutrot et al., 2022). This research included data from almost 400,000 video game players. Researchers are now also utilizing large neuroimaging datasets, such as the ABCD dataset (Barch et al., 2018), to include data from thousands of participants. The ABCD dataset also contains some environmental data about its participants in terms of where they reside. In a recent study, researchers have found that adolescents who live in neighborhoods with more air pollution had worsened developmental increases in working memory performance and that these effects were similar in terms of strength to that of parental income (Kardan et al., 2023). Furthermore, these developmental effects were mediated by weakened connectivity in brain networks that support cognitive abilities (Kardan et al., 2023). Although one needs to be careful when drawing causal inferences from observational data, these studies highlight that much information can be gleaned from these large-scale datasets and that with care causal inferences may be drawn (Pearl & Mackenzie, 2018).

Lastly, it was mentioned that Kurt Lewin had posited that human behavior is a function of the person and the environment or behavior = *f*(P,E). Environmental neuroscientists are encouraged to attempt to model and quantify these complex relationships between person-centric qualities, environmental qualities, and their interactions (Berman et al., 2019b). Such quantification and the development of new models will help the field of environmental neuroscience to test new theories and provide a new understanding of how the environment interacts with a person's qualities to produce behavior. Such modeling and theorizing have been highly fruitful in the field of urban science, complex systems, and urban scaling theory (Bettencourt, 2013; Bettencourt et al., 2007; West, 2017), and we would encourage environmental neuroscientists to also take up this complex systems approach when trying to model organism–environment systems.

References

Agorastos, A., Pervanidou, P., Chrousos, G. P., & Baker, D. G. (2019). Developmental trajectories of early life stress and trauma: A narrative review on neurobiological aspects beyond stress system dysregulation. *Frontiers in Psychiatry, 10*, 118. https://doi.org/10.3389/fpsyt.2019.00118

Asch, S. (1951). Effects of group pressure on the modification and distortion of judgments. In *Groups, leadership and men: Research in human relations*. Carnegie Press.

Barch, D. M., Albaugh, M. D., Avenevoli, S., Chang, L., Clark, D. B., Glantz, M. D., et al. (2018). Demographic, physical, and mental health assessments in the adolescent brain and cognitive development study: rationale and description. *Developmental Cognitive Neuroscience, 32*, 55–66.

Belsky, J., Jonassaint, C., Pluess, M., Stanton, M., Brummett, B., & Williams, R. (2009). Vulnerability genes or plasticity genes? *Molecular Psychiatry, 14*(8), 746–754. https://doi.org/10.1038/mp.2009.44. Epub 2009 May 19. PMID: 19455150; PMCID: PMC2834322.

Benefiel, A. C., & Greenough, W. T. (1998). Effects of experience and environment on the developing and mature brain: Implications for laboratory animal housing. *ILAR Journal, 39*, 5–11. https://doi.org/10.1093/ilar.39.1.5

Berman, M. G., Hout, M., Kardan, O., Hunter, M. R., Yourganov, G., Hanayik, T., Karimi, H., &Jonides, J. (2014). The perception of naturalness correlates with low-level visual features of environmental scenes. *PLoS ONE, 9*(12), Article e114572.

Berman, M. G., Kardan, O., Kotabe, H. P., Nusbaum, H. C., & London, S. E. (2019a). The promise of environmental neuroscience. *Nature Human Behaviour, 3*, 414–417.

Berman, M. G., Stier, A. J., & Akcelik, G. N. (2019b). Environmental neuroscience. *American Psychologist, 74*, 1039–1052.

Bettencourt, L. M. A. (2013). The origins of scaling in cities. *Science, 340*, 1438–1441. https://doi.org/10.1126/science.1235823

Bettencourt, L. M. A., Lobo, J., Helbing, D., & West, G. B. (2007). Growth, innovation, scaling, and the pace of life in cities. *PNAS, 104*(17), 7301–7306.

Blakemore, C., & Cooper, G. F. (1970). Development of the brain depends on the visual environment. *Nature, 228*, 477–478. https://doi.org/10.1038/228477a0

van den Bosch, M., & Meyer-Lindenberg, A. (2019). Environmental exposures and depression: Biological mechanisms and epidemiological evidence. *Annual Review of Public Health, 40*(1), 239–259. https://doi.org/10.1146/annurev-publhealth-040218-044106

Brancato, G., Van Hedger, K., Berman, M. G., & Van Hedger, S. (2022). Simulated nature walks improve psychological well-being along a natural to urban continuum. *Journal of Environmental Psychology, 81*(5), 101779.

Bratman, G. N., & Olvera-Alvarez, H. A. (2022). Nature and health: Perspectives and pathways. *Ecopsychology, 14*(3), 133–136. https://doi.org/10.1089/eco.2022.29007.editorial

Bratman, G. N., Daily, G. C., Levy, B. J., & Gross, J. J. (2015). The benefits of nature experience: Improved affect and cognition. *Landscape and Urban Planning, 138*, 41–50. https://doi.org/10.1016/j.landurbplan.2015.02.005

Bratman, G. N., Anderson, C. B., Berman, M. G., Cochran, B., de Vries, S., Flanders, J., Folke, C., Frumkin, H., Gross, J. J., Hartig, T., Kahn, P. H., Kuo, M., Lawler, J. J., Levin, P. S., Lindahl, T., Meyer-Lindenberg, A., Mitchell, R., Ouyang, Z., Roe, J., et al. (2019). Nature and mental health: An ecosystem service perspective. *Science Advances, 5*(7), eaax0903. https://doi.org/10.1126/sciadv.aax0903

Bratman, G. N., Olvera-Alvarez, H. A., & Gross, J. J. (2021a). The affective benefits of nature exposure. *Social and Personality Psychology Compass, 15*(8). https://doi.org/10.1111/spc3.12630

Bratman, G. N., Young, G., Mehta, A., Lee Babineaux, I., Daily, G. C., & Gross, J. J. (2021b). Affective benefits of nature contact: The role of rumination. *Frontiers in Psychology, 12*, 643866. https://doi.org/10.3389/fpsyg.2021.643866

Chang, C.-c., Cox, D. T. C., Fan, Q., Nghiem, T. P. L., Tan, C. L. Y., Oh, R. R. Y., et al. (2022). People's desire to be in nature and how they experience it are partially heritable. *PLoS Biology, 20*(2), e3001500. https://doi.org/10.1371/journal.pbio.3001500

Chun, M. M., & Jiang, Y. (1999). Top-down attentional guidance based on implicit learning of visual covariation. *Psychological Science, 10*(4), 360–365. https://doi.org/10.1111/1467-9280.00168

Coburn, A., Kardan, O., Kotabe, H. P., Steinberg, J., Hout, M. C., Robbins, A., MacDonald, J., Hayn-Leichsenring, G. U., & Berman, M. G. (2019). Psychological responses to natural patterns in architecture. *Journal of Environmental Psychology, 62*, 133–145.

Coutrot, A., Manley, E., Goodroe, S., et al. (2022). Entropy of city street networks linked to future spatial navigation ability. *Nature, 604*, 104–110. https://doi.org/10.1038/s41586-022-04486-7

Doell, K., Berman, M. G., Bratman, G. N., Knutson, B., Kühn, S., Lamm, C., Pahl, S., Sawe, N., Van Bavel, J. J., White, M., & Brosch, T. (2023). Leveraging neuroscience for climate change research. *Nature Climate Change, 13*, 1288–1297.

Frumkin, H., Bratman, G. N., Breslow, S. J., Cochran, B., Kahn, P. H., Jr., Lawler, J. J., Levin, P. S., Tandon, P. S., Varanasi, U., Wolf, K. L., & Wood, S. A. (2017, Jul 31). Nature contact and human health: A research agenda. *Environmental Health Perspectives, 125*(7), 075001. https://doi.org/10.1289/EHP1663. PMID: 28796634; PMCID: PMC5744722.

Gauthier, I., Tarr, M. J., Anderson, A. W., Skudlarski, P., & Gore, J. C. (1999 Jun). Activation of the middle fusiform 'face area' increases with expertise in recognizing novel objects. *Nature Neuroscience, 2*(6), 568–573. https://doi.org/10.1038/9224

Hartig, T. (2021). Restoration in nature: Beyond the conventional narrative. In A. R. Schutte, J. Torquati, & J. R. Stevens (Eds.), *Nature and psychology: Biological, cognitive, developmental, and social pathways to well-being (Proceedings of the 67th Annual Nebraska Symposium on Motivation)* (pp. 89–151) Springer Nature.

Hebb, D. O. (1949). *The organization of behavior: A neuropsychological theory*. John Wiley and Sons, Inc.

Hertzman, C., & Boyce, T. (2010). How experience gets under the skin to create gradients in developmental health. *Annual Review of Public Health, 31*, 329–347. 323p following 347. https://doi.org/10.1146/annurev.publhealth.012809.103538

Hubel, D. H., & Wiesel, T. N. (1962). Receptive fields, binocular interaction, and functional architecture in the cat's visual cortex. *The Journal of Physiology, 160*, 106–154.

Joye, Y., Steg, L., Unal, A. B., & Pals, R. (2016). When complex is easy on the mind: Internal repetition of visual information in complex objects is a source of perceptual fluency. *Journal of Experimental Psychology: Human Perception and Performance, 42*(1), 103–114.

Kandler, C., & Rauthmann, J. F. (2022). Conceptualizing and studying characteristics, units, and fits of persons and environments: A coherent synthesis. *European Journal of Personality, 36*(3), 293–318. https://doi.org/10.1177/08902070211048728

Kaplan, S. (1995). The restorative benefits of interacting with nature. *Journal of Environmental Psychology, 15*(3), 169–182.

Kaplan, S., & Berman, M. G. (2010). Directed attention as a common resource for executive functioning and self-regulation. *Perspectives in Psychological Science, 5*(1), 43–57.

Kardan, O., Gozdyra, P., Misic, B., Moola, F., Palmer, L., Paus, T., & Berman, M. G. (2015). Neighborhood greenspace and health in a large urban center. *Scientific Reports, 5*, 11610.

Kardan, O., Sereeyothin, C., Schertz, K. E., Angstadt, M., Weigard, A., Berman, M. G., & Rosenberg, M. D. (2023). Neighborhood air pollution is negatively associated with neurocognitive maturation in early adolescence. *bioRxiv*.

Kempermann, G. (2019). Environmental enrichment, new neurons, and the neurobiology of individuality. *Nature Reviews. Neuroscience, 20*, 235–245. https://doi.org/10.1038/s41583-019-0120-x

Korpela, K. M., Hartig, T., Kaiser, F. G., & Fuhrer, U. (2001). Restorative experience and self-regulation in favorite places. *Environment and Behavior, 33*(4), 572–589. https://doi.org/10.1177/00139160121973133

Korpela, K. M., Pasanen, T., Repo, V., Hartig, T., Staats, H., Mason, M., Alves, S., Fornara, F., Marks, T., Saini, S., Scopelliti, M., Soares, A. L., Stigsdotter, U. K., & Thompson, C. W. (2018). Environmental strategies of affect regulation and their associations with subjective well-being. *Frontiers in Psychology, 9*(APR), 1–15. https://doi.org/10.3389/fpsyg.2018.00562

Korpela, K. M., Korhonen, M., Nummi, T., Martos, T., & Sallay, V. (2020). Environmental self-regulation in favourite places of Finnish and Hungarian adults. *Journal of Environmental Psychology, 67*, 101384–101384. https://doi.org/10.1016/j.jenvp.2019.101384.

Kotabe, H. P., Kardan, O., & Berman, M. G. (2016). The order of disorder: Deconstructing visual disorder and its effect on rule-breaking. *Journal of Experimental Psychology: General, 145*(12), 1713–1727.

Kühn, S., Düzel, S., Eibich, P., Krekel, C., Wüstemann, H., Kolbe, J., et al. (2017). In search of features that constitute an "enriched environment" in humans: Associations between geographical properties and brain structure. *Scientific Reports, 7*(1), 11920.

Lederbogen, F., Kirsch, P., Haddad, L., Streit, F., Tost, H., Schuch, P., Wüst, S., Pruessner, J. C., Rietschel, M., Deuschle, M., & Meyer-Lindenberg, A. (2011). City living and urban upbringing affect neural social stress processing in humans. *Nature, 474*(7352), 498–501. https://doi.org/10.1038/nature10190

Lewin, K. (1936). *Principles of topological psychology*. McGraw-Hill.

MacKerron, G., & Mourato, S. (2013). Happiness is greater in natural environments. *Global Environmental Change, 23*(5), 992–1000.

Meidenbauer, K. L., Stenfors, C. U., Bratman, G. N., Gross, J. J., Schertz, K. E., Choe, K. W., & Berman, M. G. (2020). The affective benefits of nature exposure: What's nature got to do with it? *Journal of Environmental Psychology, 72*(2), 101498.

Mischel, W. (1968). *Personality and assessment*. Wiley.

Mitchell, R., & Popham, F. (2008, Nov 8). Effect of exposure to natural environment on health inequalities: an observational population study. *Lancet, 372*(9650), 1655–1660. https://doi.org/10.1016/S0140-6736(08)61689-X

Nusslock, R., & Miller, G. E. (2016). Early-life adversity and physical and emotional health across the lifespan: A neuroimmune network hypothesis. *Biological Psychiatry, 80*(1), 23–32. https://doi.org/10.1016/j.biopsych.2015.05.017

Oliva, A., & Torralba, A. (2007, Dec). The role of context in object recognition. *Trends in Cognitive Sciences, 11*(12), 520–527. https://doi.org/10.1016/j.tics.2007.09.009. Epub 2007 Nov 19.

Olvera-Alvarez, H. A., Appleton, A. A., Fuller, C. H., Belcourt, A., & Kubzansky, L. D. (2018). An integrated socio-environmental model of health and well-being: A conceptual framework exploring the joint contribution of environmental and social exposures to health and disease over the life span. *Current Environmental Health Reports, 5*(2), 233–243. https://doi.org/10.1007/s40572-018-0191-2

van Os, J., Kenis, G., & Rutten, B. (2010). The environment and schizophrenia. *Nature, 468*, 203–212. https://doi.org/10.1038/nature09563

OWID based on UN World Urbanization Prospects 2018 and others—processed by Our World in Data. "Urban" [dataset]. OWID based on UN World Urbanization Prospects 2018 and others [original data].

Park, C. L., Kubzansky, L. D., Chafouleas, S. M., Davidson, R. J., Keltner, D., Parsafar, P., Conwell, Y., Martin, M. Y., Hanmer, J., & Wang, K. H. (2022). Emotional well-being: What it is and why it matters. *Affective Science*. https://doi.org/10.1007/s42761-022-00163-0

Pearl, J., & Mackenzie, D. (2018). *The book of why*. Basic Books.

Poldrack, R. A. (2006). Can cognitive processes be inferred from neuroimaging data? *Trends in Cognitive Sciences, 10*(2), 59–63.

Poldrack, R. A. (2011). Inferring mental states from neuroimaging data: from reverse inference to large-scale decoding. *Neuron, 72*(5), 692–697.

Qian, K., Arichi, T., Price, A., et al. (2021). An eye tracking based virtual reality system for use inside magnetic resonance imaging systems. *Scientific Reports, 11*, 16301. https://doi.org/10.1038/s41598-021-95634-y

Richardson, M. (2019). Beyond restoration: Considering emotion regulation in natural well-being. *Ecopsychology, 11*(2), 123–129. https://doi.org/10.1089/eco.2019.0012

Sapolsky, R. M. (2015). Stress and the brain: Individual variability and the inverted-U. *Nature Neuroscience, 18*, 1344–1346.

Sawe, N., & Knutson, B. (2015). Neural valuation of environmental resources. *NeuroImage, 122*, 87–95.

Schertz, K. E., & Berman, M. G. (2019). Understanding nature and its cognitive benefits. *Current Directions in Psychological Science, 28*(5), 496–502.

Schertz, K. E., Sachdeva, S., Kardan, O., Kotabe, H. P., Wolf, K. L., & Berman, M. G. (2018). A thought in the park: The influence of naturalness and low-level visual features on expressed thoughts. *Cognition, 174*, 82–93.

Schertz, K. E., Kardan, O., & Berman, M. G. (2020). Visual features influence thought content in the absence of overt semantic information. *Attention, Perception and Psychophysics, 82* 3945–3956. https://doi.org/10.3758/s13414-020-02121-z.

Schertz, K., Saxon, J., Cardenas-Iniguez, C., Bettencourt, L., Ding, Y., Hoffman, H., & Berman, M. G. (2021). Neighborhood street activity and greenspace usage uniquely contribute to predicting crime. *NPJ: Urban Sustainability, 1*(1), 19.

Schertz, K. E., Bowman, J. E., Kotabe, H. P., Layden, E. A., Zhen, J., Lakhtakia, T., Lyu, M., Paraschos, O., Van Hedger, S. C., Vohs, K., & Berman, M. G. (2022). Environmental influences on affect and cognition: A study of natural and commercial semi-public spaces. *Journal of Environmental Psychology, 83*, 101852.

Schertz, K. E., Bowman, J. E., Kotabe, H. P., Layden, E. A., Zhen, J., Lakhtakia, T., Lyu, M., Paraschos, O., Van Hedger, S. C., Vohs, K., & Berman, M. G. (2023). Nature's path to thinking about others and the surrounding environment. *Journal of Environmental Psychology, 89*, 102046.

Schiebel, T., Gallinat, J., & Kühn, S. (2022). Testing the Biophilia theory: Automatic approach tendencies towards nature. *Journal of Environmental Psychology. 79*, 101725, https://doi.org/10.1016/j.jenvp.2021.101725

Seresinhe, C. I., Preis, T., MacKerron, G., & Moat, H. S. (2019, Mar 14). Happiness is Greater in More Scenic Locations. *Scientific Reports, 9*(1), 4498. https://doi.org/10.1038/s41598-019-40854-6. PMID: 30872776; PMCID: PMC6418136.

Slavich, G. M. (2020). Social safety theory: A biologically based evolutionary perspective on life stress, health, and behavior. *Annual Review of Clinical Psychology, 16*, 265–295. https://doi.org/10.1146/annurev-clinpsy-032816-045159

Stenfors, C. U. D., Van Hedger, S. C., Schertz, K. E., Meyer, F. A., Smith, K. E. L., Norman, G. J., et al. (2019). Positive effects of nature on cognitive performance across multiple experiments: Test order but not affect modulates the cognitive effects. *Frontiers in Psychology*. https://doi.org/10.3389/fpsyg.2019.01413

Stier, A. J., Schertz, K. E., Rim, N. W., Cardenas-Iniguez, C., Lahey, B. B., Bettencourt, L. M., & Berman, M. G. (2021). Evidence and theory for lower rates of depression in larger US urban areas. *Proceedings of the National Academy of Sciences, 118*(31), e2022472118.

Stier, A. J., Berman, M. G., & Bettencourt, L. (2021a). COVID-19 attack rate increases with city size. *NPJ: Urban Sustainability*.

Stier, A. J., Schertz, K. E., Rim, N. W., Cardenas-Iniguez, C., Lahey, B. B., Bettencourt, L. M. A., & Berman, M. G. (2021b). Rethinking depression in U.S. cities: Evidence and theory for lower rates in larger urban areas. *Proceedings of the National Academy of Sciences, 118*(31), e2022472118.

Stier, A. J., Sajjadi, S., Karimi, F., Bettencourt, L. M. A., & Berman, M. G. (2024). Implicit racial biases are lower in more populous more diverse and less segregated US cities. *Nature Communications, 15*, 961. https://doi.org/10.1038/s41467-024-45013-8

Tost, H., Reichert, M., Braun, U., Reinhard, I., Peters, R., Lautenbach, S., Hoell, A., Schwarz, E., Ebner-Priemer, U., Zipf, A., & Meyer-Lindenberg, A. (2019). Neural correlates of individual differences in affective benefit of real-life urban green space exposure. *Nature Neuroscience, 22*(9), 1389–1393. https://doi.org/10.1038/s41593-019-0451-y

Tulve, N., Ruiz, J., Lichtveld, K., Darney, S., & Quackenboss, J. (2016). Development of a conceptual framework depicting a child's total (built, natural, social) environment in order to optimize health and well-being. *Journal of Environment and Health Science, 2*(2), 1–8.

Ulrich, R. S., Simons, R. F., Losito, B. D., Fiorito, E., Miles, M. A., & Zelson, M. (1991). Stress recovery during exposure to natural and urban environments. *Journal of Environmental Psychology, 11*(3), 201–230. https://doi.org/10.1016/S0272-4944(05)80184-7

Van Hedger, S. C., Nusbaum, H. C., Heald, S. L. M., Huang, A., Kotabe, H. P., & Berman, M. G. (2019). The aesthetic preference for nature sounds depends on sound object recognition. *Cognitive Science, 43*(5), e12734.

West, G. (2017). *Scale*. Penguin Random House.

West, R. W., & Greenough, W. T. (1972). Effect of environmental complexity on cortical synapses of rats: Preliminary results. *Behavioral Biology, 7*, 279–284. https://doi.org/10.1016/S0091-6773(72)80207-4

Woo, C. C., & Leon, M. (2013). Environmental enrichment as an effective treatment for autism: a randomized controlled trial. *Behavioral Neuroscience, 127*(4), 487.

Wu, A., Yu, B., & Komiyama, T. (2020). Plasticity in olfactory bulb circuits. *Current Opinion in Neurobiology, 64*, 17–23.

Yarkoni, T., Poldrack, R. A., Nichols, T. E., Van Essen, D. C., & Wager, T. D. (2011). Large-scale automated synthesis of human functional neuroimaging data. *Nature Methods, 8*(8), 665–670. https://doi.org/10.1038/nmeth.1635. PMID: 21706013; PMCID: PMC3146590.

The Rise and Future of Environmental Neuroscience in Environmental Psychology

Lindsay J. McCunn

Abstract The rise and the future of environmental neuroscience and environmental psychology are happening now simultaneously. Both disciplines hold multivalent opportunities to understand the human condition in a variety of settings. In addition, both examine the mechanisms and extent to which humans interact with particular places—built, natural, or virtual. This chapter discusses the insights and methods native to environmental neuroscience that prompt environmental psychologists to ask questions about what the brain and the body can do with the places around us—and why and how. The answers to these questions buttress the continued merger between these fields so that we can make a positive impact on the design of our cities, our homes, our workplaces, and the natural environments around us. Some classical studies at this intersection are noted as are some of the emerging themes in today's body of research concerning attention restoration, cognitive load, creativity, mood, and various neuropsychophysiological responses to nature and architecture. A few key scholarly journals, books, and organizations that support environmental neuroscience research are also described, as is the need to augment academic programs and curricula to incorporate and build upon both fields that, together, are becoming a modernizing force in social science.

Keywords Environmental neuroscience · Environmental psychology · Interdisciplinary research · Emerging science · Applied psychology

Environmental neuroscience and environmental psychology are two specialized but wide-reaching disciplines that have much in common. Both are relatively young subfields within the larger arenas of natural and social sciences. Each holds diverse and multivalent opportunities to understand (and ultimately improve) the human

The original version of the chapter has been revised. A correction to this chapter can be found at https://doi.org/10.1007/978-3-031-64699-7_23

L. J. McCunn (✉)
Vancouver Island University, Nanaimo, BC, Canada
e-mail: lindsay.mccunn@viu.ca

S. Kühn (ed.), *Environmental Neuroscience*,
https://doi.org/10.1007/978-3-031-64699-7_2

condition in a variety of settings. Both examine the mechanisms and extent to which humans interact with particular places—whether they are built, natural, or virtual. They rest on the philosophy that a rational understanding of psychology, medicine, biology (and beyond) ought not to ignore or misrepresent the power of the environment to affect cognition, emotion, and behavior. Moreover, both have developed a mutual goal to promote human (and nonhuman) health and mental well-being by elevating environmental design for positive sensory, emotional, and behavioral outcomes as well as for optimal brain function.

Environmental psychology centers on the transactions between individuals and their physical surroundings (Gifford, 2014). Those who study it seek to understand how people sense, perceive, and remember the different spaces that they encounter. After all, our environment is everywhere, all of the time. It affects how we think, feel, and behave—and we affect environments in return. The psychological interplay at the junctures between people and places is why the word "transaction" is often used in definitions of environmental psychology. It reminds us that the human–environment relationship is complex and interdependent. This is why scholars of environmental neuroscience and environmental psychology dedicate their attention to environmental influence—rather than ignore or control it in research and experimental design.

In order to address the global problem of climate change, environmental psychology has grown to include foci concerning conservation, pro-environmental attitudes, intentions, and behaviors as well as consumer perceptions about this issue. However, many environmental psychologists continue to analyze classical topics in the field—topics that concern levels of stimulation, control, and sensory adaptation in response to a setting. In doing so, they often engage with engineers, city planners, architects, and facility managers to communicate social scientific findings about psychology, health, and design in order to make recommendations about their applications. In the first chapter of his book *Environmental Psychology: Principles and Practice* (2014), Robert Gifford includes a figure that summarizes the scope of environmental psychology (replicated here as Fig. 1). It illustrates that what people

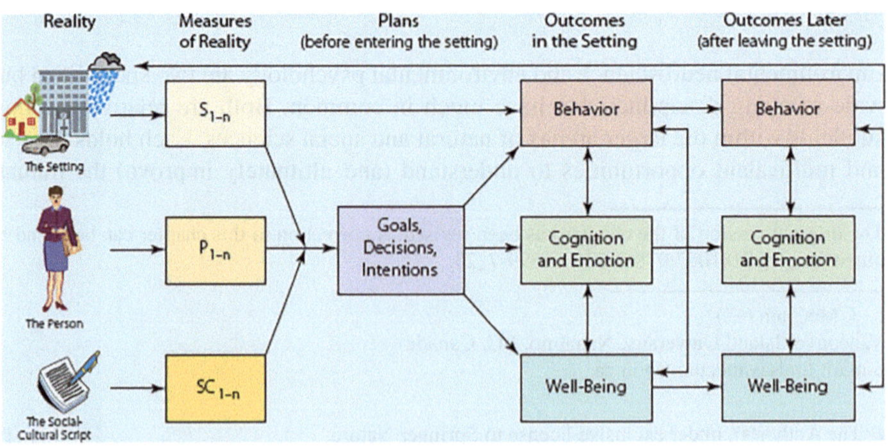

Fig. 1 An overview of environmental psychology. (Used with permission from Robert Gifford)

view, think about, plan for, and do is a function of the natural and built environments that affect us near and far. We can see that the types of research questions considered by environmental psychologists to explore the factors in the figure complement the kinds of questions asked by environmental neuroscientists. Indeed, one of the broadest goals of environmental neuroscience is to quantify the attributes of physical and social environments to systematically learn how (and to what extent) they affect the brain and, in turn, behavior (Berman et al., 2019). Thus, it is not difficult to notice the numerous entry points for environmental neuroscience in this general model of environmental psychology. As studies in environmental neuroscience become more visible in journals publishing for a readership working in the applied sciences, connections between its methods and findings with those found in environmental psychology are likely to be formed more often.

With interdisciplinarity in mind, the two leading journals in environmental psychology—the *Journal of Environmental Psychology* (*JEP*) and *Environment & Behavior*—have been open to articles on neuroscientific-related topics for a number of years. In our roles as co-editors-in-chief of *JEP*, Wesley Schultz and I receive many manuscripts with keywords related to understanding brain–environment associations. Of course, not all of these submissions withstand peer review. Over the past decade, the number of accepted papers in *JEP* with keywords such as "brain," "EEG," (➔ chapter "EEG for the Study of Environmental Neuroscience" by Grassini) and "fMRI" (➔ chapter "Mapping the Human Brain with Computational Anatomy" by Draganski, Joost and Kherif) (and so on) in their titles and abstracts has risen. A cursory online search in the journal will generate more than 20 papers published with these and similar neuroscience-related keywords (not counting the raft of literature on spatial cognition, navigation, and wayfinding or on human responses to isolated and confined environments, e.g., Suedfeld et al. (1994)).

One emerging theme within this perfunctory yet informative search of *JEP* concerns the measurement of human attention restoration, cognitive load, creativity, social functioning, brain structure, and various neuropsychophysiological responses to some facet of nature or natural stimuli. Variables such as nature perception, exposure to different amounts of urban tree cover or green roofs, nature-based hospital designs, blue (i.e., water-based) city spaces, and so on are common (see Chen et al. 2020; Grassini et al. 2019; Jung et al. 2023; Kühn et al. 2023; Lee et al. 2015; Luo et al. 2023; Mygind et al. 2023; Olszewska-Guizzo et al. 2021; Williams et al. 2018). Other notable papers concerning nature explore the virtues of birdsong and its combinational effects on physiopsychological restoration (Qi et al., 2022) and on whether viewing a flower image provides automatic recovery effects after psychological stress (Mochizuki-Kawai et al., 2020). These articles provide additional examples of an overlap between environmental neuroscience and environmental psychology that aids in our understanding of the role nature plays in enhancing health and well-being, urban living, and satisfaction indoors.

A similar theme among some of the noted articles published in *JEP* has to do with the measurement of brain activity, place immersion, mood, emotion, and judgments of beauty in response to different natural and built environments, textures, and architectural attributes (e.g., ceiling height and perceived enclosure) as well as geometries in virtual space (see Bower et al. 2019; Buttazzoni et al. 2022; Norwood et al. 2019; Shemesh et al. 2022; Vartanian et al. 2015). Publications like these indicate that environmental psychology and environmental neuroscience are both prepared and eager to inform infrastructure and design decisions to foster emotional fulfillment and health across and among real and virtual contexts.

However inspiring and instructive, these and other publications do not necessarily represent an altogether novel trajectory in research. Readers will likely be familiar with an article published in 2012 by Lengen and Kistemann titled "Sense of Place and Place Identity: Review of Neuroscientific Evidence." The authors explain that, in 1965, a report by the National Academy of Sciences and the National Research Council titled "The Science of Geography" stated that sense of place, defined as "a compound of a sense of territoriality, physical direction, and distance," is "very deeply ingrained in the human race" (National Academy of Sciences, National Research Council, 1965, p. 7). The report affirms that little is known about how humans develop a sense of place and that it should be investigated in the neuropsychological and neurobiological sciences (National Academy of Sciences, National Research Council, 1965). In response, Lengen and Kistemann's (2012) systematic meta-analysis reviews the neuroscientific literature (at the time of publication) for studies concerning sense of place. The objective was to resume the National Academy of Sciences and the National Research Council's call for more transdisciplinary research that could bring the "phenomenological sense of place approach together with current results from neuroscience, which could be significant for the understanding of sense of place, place identity, and their significance for health and well-being" (Lengen & Kistemann, 2012, p. 1163). Read today, Lengen and Kistemann's (2012) review remains helpful and promising—it confirms that place is a measurable dimension in neuronal processing and outlines studies that correlate sense of place with memory, perception, orientation, attention, emotion, and behavior at the neuronal, structural, and regional levels of the brain as well as at the network level.

In parallel, John Zeisel explains in the second edition of his book *Inquiry by Design: Environment Behaviour Neuroscience in Architecture Interiors* (2006) that social scientists ought to embrace the neuroscientific literature and methods to broaden their techniques to conceptualize and measure environment–behavior relations. Zeisel's book has alerted many environmental psychologists to neuroscientific approaches that lend themselves to understanding how wayfinding, cognitive mapping, spatial cognition (and, arguably, architecture) are reflected in our brains. It also showcases the importance of studying these topics from interdisciplinary perspectives to promote stress reduction in public settings (such as airports and hospitals) as well as for special populations (those living in memory care units, for example). Similarly, John Paul Eberhard, the founding president of the Academy of Neuroscience for Architecture (ANFA), published his book called *Brain Landscape: The Coexistence of Neuroscience and Architecture* in 2009. Eberhard contends that if our environments can alter our gene expression via epigenetic mechanisms, then the environment in general must have an undeniable and substantial effect on our bodies, brains, and behaviors alike at all times. Therefore, scholars in the academy, and decision-makers in practice, must consider the planning and design of cities, communities, buildings, and interiors seriously by respecting the connections between the brain and setting.

This suggests that encouraging additional investigation of topics that blend neuroscience, design, health, and psychology is justified in order for the field of environmental psychology to continue to mature and advance. The discipline must continue to sharpen its theories and frameworks by asking deeper questions about *why* we respond to environments in the ways we do. What is it about particular settings that make us feel or think about things in particular ways? How can those characteristics be manipulated to support better lifestyles, prosocial experiences, and eco-conscious decision-making? And, how can neuroscientific methods be straightforwardly adopted into the praxis of environmental psychology? Answering such questions requires a stronger partnership with neuroscience if environmental psychologists are to explore what is occurring in the brain when we experience places—and how we might capitalize on those processes. Examining environmental psychology's key theories through the lens of environmental neuroscience will certainly be exciting in the near future. As an example, measuring attention restoration and well-being in natural spaces (but under a variety of circumstances, amounts, and structures of naturalness, such as the number of trees, the amount of water, or sense of "wildness") will be compelling. Understanding these variables alongside neuroscientific measures of the brain in relation to lighting, biodiversity, and competing sensory stimulations (e.g., auditory, olfactory) may help translate what we already know about how humans respond to nature into design guidelines and recommendations for planners and architects who seek to create beneficial spaces at a variety of scales. As noted by Berman et al. (2019), and described by others (e.g., Gamble et al. 2014; Valtchanov et al. 2010), psychophysiological benefits can occur during and after nature exposure—even when natural stimuli are artificial or simulated. Continued research on the bounds of this influence using virtual reality (VR) and other computer-generated or artistic sources will be persuasive as applications emerge in the space exploration industry (➔ chapter "Extended Reality in Environmental Neuroscience Research" by Li et al.), for example, where mental health, quality of life, and problem-solving under stress in small, confined spaces must coincide.

In fact, many phenomenological constructs long-studied in environmental psychology, such as place attachment and place identity, are also in need of new avenues of inquiry (Lewicka, 2011). Applying environmental neuroscientific insight to research on how these and other attitudes are developed and maintained at different stages of life, and in different surroundings, is vital if environmental psychologists ask that government bodies refer to our literature as they make evidence-based decisions to enhance public life. Similarly, asking how closely eco-friendly attitudes are to brain activation patterns, shown when we are helpful or generous toward others, may be productive if we are to assert that cultivating sustainable intentions and actions is good for us as individuals and as community members. If ample similarity exists between how the brain responds to social stimuli and environmental stimuli, then encouraging people to care about our planet and the places to which we have bonded may become more impactful as we address climate change and form adaptation strategies. In an era where climate change is a relentless concern, social scientific research about how people feel (and make sustainable decisions) about spaces and places may wield more power if it is combined with environmental neuroscientific efforts to inspect the neural mechanisms behind those feelings and choices—and how to leverage them for policy development.

Moreover, people around the world are spending much of their time indoors (Klepeis et al., 2001). In response, environmental psychologists often study stimuli in the built environment that affects brain function (along with comfort, preference, control, and behavior), such as artificial lighting, thermal conditions, acoustics, and indoor air quality (➔ chapters "Aesthetics of Architecture: Emerging Insights from the Brain" by Vartanian, "The Built Environment and the Brain: Review of Emerging Methods to Investigate the Impact of Viewing Architectural Design" by Tawil and Kühn, and "Applications of Biometrics in Architectural and Environmental Design" by Lavdas and Sussman). Many environmental psychologists work alongside architects and designers to create and assess indoor and outdoor settings for their promotion of productivity, cognitive performance, and so on. It makes sense that these professional collaborations should also include environmental neuroscientists who can undoubtedly strengthen projects with knowledge and ideas that can yield more comprehensive work. For instance, amplifying human-centered design with neuroarchitecture principles is likely to be more popular as commercial mobile brain and body imaging technology devices become more accessible (Gorman et al., 2021; Metzger & Walter, 2018). In the same vein, in 2016, the United Nations estimated that 55% of the world's population currently lives in urban areas—a percentage that is estimated to increase to 68% by 2050 (United Nations, 2018). Environmental psychologists have traditionally studied the numerous behavioral and emotional effects of crowding, noise, daylighting, air pollution, spatial density, perceptions of crime, and other aspects of city living on humans. These are additional topics that lend themselves well to environmental neuroscience.

Given the growth in interest in both environmental psychology and environmental neuroscience over the last decade, and their shared goal to help develop humane environments, one cogent next step may be to encourage more postsecondary faculties and departments to ensure students' exposure to these subfields in the respective curricula of psychology and neuroscience (and in design programs too). More information about the theories, methods, applications, strengths, and challenges of environmental neuroscience could be introduced in undergraduate- and graduate-level environmental psychology textbooks to draw attention to the analogous nature of these disciplines. In turn, providing more education about the concepts, history, and utility of environmental psychology to students of neuroscience may prompt more individuals to consider this field as they build research teams in the future. In addition, inserting topics from both environmental psychology and environmental neuroscience into urban planning and architecture programs seems timely and prudent. Offering students a richer consciousness of interdisciplinary bodies of literature that can be used to check ideas against as they learn strategies to generate settings that foster climate resilience, prosocial tendencies, and better public health is rational.

Clearer awareness and access to this kind of literature may also help planning and architecture students equip themselves with evidence concerning nature's tendency to correlate with positive emotions, physical and mental health, and other advantageous outcomes. In addition, more thorough connections in education are likely to afford supplementary occasions to undertake research merging environmental psychology, environmental neuroscience, and urban planning with larger sample sizes in real-world settings—something that all three fields strive to find. I hope that the breadth of empirical work published in *JEP* can help readers from many viewpoints find inspiration.

Right now, a small but increasing number of research groups and laboratories around the world are successfully performing studies rooted in both environmental neuroscience and environmental psychology. Nonprofit organizations, such as the Environmental Design Research Association (EDRA) and ANFA, also offer pathways for scholars, practitioners, and educators to learn from each other and share resources and research. In fact, both EDRA and ANFA were established with the objective to bring these and other environment–behavior fields together to push the boundaries of science and design principles. In 2015, EDRA held its annual conference with a neuroscience-focused theme called "BrainStorm." A glance through the conference's published proceedings demonstrates how enthusiastic and informed the environment–behavior community is about topics that relate to environmental neuroscience. This aspiration to be of service to society by providing professionals working in these distinct but related disciplines opportunities to discuss ideas can be found in many other organizations centered on applied social sciences around the world, such as the International Association for People-Environment Studies (IAPS) and the International Association of Applied Psychology (IAAP).

The rise and the future of environmental neuroscience and environmental psychology are happening now simultaneously. The insights and methods native to environmental neuroscience are prompting environmental psychologists to ask questions about what the brain and the body can do with the places around us—and why and how. Answers to these questions buttress the continued merger between these fields so that we can make a positive impact on the design of our cities, our homes, our workplaces, and our lives. The relatable and reliable contributions of environmental psychology inspire the aims and the scope of environmental neuroscience in return. Together, they are becoming a modernizing force in social science.

References

Berman, M., Steir, A. J., & Akcelik, G. N. (2019). Environmental Neuroscience. *American Psychologist, 74*, 1039–1052.

Bower, I., Tucker, R., & Enticott, P. G. (2019). Impact of built environment design on emotion measured via neurophysiological correlates and subjective indicators: A systematic review. *Journal of Environmental Psychology, 66*, 101344.

Buttazzoni, A., Ellard, C., Sadiora, S., & Minaker, L. (2022). Toward conceptualizing "place immersion" as a spatial neuropsychosocial phenomenon: A multidisciplinary meta-review and -synthesis. *Journal of Environmental Psychology, 81*, 101810.

Chen, Z., He, Y., & Yu, Y. (2020). Attention restoration during environmental exposure via alpha-theta oscillations and synchronization. *Journal of Environmental Psychology, 68*, 101406.

Eberhard, J. P. (2009). *Brain landscape: The coexistence of neuroscience and architecture*. Oxford University Press.

Gamble, K. R., Howard, J. H., Jr., & Howard, D. V. (2014). Not just scenery: Viewing nature pictures improves executive attention in older adults. *Experimental Aging Research, 40*, 513–530.

Gifford, R. (2014). *Environmental psychology: Principles and practice* (5th ed.). Optimal Books.

Gorman, N., Louw, A., Craik, A., Gonzalez, J., Feng, J., & Contreras-Vidal, J. L. (2021). Design principles for mobile brain-body imaging devices with optimized ergonomics. In *Advances in usability, user experience, wearable and assistive technology* (Vol. 275, pp. 3–10). Springer.

Grassini, S., Revonsuo, A., Castellotti, S., Petrizzo, I., Benedetti, V., & Koivisto, M. (2019). Processing of natural scenery is associated with lower attentional and cognitive load compared with urban ones. *Journal of Environmental Psychology, 62*, 1–11.

Jung, D., Kim, D. E., & Kim, N. (2023). Bringing nature into hospital architecture: Machine Learning based EEG analysis of the biophilia effect in virtual reality. *Journal of Environmental Psychology, 89*, 102033.

Klepeis, N., Nelson, W., Ott, W., et al. (2001). The National Human Activity Pattern Survey (NHAPS): A resource for assessing exposure to environmental pollutants. *Journal of Exposure Science & Environmental Epidemiology, 11*, 231–252.

Kühn, S., Banaschewski, T., Bokde, A. L. W., Büchel, C., Quinlan, E. B., et al. (2023). Trees for brains: Current residential tree cover density and its association with brain structure in young adults. *Journal of Environmental Psychology, 89*, 102047.

Lee, K. E., Williams, K. J. H., Sargent, L. D., Williams, N. S. G., & Johnson, K. A. (2015). *Journal of Environmental Psychology, 42*, 182–189.

Lengen, C., & Kistemann, T. (2012). Sense of place and place identity: Review of neuroscientific evidence. *Health & Place, 18*, 1162–1171.

Lewicka, M. (2011). Place attachment: How far have we come in the last 40 years? *Journal of Environmental Psychology, 31*, 207–230.

Luo, L., Yu, P., & Jiang, B. (2023). Differentiating mental health promotion effects of various bluespaces: An electroencephalography study. *Journal of Environmental Psychology, 88*, 102010.

Metzger, C., & Walter, R. (2018). *Neuroarchitecture*. Jovis.

Mochizuki-Kawai, H., Matsuda, I., & Mochizuki, S. (2020). Viewing a flower image provides automatic recovery effects after psychological stress. *Journal of Environmental Psychology, 70*, 101445.

Mygind, L., Clark, G. M., Bigelow, F. J., Fuller-Tyszkiewicz, M., Knibbs, L. D., et al. (2023). Green enrichment for better mind readers? Residential nature and social brain function in childhood. *Journal of Environmental Psychology, 88*, 102029.

National Academy of Sciences, National Research Council. (1965). *The science of geography report of the Ad Hoc Committee on Geography*. National Academy of Sciences. National Research Council.

Norwood, M. F., Lakhani, A., Maujean, A., Zeeman, H., Creux, H., & Kendall, E. (2019). Brain activity, underlying mood and the environment: A systematic review. *Journal of Environmental Psychology, 65*, 101321.

Olszewska-Guizzo, A., Fogel, A., Escoffier, N., & Ho, R. (2021). Effects of COVID-19-related stay-at home order on neuropsychophysiological response to urban spaces: Beneficial role of exposure to nature? *Journal of Environmental Psychology, 75*, 101590.

Qi, Y., Chen, Q., Lin, F., Liu, Q., Zhang, X., et al. (2022). Comparative study on birdsong and its multisensory combinational effects on physio-psychological restoration. *Journal of Environmental Psychology, 83*, 101879.

Shemesh, A., Leisman, G., Barr, M., & Grobman, Y. J. (2022). The emotional influence of different geometries in virtual spaces: A neurocognitive examination. *Journal of Environmental Psychology, 81*, 101802.

Suedfeld, P., Steel, G. D., Wallbaum, A. B. C., Bluck, S., Livesey, N., & Capozzi, L. (1994). Explaining the effects of stimulus restriction: Testing the dynamic hemispheric asymmetry hypothesis. *Journal of Environmental Psychology, 14*, 87–100.

United Nations. (2018). *Department of Economic and Social Affairs*. Retrieved from: https://www.un.org/development/desa/en/news/population/2018-revision-of-world-urbanization-prospects.html

Valtchanov, D., Barton, K. R., & Ellard, C. (2010). Restorative effects of virtual nature settings. *Cyberpsychology, Behavior, and Social Networking, 13*, 503–512.

Vartanian, O., Navarrete, G., Chatterjee, A., Brorson Fich, L., Gonzalez-Mora, J. L., et al. (2015). Architectural design and the brain: Effects of ceiling height and perceived enclosure on beauty judgments and approach-avoidance decisions. *Journal of Environmental Psychology, 41*, 10–18.

Williams, K. J. H., Lee, K. E., Hartig, T., Sargent, L. D., Williams, N. S. G., & Johnson, K. A. (2018). Conceptualising creativity benefits of nature experience: Attention restoration and mind wandering as complementary processes. *Journal of Environmental Psychology, 59*, 36–45.

Zeisel, J. (2006). *Inquiry by design: Environment behaviour neuroscience in architecture interiors*. Norton.

Shemesh, A., Leisman, G., Bar, M., & Grobman, Y. J. (2022). The emotional influence of different geometries in virtual spaces: A neurocognitive examination. *Journal of Environmental Psychology, 81*, 101802.

[illegible] (2009). [illegible]

[illegible] (1983). [illegible]

[illegible] *Cognitive* [illegible] *Neurology*, [illegible]

[illegible]

[illegible]

Part II
The Adaptation Pathway (Environment-to-Brain): How the Environment Impacts Humans

Interplay Between the Physical Environment and the Human Brain: A Review of MRI Research

Moana Beyer, Sonja Sudimac, Maximilian O. Steininger, and Simone Kühn

Abstract Past non-human research has consistently demonstrated the profound impact of an animal's environment on its central nervous system. Over the past decade, various studies have advanced our understanding of how the human brain is linked to its physical surroundings. These studies have employed various methods of operationalising the physical environment, including population density, the outdoor environment, and pollution, as well as diverse study designs, ranging from controlled artificial viewing of pictures to walks in nature. This chapter provides a comprehensive overview of magnetic resonance imaging (MRI) studies probing the relationship between the human brain and environmental factors. Three brain regions consistently emerged across functional MRI and structural MRI investigations: the prefrontal cortex, the amygdala, and the hippocampus. Lifelong and acute exposure to more rural and greener areas may benefit amygdala and prefrontal brain activity and the cortical structure. This beneficial association might be influenced by typically lower levels of social stress in less urbanised environments compared to more urban settings. Moreover, rural environments may provide more opportunities for developing both spatial skills and the hippocampal structure. Fully understanding the causality of these effects remains a critical avenue for future inquiries, focusing on interventional and longitudinal study designs. Comprehending

M. Beyer (✉) · S. Sudimac
Max Planck Institute for Human Development, Center for Environmental Neuroscience, Berlin, Germany
e-mail: beyer@mpib-berlin.mpg.de; sudimac@mpib-berlin.mpg.de

M. O. Steininger
Social, Cognitive and Affective Neuroscience Unit, Department of Cognition, Emotion, and Methods in Psychology, Faculty of Psychology, University of Vienna, Vienna, Austria
e-mail: maximilian.steininger@univie.ac.at

S. Kühn
Max Planck Institute for Human Development, Center for Environmental Neuroscience, Berlin, Germany

University Medical Center Hamburg-Eppendorf, Department of Psychiatry and Psychotherapy, Hamburg, Germany
e-mail: kuehn@mpib-berlin.mpg.de

S. Kühn (ed.), *Environmental Neuroscience*,
https://doi.org/10.1007/978-3-031-64699-7_3

environmental factors that have beneficial or detrimental impacts on our mental and brain health is critical for effective urban planning and health policy development. This synthesis contributes to our understanding of environmental neuroscience and has significant implications for designing healthier living spaces in our increasingly urbanised world.

Keywords Environmental neuroscience · Brain structure · Brain function · Green space · Urbanicity · Pollution · Review

Introduction

In non-human neuroscience, there has been a long tradition of investigating how environments impact an organism's nervous system. Early studies on environmental enrichment in non-human animals show that exposure to stimulating environments is associated with a wide variety of robust central nervous system changes (Sale et al., 2009). The rapid development of neuroscientific methods over the past few decades has made it possible to transfer this field of research to the study of humans. By harnessing modern techniques, the emerging field of environmental neuroscience intends to investigate how the brain interacts with its surroundings (Berman et al. 2019a). Although the primary aim of the field is to shed light on the reciprocal pathways between the brain and environments, most research to date has explored the unidirectional route of how different environmental aspects impact the brain (Berman et al., 2019b). Environmental neuroscience uses various methods on different temporal and spatial scales to inform us about these pathways—from fine-grained cellular and molecular investigations to broader neuroimaging approaches. This chapter focuses on an excerpt of these findings by summarising human-based studies investigating the impact of the physical environment on the brain by employing structural magnetic resonance imaging (MRI) and functional magnetic resonance imaging (fMRI).

There is a plethora of factors in the environment that have the potential to impact both brain structure and function. In recent years, the concept of the exposome has been established to structure and organise these environmental factors (➔ chapter "Advancing the Neural Exposome" by Veasey). Like the genome's role in encapsulating all the genetic information of an organism, the exposome is intended to represent the totality of an organism's environmental influences. It can be divided into an internal, a specific, and a general external domain (Wild, 2012). While the first two domains relate to internal bodily microenvironments and aspects such as an individual's education, lifestyle, or diet, the general external domain subsumes physical and social environmental factors (Liu et al., 2023). These include pollution, temperature, green space, urbanicity, traffic, population density, or social capital. This chapter focuses on associating these factors from the general external domain with both brain structure and function. We will emphasise three aspects: urbanicity (population density), the outdoor environment, and pollution, as they provide a comprehensive overview of most research conducted to date. "Urbanicity" refers to the impact of residing in urban areas (Vlahov & Galea, 2002). Although the term is associated with many aspects, population density is an essential and the most commonly utilised characteristic of urban space in environmental neuroscience research. Population density quantifies the number of individuals within a given land area. It is a well-established general measure of urbanicity that can be used around the globe and is frequently applied for neighbourhood classifications (Melchiorri et al., 2018). "Outdoor environment" is a term used to describe a collection of areas primarily composed of natural elements typically referred to as either green or blue spaces. Although there is a lack of consensus regarding the conceptual usage of these terms, this chapter will include studies investigating areas of vegetated land or bodies of water and urban vegetation, such as parks, urban forests, or gardens (Taylor & Hochuli, 2017). Pollution encompasses air, noise, and light pollution, signifying the release of contaminants into the environment. Air pollution is a complex mixture of particulate matter (PM), gases, organic volatile compounds, and heavy metals (Barwise & Kumar, 2020). Noise pollution describes continued exposure to noise levels surpassing specific thresholds. Both air and noise pollution usually stem from similar sources, such as traffic, transportation, or industrial activity. Light pollution is typically measured as light during the night, produced, for example, from street lights, illuminated buildings, and commercial signs.

This chapter's first and second sections summarise the findings from functional and structural MRI, respectively. Each section is organised into subsections devoted to the association of population density, the outdoor environment, and pollution (air, noise, and light pollution) with MRI measures. Within these subsections, we have sorted the studies according to the type and time of exposure to the described environment. It should be noted that the time of the MRI scan may differ from the time window of exposure e.g. a study may investigate how urban upbringing is linked to the brain function or structure measured during adulthood. Other chapters in this book complement this summary by providing in-depth overviews of other relevant neuroimaging techniques (see e.g. chapter "EEG for the study of environmental neuroscience"). This chapter will conclude with a discussion on the limitations and future areas of investigation in the field.

Functional MRI Studies

Functional MRI (fMRI) tracks alterations in neural metabolism that indicate activity changes in the brain. These alterations can result from changes in information processing due to participants working on specific tasks or spontaneous fluctuations in the resting brain's activity (Chen & Glover, 2015).

On the one hand, task-based fMRI informs us about the activity or connectivity of brain areas in response to specific tasks. These tasks range from simple motor performances, such as finger tapping, to highly complex decision-making. In environmental neuroscience, past task-based fMRI studies focused on diverse experimental manipulations, such as viewing photos or videos of different environments, reward processing, and responses to negative emotions and trust games. In line with the stress recovery theory (Ulrich et al., 1991), various studies investigated how environments relate to our brain activity when we are exposed to different stressors (Lederbogen et al., 2011; Sudimac et al., 2022; Dimitrov-Discher et al., 2023).

On the other hand, resting-state fMRI examines the intrinsic networks of the brain in the absence of a specific task to estimate the spontaneous correlation between brain regions (Lee et al., 2013). The human brain comprises several resting-state in cities compared to rural environments from attentional processing to self-referential thought (Yeo et al., 2011). The interaction within and between these networks can be operationalised by different measures that characterise certain aspects of connectivity (Rubinov & Sporns, 2010). In the context of environmental neuroscience, most studies to date have investigated how these connectivity measures relate to long- or short-term environmental exposures. Although different research approaches exist, the majority of research is devoted to the association of resting-state functional connectivity and exposure with air or noise pollution (Glaubitz et al., 2022; Martínez-Vilavella et al., 2023; Pujol et al., 2016b).

Brain Function and Urban Environments

Currently, more than half of the world's population resides in urban areas, a proportion anticipated to increase to two-third by 2050 (DESA et al., 2019). Although cities offer benefits such as better access to healthcare, education, and employment, there is increasing evidence of urban living's adverse effects on mental health. The prevalence of psychiatric disorders such as mood and anxiety disorders as well as schizophrenia is up to 56% higher in cities compared to rural environments (Peen et al., 2007; Purtle et al., 2019) (➔ chapter "The CITY & the PSYCHE—A Chronology" by Mehran, Kluge and Adli). Furthermore, being raised in a town has been found to double the risk of developing schizophrenia (Krabbendam & Van Os, 2005; Van Os et al., 2010; Vassos et al., 2012).

Thus, the first studies examining the association between the brain and the environment were conducted within psychiatry research, aiming to explore the potentially detrimental effects of cities on mental health (Lederbogen et al., 2011; Haddad et al., 2015). In a seminal study exploring the link between brain activity and the environment, Lederbogen et al. (2011) examined the differences in brain activity during social stress between individuals residing in urban and rural environments. Using fMRI, they discovered that urban residents, compared to rural dwellers, showed heightened activity in the amygdala, a brain region associated with stress (Roozendaal et al., 2009). Furthermore, the intensity of this association increased stepwise from rural residents to town residents and was the highest in urban dwellers. Interestingly, this effect was lateralised to the right amygdala, previously shown to be involved in processing social and environmental stress (Mothersill & Donohoe,

2016). These findings may be interpreted within the framework of prior research that has shown an association between amygdala hyperactivity and trait anxiety (Sehlmeyer et al., 2011) and depression (Siegle et al., 2007; Yang et al., 2010). Moreover, the authors examined whether early-life urbanicity i.e. years of upbringing in a city, town, or rural environment until age 15 correlated with neural activity during the social stress task. They found that early-life urbanicity was positively associated with activity in the pregenual anterior cingulate cortex (pACC), a brain region that regulates amygdala activity and is involved in stress processing (LeDoux, 2000; Diorio et al., 1993). Thus, participants raised exclusively in cities exhibited the greatest involvement of the brain region regulating stress-related brain activity, highlighting the importance of the upbringing environment for this stress-related neural circuit. This study indicates that individuals who currently reside or were raised in urban areas may be more susceptible to social stress, as evidenced by their heightened activity in the brain regions linked to stress, compared to those living or raised in rural areas (Lederbogen et al., 2011). Alternatively, it suggests that individuals with this susceptibility to stress may move more frequently to cities. The study's authors propose that social stress may contribute to the higher incidence of mental disorders in cities, partly accounting for the detrimental effects of the urban environment on mental health (Lederbogen et al., 2013). Contrary to the frequently used explanation that urban living negatively impacts mental health, environmental psychology suggests that these findings may also reflect the protective effects of natural environments on the brain activity of people residing in less urban areas.

In addition to amygdala activity, living in a city has been associated with activity in mesocorticolimbic brain regions. An fMRI study examined brain activity in participants (n = 147) who lived in either urban or less urban areas (Krämer et al., 2017). The results revealed that urban dwellers exhibited reduced activation and modulation capability of the left ventral tegmentum area (VTA), a key component of the brain's reward circuit. Previous research has reported that VTA dopamine neurons in rodents are reactive to stress (Marinelli & McCutcheon, 2014) and that chronic mild stress can decrease VTA dopamine activity (Chang & Grace, 2014). Moreover, urban residents exhibited heightened activation in the brain regions involved in both the reward system and stress regulation, including the amygdala, the medial orbitofrontal cortex, and the pACC, compared to dwellers in less urban areas. In the literature, dysregulation of the dopamine system is associated with conditions like depression and schizophrenia (Grace, 2016). In summary, these findings suggest that chronic stress, more prevalent in urban environments, may contribute to the dysregulation of the mesolimbic dopamine system. This could result in reduced reactivity to pleasurable stimuli and increase the vulnerability to mental disorders such as depression and schizophrenia (Krämer et al., 2017).

To further explore the relationship between urbanicity and neural activity in the clinical population, Lemmers-Jansen et al. (2020) set out to investigate the association between brain activity and early-life urbanicity in a trust game, comparing psychosis patients with healthy participants. Psychosis patients are known for their heightened sensitivity to stress. In the cooperative investment phase of the task, psychosis patients who were raised in highly urban areas showed a more substantial reduction in left amygdala activity compared to those who were brought up in less urban areas. The same trend was observed in healthy participants, although it was less pronounced than in the psychosis group. These results imply that growing up in less urbanised areas might serve as a protective factor, promoting cooperative behaviour and enhancing trust related to positive feedback, as observed from the amygdala activity. Furthermore, as the correlation with early-life urbanicity was more prominent in psychosis patients than in healthy individuals, the authors concluded that psychosis patients showed an increased vulnerability to the adverse effects of urban environments during childhood.

Since environmental factors during upbringing have been linked to stress-related neural activity (Lederbogen et al., 2011; Lemmers-Jansen et al., 2020), it has been investigated whether genetic factors play a role in modulating the environmental effects on brain activity. Building on the earlier study by their group (Lederbogen et al., 2011), Streit et al. (2014) investigated whether a genotype group could affect the association between urban upbringing and stress-related neural activity. Given the association of the neuropeptide S receptor gene (*NPSR1* rs324891) with stress- and anxiety-related phenotypes (Kumsta et al., 2013; Klauke et al., 2014), it was examined whether three different genotype groups in this gene modulated the relationship between early-life urbanicity and amygdala activity during social stress. The authors showed that the participants with genotype TT exhibited a positive correlation between early-life urbanicity and stress-related neural activity. Thus, the results from the previous study by Lederbogen et al. (2011) were replicated only in participants with the TT genotype but not in those with the AT and AA genotypes. These results provided evidence that genetic variation can modulate stress-related neural activity in an interaction with early-life urbanicity and, thereby, might contribute to a higher risk of developing mental disorders. A subsequent study found a similar modulation by genetic factors in the association between early-life urbanicity and brain activity in a working memory task (Reed et al., 2018). Although task performance and reaction times did not differ between participants raised in urban environments, towns, and rural areas, those raised in urban environments showed increased activity in the dorsolateral prefrontal cortex (dlPFC) compared to their counterparts raised in rural or town settings. This indicated that participants brought up in highly urban environments were less efficient in the same working memory task than those who grew up in less urban environments.

When examining the role of genetic factors, the study revealed that different dopamine genes modulated the association between early-life urbanicity and prefrontal brain activity in the working memory task. These studies (Streit et al., 2014; Reed et al., 2018) imply that gene–environment interactions may partially mediate the increased risk of developing mental disorders in an urban environment through the expression of stress-related and dopamine genes. They emphasise the importance of considering genetic factors combined with urban upbringing for understanding alterations in stress- and cognition-related brain activity in adulthood, which might be associated with an increased risk of developing mental disorders.

Besides task-based fMRI approaches, an extensive study involving adolescents and young adults from Europe and China, with a substantial study size of 3867 participants, investigated how population density is related to resting-state connectivity (Xu et al., 2022). Population density across each participant's lifespan was operationalised with a satellite-based measure using lifetime residential geographies. This measure was then associated with between- and within-network connectivity. The study revealed replicable and robust associations between population density and changes in within- and between-network connectivity. Specifically, higher density was correlated with lower and higher within-network connectivity of the anterior default mode network (aDMN) and cerebellar network, respectively. Moreover, population density was related to changes in the between-network connectivity of the aDMN with the cerebellar network, the executive control network (ECN), and the right frontoparietal network (FPN). Notably, these connectivity changes were predominantly observed in the aDMN which is interesting for several reasons. Firstly, the medial prefrontal cortex (mPFC), a core region of the aDMN, showed a decrease in grey matter volume with increased levels of population density. Furthermore, the mPFC and aDMN served as mediators in the association between population density and increased symptoms of depression in this sample. Lastly, decreases in mPFC volume have been correlated with stress exposure (Belleau et al., 2019). As discussed above, alterations in stress levels and associated brain activity present a promising candidate mechanism for explaining how environments with varying degrees of urbanicity impact the brain and mental health.

In summary, living in an urban environment can modulate the activity and connectivity in brain regions involved in stress and reward processing (Lederbogen et al., 2011; Krämer et al., 2017; Xu et al., 2022). As dysregulation of these systems has been linked to mental disorders like depression and schizophrenia (Siegle et al., 2007; Yang et al., 2010; Rauch et al., 2010; Pankow et al., 2013; Grace, 2016), they may constitute a neural mechanism through which urban living increases the risk of mental disorders in cities. Growing up in a highly urban environment has been associated with heightened neural activity during stress processing (Lederbogen et al., 2011) and altered brain activity in a trust game task with the latter effect being particularly pronounced in psychosis patients (Lemmers-Jansen et al., 2020). This may imply a heightened vulnerability to the detrimental effects of urban environments or, conversely, the beneficial effects of less urban influences during upbringing. Furthermore, the link between urban upbringing and brain activity may vary across different genotypes (Streit et al., 2014; Reed et al., 2018), suggesting why certain individuals are more prone to urban-related mental health risks. Lastly, different levels of urbanicity have been related to alterations in the functional connectivity patterns in brain areas and networks previously associated with depression, rumination, and stress (Xu et al., 2022).

Brain Function and Outdoor and Green Environments

Besides exploring brain activity and urbanicity, recent research has begun to examine which urban environmental features drive this association given that an understanding of the same is crucial for informing concrete steps for urban planning to improve mental health and brain development. Two studies have examined the association between brain activity involved in processing negative emotions and urban features such as green spaces and street network integration (Tost et al., 2019; Dimitrov-Discher et al., 2023).

The first study (n = 60) examined the association between real-time emotional responses while in urban green spaces and brain activity during a task evoking negative emotions (Tost et al., 2019). The authors initially demonstrated that participants who spent more time in urban green spaces during their daily activities showed more positive emotional responses. Furthermore, positive emotions when being in urban green spaces were negatively associated with activity in the dlPFC and ventromedial prefrontal cortex (vmPFC), brain regions linked to emotion regulation. This suggests that individuals exhibiting diminished activity in these regulatory prefrontal regions experienced greater emotional benefits from green spaces. In accordance with this finding, hypoactivation in the dlPFC has been observed in individuals with an increased risk of depression and anxiety disorder (Opel et al., 2017). Furthermore, in this study, dlPFC hypoactivation was observed in a task eliciting negative emotions. This implies that green spaces may act as a protective factor by compensating for diminished regulatory prefrontal capacity. The study also revealed a negative correlation between green space exposure and emotional benefits from green spaces, specifically in neighbourhoods with high psychiatric risk. This suggests that those city dwellers who habitually spend time in environments with less green space and with a heightened risk of mental illness may particularly benefit from urban green spaces. These findings emphasise the potential value of green areas in nature-based interventions aimed at enhancing mental health, especially in urban settings with limited natural environments.

Another study (n = 42) explored the correlation between street integration for the street closest to the participant's address, as well as the participant's street integration in their neighbourhood, and brain activity during a social stress task (Dimitrov-Discher et al., 2023). Although the authors observed no association between brain activity and street integration of the closest street, they found that more integrated street networks in the participants' neighbourhoods were linked to lower neural activity during social stress. The most involved brain regions (dorsal anterior cingulate cortex (dACC), insula, thalamus, caudate putamen, precentral gyrus) are involved in the salience network (SN), active during the detection of salient stimuli and threats. Furthermore, the authors found activity differences in the precuneus, supramarginal gyrus, and parahippocampal cortex, all associated with the default mode network (DMN). Although the DMN is related to many functions, a major function ascribed to it is self-referential thought and rumination (Muscatell et al., 2015). Thus, the authors proposed that strong social support, which is assumed to be higher in neighbourhoods with highly integrated streets, might attenuate reactivity to threats, resulting in decreased activity during the stress task, including brain regions within the DMN and the salience network, with the former linked to rumination and rejection and the latter to threat-related stimulus responses. Alternatively, the authors suggested that living in an area with integrated streets may include everyday exposure to many different stimuli (such as busy streets with people and traffic). This could lead to habituation and less responsivity to threat-related stimuli, mirrored in the attenuated activity of the reported regions in those participants who live in highly integrated streets (Dimitrov-Discher et al., 2023).

Studies on urban environment features have found that green spaces in cities and highly integrated street networks may be beneficial for neural activity while processing negative emotions (Tost et al., 2019; Dimitrov-Discher et al., 2023). However, these studies, alongside those exploring the link between brain activity and urban living, are cross-sectional. Consequently, they lack the capacity to establish a cause-and-effect relationship. The results should be interpreted cautiously, considering the possibility of a reverse relationship, namely, that people with particular neurological characteristics choose to live in specific environments (a phenomenon summarised under the term "selective migration").

Urban and Natural Environments in fMRI Lab Studies

To achieve greater methodological control and to facilitate the inference of causality, several studies moved to empirical investigations within laboratory settings. The initial studies in these controlled settings focused on differences in brain activity while being exposed to visual stimuli, primarily photos, depicting different environments, with a focus on natural and urban settings.

One of the initial studies demonstrated that while viewing rural versus urban photos, the anterior cingulate cortex (ACC) and the basal ganglia were activated among other regions (Kim et al., 2010). The authors proposed that ACC activity, while viewing rural photos, was related to the cognitive and emotional processing of the pictures, whereas the activity in the basal ganglia may be linked to the processing of positive emotions. In contrast, viewing urban photos was predominantly linked to activity in the temporal cortex (hippocampus and parahippocampal gyrus), the lingual gyrus, and the amygdala. The authors suggested that when viewing urban photos, the hippocampus and parahippocampal gyrus, the lingual gyrus, and the amygdala were activated since they are involved in processing negative and threat-related stimuli. A follow-up study showed that when viewing uncomfortable urban photos, neural activity was observed in the hippocampus, the parahippocampal gyrus, and the amygdala (Kim & Jeong, 2014), consistent with the previous findings in response to urban photos (Kim et al., 2010). Furthermore, the authors interpreted heightened activity in the mPFC, the ventrolateral prefrontal cortex (vlPFC), the ACC, and the insula as cognitive control of emotions during uncomfortable stimuli. Conversely, comfortable scenes were related only to the activation in the calcarine gyrus, a region related to visual processing. However, these findings were constrained by the small sample size of only 13 young adults, warranting cautious interpretation. Interestingly, similar brain activity patterns emerged when only imagining non-beautiful environments versus beautiful environments (Vedder et al., 2015). Namely, during the mentalisation of unpleasant scenes, heightened activity was detected in the hippocampus, amygdala, mPFC, and ACC, those brain regions that were previously involved in processing uncomfortable urban photos (Kim & Jeong, 2014). Additionally, when participants were asked to visualise non-beautiful stimuli, their orbitofrontal and insular cortex were involved. The authors proposed that when responding to aversive stimuli, additional neural engagement is needed in these brain regions for emotion processing and cognitive control (Vedder et al., 2015). Nonetheless, the study's sample size was relatively small, comprising only 16 participants, suggesting that the conclusions drawn from this study should be handled with caution.

Some studies suggested that viewing urban and natural environments is related to not only distinct brain activity but also distinct neural pathways. In a study in which the participants ($n = 31$) observed pleasant, neutral, and unpleasant scenery photos (Seiyama et al., 2018), viewing pleasant environments was associated with activity changes in the dorsal pathway, whereas the ventral pathway was implicated when watching unpleasant scenery. Since the dorsal pathway is related to spatial location processing and the ventral pathway to object recognition, the authors proposed that pleasant landscapes may involve spatial processing (e.g. depth and location), whereas participants may be more focused on object recognition (e.g. colour and shape) when viewing unpleasant scenes. Additionally, the authors suggest that the observed increase in hippocampus activity when viewing unpleasant, urban photos, which were familiar to the participants, may reflect memory processes. In another study ($n = 24$), viewing low restorative photos was linked to activity in the cortical areas implicated in top-down cognitive control, whereas viewing highly restorative photos was linked to activated areas important for bottom-up, involuntary attention (Martínez-Soto et al., 2013). The authors interpreted activity in these brain regions as indicators of top-down, voluntary attention when viewing photographs of man-made structures and of involuntary attention when observing highly restorative photos of nature. The authors further suggested that these results align with attention restoration theory (ART), which states that urban environments elicit voluntary attention, whereas natural environments restore attentional capacities by capturing involuntary attention (Kaplan, 1995).

To further understand neural responses to features of natural environments, Tang et al. (2017) examined brain activity in 31 participants while presenting photos of an urban environment and different natural environments. When viewing urban photos, the regions involved in basic visual processing (cuneus) and attention (precuneus and right cingulate gyrus) were more activated. The authors proposed that this might be the case because urban sceneries may demand higher attentional focus, as previously suggested by Martínez-Soto et al. (2013) and Vedder et al. (2015) when imagining aversive photos and in line with the ART. In another study, the participants ($n = 44$) observed photos with different urban green densities. They showed activity in the regions involved in spatial processing (superior parietal lobule and superior temporal gyrus), executive attention (superior parietal lobule, precentral gyrus, and medial frontal gyrus), and early visual processing (cuneus) (Chang et al., 2021). The authors also interpreted the involvement of attention-related brain regions when viewing urban green photos in the scope of ART, thus highlighting nature's capacity to induce attention restoration. After accounting for low-level visual features, by contrasting the photos with their scrambled versions, green density was associated with brain activity in the cuneus and ventral posterior cingulate cortex (PCC) as well as with the stress ratings of the photos. Since the ventral PCC (vPCC) is connected to the limbic structures and is associated with spatial memory and stress modulation, the authors proposed that it is the key region for moderating stress-related response to urban photos. Furthermore, the early engagement of the vPCC when viewing photos with less greenery was interpreted by the authors in the context of ART, thus suggesting that environments with lower green density demand more attentional resources (Kaplan, 1995).

While the aforementioned research employed task-based designs, three notable studies adopted resting-state approaches to examine neural responses during exposure to natural stimuli. A study from our group on exposure to nature and urban stimuli, matched for ratings of scenicness, revealed changes in functional connectivity corresponding to the two sets of photos (Kühn et al., 2021b). Viewing natural photos resulted in heightened connectivity between the DMN and the dorsal attention and somatomotor network as well as higher correlations between the ventral attention network (VAN) and dorsal attention network (DAN). Interestingly, city upbringing moderated the effects of nature. Reduced connectivity between these networks while watching nature correlated with a longer duration of participants' upbringing in cities. These findings resonate with cross-sectional findings linking differences in lifetime exposures to urbanicity with changes in between-network connectivity involving the DMN (Xu et al., 2022). However, the pattern of results is still inconclusive. For instance, another study examining alterations in connectivity patterns within and between networks (including the DMN) did not find supporting results (Gould van Praag et al., 2017). Participants were presented with naturalistic versus artificial soundscapes at rest. Although no changes in between-network connectivity were observed, listening to naturalistic soundscapes was associated with alterations in within-network connectivity in the DMN. The relatively stable involvement of the DMN across the studies indicates that exposure to natural stimuli might impact processes related to self-referential thought, mind-wandering, or attention. However, further research is needed to shed light on the robustness and underlying mechanisms of these results. Another study from our group also examined brain functional connectivity when listening to natural and urban soundscapes (Stobbe et al., 2023). Listening to urban sounds led to higher brain entropy in the posterior cingulate gyrus, cuneus, precuneus, and occipital lobe compared to natural sounds, possibly reflecting the unpredictability of noise in an urban environment. Furthermore, higher brain entropy when listening to urban sounds was linked to a stronger decrease in positive mood, suggesting that increased brain entropy may be a potential negative consequence of exposure to urban sounds. Additionally, higher functional connectivity was found between areas in the auditory, cinguloopercular, and somatomotor head and mouth networks when listening to natural compared to urban sounds, which correlated positively with improved working memory capacity during exposure to natural sounds. These findings imply that the gain in working memory performance is associated with higher connectivity with, among others, auditory functional brain networks while listening to nature sounds. However, it is important to note that the correlations between functional connectivity and behavioural data do not withstand multiple comparison corrections, and the results should, therefore, be treated with caution.

In summary, the above-mentioned studies have shown distinct neural activity when individuals merely viewed photos of urban versus less urban environments (Kim et al., 2010; Kim & Jeong, 2014). Moreover, research has suggested that viewing urban and natural environments is related to distinct neural pathways (Seiyama et al., 2018; Martínez-Soto et al., 2013). Several studies interpreted heightened neural activity in regions related to attention processing within the scope of the ART (Martínez-Soto et al., 2013; Tang et al., 2017; Chang & Grace, 2014; Vedder et al., 2015). Additionally, viewing natural stimuli enhanced the connectivity between several brain networks, and this effect was moderated by urban upbringing, with less time spent in urban environments during upbringing linked to higher connectivity (Kühn et al., 2021b).

The studies reviewed have revealed the diverse effects of short-term exposure to natural stimuli on brain activity and connectivity. However, the reported findings are highly heterogeneous, likely due to the considerable variation in the stimuli across studies. Methodological considerations, such as choosing adequate comparisons (e.g. by controlling for low-level properties of the different stimuli), are essential to further assess which mechanisms are responsible for the observed differences. In addition to the heterogeneity of the results, the interpretability of these findings is further limited by the elevated risk of reverse inference (Poldrack, 2006). When exposed to natural stimuli, changes in a certain brain region do not automatically imply that a particular psychological process is at play. In conjunction with the experimental tasks associated with specific psychological processes, interventional approaches provide a solution to overcome these shortcomings (Roberts & Christopoulos, 2018). For instance, demonstrating that a nature walk is linked to decreased amygdala activity during a stress task enables a more confident conclusion that nature diminishes the stress-related response compared to simply observing changes in "stress-related" brain regions while viewing pictures of nature.

Urban and Natural Environments in Experimental fMRI Intervention Studies

Although the MRI methodology offers high spatial resolution and can measure activity in deeper brain structures, its immobility prevents the measurement of brain activity in a natural or an urban environment. However, to investigate the causal link between the environment and brain activity, researchers have begun conducting fMRI intervention studies, in which brain activity is measured both before and after the intervention, namely, the exposure to natural and urban settings.

The first fMRI intervention study examined neural correlates of rumination after a 90-minute walk in an urban (a busy street in Stanford) versus a natural environment (an urban green space; Bratman et al., 2015). To explore how urban living may be connected with the onset of mental disorders, Bratman an colleagues first examined whether rumination, a harmful self-focused thought pattern linked to an increased risk of depression and other mental health issues, would decrease after a walk in nature. Indeed, the self-reported levels of state rumination ($n = 38$) showed a reduction in negative self-referential thoughts following the nature walk. The authors used a neuroimaging method known as arterial spin labelling (ASL) to measure brain activity. ASL measures regional cerebral blood flow in specific regions of interest. This method enables the measurement of brain activity associated with enduring mental states like rumination. Notably, the authors observed diminished activity in the subgenual prefrontal cortex (sgPFC), a brain area associated with self-focused behavioural withdrawal linked to rumination. These findings imply that the reduction in rumination, associated with sgPFC activity, might be a key neural mechanism through which exposure to nature contributes to mental health or, conversely, through which urbanisation, by limiting access to natural environments, may be linked to a higher prevalence of mental disorders.

Another proposed connection between urban environments and a heightened risk of mental disorders in cities is social stress, which has been referred to as a typical characteristic of urban settings (Lederbogen et al., 2011; Lederbogen et al., 2013; Haddad et al., 2015). Therefore, in a different fMRI intervention study ($n = 63$), we explored how short-term exposure to a natural versus urban environment affects stress-related brain activity (Sudimac et al., 2022). We employed an fMRI paradigm that elicits activity of the amygdala, a brain region related to stress processing and urbanicity (Lederbogen et al., 2011), both before and after a 60-minute walk in either an urban (a busy street in Berlin) or a natural environment (an urban forest). We found that amygdala activity remained stable after the walk in the urban environment. However, it decreased after the walk in nature. These results suggest that exposure to nature—even for only an hour—can be beneficial for stress-related activity. The findings align with Bratman et al.'s (2015) study in which the walk in nature was beneficial for rumination and brain activity linked to rumination, but no changes were observed after the walk in the urban environment (Bratman et al., 2015). Notably, the effect was driven by the activity in the right amygdala, aligning with a previous study showing that the association between amygdala activity during social stress and living in a city was also lateralised to the right amygdala (Lederbogen et al., 2011). Interestingly, reduced amygdala activity after the walk in nature was more prominent in women (Sudimac & Kühn, 2022), highlighting the importance of examining interindividual differences when investigating the impact of the environment on the brain. The results of this study may indicate a potential mechanism behind the previously reported association between urbanicity and stress-related activity (Lederbogen et al., 2011). As a 60-minute walk can decrease amygdala activity, repeated exposures to the natural environment may heighten the amygdala's activation threshold. As a result, people who live in rural areas, typically surrounded by nature, may show lower amygdala activity during social stress than urban dwellers, as previously reported by Lederbogen et al. (2011).

In summary, studies examining the causal relationship between in situ exposure to environments and brain activity are scarce. Current evidence indicates that even one walk in nature can decrease the neural activity associated with rumination and stress, whereas walking in the urban environment does not increase activity in these brain regions (Bratman et al., 2015; Sudimac et al., 2022). This implies that the effect is driven by the beneficial effects of nature rather than by the detrimental impact of an urban environment. Nonetheless, it is important to consider that both studies were conducted in an urban setting and in participants who reside in a city, which may explain the lack of heightened neural activity after exposure to a walk in the urban environment. Future studies should, therefore, explore the impact of natural and urban environments on individuals from diverse backgrounds, ranging from rural to highly urbanised areas. Moreover, the duration of nature's beneficial impact on neural activity and the specific environmental features responsible for this effect remain to be established.

Brain Function and Polluted Environments

Air and noise pollution are associated with various health factors, including structural and functional changes in the brain. For example, chronic noise exposure has been demonstrated to affect the neurohormonal stress response both directly and indirectly through changes in sleep and activity patterns or elevated levels of depression (Hahad et al., 2022). Furthermore, specific air pollutants can directly cross the blood–brain barrier, causing neurotoxic effects by elevating the levels of pro-inflammatory cytokines (Kang et al., 2021). These are just two of many possible mechanisms by which pollutants and brain function might be related (Hahad et al., 2022; Block et al., 2012). Neuroimaging studies are only beginning to uncover these associations at the level of altered brain function and structure (see also section "Brain Structure and Polluted Environments"). Most studies using fMRI to relate air and noise pollution to the brain investigated alterations in resting-state connectivity. This research focuses on the extent to which pollution alters connectivity within and between resting-state cortical networks.

Brain Function and Air Pollution

One of the first studies investigating the association between air pollution and brain function using resting-state analysis was conducted in school-aged children (Pujol et al., 2016b). The authors computed a general indicator of traffic pollution by combining measurements of NO_2 (nitrogen dioxide) and $PM_{2.5}$ (particulate matter) assessed through aerosol sampling in children's classrooms and playgrounds and analysed its relation to brain structure and functional connectivity during rest. Both substances represent pollutants that are monitored and regulated by air quality standards in various countries. Testing 263 school-aged children (8–12 years of age), they found that higher air pollution was associated with decreased within-network connectivity and increased between-network connectivity of the DMN with the lateral operculum, which the authors interpreted as an indicator of lower network segregation. Increases in connectivity between networks associated with higher air pollution were confirmed by a second study investigating children from a similar age group (Pérez-Crespo et al., 2022). The study estimated levels of NO_x (nitrogen oxide), NO_2, and PM from the home addresses of 2197 children for different time periods. It revealed that higher levels of air pollutants were associated with elevated functional connectivity between various networks. The majority of increased between-network connections included the DMN and the task-positive network, a network usually active during tasks demanding high levels of attention. These differential alterations in between- and within-network patterns are interesting because they mirror the changes associated with neurotypical development. For instance, it has been shown that within-network connectivity from childhood to early adolescence increases for networks such as the DMN, FPN, or dorsal attention network (DAN). Furthermore, for the same developmental period, between-network patterns involving the DMN are characterised by decreased connectivity (Sanders et al., 2023). These age-related changes in brain maturation oppose the impact of air pollution reported above. Corroborating this interpretation, Pujol et al. (2016b) showed that their participant's age had an opposite effect on within- and between-network connectivity patterns compared to air pollution, thus suggesting a potential association between air pollution and disrupted brain maturation.

However, these results and the interpretation that they inversely resemble age-related trajectories in brain development have been challenged as being too coarse by a current longitudinal study (Cotter et al., 2023). The authors analysed data from 9497 children to investigate how exposure to varying levels of air pollution over the period of 1 year relates to changes in resting-state connectivity in a 2-year follow-up scan. They found that exposure to air pollutants was associated with both increases and decreases in within- and between-network connectivity and that the patterns depended on the specific pollutant and network investigated. For instance, while higher levels of O_3 (ground-level ozone) were related to an increase in within-network connectivity of the DMN, higher NO_2 was associated with decreased connectivity within the FPN. Similarly, higher $PM_{2.5}$ levels were related to elevations of between-network connectivity of the DMN–FPN and the DMN and salience network (SN), whereas increased levels of NO_2 showed decreased connectivity between the FPN–SN and FPN–DMN. The authors concluded that the relationship should be interpreted more nuancedly and that factors such as timing of exposure and chemical composition of the pollutants must be considered. Due to the age-dependent nature of the periods of plasticity in different brain regions and neurodevelopmental processes, air pollutants might have distinct effects on the maturing brain depending on the specific exposure time window across an individual's lifespan.

The previously mentioned observational studies have explored the impact of estimated or directly measured pollution levels over extended periods in different environments. Nevertheless, air pollution is correlated with several additional environmental factors that might have contributed to the reported associations. To increase the magnitude of experimental control, a recent study has exposed participants to diesel exhaust (composed of $PM_{2.5}$) and compared its impact on resting-state connectivity with a control condition using filtered air for the same duration (Gawryluk et al., 2023). Before and after the exposure, during which participants exerted light physical effort for 2 hours, within-network connectivity was assessed in the DMN. The study revealed no change in the within-network connectivity patterns before and after the exposure to air pollution but an increase in connectivity within the DMN in the filtered air condition. Since increased functional connectivity in the DMN is typically understood as a beneficial effect of exercise, the authors concluded that the benefits associated with light physical effort are disrupted when exposed to polluted air. However, whether and how these results might relate to long-term chronic exposure to air pollution remains to be tested.

Brain Function and Noise Pollution

Like the studies on air pollution, most fMRI studies on noise pollution have investigated the relationship with functional connectivity measures. While environmental noise can originate from various sources, traffic is the main contributor to noise pollution in urban areas (Morillas et al., 2018). Therefore, it is not surprising that the few existing studies to date have focused on the association of traffic noise with brain function. A study analysing exposure to chronic traffic noise in 574 older adult participants reported small associations between noise and connectivity measures (Glaubitz et al., 2022). The authors found that higher noise levels were associated with decreases in within-network connectivity of the visual network and elevations in the between-network connectivity involving the DAN. Furthermore, more noise was related to lower network segregation in the FPN and visual network, indicating that the results were driven by a pattern of low within- and high between-network connections. Similar to the interpretation of air pollution described above, the authors hypothesised that this rise in network segregation and the decline in network integration mirrors age-related changes in functional brain organisation. Thus, according to the study, exposure to noise (and air) pollution might contribute to accelerated neural and cognitive ageing in older adults.

An accelerated brain maturation pattern linked to increased noise exposure was also observed in research investigating 229 children (Martínez-Vilavella et al., 2023). In this study, the noise levels in classrooms of school-aged children (8–12 years) were measured and associated with connectivity data from the inferior colliculus, the medial geniculate body of the thalamus, and the primary auditory cortex, all of which are part of the auditory pathway. Focusing on distinct elements of the auditory pathway, the study aimed to gain a more nuanced understanding of the influence of noise on the brain. Higher noise was associated with increased connectivity within the subcortical regions of the pathway (i.e. the inferior colliculus and thalamus). As this increased functional coupling is linked to age-related brain maturation, the authors concluded that increased noise speeds up the maturation process, potentially resulting in faster but less complete development of the auditory system. These effects were specific to changes in subcortical regions and could not be found for cortical areas. This absence of changes in cortical functional coupling between brain regions is in line with an investigation that found no differences in the connectivity patterns of school-aged children in relation to noise levels in their homes (Pérez-Crespo et al., 2022).

Taken together, the studies on air and noise pollution suggest that both factors influence connectivity patterns in children and adults. Some of the studies showed that estimates of noise and air pollution are characterised by within- and between-network connectivity alterations that resemble age-related trajectories, suggesting that pollution might modify maturational processes (Martínez-Vilavella et al., 2023; Glaubitz et al., 2022; Pujol et al., 2016b; Pérez-Crespo et al., 2022). However, this conclusion has been challenged by other findings showing a more complex or absent pattern of results (Cotter et al., 2023; Pérez-Crespo et al., 2022). Although the number of studies is still limited, it has been suggested that these discrepancies might be due to differences in the timing of exposure to and composition of pollutants (Cotter et al., 2023). Especially regarding children and adolescents who are characterised by complex and dynamic changes in brain structure and function, the timing of exposure might be vital to gaining more detailed insights into the mechanisms at play. Furthermore, apart from one study (Gawryluk et al., 2023), all research conducted so far used correlational approaches. Since noise and air pollution are inherently linked and associated with additional confounding variables, great care should be taken to control for their influence in future studies.

Summary of Brain Function and the Physical Environment

To summarise, existing research indicates that our surrounding environment impacts brain activity. Growing up and living in an urban environment, the presence of green spaces and street integration in a city seems to moderate neural activity (Lederbogen et al., 2011; Krämer et al., 2017; Lemmers-Jansen et al., 2020; Tost et al., 2019; Dimitrov-Discher et al., 2023). Additionally, specific characteristics of the physical environment, such as air and noise pollution, are associated with the connectivity of several brain networks (Pujol et al., 2016b; Cotter et al., 2023; Glaubitz et al., 2022; Martínez-Vilavella et al., 2023). Furthermore, visual stimulation with photos of urban and natural environments can elicit distinct patterns of brain function (Kim et al., 2010; Kim & Jeong, 2014; Seiyama et al., 2018; Martínez-Soto et al., 2013; Tang et al., 2017). Studies examining the causal link between the physical environment and the brain showed that short-term exposure to nature, such as a walk, can decrease neural activity related to rumination and stress (Bratman et al., 2015; Sudimac et al., 2022).

Structural MRI Studies

Beyond its capacity to scrutinise brain activation, an MRI allows for the non-invasive and in vivo examination of the brain's anatomy (➔ chapter "Mapping the Human Brain with Computational Anatomy" by Draganski, Joost and Kherif), thus facilitating the investigation of how our brain is structurally associated with our physical environment. The brain encompasses two primary tissue types that can be visualised by MRI. Grey matter (GM) consists of nerve cell bodies, dendrites, and unmyelinated axons and receives its name from the greyish colour when dissected. To further investigate the shape of the cortex, its surface is frequently described by means of thickness, surface area, and convolution i.e. gyrification. White matter (WM) primarily consists of nerve fibres. To delve into its microstructural properties, specialised MRI techniques like diffusion tensor imaging (DTI) come into play. DTI measures the magnitude and directionality of water diffusion within white matter. The characterisation of white matter's diffusion profile employs metrics such as fractional anisotropy (FA) for overall directionality and mean diffusivity (MD) to quantify the magnitude of water diffusion within the brain tissue. Structural MRI scans further provide the means to investigate ventricles, the fluid-filled cavities within the brain.

Brain Structure and Urban Environments

Haddad et al. (2015) were the first to investigate how being raised in an urban environment may affect brain structure after research had previously only focused on the effects of brain function. They defined early-life urbanicity as the reported years that participants lived in cities, towns, and rural regions for the first 15 years, employing population density as a proxy for urbanicity. Their findings unveiled an inverse association between early-life urbanicity and grey matter volume (GMV) in the dlPFC and pACC (observed exclusively in men; $n = 110$).

This finding was replicated in a larger cohort ($n = 290$), corroborating that adults raised in an urban environment exhibited reduced grey matter volume in the bilateral dlPFC (Lammeyer et al., 2019). Furthermore, authors from this study uncovered an inverse correlation within the right inferior parietal lobe (IPL), accompanied by a positive association with grey matter volume in the left precuneus. Therefore, the authors concluded that being raised in an urban setting has adverse effects on frontal brain structure and that this may be caused by the social stress experienced due to cities' complex environments, as the dlPFC is highly sensitive to stress exposure (Luethi et al., 2009; Qin et al., 2009). However, these types of cross-sectional studies can only hint at the potential underlying causal effects but do not provide evidence.

In contrast, a recent study by Zhang et al. (2023) has found a negative relationship between early-life urbanicity and dlPFC volume, concluding that urban upbringing benefits brain structure. They further reported that dlPFC volume mediated between childhood urbanicity and cognition.

Moreover, a cross-continental study (Asian adults: $n = 2176$; European adults: $n = 415$) revealed a negative correlation between mean urbanicity before the age of 18 and mPFC volume, alongside a positive association with cerebellar volume (Xu et al., 2022). These results are not only in accordance with previous studies but further suggest that the association between urbanicity and brain morphology may be robust across geographies and sociocultural conditions. This finding could be replicated in another Asian study ($n = 490$, Zhang et al., 2021). Moreover, Xu et al. (2022) reported that these differences in mPFC volume were primarily determined by cortical surface area and not cortical thickness (the shortest distance between white matter and the pial surface). However, this contradicts a study that revealed that adults ($n = 85$) who had been raised in urban or rural settings during their first 15 years of life had thinner cortices in multiple regions, including the left dlPFC, bilateral mPFC, left superior temporal cortex, and left parahippocampal cortex (Besteher et al., 2017). More studies must be conducted to understand whether differences in grey matter volume stem from alterations in cortical thickness or surface area.

While our lab did not replicate either of these findings, we found that 14-year-old adolescents ($n = 106$) raised in cities displayed reduced volumes in the left hippocampal formation, coupled with worse performances in spatial processing (Kühn et al., 2020). This may suggest that rural environments offer more opportunities for developing spatial skills compared to urban environments, mediated by the hippocampal formation.

As of now, the exploration into the correlation between overall urban living and the second type of brain tissue—white matter structure—remains relatively limited. One DTI study revealed inverse associations between early-life urbanicity and white matter anisotropy of the left superior longitudinal fasciculus (SLF), the left corticospinal tract, and the left superior corona radiata (Lammeyer et al., 2019). In contrast, another study found no significant correlations between brain fractional anisotropy, a measure of white matter integrity, and the current urbanicity level (Xu et al., 2022).

The collective findings from these studies suggest a potential negative impact of being raised in urban environments on the volume of the dorsolateral and medial prefrontal cortices (PFCs). While the lateral PFC likely plays a crucial role in self-regulation (Cohen & Lieberman, 2010), the medial PFC may exert a monitoring function, such as action conflict, decision uncertainty, and reward expectations (Botvinick et al., 2001; Duverne & Koechlin, 2017; Ridderinkhof et al., 2004). Research suggests that the bigger the prefrontal cortices, the better an individual's cognitive abilities (Basten et al., 2015). At this point, whether there are also volume differences of other grey matter structures, such as the hippocampus, cortical surface characteristics, or white matter, is uncertain. Moreover, the research by Xu et al. (2022) highlights an important aspect: the relationship between rural/urban upbringing and brain structure might not solely hinge on how densely populated an area is. They reported that brain differences were driven by the proportion of built-up, agricultural land, and night-time light.

Brain Structure and Outdoor and Green Environments

One way to characterise the appearance of a city is to investigate its characteristics such as green spaces, forest density, water, and neighbourhood walkability. Although there is accumulating evidence of the positive effects of green spaces on mental health, the investigation of their effects on the brain has only recently attracted attention. The first to study how exposure to greenness during upbringing is associated with brain structure were Dadvand et al. (2018). They investigated how lifelong green exposure i.e. the satellite-based normalised difference vegetation index (NDVI) was associated with grey matter volume in primary school children living in Barcelona (n = 253). They reported a positive association with the prefrontal cortex and the left premotor cortex, overlapping with brain clusters associated with working memory and attentiveness. Additionally, adjacent white matter regions were also affected. In accordance with the findings from Barcelona (Dadvand et al., 2018), our lab observed a positive association between lifelong open green space coverage and grey matter volume in the bilateral PFC in young adolescents (n = 95, Kühn et al., 2023). Additionally, tree cover density was negatively associated with multiple clusters in bilateral lateral and medial PFCs. Lastly, we evaluated the visibility of the sky i.e. how much sky one can see from an eye-level perspective and found a positive association with clusters in the lateral and medial PFCs. All three analyses overlapped in one region: the medial PFC.

While the studies mentioned above focused on upbringing in a particular environment, other studies set out to understand the link between brain structure and an individual's acute living environment i.e. the environment in which they reside or spend their time, independent of their upbringing. Overall, the majority of these types of studies examine older cohorts. This allows for a more reliable investigation of long-term exposure to a certain environment as older adults tend to reside longer at one location and also spend more time in their residential areas compared to younger adults. For example, we found that older adults (n = 341) who lived near forests exhibited higher amygdala integrity (Kühn et al., 2017). Amygdala integrity was defined as a latent factor built from grey matter volume, mean diffusivity, and the magnetisation transfer ratio. Unlike studies exploring urban upbringing and brain structure, no significant connections emerged for prefrontal regions nor did other land categories such as urban green, water, and wasteland show notable associations. Within the same sample, we examined urban landscapes, delineating areas tied to industry, commerce, transportation, and urban green spaces like parks and gardens, around the participants' home addresses (n = 207, Kühn et al., 2021a). We found that participants living in a part of their town with more urban areas exhibited reduced grey matter volume in the pregenual ACC, extending into the subgenual ACC, a region within the prefrontal cortex. Likewise, an overlapping brain cluster indicated higher grey matter volume among those surrounded by more urban greenery. This study revealed that urban greenery could explain variance in grey matter volume above and beyond urbanicity. Volume reductions in the amygdala and inferior frontal regions were also found in participants who lived in an environment dominated by green and blue spaces and at long distances from waste and energy facilities (Xu et al., 2023). These volume reductions appeared to play a role in mediating anxiety symptoms among the participants. The authors hypothesised that emotional regulation mechanisms might underpin this association.

While all studies previously mentioned in this section investigated the acute environment as a function of land use around individuals' home addresses, Cerin et al. (2017) assessed how neighbourhood walkability was associated with levels and brain volume changes in older adults (n = 146). They reported that the more walkable the residential environment, the larger the participants' right hippocampal volume. The authors concluded that long-term exposure to spatially and visually complex surroundings may spark neurogenesis in hippocampal regions due to the cognitive processes required to navigate these environments. Additionally, they found an association between walkable neighbourhoods and smaller ventricular volume. Ventricular enlargement is considered a measure reflecting global brain atrophy in older individuals. However, these findings could not or only borderline be replicated in two similar-aged samples (n = 1125 and n = 228) for neighbourhood greenness—neither for hippocampal nor ventricular volume (Besser et al., 2021; Crous-Bou et al., 2020).

Except for one (Xu et al., 2022), all aforementioned studies investigating green and outdoor environments focused on older participants. However, we were also interested in how the outdoors and natural environments may be linked to brain structure in young adults. To that aim, we implemented a longitudinal design, scanning 6 young adults 40–50 times over 6–8 months (Kühn et al., 2022). We found that the time spent outdoors 24 hours before the MRI acquisition was positively associated with grey matter volume in the right dlPFC. The results remained significant after controlling for physical activity, fluid intake, free time, and hours of sunshine. Furthermore, preliminary evidence from an intervention study shows that as little as a 1-hour walk in nature can increase the volume in the subiculum, a hippocampal subfield involved in stress response inhibition (n = 60, Sudimac & Kühn, 2024). This volume change was associated with a decrease in self-reported rumination. This may support the potential role of nature-induced changes in brain structure in influencing cognitive processes related to stress regulation and mental well-being.

To our knowledge, only two studies have investigated how acute greenness could be linked to cortical thickness. They reported that the presence of more green spaces around participants' residences was linked to thicker cortices in areas that have been shown to be involved in Alzheimer's disease (Crous-Bou et al., 2020) and in the parietal and occipital regions in an urban population (Min et al., 2021). These findings hint at a potential protective role of green surroundings in cortical thickness, potentially explained by reduced chronic stress (Lucassen et al., 2014; Radley et al., 2015).

In summary, although these studies used vastly different definitions of greenness and the outdoors, multiple studies have reported variations in prefrontal cortices' grey matter volumes (e.g. Dadvand et al., 2018; Kühn et al., 2023). This could mean that the link between urban upbringing and reduced prefrontal volume may be better explained by the lack of (open) green spaces in cities and may not only be a function of a densely populated area. Furthermore, walking in nature and residing in walkable neighbourhoods may increase hippocampal volume (Cerin et al., 2017; Sudimac & Kühn, 2024). Merely having a green neighbourhood will not suffice to trigger volume changes in the hippocampus (Besser et al., 2021; Crous-Bou et al., 2020). This is in accordance with the general role of the hippocampus involving spatial memory and representation (Lisman et al., 2017). In general, the brain regions from this body of research overlap with the brain areas identified in research on global urbanicity measured with population density. In conclusion, living in greener, walkable cities and spending time outdoors appear to have a positive impact on brain structure.

Brain Structure and Polluted Environments

Brain Structure and Air Pollution

Inversely associated with residential green and natural environments is air pollution (Barwise & Kumar, 2020). Until now, the impact of air pollution on brain structure has received the most comprehensive research attention within the field of urban environmental studies. It was established that air pollution may affect brain structure as early as in the womb. Specific components of the air can cross the placenta and blood–brain barrier (reviewed by Brown et al., 2007), and, as the brain is still rapidly developing during pregnancy and the first years of life, even low levels of exposure may cause permanent changes in the brain (Grandjean & Landrigan, 2014; Rice & Barone Jr, 2000). Therefore, prenatal and early postnatal periods are considered the vulnerable windows for effects of environmental exposures (Grandjean & Landrigan, 2014).

White Matter

One of the toxic components of $PM_{2.5}$ are polycyclic aromatic hydrocarbons (PAHs), which are released into the air during incomplete combustion of fossil fuels, tobacco, and other organic materials (Bostrom et al., 2002). They are considered neurotoxic, particularly for the developing foetus. Peterson et al. (2015) were the first to investigate the effects of prenatal exposure in humans based on a neuroimaging investigation of 40 children from New York City. They found a dose–response relationship between increased prenatal PAH exposure (measured from personal air samples of $PM_{2.5}$ during pregnancy) and white matter surface reductions. These reductions globally affected almost the entire left hemisphere. These findings were independent of the levels of air pollution encountered during childhood. After this initial study, more publications on prenatal exposure to air pollutants followed. However, multiple studies failed to find white matter variations following prenatal exposure to air pollutants—both in childhood (Guxens et al., 2018) and during the first year of life (Bos et al., 2023). Nevertheless, volume reductions in one white matter structure have been highlighted in children (Mortamais et al., 2019; Guxens et al., 2022) and adolescents (Lubczyńska et al., 2021) after $PM_{2.5}$ exposure during the final trimester of pregnancy: the corpus callosum. The corpus callosum is the largest WM bundle in the brain, consisting of mostly myelinated axons and connecting our two hemispheres (Banich, 2003). Even when the exposure levels did not exceed the limit target values established in the European Union, reduced callosum volumes were recorded. White matter disruption partially mediated the association between fine particle exposure and impaired inhibitory control (Guxens et al., 2018) and slower information processing speed in children (Peterson et al., 2015), potentially giving rise to mental health and cognitive delays later in life.

As the brain undergoes ongoing development, exposure to air pollution may continue to be especially harmful throughout childhood. Previous studies reported diverse structural changes in association with childhood exposure, including changes in white matter volumes in the right parietal and bilateral temporal areas (Calderón-Garcidueñas et al., 2011), smaller corpus callosum volume (in line with prenatal exposure; Lubczyńska et al., 2021; Guxens et al., 2022), prefrontal white matter hyperintense lesions (Calderón-Garcidueñas et al., 2008), and increases in cellular barriers in white matter (Burnor et al., 2021). Peterson et al. (2015), who studied both prenatal and childhood exposures, found that while prenatal pollutant exposure affected large areas of the left hemisphere white matter, the effect was less severe for postnatal exposures at the age of 5 years. However, they reported distinct effects of later childhood exposure to PAHs in the dorsal prefrontal white matter, independently of prenatal PAH exposure. Additional studies employing DTI reinforced the findings of a link between white matter and childhood exposure to air pollution: Tailpipe and non-tailpipe emissions were associated with lower FA and higher MD in the white matter of pre-adolescents (Lubczyńska et al., 2020), and airborne elemental copper was linked to differences in the white matter microstructure adjacent to the caudate nucleus (Pujol et al., 2016a). In summary, childhood exposure to air pollution seems to be associated with widespread changes in the white matter structure.

In contrast, the scientific exploration of acute exposure to specific air pollutants or cumulative exposure throughout one's lifetime i.e. in (late) adulthood is comparatively less comprehensively understood than research focusing on exposure during pregnancy or childhood. However, there is some evidence from older samples that long-term ambient exposure to particulate matter may ultimately be associated with reduced white matter volume in the frontal lobe, smaller clusters in the temporal, parietal, and occipital lobes (Casanova et al., 2016; Chen et al., 2015), and smaller corpus callosum volume in women (Chen et al., 2017).

Grey Matter

Research has not only found changes in white matter associated with air pollution but also in grey matter. For example, prenatal exposure was reported to affect various brain regions. Studies reported smaller hippocampal volume (Lubczyńska et al., 2021; Guxens et al., 2022) and amygdala volume (Lubczyńska et al., 2021; Guxens et al., 2022), nucleus accumbens (Guxens et al., 2022), and cerebellum (Lubczyńska et al., 2021; Guxens et al., 2022) in pre-adolescents (9- to 12-year-olds). Such differences in grey matter were already noticeable in neonates, whose brains were scanned at around 40 weeks of age ($n = 469$, Bos et al., 2023). Furthermore, an increasing number of studies have reported that higher exposure to several air pollutants during foetal life was associated with thinner cortices in the frontal and parietal regions in both school-aged children (Guxens et al., 2018, 2022; Lubczyńska et al., 2021; Peterson et al., 2022) and pre-adolescents (Guxens et al., 2022). Such findings were also found in 12-year-olds whose exposure to air pollutants during their first year of life was investigated (Beckwith et al., 2020). Contrarily, some brain regions also showed thicker cortices in several regions (Peterson et al., 2022) or no associations at all (Peterson et al., 2015).

Moreover, several studies investigated childhood exposure to air pollution (mostly focusing on the oxidative potential of $PM_{2.5}$), linking it to alterations in various brain regions among pre-adolescents. Amongst other, findings included reduced hippocampal volume (Lubczyńska et al., 2021; Guxens et al., 2022), larger nucleus accumbens (Guxens et al., 2022; Lubczyńska et al., 2021), and smaller caudate nucleus volumes (Mortamais et al., 2017). Moreover, hemisphere-specific differences in grey matter and cortical thickness across global cortical regions in pre-adolescents were reported (Cserbik et al., 2020) as well as a larger cortical surface area in several areas (Lubczyńska et al., 2021; Guxens et al., 2022). In contrast, some studies reported no differences in grey matter at all (e.g. Pujol et al., 2016b).

Lastly, studies on long-term exposure to certain air pollutants in older adults revealed that ambient exposure to particulate matter may ultimately be associated with reduced grey matter volume in the (pre)frontal cortex (Casanova et al., 2016; Power et al., 2018; Gale et al., 2020; Xu et al., 2023), the left amygdala (Xu et al., 2023), and the left hippocampus (Hedges et al., 2019) in middle- to older-aged adults. Furthermore, higher exposure to air pollutants, particularly NO_2 and PM_{10}, was associated with reduced cortical thickness of the brain areas affected by Alzheimer's disease (Crous-Bou et al., 2020). Additionally, participants who were faced with more air pollution showed more global and local brain atrophy, including lower total brain volume (Wilker et al., 2015; Chen et al., 2015), lower total grey matter volume (Furlong et al., 2022), greater ventricle volumes (Crous-Bou et al., 2020), and less local gyrification within the right frontoparietal network (Nußbaum et al., 2020).

There is mounting evidence that exposure to air pollution is likely to affect the human brain as early as in the womb. Most air pollutants have traffic as their primary source (often referred to as TRAP; traffic-related air pollution) and are, therefore, highly ubiquitous in urban settings. In accordance with the studies on population density and green outdoor environments, the results converge in their finding of grey matter abnormalities in the prefrontal cortices (e.g. Casanova et al., 2016; Gale et al., 2020). The contrary effects of green environments and air pollution may both explain the association between urban areas and brain structure. In general, air pollution has been associated with a decline in cognitive function, such as episodic memory, and possibly dementia (Power et al., 2016; Ailshire & Crimmins, 2014; Tonne et al., 2014). In conclusion, air pollution exposure throughout the lifespan may lead to changes in frontal grey matter and these may, in turn, affect cognitive decline. However, more studies using mediation analyses are necessary to fully understand this relationship. Additionally, independent of the time of exposure, air pollution has been associated with reduced corpus callosum volume (e.g. Lubczyńska et al., 2021; Guxens et al., 2022). These findings are supported by findings from experimental studies, which report an association between air pollution exposure and increased inflammatory reactions within the corpus callosum, accompanied by alterations in the size of the structure (Allen et al., 2017; Klocke et al., 2017; Morris-Schaffer et al., 2019). Mice exposed to airborne particles during gestation even showed an increased number of metal particles in the corpus callosum (Allen et al., 2014). Moreover, reductions in corpus callosum size have been associated with early-life stress in experimental (Jackowski et al., 2011) and epidemiological works (Teicher et al., 2004). Although changes in subcortical structures such as the amygdala, hippocampus, and nucleus accumbens in conjunction with air pollution have been reported (e.g. Hedges et al., 2019; Guxens et al., 2022), the majority of studies did not find significant associations (Chen et al., 2015; Calderón-Garcidueñas et al., 2011; Wilker et al., 2015; Casanova et al., 2016; Crous-Bou et al., 2020; Power et al., 2018). Nonetheless, in summary, these findings call for further improvements in air quality in cities to protect the developing brain.

Brain Structure and Noise Pollution

Like air pollution, noise pollution is a product of traffic, industry, and construction. Not much evidence points towards a link between noise pollution and brain structure. One study reported that participants who lived between 100 and 200 metres from a heavily trafficked road showed less gyrification in the right dlPFC and the PCC/precuneus compared to those who lived more than 200 metres from a heavily trafficked road (Nußbaum et al., 2020). In contrast, another study reported no association between exposure to noise levels and cortical thickness (Crous-Bou et al., 2020).

Brain Structure and Light Pollution

Another environmental issue that accompanies increased urbanisation is light pollution. A world atlas of artificial sky luminance shows that more than 80% of the world population and more than 99% of US and European residents live under light-polluted skies (Falchi et al., 2016). The potential health impacts of such light pollution range from sleep disorders to cancers (reviewed by Cao et al., 2023). Nocturnal light pollution may act by disrupting our circadian rhythmicity, which has been reliably associated with a range of adverse mental health and well-being outcomes (Lyall et al., 2018). Nevertheless, to our current understanding, only one study has investigated the effect of light pollution on brain structure. The main objective of this study was to examine the association between lifelong global urbanicity and brain structure (Xu et al., 2022). They reported that the correlation between urbanicity and grey matter volume within the medial PFC was driven by night-time light and the percentage of built-up and that the correlation with cerebellar GMV was determined by night-time light and the percentage of cropland (Xu et al., 2022). More research is needed to better understand whether light pollution may also be associated with changes in brain structure and function.

Summary of Brain Structure and the Physical Environment

In conclusion, when it comes to the brain structure, what matters is which environment one grew up in, spend their daily lives in and where their mothers lived during their pregnancy. Moreover, this research underscores that this is not only due to living in a densely populated area but also due to specific exposures such as exposure to air pollution, exposure to and the type of green space, and the time spent outdoors. In summary, this body of research suggests that participants who were raised in an urban, air-polluted, densely populated area show reduced volume and thickness of the dorsolateral (Haddad et al., 2015; Besteher et al., 2017; Lammeyer et al., 2019; Kühn et al., 2020) and medial (Xu et al., 2022; Zhang et al., 2023) prefrontal cortices and the hippocampus (Kühn et al., 2020; Besteher et al., 2017; Lubczyńska et al., 2021; Guxens et al., 2022). Moreover, living in a green, unpolluted environment and spending time outdoors likewise tend to be associated with the prefrontal cortices (dlPFC: Kühn et al., 2022, Kühn et al., 2023; mPFC: Besteher et al., 2017; Xu et al., 2022; Kühn et al., 2023; Casanova et al., 2016), whereas taking walks in nature and having a walkable neighbourhood have been associated with greater hippocampal volume (Sudimac & Kühn, 2024; Cerin et al., 2017). Lastly, air pollution has been associated with reduced corpus callosum volume (Mortamais et al., 2019, Guxens et al., 2022; Lubczyńska et al., 2021; Chen et al., 2017), reduced amygdala volume (Lubczyńska et al., 2021; Guxens et al., 2022; Xu et al., 2023), and potentially more global and local atrophy (Nußbaum et al., 2020; Crous-Bou et al., 2020; Wilker et al., 2015; Chen et al., 2015).

Furthermore, when being in urban green spaces, positive emotions were negatively associated with activity in the dlPFC and vmPFC, the brain regions linked to emotion regulation. This suggests that individuals exhibiting diminished activity in these regulatory prefrontal regions experienced greater emotional benefits from green spaces. In accordance with this finding, hypoactivation in the dlPFC was likewise demonstrated in individuals with an increased risk of depression and anxiety disorder (Opel et al., 2017). Furthermore, in this study, dlPFC hypoactivation was observed in a task eliciting negative emotions. This implies that green spaces may act as a protective factor by compensating for diminished regulatory prefrontal capacity. Moreover, the study revealed a negative correlation between green space exposure and emotional benefits from green spaces, specifically in neighbourhoods at a high psychiatric risk. This suggests that those city dwellers who habitually spend time in environments with less green space and with a heightened risk of mental illness may particularly benefit from urban green spaces. These findings emphasise the potential importance of green areas in nature-based interventions aimed at enhancing mental health, especially in urban settings with limited natural environments.

Furthermore, reduced activity in the dlPFC and mPFC, associated with experiencing positive emotions in urban green spaces, implies that natural areas within cities may serve as protective factors, potentially offsetting the decrease in regulatory prefrontal activity.

Conclusions

To comprehend why a high incidence of mental disorders is linked to residing and growing up in urban areas and to understand the impact of one's environment in general, researchers began to explore the relationship between the urban environment and brain function and structure. By exploring the diverse aspects of urban and rural environments, such as population density, green space, walkability, and pollution, specific brain regions were repeatedly highlighted across studies, across environmental features, across the time and duration of exposure, across the time of the brain scan, and across brain measures.

In the following conclusion, we summarise the brain regions that have demonstrated repeated associations with different environmental factors while elucidating the challenges and limitations of the conducted research. Firstly, convergent evidence points towards the effects on the structure and function of the prefrontal cortex, particularly in the dorsolateral and medial PFCs. Individuals raised in cities exhibited increased dlPFC activity during working memory tasks compared to those raised in towns or rural areas (Reed et al., 2018). Additionally, diminished activity in the dlPFC and mPFC, associated with experiencing positive emotions in urban green spaces, implies that natural areas within cities may serve as protective factors, potentially compensating for decreased regulatory prefrontal activity (Tost et al., 2019). Moreover, ruminative thoughts and activity in the sgPFC, previously related to rumination, decreased after a walk in nature (Bratman et al., 2015). Furthermore, studies indicate that various environmental factors, such as population density or air pollution, are associated with altered within- and between-network connectivity patterns of the anterior part of the DMN, which encompasses the mPFC (Pérez-Crespo et al., 2022; Pujol et al., 2016a; Xu et al., 2022). Investigations into brain structure have revealed reduced grey matter volume in the dlPFC (Haddad et al.,

2015; Besteher et al., 2017; Lammeyer et al., 2019) and mPFC (Xu et al., 2022) among individuals exposed to urban environments during their early years. In accordance, urban upbringing also correlated with thinner cortices in the dlPFC and mPFC (Besteher et al., 2017) and lower cortical surface area in the mPFC (Xu et al., 2022). These findings suggest that lifelong and acute exposure to more rural—and most likely more green areas—has beneficial effects on PFC activity and cortical structure. Conversely, a frequently proposed mechanism through which cities may act on the brain is social stress: Cities are highly complex and stressful environments, requiring a high degree of adaption (Lederbogen et al., 2011). This may be experienced as uncontrollable stress, which, in turn, may also result in functional and morphological changes—especially during upbringing—and predominantly in the prefrontal cortex (Arnsten, 2009). Taking this one step further, this may have adverse effects on mental health (Nuechterlein & Dawson, 1984; Os et al., 2005), given that deficits in the prefrontal cortex have been consistently associated with the presence of psychiatric disease (Goodkind et al., 2015). However, we cannot draw definite conclusions about the causality between early-life urbanicity and brain function and structure in the PFC, as all but one study (Bratman et al., 2015) are non-interventional. However, these individuals did not choose the environment they were raised in, thus counteracting the argument of selective migration, which refers to people with specific neurological characteristics choosing to live in certain environments. We advocate for more interventional studies to gather additional evidence to substantiate causal interpretations.

Secondly, previous functional MRI investigations have revealed decreased amygdala activity in individuals who resided in less urban environments (Lederbogen et al., 2011; Krämer et al., 2017) and in those who took a walk in nature (Sudimac et al., 2022). Interestingly, these findings emphasise the involvement of the right amygdala, which was previously found to be involved in the processing of social environmental stress (Mothersill & Donohoe, 2016) and heightened during processing negative emotions in patients with schizophrenia (Pankow et al., 2013). The amygdala has been elicited when observing pictures of urban environments (Kim et al., 2010; Kim & Jeong, 2014) and imagining unpleasant stimuli (Vedder et al., 2015). Moreover, structural MRI studies in children have linked prenatal and acute exposure to air pollution to increased amygdala volumes (Lubczyńska et al., 2021; Guxens et al., 2022), aligning with previous research on chronic and acute stress responses. In contrast, research involving older samples reported smaller amygdala volume and reduced integrity in deprived, air-polluted areas that lacked green spaces (Xu et al., 2023) and low forest coverage (Kühn et al., 2017), respectively. Amygdala activity has been associated with physical proximity to another person, and damage to the amygdalae has led to the absence of personal space boundaries towards others (Kennedy et al., 2009). Thus, the amygdala's involvement in urban environments may represent a higher perceived threat because of personal space violations due to high population density in cities. Moreover, Xu et al. (2023) reported that the correlation between anxiety symptoms and deprived, air-polluted, non-green environments was mediated by volume reductions in the amygdala and other brain regions involved in reward processing. Taken together, these results suggest that both amygdala activity and structure may be affected by heightened social stress, present in highly urban environments, potentially serving as an explanation for why the prevalence rates of mood and anxiety disorders are up to 56% higher in urban compared to rural areas (Peen et al., 2007; Van Os et al., 2010; Lederbogen et al., 2011; Lederbogen et al., 2013).

Thirdly, besides the amygdala, the hippocampus has been implicated in the processing of urban photos (Kim et al., 2010; Kim & Jeong, 2014; Seiyama et al., 2018) and in imagining unpleasant scenarios (Vedder et al., 2015). The hippocampus is involved in emotional processing (Kim & Jeong, 2014), but its activity may also reflect recognition of the shown stimuli (Seiyama et al., 2018), considering it is a central brain region for memory processing. Moreover, structural alterations in the hippocampus have been associated with activities involving walking, wayfinding, and overall spatial processing (Sudimac & Kühn, 2024; Cerin et al., 2017; Kühn et al., 2020). Consistently, the hippocampus is essential for memory, spatial navigation, and cognition. More precisely, it plays a pivotal role in spatial memory and spatial representation of environments, skills that help us to remember and navigate in our surroundings (Lisman et al., 2017). Rural environments may offer more opportunities for developing spatial skills compared to urban environments—mediated by hippocampal volume. This might elucidate the reduced hippocampal grey matter volume (Kühn et al., 2020) and cortical thickness (Besteher et al., 2017) in individuals raised in cities.

Lastly, another region demonstrated to be negatively associated with urban and natural environments is the ACC. Increased activity of the ACC has been linked to current city dwelling (Krämer et al., 2017) and urban upbringing (Lederbogen et al., 2011). Additionally, in a social stress task, urban upbringing has been associated with increased activity in the pACC, a region that regulates the amygdala and is involved in stress processing (Lederbogen et al., 2011). Moreover, its activity was decreased during social stress in participants who lived in a neighbourhood with highly integrated streets (Dimitrov-Discher et al., 2023). Thus, the results imply that the ACC may be responsive to being raised and currently living in urban environments, potentially reflecting the social stress typical of such settings. While the structural findings show less variability than the results from the functional MRI studies, only a few studies have identified effects in the ACC. These studies reported that living close to urban areas and early-life urbanicity (exclusively in men) was linked to reduced GMV in the ACC (Kühn et al., 2021a; Haddad et al., 2015), whereas increased GMV within this region was associated with urban greenery (Kühn et al., 2021a). In summary, the ACC plays a crucial role in regulating negative emotions, stress, and amygdala function and has been suggested to mediate the relationship between mPFC activity and affective symptoms (Yoshimura et al., 2010; LeDoux, 2000; Diorio et al., 1993).

Overall, this research field faces several challenges, including reverse inference, issues of causal interpretations, gene–environment interactions, environmental complexity, focus on Western societies, and selective migration.

Firstly, to address reverse inference and to advance beyond associative results, a shift towards understanding causal effects can be achieved by focusing on interventional designs, as has already been done by some research groups (e.g. Bratman et al., 2015; Sudimac & Kühn, 2022, 2024). For example, future investigations could explore how neural activity and brain structure change when moving from a rural to an urban neighbourhood, or vice versa. Comprehending the cause-and-effect dynamics between the environment and the brain would enable us to design environments that are the most conducive to our mental and brain health.

Secondly, our environments are tremendously complex and interact with factors, such as personality, lifestyle, and genetic predispositions. In addition to the physical features of one's environment, social aspects differ between urban and rural areas. For example, living in a populated area may have certain benefits, such as better access to healthcare, infrastructure, education, and occupational opportunities. Still, it may also relate to social, economic, and material deprivation. Moreover, prospective studies ought to combine information from different spatiotemporal resolutions such as including ecological momentary assessment and satellite-based measures. Issues of complexity and confounding factors can be tackled using two approaches: Firstly, studies could adopt approaches encompassing a broader spectrum of environmental and individual-specific demographic and personality-related variables rather than focusing solely on isolated factors such as population density or green space. Feature selection methods may help overcome statistical challenges. Secondly, another approach that can mitigate gene–environment interactions is studying identical twins, as they not only share the same genome but typically also their early-life and family environment until they move to different geographical and social areas. In general, interdisciplinary work will help combine data from different fields and face other issues linked to these complex data.

Thirdly, while interventional and longitudinal approaches will likely provide the most convincing evidence within the next decade, these study designs, in combination with MRI, do not yet allow for simultaneous measurement of experiencing an environment and brain function (or structure). Although the settings in MRI research allow for high experimental control, they also suffer from limitations related to ecological validity. Establishing a multimodal environmental experience in the scanner has proven challenging, and further approaches are needed to tackle this issue. One possible direction is the use of magnetic resonance–virtual reality (MR-VR) studies that may be able to bridge this gap between using unrealistic, environmental stimuli while being able to measure brain activity and exposure to particular surroundings simultaneously but sacrificing the measurement of live brain activity. Furthermore, MR-VR would allow for manipulating the virtual environment and measuring specific features within a realistic natural setting while concurrently evaluating brain functions in response to these features. Alongside using MRI, other neuroimaging techniques, such as EEG or functional near-infrared spectroscopy (fNIRS), should be harnessed to investigate exposure to environments in real-world contexts. Although these methods suffer from drawbacks in spatial resolution, they allow measuring brain activity in more realistic contexts.

Fourthly, related to the issue of the temporal gap between exposure to the environment and the measurements of the brain, future studies ought to investigate the developmental trajectories of individuals and the potential critical time windows of exposure.

Lastly, studies should diversify, including different age groups, genders, clinical groups, and individuals at risk of mental disorders, to identify those who might benefit most from exposure to nature. Moreover, comparing groups to individuals will further clarify how important the social component is. Prospective research should also include samples from societies that are not WEIRD (Western, educated, industrialised, rich, and democratic) while ensuring that sample sizes are sufficiently large. Moreover, the vast differences in natural environments across countries, ranging from deserts and beaches to mountains and forests, along with varying cultural perceptions and values of nature, should be considered. These variations are crucial for studying nature's effects on mental health and brain function and structure and for applying these findings to medical or urban design policies across different cultural contexts.

In conclusion, our surrounding environment shapes our brain function and structure. This research provides preliminary evidence that living in and growing up in urban, densely populated and polluted areas may have detrimental effects on our mental health and the brain, whereas exposure to nature offers salutogenic effects. However, there is still a gap in understanding the duration and lasting impact of nature's positive effects, the optimal length and frequency of exposure, and how individual differences influence these outcomes. Nonetheless, the field of research underscores the urgent need for future urban planning to prioritise integrating more green and natural spaces while concurrently reducing exposure to air pollution, especially among pregnant women and children. As we continue to unravel the complex relationships between our environments and our brains, this research advanced the concept of "green prescriptions"—health interventions where medical professionals prescribe nature-based activities as part of treatment plans (see Van den Berg, 2017). In summary, these initiatives hold promise in fostering healthier environments and mitigating potential detrimental impacts on brain development and overall well-being.

Acknowledgements During the work on her dissertation, MB was a pre-doctoral fellow of the International Max Planck Research School on Computational Methods in Psychiatry and Ageing Research (COMP2PSYCH, https://www.mps-ucl-centre.mpg.de/comp2psych; participating institutions: Max Planck Institute for Human Development, University College London).

During her dissertation, SS was a pre-doctoral fellow of the International Max Planck Research School on the Life Course (LIFE, www.imprs-life.mpg.de; participating institutions: Max Planck Institute for Human Development, Freie Universität Berlin, Humboldt-Universität zu Berlin, University of Michigan, University of Virginia, University of Zurich).

MS's time on this project was funded by the Austrian Science Fund (FWF: W1262-B29).

The work has been partially funded by the European Union (ERC-2022-CoG-BrainScape-101086188 awarded to SK). Views and opinions expressed are, however, those of the author(s) only and do not necessarily reflect those of the European Union or the European Research Council Executive Agency (ERCEA). Neither the European Union nor the granting authority can be held responsible for them.

References

Ailshire, J. A., & Crimmins, E. M. (2014). Fine particulate matter air pollution and cognitive function among older US adults. *American Journal of Epidemiology, 180*(4), 359–366.

Allen, J. L., Liu, X., Pelkowski, S., Palmer, B., Conrad, K., Oberdörster, G., Weston, D., Mayer-Pröschel, M., & Cory-Slechta, D. A. (2014). Early postnatal exposure to ultrafine particulate matter air pollution: Persistent ventriculomegaly, neurochemical disruption, and glial activation preferentially in male mice. *Environmental Health Perspectives, 122*(9), 939–945.

Allen, J. L., Oberdorster, G., Morris-Schaffer, K., Wong, C., Klocke, C., Sobolewski, M., Conrad, K., Mayer-Proschel, M., & Cory-Slechta, D. A. (2017). Developmental neurotoxicity of inhaled ambient ultrafine particle air pollution: Parallels with neuropathological and behavioral features of autism and other neurodevelopmental disorders. *Neurotoxicology, 59*, 140–154.

Arnsten, A. F. T. (2009). Stress signalling pathways that impair prefrontal cortex structure and function. *Nature Reviews Neuroscience, 10*(6), 410–422.

Banich, M. T. (2003). Interaction between the hemispheres and its implications for the processing capacity of the brain. The asymmetrical brain. *Hugdahl K& Davidson RJ*, 261–302.

Barwise, Y., & Kumar, P. (2020). Designing vegetation barriers for urban air pollution abatement: A practical review for appropriate plant species selection. *npj Climate and Atmospheric Science, 3*(1), 12.

Basten, U., Hilger, K., & Fiebach, C. J. (2015). Where smart brains are different: A quantitative meta-analysis of functional and structural brain imaging studies on intelligence. *Intelligence, 51*, 10–27.

Beckwith, T., Cecil, K., Altaye, M., Severs, R., Wolfe, C., Percy, Z., Maloney, T., et al. (2020). Reduced gray matter volume and cortical thickness associated with traffic-related air pollution in a longitudinally studied pediatric cohort. *PLoS One, 15*(1), e0228092.

Belleau, E. L., Treadway, M. T., & Pizzagalli, D. A. (2019). The impact of stress and major depressive disorder on hippocampal and medial prefrontal cortex morphology. *Biological Psychiatry, 85*(6), 443–453.

Berman, M. G., Kardan, O., Kotabe, H. P., Nusbaum, H. C., & London, S. E. (2019a). The promise of environmental neuroscience. *Nature Human Behaviour, 3*(5), 414–417.

Berman, M. G., Stier, A. J., & Akcelik, G. N. (2019b). Environmental neuroscience. *American Psychologist, 74*(9), 1039.

Besser, L. M., Lovasi, G. S., Michael, Y. L., Garg, P., Hirsch, J. A., Siscovick, D., Hurvitz, P., et al. (2021). Associations between neighborhood greenspace and brain imaging measures in non-demented older adults: The cardiovascular health study. *Social Psychiatry and Psychiatric Epidemiology, 56*, 1575–1585.

Besteher, B., Gaser, C., Spalthoff, R., & Nenadić, I. (2017). Associations between urban upbringing and cortical thickness and gyrification. *Journal of Psychiatric Research, 95*, 114–120.

Block, M. L., Elder, A., Auten, R. L., Bilbo, S. D., Chen, H., Chen, J.-C., Cory-Slechta, D. A., et al. (2012). The outdoor air pollution and brain health workshop. *Neurotoxicology, 33*(5), 972–984.

Bos, B., Barratt, B., Batalle, D., Gale-Grant, O., Hughes, E. J., Beevers, S., Cordero-Grande, L., et al. (2023). Prenatal exposure to air pollution is associated with structural changes in the neonatal brain. *Environment International, 174*, 107921.

Bostrom, C.-E., Gerde, P., Hanberg, A., Jernstrom, B., Johansson, C., Kyrklund, T., Rannug, A., Tornqvist, M., Victorin, K., & Westerholm, R. (2002). Special report: Cancer risk assessment for polycyclic aromatic hydrocarbons-Cancer risk assessment, indicators, and guidelines for polycyclic aromatic hydrocarbons in the ambient air. *Environmental Health Perspectives Supplement, 110*(3), 451–488.

Botvinick, M. M., Braver, T. S., Barch, D. M., Carter, C. S., & Cohen, J. D. (2001). Conflict monitoring and cognitive control. *Psychological Review, 108*(3), 624.

Bratman, G. N., Paul Hamilton, J., Hahn, K. S., Daily, G. C., & Gross, J. J. (2015). Nature experience reduces rumination and subgenual prefrontal cortex activation. *Proceedings of the National Academy of Sciences, 112*(28), 8567–8572.

Brown, L.'. N. A., Khousbouei, H., Goodwin, J. S., Irvin-Wilson, C. V., Ramesh, A., Sheng, L., McCallister, M. M., Jiang, G. C. T., Aschner, M., & Hood, D. B. (2007). Down-regulation of early ionotrophic glutamate receptor subunit developmental expression as a mechanism for observed plasticity deficits following gestational exposure to Benzo (a) pyrene. *Neurotoxicology, 28*(5), 965–978.

Burnor, E., Cserbik, D., Cotter, D. L., Palmer, C. E., Ahmadi, H., Eckel, S. P., Berhane, K., et al. (2021). Association of outdoor ambient fine particulate matter with intracellular white matter microstructural properties among children. *JAMA Network Open, 4*(12), e2138300.

Calderón-Garcidueñas, L., Engle, R., Mora-Tiscareño, A., Styner, M., Gómez-Garza, G., Zhu, H., Jewells, V., et al. (2011). Exposure to severe urban air pollution influences cognitive outcomes, brain volume and systemic inflammation in clinically healthy children. *Brain and Cognition, 77*(3), 345–355.

Calderón-Garcidueñas, L., Mora-Tiscareño, A., Ontiveros, E., Gómez-Garza, G., Barragán-Mejía, G., Broadway, J., Chapman, S., et al. (2008). Air pollution, cognitive deficits and brain abnormalities: A pilot tudy with children and dogs. *Brain and Cognition, 68*(2), 117–127.

Cao, M., Xu, T., & Yin, D. (2023). Understanding light pollution: Recent advances on its health threats and regulations. *Journal of Environmental Sciences, 127*, 589–602.

Casanova, R., Wang, X., Reyes, J., Akita, Y., Serre, M. L., Vizuete, W., Chui, H. C., et al. (2016). A voxel-based morphometry study reveals local brain structural alterations associated with ambient fine particles in older women. *Frontiers in Human Neuroscience, 10*, 495.

Cerin, E., Rainey-Smith, S. R., Ames, D., Lautenschlager, N. T., Lance Macaulay, S., Fowler, C., Robertson, J. S., et al. (2017). Associations of neighborhood environment with brain imaging outcomes in the Australian imaging, biomarkers and lifestyle cohort. *Alzheimer's & Dementia, 13*(4), 388–398.

Chang, C.-h., & Grace, A. A. (2014). Amygdala-ventral pallidum pathway decreases dopamine activity after chronic mild stress in rats. *Biological Psychiatry, 76*(3), 223–230.

Chang, D. H. F., Jiang, B., Wong, N. H. L., Wong, J. J., Webster, C., & Lee, T. M. C. (2021). The human posterior cingulate and the stress-response benefits of viewing green urban landscapes. *NeuroImage, 226*, 117555.

Chen, J. E., & Glover, G. H. (2015). Functional magnetic resonance imaging methods. *Neuropsychology Review, 25*, 289–313.

Chen, J.-C., Wang, X., Serre, M., Cen, S., Franklin, M., & Espeland, M. (2017). Particulate air pollutants, brain structure, and neurocognitive disorders in older women. *Research Reports: Health Effects Institute, 2017*(193), 1–65.

Chen, J.-C., Wang, X., Wellenius, G. A., Serre, M. L., Driscoll, I., Casanova, R., McArdle, J. J., Manson, J. A. E., Chui, H. C., & Espeland, M. A. (2015). Ambient air pollution and neurotoxicity on brain structure: Evidence from women's health initiative memory study. *Annals of Neurology, 78*(3), 466–476.

Cohen, J. R., & Lieberman, M. D. (2010). The common neural basis of exerting self-control in multiple domains. *Self Control in Society, Mind, and Brain, 1*, 141–162.

Cotter, D. L., Campbell, C. E., Sukumaran, K., McConnell, R., Berhane, K., Schwartz, J., Hackman, D. A., Ahmadi, H., Chen, J.-C., & Herting, M. M. (2023). Effects of ambient fine particulates, nitrogen dioxide, and ozone on maturation of functional brain networks across early adolescence. *Environment International, 177*, 108001.

Crous-Bou, M., Gascon, M., Gispert, J. D., Cirach, M., Sánchez-Benavides, G., Falcon, C., Arenaza-Urquijo, E. M., et al. (2020). Impact of urban environmental exposures on cognitive performance and brain structure of healthy individuals at risk for Alzheimer's dementia. *Environment International, 138*, 105546.

Cserbik, D., Chen, J.-C., McConnell, R., Berhane, K., Sowell, E. R., Schwartz, J., Hackman, D. A., Kan, E., Fan, C. C., & Herting, M. M. (2020). Fine particulate matter exposure during childhood relates to hemispheric-specific differences in brain structure. *Environment International, 143*, 105933.

Dadvand, P., Pujol, J., Macià, Dídac, Martínez-Vilavella, G., Blanco-Hinojo, L., Mortamais, M., Alvarez-Pedrerol, M., et al. (2018). The association between lifelong greenspace exposure and 3-dimensional brain magnetic resonance imaging in Barcelona Schoolchildren. *Environmental Health Perspectives, 126*(2), 027012.

DESA, UN, et al. (2019). *World urbanization prospects: The 2018 revision (St/Esa/Ser. A/420)*. United Nations.

Dimitrov-Discher, A., Lanqing, G., Pandit, L., Veer, I. M., Walter, H., Adli, M., & Knöll, M. (2023). Stress and streets: How the network structure of streets is associated with stress-related brain activation. *Journal of Environmental Psychology, 91*, 102142.

Diorio, D., Viau, V., & Meaney, M. J. (1993). The role of the medial prefrontal cortex (Cingulate Gyrus) in the regulation of hypothalamic-pituitary-adrenal responses to stress. *Journal of Neuroscience, 13*(9), 3839–3847.

Duverne, S., & Koechlin, E. (2017). Rewards and cognitive control in the human prefrontal cortex. *Cerebral Cortex, 27*(10), 5024–5039.

Falchi, F., Cinzano, P., Duriscoe, D., Kyba, C. C. M., Elvidge, C. D., Baugh, K., Portnov, B. A., Rybnikova, N. A., & Furgoni, R. (2016). The new world atlas of artificial night sky brightness. *Science Advances, 2*(6), e1600377.

Furlong, M. A., Alexander, G. E., Klimentidis, Y. C., & Raichlen, D. A. (2022). Association of air pollution and physical activity with brain volumes. *Neurology, 98*(4), e416–e426.

Gale, S. D., Erickson, L. D., Anderson, J. E., Brown, B. L., & Hedges, D. W. (2020). Association between exposure to air pollution and prefrontal cortical volume in adults: A cross-sectional study from the UK Biobank. *Environmental Research, 185*, 109365.

Gawryluk, J. R., Palombo, D. J., Curran, J., Parker, A., & Carlsten, C. (2023). Brief diesel exhaust exposure acutely impairs functional brain connectivity in humans: A randomized controlled crossover study. *Environmental Health, 22*(1), 1–7.

Glaubitz, L., Stumme, J., Lucht, S., Moebus, S., Schramm, S., Jockwitz, C., Hoffmann, B., & Caspers, S. (2022). Association between long-term air pollution, chronic traffic noise, and resting-state functional connectivity in the 1000BRAINS Study. *Environmental Health Perspectives, 130*(9), 097007.

Goodkind, M., Eickhoff, S. B., Oathes, D. J., Jiang, Y., Chang, A., Jones-Hagata, L. B., Ortega, B. N., et al. (2015). Identification of a common neurobiological substrate for mental illness. *JAMA Psychiatry, 72*(4), 305–315.

van Praag, G., Cassandra, D., Garfinkel, S. N., Sparasci, O., Mees, A., Philippides, A. O., Ware, M., Ottaviani, C., & Critchley, H. D. (2017). Mind-wandering and alterations to default mode network connectivity when listening to naturalistic versus artificial sounds. *Scientific Reports, 7*(1), 45273.

Grace, A. A. (2016). Dysregulation of the dopamine system in the pathophysiology of schizophrenia and depression. *Nature Reviews Neuroscience, 17*(8), 524–532.

Grandjean, P., & Landrigan, P. J. (2014). Neurobehavioural effects of developmental toxicity. *The Lancet Neurology, 13*(3), 330–338.

Guxens, M., Lubczyńska, M. J., Muetzel, R. L., Dalmau-Bueno, A., Jaddoe, V. W. V., Hoek, G., van der Lugt, A., et al. (2018). Air pollution exposure during fetal life, brain morphology, and cognitive function in school-age children. *Biological Psychiatry, 84*(4), 295–303.

Guxens, M., Lubczyńska, M. J., Pérez-Crespo, L., Muetzel, R. L., El Marroun, H., Basagaña, X., Hoek, G., & Tiemeier, H. (2022). Associations of Air Pollution on the Brain in Children: A Brain Imaging Study. *Research Reports: Health Effects Institute, 2022*(209), 1–61.

Haddad, L., Schäfer, A., Streit, F., Lederbogen, F., Grimm, O., Wüst, S., Deuschle, M., Kirsch, P., Tost, H., & Meyer-Lindenberg, A. (2015). Brain structure correlates of urban upbringing, an environmental risk factor for schizophrenia. *Schizophrenia Bulletin, 41*(1), 115–122.

Hahad, O., Jimenez, M. T. B., Kuntic, M., Frenis, K., Steven, S., Daiber, A., & Muenzel, T. (2022). Cerebral consequences of environmental noise exposure. *Environment International, 165*, 107306.

Hedges, D. W., Erickson, L. D., Kunzelman, J., Brown, B. L., & Gale, S. D. (2019). Association between exposure to air pollution and hippocampal volume in adults in the UK Biobank. *Neurotoxicology, 74*, 108–120.

Jackowski, A., Perera, T. D., Abdallah, C. G., Garrido, G., Tang, C. Y., Martinez, J., Mathew, S. J., et al. (2011). Early-life stress, corpus callosum development, hippocampal volumetrics, and anxious behavior in male nonhuman primates. *Psychiatry Research: Neuroimaging, 192*(1), 37–44.

Kang, Y. J., Tan, H.-Y., Lee, C. Y., & Cho, H. (2021). An air particulate pollutant induces neuroinflammation and neurodegeneration in human brain models. *Advanced Science, 8*(21), 2101251.

Kaplan, S. (1995). The restorative benefits of nature: Toward an integrative framework. *Journal of Environmental Psychology, 15*(3), 169–182.

Kennedy, D. P., Jan Gläscher, J., Tyszka, M., & Adolphs, R. (2009). Personal space regulation by the human amygdala. *Nature Neuroscience, 12*(10), 1226–1227.

Kim, G.-W., & Jeong, G.-W. (2014). Brain activation patterns associated with the human comfortability of residential environments: 3.0-t functional MRI. *Neuroreport, 25*(12), 915–920.

Kim, T.-H., Jeong, G.-W., Baek, H.-S., Kim, G.-W., Sundaram, T., Kang, H.-K., Lee, S.-W., Kim, H.-J., & Song, J.-K. (2010). Human brain activation in response to visual stimulation with rural and urban scenery pictures: A functional magnetic resonance imaging study. *Science of the Total Environment, 408*(12), 2600–2607.

Klauke, B., Deckert, J., Zwanzger, P., Baumann, C., Arolt, V., Pauli, P., Reif, A., & Domschke, K. (2014). Neuropeptide S receptor gene (NPSR) and life events: G × E effects on anxiety sensitivity and its subdimensions. *The World Journal of Biological Psychiatry, 15*(1), 17–25.

Klocke, C., Allen, J. L., Sobolewski, M., Mayer-Pröschel, M., Blum, J. L., Lauterstein, D., Zelikoff, J. T., & Cory-Slechta, D. A. (2017). Neuropathological consequences of gestational exposure to concentrated ambient fine and ultrafine particles in the mouse. *Toxicological Sciences, 156*(2), 492–508.

Krabbendam, L., & Van Os, J. (2005). Schizophrenia and urbanicity: A major environmental influence—conditional on genetic risk. *Schizophrenia Bulletin, 31*(4), 795–799.

Krämer, B., Diekhof, E. K., & Gruber, O. (2017). Effects of city living on the mesolimbic reward system—an fMRI study. *Human Brain Mapping, 38*(7), 3444–3453.

Kühn, S., Banaschewski, T., Bokde, A. L. W., Büchel, C., Quinlan, E. B., Desrivières, S., Flor, H., et al. (2020). Brain structure and habitat: Do the brains of our children tell us where they have been brought up? *NeuroImage, 222*, 117225.

Kühn, S., Düzel, S., Eibich, P., Krekel, C., Wüstemann, H., Kolbe, J., Martensson, J., et al. (2017). In search of features that constitute an 'enriched environment' in humans: Associations between geographical properties and brain structure. *Scientific Reports, 7*(1), 11920.

Kühn, S., Düzel, S., Mascherek, A., Eibich, P., Krekel, C., Kolbe, J., Goebel, J., Gallinat, J., Wagner, G. G., & Lindenberger, U. (2021a). Urban green is more than the absence of city: structural and functional neural basis of urbanicity and green space in the neighbourhood of older adults. *Landscape and Urban Planning, 214*, 104196.

Kühn, S., Forlim, C. G., Lender, A., Wirtz, J., & Gallinat, J. (2021b). Brain functional connectivity differs when viewing pictures from natural and built environments using fMRI resting state analysis. *Scientific Reports, 11*(1), 4110.

Kühn, S., Mascherek, A., Filevich, E., Lisofsky, N., Becker, M., Butler, O., Lochstet, M., et al. (2022). Spend time outdoors for your brain–an in-depth longitudinal MRI study. *The World Journal of Biological Psychiatry, 23*(3), 201–207.

Kühn, S., Schmalen, K., Beijers, R., Tyborowska, A., Roelofs, K., & de Weerth, C. (2023). Green is not the same as green: Differentiating between the association of trees and open green spaces with children's brain structure in The Netherlands. *Environment and Behavior, 55*(5), 311–334.

Kumsta, R., Chen, F. S., Pape, H.-C., & Heinrichs, M. (2013). Neuropeptide s receptor gene is associated with cortisol responses to social stress in humans. *Biological Psychology, 93*(2), 304–307.

Lammeyer, S., Dietsche, B., Dannlowski, U., Kircher, T., & Krug, A. (2019). Evidence of brain network aberration in healthy subjects with urban upbringing–a multimodal DTI and VBM study. *Schizophrenia Research, 208*, 133–137.
Lederbogen, F., Haddad, L., & Meyer-Lindenberg, A. (2013). Urban social stress–Risk factor for mental disorders. The case of schizophrenia. *Environmental Pollution, 183*, 2–6.
Lederbogen, F., Kirsch, P., Haddad, L., Streit, F., Tost, H., Schuch, P., Wüst, S., et al. (2011). City living and urban upbringing affect neural social stress processing in humans. *Nature, 474*(7352), 498–501.
LeDoux, J. E. (2000). Emotion circuits in the brain. *Annual Review of Neuroscience, 23*(1), 155–184.
Lee, M. H., Smyser, C. D., & Shimony, J. S. (2013). Resting-state fMRI: A review of methods and clinical applications. *American Journal of Neuroradiology, 34*(10), 1866–1872.
Lemmers-Jansen, I. L. J., Fett, A.-K. J., van Os, J., Veltman, D. J., & Krabbendam, L. (2020). Trust and the city: Linking urban upbringing to neural mechanisms of trust in psychosis. *Australian & New Zealand Journal of Psychiatry, 54*(2), 138–149.
Lisman, J., Buzsáki, G., Eichenbaum, H., Nadel, L., Ranganath, C., David, A., & Redish. (2017). Viewpoints: How the hippocampus contributes to memory, navigation and cognition. *Nature Neuroscience, 20*(11), 1434–1447.
Liu, F., Jiayuan, X., Guo, L., Qin, W., Liang, M., Schumann, G., & Chunshui, Y. (2023). Environmental neuroscience linking exposome to brain structure and function underlying cognition and behavior. *Molecular Psychiatry, 28*(1), 17–27.
Lubczyńska, M. J., Muetzel, R. L., El Marroun, H., Basagaña, X., Strak, M., Denault, W., Jaddoe, V. W. V., et al. (2020). Exposure to air pollution during pregnancy and childhood, and white matter microstructure in preadolescents. *Environmental Health Perspectives, 128*(2), 027005.
Lubczyńska, M. J., Muetzel, R. L., El Marroun, H., Hoek, G., Kooter, I. M., Thomson, E. M., Hillegers, M., et al. (2021). Air pollution exposure during pregnancy and childhood and brain morphology in preadolescents. *Environmental Research, 198*, 110446.
Lucassen, P. J., Pruessner, J., Sousa, N., Almeida, O. F. X., Van Dam, A. M., Rajkowska, G., Swaab, D. F., & Czéh, B. (2014). Neuropathology of stress. *Acta Neuropathologica, 127*, 109–135.
Luethi, M., Meier, B., & Sandi, C. (2009). Stress effects on working memory, explicit memory, and implicit memory for neutral and emotional stimuli in healthy men. *Frontiers in Behavioral Neuroscience, 2*, 5.
Lyall, L. M., Wyse, C. A., Graham, N., Ferguson, A., Lyall, D. M., Cullen, B., Celis Morales, C. A., et al. (2018). Association of disrupted circadian rhythmicity with mood disorders, subjective wellbeing, and cognitive function: A cross-sectional study of 91 105 participants from the UK Biobank. *The Lancet Psychiatry, 5*(6), 507–514.
Marinelli, M., & McCutcheon, J. E. (2014). Heterogeneity of dopamine neuron activity across traits and states. *Neuroscience, 282*, 176–197.
Martínez-Soto, J., Gonzales-Santos, L., Pasaye, E., & Barrios, F. A. (2013). Exploration of neural correlates of restorative environment exposure through functional magnetic resonance. *Intelligent Buildings International, 5*(sup1), 10–28.
Martínez-Vilavella, G., Pujol, J., Blanco-Hinojo, L., Deus, J., Rivas, I., Persavento, C., Sunyer, J., & Foraster, M. (2023). The effects of exposure to road traffic noise at school on central auditory pathway functional connectivity. *Environmental Research, 226*, 115574.
Melchiorri, M., Florczyk, A. J., Freire, S., Schiavina, M., Pesaresi, M., & Kemper, T. (2018). Unveiling 25 years of planetary urbanization with remote sensing: Perspectives from the global human settlement layer. *Remote Sensing, 10*(5), 768.
Min, K.-D., Kim, J. S., Park, Y. H., Shin, H. Y., Kim, C., Seo, S. W., & Kim, S.-Y. (2021). New assessment for residential greenness and the association with cortical thickness in cognitively healthy adults. *Science of the Total Environment, 778*, 146129.
Morillas, J. M., Barrigón, G. R., Gozalo, D. M., González, P. A., & Moraga, and Rosendo Vílchez-Gómez. (2018). Noise pollution and urban planning. *Current Pollution Reports, 4*(3), 208–219.

Morris-Schaffer, K., Merrill, A. K., Wong, C., Jew, K., Sobolewski, M., & Cory-Slechta, D. A. (2019). Limited developmental neurotoxicity from neonatal inhalation exposure to diesel exhaust particles in C57BL/6 mice. *Particle and Fibre Toxicology, 16*(1), 1–14.

Mortamais, M., Pujol, J., Martínez-Vilavella, G., Fenoll, R., Reynes, C., Sabatier, R., Rivas, I., et al. (2019). Effects of prenatal exposure to particulate matter air pollution on corpus callosum and behavioral problems in children. *Environmental Research, 178*, 108734.

Mortamais, M., Pujol, J., Van Drooge, B. L., Macià, D., Martínez-Vilavella, G., Reynes, C., Sabatier, R., et al. (2017). Effect of exposure to polycyclic aromatic hydrocarbons on basal Ganglia and attention-deficit hyperactivity disorder symptoms in primary school children. *Environment International, 105*, 12–19.

Mothersill, O., & Donohoe, G. (2016). Neural effects of social environmental stress–An activation likelihood estimation meta-analysis. *Psychological Medicine, 46*(10), 2015–2023.

Muscatell, K. A., Dedovic, K., Slavich, G. M., Jarcho, M. R., Breen, E. C., Bower, J. E., Irwin, M. R., & Eisenberger, N. I. (2015). Greater amygdala activity and dorsomedial prefrontal–amygdala coupling are associated with enhanced inflammatory responses to stress. *Brain, Behavior, and Immunity, 43*, 46–53.

Nuechterlein, K. H., & Dawson, M. E. (1984). A heuristic vulnerability/stress model of schizophrenic episodes. *Schizophrenia Bulletin, 10*(2), 300.

Nußbaum, R., Lucht, S., Jockwitz, C., Moebus, S., Engel, M., Jöckel, K.-H., Caspers, S., & Hoffmann, B. (2020). Associations of air pollution and noise with local brain structure in a cohort of older adults. *Environmental Health Perspectives, 128*(6), 067012.

Opel, N., Redlich, R., Grotegerd, D., Dohm, K., Zaremba, D., Meinert, S., Bürger, C., et al. (2017). Prefrontal brain responsiveness to negative stimuli distinguishes familial risk for major depression from acute disorder. *Journal of Psychiatry and Neuroscience, 42*(5), 343–352.

Os, J. v., Krabbendam, L., Myin-Germeys, I., & Delespaul, P. (2005). The schizophrenia envirome. *Current Opinion in Psychiatry, 18*(2), 141–145.

Pankow, A., Friedel, E., Sterzer, P., Seiferth, N., Walter, H., Heinz, A., & Schlagenhauf, F. (2013). Altered amygdala activation in schizophrenia patients during emotion processing. *Schizophrenia Research, 150*(1), 101–106.

Peen, J., Dekker, J., Schoevers, R. A., ten Have, M., de Graaf, R., & Beekman, A. T. (2007). Is the prevalence of psychiatric disorders associated with urbanization? *Social Psychiatry and Psychiatric Epidemiology, 42*, 984–989.

Pérez-Crespo, L., Kusters, M. S. W., López-Vicente, M., Lubczyńska, M. J., Foraster, M., White, T., Hoek, G., Tiemeier, H., Muetzel, R. L., & Guxens, M. (2022). Exposure to traffic-related air pollution and noise during pregnancy and childhood, and functional brain connectivity in preadolescents. *Environment International, 164*, 107275.

Peterson, B. S., Bansal, R., Sawardekar, S., Nati, C., Elgabalawy, E. R., Hoepner, L. A., Garcia, W., et al. (2022). Prenatal exposure to air pollution is associated with altered brain structure, function, and metabolism in childhood. *Journal of Child Psychology and Psychiatry, 63*(11), 1316–1331.

Peterson, B. S., Rauh, V. A., Bansal, R., Hao, X., Toth, Z., Nati, G., Walsh, K., et al. (2015). Effects of prenatal exposure to air pollutants (polycyclic aromatic hydrocarbons) on the development of brain white matter, cognition, and behavior in later childhood. *JAMA Psychiatry, 72*(6), 531–540.

Poldrack, R. A. (2006). Can cognitive processes be inferred from neuroimaging data? *Trends in Cognitive Sciences, 10*(2), 59–63.

Power, M. C., Adar, S. D., Yanosky, J. D., & Weuve, J. (2016). Exposure to air pollution as a potential contributor to cognitive function, cognitive decline, brain imaging, and dementia: A systematic review of epidemiologic research. *Neurotoxicology, 56*, 235–253.

Power, M. C., Lamichhane, A. P., Liao, D., Xiaohui, X., Jack, C. R., Gottesman, R. F., Mosley, T., Stewart, J. D., Yanosky, J. D., & Whitsel, E. A. (2018). The association of long-term exposure to particulate matter air pollution with brain MRI findings: The ARIC study. *Environmental Health Perspectives, 126*(2), 027009.

Pujol, J., Fenoll, R., Macià, D́idac, Martínez-Vilavella, G., Alvarez-Pedrerol, M., Rivas, I., Forns, J., et al. (2016a). Airborne copper exposure in school environments associated with poorer motor performance and altered basal ganglia. *Brain and Behavior, 6*(6), e00467.

Pujol, J., Martínez-Vilavella, G., Macià, D́idac, Fenoll, R., Alvarez-Pedrerol, M., Rivas, I., Forns, J., et al. (2016b). Traffic pollution exposure is associated with altered brain connectivity in school children. *NeuroImage, 129*, 175–184.

Purtle, J., Nelson, K. L., Yang, Y., Langellier, B., Stankov, I., Diez, A. V., & Roux. (2019). Urban–Rural differences in older adult depression: A systematic review and meta-analysis of comparative studies. *American Journal of Preventive Medicine, 56*(4), 603–613.

Qin, S., Hermans, E. J., Van Marle, H. J. F., Luo, J., & Fernández, G. (2009). Acute psychological stress reduces working memory-related activity in the dorsolateral prefrontal cortex. *Biological Psychiatry, 66*(1), 25–32.

Radley, J., Morilak, D., Viau, V., & Campeau, S. (2015). Chronic stress and brain plasticity: Mechanisms underlying adaptive and maladaptive changes and implications for stress-related CNS disorders. *Neuroscience & Biobehavioral Reviews, 58*, 79–91.

Rauch, A. V., Reker, M., Ohrmann, P., Pedersen, A., Bauer, J., Dannlowski, U., Harding, L., et al. (2010). Increased amygdala activation during automatic processing of facial emotion in schizophrenia. *Psychiatry Research: Neuroimaging, 182*(3), 200–206.

Reed, J. L., D'Ambrosio, E., Marenco, S., Ursini, G., Zheutlin, A. B., Blasi, G., Spencer, B. E., et al. (2018). Interaction of childhood urbanicity and variation in dopamine genes alters adult prefrontal function as measured by functional magnetic resonance imaging (fMRI). *PLoS One, 13*(4), e0195189.

Rice, D., & Barone, S., Jr. (2000). Critical periods of vulnerability for the developing nervous system: Evidence from humans and animal models. *Environmental Health Perspectives, 108*(suppl 3), 511–533.

Ridderinkhof, K. R., Ullsperger, M., Crone, E. A., & Nieuwenhuis, S. (2004). The role of the medial frontal cortex in cognitive control. *Science, 306*(5695), 443–447.

Roberts, A. C., & Christopoulos, G. I. (2018). Comment on 'using functional magnetic resonance imaging (fMRI) to analyze brain region activity when viewing landscapes'. *Landscape and Urban Planning, 172*, 25–28.

Roozendaal, B., McEwen, B. S., & Chattarji, S. (2009). Stress, memory and the amygdala. *Nature Reviews Neuroscience, 10*(6), 423–433.

Rubinov, M., & Sporns, O. (2010). Complex network measures of brain connectivity: Uses and interpretations. *NeuroImage, 52*(3), 1059–1069.

Sale, A., Berardi, N., & Maffei, L. (2009). Enrich the environment to empower the brain. *Trends in Neurosciences, 32*(4), 233–239.

Sanders, A. F. P., Harms, M. P., Kandala, S., Marek, S., Somerville, L. H., Bookheimer, S. Y., Dapretto, M., et al. (2023). Age-related differences in resting-state functional connectivity from childhood to adolescence. *Cerebral Cortex, 33*(11), 6928–6942.

Sehlmeyer, C., Dannlowski, U., Schöning, S., Kugel, H., Pyka, M., Pfleiderer, B., Zwitserlood, P., et al. (2011). Neural correlates of trait anxiety in fear extinction. *Psychological Medicine, 41*(4), 789–798.

Seiyama, A., Yamada, K., Osaki, K., Nakai, R., Matsumoto, J., & Yoshimura, A. (2018). Neural bases on cognitive aspect of landscape evaluation: a study using functional magnetic resonance imaging. *Journal of Neuroscience, 9*(263), 10-21767.

Siegle, G. J., Thompson, W., Carter, C. S., Steinhauer, S. R., & Thase, M. E. (2007). Increased amygdala and decreased dorsolateral prefrontal BOLD responses in unipolar depression: Related and independent features. *Biological Psychiatry, 61*(2), 198–209.

Stobbe, E., Forlim, C. G., & Kühn, S. (2023). Impact of exposure to natural versus urban soundscapes on brain functional connectivity, BOLD entropy and behavior. *Environmental Research, 244*, 117788.

Streit, F., Haddad, L., Paul, T., Frank, J., Schäfer, A., Nikitopoulos, J., Akdeniz, C., et al. (2014). A functional variant in the neuropeptide S receptor 1 gene moderates the influence of urban upbringing on stress processing in the amygdala. *Stress, 17*(4), 352–361.

Sudimac, S., & Kühn, S. (2022). A one-hour walk in nature reduces amygdala activity in women, but not in men. *Frontiers in Psychology, 13*, 931905.

Sudimac, S., & Kühn, S. (2024). Can a nature walk change your brain? Investigating Hippocampal Brain Plasticity After One Hour in a Forest. Environmental Research;262(Pt 1):119813.

Sudimac, S., Sale, V., & Kühn, S. (2022). How nature nurtures: Amygdala activity decreases as the result of a one-hour walk in nature. *Molecular Psychiatry, 27*(11), 4446–4452.

Tang, I.-C., Tsai, Y.-P., Lin, Y.-J., Chen, J.-H., Hsieh, C.-H., Hung, S.-H., Sullivan, W. C., Tang, H.-F., & Chang, C.-Y. (2017). Using functional magnetic resonance imaging (fMRI) to analyze brain region activity when viewing landscapes. *Landscape and Urban Planning, 162*, 137–144.

Taylor, L., & Hochuli, D. F. (2017). Defining greenspace: Multiple uses across multiple disciplines. *Landscape and Urban Planning, 158*, 25–38.

Teicher, M. H., Dumont, N. L., Ito, Y., Vaituzis, C., Giedd, J. N., & Andersen, S. L. (2004). Childhood neglect is associated with reduced corpus callosum area. *Biological Psychiatry, 56*(2), 80–85.

Tonne, C., Elbaz, A., Beevers, S., & Singh-Manoux, A. (2014). Traffic-related air pollution in relation to cognitive function in older adults. *Epidemiology (Cambridge, Mass.), 25*(5), 674.

Tost, H., Reichert, M., Braun, U., Reinhard, I., Peters, R., Lautenbach, S., Hoell, A., et al. (2019). Neural correlates of individual differences in affective benefit of real-life urban green space exposure. *Nature Neuroscience, 22*(9), 1389–1393.

Ulrich, R. S., Simons, R. F., Losito, B. D., Fiorito, E., Miles, M. A., & Zelson, M. (1991). Stress recovery during exposure to natural and urban environments. *Journal of Environmental Psycholo.gy, 11*(3), 201–230.

Van den Berg, A. E. (2017). From green space to green prescriptions: Challenges and opportunities for research and practice. *Frontiers in Psychology, 8*, 268.

Van Os, J., Kenis, G., & Rutten, B. P. F. (2010). The environment and schizophrenia. *Nature, 468*(7321), 203–212.

Vassos, E., Pedersen, C. B., Murray, R. M., Collier, D. A., & Lewis, C. M. (2012). Meta-analysis of the association of urbanicity with schizophrenia. *Schizophrenia Bulletin, 38*(6), 1118–1123.

Vedder, A., Smigielski, L., Gutyrchik, E., Bao, Y., Blautzik, J., Pöppel, E., Zaytseva, Y., & Russell, E. (2015). Neurofunctional correlates of environmental cognition: An fMRI study with images from episodic memory. *PLoS One, 10*(4), e0122470.

Vlahov, D., & Galea, S. (2002). Urbanization, urbanicity, and health. *Journal of Urban Health, 79*, S1–S12.

Wild, C. P. (2012). The exposome: From concept to utility. *International Journal of Epidemiology, 41*(1), 24–32.

Wilker, E. H., Preis, S. R., Beiser, A. S., Wolf, P. A., Rhoda, A., Kloog, I., Li, W., et al. (2015). Long-term exposure to fine particulate matter, residential proximity to major roads and measures of brain structure. *Stroke, 46*(5), 1161–1166.

Xu, J., Liu, N., Polemiti, E., Garcia-Mondragon, L., Tang, J., Liu, X., ... & Schumann, G. (2023). Effects of urban living environments on mental health in adults. *Nature Medicine, 29*(6), 1456–1467.

Xu, J., Liu, X., Li, Q., Goldblatt, R., Qin, W., Liu, F., Chu, C., et al. (2022). Global urbanicity is associated with brain and behaviour in young people. *Nature Human Behaviour, 6*(2), 279–293.

Yang, T. T., Simmons, A. N., Matthews, S. C., Tapert, S. F., Frank, G. K., Max, J. E., Bischoff-Grethe, A., et al. (2010). Adolescents with major depression demonstrate increased amygdala activation. *Journal of the American Academy of Child & Adolescent Psychiatry, 49*(1), 42–51.

Yeo, B. T. T., Krienen, F. M., Sepulcre, J., Sabuncu, M. R., Lashkari, D., Hollinshead, M., Roffman, J. L., et al. (2011). The organization of the human cerebral cortex estimated by intrinsic functional connectivity. *Journal of Neurophysiology, 106*(3), 1125–1165.

Yoshimura, S., Okamoto, Y., Onoda, K., Matsunaga, M., Ueda, K., Suzuki, S.-i., et al. (2010). Rostral anterior cingulate cortex activity mediates the relationship between the depressive symptoms and the medial prefrontal cortex activity. *Journal of Affective Disorders, 122*(1–2), 76–85.

Zhang, X., Yan, H., Hao, Y., Zhang, Y., Tan, H. Y., Zhang, D., & Yue, W. (2023). The effects of environmental factors associated with childhood urbanicity on brain structure and cognition. *BMC Psychiatry, 23*(1), 1–11.

Zhang, X., Yan, H., Hao, Y., Zhao, X., Shah, S., Dong, Z., Yang, G., et al. (2021). Childhood urbanicity interacts with polygenic risk for depression to affect stress-related medial prefrontal function. *Translational Psychiatry, 11*(1), 522.

Human Brain and Behavior in the Geospatial Context: Why and How

Tomas Paus

Abstract This chapter introduces the concept of "digital ethology," as envisioned during the Ernst Strüngmann Forum convened in July 2022. By "digital ethology," we mean the observation of human behavior in the "wild," that is behavior unfolding in its natural context and without interference by the observer. We focus on the "digital footprints" left by our behavior at the area level (neighborhoods, cities, counties) and on the constraints of human behavior by the surrounding physical and built environments modeled from digital sources. We point out the potential of linkages between the area-level characteristics of our physical, built, and social environments and individual-level (health) outcomes, with the latter captured in administrative health databases and large community-based cohorts. In this way, we see digital ethology as a conceptual and practical framework complementing population neuroscience.

Keywords Population neurosciences · Digital ethology · Geospatial mapping · Brain development · Mental health

This chapter was first published in *Digital Ethology: Human Behavior in Geospatial Context*, edited by Tomáš Paus and Hye-Chung Kum, reprinted courtesy of The MIT Press.

T. Paus (✉)
Departments of Psychiatry and Neuroscience, Faculty of Medicine, University of Montreal, Montreal, QC, Canada

S. Kühn (ed.), *Environmental Neuroscience*,
https://doi.org/10.1007/978-3-031-64699-7_4

General Background

From conception onward, the individual is developing, maturing, working, playing, and aging in their[1] context. As illustrated in Fig. 1, multiple layers of environment (context) surround an individual across space and time: from the uteroplacental circulation connecting the fetus and their mother before birth to the influence of their caregivers, extended family, and peers during childhood, adolescence, and adulthood. This "proximal" context (in green) is embedded in larger geospatial units, such as specific neighborhoods, cities, or countries (in blue). All environmental influences unfold in time throughout the individual's lifespan. Needless to say, the different layers interact, in a bidirectional manner, with each other. Thus, for instance, a pregnant person responds to signals generated by the fetus, and vice versa (Fowden et al., 2022; Kolle et al., 2020; Menon, 2019), a pregnant person interacts with their partner, and vice versa (Khaled et al., 2021; Saxbe et al., 2018), and the caregiver interacts with the child, and vice versa (Carollo et al., 2023; Paquette & St. George, 2023). At the same time, the individual and those in their proximal context (e.g., caregivers and peers) act as both recipients and co-creators of their area-level environment along all its dimensions, including the physical environment (e.g., air quality), built environment (e.g., parks and transportation networks), and social environment (e.g., social cohesion). Different aspects of the environment change over time in an interdependent manner (e.g., air quality, vehicular traffic, lack of green spaces, demographic characteristics), often reflecting the resources and policies in place at different levels of geospatial granularity (e.g., country, city, neighborhood). Both within and across countries, the lack of

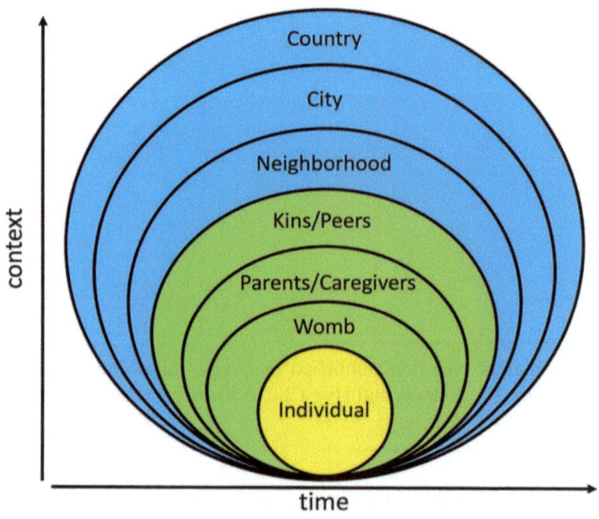

Fig. 1 Conceptualization of the multiple layers that comprise the contextual environment of an individual across space and time

[1] Throughout this chapter, "they" (and its derivations) is used as a gender-neutral third-person pronoun.

environmental justice is reflected in the disproportional exposures of marginalized communities to various combinations of adverse environments and, in turn, their *combined* health effects (Van Horne et al., 2023).

For those of us interested in understanding the forces that shape the human brain and behavior, from conception onward, the complexity of this multilayered "exposome" (Wild, 2005; Munzel et al., 2023) (➔ chapter "Advancing the Neural Exposome" by Veasey) is staggering. The field of population neuroscience emerged to face this challenge; it brings together epidemiology, genetics, and neuroscience to gain insights into the factors underpinning the interindividual variability in the structure and function of the human brain (Paus, 2010, 2013, 2016). Owing to the ease of characterizing the individual's genome and the advances in our understanding of related biological processes, initial studies focused on the genetic side of the equation. Working mostly in the context of international consortia, such as ENIGMA (Enhancing NeuroImaging Genetics Through Meta Analysis) (Thompson et al., 2014) and CHARGE (Cohorts for Heart and Aging Research in Genomic Epidemiology) (Psaty et al., 2009), we have learned a great deal about the molecular architecture of various quantitative traits derived from magnetic resonance images of the human brain (Grasby et al., 2020; Shin et al., 2020; Satizabal et al., 2019), but efforts on the environmental front lag behind. This is understandable given the difficulty of characterizing an individual's environment. Published studies in this area have addressed a handful of factors—one at the time—from the different context layers illustrated in Fig. 1, such as an intrauterine environment (e.g., exposure to maternal cigarette smoking during pregnancy) (Toro et al., 2008; Muller et al., 2013), a family environment (e.g., family socioeconomic status) (Noble et al., 2015), population density (Xu et al., 2022a), and variations in the physical (e.g., air pollution) (Sukumaran et al., 2023), built (e.g., green space) (Kardan et al., 2015)), and social (e.g., income inequality) (Parker et al., 2017), environments across neighborhoods, cities, and/or countries. Although encouraging, major gaps remain. The Ernst Strüngmann Forum on Digital Ethology, convened in Frankfurt am Main, Germany in July 2022, brought together scholars and experts to address a number of conceptual and practical gaps in this area.

As pointed out above, the scarcity of multidimensional data that can be used to characterize an individual's environment in an integrated manner presents the key challenge to studying relationships between the multilayered, multi-domain environment and individual-level outcomes, such as brain development and aging. The Forum addressed this challenge in two ways: *Conceptually*, it called for adopting an ethological approach whereby human behavior is observed, or inferred, in the "wild," that is, without influencing the observed individual (e.g., by asking them questions). *Practically*, it called for focusing on data sources that either exist or can be readily harnessed at an aggregate level, with different area-level (spatial) granularity (e.g., neighborhood, city, country). The ethological framework presented in Dumas et al. (chapter "The Rise and Future of Environmental Neuroscience in Environmental Psychology") underpins the name to this Forum. By "digital ethology," we mean the observation of human behavior through its digital manifestations, such as the use of a search engine, a payment card, or through posting on social media. This behavior leaves "digital footprints" that are particularly relevant for characterizing the social environment of a given area-level unit, as discussed by Weigle et al. (chapter "Extended Reality in Environmental Neuroscience Research"). Human behavior is also reflected and constrained by the surrounding physical and built environments, as outlined by Lovasi et al. (chapter "Interplay Between the Physical Environment and the Human Brain: Insights from MRI Research"). Finally, variations in individual-level outcomes as a function of the multidimensional area-level environment can best be studied using large datasets; the practicalities as well as legal and ethical considerations are addressed by Bauzer Medeiros et al. (chapter "Advancing the Neural Exposome"). Finally, chapters "Leveraging the Social Neuroscience of Prosocial Behavior to Advance Our Understanding of Pro-environmental Behavior", "Neuroscientific Approaches to Understand Barriers and Promotors of Pro-environmental Behavior", "Aesthetics of Architecture: Emerging Insights from the Brain", "The Built Environment and the Brain: Emerging Methods to Investigate the Impact of Viewing Architectural Design", "Application of Biometrics in Architectural and Environmental Design", "The Potential of Biophilic Design and Nature to Improve Health, Creativity, and Wellbeing", and "Environmental Exposures and Child Neurodevelopment" provide primers to many of the concepts and strategies that underpin digital ethology.

A Case Study: Inequalities in Area-Level Environment and Brain Health[2]

Social, economic, and political conditions produce *health inequalities* within and across countries (Metzl & Hansen, 2018; Scambler, 2012; Stuart & Soulsby, 2011). In high-income countries, for instance, individuals are more likely to experience poor mental health if they grow up in households with low income (Bjorkenstam et al., 2017) or affluence (Elgar et al., 2015; Rajmil et al., 2014), live in areas with high deprivation (Kivimaki et al., 2020), or experience inequalities in income distribution (Mangalore et al., 2007). Certain communities are disadvantaged more than others (Waldron, 2018). This is especially true for Indigenous (Ogilvie et al., 2021) and racialized (Castro-Ramirez et al., 2021) communities, which are at a higher risk of mental health problems and simultaneously experience a lower likelihood of receiving evidence-based treatment (Castro-Ramirez et al., 2021). At the area level, our physical, built, and social environments combine to create ecosystems in which we live and work. Together, these ecosystems, as well as the structures and systems that produce them, contribute to what has been termed the "social and structural determinants of health" (Vandenbroucke, 1990; Diderichsen et al., 2001).

[2] This section is a modified version of an article published in Paus et al. (2022).

As described elsewhere (Paus, 2016), there are countless permutations of the physical, built, and social environments that surround us in space and time. We both "receive" and "create" our environments (Kendler et al., 2003), thus co-determining what air we breathe, how many steps we take, how hot or cold we are, and what and who we see, hear, and interact with during our commutes. Together with our genes, these "external exposures" contribute to "internal" environments that exist in our body: on body surfaces (e.g., microbes on our skin and in the gut), in the lungs (e.g., particulate matter), circulating blood (e.g., toxins, micronutrients, inflammatory molecules), and the brain (e.g., stress- and reward-related neurotransmitters, cumulative engagement of specific neural circuits).

As pointed out above, the use of aggregate-level (spatial) data, produced from multiple locations and time points, is one strategy for characterizing the physical, built, and social environments surrounding the individual. In turn, linking such aggregate-level data to individual-level information about a person's health in general, and brain health in particular, is the first step toward understanding these relationships. Below, the basic steps in this process are reviewed, which are covered in depth in chapters "Leveraging the Social Neuroscience of Prosocial Behavior to Advance Our Understanding of Pro-environmental Behavior", "Neuroscientific Approaches to Understand Barriers and Promotors of Pro-environmental Behavior", "Aesthetics of Architecture: Emerging Insights from the Brain", "The Built Environment and the Brain: Emerging Methods to Investigate the Impact of Viewing Architectural Design", "Application of Biometrics in Architectural and Environmental Design", "The Potential of Biophilic Design and Nature to Improve Health, Creativity, and Wellbeing", and "Environmental Exposures and Child Neurodevelopment".

Geospatial Mapping of Area-Level Environments

Geospatial science and related tools enable spatial analysis and visualization of the external environments in which we spend a considerable amount of our lives (e.g., our residence, place of work, school, recreation, or a commute path) and an evaluation of their impact on our health (for an example see chapter "Neuroscience and Geographic Information Systems to Investigate the Impact of Global Warming on Mood Disorders and Brain Plasticity in Urban Areas"). Datasets can be created at different levels of spatial granularity matching the goals of a given study and availability of relevant data. In Canada, for example, geographic units include six-digit postal codes, Canadian Census geographic units such as dissemination areas (400–700 persons), and census tracts (2,500–8,000 persons), or larger areas such as city districts. The spatial unit used to link geospatial datasets to health data varies; depending on the study and actions necessary to protect the confidentiality of study participants, this can be as precise as the exact street address or a postal code (half of a city block in a dense urban area) or as coarse as a city district, a county, a province/state, or a country. The temporal dimension depends on the type of data; it may range from data sampled monthly (e.g., air quality), annually (e.g., public transportation), or up to every 5 years (e.g., the Canadian Census).

Spatiotemporal datasets can be created using existing tools and databases provided by large GIS (Geographic Information System)-based organization and companies, such as ESRI, DMTI Spatial, and Google Earth Engine, as well as open sources (e.g., Open Street Map), government sources (e.g., Statistics Canada), and academic organizations. In Canada, we have acquired, curated, and disseminated geospatially coded information about the physical and built environments through the Canadian Urban Environmental Health Research Consortium (CANUE) (Brook et al., 2018). Metrics derived from different sources can be combined to ask, for example, questions about the relationship between socioeconomic indicators (e.g., household income) and the built environment (e.g., access to parks), and thereby used to assess inequity in the spatial distribution of environmental good or hazards. Figure 2 illustrates inequality in the access to parks and recreation (derived from Open Street Map data) across areas with high levels (top 20%) of material deprivation (derived from Canadian Census data; Pampalon et al., 2012).

In addition to sourcing and creating data about physical and built environments from existing databases (see Table 1 in Paus (2016), one can also derive relevant measures from new data streams such as high-resolution satellite and street-level imagery combined with machine learning techniques (Weichenthal et al., 2019). For example, Google Street View allows investigators to assess different features of the

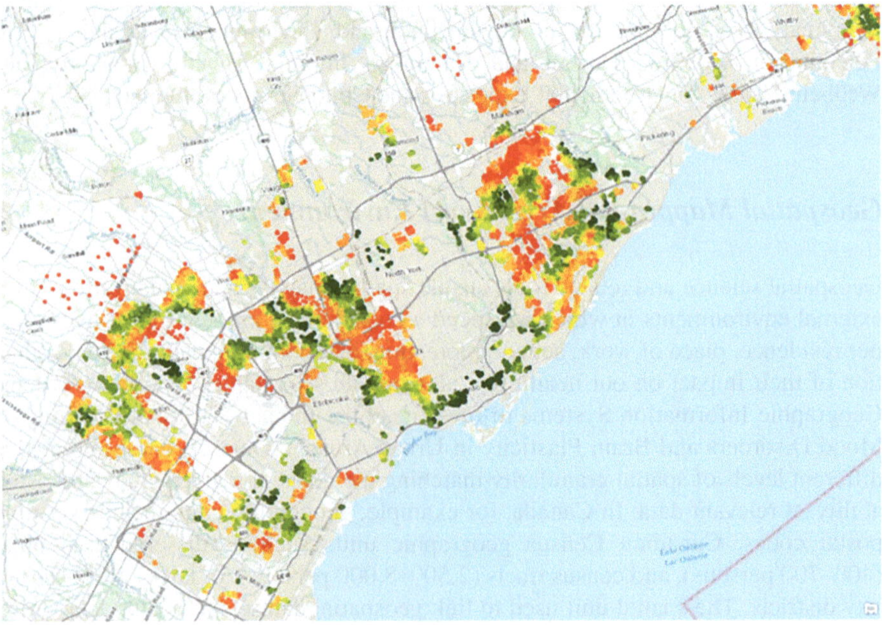

Fig. 2 Material deprivation and access to parks and recreation, Greater Toronto Area. All colored areas represent postal codes characterized by high levels (top 20%) of material deprivation (Pampalon et al., 2012). Green indicates postal codes in the highest 10% density of park and recreational amenity within 1 km; red indicates postal codes in the lowest 10%. (Source: Open Street Map)

Table 1 Examples of measures, with the corresponding sources of raw geospatially coded data and examples of the new types of data to be derived

Physical and built environments	Social environment
Air quality (NO_2, O_3, SO_2, PM2.5)[a]	(b) Demographic[f]
Greenness (greenest pixel, tree canopy)[a]	(c) Households[f]
Nighttime light[a]	(d) Socioeconomic[f]
Noise[b]	Water quality concerns[g]
Public transportation[c]	Composting and recycling behavior[g]
Proximity to roads[d]	Involvement in outdoor activities[g]
Proximity to retail outlets and sales for alcohol, tobacco, cannabis, gambling[d]	Caregiving and care receiving[h]
Green roads[e]	Social identity[h]
Facility index[e]	Giving, volunteering, and participating[h]
(a) Cumulative opportunities[e]	Victimization[h]
	Social media and search engine use by youth: frequency and time of day[i]
	Social media and search engine use by youth: content[j]
	Built environment predictors of psychosocial stress[j]
	Built environment predictors of social cohesion[j]

[a]Landsat
[b]CANUE
[c]OpenStreetMap (OSM)
[d]DMTI Spatial
[e]OSM and CANUE
[f]Census
[g]Household and the Environment Survey (Canada)
[h]The General Social Survey (Canada)
[i]Newly derived measures from raw data streams (e.g., Twitter, Google search engines)
[j]Newly derived measures from raw data streams (satellite and street-view imagery). (a) Travel times (walking, public transport) to jobs, leisure, and shopping as well as health, medical, and social services; (b) population (total and densities), proportions (by age, sex, ethnicity, marital status, mobility/migration status, religion, mother tongue); (c) household size, total housing units, proportion rented, type of dwelling; (d) household income, unemployment rate, proportion below poverty line, proportion (by age/sex) in labor force

built environment using panoramic street-level images taken mostly by camera-equipped cars, whereas recent satellite technology has provided daily coverage of the most inhabited areas on Earth at a resolution of only a few meters. These geo-coded images can be rated for various features, such as signs of physical disorder (e.g., litter, graffiti), physical decay (e.g., poor conditions of sidewalks), types of stores, traffic, or street walkability (Odgers et al., 2012; Less et al., 2015); this approach does have, however, some limitations (Curtis et al., 2013). In turn, these computer vision and machine learning algorithms can use these image data to generate indirect indices of the social environment (e.g., psychosocial stress) and physical environment (e.g., air or noise pollution) in a manner similar to that used by others to derive measures characterizing living environment, health, and crime (Suel et al., 2019).

As summarized in Table 1 (social environment), a wealth of data speak to the basic (often self-reported) measures of socioeconomic factors (e.g., education, employment, immigration, household spending habits, volunteering, and giving) collected by governmental agencies (e.g., census) and national surveys. One can, however, also use data from digital streams (e.g., search engines, social media) to generate new measures of the social environment that are relevant for attitudes vis-à-vis health and health interventions (e.g., vaccination), as well as social cohesion, social support, and role models, and, most recently, for the emerging issues related to environmental anxiety (To et al., 2021; Soutar & Wand, 2022; Usher, 2022; Hickman et al., 2021).

Once properly curated, all aggregate-level data (e.g., see Table 1) should be described using comprehensive metadata and coded to different geographic units (e.g., postal codes, dissemination areas, and other census geographies), as has been previously done by CANUE.

Linkage to Individual-Level Data

Ultimately, what we are interested in doing is to link the aggregate-level "exposures" described above to individual-level "outcomes." In this section, two examples illustrate how this can be achieved using administrative health databases and data acquired in research cohorts.

Administrative Data

In the recent past, we have all seen the power of mapping administrative data related to COVID-19 (across countries, provinces/states, or cities) and communicating these numbers to the public. In Canada, administrative health data (i.e., data captured during the course of providing services or running programs) are made available for research use by provincial governments and other agencies, often in close partnership with academic organizations (Lucyk et al., 2015). In all provinces, these data are longitudinal and population-based, covering all residents who have received health-care and social services (e.g., education) from birth onward. This creates comprehensive and important data for the population of interest, such as youth.

In the province of Ontario, for example, administrative health data are curated and made available for research by the Institute for Clinical Evaluative Sciences (ICES), a not-for-profit research institute made up of a community of research, data, and clinical experts that provide a secure and accessible inventory of Ontario's health-related data. Behind a firewall, ICES provides access to coded and linkable databases containing, for example, the Ontario Mental Health Reporting System. Just in the City of Toronto, these data are available for about 270,000 adolescents and youth (12–22 years of age). In addition to health data, many of the provincial custodians of administrative data provide access to other linked datasets, such as education, workplace, or justice data (e.g., Population Data BC). When linking administrative data to geospatial datasets containing area-level characteristics of the physical, built, and social environments, one would typically use the residential six-digit postal codes (Canada) and relevant geographies (e.g., dissemination blocks) reported in the administrative data for each individual. Postal code-indexed geospatial datasets are linked in the secure environments controlled by the custodian of the individual-level health data (Boyd et al., 2012; Kum & Ahalt, 2013; Pencarrick Hertzman et al., 2013). Here, ethical and legal guidance is necessary to provide assurance to data stewards that this form of data linkage and access can be done in a privacy-preserving and transparent manner that respects all applicable legal, regulatory, and ethical requirements. Ongoing efforts address issues relevant for ensuring public trust, such as transparency of the current practices and systems of governance, and for understanding public opinion regarding the use of "big data" in the service of population health (Aitken et al., 2016) (O'Brien et al., 2019; Schmit et al., 2021).

Cohort Studies

One of the key advantages of administrative health data is their population-wide coverage; by definition, these data show only the tip of the "health iceberg," namely, individuals with health issues significant enough to enter the health-care system. This is where community-based cohort studies come in as a complementary source of information, with longitudinal birth cohorts being the most valuable. For example, birth cohorts are well suited for investigating the relationships between brain health (individual-level data) and context (aggregate-level characteristics of the environment) for several reasons:

1. Many birth cohorts, such as ALSPAC (Avon Longitudinal Study of Parents and Children) (Boyd et al., 2013), Generation R (Tiemeier et al., 2012), and Northern Finland Birth Cohorts (Rantakallio, 1988), ascertained their participants (pregnant women) in a relatively small geographic area.
2. Each cohort includes a relatively large sample size of individuals (~10,000).
3. The brain (e.g., mental) health of cohort members is assessed using a number of instruments, often on a continuous scale.

The combination of the first two features makes it likely that a reasonable number of participants live in each geospatial unit, hence providing sufficient statistical power to investigate these relationships. The third feature (assessment) permits the capture of "subclinical" mental health problems. Finally, additional deep phenotyping of cohort members through, for example, cognitive assessment, neuroimaging, blood-based biomarkers (e.g., inflammation), genotyping and epigenotyping provides rich information suitable for detailed modeling of exposure–outcome relationships and their mediators and moderators (Paus, 2013).

Social Inequality and Mental Health

To close this section, let us consider a hypothetical example illustrating how one can use aggregate-level information about the physical, built, and social environments to unpack the relationship between poverty and mental health. As pointed out by Diderichsen et al. (2001), and represented in Fig. 3, social stratification—with poverty being but one example of social, economic, and political inequalities—generates a vicious circle: Disadvantaged persons are more likely to be exposed to harmful or deprived physical (e.g., air pollution), built (e.g., access to food stores), and social (e.g., lack of social support) environments as well as to population-level challenges (e.g., heat waves, SARS-CoV-2). These exposures lead to an increased vulnerability to other exposures (e.g., victimization), and both the exposures and

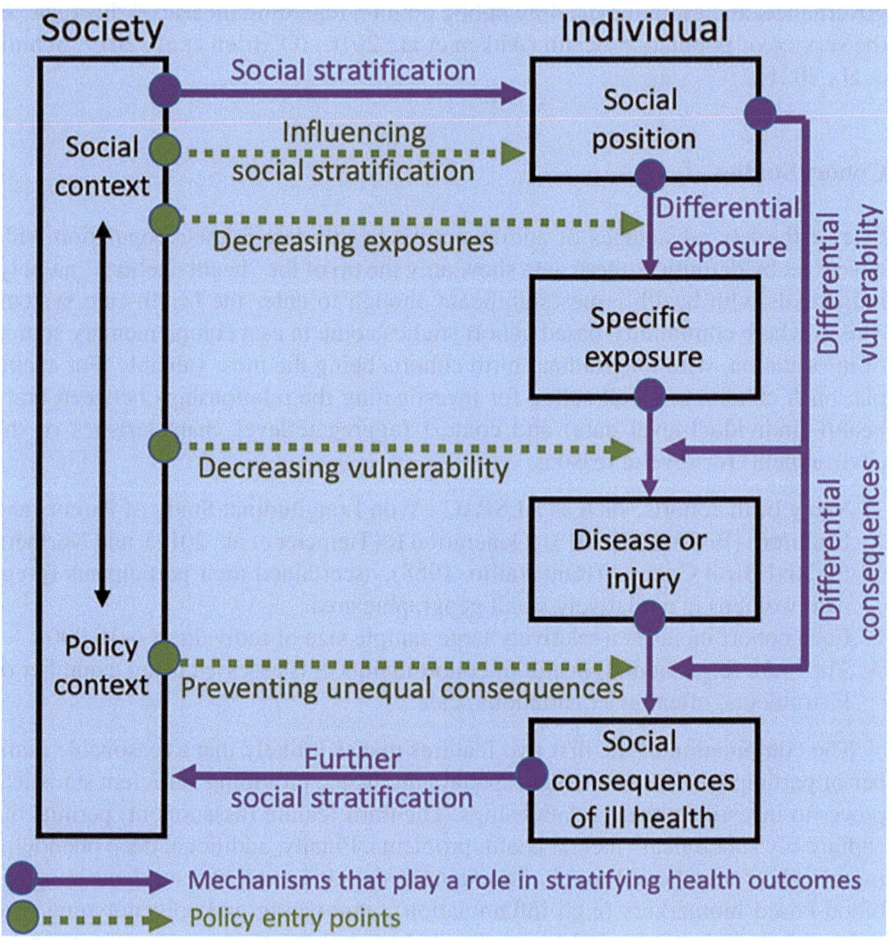

Fig. 3 From structural inequalities to ill health. (From Diderichsen et al. (2001))

vulnerabilities combined precipitate (mental) illness. This vicious circle is closed by the illness leading to further social stratification (e.g., lost educational and employment opportunities). Having extensive multi-domain area-level datasets that can be used to characterize the physical, built, and social environments would enable us to test a variety of possible pathways (and their combination) leading from social stratification to brain health; decomposition analysis is but one method that can be used to quantify the contributions of various factors to the observed outcomes (O'Donnell et al., 2008).

Looking Forward

As discussed by Lovasi et al. and outlined in the environmental justice framework for exposure science (Van Horne et al., 2023), complexities of the multilayered relationships between the individual and their environment require not only top-quality data and conceptual and analytical approaches but also meaningful engagement with communities and their policymakers as well as development and implementation of adequate strategies by funders, academic institutions, and journal editors working in this research field. Innovative methods should be employed to address one of the main limitations of observational strategies, namely, the difficulty of making causal inferences (discussed in this book by Dumas et al. (chapter "How Can Concepts of Ethology Be Applied to Large-Scale Digital Data?"). For example, dense time series of multiple exposures and outcomes offer an opportunity for estimating Granger causality (Imran et al., 2023). Pseudo-experimental design can explore the causality and directionality in cases of discrete events that affect the local environment; it should be noted that events such as forest fires may impact not only physical (air quality, Khraishah et al., 2022) but also built (loss of infrastructure) and social (evacuation) environments. To begin to address this issue, at least for certain environments, Mendelian randomization (Smith & Ebrahim, 2003) could be used. For example, using genetic variants associated with biomarkers of low-grade inflammation (Xu et al., 2022b; Liu et al., 2019), one can test a mechanistic path by which air pollution affects brain-related outcomes (Fani et al., 2021). Finally, as reflected in the diversity of the Forum participants, this enterprise requires experts from wide-ranging domains, including geospatial and data science, behavioral and brain science, epidemiology and public health, ethics and law, and urban planning. Finding a common language and purpose will allow us to work together toward the understanding of how humans transform their environments and how environments shape human brain and behavior.

References

Aitken, M., De St Jorre, J., Pagliari, C., Jepson, R., & Cunningham-Burley, S. (2016). Public responses to the sharing and linkage of health data for research purposes: A systematic review and thematic synthesis of qualitative studies. *BMC Medical Ethics, 17*, 73.

Bjorkenstam, E., Cheng, S., Burstrom, B., Pebley, A. R., Bjorkenstam, C., & Kosidou, K. (2017). Association between income trajectories in childhood and psychiatric disorder: A Swedish population-based study. *Journal of Epidemiology and Community Health, 71*, 648–654.

Boyd, J. H., Ferrante, A. M., O'Keefe, C. M., Bass, A. J., Randall, S. M., & Semmens, J. B. (2012). Data linkage infrastructure for cross-jurisdictional health-related research in Australia. *BMC Health Services Research, 12*, 480.

Boyd, A., Golding, J., Macleod, J., Lawlor, D. A., Fraser, A., Henderson, J., Molloy, L., Ness, A., Ring, S., & Davey Smith, G. (2013). Cohort Profile: The "children of the 90s"—The index offspring of the Avon Longitudinal Study of Parents and Children. *International Journal of Epidemiology, 42*, 111–127.

Brook, J. R., Setton, E. M., Seed, E., Shooshtari, M., Doiron, D., & Consortium, C. T. C. U. E. H. R. (2018). The Canadian Urban Environmental Health Research Consortium: A protocol for building a national environmental exposure data platform for integrated analyses of urban form and health. *BMC Public Health, 18*, 114.

Carollo, A., Montefalcone, P., Bornstein, M. H., & Esposito, G. (2023). A Scientometric review of infant cry and caregiver responsiveness: Literature trends and research gaps over 60 years of developmental study. *Children (Basel), 10*, 1042.

Castro-Ramirez, F., Al-Suwaidi, M., Garcia, P., Rankin, O., Ricard, J. R., & Nock, M. K. (2021). Racism and poverty are barriers to the treatment of youth mental health concerns. *Journal of Clinical Child and Adolescent Psychology, 50*, 534–546.

Curtis, J. W., Curtis, A., Mapes, J., Szell, A. B., & Cinderich, A. (2013). Using Google Street View for systematic observation of the built environment: Analysis of spatio-temporal instability of imagery dates. *International Journal of Health Geographics, 12*, 53.

Diderichsen, F., Evans, T., & Whitehead, M. (2001). The social basis of disparities in health. In M. Whitehead, T. Evans, F. Diderichsen, A. Bhuiya, & M. Wirth (Eds.), *Challenging inequities in health: From ethics to action*. Oxford University Press.

Elgar, F. J., Pfortner, T. K., Moor, I., De Clercq, B., Stevens, G. W., & Currie, C. (2015). Socioeconomic inequalities in adolescent health 2002-2010: A time-series analysis of 34 countries participating in the Health Behaviour in School-aged Children study. *Lancet, 385*, 2088–2095.

Fani, L., Georgakis, M. K., Ikram, M. A., Ikram, M. K., Malik, R., & Dichgans, M. (2021). Circulating biomarkers of immunity and inflammation, risk of Alzheimer's disease, and hippocampal volume: A Mendelian randomization study. *Translational Psychiatry, 11*, 291.

Fowden, A. L., Vaughan, O. R., Murray, A. J., & Forhead, A. J. (2022). Metabolic consequences of glucocorticoid exposure before birth. *Nutrients, 14*, 2072–6643.

Grasby, K. L., Jahanshad, N., Painter, J. N., Colodro-Conde, L., Bralten, J., Hibar, D. P., Lind, P. A., Pizzagalli, F., Ching, C. R. K., Mcmahon, M. A. B., Shatokhina, N., Zsembik, L. C. P., Thomopoulos, S. I., Zhu, A. H., Strike, L. T., Agartz, I., Alhusaini, S., Almeida, M. A. A., Alnaes, D., Amlien, I. K., Andersson, M., Ard, T., Armstrong, N. J., Ashley-Koch, A., Atkins, J. R., Bernard, M., Brouwer, R. M., Buimer, E. E. L., Bulow, R., Burger, C., Cannon, D. M., Chakravarty, M., Chen, Q., Cheung, J. W., Couvy-Duchesne, B., Dale, A. M., Dalvie, S., De Araujo, T. K., De Zubicaray, G. I., De Zwarte, S. M. C., Den Braber, A., Doan, N. T., Dohm, K., Ehrlich, S., Engelbrecht, H. R., Erk, S., Fan, C. C., Fedko, I. O., Foley, S. F., Ford, J. M., Fukunaga, M., Garrett, M. E., Ge, T., Giddaluru, S., Goldman, A. L., Green, M. J., Groenewold, N. A., Grotegerd, D., Gurholt, T. P., Gutman, B. A., Hansell, N. K., Harris, M. A., Harrison, M. B., Haswell, C. C., Hauser, M., Herms, S., Heslenfeld, D. J., Ho, N. F., Hoehn, D., Hoffmann, P., Holleran, L., Hoogman, M., Hottenga, J. J., Ikeda, M., Janowitz, D., Jansen, I. E., Jia, T., Jockwitz, C., Kanai, R., Karama, S., Kasperaviciute, D., Kaufmann, T., Kelly,

S., Kikuchi, M., Klein, M., Knapp, M., Knodt, A. R., Kramer, B., Lam, M., Lancaster, T. M., Lee, P. H., Lett, T. A., Lewis, L. B., Lopes-Cendes, I., Luciano, M., Macciardi, F., Marquand, A. F., Mathias, S. R., Melzer, T. R., Milaneschi, Y., et al. (2020). The genetic architecture of the human cerebral cortex. *Science, 367*. https://doi.org/10.1126/science.aay6690

Hickman, C., Marks, E., Pihkala, P., Clayton, S., Lewandowski, R. E., Mayall, E. E., Wray, B., Mellor, C., & Van Susteren, L. (2021). Climate anxiety in children and young people and their beliefs about government responses to climate change: A global survey. *Lancet Planetary Health, 5*, e863–e873.

Imran, M., Khan, S., Nassani, A. A., Haffar, M., Khan, H. U. R., & Zaman, K. (2023). Access to sustainable healthcare infrastructure: A review of industrial emissions, coal fires, and particulate matter. *Environmental Science and Pollution Research International, 30*, 69080–69095.

Kardan, O., Gozdyra, P., Misic, B., Moola, F., Palmer, L. J., Paus, T., & Berman, M. G. (2015). Neighborhood greenspace and health in a large urban center. *Scientific Reports, 5*, 11610.

Kendler, K. S., Gardner, C. O., & Prescott, C. A. (2003). Personality and the experience of environmental adversity. *Psychological Medicine, 33*, 1193–1202.

Khaled, M., Corner, G. W., Morris, A., Havaldar, S., Luo, E., & Saxbe, D. E. (2021). Physiological linkage in pregnancy: Couples' cortisol, negative conflict behavior, and postpartum depression. *Biological Psychology, 161*, 108075.

Khraishah, H., Alahmad, B., Ostergard, R. L., Jr., Alashqar, A., Albaghdadi, M., Vellanki, N., Chowdhury, M. M., Al-Kindi, S. G., Zanobetti, A., Gasparrini, A., & Rajagopalan, S. (2022). Climate change and cardiovascular disease: Implications for global health. *Nature Reviews. Cardiology, 19*, 798–812.

Kivimaki, M., Batty, G. D., Pentti, J., Shipley, M. J., Sipila, P. N., Nyberg, S. T., Suominen, S. B., Oksanen, T., Stenholm, S., Virtanen, M., Marmot, M. G., Singh-Manoux, A., Brunner, E. J., Lindbohm, J. V., Ferrie, J. E., & Vahtera, J. (2020). Association between socioeconomic status and the development of mental and physical health conditions in adulthood: A multi-cohort study. *The Lancet Public Health, 5*, e140–e149.

Kolle, S., Hughes, B., & Steele, H. (2020). Early embryo-maternal communication in the oviduct: A review. *Molecular Reproduction and Development, 87*, 650–662.

Kum, H. C., & Ahalt, S. (2013). Privacy-by-design: Understanding data access models for secondary data. *AMIA Summits on Translational Science Proceedings, 2013*, 126–130.

Less, E. L., Mckee, P., Toomey, T., Nelson, T., Erickson, D., Xiong, S., & Jones-Webb, R. (2015). Matching study areas using Google Street View: A new application for an emerging technology. *Evaluation and Program Planning, 53*, 72–79.

Liu, Q., Gu, X., Deng, F., Mu, L., Baccarelli, A. A., Guo, X., & Wu, S. (2019). Ambient particulate air pollution and circulating C-reactive protein level: A systematic review and meta-analysis. *International Journal of Hygiene and Environmental Health, 222*, 756–764.

Lucyk, K., Lu, M., Sajobi, T., & Quan, H. (2015). Administrative health data in Canada: Lessons from history. *BMC Medical Informatics and Decision Making, 15*, 69.

Mangalore, R., Knapp, M., & Jenkins, R. (2007). Income-related inequality in mental health in Britain: The concentration index approach. *Psychological Medicine, 37*, 1037–1045.

Menon, R. (2019). Initiation of human parturition: Signaling from senescent fetal tissues via extracellular vesicle mediated paracrine mechanism. *Obstetrics & Gynecology Science, 62*, 199–211.

Metzl, J. M., & Hansen, H. (2018). Structural competency and psychiatry. *JAMA Psychiatry, 75*, 115–116.

Muller, K. U., Mennigen, E., Ripke, S., Banaschewski, T., Barker, G. J., Buchel, C., Conrod, P., Fauth-Buhler, M., Flor, H., Garavan, H., Heinz, A., Lawrence, C., Loth, E., Mann, K., Martinot, J. L., Pausova, Z., Rietschel, M., Strohle, A., Struve, M., Walaszek, B., Schumann, G., Paus, T., Smolka, M. N., & Consortium, I. (2013). Altered reward processing in adolescents with prenatal exposure to maternal cigarette smoking. *JAMA Psychiatry, 70*, 847–856.

Munzel, T., Sorensen, M., Hahad, O., Nieuwenhuijsen, M., & Daiber, A. (2023). The contribution of the exposome to the burden of cardiovascular disease. *Nature Reviews. Cardiology*. https://doi.org/10.1038/s41569-023-00873-3

Noble, K. G., Houston, S. M., Brito, N. H., Bartsch, H., Kan, E., Kuperman, J. M., Akshoomoff, N., Amaral, D. G., Bloss, C. S., Libiger, O., Schork, N. J., Murray, S. S., Casey, B. J., Chang, L., Ernst, T. M., Frazier, J. A., Gruen, J. R., Kennedy, D. N., Van Zijl, P., Mostofsky, S., Kaufmann, W. E., Kenet, T., Dale, A. M., Jernigan, T. L., & Sowell, E. R. (2015). Family income, parental education and brain structure in children and adolescents. *Nature Neuroscience, 18*, 773–778.

O'Brien, E. C., Rodriguez, A. M., Kum, H. C., Schanberg, L. E., Fitz-Randolph, M., O'Brien, S. M., & Setoguchi, S. (2019). Patient perspectives on the linkage of health data for research: Insights from an online patient community questionnaire. *International Journal of Medical Informatics, 127*, 9–17.

O'Donnell, O., Van Doorslaer, E., Wagstaff, A., & Lindelow, M. (2008). *Analyzing health equity using household survey data*. World Bank.

Odgers, C. L., Caspi, A., Bates, C. J., Sampson, R. J., & Moffitt, T. E. (2012). Systematic social observation of children's neighborhoods using Google Street View: A reliable and cost-effective method. *Journal of Child Psychology and Psychiatry, 53*, 1009–1017.

Ogilvie, J. M., Tzoumakis, S., Allard, T., Thompson, C., Kisely, S., & Stewart, A. (2021). Prevalence of psychiatric disorders for Indigenous Australians: A population-based birth cohort study. *Epidemiology and Psychiatric Sciences, 30*, e21.

Pampalon, R., Hamel, D., Gamache, P., Philibert, M. D., Raymond, G., & Simpson, A. (2012). An area-based material and social deprivation index for public health in Quebec and Canada. *Canadian Journal of Public Health, 103*, S17–S22.

Paquette, D., & St. George, J. M. (2023). Proximate and ultimate mechanisms of human father-child rough-and-tumble play. *Neuroscience and Biobehavioral Reviews, 149*, 105151.

Parker, N., Wong, A. P., Leonard, G., Perron, M., Pike, B., Richer, L., Veillette, S., Pausova, Z., & Paus, T. (2017). Income inequality, gene expression, and brain maturation during adolescence. *Scientific Reports, 7*, 7397.

Paus, T. (2010). Population neuroscience: Why and how. *Human Brain Mapping, 31*, 891–903.

Paus, T. (2013). *Population neuroscience*. Springer-Verlag.

Paus, T. (2016). Population neuroscience. *Handbook of Clinical Neurology, 138*, 17–37.

Paus, T., Brook, J., & Doiron, D. (2022). Mapping inequalities in the physical, built and social environment in population-based studies of brain health. *Frontiers in Neuroimaging, 1*, 884191.

Pencarrick Hertzman, C., Meagher, N., & Mcgrail, K. M. (2013). Privacy by Design at Population Data BC: A case study describing the technical, administrative, and physical controls for privacy-sensitive secondary use of personal information for research in the public interest. *Journal of the American Medical Informatics Association, 20*, 25–28.

Psaty, B. M., O'Donnell, C. J., Gudnason, V., Lunetta, K. L., Folsom, A. R., Rotter, J. I., Uitterlinden, A. G., Harris, T. B., Witteman, J. C., Boerwinkle, E., & Consortium, C. (2009). Cohorts for Heart and Aging Research in Genomic Epidemiology (CHARGE) Consortium: Design of prospective meta-analyses of genome-wide association studies from 5 cohorts. *Circulation. Cardiovascular Genetics, 2*, 73–80.

Rajmil, L., Herdman, M., Ravens-Sieberer, U., Erhart, M., Alonso, J., & European, K. G. (2014). Socioeconomic inequalities in mental health and health-related quality of life (HRQOL) in children and adolescents from 11 European countries. *International Journal of Public Health, 59*, 95–105.

Rantakallio, P. (1988). The longitudinal study of the northern Finland birth cohort of 1966. *Paediatric and Perinatal Epidemiology, 2*, 59–88.

Satizabal, C. L., Adams, H. H. H., Hibar, D. P., White, C. C., Knol, M. J., Stein, J. L., Scholz, M., Sargurupremraj, M., Jahanshad, N., Roshchupkin, G. V., Smith, A. V., Bis, J. C., Jian, X., Luciano, M., Hofer, E., Teumer, A., Van Der Lee, S. J., Yang, J., Yanek, L. R., Lee, T. V., Li, S., Hu, Y., Koh, J. Y., Eicher, J. D., Desrivieres, S., Arias-Vasquez, A., Chauhan, G., Athanasiu, L., Renteria, M. E., Kim, S., Hoehn, D., Armstrong, N. J., Chen, Q., Holmes, A. J., Den Braber,

A., Kloszewska, I., Andersson, M., Espeseth, T., Grimm, O., Abramovic, L., Alhusaini, S., Milaneschi, Y., Papmeyer, M., Axelsson, T., Ehrlich, S., Roiz-Santianez, R., Kraemer, B., Haberg, A. K., Jones, H. J., Pike, G. B., Stein, D. J., Stevens, A., Bralten, J., Vernooij, M. W., Harris, T. B., Filippi, I., Witte, A. V., Guadalupe, T., Wittfeld, K., Mosley, T. H., Becker, J. T., Doan, N. T., Hagenaars, S. P., Saba, Y., Cuellar-Partida, G., Amin, N., Hilal, S., Nho, K., Mirza-Schreiber, N., Arfanakis, K., Becker, D. M., Ames, D., Goldman, A. L., Lee, P. H., Boomsma, D. I., Lovestone, S., Giddaluru, S., Le Hellard, S., Mattheisen, M., Bohlken, M. M., Kasperaviciute, D., Schmaal, L., Lawrie, S. M., Agartz, I., Walton, E., Tordesillas-Gutierrez, D., Davies, G. E., Shin, J., Ipser, J. C., Vinke, L. N., Hoogman, M., Jia, T., Burkhardt, R., Klein, M., Crivello, F., Janowitz, D., Carmichael, O., Haukvik, U. K., Aribisala, B. S., Schmidt, H., et al. (2019). Genetic architecture of subcortical brain structures in 38,851 individuals. *Nature Genetics, 51*, 1624–1636.

Saxbe, D., Corner, G. W., Khaled, M., Horton, K., Wu, B., & Khoddam, H. L. (2018). The weight of fatherhood: Identifying mechanisms to explain paternal perinatal weight gain. *Health Psychology Review, 12*, 294–311.

Scambler, G. (2012). Health inequalities. *Sociology of Health & Illness, 34*, 130–146.

Schmit, C., Giannouchos, T., Ramezani, M., Zheng, Q., Morrisey, M. A., & Kum, H. C. (2021). US privacy laws go against public preferences and impede public health and research: Survey study. *Journal of Medical Internet Research, 23*, e25266.

Shin, J., Ma, S., Hofer, E., Patel, Y., Vosberg, D. E., Tilley, S., Roshchupkin, G. V., Sousa, A. M. M., Jian, X., Gottesman, R., Mosley, T. H., Fornage, M., Saba, Y., Pirpamer, L., Schmidt, R., Schmidt, H., Carrion-Castillo, A., Crivello, F., Mazoyer, B., Bis, J. C., Li, S., Yang, Q., Luciano, M., Karama, S., Lewis, L., Bastin, M. E., Harris, M. A., Wardlaw, J. M., Deary, I. E., Scholz, M., Loeffler, M., Witte, A. V., Beyer, F., Villringer, A., Armstrong, N. J., Mather, K. A., Ames, D., Jiang, J., Kwok, J. B., Schofield, P. R., Thalamuthu, A., Trollor, J. N., Wright, M. J., Brodaty, H., Wen, W., Sachdev, P. S., Terzikhan, N., Evans, T. E., Adams, H., Ikram, M. A., Frenzel, S., Auwera-Palitschka, S. V., Wittfeld, K., Bulow, R., Grabe, H. J., Tzourio, C., Mishra, A., Maingault, S., Debette, S., Gillespie, N. A., Franz, C. E., Kremen, W. S., Ding, L., Jahanshad, N., Consortium, E., Sestan, N., Pausova, Z., Seshadri, S., Paus, T., & Neuro, C. W. G. (2020). Global and regional development of the human cerebral cortex: Molecular architecture and occupational aptitudes. *Cerebral Cortex, 30*, 4121–4139.

Smith, G. D., & Ebrahim, S. (2003). 'Mendelian randomization': Can genetic epidemiology contribute to understanding environmental determinants of disease? *International Journal of Epidemiology, 32*, 1–22.

Soutar, C., & Wand, A. P. F. (2022). Understanding the spectrum of anxiety responses to climate change: A systematic review of the qualitative literature. *International Journal of Environmental Research and Public Health, 19*, 990.

Stuart, K., & Soulsby, E. J. (2011). Reducing global health inequalities. Part 1. *Journal of the Royal Society of Medicine, 104*, 321–326.

Suel, E., Polak, J. W., Bennett, J. E., & Ezzati, M. (2019). Measuring social, environmental and health inequalities using deep learning and street imagery. *Scientific Reports, 9*, 6229.

Sukumaran, K., Cardenas-Iniguez, C., Burnor, E., Bottenhorn, K. L., Hackman, D. A., Mcconnell, R., Berhane, K., Schwartz, J., Chen, J. C., & Herting, M. M. (2023). Ambient fine particulate exposure and subcortical gray matter microarchitecture in 9- and 10-year-old children across the United States. *iScience, 26*, 106087.

Thompson, P. M., Stein, J. L., Medland, S. E., Hibar, D. P., Vasquez, A. A., Renteria, M. E., Toro, R., Jahanshad, N., Schumann, G., Franke, B., Wright, M. J., Martin, N. G., Agartz, I., Alda, M., Alhusaini, S., Almasy, L., Almeida, J., Alpert, K., Andreasen, N. C., Andreassen, O. A., Apostolova, L. G., Appel, K., Armstrong, N. J., Aribisala, B., Bastin, M. E., Bauer, M., Bearden, C. E., Bergmann, O., Binder, E. B., Blangero, J., Bockholt, H. J., Boen, E., Bois, C., Boomsma, D. I., Booth, T., Bowman, I. J., Bralten, J., Brouwer, R. M., Brunner, H. G., Brohawn, D. G., Buckner, R. L., Buitelaar, J., Bulayeva, K., Bustillo, J. R., Calhoun, V. D., Cannon, D. M., Cantor, R. M., Carless, M. A., Caseras, X., Cavalleri, G. L., Chakravarty, M. M., Chang, K. D.,

Ching, C. R., Christoforou, A., Cichon, S., Clark, V. P., Conrod, P., Coppola, G., Crespo-Facorro, B., Curran, J. E., Czisch, M., Deary, I. J., De Geus, E. J., Den Braber, A., Delvecchio, G., Depondt, C., De Haan, L., De Zubicaray, G. I., Dima, D., Dimitrova, R., Djurovic, S., Dong, H., Donohoe, G., Duggirala, R., Dyer, T. D., Ehrlich, S., Ekman, C. J., Elvsashagen, T., Emsell, L., Erk, S., Espeseth, T., Fagerness, J., Fears, S., Fedko, I., Fernandez, G., Fisher, S. E., Foroud, T., Fox, P. T., Francks, C., Frangou, S., Frey, E. M., Frodl, T., Frouin, V., Garavan, H., Giddaluru, S., Glahn, D. C., Godlewska, B., Goldstein, R. Z., Gollub, R. L., Grabe, H. J., et al. (2014). The ENIGMA Consortium: Large-scale collaborative analyses of neuroimaging and genetic data. *Brain Imaging and Behavior, 8*, 153–182.

Tiemeier, H., Velders, F. P., Szekely, E., Roza, S. J., Dieleman, G., Jaddoe, V. W., Uitterlinden, A. G., White, T. J., Bakermans-Kranenburg, M. J., Hofman, A., Van Ijzendoorn, M. H., Hudziak, J. J., & Verhulst, F. C. (2012). The Generation R Study: A review of design, findings to date, and a study of the 5-HTTLPR by environmental interaction from fetal life onward. *Journal of the American Academy of Child and Adolescent Psychiatry, 51*(1119–1135), e7.

To, P., Eboreime, E., & Agyapong, V. I. O. (2021). The impact of wildfires on mental health: A scoping review. *Behavioral Sciences (Basel), 11*, 126.

Toro, R., Leonard, G., Lerner, J. V., Lerner, R. M., Perron, M., Pike, G. B., Richer, L., Veillette, S., Pausova, Z., & Paus, T. (2008). Prenatal exposure to maternal cigarette smoking and the adolescent cerebral cortex. *Neuropsychopharmacology, 33*, 1019–1027.

Usher, C. (2022). Eco-anxiety. *Journal of the American Academy of Child and Adolescent Psychiatry, 61*, 341–342.

Van Horne, Y. O., Alcala, C. S., Peltier, R. E., Quintana, P. J. E., Seto, E., Gonzales, M., Johnston, J. E., Montoya, L. D., Quiros-Alcala, L., & Beamer, P. I. (2023). An applied environmental justice framework for exposure science. *Journal of Exposure Science & Environmental Epidemiology, 33*, 1–11.

Vandenbroucke, J. P. (1990). Epidemiology in transition: A historical hypothesis. *Epidemiology, 1*, 164–167.

Waldron, I. (2018). *There's something in the water: Environmental racism in indigenous & black communities*. Fernwood Publishing.

Weichenthal, S., Hatzopoulou, M., & Brauer, M. (2019). A picture tells a thousand...exposures: Opportunities and challenges of deep learning image analyses in exposure science and environmental epidemiology. *Environment International, 122*, 3–10.

Wild, C. P. (2005). Complementing the genome with an "exposome": The outstanding challenge of environmental exposure measurement in molecular epidemiology. *Cancer Epidemiology, Biomarkers & Prevention, 14*, 1847–1850.

Xu, J., Liu, X., Li, Q., Goldblatt, R., Qin, W., Liu, F., Chu, C., Luo, Q., Ing, A., Guo, L., Liu, N., Liu, H., Huang, C., Cheng, J., Wang, M., Geng, Z., Zhu, W., Zhang, B., Liao, W., Qiu, S., Zhang, H., Xu, X., Yu, Y., Gao, B., Han, T., Cui, G., Chen, F., Xian, J., Li, J., Zhang, J., Zuo, X. N., Wang, D., Shen, W., Miao, Y., Yuan, F., Lui, S., Zhang, X., Xu, K., Zhang, L., Ye, Z., Banaschewski, T., Barker, G. J., Bokde, A. L. W., Flor, H., Grigis, A., Garavan, H., Gowland, P., Heinz, A., Bruhl, R., Martinot, J. L., Artiges, E., Nees, F., Orfanos, D. P., Lemaitre, H., Paus, T., Poustka, L., Robinson, L., Hohmann, S., Frohner, J. H., Smolka, M. N., Walter, H., Whelan, R., Winterer, J., Patrick, K., Calhoun, V., Li, M. J., Liang, M., Gong, P., Barker, E. D., Clinton, N., Marquand, A., Yu, L., Yu, C., Schumann, G., & Chimgen, & Consortia, I. (2022a). Global urbanicity is associated with brain and behaviour in young people. *Nature Human Behaviour, 6*, 279–293.

Xu, Z., Wang, W., Liu, Q., Li, Z., Lei, L., Ren, L., Deng, F., Guo, X., & Wu, S. (2022b). Association between gaseous air pollutants and biomarkers of systemic inflammation: A systematic review and meta-analysis. *Environmental Pollution, 292*, 118336.

Advancing the Neural Exposome

Sigrid C. Veasey

Abstract Epidemiologic genetics studies have revealed key roles for genes in human disease but have also reinforced the need to include environmental exposures in risk assessment for most disorders and conditions. The concept of exposome was developed over two decades ago to include the lifetime composite of chemicals and other exposures that negatively impact health. The brain is uniquely susceptible to a multitude of chemical exposures, especially across specific windows of neurodevelopment, yet may influence the development of neurodegenerative processes as well. Here, we highlight several categories of the best-studied environmental neurotoxicants. These studies, however, also serve to emphasize that so much more must be learned about individual chemical exposures and combinations of chemical exposures to understand their roles in human disease. Newer techniques have been developed to characterize the exposome in large populations, by taking advantage of existing big data collections and incorporating newer tools to assess dynamic aspects of exposures and will also be described below. Finally, high-throughput screens are being developed to rapidly assess relative neurotoxicity for one or more chemicals. Genes identified may be used to the focus laboratory animal studies on select chemicals and to assess in human studies. It is the merging of all these techniques that will move this field most efficiently.

Keywords Exposome · Environmental neurotoxicity · Endocrine-disrupting chemicals · Neurodevelopment · Neurodegeneration

This work was supported in part by NIH NIEHS P30 ES013508

S. C. Veasey (✉)
Chronobiology and Sleep Institute, Department of Medicine, Perelman School of Medicine, University of Pennsylvania, Philadelphia, PA, USA
e-mail: veasey@pennmedicine.upenn.edu

S. Kühn (ed.), *Environmental Neuroscience*,
https://doi.org/10.1007/978-3-031-64699-7_5

Exposome: Definition and Significance

Successful sequencing and mapping of the human genome helped deliver a wealth of innovative molecular and analytical tools that, in turn, have fostered a vast number of genomic studies identifying genes associated with human disease. However, as use of the new tools led to an explosion in genome-wide association studies, researchers soon realized that most genes associated with specific diseases are of high prevalence and low penetrance, meaning that other genes and/or specific environment exposures must also contribute to risk of disease. The pace of study of environmental risk factors for human disease, which typically occurs one exposure risk at a time, has developed more slowly, yet research has delivered the neural impact of increasing numbers of distinct environmental exposures that contribute to specific human disease. These efforts and results have prompted consideration of how best to analyze environmental risks in addition to genetic risk assessment. The term "exposome" was developed for the complimentary consideration of environmental exposures that might also contribute to disease risk or influence disease phenotype (Wild, 2005).

An exposome is defined as the comprehensive set of cumulative exposures for an individual from conception to the point in life when disease or a phenotype is defined, or up until death. Inherently, every living thing has its own distinct exposome. Three domains of exposures contribute to this collective exposome. The first is termed general external, which refers to the broader physical and psychosocial exposures plaguing a regional environment (e.g., ambient temperatures, noise, air/water/soil pollution, socioeconomic stress, education access). The second domain has been termed specific external exposures, as these can be modified in part through individual actions (e.g., radiation/ultraviolet exposure, tobacco use, alcohol and drug use, home pesticides, and diet), and the third domain is comprised of internal exposure modifiers (e.g., development, aging, metabolism, microbiota, infection exposures, hormone activity/responses, immune/inflammatory responses, genetic predisposition to inflammatory, or detrimental responses to exposures). An additional concept for the exposome is needed, however, and that is for a relevant exposome, meaning the exposures within any of the three domains that induce a biological reaction that threatens health. While described as three domains, all domains should be considered collectively, and there is clearly overlap across domains. For example, government spraying of pesticides regionally would fall under general external, while pesticide use on one's lawn or in the home would be included in the specific external domain. Moreover, responses to general external or specific external domain exposures may then influence internal exposure responses (e.g., shorten telomers, induce oxidative stress, change skin, and gut microbiome), and some exposures, for example socioeconomic stress could be considered in both the general and specific external exposomes. Overall, it is less important to consider to which domain an exposure belongs, as it is to consider all types of exposures that might contribute to one's collective exposome. An example of a lifetime of exposures across all three domains is provided in Fig. 1. It is important to consider that some exposures such as nature, healthy diet, and exercise may be protective.

In contrast with genetics, the exposome is far more dynamic as it is continuously changing across the lifespan (➔ chapters "Environmental Exposures and Child Neurodevelopment" by Peuters et al. and "Neighbourhood Urban Environments and Cognitive Health in Ageing Populations" by Soloveva and Cerin et al.). Important considerations include age at exposure, duration-and dose-dependent responses, the consequences of additional exposures that impart additive or synergistic effects, and genetic determinants of response to exposure to individual exposures across the ages that will then impact disease phenotype.

It is important to appreciate how significantly environmental exposures contribute to illness and disability. The Global Burden of Disease Project examined the

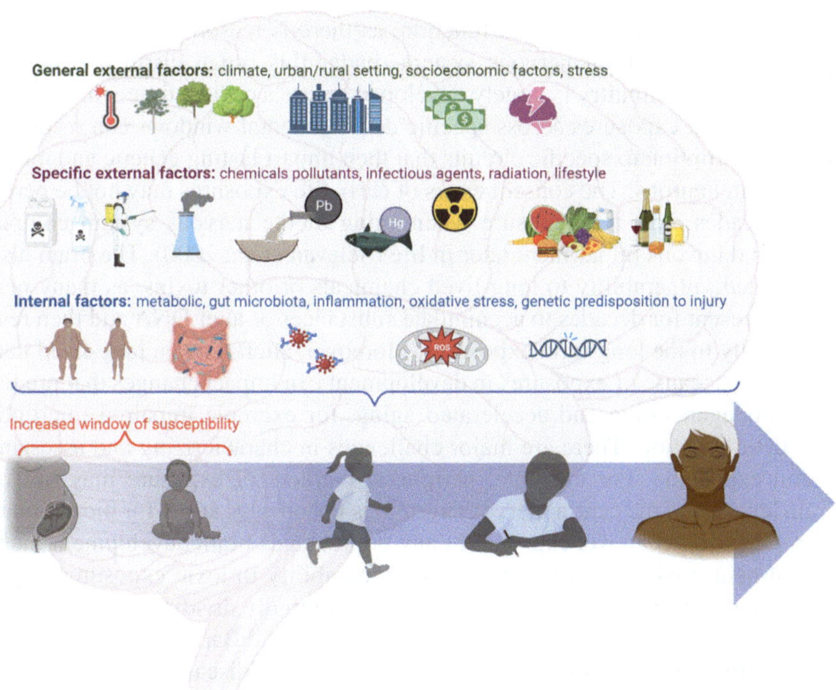

Fig. 1 Example of the neural exposome in an older woman. Exposures occur continuously across the lifetime (large blue arrow) but are perhaps more likely to induce lasting effects when exposures occur before the blood–brain barrier develops in utero or as the brain most rapidly develops (red bracketed ages). Exposures may be categorized as general external exposures (upper, green text) defined as exposures experienced by all in the vicinity). Images directly below represent some of examples of general external factors. A second category is a group of specific external factors (purple text), defined as exposures incurred by only a subset of individuals in the vicinity because of differences in lead pipes into home, choosing fish with higher mercury (Hg) levels, or household use of insecticides, etc. Images below provide some examples. The third category, internal factors (blue text), may be considered as an individual's intrinsic predisposition or resilience to neurotoxicity from exposures. Here alterations in underlying inflammatory responses from obesity, or oxidative stress from other disorders and genetic predispositions to detrimental responses may influence overall neural injury from exposome. The figure was generated with Biorender

impact of over 80 metabolic and behavioral environmental exposures worldwide and determined that these risk factors contributed to more than half of the deaths in the 195 countries studied. A fundamental tragedy, thus, is that most of the environmental exposures that contribute to illness, disability, and death are presumably modifiable.

Neural-Specific Considerations Regarding the Exposome Vulnerability

The term, "neuroexposome," has been proposed to encapsulate all exposures that may influence brain health and function, as there is reason to believe that unique qualities of the central nervous system render this organ distinctly susceptible. Notably, neural circuitry is largely developed across a critical time window (now or never), so that exposures across specific developmental windows can result in permanent disruption to specific circuits that then impart lasting genetic and metabolic changes in neurons. The consequences of early life exposures may not be perceived until decades after the exposure when aging of the nervous system compounds injury and unveils impairment later in life (Selevan et al., 2000). The brain also has heightened vulnerability to long-lived chemicals or other toxins, as many neurons can be present for decades to accumulate substances or alter DNA and then respond perpetually to the long-term exposures. Moreover, alterations in long-lived neurons that occur because of exposures in development can impact changes that predispose to neurodegeneration and accelerated aging, for example impaired microglial or astrocytic function. There are major challenges in characterizing and assessing the neural exposome. For example, peripheral markers of exposure may not reflect brain levels, and the central nervous system is not an easy target for biopsy material.

As mentioned above, exposure at any point across brain development increases risk of neurotoxicity, yet the increased vulnerability to toxic exposures is present even prior to brain development when cells are rapidly dividing and there is effectively no blood–brain barrier. Three decades ago, David Barker proposed that alterations in the fetal environment could lead to chronic disease later in life (Barker et al., 1993) (➔ chapter "Environmental Exposures and Child Neurodevelopment" by Peuters et al.). Later Lahiri proposed that prenatal epigenetic changes could induce late in life neurological disease (Lahiri et al., 2009) and set the stage for heightened vulnerability to additional hits later in life that collectively result in a specific neurologic disorder (Heindel et al., 2015).

In light of the heightened vulnerability of the brain to various environmental exposures and the challenges in accurately characterizing exposure doses and timing, it is not surprising that we will need to develop new methods to identify the most impactful exposure risks and the individuals at heightened neuroexposome risk, and in turn prioritize international collaborative efforts to reduce our greatest threats in the environment.

The Study of Select Neurotoxicants from Bench to Bedside

Neurotoxicants are chemicals and substances that can alter brain health, cognitive function, and/or behavior. Neural effects of toxicants may be temporary or long-lived, and effects may be observed acutely or after a delayed period, and for many individuals, the consequences are never considered or appreciated. Notably, within some developed countries prevalences for both autism and attention deficit disorder have increased significantly in the past two decades (Landrigan et al., 2012). Increased occurrences of these disorders are unlikely to be genetic. Thus, in addition to increased recognition of the conditions, altered exposomes across the time of increase in given populations must be considered. There are, however, over 350,000 chemicals that are registered for use in developed countries (Wang et al., 2020), and while only 5000 may be used widely enough to be considered universal threats, individuals in specific regions of the world are still at risk of other chemicals through occupational or local exposures. To date only 1000 neurotoxicants have been substantiated using animal models, and human epidemiology studies suggest increased risk for neural injury for just over 200 chemicals (Grandjean & Landrigan, 2006, 2014). However, most studies have examined the effects of one chemical on one phenotype. Clearly, there is still an enormous amount of work to do to define precisely who is at risk where, at what age, what are the consequences (acute and delayed). Research has focused on particular neurotoxicants, including lead exposure, the particulate matter in air pollution, and endocrine-disrupting chemicals. For each of these emphasized areas, findings in animal models from exposures and findings in human epidemiology studies are discussed.

Lead (Pb) Exposure

Neurotoxic effects of Pb have been recognized for over two millennia, yet the metal has been continuously used in diverse arrays of man-made products, such that exposures to this toxin exist to this day from existing Pb products, and occupational Pb exposure in developing countries persists, as Pb use is only partially regulated. Human exposure to Pb generally occurs through contact with soil, air, water, or food. Major sources of Pb across the last 100 years have been older paints, lead-containing pipes carrying drinking water, and sites enriched in Pb-containing gasoline. The banning of producing Pb-based paints and gasoline helped reduce the average Pb levels in children's blood from 13.7ug/dl to 3.2 ug/dl but not to 0 ug/dl. Unfortunately, to this day, buildings, plumbing systems, and roads in older cities contain sufficient Pb that drinking water piped into these buildings exposes people to Pb, and when older buildings are demolished, Pb may acutely increase in proximal air and water to acutely increase exposure. Occupations including refining, battery manufacturing, and smelting continue to expose workers to higher levels of Pb. Pottery, various homeopathic remedies, and cosmetics may also expose individuals to Pb (Sanborn et al., 2002). Notably, absorption of Pb may be increased by deficiencies in nutrient metals, zinc, iron and/or calcium (Karri et al., 2008), such that disadvantaged children and adults are at higher risk for a given exposure. Children of parents working with Pb can be exposed when parents working in Pb-product industries bring home Pb dust on clothing, and fetuses may be exposed when Pb in the mother crosses the placenta. Pregnancy promotes the release of Pb from bones into the bloodstream, increasing Pb exposure to both the mother and unborn baby.

Pb can interfere with neurotransmitter systems and neural organelle function to have diverse effects in both the central and peripheral nervous systems. Once the blood–brain barrier is developed, Pb will compete with calcium for crossing the blood–brain barrier; in turn reductions in calcium will interfere with many neuronal functions including neurotransmission. Either through neurotransmitter excitatory effects or through mitochondrial injury, neuronal injury, neurotransmission dysfunction, and neuron loss may occur. Brain imaging studies have revealed loss of cortical gray matter in individuals with higher blood Pb levels. Dopaminergic neurons may be particularly susceptible to the effects of Pb exposure, resulting in increased dopaminergic signaling (Zuch et al., 1998; Szczerbak et al., 2007), which may then influence the risk of aggression, impulsivity, excessive motor behaviors, poor sleep, and/or addiction. Pb exposure may also disturb cholinergic signaling, as Pb impairs acetylcholinesterase release and can result in acetylcholine accumulation (Radulescu & Lundgren, 2019). Blood Pb levels at ages 3–5 predict poorer memory and attention at age of 12 years (Halabicky et al., 2022). Effects of higher blood Pb levels on behavior have been established for children, including behavioral disorders, attention deficit disorders, psychological conditions, and lower intelligence quotients (IQ). Very high levels (>80 ug.dl) are associated with encephalopathy, seizures ataxia, drowsiness, and even coma. A study of over 12,000 babies in the United States found that umbilical cord Pb levels predict the intelligence quotient (IQ) of children at 24 months. Pb exposure may also contribute to hearing loss and peripheral neuropathy.

Pb exposure increases the risk of brain injury across the lifespan. Whether Pb exposure increases the risk of autism is not known. Several studies have found elevated blood Pb levels in some children with autism disorders but not in all children (Hauptman et al., 2019). It is also possible that early life Pb exposure could result in mutagenicity or genotoxicity to promote the development of autism disorders (Pugsley et al., 2022). Early life Pb exposure is a risk for Alzheimer's disease (AD) and related dementias late-in-life, suggesting important delayed effects of Pb toxicity. Studies in animal models support a direct connection by showing that Pb exposure from conception to postnatal day 20 in mice resulted in increased amyloid plaque formation at 20 months of age. Similar findings have been shown in nonhuman primates, exposed to Pb from birth until 1 year of age, where increased amyloid plaques were present in aged (20-year-old monkeys). In animal models, Pb exposure has its greatest effects on AD risk in the developing nervous system. While effects of two other metals arsenic and cadmium had no apparent effect on AD-like pathology, exposure to all three metals markedly increased AD pathology supporting a synergistic effect and the importance of comprehensively assessing metal exposures. In summary, a large body of data supports Pb as an important neurotoxicant, and any exposure to Pb may be detrimental to brain health.

An important unanswered question regarding Pb toxicity is whether the chronic effects can be lessened or reversed in any way. There is some evidence that environmental enrichment reduced stress and improved nutrition, and exercise could help promote brain health to lessen effects. An equally important question to be answered is whether there are genes or other exposures that act synergistically to accentuate Pb effect on brain function and health.

Particulate Matter Exposure

Global industrialization has substantially increased health risk to air pollution exposures. Air pollution is chemically diverse and can include any combination of particulate matter (PM), which is largely comprised of nitrogen dioxide, sulfur dioxide, and polycyclic aromatic hydrocarbons. Particulate matter (PM), in turn, is composed of metals (Pb, arsenic, iron, titanium), elemental carbon, organic carbons, sulfate and nitrates, and various composites of PM are categorized by their size. Common sources include automobile exhaust, industrial waste, construction, and demolition dust. PM10 has diameters of 2.5–10 μm; Fine PM or PM2.5 has diameters 100ηm–2.5 μm, and ultrafine PM has diameters <100ηm. PM2.5 and ultrafine are sufficiently small to enter the lungs, be absorbed into the bloodstream, and then cross the blood–brain barrier. Thus, PM2.5 and ultrafine PM may carry a higher threat to brain health and function (Calderon-Garciduenas et al., 2020). Additionally, PM may alter gut microbiota to induce neuroinflammation (Panda et al., 2023).

The mechanisms by which PM injures the brain are just beginning to be understood. Studies in animal models of exposure to PM2.5 have found increased glial reactivity in several brain regions including the frontal cortices, hippocampus, cerebellum, and midbrain (Di Domenico et al., 2020; Gomez-Budia et al., 2020). In animal models, PM2.5 exposures depending on concentration and duration of exposure can result in anxiety-like behaviors, poorer learning, and memory, where lower doses may result in anxiety and higher doses in learning impairments (Liu et al., 2023). PM exposure in animal models also increases DNA methylation and oxidative stress in the brain (Wei et al., 2017). How specifically PM2.5 results in oxidative stress is not known, but antioxidant enzyme levels are reduced in response to PM2.5 (Fagundes et al., 2015; Cole et al., 2016). PM exposure also increases brain levels of N-methyl-D-aspartate (NMDA), including in the hippocampus and increases the levels of specific NMDA receptor subtypes in the hippocampus that promote excitotoxicity (Allen et al., 2014; Kim et al., 2020). Finally, there is newer evidence that PM may promote iron toxicity by disrupting iron metabolism in a manner that also involves excitotoxicity (Mei et al., 2024). In an AD knock-in mouse model, ultrafine PM exposure increased amyloid plaque deposition and glial reactivity (Kilian et al., 2023). PM2.5 exposures can injure dopaminergic neurons and promote a-synuclein fibrillization and neurodegeneration in a transgenic

Parkinson's mouse model (Choi et al., 2022; Yuan et al., 2022). From the above animal model studies, it is evident that PM exposures place the brain at significant risk.

Epidemiologic studies in humans support roles for PM2.5 in both neurodevelopmental and neurodegenerative disorders. Children exposed to PM2.5 in utero have reduced volumes of the corpus collosum and evidence behavioral disorders (Mortamais et al., 2019). Higher exposures to PM2.5 across the first 3 years of life increase the risk for autism spectrum disorders (Raz et al., 2015). PM2.5 levels are associated with neuropathologic hallmarks of Alzheimer's disease including amyloid plaques and tau hyperphosphorylated tangles (Thiankhaw et al., 2022). Similarly, higher PM2.5 exposures were associated with Parkinson's risk, in a region-specific way suggesting that specific sub-particles in air pollution may differentially influence risk of Parkinson's (Krzyzanowski et al., 2023). Major challenges for advancing this science are that the composition of PM at all sizes is highly variable such that toxicities vary even from car exhaust in one city to the next. High-throughput screens of simpler life forms, for example planarians, c. elegans, drosophila, and/or zebrafish are needed to identify which specific particle components impose the greatest toxicity for the nervous system.

Endocrine-Disrupting Chemicals

Endocrine-disrupting chemicals (EDCs) are a diverse group of environmental chemicals that have in common effects on hormone synthesis or hormone receptor signaling (Rosenfeld & Cooke, 2019). However, it is now understood that these chemicals also exert neurotoxic effects at nonsteroid receptors, act as transcriptional coactivators or suppressers, and some of the ECDs act, in part, through interference with enzymatic pathways (Diamanti-Kandarakis et al., 2009). EDCs have also been found to have direct effects on genes, and the chemicals can perturb cellular organelles, including the endoplasmic reticulum (Kusunoki et al., 2008). EDCs are present in a wide-ranging group of exposures, including pesticides, industrial solvents and lubricants, flame retardants, cosmetics, and plastic containers and plasticizers. Indeed, these chemicals are ubiquitous such that most humans are exposed to multiple EDCs every day, particularly EDCs used in plastic containers, flame retardants on furniture and clothing and pesticides on lawns and in soil and water, and thereby in fish, animals, and dairy products. Biological half-lives of these chemicals vary across chemical families. Parabens, phthalates, benzophenones, and bisphenols are quite short (1–20 hours), and in contrast, the half-lives of organotins (perfluoroalkanes, dioxins, and PCB's) range from 3 to 40 years (Denuziere & Ghersi-Egea, 2022).

Bisphenol A (BPA) is an EDC commonly used in food and beverage containers. Mice exposed to low-level BPA across gestation show structural alterations in brain nuclei, impaired spatial memory, and increased aggression and anxiety (Raja et al., 2022; Zheng et al., 2022). BPA has also been shown to induce oxidative stress in hypothalamic neurons (in vitro) that contributes to increased neuropeptide Y synthesis, which may modulate both food intake and reproduction (Loganathan et al., 2020). Many pesticide ECDs (organochlorines and organophosphates) have neurotoxic effects, where dopaminergic neurons have heightened susceptibility to α-synuslein misfolding and degeneration (Rocha et al., 2022). Organochlorine pesticides act in part by reducing anticholinesterase levels but can have additional neurotoxic effects (McKnight & Hack, 2020). Specific mixtures of EDCs have been shown to impair social interaction in male mice (Repouskou et al., 2020). Effects on the brain and peripheral nervous system vary across ECDs, as the binding affinities for various hormonal and nonhormonal receptors and additional effects differ with chemicals. Gestational exposure to organophosphates in mice results in reduced neuritic (axonal and dendritic) complexity and reduced neurogenesis (Tran et al., 2020). Normal maternal and fetal thyroid hormone levels are essential for brain development, and EDC exposure (especially polybrominated diphenyl ethers, in specific flame retardants) increases the risk of hypothyroidism-induced neurodevelopmental disorders (Oulhote et al., 2016). EDCs may also disrupt the blood–brain barrier and increase neuroinflammation (Denuziere & Ghersi-Egea, 2022).

As expected from such a diverse group of chemicals, neurotoxicities from exposures to EDCs are extensive and include acute toxicity syndromes, neurodevelopmental disorders, neuropsychiatric disorders, and neurodegenerative processes. A large epidemiology study in France examined prenatal exposure to pyrethroid insecticides on behavior at age of 6 years. Increased maternal levels of pyrethroid metabolites increased the risk for borderline social behavior disorder and difficulties externalizing (Viel et al., 2017). EDC exposure in the home is associated with poorer academic performance in grade school (Clark-Reyna et al., 2016). Maternal third trimester levels of phthalates and BPA predict social difficulties, specifically poorer social awareness, and social communication skills (Miodovnik et al., 2011). However, comprehensive examination of all epidemiological studies assessing ECD risk for autism spectrum disorders does not support an overall elevated risk (Cunha et al., 2023). Nonetheless, given the variability in chemicals, much more work must be done before concluding no effects. The increased risk of Parkinson's disease from ECD pesticide exposures is well established and there are important environment–gene interactions, where gene mutations alter the metabolism of pesticides to increase neurotoxicity (McCann et al., 1997; Ascherio et al., 2006). Associations between EDC exposures and AD and related dementias are weaker, with conflicting results across studies (Killin et al., 2016).

High-Throughput Screening for Neurotoxicants

With over 5000 chemicals to test as important neurotoxicants, the present approach of testing one or even several chemicals at the same time in laboratory rodents to assess behavior and pathological effects will take decades to complete; thus, higher throughput approaches are needed. Three-dimensional (3D) cell cultures of neural stem cells allow assessment of toxicant effects on cell–cell communication and effects on the extracellular matrix (Kang et al., 2021). It is now possible to develop neural organoids using tissue from specific brain regions, for example, the hippocampus. Organ-on-a-chip techniques may provide adequate nutrients and oxygenation. A clear advantage of these tissue models is that they may be created with human-derived cells. Now that more and more labs can maintain brain organoids for 1–2 months, we can expect that high-throughput human neural 3-D tissue screens for environmental toxins will soon unveil new neurotoxicants. Further, because these methods are high-throughput, the possibility of testing multiple combinations is now more realistic. Planarians have been used for a while for the study of environmental toxicology. The animals have central neural tissue and allow for the study of neurodevelopment, behavior, and neurodegeneration (Bayingana et al., 2023; Rabeler et al., 2023). Recent screens have examined numerous chemicals at multiple doses, assessing effects on behavior, and neurodevelopment (Zhang et al., 2007; Zhang et al., 2019). Zebrafish are an excellent model in which to test water-soluble chemicals, and successful screens have been implemented to assess the effects of over 10 polyfluoroalkyl chemicals across approximately 10 exposure concentrations to assess developmental and behaviors, for example motor activity (Gaballah et al., 2020). *Drosophila melanogaster* and *c.elegans* may provide high-throughput models for PM exposure and hydrophobic chemical exposures, respectively. Efficiency with these screens could be further enhanced if researchers in the field develop a registry to specific what screens are being performed and what tools are needed.

Measuring the Exposome

To capture the complete exposome would require continuous measurement of thousands of chemicals and other biological stressors continuously in the environment and in all body tissues. Admittedly, measuring the exposome will always be incomplete, yet there are newer approaches that may be used to shift from the conceptual exposome to providing key exposure information. Large cohorts with associated diagnoses and biobanks that were created for genetic analyses can now be reevaluated for exposure analyses, including plasma biomarkers. Clearly, the optimal approach will involve integrating genetic and exposure information, and this effort will require international collaborations (Wild, 2005, 2012). We should also take advantage of other large datasets from geographic information systems, which collect mapped data from sensors. This data can be used to provide spatial information for green space, air pollution, light and noise pollution, water contamination, and soil contamination, and because of the ubiquitous use of cellular phones with global positioning systems, mobility of people may also be analyzed. Maps like this exist for ambient temperatures around the world, nighttime light levels, and for PM2.5 levels globally. All of these external exposome data may be analyzed now and related to the prevalence of specific neurodevelopmental, behavioral, and neurodegenerative processes. Portable sensors are increasingly being used to detect additional exposures. A silicone wristband has been developed to detect chemical exposures across a 30-day period. This device has been implemented in a study of over 1000 participants, where the band detected approximately 50 environmental chemicals with known neurotoxicity, including phthalates, polycyclic aromatic hydrocarbons, and select pesticides (O'Connell et al., 2014). Personal sampling devices may also be used to detect physiological responses (e.g., blood pressure, heart rate, oxygenation). Smart phones routinely capture use, location, and ambient light but optical add-ons can be used to measure ambient aerosols. Similarly, newer techniques can be used to collect questionnaire data from far larger populations globally by using smartphones, social media, and/or other heavily trafficked websites.

Concluding Remarks

At the time of the initial concept of exposome, the task of identifying and detailing chemicals that negatively impact neurobiology seemed insurmountable. Animal model studies have begun to characterize neurodevelopmental and neurobehavioral effects of select chemicals, yet the number of chemicals in need of study is formidable. With the combined advancement, however, of tools to detail geospatial risks influencing human disease and behavior and the development and refinement of several excellent high-throughput screens for neurotoxicants, it is likely that just as the genome was sequenced, we will soon have the complimentary neuroexposome date to effectively complement genetic studies.

References

Allen, J. L., Liu, X., Pelkowski, S., Palmer, B., Conrad, K., Oberdorster, G., Weston, D., Mayer-Proschel, M., & Cory-Slechta, D. A. (2014). Early postnatal exposure to ultrafine particulate matter air pollution: persistent ventriculomegaly, neurochemical disruption, and glial activation preferentially in male mice. *Environmental Health Perspectives, 122*, 939–945.

Ascherio, A., Chen, H., Weisskopf, M. G., O'Reilly, E., McCullough, M. L., Calle, E. E., Schwarzschild, M. A., & Thun, M. J. (2006). Pesticide exposure and risk for Parkinson's disease. *Annals of Neurology, 60*, 197–203.

Barker, D. J., Gluckman, P. D., Godfrey, K. M., Harding, J. E., Owens, J. A., & Robinson, J. S. (1993). Fetal nutrition and cardiovascular disease in adult life. *Lancet, 341*, 938–941.

Bayingana, K., Ireland, D., Rosenthal, E., Rabeler, C., & Collins, E. S. (2023). Adult and regenerating planarians respond differentially to chronic drug exposure. *Neurotoxicology and Teratology, 96*, 107148.

Calderon-Garciduenas, L., Herrera-Soto, A., Jury, N., Maher, B. A., Gonzalez-Maciel, A., Reynoso-Robles, R., Ruiz-Rudolph, P., van Zundert, B., & Varela-Nallar, L. (2020). Reduced repressive epigenetic marks, increased DNA damage and Alzheimer's disease hallmarks in the brain of humans and mice exposed to particulate urban air pollution. *Environmental Research, 183*, 109226.

Choi, D., Lee, G., Kim, K. H., & Bae, H. (2022). Particulate matter exacerbates the death of dopaminergic neurons in Parkinson's disease through an inflammatory response. *International Journal of Molecular Sciences, 23*(12), 6487.

Clark-Reyna, S. E., Grineski, S. E., & Collins, T. W. (2016). Ambient concentrations of metabolic disrupting chemicals and children's academic achievement in El Paso, Texas. *International Journal of Environmental Research and Public Health, 13*(9), 874.

Cole, T. B., Coburn, J., Dao, K., Roque, P., Chang, Y. C., Kalia, V., Guilarte, T. R., Dziedzic, J., & Costa, L. G. (2016). Sex and genetic differences in the effects of acute diesel exhaust exposure on inflammation and oxidative stress in mouse brain. *Toxicology, 374*, 1–9.

Cunha, Y. G. O., do Amaral, G. C. B., Felix, A. A., Blumberg, B., & Amato, A. A. (2023). Early-life exposure to endocrine-disrupting chemicals and autistic traits in childhood and adolescence: A systematic review of epidemiological studies. *Frontiers in Endocrinology (Lausanne), 14*, 1184546.

Denuziere, A., & Ghersi-Egea, J. F. (2022). Cerebral concentration and toxicity of endocrine disrupting chemicals: The implication of blood-brain interfaces. *Neurotoxicology, 91*, 100–118.

Di Domenico, M., Benevenuto, S. G. M., Tomasini, P. P., Yariwake, V. Y., de Oliveira, A. N., Rahmeier, F. L., da Cruz, F. M., Moura, D. J., Nascimento Saldiva, P. H., & Veras, M. M. (2020). Concentrated ambient fine particulate matter (PM(2.5)) exposure induce brain damage in pre and postnatal exposed mice. *Neurotoxicology, 79*, 127–141.

Diamanti-Kandarakis, E., Bourguignon, J. P., Giudice, L. C., Hauser, R., Prins, G. S., Soto, A. M., Zoeller, R. T., & Gore, A. C. (2009). Endocrine-disrupting chemicals: An Endocrine Society scientific statement. *Endocrine Reviews, 30*, 293–342.

Fagundes, L. S., Fleck Ada, S., Zanchi, A. C., Saldiva, P. H., & Rhoden, C. R. (2015). Direct contact with particulate matter increases oxidative stress in different brain structures. *Inhalation Toxicology, 27*, 462–467.

Gaballah, S., Swank, A., Sobus, J. R., Howey, X. M., Schmid, J., Catron, T., McCord, J., Hines, E., Strynar, M., & Tal, T. (2020). Evaluation of developmental toxicity, developmental neurotoxicity, and tissue dose in Zebrafish exposed to GenX and other PFAS. *Environmental Health Perspectives, 128*, 47005.

Gomez-Budia, M., Konttinen, H., Saveleva, L., Korhonen, P., Jalava, P. I., Kanninen, K. M., & Malm, T. (2020). Glial smog: Interplay between air pollution and astrocyte-microglia interactions. *Neurochemistry International, 136*, 104715.

Grandjean, P., & Landrigan, P. J. (2006). Developmental neurotoxicity of industrial chemicals. *Lancet, 368*, 2167–2178.

Grandjean, P., & Landrigan, P. J. (2014). Neurobehavioural effects of developmental toxicity. *The Lancet Neurology, 13*, 330–338.

Halabicky, O. M., Ji, X., Gur, R. E., Gur, R. C., Yan, C., Chen, A., & Liu, J. (2022). Childhood lead exposure and sex-based neurobehavioral functioning in adolescence. *Neurotoxicology, 93*, 1–8.

Hauptman, M., Stierman, B., & Woolf, A. D. (2019). Children with autism spectrum disorder and lead poisoning: Diagnostic challenges and management complexities. *Clinical Pediatrics (Phila), 58*, 605–612.

Heindel, J. J., Balbus, J., Birnbaum, L., Brune-Drisse, M. N., Grandjean, P., Gray, K., Landrigan, P. J., Sly, P. D., Suk, W., Cory Slechta, D., Thompson, C., & Hanson, M. (2015). Developmental origins of health and disease: Integrating environmental influences. *Endocrinology, 156*, 3416–3421.

Kang, S. Y., Joshi, P., & Lee, M. Y. (2021). High-throughput screening of compound neurotoxicity using 3D-cultured neural stem cells on a 384-pillar plate. *Current Protocols, 1*, e107.

Karri, S. K., Saper, R. B., & Kales, S. N. (2008). Lead encephalopathy due to traditional medicines. *Current Drug Safety, 3*, 54–59.

Kilian, J. G., Mejias-Ortega, M., Hsu, H. W., Herman, D. A., Vidal, J., Arechavala, R. J., Renusch, S., Dalal, H., Hasen, I., Ting, A., Rodriguez-Ortiz, C. J., Lim, S. L., Lin, X., Vu, J., Saito, T.,

Saido, T. C., Kleinman, M. T., & Kitazawa, M. (2023). Exposure to quasi-ultrafine particulate matter accelerates memory impairment and Alzheimer's disease-like neuropathology in the AppNL-G-F knock-in mouse model. *Toxicological Sciences: An Official Journal of the Society of Toxicology, 193*, 175–191.

Killin, L. O., Starr, J. M., Shiue, I. J., & Russ, T. C. (2016). Environmental risk factors for dementia: A systematic review. *BMC Geriatrics, 16*, 175.

Kim, H., Kim, W. H., Kim, Y. Y., & Park, H. Y. (2020). Air pollution and central nervous system disease: A review of the impact of fine particulate matter on neurological disorders. *Frontiers in Public Health, 8*, 575330.

Krzyzanowski, B., Searles Nielsen, S., Turner, J. R., & Racette, B. A. (2023). Fine particulate matter and Parkinson disease risk among medicare beneficiaries. *Neurology, 101*, e2058–e2067.

Kusunoki, T., Shimoke, K., Komatsubara, S., Kishi, S., & Ikeuchi, T. (2008). p-Nonylphenol induces endoplasmic reticulum stress-mediated apoptosis in neuronally differentiated PC12 cells. *Neuroscience Letters, 431*, 256–261.

Lahiri, D. K., Maloney, B., & Zawia, N. H. (2009). The LEARn model: An epigenetic explanation for idiopathic neurobiological diseases. *Molecular Psychiatry, 14*, 992–1003.

Landrigan, P. J., Lambertini, L., & Birnbaum, L. S. (2012). A research strategy to discover the environmental causes of autism and neurodevelopmental disabilities. *Environmental Health Perspectives, 120*, a258–a260.

Liu, F., Liu, C., Liu, Y., Wang, J., Wang, Y., & Yan, B. (2023). Neurotoxicity of the air-borne particles: From molecular events to human diseases. *Journal of Hazardous Materials, 457*, 131827.

Loganathan, N., McIlwraith, E. K., & Belsham, D. D. (2020). BPA differentially regulates NPY expression in hypothalamic neurons through a mechanism involving oxidative stress. *Endocrinology, 161*(11), bqaa170.

McCann, S. J., Pond, S. M., James, K. M., & Le Couteur, D. G. (1997). The association between polymorphisms in the cytochrome P-450 2D6 gene and Parkinson's disease: a case-control study and meta-analysis. *Journal of the Neurological Sciences, 153*, 50–53.

McKnight, S., & Hack, N. (2020). Toxin-induced Parkinsonism. *Neurologic Clinics, 38*, 853–865.

Mei, H., Wu, D., Yong, Z., Cao, Y., Chang, Y., Liang, J., Jiang, X., Xu, H., Yang, J., Shi, X., Xie, R., Zhao, W., Wu, Y., & Liu, Y. (2024). PM(2.5) exposure exacerbates seizure symptoms and cognitive dysfunction by disrupting iron metabolism and the Nrf2-mediated ferroptosis pathway. *The Science of the Total Environment, 910*, 168578.

Miodovnik, A., Engel, S. M., Zhu, C., Ye, X., Soorya, L. V., Silva, M. J., Calafat, A. M., & Wolff, M. S. (2011). Endocrine disruptors and childhood social impairment. *Neurotoxicology, 32*, 261–267.

Mortamais, M., Pujol, J., Martinez-Vilavella, G., Fenoll, R., Reynes, C., Sabatier, R., Rivas, I., Forns, J., Vilor-Tejedor, N., Alemany, S., Cirach, M., Alvarez-Pedrerol, M., Nieuwenhuijsen, M., & Sunyer, J. (2019). Effects of prenatal exposure to particulate matter air pollution on corpus callosum and behavioral problems in children. *Environmental Research, 178*, 108734.

O'Connell, S. G., Kincl, L. D., & Anderson, K. A. (2014). Silicone wristbands as personal passive samplers. *Environmental Science & Technology, 48*, 3327–3335.

Oulhote, Y., Chevrier, J., & Bouchard, M. F. (2016). Exposure to polybrominated diphenyl ethers (PBDEs) and hypothyroidism in Canadian women. *The Journal of Clinical Endocrinology and Metabolism, 101*, 590–598.

Panda, S. R., Chaudhari, V. B., Ahmed, S., Kwatra, M., Jala, A., Ponneganti, S., Pawar, S. D., Borkar, R. M., Sharma, P., & Naidu, V. G. M. (2023). Ambient particulate matter (PM(2.5)) exposure contributes to neurodegeneration through the microbiome-gut-brain axis: Therapeutic role of melatonin. *Environmental Toxicology and Pharmacology, 101*, 104183.

Pugsley, K., Scherer, S. W., Bellgrove, M. A., & Hawi, Z. (2022). Environmental exposures associated with elevated risk for autism spectrum disorder may augment the burden of deleterious de novo mutations among probands. *Molecular Psychiatry, 27*, 710–730.

Rabeler, C., Gong, T., Ireland, D., Cochet-Escartin, O., & Collins, E. S. (2023). Acetylcholinesterase activity staining in freshwater planarians. *Current Protocols, 3*, e674.

Radulescu, A., & Lundgren, S. (2019). A pharmacokinetic model of lead absorption and calcium competitive dynamics. *Scientific Reports, 9*, 14225.

Raja, G. L., Subhashree, K. D., & Kantayya, K. E. (2022). In utero exposure to endocrine disruptors and developmental neurotoxicity: Implications for behavioural and neurological disorders in adult life. *Environmental Research, 203*, 111829.

Raz, R., Roberts, A. L., Lyall, K., Hart, J. E., Just, A. C., Laden, F., & Weisskopf, M. G. (2015). Autism spectrum disorder and particulate matter air pollution before, during, and after pregnancy: A nested case-control analysis within the Nurses' Health Study II Cohort. *Environmental Health Perspectives, 123*, 264–270.

Repouskou, A., Papadopoulou, A. K., Panagiotidou, E., Trichas, P., Lindh, C., Bergman, A., Gennings, C., Bornehag, C. G., Ruegg, J., Kitraki, E., & Stamatakis, A. (2020). Long term transcriptional and behavioral effects in mice developmentally exposed to a mixture of endocrine disruptors associated with delayed human neurodevelopment. *Scientific Reports, 10*, 9367.

Rocha, S. M., Bantle, C. M., Aboellail, T., Chatterjee, D., Smeyne, R. J., & Tjalkens, R. B. (2022). Rotenone induces regionally distinct alpha-synuclein protein aggregation and activation of glia prior to loss of dopaminergic neurons in C57Bl/6 mice. *Neurobiology of Disease, 167*, 105685.

Rosenfeld, C. S., & Cooke, P. S. (2019). Endocrine disruption through membrane estrogen receptors and novel pathways leading to rapid toxicological and epigenetic effects. *The Journal of Steroid Biochemistry and Molecular Biology, 187*, 106–117.

Sanborn, M. D., Abelsohn, A., Campbell, M., & Weir, E. (2002). Identifying and managing adverse environmental health effects: 3. Lead exposure. *Canadian Medical Association Journal, 166*, 1287–1292.

Selevan, S. G., Kimmel, C. A., & Mendola, P. (2000). Identifying critical windows of exposure for children's health. *Environmental Health Perspectives, 108*(Suppl 3), 451–455.

Szczerbak, G., Nowak, P., Kostrzewa, R. M., & Brus, R. (2007). Maternal lead exposure produces long-term enhancement of dopaminergic reactivity in rat offspring. *Neurochemical Research, 32*, 1791–1798.

Thiankhaw, K., Chattipakorn, N., & Chattipakorn, S. C. (2022). PM2.5 exposure in association with AD-related neuropathology and cognitive outcomes. *Environmental Pollution, 292*, 118320.

Tran, D. N., Jung, E. M., Yoo, Y. M., & Jeung, E. B. (2020). 4-tert-Octylphenol exposure disrupts brain development and subsequent motor, cognition, social, and behavioral functions. *Oxidative Medicine and Cellular Longevity, 2020*, 8875604.

Viel, J. F., Rouget, F., Warembourg, C., Monfort, C., Limon, G., Cordier, S., & Chevrier, C. (2017). Behavioural disorders in 6-year-old children and pyrethroid insecticide exposure: The PELAGIE mother-child cohort. *Occupational and Environmental Medicine, 74*, 275–281.

Wang, Z., Walker, G. W., Muir, D. C. G., & Nagatani-Yoshida, K. (2020). Toward a global understanding of chemical pollution: A first comprehensive analysis of national and regional chemical inventories. *Environmental Science & Technology, 54*, 2575–2584.

Wei, H., Feng, Y., Liang, F., Cheng, W., Wu, X., Zhou, R., & Wang, Y. (2017). Role of oxidative stress and DNA hydroxymethylation in the neurotoxicity of fine particulate matter. *Toxicology, 380*, 94–103.

Wild, C. P. (2005). Complementing the genome with an "exposome": The outstanding challenge of environmental exposure measurement in molecular epidemiology. *Cancer Epidemiology, Biomarkers & Prevention, 14*, 1847–1850.

Wild, C. P. (2012). The exposome: From concept to utility. *International Journal of Epidemiology, 41*, 24–32.

Yuan, X., Yang, Y., Liu, C., Tian, Y., Xia, D., Liu, Z., Pan, L., Xiong, M., Xiong, J., Meng, L., Zhang, Z., Ye, K., Jiang, H., & Zhang, Z. (2022). Fine particulate matter triggers alpha-synuclein fibrillization and Parkinson-like neurodegeneration. *Movement Disorders, 37*, 1817–1830.

Zhang, S., Lin, L., Kaur, S., Thankachan, S., Blanco-Centurion, C., Yanagisawa, M., Mignot, E., & Shiromani, P. J. (2007). The development of hypocretin (orexin) deficiency in hypocretin/ataxin-3 transgenic rats. *Neuroscience, 148*, 34–43.

Zhang, S., Ireland, D., Sipes, N. S., Behl, M., & Collins, E. S. (2019). Screening for neurotoxic potential of 15 flame retardants using freshwater planarians. *Neurotoxicology and Teratology, 73*, 54–66.

Zheng, J., Reynolds, J. E., Long, M., Ostertag, C., Pollock, T., Hamilton, M., Dunn, J. F., Liu, J., Martin, J., Grohs, M., Landman, B., Huo, Y., Dewey, D., Kurrasch, D., & Lebel, C. (2022). The effects of prenatal bisphenol A exposure on brain volume of children and young mice. *Environmental Research, 214*, 114040.

Zuch, C. L., O'Mara, D. J., & Cory-Slechta, D. A. (1998). Low-level lead exposure selectively enhances dopamine overflow in nucleus accumbens: an in vivo electrochemistry time course assessment. *Toxicology and Applied Pharmacology, 150*, 174–185.

Part III
The Mitigation Pathway (Brain-to-Environment): How Humans Act on the Environment

Leveraging the Social Neuroscience of Prosocial Behavior to Advance Our Understanding of Pro-environmental Behavior

Claus Lamm, Annika Wyss, Tobias Brosch, Boryana Todorova, and Kimberly C. Doell

Abstract This chapter explores the intersection of social neuroscience and pro-environmental behavior to understand how prosocial neuroscientific approaches can inform and enhance climate change mitigation efforts. It emphasizes the critical role of understanding cognitive, affective, and motivational antecedents of behavior to address climate change effectively. By examining the neural underpinnings of prosocial behavior, we highlight how insights from social neuroscience may be used to foster pro-environmental decisions and actions. We first discuss methodologies like functional magnetic resonance imaging and transcranial magnetic stimulation, underscoring their potential to reveal determinants of pro-environmental behavior. Then, we discuss the importance of empathy, social cognition, and the neural mechanisms underlying prosociality, suggesting that these can provide a foundation for

C. Lamm · B. Todorova
Department of Cognition, Emotion, and Methods in Psychology, Faculty of Psychology, University of Vienna, Vienna, Austria
e-mail: claus.lamm@univie.ac.at

A. Wyss
Department of Social Neuroscience and Social Psychology, University of Bern, Bern, Switzerland

T. Brosch
Swiss Center for Affective Sciences, University of Geneva, Geneva, Switzerland

Consumer Decision and Sustainable Behavior Lab, Department of Psychology, University of Geneva, Geneva, Switzerland

K. C. Doell (✉)
Department of Cognition, Emotion, and Methods in Psychology, Faculty of Psychology, University of Vienna, Vienna, Austria

Centre for the Advanced Study of Collective Behaviour, University of Konstanz, Konstanz, Germany

Department of Collective Behaviour, Max-Planck Institute of Animal Behavior, Kontstanz, Germany
e-mail: kimberly.doell@univie.ac.at

S. Kühn (ed.), *Environmental Neuroscience*,
https://doi.org/10.1007/978-3-031-64699-7_6

investigating and encouraging pro-environmental behavior. In addition, the chapter aims to pave the way for a new subdiscipline, environmental social neuroscience, focusing on the cognitive, affective, and social drivers of environmental decision-making and behavior. Integrating neuroscientific and environmental psychology approaches requires innovative research to understand and mitigate the impacts of climate change, ultimately contributing to a sustainable future.

Keywords Prosocial behavior · Sustainability neuroscience · Proenvironmental behavior · Social neuroscience

Man-made climate change and its mitigation are major pressing challenges for humanity. As the term man-made suggests, humans are both the source and the possible remedy for the climate changes we have been inducing over the centuries. Mitigation requires the coordination of system-level and individual-level strategies and changes in decisions and behaviors (IPCC, 2023). In order to be targeted and effective, we need to better understand the reasons that brought us into this situation and the mechanisms by which we may get out of this, or at least slow it down to a considerable extent. In this, the fields of psychological and behavioral sciences can play a major role, as one of their main foci is to advance our understanding of the drivers and obstacles of individual and collective behaviors. Within these fields, knowledge of the neural underpinnings of psychological and behavioral phenomena has become highly influential over the last decades.

Driven by the advent of modern brain imaging and brain stimulation methodologies, such as functional magnetic resonance imaging (fMRI) or transcranial magnetic stimulation (TMS), and their recent more pervasive combination with neuro-computational approaches, fields such as cognitive and affective neuroscience complement the behavioral approaches and provide key insights into the cognitive, affective and motivational antecedents of behavior both at the individual, group, and collective-societal levels. It is thus surprising that these approaches have rarely been exploited to advance our understanding of the cognitive, affective, and motivational factors that drive pro-environmental decision-making and behavior, as well as their harmful counterparts (but see Doell et al., 2023 for a recent review).

One particular subdomain where this lack of research seems a particularly missed opportunity is the domain of social neuroscience. Humans, after all, are often labeled as “social animals,” and large parts of evolution may have driven neural adaptations geared toward expanding our “social brains” and fostering ever more complex social cognition and behavior (Adolphs, 2009). Moreover, man-made climate change not only has strong social causes but has and will have increasingly stronger effects on our societies and social cohesion. Finally, fostering system- and individual-level changes geared towards more pro-environmental and sustainable behavior shares many features, on a conceptual as well as applied level, with fostering prosocial behavior.

The analogies between prosocial and pro-environmental behavior are obvious: both are behaviors that require weighing costs and benefits for the individual against costs and benefits for others or for the (local or global) environment (here defined as the ecological aspects of the environment). Moreover, the latter will naturally impact other humans, in the sense that climate change mitigation will also benefit others, including the individuals engaging in the pro-environmental behaviors. Costs are broadly defined, including monetary, effort and time costs, for example helping a colleague move house or bringing recycling goods to the recycling containers. Also, prosocial and pro-environmental behaviors are at their core social dilemmas of a similar structure. They often create a conflict between acting to maximize the self-interests of the individual versus foregoing some benefits or incurring costs for the self on behalf of benefitting others, or the environment, respectively.

This weighing of self-focused against other-focused ("altruistic") interests has fascinated scholars of various disciplines for centuries, with (possibly ill-posed) questions such as whether we are genuinely selfish or altruistic being heavily researched and debated up to today. It will come as no big surprise, then, that social neuroscience, since its establishment not much more than 30 years ago (e.g., Cacioppo & Berntson, 1992), has devoted considerable thought and research to prosociality as well (e.g., Lamm & Forbes, 2022 for recent review). While this has, of course, not been a magic bullet that all of sudden allowed us to answer all open questions, it has yielded conceptually and practically relevant insights into the drivers, obstacles, and moderators of prosocial cognition, emotion, motivation, and behavior. Of central importance for the present context, it has provided us with key insights into how prosociality can be investigated with methods from human neuroscience, and how their integration with cognitive and behavioral methods may yield better knowledge and insights into what makes us more or less prosocial.

The aim of this book chapter is, thus, to outline and reflect on the analogies between these approaches and the investigation of pro-environmental behaviors. Here, pro-environmental behaviors are defined as a class of behaviors that help to mitigate environmental issues, particularly climate change (Lange & Dewitte, 2019). Importantly, such behaviors can include the commission of acts that benefit the environment (e.g., recycling), and the omission of acts that harm the environment (e.g., opting out of driving to work). Similarly, environmentally harmful behaviors are defined as those that specifically worsen the environment, either directly (e.g., air travel) or via omission (e.g., opting not to recycle; Thøgersen, 2014).

In drawing such analogies, we will also review the empirical research that has already been performed in that respect. Considering that such research is rather scant, though, our main intention is to highlight future research avenues and to provide a possible framework, which will foster research and ultimately establish a new subdiscipline we propose to call *Environmental Social Neuroscience*.

The chapter is structured as follows. We will start with an overview of the current methodologies available and previously used to investigate the neural bases of human prosocial behavior and we will describe their basic advantages and limitations for investigating pro-environmental decision-making and behavior. This will be followed by a reflection on overall principles driving (social) decision-making and their specific relevance for research on sustainable decision-making and pro-environmental behavior. We will then use a selective review of the social neuroscience literature on the drivers and barriers of prosocial behavior to outline specifically how the conceptual, experimental, and methodological approaches in this domain may inform the neuroscientific investigation of pro-environmental decision-making and behavior, outlining wherever available already existing research, or possible concrete examples for such research (➔ chapter "Neuroscientific Approaches to Understand Barriers and Promotors of Pro-environmental Behavior" by Krebs and Sawe).

Neuroscientific Approaches for Studying Prosocial and Pro-environmental Behavior

The integration of neuroscientific insights into the study of human decision-making has gained popularity in recent years. In prosociality research, adding a neural level to the analysis has enabled inferences that help evaluate competing theories and make more accurate predictions about the proximate causes of behavior (e.g., Wyss & Knoch, 2022, for a review). However, the potential of neuroscience for shedding light on the determinants of pro-environmental behavior has remained largely untapped (Doell et al., 2023). Therefore, we elaborate on three neuroscientific approaches that have the potential to further expand our understanding of pro-environmental decision-making.

One way to investigate the processes that underlie human prosociality has been task-based brain imaging. Here, brain activity is recorded while participants are engaging in experimental tasks designed in such a way that they recruit the psychological and brain processes of interest. The aim is, then, to analyze the data in a way that allows to associate the brain data to cognitive, affective, or behavioral processes relevant for the task performance. In terms of the brain imaging techniques, fMRI has been particularly popular. It provides images showing functional activity within or connectivity between brain regions during the execution of a task. While the use of fMRI has yielded many insights into the mechanisms underlying prosocial decision-making (e.g., Bellucci et al., 2020; Cutler & Campbell-Meiklejohn, 2019; Rhoads et al., 2021) and extended theoretical debates about the nature of cooperation (e.g., Hackel et al., 2020; Pärnamets et al., 2020), its application to understand pro-environmental behavior is still in its infancy, with a few notable exceptions (e.g., Baumgartner et al., 2023; Brevers et al., 2021; Sawe et al., 2022; Sawe & Knutson, 2015; Vezich et al., 2016). For instance, Baumgartner et al. (2023) combined fMRI with a consequential intergenerational sustainability paradigm to assess task-dependent functional activity and connectivity. They found that decision-making was linked to activity and connectivity within and between brain areas of the mentalizing and cognitive control network, suggesting that differences in perspective-taking and self-control mechanisms may be associated with individual differences in intergenerationally sustainable behavior. This study also illustrates how novel-innovative experimental paradigms adapted for their use in fMRI scanners may advance our insights into the processes involved in sustainable decision-making (see also below).

Research not only in the area of prosociality but also in environmental sustainability has underscored the importance of internal values and contextual constraints in shaping decision-making (e.g., Berger & Wyss, 2021; Thielmann et al., 2020; Wyss et al., 2022). An addition to task-based neuroimaging is, thus, the so-called neural trait approach that does not analyze activation during task performance but exploits brain-based measures of individual differences connected to a phenomenon of interest. Exploring sources of interindividual differences in pro-environmental motives may significantly contribute to our understanding of environmentally sustainable behavior and its determinants. The neural trait approach provides a valuable tool to explain such interindividual variation. More specifically, it seeks to explain behavioral heterogeneity with task-independent neural characteristics—so-called neural traits (Nash et al., 2015). These traits have so far been mainly identified using electroencephalography (EEG)(➔ chapter "EEG for the Study of Environmental Neuroscience" by Grassini) while participants are at rest, or structural magnetic resonance imaging (sMRI) (➔ chapter "Mapping the Human Brain with Computational Anatomy" by Draganski, Joost and Kherif) , but other methods (such as magnetoencephalography or functional near infrared spectroscopy) are conceivable as well.

sMRI provides information about neuroanatomical differences in white matter (i.e., structural connections) or gray matter (i.e., cortical thickness and volume). As a pioneering example, Guizar Rosales et al. (2022) investigated trait-like neural markers (i.e., cortical thickness) of interindividual differences in intergenerational sustainability. They found that individuals displaying more sustainable behavior in an intergenerational sustainability dilemma game showed greater cortical thickness of the dorsomedial prefrontal cortex (dmPFC) and dlPFC. Supported by further mediation analyses, the authors argue that greater cortical thickness of these brain areas might help individuals to better adopt the perspective of future others and to resist the temptation of their own personal gains for the benefit of said generations.

Brain imaging techniques have certainly advanced our understanding of the processes involved in prosocial behavior and provide considerable potential to similarly do so in pro-environmental decision-making. Their nature as so-called "neural correlate" measures represents a major limitation toward our aim to establish more causal-mechanistic insights, such as which brain processes are the proximate drivers of certain behaviors. To overcome this limitation, research on prosociality has been relying on noninvasive brain stimulation methods, such as transcranial magnetic stimulation (TMS) and transcranial electrical stimulation (TES). These methods allow the safe modulation of brain activity, which is, however, limited to areas a few centimeters beneath the scalp (Brunyé, 2021; Polanía et al., 2018; but see also Luber et al., 2022). More recently, a novel approach based on functional ultrasound stimulation has also emerged, which allows for stimulation of subcortical areas involved in emotion processing, such as the amygdala or nucleus accumbens (Yaakub et al., 2023). In the prosocial domain, combining these methods with neuroimaging has provided key insights into the specific role(s) of certain brain areas and their interplay (e.g., Bukowski et al., 2020; Silani et al., 2013a; see also below). Like brain imaging methods, brain stimulation has rarely been applied in the context of environmental sustainability, with two notable exceptions. In one study, Langenbach et al. (2022) investigated whether intergenerational mentalizing (i.e., taking the perspective of people in the future) is causally involved in sustainable behavior. They applied high-definition transcranial current stimulation (HD-tDCS) to the temporoparietal junction (TPJ) while participants engaged in a consequential intergenerational dilemma. Excitatory HD-tDCS over the TPJ increased sustainable decision-making, suggesting that reduced engagement in mentalizing might be one of the drivers of inadequate behavioral responses to environmental issues.

Taken together, this brief overview of methodologies illustrates how studying human behavior using neuroscientific approaches may provide valuable insights into the determinants pro-environmental behavior. Similar to the domain of prosociality, it will be key not only to develop the technological or analytic side of the neuroscientific measurements themselves, but to also development and pursue further innovation regarding the experimental paradigms. In fact, one reason for the scarcer use of neuroscientific methods in research on pro-environmental behavior so far may stem from a lack of suitable task paradigms. Such paradigms have been a cornerstone of prosociality research (see Van Dijk & De Dreu, 2021, for an overview), where they mirror some of the critical contingencies of real-life situations but are more abstract and stylized and thus provide more control over key parameters. Within pro-environmental behavior research, the development of similar tasks is still in its infancy (e.g., Berger & Wyss, 2021; Lange, 2022, for a review). However, as methodologies continue to evolve and the interdisciplinary efforts intensify, the prospect of harnessing neuroscientific approaches to shed more light on the determinants of pro-environmental decision-making appears increasingly encouraging, and we hope this chapter will instill some more efforts and avenues in that direction.

Humans as (Ir)rational Decision-Makers

Research on the mechanisms underlying human decision-making and behavior has identified several psychological barriers that may prevent sustainable behavior, thus contributing to climate change (Gifford, 2011). Rational-agent models of decision-making assume that people base their decisions on consistent preferences while aiming to maximize the utility of their actions (Becker, 1976). When determining their preferences, people are thought to have complete access to all the information that is relevant to the decision, to engage in an exhaustive analysis of the potential costs and benefits of the available outcomes as well as the probability they will occur, and to identify and choose the most favorable behavioral option (Ajzen, 1991). Preferences can be informed by diverse factors such as people's belief structures (e.g., whether they believe in man-made climate change or not; van Valkengoed et al., 2021), core values (e.g., whether they think that protecting the planet should be an important priority in one's life; Steg, 2016), or perceived social norms (e.g., whether they think that other people engage in climate action (Constantino et al., 2022).

When it comes to the capacity to make rational, utility-maximizing decisions in the face of climate change, several complicating factors come into play. Perceiving the threat that is posed by climate change can be cognitively as well as emotionally challenging. Climate change is an abstract and complex phenomenon. It is composed of disparate and seemingly incongruous events (e.g., increased rainfall in one region and increased droughts in others), and it is often communicated using probabilities and increases in average temperature that can appear relatively minor (e.g., attempts to keep warming to 1.5 °C). At the same time, climate change information and communication often appeal to affect and emotions, with climate change anxiety being one prevalent factor that leads to changes in pro-environmental beliefs and behaviors (Chapman et al., 2017). In the absence of direct personal experience with the consequences of climate change, however, it can be difficult to understand the gravity of the problem and to infer the utility related to changing one's way of life (Markowitz & Shariff, 2012; Weber, 2006). Moreover, many competing and conflicting motivations can drive behavior in the context of climate change. Successful and impactful climate action will be costly and will require substantial shifts of consumption patterns resulting in less greenhouse gas emissions (requiring, e.g., much less flying and meat consumption; Ivanova et al., 2020). Giving up currently held comforts and securities to achieve a more livable future stands in conflict with many self-focused values and motivations (Steg, 2016).

Over the last few decades, an alternative to the rational-agent model of decision-making has gained influence, and this model is likely to be informative in the context of the thoughts outlined in the present chapter as well. This approach acknowledges that people are limited in their ability to make rational decisions. Rooted in dual-process theories of cognition and information processing (Evans, 2008), it recognizes that human behavior is usually not driven by the elaborate and rational thought processes assumed by rational-agent models. Instead, it relies on more automatic and computationally less intensive forms of decision-making that allow people to navigate the countless demands of everyday life in the face of limited time, available information, and computational power (Gigerenzer & Gaissmaier, 2011). So-called "boundedly rational" decision-makers tend to construct their preference in an ad hoc fashion, based on cognitive shortcuts and biases, which makes decision outcomes highly susceptible to contextual influences (Payne et al., 1992). This explains for example why climate change beliefs may vary depending on the season and the occurrence of some major weather events. Moreover, decisions and display of behaviors are often based on automatic processes, reflecting an adaptation of underlying social preferences to a context where they may or may not be appropriate, and where behaviors may thus seem more or less "(ir)rational." This will be the case, especially in situations or contexts where information or prior experience is scarce.

One such preference is whether humans are genuinely selfish or prosocial-altruistic. While the field of economics has pervasively argued for the former, and advocated views of a "homo oeconomicus" that maximizes self-benefit, research in behavioral economics and neuroeconomics has consistently indicated that humans can be strongly driven by prosocial motives and preferences (e.g., Fehr & Schmidt, 1999;

Morishima et al., 2012). The "default" preference and thus the one selected in automatic decision-making and behavior has even been suggested to act cooperatively prosocial rather than selfishly. This has also led to the proposal that humans may rather be "intuitively prosocial" than "intuitively selfish" (e.g., Zaki & Mitchell, 2013).

More recent research on contextual factors and moderators, however, indicates that no matter what our preferences ultimately may be, they seem much less automatic and universal than previously thought. For instance, when targeting implicit prosocial learning processes using a reinforcement learning approach, somewhat discrepant findings have been reported (Lengersdorff et al., 2020; Lockwood et al., 2016). This was also associated with a difference in neural networks involved in the learning and decision-making processes: when money was at stake, participants showed behavior more in line with selfish preferences, while prosocial preferences prevailed when participants had to learn to avoid harm (painful shocks) being inflicted on another person. Findings such as these fit well into the overall insight that research should not focus on whether we are either genuinely selfish or altruistic, but that we should rather investigate the contextual, situational, and person-level factors that determine whether or not our preferences come more or less into play. In analogy, we should also not investigate whether humans are genuinely pro-environmental but rather which factors influence our pro-environmental decision-making and behaviors.

From Prosocial to Pro-environmental Behavior: Hindrances, Obstacles, and Barriers

Prosocial behavior is not an automatism, and there are many factors that can prevent or suppress its occurrence. We now turn to discuss a few of them, against the backdrop of how this may inform research on obstacles for pro-environmental and sustainable decision-making and behavior, especially in the context of climate change.

Diffusion of Responsibility and Social Discounting

We start with the bystander effect when the presence of others deters an individual from acting during an emergency. Research on this phenomenon was strongly influenced by the case of Kitty Genovese in 1968; she was murdered, and even though there were 38 witnesses that saw or heard the attack, no one effectively intervened. This tragic case spurred the creation of entire lines of research that helped identify factors such as diffusion of responsibility (i.e., when people who need to make a decision wait for someone else to act first) or pluralistic ignorance (i.e., when people mistakenly believe that everyone else holds an opinion different from their own; Latané & Darley, 1970). These factors prevent individuals from acting because they think that others are more qualified to help, or where they think that others are not helping either and that this justifies their own inaction.

While no direct neuroscientific investigation of the bystander effects exists, to our knowledge, a neuroimaging study by Cui et al. (2015) provided interesting insights into the neural networks that underpin how we perceive being responsible for another's pain, and the possible diffusion of such responsibility. When participants were solely responsible for pain being inflicted on another person, they showed more activation in pain affect sharing neural regions (including the anterior insular and mid-cingulate cortex), than when they shared the responsibility for causing the pain with another participant. This suggests that the affective response to another person's suffering, and thus the likelihood to engage in actions that alleviate that suffering, is reduced when the supposed or actual responsibility is socially distributed.

It seems obvious, then, to use a similar design to test whether thinking of climate change as something for which we are all collectively responsible, as compared to which we all individually contribute could lead to a weakened aversive response and thus a reduced likelihood to engage in pro-environmental action. Importantly, such a study would enable insights into whether similar affective factors (as measured by activation of affect-related areas) account for diffusion of responsibility regarding pro-environmental action, or whether more cognitive-evaluative factors (such as e.g., pluralistic ignorance) play a role instead, or whether these factors may even operate in parallel.

Diffusion of responsibility and pluralistic ignorance are cases that demonstrate a weakening of affective responses and a related reduced tendency to act. The underlying mechanism seems to be predominantly cognitive-evaluative, in the sense that attributions as to who should help (others rather than myself) reduce empathy and related prosocial motivations and behavior. Related, but possibly more affect-driven mechanisms are social discounting and compassion collapse to which we turn to next (Hagman et al., 2022). Social discounting describes the fact that the farther away a person is perceived to us (in terms of our social-affective closeness to them), the less we seem to care about and behave prosocially toward them (note that this is related to, but not identical to ingroup/outgroup mechanisms, which we will discuss further below). Social discounting has been investigated in a series of studies, using a variety of methodological approaches from hormone administration to neuroimaging (e.g., Forbes et al., 2023a; Sellitto et al., 2021; Sellitto & Kalenscher, 2022). Their findings indicate that areas such as the temporoparietal junction (TPJ) and the anterior insula, and thus what seems to be an interaction between supposedly "cognitive" and "affective" representations, underpin and shape the discounting of the needs of more distant others. More specifically, when disruptively stimulating the right TPJ using TMS, increases in the discounting of delayed and prosocial rewards were observed (Soutschek et al., 2016). Moreover, this effect was connected to decreases in perspective taking and thus interpreted as a suppression of the ability to take a less self-centered (and thus more other-oriented) perspective. This finding and experimental logic could be easily exploited in research on pro-environmental behavior, with distance being varied either in terms of time, geographic, cultural, or other aspects that are relevant drivers of pro-environmentalism and sustainable behavior. For instance, one would predict that pro-environmental behavior would be diminished by disruptive stimulation of the right TPJ or other areas connected to perspective taking and mentalizing (Lamm et al., 2016, for review). The research on social discounting could also be used to address questions related to intergenerational perceptions and behaviors, such as when framing decisions to be on behalf of either future generations or of different generations currently alive.

Compassion collapse or compassion fade (Västfjäll et al., 2014) is another intriguing phenomenon with direct relevance for understanding pro-environmental behaviors and collective action. It describes the decrease of feelings of compassion and of prosocial acts with increasing numbers of persons in need of help. Such a decrease seems paradoxical, as from a moral-utilitarian perspective, it would be considered rational that helping proportionally increases with the number of people that need help. As discussed above, humans do not always follow rational principles, and compassion collapse may be "bounded" by perceptual and attributional factors. For instance, being able to single out individuals from a larger mass of people and attending to their concrete emotional expressions seems to be more effective in causing empathic responses than the more abstract and necessarily mediated representations of the suffering of thousands (Kogut & Ritov, 2005; Västfjäll et al., 2014).

The theoretical insights and empirical approaches from research on compassion fade seem highly promising and relevant to inform and better understand what one could call pro-environmental or sustainable behavior loss or fade. It is indeed a well-documented finding that we tend to engage in particular circumscribed pro-environmental actions (such as e.g., recycling, switching off the lights) that may not always be the most effective ones, while at the same showing a seemingly irrational reluctance to address challenges that are much larger in scale and putative effect (such as changing our mobility and consumption behavior, or more systemic changes of the economic system; Nielsen, et al., 2021a; Nielsen, et al., 2021b). Using the theoretical framework of compassion fade and a neuroscientific approach, it would be highly interesting to investigate which neural networks and associated psychological processes may explain such phenomena, and thus which angles of interventions—that is, cognitive, affective, or motivation—should be used to remedy them. Hypotheses to be tested could be, for example, whether computations involved in numeracy, such as in the parietal cortex, underpin the seeming failure to respond to different levels of impact or whether it is rather (as suggested by the literature on social discounting) networks linked to affect and closeness, including insular and cingulate cortex, or to cognitive appraisal and perspective taking, such as the temporoparietal junction and dorsomedial PFC (see e.g., Majdandžić et al., 2012; Ugazio et al., 2021; Ye et al., 2020).

Social Groups and Identities

Other aspects with a major role in the social psychology and neuroscience of prosocial behavior are social identities and ingroup versus outgroup phenomena (see Doell et al., 2021b). In brief, whether someone is perceived as being or not being part of the group I associate myself with, that is, an in- or outgroup member, can have a major impact on empathy and prosocial behavior toward that person (see e.g., Hein et al., 2010, for neuroimaging evidence).

Notably, it has been argued that our tendency to engage in- versus out-group categorizations is strongly biologically rooted and a highly prevalent feature that affects many aspects of social cognition and behavior. One such root may be the evolution of mammalian rearing and parental behavior, where the ingroup is one's own offspring or other genetically related individuals (Decety & Svetlova, 2012). It has thus been intensely debated how the resulting biases, stereotypes, prejudices, and preferential behavior of some individuals and groups speak for a "dark side" of empathy (Bloom, 2017), in the sense that empathy per se is not a moral or ethical emotion, but that it needs to be combined with virtues, moral, and ethical stances that counteract or even eliminate such biases (such as, e.g., that all human beings should be treated equal; see also Decety & Cowell, 2014; Lamm & Majdandžić, 2015, for reviews). The research on in-/out-group biases in prosociality as well as the debate of how this may induce "dark" aspects of social cognition, emotion, and behaviors is of high and direct relevance for advancing our understanding of pro-environmentalism and sustainability.

Whether we perceive ourselves as part of nature and identify nature or elements of it (such as certain habitats, or specific types of flora and fauna) as belonging to our ingroup, or instead perceive nature as something "out there" that may even need to be fought or controlled (a view held extensively for extended periods in human history, Cronon, 1996) will certainly have effects on pro-environmental concern and behavior. While this aspect, to our knowledge, has not yet been investigated explicitly, indirect insights can be gained from an fMRI study on how vegetarians, vegans, and omnivores respond to pictures of animals in negative situations (e.g., being wounded; Filippi et al., 2010). This revealed that vegans and vegetarians showed neural responses indicative of a stronger engagement of (partially distinct) affect-related (insula or amygdala), visual-perceptual, as well as prefrontal areas than omnivores. Of course, such findings may be explained by a variety of other factors, unrelated to in-/out-group mechanisms, and they also do not align fully with the modulation of responses to human in- vs. out-group members and their suffering. Moreover, we cannot know whether vegans and vegetarians already showed such different responses before becoming vegan or vegetarian, or whether they are a consequence of their dietary choices.

Future research in this domain could try to manipulate in-versus-out-group perceptions experimentally, geared toward reducing intergroup boundaries. In that respect, it is interesting to note that Baumgartner et al. (2014) show that TMS of the right TPJ reduces parochialist punishment, and this logic could be easily extended toward investigations of how this right TPJ disruption increases decisions to protect the environment. Ideally, such an approach would also include elements of social–psychological manipulations, such as framing paradigms, vignettes, or other types of interventions used in research on inter-group conflict (De Dreu et al., 2023). This should also include a focus on how anthropomorphization and humanization (e.g., Kteily et al., 2015; Majdandžić et al., 2012; Waytz et al., 2010) could be exploited to foster pro-environmental and nature-protective actions. It would, for instance, be intriguing to investigate how such interventions could influence the well-documented preferences to be concerned by the needs of other mammals, and in particular for the ones that by means of their physical appearance and visual display activate stronger feelings of care and concern (Mishra & Mehta, 2023). Ideally, such interventions would help us extend our in-group circle to animals such as insects or mollusks that often are much more relevant for biodiversity but are usually met with disgust, rejection, or just plain disinterest. Neural measures would allow us, in this case, to pinpoint whether such manipulations engage similar networks and processes as those when breaking up human-to-human boundaries, or whether human-to-animal boundaries engage distinct mechanisms (see Boch et al., 2023, for research showing distinct as well as similar brain areas involved in human–human and human–animal interaction).

Stress and Prosociality

Closing up our selective review on hindrances and obstacles, we turn to the role of stress as an example of how challenging aversive situations may influence prosociality and pro-environmental behavior. Despite a clearly bad reputation in terms of its folk psychological understanding, stress is scientifically understood as an adaptive response of the organism, triggered by challenging situations that require the mobilization of extra efforts and energy to cope with them. The literature on how stress influences social cognition and prosocial behavior is surprisingly scarce, and controversial. A recent meta-analysis (Nitschke et al., 2022; see also Faber & Häusser, 2022; von Dawans et al., 2021, for reviews) found no consistent differences between stressed and control groups in their prosocial behaviors. Stress, depending on the context or individual factors, may thus have beneficial and adverse effects alike on prosocial behavior. In a series of neuroimaging and psychoendocrinological studies, we have extensively investigated how stress influences empathy and prosocial behavior (Forbes et al., 2023b; Tomova et al., 2017, 2019, 2020). This, again, has painted a mixed picture. While one fMRI study showed that stress increases prosocial behavior (Tomova et al., 2017; see also Tomova et al., 2020), another fMRI study showed that stress made us more self-focused (Forbes et al., 2023b). The studies differed in the way in which prosocial behavior was induced and measured. While in one case, a dictator game and thus the effortless sharing of money from a windfall was used, the other study required participants to make decisions as to whether they wanted to invest physical effort to help another person. Aversive experiences such as stress can thus have a decisive impact on behaviors that are costly and effortful, such as taking more sustainable means of transportation, or buying and procuring the more expensive organic produce.

Antecedents and Ways to Foster Prosocial and Pro-environmental Behavior

Depending on the scientific discipline, different aspects of the antecedents of prosocial behavior will be highlighted. Our review is guided by social psychological approaches. We start by discussing antecedents that pertain to social cognition, that is, empathy, (empathic) concern, and perspective taking, then review research on more behavioral–motivational factors such as prosocial preferences and then focus on antecedents that connect behaviors to social norms and morality.

Social-Cognitive Aspects: Empathy

Emotions play a strong role in driving human behavior, and prosocial and pro-environmental behaviors are no exception. As the origin of the term "emotion" (from the Latin "e-movere") indicates, emotions' primary function is to set things in motion. In the case of prosocial behavior, empathy, which according to one definition is the ability to share and understand the affective state of others (e.g., Lamm et al., 2019, for review), may act as a strong driver to motivate prosocial behavior. For instance, sharing the affective concomitants of another person's pain may move and motivate us—in the very sense of the "e-movere"—to help that person by reducing their pain (e.g., Crockett et al., 2014; Hartmann et al., 2022).

The neural networks that have been associated with empathy are closely connected to affect and emotion processing, as well as the awareness of the affective states by means of interoceptive processes. They include, foremost and in particular when negative emotions are shared, the anterior insular cortex and the mid-cingulate cortex (Lamm et al., 2019, for review; Lamm et al., 2011; Schurz et al., 2021, for meta-analyses), while the sharing of positive affect and emotions has been connected to areas of the reward system and hedonic experiences (Lamm et al., 2015; Morelli et al., 2015). This pattern of findings is highly relevant in the present context as it suggests that vicarious affective processes similar to first-hand effect play a key role in empathizing with others. This is how it may set helping behavior "in motion."

The affective underpinnings of empathy also bring into play another important aspect, which is that empathy sensu affect sharing is not the same as sympathy, compassion, or concern (Singer & Lamm, 2009 for review). The latter three are related, yet distinct terms and phenomena, although their occurrence may be driven by affect sharing; however, all three have to be distinguished from a decisive additional element, which is that they care a component of care or concern for the other person. As such, they have much better leverage to instigate prosocial behaviors, as care and concern are even more direct drivers of behavior than empathy. This, at first, may be counterintuitive, as the everyday usage of the term empathy usually implicitly suggests to carry a care and concern element.

From a scientific viewpoint, however, we need to distinguish the "feeling as" (i.e., affect felt as the affect of the other) from the "feeling for" (i.e., feelings of care and concern for the other) component—with the latter being more instrumental for prosocial behavior (Batson, 2009 for review). In this context, we also need to delineate affect sharing and concern from feelings of personal distress, which is the experience of negative affect that may occur in response to the plight of others. This distinction is particularly important when it comes to understanding pro-environmental behavior, as personal distress may result in self-centered rather than prosocial responses. More specifically, it entails a strong focus on re-establishing a neutral or positive affective state for oneself and thus may detract from an other-oriented focus and attempts to help them overcome their plight (Decety & Lamm, 2011). Personal distress often leads to disengagement from an upsetting situation, such as when switching off the overwhelming bad news or when quickly passing and expressly ignoring a person in clear need of help. Interestingly, these related yet distinct experiences in response to the suffering of others are associated with different neural networks. While personal distress and negative affect sharing have predominantly been connected to the insular and cingulate cortex, care and concern are associated with the mesocorticolimbic motivational and reward system as well as areas that underpin "warm and tender" compassion-related feelings (Ashar et al., 2017; Klimecki et al., 2013; Singer & Klimecki, 2014, for review).

How may these findings inform our quest for understanding the antecedents of pro-environmental behavior? We argue they highlight that when exposing research participants to the "plight" of the natural-ecological environment (the eco- and biosphere), as induced by climate change or other ways of environmental pollution or destruction, we need to consider that this may result in different and nuanced responses, differentially activate the associated neural networks, and in this way also influence pro-environmental (and environmentally harmful) decisions and behaviors in varying ways. For instance, portrayals of environmental destruction or "doom and gloom" styled negative emotion communications may overwhelm and personally distress people, ultimately leading to disengagement and inaction (Chapman et al., 2017).

Meanwhile, care- and concern-based portrayals may lead to a focus on support and action towards protecting the environment. However, interindividual differences in specific traits also need to be considered. For example, in participants with low levels of environmentally relevant trait affect (i.e., one's predisposition to experience positive emotions in an environmental context), positive emotional communications resulted in the commission of fewer pro-environmental messages throughout the day (compared to neutral and even negative environmental messages; Doell et al., 2021a, b). Thus, while concern-based or emotionally positive communication strategies may result in the commission of more pro-environmental behaviors in some, these same interventions may backfire in others.

Regardless, consistent empirical and meta-analytic evidence has shown that emotions about climate change are some of the strongest predictors of climate change-related decisions and behaviors (e.g., van der Linden, 2015; Xie et al., 2019). Generally, it will be important to leverage emotional responses and not just fact- or knowledge-based responses, if the aim is to motivate changes in pro-environmental and sustainable behaviors (for a review, see Brosch, 2021). Here, neuroscientific methodologies can help to better understand and characterize these often-conflicting results and may even contribute to the development of better targeted interventions.

In fact, one of the few studies in environmental social neuroscience looked at neural responses to the harm of people, animals, and nature (Mathur et al., 2016). More specifically, the authors exposed participants in the fMRI scanner to images of humans, non-human animals, and nature (mostly scenes of landscapes or patches of land) in situations that could be perceived as harmful, threatening, or damaging, or in neutral control conditions. The findings revealed largely overlapping networks for the three categories, including the anterior insular and the mid-cingulate cortex. This suggests that harm to nature or nonhuman animals may recruit similar networks as harm being inflicted in humans, a finding that may not only be exploited to foster pro-environmental action but also confirms the conceptual analogies between prosocial and pro-environmental responses the present paper focuses on. However, further research is definitely needed. Neuroimaging studies should try to pinpoint how distinct aspects such as personal distress, "shared" affective responses, and feelings of concern may rely on representations that are similar for humans and the environment and whether they may predict individual differences in pro-environmental behaviors.

Social-Cognitive Aspects: Perspective Taking

Perspective taking, that is, the capacity to view (physical or mental) situations from different stances, is a key skill for social cognition and behavior in general, but it also plays a special role in the context of empathy and prosociality. It enables us to overcome egocentric tendencies and to gain access to inner states of others that otherwise may not be accessible. Perspective taking can be more or less implicit or explicit ("automatic" vs. "deliberate"), and it often needs to be accompanied by another process that is referred to as self-other distinction, which avoids the confusion between one's own views and inner states (thoughts, feelings, motives) from those of the other person.

Self-other distinction is of particular importance when taking the perspective of others, where it prevents that we mistakenly not just "step into someone else's shoes," but that we falsely start assuming those shoes are actually our own. In the context of prosociality and helping others, it should inform a helper whether another person is really in need of help, or whether we mis-evaluate their needs due to an egocentric stance. Adequate self-other distinction also helps to avoid becoming overwhelmed by confusing the aversive feelings of the other with one's own affective state, which is it plays a particular role in psychotherapeutic interactions (Silani et al., 2013b for review).

Perspective taking as been shown to act as a key driver and a moderator of empathy and empathic concern. While earlier accounts had conceptualized empathy as an automatic and mainly bottom-up driven phenomenon, later findings and approaches indicated that empathy is by no means automatic but rather malleable by a variety of factors. One prevalent aspect is the stance from which stimuli about others' pain or suffering are perceived. In a study by Lamm et al., 2007, participants were instructed to adopt a more involved versus a more detached perspective, and this recruited different areas involved in personal distress, affect sharing, or self-other distinction, suggestive of an inverted U-shape between affective engagement and empathy as well as concern. In other words, engaging in a perspective that led to overwhelmingly strong emotional responses driven by personal distress, and not or very little engagement with the plight of the other person leading to a lack of emotional involvement, led to lowest empathy and concern. What seems optimal then is a level of involvement that is neither too intense nor too detached. This logic could also be used for understanding better which areas and processes drive pro-environmental behavior, and where the peak of the inverted U is situated so that behavior can be optimally fostered.

For instance, researchers have recently started to investigate the relationships between climate anxiety and climate action, speculating that too much anxiety might lead to "eco-paralysis" (Clayton, 2020). However, this has not yet been fully supported in the literature. Indeed, multiple studies have found a positive relationship between climate anxiety and climate action (Hogg et al., 2021; Whitmarsh et al., 2022; Wullenkord et al., 2021). It should be noted that the operationalization of how climate anxiety is measured and whether it differs from worry has been unclear and often intertwined. In contrast, one study using a more stringent definition of climate anxiety (using a scale inspired by the clinical domain) found no relationship between climate anxiety and behavior, potentially reflecting a state of tension between motivating and paralyzing effects (Clayton & Karazsia, 2020). Thus, it is not entirely clear when anxiety fosters climate action and whether too much anxiety might be detrimental and overwhelming (thus leading to either a freezing or a withdrawal behavioral response).

Here, a neuroscientific approach may complement studies using self-reports of anxiety and be gauged to predict pro-environmental behaviors based on the engagement of distinct networks related to threat and threat coping. It would be particularly interesting to exploit recent approaches investigating the role and mechanisms of social support, which may have buffering effects on stress and anxiety (e.g., Goldstein et al., 2018; Shamay-Tsoory & Levy-Gigi, 2021) but also of stress and anxiety contagion, which may have detrimental effects (e.g., Nitschke & Bartz, 2023).

Interestingly, and somewhat surprisingly, a recent behavioral (online) study was not able to reveal an effect of perspective taking for others on climate change mitigation measures, such as donations, signing a petition, or approving of mitigation policies (Koessler et al., 2023). A neuroscientific approach could be particularly helpful to pinpoint the cognitive, affective, motivational, and other potentially relevant processes that may have been involved in participants when engaging in perspective taking and to use that information to finetune either the perspective taking instructions or to connect them better to the investigated mitigation measures. Another worthwhile research avenue would be to exploit the phenomenon that perspective taking cannot only be applied to better understand other human beings or view things from their perspective but to pretty much any real or even hypothetical entities. One could, for instance, ask participants to adopt the perspective of "mother nature," or of parts of the ecosystem for which humans are usually less prone to show empathic or concern responses. This could provide key insights into how to fight the global decline in biodiversity, which would require much more protection of animals and ecosystems to which we naturally feel less connected to (such as insects, or swamps). In this respect, a neuroscientific approach may inform us whether perspective taking for such animals would increase empathy and concern-related responses in similar neural networks as those engaged in response to animals, we usually feel closer to or whether it engages distinct networks and affective responses. Depending on the findings, one would then decide on optimal intervention strategies, such as whether they should involve some elements of anthropomorphisation (i.e., conveying the animals or plants as having similar attributes as humans, which, however, may come at its own risk, Williams et al., 2020), or whether other types of affect-inducing strategies (such as focusing on the benefits of the distant animals for biodiversity) may be more effective.

Perspective Taking for Future Selves

Perspective taking can also be exploited to better understand one's future selves, or the future selves of generations after us that will be most affected by climate change. Indeed, using brain stimulation methodology, Soutschek et al. (2016) could show that the right temporoparietal junction (rTPJ) plays a key role in discounting future rewards for an imagined future self but also for imagined future individuals. Moreover, these effects on discounting were linked to deficits in perspective-taking. When combining brain stimulation with neuroimaging, the authors moreover pinpointed the orientation toward future rewards (and selves, as well as others) as a mechanism that enables the delay of gratification (Soutschek et al., 2020). This is an exciting finding, as it suggests that similar mechanisms could be leveraged to foster future-orientation, an important element of sustainable choices and behavior and thus buffer the negative effects of temporal discounting (see above).

One example where related research has been performed in the domain of sustainability research comes from Brosch et al. (2018), who exposed participants to different consequences of climate change while undergoing functional MRI. The authors manipulated the timescale from which these consequences would be experienced, such that they were either within the participants' lifetime or several decades later (and thus presumably affected only future generations). Participants experienced increased activation in a region involved in the encoding of the personal significance of future events (i.e., a specific portion of the vmPFC) when envisioning far future events. Importantly, the degree of activation in this region was highest for participants with high levels of self-transcendent values (i.e., high levels of benevolence and care for the welfare of others). The highly self-transcendent participants also reported higher subjective feelings of concern about the future events. Thus, this work outlines the neural substrates by which people with differing value structures feel concern for the plight of future generations and highlights the utility of combining neuroimaging approaches with more traditional environmental psychology approaches.

Another study where prospective thinking has been investigated is the fMRI study by Brevers et al. (2021). These authors could demonstrate that asking participants to engage in prospective thinking about sustainable or unsustainable behaviors activated areas involved in valuation and memory, such as the ventro-medial prefrontal cortex and hippocampus. Interestingly, areas of the social brain, such as the rTPJ, did not show up in their study. This suggests that this area may play a role only when used as framing of the future sustainable behavior as a prosocial or other-oriented act.

Conclusions and Outlook

Man-made climate change poses a major threat to humanity. It not only affects the geo- and biosphere of our planet but will globally disrupt societies and the socioeconomic conditions under which we lived so far. While there is wide consensus by now that we can only tackle this challenge together and across national, political, and economic boundaries, how this joint effort should be organized and implemented is a matter of intense debate. Science and academia not only have the knowledge and tools but also the obligation to contribute to this discourse. While the natural and life sciences have been central in documenting and understanding the causes and effects, we need concerted efforts by the social sciences to better understand mitigation of as well as adaptation to climate change and its downstream effects.

Besides behavioral science-based research approaches documenting and explaining meso- and macro-level cognition and behavior, we need to tap the potential of micro-level investigations and the insights they provide in the putative mechanisms driving individual-level processes, decision-making, and behaviors. Social neuroscience as a discipline that has emerged at the intersection of social psychology and behavioral sciences with neuroscience seems optimally positioned to lead on such efforts. In the present chapter, we have thus attempted to outline some analogies between the research on prosocial and pro-environmental behavior that we may want to exploit. We hope this will motivate research efforts that bridge across and integrate all relevant levels of observation, from molecules to collective behaviors, just as social neuroscience has attempted to for several decades now. Importantly, these efforts should not only aim at basic research on the drivers and obstacles of pro-environmental behavior but rather be geared toward understanding of how these insights may be used to ensure a sustainable future for us all and the generations after us on our planet (➔ chapter "Neuroscientific Approaches to Understand Barriers and Promotors of Pro-environmental Behavior" by Krebs and Sawe).

References

Adolphs, R. (2009). The social brain: Neural basis of social knowledge. *Annual Review of Psychology, 60*(1), 693–716. https://doi.org/10.1146/annurev.psych.60.110707.163514

Ajzen, I. (1991). The theory of planned behavior. *Organizational Behavior and Human Decision Processes, 50*(2), 179–211. https://doi.org/10.1016/0749-5978(91)90020-T

Ashar, Y. K., Andrews-Hanna, J. R., Dimidjian, S., & Wager, T. D. (2017). Empathic care and distress: Predictive brain markers and dissociable brain systems. *Neuron, 94*(6), 1263–1273.e4. https://doi.org/10.1016/j.neuron.2017.05.014

Batson, C. D. (2009). These things called empathy: Eight related but distinct phenomena. In *The social neuroscience of empathy* (pp. 3–15). Boston Review. https://doi.org/10.7551/mitpress/9780262012973.003.0002

Baumgartner, T., Schiller, B., Rieskamp, J., Gianotti, L. R. R., & Knoch, D. (2014). Diminishing parochialism in intergroup conflict by disrupting the right temporo-parietal junction. *Social Cognitive and Affective Neuroscience, 9*(5), 653–660. https://doi.org/10.1093/scan/nst023

Baumgartner, T., Guizar Rosales, E., & Knoch, D. (2023). Neural mechanisms underlying inter-individual differences in intergenerational sustainable behavior. *Scientific Reports, 13*(1), 1. https://doi.org/10.1038/s41598-023-44250-z

Becker, G. S. (1976). *The economic approach to human behavior*. University of Chicago Press.

Bellucci, G., Camilleri, J. A., Eickhoff, S. B., & Krueger, F. (2020). Neural signatures of prosocial behaviors. *Neuroscience & Biobehavioral Reviews, 118*, 186–195. https://doi.org/10.1016/j.neubiorev.2020.07.006

Berger, S., & Wyss, A. M. (2021). Measuring pro-environmental behavior using the carbon emission task. *Journal of Environmental Psychology, 75*, 101613. https://doi.org/10.1016/j.jenvp.2021.101613

Bloom, P. (2017). Empathy and its discontents. *Trends in Cognitive Sciences, 21*(1), 24–31. https://doi.org/10.1016/j.tics.2016.11.004

Boch, M., Wagner, I. C., Karl, S., Huber, L., & Lamm, C. (2023). Functionally analogous body- and animacy-responsive areas are present in the dog (Canis familiaris) and human occipito-temporal lobe. *Communications Biology, 6*(1), 1. https://doi.org/10.1038/s42003-023-05014-7

Brevers, D., Baeken, C., Maurage, P., Sescousse, G., Vögele, C., & Billieux, J. (2021). Brain mechanisms underlying prospective thinking of sustainable behaviours. *Nature Sustainability, 4*(5), 5. https://doi.org/10.1038/s41893-020-00658-3

Brosch, T. (2021). Affect and emotions as drivers of climate change perception and action: A review. *Current Opinion in Behavioral Sciences, 42*, 15–21. https://doi.org/10.1016/j.cobeha.2021.02.001

Brosch, T., Stussi, Y., Desrichard, O., & Sander, D. (2018). Not my future? Core values and the neural representation of future events. *Cognitive, Affective, & Behavioral Neuroscience, 18*(3), 476–484. https://doi.org/10.3758/s13415-018-0581-9

Brunyé, T. (2021). Non-invasive brain stimulation effects on the perceptual and cognitive processes underlying decision-making: A mini review. *Journal of Cognitive Enhancement, 5*. https://doi.org/10.1007/s41465-020-00186-0

Bukowski, H., Tik, M., Silani, G., Ruff, C. C., Windischberger, C., & Lamm, C. (2020). When differences matter: rTMS/fMRI reveals how differences in dispositional empathy translate to distinct neural underpinnings of self-other distinction in empathy. *Cortex, 128*, 143–161. https://doi.org/10.1016/j.cortex.2020.03.009

Cacioppo, J. T., & Berntson, G. G. (1992). Social psychological contributions to the decade of the brain. Doctrine of multilevel analysis. *The American Psychologist, 47*(8), 1019–1028. https://doi.org/10.1037/0003-066x.47.8.1019

Chapman, D. A., Lickel, B., & Markowitz, E. M. (2017). Reassessing emotion in climate change communication. *Nature Climate Change, 7*(12), 12. https://doi.org/10.1038/s41558-017-0021-9

Clayton, S. (2020). Climate anxiety: Psychological responses to climate change. *Journal of Anxiety Disorders, 74*, 102263. https://doi.org/10.1016/j.janxdis.2020.102263

Clayton, S., & Karazsia, B. T. (2020). Development and validation of a measure of climate change anxiety. *Journal of Environmental Psychology, 69*, 101434. https://doi.org/10.1016/j.jenvp.2020.101434

Constantino, S. M., Sparkman, G., Kraft-Todd, G. T., Bicchieri, C., Centola, D., Shell-Duncan, B., Vogt, S., & Weber, E. U. (2022). Scaling up change: A critical review and practical guide to harnessing social norms for climate action. *Psychological Science in the Public Interest, 23*(2), 50–97. https://doi.org/10.1177/15291006221105279

Crockett, M. J., Kurth-Nelson, Z., Siegel, J. Z., Dayan, P., & Dolan, R. J. (2014). Harm to others outweighs harm to self in moral decision making. *Proceedings of the National Academy of Sciences of the United States of America, 111*(48), 17320–17325. https://doi.org/10.1073/pnas.1408988111

Cronon, W. (1996). *Uncommon ground: Rethinking the human place in nature*. W. W. Norton.

Cui, F., Abdelgabar, A.-R., Keysers, C., & Gazzola, V. (2015). Responsibility modulates pain-matrix activation elicited by the expressions of others in pain. *NeuroImage, 114*, 371–378. https://doi.org/10.1016/j.neuroimage.2015.03.034

Cutler, J., & Campbell-Meiklejohn, D. (2019). A comparative fMRI meta-analysis of altruistic and strategic decisions to give. *NeuroImage, 184*, 227–241. https://doi.org/10.1016/j.neuroimage.2018.09.009

von Dawans, B., Strojny, J., & Domes, G. (2021). The effects of acute stress and stress hormones on social cognition and behavior: Current state of research and future directions. *Neuroscience and Biobehavioral Reviews, 121*, 75–88. https://doi.org/10.1016/j.neubiorev.2020.11.026

De Dreu, C. K. W., Gross, J., & Romano, A. (2023). Group formation and the evolution of human social organization. *Perspectives on Psychological Science: A Journal of the Association for Psychological Science*, 17456916231179156. https://doi.org/10.1177/17456916231179156

Decety, J., & Cowell, J. M. (2014). Friends or foes: Is empathy necessary for moral behavior? *Perspectives on Psychological Science: A Journal of the Association for Psychological Science, 9*(5), 525–537. https://doi.org/10.1177/1745691614545130

Decety, J., & Lamm, C. (2011). 15 Empathy versus personal distress: Recent evidence from social neuroscience. In *The social neuroscience of empathy* (Vol. 199). MIT Press.

Decety, J., & Svetlova, M. (2012). Putting together phylogenetic and ontogenetic perspectives on empathy. *Developmental Cognitive Neuroscience, 2*(1), 1–24. https://doi.org/10.1016/j.dcn.2011.05.003

Doell, K. C., Conte, B., & Brosch, T. (2021a). Interindividual differences in environmentally relevant positive trait affect impacts sustainable behavior in everyday life. *Scientific Reports, 11*(1), 1. https://doi.org/10.1038/s41598-021-99438-y

Doell, K. C., Pärnamets, P., Harris, E. A., Hackel, L. M., & Van Bavel, J. J. (2021b). Understanding the effects of partisan identity on climate change. *Current Opinion in Behavioral Sciences, 42*, 54–59. https://doi.org/10.1016/j.cobeha.2021.03.013

Doell, K. C., Berman, M. G., Bratman, G. N., Knutson, B., Kühn, S., Lamm, C., Pahl, S., Sawe, N., Van Bavel, J. J., White, M. P., & Brosch, T. (2023). Leveraging neuroscience for climate change research. *Nature Climate Change*, 1–10. https://doi.org/10.1038/s41558-023-01857-4

Evans, J. S. B. T. (2008). Dual-processing accounts of reasoning, judgment, and social cognition. *Annual Review of Psychology, 59*, 255–278. https://doi.org/10.1146/annurev.psych.59.103006.093629

Faber, N. S., & Häusser, J. A. (2022). Why stress and hunger both increase and decrease prosocial behaviour. *Current Opinion in Psychology, 44*, 49–57. https://doi.org/10.1016/j.copsyc.2021.08.023

Fehr, E., & Schmidt, K. M. (1999). A theory of fairness, competition, and cooperation. *The Quarterly Journal of Economics, 114*(3), 817–868.

Filippi, M., Riccitelli, G., Falini, A., Di Salle, F., Vuilleumier, P., Comi, G., & Rocca, M. A. (2010). The brain functional networks associated to human and animal suffering differ among omnivores, vegetarians and vegans. *PLoS One, 5*(5), e10847. https://doi.org/10.1371/journal.pone.0010847

Forbes, P. A., Chaliani, I., Schilbach, L., & Kalenscher, T. (2023a). Autistic adults show enhanced generosity to socially distant others. *Autism: The International Journal of Research and Practice*, 13623613231190674. https://doi.org/10.1177/13623613231190674

Forbes, P. A., Aydogan, G., Braunstein, J., Todorova, B., Wagner, I. C., Lockwood, P. L., Apps, M. A. J., Ruff, C. C., & Lamm, C. (2023b). Acute stress reduces effortful prosocial behaviour. *eLife, 12*. https://doi.org/10.7554/eLife.87271

Gifford, R. (2011). The dragons of inaction: Psychological barriers that limit climate change mitigation and adaptation. *American Psychologist, 66*(4), 290–302. https://doi.org/10.1037/a0023566

Gigerenzer, G., & Gaissmaier, W. (2011). Heuristic decision making. *Annual Review of Psychology, 62*, 451–482. https://doi.org/10.1146/annurev-psych-120709-145346

Goldstein, P., Weissman-Fogel, I., Dumas, G., & Shamay-Tsoory, S. G. (2018). Brain-to-brain coupling during handholding is associated with pain reduction. *Proceedings of the National Academy of Sciences of the United States of America, 115*(11), E2528–E2537. https://doi.org/10.1073/pnas.1703643115

Guizar Rosales, E., Baumgartner, T., & Knoch, D. (2022). Interindividual differences in intergenerational sustainable behavior are associated with cortical thickness of the dorsomedial and dorsolateral prefrontal cortex. *NeuroImage, 264*, 119664. https://doi.org/10.1016/j.neuroimage.2022.119664

Hackel, L., Wills, J., & Van Bavel, J. (2020). Shifting prosocial intuitions: Neurocognitive evidence for a value-based account of group-based cooperation. *Social Cognitive and Affective Neuroscience, 15*(4), 371–381. https://doi.org/10.1093/scan/nsaa055

Hagman, W., Tinghög, G., Dickert, S., Slovic, P., & Västfjäll, D. (2022). Motivated down-regulation of emotion and compassion collapse revisited. *Frontiers in Psychology, 13*, 801150. https://doi.org/10.3389/fpsyg.2022.801150

Hartmann, H., Forbes, P. A. G., Rütgen, M., & Lamm, C. (2022). Placebo analgesia reduces costly prosocial helping to lower another person's pain. *Psychological Science, 33*(11), 1867–1881. https://doi.org/10.1177/09567976221119727

Hein, G., Silani, G., Preuschoff, K., Batson, C. D., & Singer, T. (2010). Neural responses to ingroup and outgroup members' suffering predict individual differences in costly helping. *Neuron, 68*(1), 149–160. https://doi.org/10.1016/j.neuron.2010.09.003

Hogg, T. L., Stanley, S. K., O'Brien, L. V., Wilson, M. S., & Watsford, C. R. (2021). The Hogg Eco-Anxiety Scale: Development and validation of a multidimensional scale. *Global Environmental Change, 71*, 102391. https://doi.org/10.1016/j.gloenvcha.2021.102391

IPCC, 2023: Climate change 2023: Synthesis report. A report of the intergovernmental panel on climate change. Contribution of working groups I, II and III to the sixth assessment report of the intergovernmental panel on climate change [Core Writing Team, H. Lee and J. Romero (eds.)]. (2023). IPCC, Geneva, Switzerland. https://www.ipcc.ch/report/ar6/syr/downloads/report/IPCC_AR6_SYR_LongerReport.pdf

Ivanova, D., Barrett, J., Wiedenhofer, D., Macura, B., Callaghan, M., & Creutzig, F. (2020). Quantifying the potential for climate change mitigation of consumption options. *Environmental Research Letters.* https://doi.org/10.1088/1748-9326/ab8589

Klimecki, O. M., Leiberg, S., Lamm, C., & Singer, T. (2013). Functional neural plasticity and associated changes in positive affect after compassion training. *Cerebral Cortex (New York, N.Y.: 1991), 23*(7), 1552–1561. https://doi.org/10.1093/cercor/bhs142

Koessler, A.-K., Heinz, N., & Engel, S. (2023). Perspective-taking with affected others to promote climate change mitigation. *Frontiers in Psychology, 14*. https://www.frontiersin.org/articles/10.3389/fpsyg.2023.1225165

Kogut, T., & Ritov, I. (2005). The "identified victim" effect: An identified group, or just a single individual? *Journal of Behavioral Decision Making, 18*(3), 157–167. https://doi.org/10.1002/bdm.492

Kteily, N., Bruneau, E., Waytz, A., & Cotterill, S. (2015). The ascent of man: Theoretical and empirical evidence for blatant dehumanization. *Journal of Personality and Social Psychology, 109*(5), 901–931. https://doi.org/10.1037/pspp0000048

Lamm, C., & Forbes, P. (2022). Neurobiology of prosociality: Investigating the link between empathy and prosocial behavior in the brain. https://doi.org/10.1017/9781108876681.005.

Lamm, C., & Majdandžić, J. (2015). The role of shared neural activations, mirror neurons, and morality in empathy–A critical comment. *Neuroscience Research, 90*, 15–24.

Lamm, C., Batson, C. D., & Decety, J. (2007). The neural substrate of human empathy: Effects of perspective-taking and cognitive appraisal. *Journal of Cognitive Neuroscience, 19*(1), 42–58.

Lamm, C., Decety, J., & Singer, T. (2011). Meta-analytic evidence for common and distinct neural networks associated with directly experienced pain and empathy for pain. *NeuroImage, 54*(3), 2492–2502. https://doi.org/10.1016/j.neuroimage.2010.10.014

Lamm, C., Silani, G., & Singer, T. (2015). Distinct neural networks underlying empathy for pleasant and unpleasant touch. *Cortex; a Journal Devoted to the Study of the Nervous System and Behavior, 70*, 79–89. https://doi.org/10.1016/j.cortex.2015.01.021

Lamm, C., Bukowski, H., & Silani, G. (2016). From shared to distinct self-other representations in empathy: Evidence from neurotypical function and socio-cognitive disorders. *Philosophical Transactions of the Royal Society of London. Series B, Biological Sciences, 371*(1686), 20150083. https://doi.org/10.1098/rstb.2015.0083

Lamm, C., Rütgen, M., & Wagner, I. C. (2019). Imaging empathy and prosocial emotions. *Neuroscience Letters, 693*, 49–53. https://doi.org/10.1016/j.neulet.2017.06.054

Lange, F. (2022). Behavioral paradigms for studying pro-environmental behavior: A systematic review. *Behavior Research Methods, 55*. https://doi.org/10.3758/s13428-022-01825-4

Lange, F., & Dewitte, S. (2019). Measuring pro-environmental behavior: Review and recommendations. *Journal of Environmental Psychology, 63*, 92–100. https://doi.org/10.1016/j.jenvp.2019.04.009

Langenbach, B. P., Savic, B., Baumgartner, T., Wyss, A. M., & Knoch, D. (2022). Mentalizing with the future: Electrical stimulation of the right TPJ increases sustainable decision-making. *Cortex, 146*, 227–237. https://doi.org/10.1016/j.cortex.2021.11.006

Latané, B., & Darley, J. M. (1970). *The unresponsive bystander: Why doesn't he help?* Appleton-Century Crofts.

Lengersdorff, L. L., Wagner, I. C., Lockwood, P. L., & Lamm, C. (2020). When implicit prosociality trumps selfishness: The neural valuation system underpins more optimal choices when learning to avoid harm to others than to oneself. *Journal of Neuroscience, 40*(38), 7286–7299.

van der Linden, S. (2015). The social-psychological determinants of climate change risk perceptions: Towards a comprehensive model. *Journal of Environmental Psychology, 41*, 112–124. https://doi.org/10.1016/j.jenvp.2014.11.012

Lockwood, P. L., Apps, M. A. J., Valton, V., Viding, E., & Roiser, J. P. (2016). Neurocomputational mechanisms of prosocial learning and links to empathy. *Proceedings of the National Academy of Sciences, 113*(35), 9763–9768. https://doi.org/10.1073/pnas.1603198113

Luber, B., Davis, S. W., Deng, Z.-D., Murphy, D., Martella, A., Peterchev, A. V., & Lisanby, S. H. (2022). Using diffusion tensor imaging to effectively target TMS to deep brain structures. *NeuroImage, 249*, 118863. https://doi.org/10.1016/j.neuroimage.2021.118863

Majdandžić, J., Bauer, H., Windischberger, C., Moser, E., Engl, E., & Lamm, C. (2012). The human factor: Behavioral and neural correlates of humanized perception in moral decision making. *PLoS One, 7*(10), e47698. https://doi.org/10.1371/journal.pone.0047698

Markowitz, E. M., & Shariff, A. F. (2012). Climate change and moral judgement. *Nature Climate Change, 2*(4), 4. https://doi.org/10.1038/nclimate1378

Mathur, V. A., Cheon, B. K., Harada, T., Scimeca, J. M., & Chiao, J. Y. (2016). Overlapping neural response to the pain or harm of people, animals, and nature. *Neuropsychologia, 81*, 265–273. https://doi.org/10.1016/j.neuropsychologia.2015.12.025

Mishra, R., & Mehta, R. (2023). The effects of food anthropomorphism on consumer behavior: A systematic literature review with integrative framework and future research directions. *Appetite, 190*, 107035. https://doi.org/10.1016/j.appet.2023.107035

Morelli, S. A., Sacchet, M. D., & Zaki, J. (2015). Common and distinct neural correlates of personal and vicarious reward: A quantitative meta-analysis. *NeuroImage, 112*, 244–253. https://doi.org/10.1016/j.neuroimage.2014.12.056

Morishima, Y., Schunk, D., Bruhin, A., Ruff, C. C., & Fehr, E. (2012). Linking brain structure and activation in temporoparietal junction to explain the neurobiology of human altruism. *Neuron, 75*(1), 73–79. https://doi.org/10.1016/j.neuron.2012.05.021

Nash, K., Gianotti, L. R. R., & Knoch, D. (2015). A neural trait approach to exploring individual differences in social preferences. *Frontiers in Behavioral Neuroscience, 8*. https://www.frontiersin.org/articles/10.3389/fnbeh.2014.00458

Nielsen, K., Cologna, V., Lange, F., Brick, C., & Stern, P. (2021a). The case for impact-focused environmental psychology. *Journal of Environmental Psychology*. https://doi.org/10.31234/osf.io/w39c5

Nielsen, K., Nicholas, K., Creutzig, F., Dietz, T., & Stern, P. (2021b). The role of high-socioeconomic-status people in locking in or rapidly reducing energy-driven greenhouse gas emissions. *Nature Energy, 6*. https://doi.org/10.1038/s41560-021-00900-y

Nitschke, J. P., & Bartz, J. A. (2023). The association between acute stress & empathy: A systematic literature review. *Neuroscience & Biobehavioral Reviews, 144*, 105003. https://doi.org/10.1016/j.neubiorev.2022.105003

Nitschke, J. P., Forbes, P. A. G., & Lamm, C. (2022). Does stress make us more-or less-prosocial? A systematic review and meta-analysis of the effects of acute stress on prosocial behaviours using economic games. *Neuroscience and Biobehavioral Reviews, 142*, 104905. https://doi.org/10.1016/j.neubiorev.2022.104905

Pärnamets, P., Shuster, A., Reinero, D. A., & Van Bavel, J. J. (2020). A value-based framework for understanding cooperation. *Current Directions in Psychological Science, 29*(3), 227–234. https://doi.org/10.1177/0963721420906200

Payne, J. W., Bettman, J. R., & Johnson, E. J. (1992). Behavioral decision research: A constructive processing perspective. *Annual Review of Psychology, 43*, 87–131. https://doi.org/10.1146/annurev.ps.43.020192.000511

Polanía, R., Nitsche, M. A., & Ruff, C. C. (2018). Studying and modifying brain function with non-invasive brain stimulation. *Nature Neuroscience, 21*(2), 2. https://doi.org/10.1038/s41593-017-0054-4

Rhoads, S., Cutler, J., & Marsh, A. (2021). A feature-based network analysis and fMRI meta-analysis reveal three distinct types of prosocial decisions. *Social Cognitive and Affective Neuroscience, 16*. https://doi.org/10.1093/scan/nsab079

Sawe, N., & Knutson, B. (2015). Neural valuation of environmental resources. *NeuroImage, 122*, 87–95. https://doi.org/10.1016/j.neuroimage.2015.08.010

Sawe, N., Srirangarajan, T., Sahoo, A., Tang, G. S., & Knutson, B. (2022). Neural responses clarify how ecolabels promote sustainable purchases. *NeuroImage, 263*, 119668. https://doi.org/10.1016/j.neuroimage.2022.119668

Schurz, M., Radua, J., Tholen, M. G., Maliske, L., Margulies, D. S., Mars, R. B., Sallet, J., & Kanske, P. (2021). Toward a hierarchical model of social cognition: A neuroimaging meta-analysis and integrative review of empathy and theory of mind. *Psychological Bulletin, 147*(3), 293–327. https://doi.org/10.1037/bul0000303

Sellitto, M., & Kalenscher, T. (2022). Variations in progesterone and estradiol across the menstrual cycle predict generosity toward socially close others. *Psychoneuroendocrinology, 140*, 105720. https://doi.org/10.1016/j.psyneuen.2022.105720

Sellitto, M., Neufang, S., Schweda, A., Weber, B., & Kalenscher, T. (2021). Arbitration between insula and temporoparietal junction subserves framing-induced boosts in generosity during social discounting. *NeuroImage, 238*, 118211. https://doi.org/10.1016/j.neuroimage.2021.118211

Shamay-Tsoory, S. G., & Levy-Gigi, E. (2021). You name it: Interpersonal affect labeling diminishes distress in romantic couples. *Behavior Therapy, 52*(2), 455–464. https://doi.org/10.1016/j.beth.2020.06.004

Silani, G., Lamm, C., Ruff, C. C., & Singer, T. (2013a). Right supramarginal gyrus is crucial to overcome emotional egocentricity bias in social judgments. *Journal of Neuroscience, 33*(39), 15466–15476.

Silani, G., Zucconi, A., & Lamm, C. (2013b). Carl Rogers meets the neurosciences: Insights from social neuroscience for client-centered therapy. In J. H. D. Cornelius-White, R. Motschnig-Pitrik, & M. Lux (Eds.), *Interdisciplinary handbook of the person-centered approach* (pp. 63–78). Springer. https://doi.org/10.1007/978-1-4614-7141-7_5

Singer, T., & Klimecki, O. M. (2014). Empathy and compassion. *Current Biology: CB, 24*(18), R875–R878. https://doi.org/10.1016/j.cub.2014.06.054

Singer, T., & Lamm, C. (2009). The social neuroscience of empathy. *Annals of the New York Academy of Sciences, 1156*, 81–96. https://doi.org/10.1111/j.1749-6632.2009.04418.x

Soutschek, A., Ruff, C. C., Strombach, T., Kalenscher, T., & Tobler, P. N. (2016). Brain stimulation reveals crucial role of overcoming self-centeredness in self-control. *Science Advances, 2*(10), e1600992. https://doi.org/10.1126/sciadv.1600992

Soutschek, A., Moisa, M., Ruff, C. C., & Tobler, P. N. (2020). The right temporoparietal junction enables delay of gratification by allowing decision makers to focus on future events. *PLoS Biology, 18*(8), e3000800.

Steg, L. (2016). Values, norms, and intrinsic motivation to act proenvironmentally. *Annual Review of Environment and Resources, 41*. https://doi.org/10.1146/annurev-environ-110615-085947

Thielmann, I., Spadaro, G., & Balliet, D. (2020). Personality and prosocial behavior: A theoretical framework and meta-analysis. *Psychological Bulletin, 146*(1), 30–90. https://doi.org/10.1037/bul0000217

Thøgersen, J. (2014). Unsustainable consumption. *European Psychologist, 19*(2), 84–95. https://doi.org/10.1027/1016-9040/a000176

Tomova, L., Majdandžic, J., Hummer, A., Windischberger, C., Heinrichs, M., & Lamm, C. (2017). Increased neural responses to empathy for pain might explain how acute stress increases prosociality. *Social Cognitive and Affective Neuroscience, 12*(3), 401–408. https://doi.org/10.1093/scan/nsw146

Tomova, L., Heinrichs, M., & Lamm, C. (2019). The Other and Me: Effects of oxytocin on self-other distinction. *International Journal of Psychophysiology: Official Journal of the International Organization of Psychophysiology, 136*, 49–53. https://doi.org/10.1016/j.ijpsycho.2018.03.008

Tomova, L., Saxe, R., Klöbl, M., Lanzenberger, R., & Lamm, C. (2020). Acute stress alters neural patterns of value representation for others. *NeuroImage, 209*, 116497. https://doi.org/10.1016/j.neuroimage.2019.116497

Ugazio, G., Grueschow, M., Polania, R., Lamm, C., Tobler, P., & Ruff, C. (2021). Neurocomputational foundations of moral preferences. *Social Cognitive and Affective Neuroscience, 17*(3), 253–265. https://doi.org/10.1093/scan/nsab100

van Valkengoed, A. M., Steg, L., & Perlaviciute, G. (2021). Development and validation of a climate change perceptions scale. *Journal of Environmental Psychology, 76*, 101652. https://doi.org/10.1016/j.jenvp.2021.101652

Van Dijk, E., & De Dreu, C. K. W. (2021). Experimental games and social decision making. *Annual Review of Psychology, 72*(1), 415–438. https://doi.org/10.1146/annurev-psych-081420-110718

Västfjäll, D., Slovic, P., Mayorga, M., & Peters, E. (2014). Compassion fade: Affect and charity are greatest for a single child in need. *PLoS One, 9*(6), e100115. https://doi.org/10.1371/journal.pone.0100115

Vezich, I., Gunter, B., & Lieberman, M. (2016). The mere green effect: An fMRI study of pro-environmental advertisements. *Social Neuroscience, 12*. https://doi.org/10.1080/17470919.2016.1182587

Waytz, A., Morewedge, C. K., Epley, N., Monteleone, G., Gao, J.-H., & Cacioppo, J. T. (2010). Making sense by making sentient: Effectance motivation increases anthropomorphism. *Journal of Personality and Social Psychology, 99*(3), 410–435. https://doi.org/10.1037/a0020240

Weber, E. (2006). Experience-based and description-based perceptions of long-term risk: Why global warming does not scare us (yet). *Climatic Change, 77*, 103–120. https://doi.org/10.1007/s10584-006-9060-3

Whitmarsh, L., Player, L., Jiongco, A., James, M., Williams, M., Marks, E., & Kennedy-Williams, P. (2022). Climate anxiety: What predicts it and how is it related to climate action? *Journal of Environmental Psychology, 83*, 101866. https://doi.org/10.1016/j.jenvp.2022.101866

Williams, L. A., Brosnan, S. F., & Clay, Z. (2020). Anthropomorphism in comparative affective science: Advocating a mindful approach. *Neuroscience & Biobehavioral Reviews, 115*, 299–307.

Wullenkord, M. C., Tröger, J., Hamann, K. R. S., Loy, L. S., & Reese, G. (2021). Anxiety and climate change: A validation of the Climate Anxiety Scale in a German-speaking quota sample and an investigation of psychological correlates. *Climatic Change, 168*(3), 20. https://doi.org/10.1007/s10584-021-03234-6

Wyss, A. M., & Knoch, D. (2022). Neuroscientific approaches to study prosociality. *Current Opinion in Psychology, 44*, 38–43. https://doi.org/10.1016/j.copsyc.2021.08.028

Wyss, A. M., Knoch, D., & Berger, S. (2022). When and how pro-environmental attitudes turn into behavior: The role of costs, benefits, and self-control. *Journal of Environmental Psychology, 79*, 101748. https://doi.org/10.1016/j.jenvp.2021.101748

Xie, B., Brewer, M. B., Hayes, B. K., McDonald, R. I., & Newell, B. R. (2019). Predicting climate change risk perception and willingness to act. *Journal of Environmental Psychology, 65*, 101331. https://doi.org/10.1016/j.jenvp.2019.101331

Yaakub, S. N., White, T. A., Roberts, J., Martin, E., Verhagen, L., Stagg, C. J., Hall, S., & Fouragnan, E. F. (2023). Transcranial focused ultrasound-mediated neurochemical and functional connectivity changes in deep cortical regions in humans. *Nature Communications, 14*(1), 5318. https://doi.org/10.1038/s41467-023-40998-0

Ye, Z., Heldmann, M., Slovic, P., & Münte, T. F. (2020). Brain imaging evidence for why we are numbed by numbers. *Scientific Reports, 10*(1), 1. https://doi.org/10.1038/s41598-020-66234-z

Zaki, J., & Mitchell, J. P. (2013). Intuitive prosociality. *Current Directions in Psychological Science, 22*(6), 466–470. https://doi.org/10.1177/0963721413492764

Neuroscientific Approaches to Understand Barriers and Promotors of Pro-environmental Behavior

Ruth M. Krebs and Nik Sawe

Abstract Despite the mounting impacts of climate change, shrinking biodiversity, and related environmental challenges, individuals find it difficult to adopt more sustainable behaviors. This is often a motivational problem transcending lack of knowledge or structural barriers, in that even individuals with pro-environmental attitudes often fail to act accordingly. Environmental psychology and related disciplines have identified factors that contribute to this so-called intention–behavior gap, including low-level cognitive biases, worldviews, social norms, risk perception, and perceived behavioral control. These observations are largely based on self-report data or choice behavior in experimental settings, both featuring particular advantages and disadvantages. More recently, neuroscientific approaches have been adopted to the pro-environmental behavior domain, as they are able to assess more implicit processes related to outcome evaluation, decision-making, and motivation in general that guide our actions. In turn understanding these processes can be valuable for promoting more sustainable behaviors to mitigate climate change and counteract biodiversity decline. In this chapter, we highlight two lines of work at the intersection of cognitive neuroscience and environmental psychology in detail, discuss related research and future avenues, as well as implications and critical issues in this emerging field.

Keywords Pro-environmental behavior · Neuroeconomics · Cognitive effort · Electroencephalography · Functional magnetic resonance imaging

R. M. Krebs (✉)
Department of Experimental Psychology, Faculty of Psychology and Educational Sciences, Ghent University, Ghent, Belgium

N. Sawe
Emmett Interdisciplinary Program in Environment and Resources, Stanford University, Stanford, CA, USA

S. Kühn (ed.), *Environmental Neuroscience*,
https://doi.org/10.1007/978-3-031-64699-7_7

Introduction

As the climate crisis has grown in urgency and prominence in both the political and public arenas, an increasingly diverse array of research fields have applied their methods to examine the issue, from engineering and agriculture to economics and psychology. Large-scale contextual factors play an important role in addressing the crisis and providing sustainable solutions, such as providing the necessary financial resources and infrastructure, access to relevant information and sustainable alternatives, and navigating social identity (Amel et al., 2017; Fritsche et al., 2018; Nielsen et al., 2021). However, changes in environmental attitudes and behavior at the individual scale are also of great importance and can in turn contribute to wide-reaching changes as the societal consensus shifts. One pronounced divide lies between the knowledge and awareness of climate-related and environmental problems (global warming impacts, air and sea pollution, and shrinking biodiversity) on the one hand, and changes in individual behavior and policy that are needed to counteract the negative development on the other hand. This is referred to as the intention-behavior (or attitude-behavior) gap—a phenomenon observed in many domains, ranging from pro-environmental and health behavior to voting behavior (Burg et al., 2006; Holland et al., 2002; Kollmuss & Agyeman, 2002; Pieters & Verplanken, 1995). Of note, while health behaviors typically benefit the individual first and foremost, the effects on society are often secondary (with pandemic response being a notable exception). This is reversed for most pro-environmental actions, rendering behavior change for the climate even less "attractive." Researchers try to understand the factors that drive certain people to engage in specific pro-environmental behaviors or promote pro-environmental goals, potential mechanisms to facilitate those behaviors (Spurlock et al., 2020), as well as the barriers that prevent individuals from doing so (Gifford, 2011). In addition to structural barriers and lack of knowledge, numerous psychological barriers can prevent us from adopting more sustainable behaviors (including psychological biases, restricted worldviews, social norms, risk perception, and perceived behavioral control). One set of barriers relates to reactions to the *perceived risk* of climate-related outcomes: specifically, that these have to be both geographically and temporally proximal, certain to occur, and affect the individual directly in order to be perceived as relevant and high priority (Gattig & Hendrickx, 2007). In turn, climate-related outcomes that are distant, uncertain, and not affecting us directly are associated with a low level of *perceived behavioral control*, that is, the feeling that your actions are not instrumental in mitigating the negative outcomes. The concept of behavioral control has been featured in early theories of pro-environmental behavior (Ajzen, 1991) and is considered a key driver for behavior in reward-based motivation (here termed action-outcome contingency or efficacy (see, Frober & Dreisbach, 2014; Fromer et al., 2021)). Low perceived risk and low perceived behavioral control are unfortunately inherent properties of many pro-environmental behaviors and associated outcomes, rendering research into factors that could modulate these perceptions or promote pro-environmental behavior via other routes even more important.

Over the past decade, topics related to understanding and promoting pro-environmental behavior that were traditionally situated in environmental psychology and related social science disciplines have been picked up by the broader field of behavioral sciences, including (neuro)economics and neuroscience (e.g., Berman et al., 2019; Doell et al., 2023; Leeuwis et al., 2022; Sawe, 2019). In this chapter, we will showcase research that applies neuroscientific approaches to better understand pro-environmental behavior. Building on the wealth of neuroscientific frameworks in the domain of motivation, effort-based decision-making, and neuroeconomics (Braver et al., 2014; Glimcher & Rustichini, 2004; Shenhav et al., 2021; Westbrook & Braver, 2015), neuroscientific techniques including functional magnetic resonance imaging (fMRI) and electroencephalography (EEG) can provide valuable insights into the underlying processes that give rise to pro-environmental behavior (➔ Grassini EEG chapter, ➔ Draganski MRI chapter). More specifically, they can help to elucidate the factors that increase or decrease the likelihood of making pro-environmental choices. Pro-environmental behaviors and attitudes have traditionally been assessed using self-report scales and field studies, and only more recently via experimental paradigms (Lange, 2022), including studies that utilize neuroimaging data (Doell et al., 2023). All these approaches each entail unique advantages and should be considered largely complementary. The distinctive contribution of neuroscience here is to provide insights into implicit processes during stimulus encoding and decision-making, lending nuance to the ensuing behavior.

In this chapter, we will focus on neural signatures broadly related to the *value* of environmental outcomes, as well as factors that render pro-environmental decisions and behaviors *costly*. Both the expected outcome value and the costs associated with securing this outcome determine the probability to engage in pro-environmental behaviors (as well as other effortful everyday behaviors). In keeping with contemporary neuroscientific frameworks, the calculation of these value and cost components and their subsequent integration both rely on neural processes related to subjective evaluation, uncertainty, and conflict resolution (Shenhav et al., 2021; Verguts et al., 2015; Westbrook et al., 2019). These can in turn be traced using neuroimaging (first and foremost fMRI and EEG) and complementary techniques (such as eye-tracking) without explicit inquisition (e.g., "How much do you value this outcome?", "How difficult was this decision?")—which entails several disadvantages such as distraction from the task and influences of social desirability (e.g., Arnold & Feldman, 1981). As such, neuroimaging data can complement the behavioral data to gain a more comprehensive picture about how pro-environmental decisions and actions come about. Moreover, combining data from different sources (e.g., self-report, performance measures, and neural signatures) is highly valuable when exploring interindividual differences, which have proven to be relevant in the domain of motivation (Costumero et al., 2013; Krebs et al., 2011) as well as pro-environmental choice behavior (Sawe & Knutson, 2015; Sawe et al., 2022).

Below we showcase two lines of research at the intersection between cognitive neuroscience and environmental psychology in detail, followed by discussing related ongoing research and future avenues, as well as implications and critical issues. Note that although we use "pro-environmental behavior" as a general construct throughout the chapter, each experimental study employs a specific operationalization thereof within a unique study context.

A Neuroeconomic Approach to Understand Pro-environmental Behavior

One avenue to better understand pro-environmental behavior through neuroscience is to apply experimental techniques from the emerging field of neuroeconomics, which combines brain imaging with behavioral economics experiments to deconstruct the decision-making process.

In this section, we will provide the key objectives of this approach as well as concrete examples to highlight how neuroscientific data can create added value in understanding the underlying mechanisms of certain choices, considering mediating factors such as affective evaluation and cost-benefit assessment. Very broadly, economic decisions, including those outside of the environmental domain, are influenced by counterbalancing the financial costs or investment with the expected value of the good or goal for the individual. These two features are considered in a cost–benefit analysis. One way to investigate these considerations for different goods and in different contexts is to assess an individual's willingness to pay (Kahneman et al., 1993) through various elicitation methods. Combining this willingness-to-pay measure with neuroscientific data has been leveraged to understand economic decisions in a variety of contexts, including food preferences (Plassmann et al., 2007), social charity (Genevsky et al., 2013), microlending (Genevsky & Knutson, 2015), as well as for environmental outcomes (Sawe & Knutson, 2015; Sawe et al., 2022). The latter studies aimed to isolate neural signatures associated with the affective response to environmental goods and relate them back to facets of the decision process, even predicting downstream behavior using neural data.

For instance, in the study by Sawe and Knutson (2015), participants engaged in a willingness-to-pay task while fMRI data were recorded. Specifically, participants were asked to decide from trial-to-trial whether they would be willing to donate varying amounts of money to protect natural park lands in various iconic or non-iconic settings from proposed conservative or destructive repurposed land uses. On the behavioral level, destructive land uses motivated participants to donate more than iconic features of the landscapes themselves. On the neural level, encountering iconic parks was associated with increased activity in the ventral striatum (specifically the nucleus accumbens): a region that is sensitive to positive stimulus value and associated with approach behavior (Knutson & Greer, 2008). During the subsequent display of the proposed land use, destructive uses elicited increased activity in anterior insula, a region that is implicated in processing losses and negative affect and associated with avoidance behavior (Knutson & Greer, 2008). These opposing affective signals appeared to be integrated in medial prefrontal cortex, which is implicated in cost–benefit assessments. Importantly, anterior insula activity predicted increased willingness to donate to preserve the parks from the proposed destructive land uses. In contrast, medial prefrontal activity was associated with reduced willingness to donate, which seems to suggest that activity modulation in this region reflects the cost–benefit integration of personal gains (rather than the environmental outcome).

In a more recent study, a similar approach was adopted to investigate how ecolabels denoting energy efficiency might influence consumer appliance purchasing choices (Sawe et al., 2022). Here, increased nucleus accumbens activity was associated positively with purchasing choices, and increased in the presence of the Energy Star ecolabel, suggesting that ecolabels can promote environmental choices via positive affect. A similar observation was featured in one of the first neuroimaging studies on pro-environmental behavior by Linder and colleagues (Linder et al., 2010). They showed that food items with a well-known organic label were associated with increased ventral striatum activity, likely reflecting an overall positive connotation.

The above observations globally align with the affect–integration–motivation (AIM) framework proposed by Samanez-Larkin and Knutson (Samanez-Larkin & Knutson, 2015). While the nucleus accumbens is highly sensitive to (the anticipation of) positive, rewarding events, and approach behavior (Bartra et al., 2013; Knutson & Greer, 2008; Parro et al., 2018), anterior insula activation is often associated with negative signals and avoidance behavior (Labrakakis, 2023; Palminteri et al., 2012; Preuschoff et al., 2008). According to the AIM framework, the initial signals that might trigger approach or avoidance behavior are thought to be integrated with other relevant information in medial prefrontal cortex in order to guide decisions and actions (Samanez-Larkin & Knutson, 2015). Note that while these value-related signals are most relevant in the present context, the roles of nucleus accumbens and anterior insula are much more diverse than merely processing positive and negative valence, respectively (Floresco, 2015; Molnar-Szakacs & Uddin, 2022).

Together the above neuroimaging studies on pro-environmental decisions highlight that both positive and negative affective responses to environmental stimuli can motivate pro-environmental actions. This contrasts with the observation that negative environmental scenarios are less effective in capturing attention and might even hamper action (Brosch, 2021; Carlson et al., 2020). The extent to which positive and negative stimuli can lead to modification of behavior likely depends on the strength of these signals and the task context (e.g., extreme catastrophic scenarios can lead to aversion behavior and thus inaction), as well as individual differences such as environmental attitudes. For instance, anterior insula response toward destructive land use was elevated in those with strong pro-environmental attitudes in the parks donation study (Sawe & Knutson, 2015). Individual differences can even cause differential processing of environmental stimuli, leading to discrete types of thinkers where different brain areas predict their behavior. For instance, in the eco-labeling study described above (Sawe et al., 2022), the purchasing behavior of consumers with high discount rates (e.g., more financially impulsive) was predicted positively by nucleus accumbens activity, while the purchasing decisions of those with low discount rates (e.g., more patient financially) were instead negatively predicted by medial prefrontal cortex activity, which appeared to hold individuals back from purchasing, similar to the inhibitory role of the medial prefrontal cortex in the parks donation study (see, Sawe & Knutson, 2015).

A related study focusing on the subjective value of environmental public goods (e.g., a clean city canal) also featured intriguing interindividual differences (Khaw et al., 2015); here the authors assessed fMRI activity in response to viewing specific environmental goods alongside other consumer goods and leisure activities. After the fMRI session, the subjective value of these goods was assessed with different contingent valuation approaches. In contrast to the other consumer goods, the subjective evaluation of environmental goods (as assessed by the willingness-to-pay taxes to protect this good) did not correlate with brain activity in typical value-related regions (e.g., ventral striatum and ventromedial prefrontal cortex). Instead, and only in a subgroup of participants, subjective value was associated with dorsomedial prefrontal cortex activity when processing environmental goods, which is indicative of higher cognitive control demand.

The above studies focus on assessing underlying affective responses that mediate environmental decisions. Other factors that influence economic decisions and choice preference more broadly include uncertainty (Platt & Huettel, 2008), temporal discounting (e.g., Scheres et al., 2013), and social context (e.g., Fehr & Camerer, 2007; Mailath & Postlewaite, 2003). To date, only few neuroimaging studies have examined these factors in the context of environmental decisions (see, Munuera & Burguière, 2022; Sawe & Chawla, 2021; Sawe et al., 2022). We will discuss several examples in section "Related (Future) Research" below. Finally, existing work grounded in environmental psychology that explores how individuals consider the different costs and benefits of pro-environmental goals and actions may lend themselves to being complemented by neuroimaging techniques. One example is the carbon emission task (Berger & Wyss, 2021), in which participants have to decide to accept or forego varying monetary incentives that are associated with varying levels of real carbon emissions. A related approach has been implemented by Lange and colleagues, in one case trading-off carbon emissions with personal waiting time (Lange et al., 2018), and in another trading-off pro-environmental donations with personal effort investment (Lange & Dewitte, 2022). All these approaches provide an experimental measure of an individual's pro-environmental behavior propensity, based on the integration of the specific costs and outcomes at stake.

The Cost and Value of Cognitive Effort in Pro-environmental Behavior

While the previous section focused on the value of pro-environmental outcomes in the context of economic decision-making by considering the amount that participants are willing to pay or donate, the current section features a different type of resource that can be invested by an individual, namely *cognitive effort*. In addition to financial investments (e.g., buying solar panels or supporting an environmental organization) and physical effort (e.g., commuting via bicycle), implementing actions that are environmentally relevant entails the allocation of cognitive resources at some point in the process. Even if a more sustainable action is equally effortful as a less sustainable option (e.g., selecting the organic version of a product from the same shelf), it is often preceded or accompanied by additional cognitive costs such as setting a goal in advance or replacing a habitual action in the moment.

Here, we focus on the role of cognitive effort in pro-environmental behavior, and more specifically, considering effort allocation as a general barrier for behavior change (Krebs et al., 2023). In keeping with contemporary frameworks, cognitive effort can be conceptualized as the deliberate allocation of cognitive resources to carry out a certain task or achieve a certain goal (see, Kool & Botvinick, 2018; Massin, 2017; Westbrook & Braver, 2015). Although investing cognitive effort is an omnipresent necessity in modern everyday life, humans typically tend to follow the law of least effort wherever possible (Hull, 1943). Since cognitive effort is considered costly in most contexts, the willingness to allocate effort depends on the expected outcomes (Hughes et al., 2015; Krebs et al., 2012; Schevernels et al., 2014; Treadway et al., 2009; Westbrook & Braver, 2015). The relationship between expected cognitive effort and outcome has been formalized in the Expected Value of Control framework (Shenhav et al., 2013; Shenhav et al., 2021). It posits that cognitive control (which is directly associated with effort) depends on the expected value of exerting control, the amount of control required, and its costs.

In a recent study, Krebs and colleagues have tested the basic assumption that the willingness to invest cognitive effort is diminished for pro-environmental outcomes compared to personal outcomes (Krebs et al., 2023). Specifically, personal outcomes and pro-environmental outcomes were embedded in a conflict task that is known to probe cognitive effort (i.e., counting Stroop, Bush et al., 2006). Task performance in personal and pro-environmental outcome trials was linked to personal incentives and donations to a pro-environmental organization, respectively. The data revealed that individuals invested less cognitive effort in trials featuring pro-environmental compared to personal outcomes, as indexed by inferior task performance. Notably, the objective outcome value as well as task demands were identical for both outcome types. This suggests that a lower *subjective* value of pro-environmental outcomes is tipping the cost–benefit assessment toward the costs associated with the cognitive task.

The above discussed data reflect the behavioral consequences of considering certain reward and effort contingencies. However, behavioral output reflects only one point in time, that is the end of the decision process rather than the processes that lead up to this output. In order to gain deeper insights into the mechanisms underlying pro-environmental behavior, and in particular the temporal dynamics, we can utilize EEG (Wang & van den Berg, 2021) (➔ chapter "EEG for the Study of Environmental Neuroscience" by Grassini). This method has superior temporal resolution to the aforementioned fMRI approach, tracking cognitive processes on the order of milliseconds. While the spatial resolution is inferior to fMRI

(➔ chapter "Mapping the Human Brain with Computational Anatomy" by Draganski, Joost and Kherif), previous work has identified EEG components, that is, event-related potentials (ERPs), that are linked to specific cognitive processes. In the present context, we focus on ERP components related to cognitive effort and motivation, including indices of subjective value, preparatory control, and feedback evaluation (Carsten et al., 2020; Fromer et al., 2021; Schevernels et al., 2014).

In order to assess processes related to the integration of expected value and costs of effort allocation, the above pro-environmental reward paradigm is modified to also include advance information about the upcoming effort level. Participants first see a cue providing information about the expected effort (high vs. low difficulty of the target display) and expected reward magnitude (high vs. low). The subsequent target varies in difficulty accordingly, and the feedback varies according to the expected reward (given fast and correct performance). In addition to the replication on the behavioral level, with reduced effort investment for pro-environmental compared to personal outcomes (Krebs et al., 2023), ERPs will be assessed during cue, target, and feedback processing. In keeping with previous work on effort and motivation (Schevernels et al., 2014), the most relevant components are the cue-related P3, which is sensitive to outcome value, and the cue-related contingent negative variation (CNV), which indexes differential preparatory activity based on reward and task difficulty. With the paradigm described above, it can be tested whether pro-environmental outcomes are indeed inferior in terms of their subjective value at the cue level, or whether differences emerge only later during the preparatory phase when effort needs to be invested. In the same vein, differential preparatory activity in response to reward cues can be assessed using a time–frequency approach. For instance, recent work has demonstrated that reward cues lead to an increase in mid-frontal theta activity (Arnau et al., 2023). The discussed cue-related modulations are expected to align with subsequent target processing, featuring larger P3 components for targets associated with a higher reward value (Krebs et al., 2013). The pro-environmental reward paradigm, also allows to explore putative differences in *effort discounting* (i.e., reward value decreases with increasing expected effort, Apps et al., 2015), in the sense that the reward value of pro-environmental outcomes might be reduced more strongly by expected effort. Based on the trial structure (cue-interval-target), we can assess at which moment effort discounting diverges between different outcome types—if at all. Finally, feedback-locked activity (Feedback-Related Negativity, FRN) is also expected to vary with reward level (Carsten et al., 2020) as well as based on actual effort investment (Palidis & Gribble, 2020). In keeping with the notion that pro-environmental outcomes have a lower subjective value, the FRN might be overall smaller for these outcomes but could be differentially increased in high effort trials in which positive feedback is less frequent. This latter pattern would be reminiscent of a positive reward prediction error (Fischer & Ullsperger, 2013).

While the above approach probes continuous effort investment ("how much effort does the participant invest during a trial with specific effort and outcome contingencies?"), an alternative approach concerns the decision of whether to invest effort at all ("will the participant perform a trial given specific effort and outcome contingencies or not?"). This bridges conceptually to section "A Neuroeconomic Approach to Understand Pro-environmental Behavior" in which mostly binary yes/no decisions are made based on expected financial gains. To this end, explicit choices can be embedded in the pro-environmental reward paradigm that vary with regard to the involved effort as well as the outcome magnitude (for a similar design see, Lange & Dewitte, 2022). In this context, EEG data can provide insights into different parts of the decision process. Specifically, mirroring previous studies with classic gambling tasks (Cui et al., 2013), ERP components can be assessed during different phases, that is, option evaluation, response selection, and feedback processing. Each option features a unique combination of effort level and outcome magnitude, requiring a binary choice of whether to accept the offer or not. In case an option is accepted, participants will engage in a task of the accordant effort level. Note that this is an extension to classic gambling tasks that do not require performance in a cognitive task (e.g., Cui et al., 2013). This setup allows to assess differences between pro-environmental and personal outcomes with a special focus on difficult decisions during the evaluation and response selection phase. For instance, the P3 component during option evaluation is expected to scale with expected outcome magnitude but is discounted by effort level, especially for pro-environmental outcomes. This would mirror *disadvantageous options* in a gambling context (Cui et al., 2013). These difficult decisions are expected to be accompanied by more negative deflections in the time window directly preceding the response, even if the decision is positive. Hence, positive pro-environmental decisions may not only have a lower probability compared to personal ones but also require more cognitive resources when they are disadvantageous. The reduced willingness to invest effort for pro-environmental outcomes due to lower subjective value conceptually relates to work on pro-social behavior, where Lockwood and colleagues (Lockwood et al., 2017) have shown that participants are less willing to invest *physical* effort for pro-social compared to personal outcomes. Notably, in this task realm, choosing for personal compared to pro-social outcomes was associated with more pronounced value representation on the neural level as assessed with fMRI (Lockwood et al., 2022).

Related (Future) Research

In this section, we will highlight other existing studies as well as future research utilizing neuroscience methods to gain insights into individuals' motivation to engage in pro-environmental behavior. These include processes related to learning and memory, as well as influences from social factors.

One study explored differences between implementing and reducing certain behaviors in the future to benefit the environment (Brevers et al., 2021). The authors collected fMRI data while participants engaged in a task with behavior cues that either signaled an action that individuals might want to implement or increase (e.g., a picture of a train) or an action they might want to reduce or avoid (e.g., a picture of a disposable cup)—both with the goal to decrease environmental impact. After passive viewing of several cues from one category, they were asked to select the behavior that would seem most feasible to undertake in the future. In general, "implement" behaviors were considered more feasible than "reduce" behaviors. The neural data revealed that picturing "implement" actions was associated with increased activity in ventromedial prefrontal regions as well as hippocampus, which may be indicative of forming new representations of known behaviors during the prospective thinking task. In contrast, picturing "reduce" actions was accompanied by right dorsolateral prefrontal cortex activity, which was, moreover, negatively correlated with hippocampus activity. This could be indicative of a downregulation of retrieval-related activity for these behaviors via dorsolateral prefrontal cortex. Together these results provide initial evidence for an involvement of episodic memory processes for implementing novel sustainable behaviors and forming long-term habits. Inspired by this, future research may be directed toward exploring how climate-related information is encoded and stored in, as well as retrieved from, episodic memory, such as the approach known as *differential neural activity based on memory* (Paller & Wagner, 2002). Here, successful versus unsuccessful memory retrieval is traced back to differences in encoding-related brain activity. Moreover, such research could delineate factors that contribute to robust memory representations, which in turn serve the implementation of future behavior.

Next to episodic memory formation, learning processes, in particular reinforcement learning, play a crucial role in shaping behavior. Processes related to reinforcement learning, which heavily relies on the dopaminergic system, have been extensively studied using neuroscience approaches and computational modeling (Daw & Tobler, 2014; Westbrook & Braver, 2016). In a recent review paper, it has been proposed that dopamine-dependent reinforcement learning might be a promising route to promote pro-environmental behavior (Munuera & Burguière, 2022). In particular, the authors elaborate on the phenomenon of *temporal discounting* (i.e., the reduction of reward value for delayed outcomes). This discounting affects economic decisions as well as reward-based motivation in general and is directly related to dopaminergic neurotransmission in that the system favors immediate over long-term outcomes (Kobayashi & Schultz, 2008). As discussed above, one crucial issue with pro-environmental outcomes is that they are inherently distant in time and space. It is argued that in order to increase motivation for such actions via reinforcement learning principles, they would need to yield short-term outcomes. Intriguingly, while the majority of people favor immediate over delayed outcomes, there is evidence for interindividual differences in the sense that individuals who display enhanced reward processing, as indexed by EEG signatures, are more willing to wait for larger rewards (Pornpattananangkul & Nusslock, 2016). This observation seems highly relevant for promoting pro-environmental behaviors in that it might be possible to reduce temporal discounting of pro-environmental goals by strengthening reward-based motivation. Related to the above processes, dopamine is also involved in the transition from goal-directed effortful behavior to more habitual responses (Graybiel, 2008), which can be leveraged in the context of pro-environmental behavior in both directions, that is implementing novel sustainable habits and breaking unsustainable ones (Verplanken & Whitmarsh, 2021).

Other studies have implemented interindividual difference approaches involving neuroimaging data to identify "green" individuals. One study implemented a consumer-choice task while recording EEG and found that individuals that made more sustainable choices exhibited larger frontal theta activity as compared to consumers with less sustainable choices on average (Lee et al., 2014). Intriguingly, the differential theta activity was specific to the processing of product information rather than price information, suggesting that this activity pattern is related to cognitive engagement when thinking about the product rather than merely financial considerations. In a similar vein, another group reports correlations between baseline EEG activity in lateral prefrontal cortex and frequency of self-reported pro-environmental behavior (Baumgartner et al., 2019)—which was, however, the result of an exploratory analysis and warrants further corroboration. The same group more recently investigated the relationship between structural properties of the brain and pro-environmental behavior (Rosales et al., 2022). They found that cortical thickness in prefrontal cortex regions was correlated with sustainable versus unsustainable behaviors during an intergenerational dilemma paradigm, which reflects the willingness to implement sustainable actions that would benefit future generations. This relationship was mediated by interindividual differences in perspective taking. These findings highlight the added value of interindividual difference approaches with multiple methods. Here, structural brain properties are considered to be reflective of traits related to pro-environmental behavior and perspective taking. More generally, the latter study also illustrates the contribution of social psychology perspectives and paradigms to understanding behavior that is relevant for climate change mitigation (Hu et al., 2023). Expanding on this, applying hyperscanning, which allows the tracking of brain activity (e.g., EEG) simultaneously across multiple participants, would seem highly relevant. In turn, this data can be related to study the underlying mechanisms of dynamic social interactions such as in social dilemma games (for review see, Liu et al., 2018).

Finally, we will briefly discuss the possibilities to utilize neuroscience to predict behavior at the population level. Several studies in the field of neuroeconomics and neuromarketing have demonstrated how neural data can be utilized not only to predict choices of an individual but also to scale and predict choices of an independent sample and even large-scale market data (Knutson & Genevsky, 2018; Sawe et al., 2022). Examples of such population-level forecasts range from microlending (Genevsky & Knutson, 2015), to music and food preferences (Berns & Moore, 2012; Kühn et al., 2016), to health campaigns (Falk et al., 2016). In the context of sustainable behavior, such predictions would be highly valuable in the context of influencing consumer behavior and advising policy regarding successful strategies or campaigns (see, Sawe, 2019; Sawe & Chawla, 2021).

To illustrate this approach, we highlight one study that assessed fMRI in smokers while watching different anti-smoking campaign spots, along with self-reported potential effectiveness of these campaigns (Falk et al., 2016). Both neural and self-report data were used as predictors for call volume to the advertised quit-smoking help center. Intriguingly, neural activity modulations in medial prefrontal cortex were predictive of behavior change intention, while the self-reports were not. In addition to offering clear recommendations for campaign content, this study again illustrates that neural measures can reflect implicit processes above and beyond individual's overt responses. A similar forecasting approach has been adopted in the eco-labeling study discussed above (Sawe et al., 2022). Here, neural activity in the study sample that underwent fMRI scanning was used to predict purchase decisions in an independent sample. Specifically, averaged activity from the nucleus accumbens associated with selecting Energy Star ecolabel products did forecast decisions about similar products in a national survey including 1550 households.

Note that this chapter is not meant to provide a comprehensive overview of all the work that has been conducted at the intersection between cognitive neuroscience and environmental psychology but aims to illustrate possible research avenues to utilize neural data to better understand barriers and promotors of pro-environmental behavior.

Implications and Critical Issues

The insights from combining behavioral and neuroscientific approaches can contribute to a better understanding of pro-environmental decisions and behaviors, beyond self-report and choice data. As such, the focus lies on more implicit processes related to the subjective value of certain outcomes, the expected costs, and the dynamics of the decision process. Apart from insights in the underlying processes of engaging or not engaging in pro-environmental behaviors of individuals and groups, this research can be utilized to predict decisions of larger samples. In turn, it holds promise in devising behavioral interventions (e.g., to increase subjective value of certain pro-environmental outcomes) and inform policy (e.g., on energy saving or wildlife conservation campaigns). In this context, it is worth noting that a recent large-scale project exposed limitations of existing intervention approaches (Vlasceanu et al., 2024), which resonates with the call for a more holistic approach to better formulate general recommendations (Lange & Brick, 2021). Moreover, being grounded in established (neuroscientific) frameworks, this line of work answers to the call for more rigorous experimental approaches in the domain of pro-environmental behavior in general and might in particular circumvent cognitive biases and influences of social desirability that are more pronounced in explicit choice and self-report data (Arnold & Feldman, 1981; Kormos & Gifford, 2014; Lange, 2023).

Another great potential area of research, evidenced in some of the aforementioned studies, lies in combining different types of data from the same participants to explore the contribution of interindividual differences in pro-environmental behavior. There are manifold opportunities, including the combination of behavioral, self-report, and neural data from one source (see examples above) but also combining different neuroimaging or stimulation techniques in one study (e.g.,

EEG, fMRI, magneto-encephalography (MEG), transcranial magnetic stimulation (TMS)), alongside structural data (e.g., gray matter volume and structural connectivity). Finally, the experimental approaches to understand and predict pro-environmental behavior featured in this chapter have been partly adopted from—or can in turn be adopted to—other applied domains, such as health and pro-social behavior.

We conclude the chapter by considering critical issues in the context of utilizing neuroscience methods to contribute to climate change mitigation. From a very general perspective, researchers have to clearly define which type of pro-environmental behaviors is measured and acknowledge that these may not represent the wide range of pro-environmental behaviors in real life. In the examples discussed above, the focus is on specific pro-environmental behavior related to conserving resources, reducing CO2 emissions, and promoting biodiversity, which are considered as model examples for pro-environmental behavior but might not apply universally across pro-environmental behaviors (Lange & Dewitte, 2019). Further, the way in which pro-environmental outcomes are presented to the participant matters; for instance, monetary donations lend themselves very well to repeated-trial manipulations that are crucial for the robust assessment of behavioral and neural responses. Related to this, it is important to acknowledge that neuroimaging experiments might pose some limitations in their design, as complex stimuli and narratives—which are often used in economic decision tasks or environmental paradigms—can confound the neuroimaging data, hampering inferences regarding the manipulation of interest. It is hence advised to simplify the elements of the tasks while still ensuring that the outcomes and associated contingencies are relevant for the participants.

Finally, when trying to understand and ultimately promote pro-environmental behaviors, it seems relevant to consider the environmental impact of the research practices themselves (Aron et al., 2020; Rae, 2023). This first and foremost concerns energy consumption associated with operating the devices as well as computing power during data processing. Depending on the research question at hand, researchers may consider the potential added value of the research against the associated environmental costs and try to reduce these costs if possible by adopting more efficient computing and storing pipelines, engage in data sharing, use alternative methods, and apply "reduce, reuse, recycle" principles in the lab in general. One technical example is the installation of a helium recycling tank for MEG devices, or switching to devices that do not require liquid helium (Laine et al., 2019). Another avenue is related to data sharing, which may allow to answer new research questions using existing data from a depository (e.g., Human Connectome Project; Open Science Framework). To give an example for alternative methods, functional near-infrared spectroscopy (fNIRS) can be a low-energy consuming alternative for fMRI if the research focuses on hemodynamic modulations in prefrontal cortex and other regions on the surface of the brain (e.g., Fishburn et al., 2014).

References

Ajzen, I. (1991). The theory of planned ehavior. *Organizational Behavior and Human Decision Processes, 50*(2), 179–211. https://doi.org/10.1016/0749-5978(91)90020-T

Amel, E., Manning, C., Scott, B., & Koger, S. (2017). Beyond the roots of human inaction: Fostering collective effort toward ecosystem conservation. *Science, 356*(6335), 275–278. https://doi.org/10.1126/science.aal1931

Apps, M. A. J., Grima, L. L., Manohar, S., & Husain, M. (2015). The role of cognitive effort in subjective reward devaluation and risky decision-making. *Scientific Reports, 5*, 16880. https://doi.org/10.1038/srep16880

Arnau, S., Liegel, N., & Wascher, E. (2023). Frontal midline theta power during the cue-target-interval reflects increased cognitive effort in rewarded task-switching. bioRxiv, 2023.2009.2025.559275. https://doi.org/10.1101/2023.09.25.559275

Arnau, S., Liegel, N., & Wascher, E. (2023). Frontal midline theta power during the cue-target-interval reflects increased cognitive effort in rewarded task-switching.

Arnold, H. J., & Feldman, D. C. (1981). Social desirability response bias in self-report choice situations. *Academy of Management Journal, 24*(2), 377–385. https://doi.org/10.5465/255848

Aron, A. R., Ivry, R. B., Jeffery, K. J., Poldrack, R. A., Schmidt, R., Summerfield, C., & Urai, A. E. (2020). How can neuroscientists respond to the climate emergency? *Neuron, 106*(1), 17–20. https://doi.org/10.1016/j.neuron.2020.02.019

Bartra, O., McGuire, J. T., & Kable, J. W. (2013). The valuation system: A coordinate-based meta-analysis of BOLD fMRI experiments examining neural correlates of subjective value. *NeuroImage, 76*(1), 412–427. https://doi.org/10.1016/j.neuroimage.2013.02.063

Baumgartner, T., Langenbach, B. P., Gianotti, L. R. R., Müri, R. M., & Knoch, D. (2019). Frequency of everyday pro-environmental behaviour is explained by baseline activation in lateral prefrontal cortex. *Scientific Reports, 9*, 9. https://doi.org/10.1038/s41598-018-36956-2

Berger, S., & Wyss, A. M. (2021). Measuring pro-environmental behavior using the carbon emission task. *Journal of Environmental Psychology, 75*(22), 101613. https://doi.org/10.1016/j.jenvp.2021.101613

Berman, M. G., Kardan, O., Kotabe, H. P., Nusbaum, H. C., & London, S. E. (2019). The promise of environmental neuroscience comment. *Nature Human Behaviour, 3*(5), 414–417. https://doi.org/10.1038/s41562-019-0577-7

Berns, G. S., & Moore, S. E. (2012). A neural predictor of cultural popularity. *Journal of Consumer Psychology, 22*(1), 154–160. https://doi.org/10.1016/j.jcps.2011.05.001

Braver, T. S., Krug, M. K., Chiew, K. S., Kool, W., Westbrook, J. A., Clement, N. J., et al. (2014). Mechanisms of motivation-cognition interaction: challenges and opportunities. *Cognitive, Affective, & Behavioral Neuroscience, 14*(2), 443–472. https://doi.org/10.3758/s13415-014-0300-0

Brevers, D., Baeken, C., Maurage, P., Sescousse, G., Vögele, C., & Billieux, J. (2021). Brain mechanisms underlying prospective thinking of sustainable behaviours. *Nature Sustainability, 4*(5), 433–439. https://doi.org/10.1038/s41893-020-00658-3

Brosch, T. (2021). Affect and emotions as drivers of climate change perception and action: A review. *Current Opinion in Behavioral Sciences, 42*, 15–21. https://doi.org/10.1016/j.cobeha.2021.02.001

Burg, J., de Vet, E., de Nooijer, J., & Verplanken, B. (2006). Predicting fruit consumption: Cognitions, intention, and habits. *Journal of Nutrition Education and Behavior, 38*(2), 73–81. https://doi.org/10.1016/j.jneb.2005.11.027

Bush, G., Whalen, P. J., Shin, L. M., & Rauch, S. L. (2006). The counting Stroop: A cognitive interference task. *Nature Protocols, 1*(1), 230–233. https://doi.org/10.1038/nprot.2006.35

Carlson, J. M., Kaull, H., Steinhauer, M., Zigarac, A., & Cammarata, J. (2020). Paying attention to climate change: Positive images of climate change solutions capture attention. *Journal of Environmental Psychology, 71*, 101477. https://doi.org/10.1016/j.jenvp.2020.101477

Carsten, T., Kostandyan, M., Boehler, C. N., & Krebs, R. M. (2020). Comparing the motivational value of rewards and losses in an EEG-pupillometry study. *European Journal of Neuroscience*. https://doi.org/10.1111/ejn.15064

Costumero, V., Barros-Loscertales, A., Bustamante, J. C., Ventura-Campos, N., Fuentes, P., & Avila, C. (2013). Reward sensitivity modulates connectivity among reward brain areas during processing of anticipatory reward cues. *European Journal of Neuroscience, 38*(3), 2399–2407. https://doi.org/10.1111/ejn.12234

Cui, J. F., Chen, Y. H., Wang, Y., Shum, D. H. K., & Chan, R. C. K. (2013). Neural correlates of uncertain decision making: ERP evidence from the Iowa Gambling Task. *Frontiers in Human Neuroscience, 7*, 776. https://doi.org/10.3389/fnhum.2013.00776

Daw, N. D., & Tobler, P. N. (2014). Value learning through reinforcement: The basics of dopamine and reinforcement learning. In P. W. Glimcher & E. Fehr (Eds.), *Neuroeconomics (second edition) decision making and the brain* (pp. 283–298). Elsevier.

Doell, K. C., Berman, M. G., Bratman, G. N., Knutson, B., Kühn, S., Lamm, C., et al. (2023). Leveraging neuroscience for climate change research. *Nature Climate Change*. https://doi.org/10.1038/s41558-023-01857-4

Falk, E. B., O'Donnell, M. B., Tompson, S., Gonzalez, R., Dal Cin, S., Strecher, V., et al. (2016). Functional brain imaging predicts public health campaign success. *Social Cognitive and Affective Neuroscience, 11*(2), 204–214. https://doi.org/10.1093/scan/nsv108

Fehr, E., & Camerer, C. F. (2007). Social neuroeconomics: The neural circuitry of social preferences. *Trends in Cognitive Sciences, 11*(10), 419–427. https://doi.org/10.1016/j.tics.2007.09.002

Fischer, A. G., & Ullsperger, M. (2013). Real and fictive outcomes are processed differently but converge on a common adaptive mechanism. *Neuron, 79*(6), 1243–1255. https://doi.org/10.1016/j.neuron.2013.07.006

Fishburn, F. A., Norr, M. E., Medvedev, A. V., & Vaidya, C. J. (2014). Sensitivity of fNIRS to cognitive state and load. *Frontiers in Human Neuroscience, 8*, 76. https://doi.org/10.3389/fnhum.2014.00076

Floresco, S. B. (2015). The nucleus accumbens: An interface between cognition, emotion, and action. *Annual Review of Psychology, 66*, 25–52. https://doi.org/10.1146/annurev-psych-010213-115159

Fritsche, I., Barth, M., Jugert, P., Masson, T., & Reese, G. (2018). A Social Identity Model of Pro-Environmental Action (SIMPEA). *Psychological Review, 125*(2), 245–269. https://doi.org/10.1037/rev0000090

Frober, K., & Dreisbach, G. (2014). The differential influences of positive affect, random reward, and performance-contingent reward on cognitive control. *Cognitive, Affective, & Behavioral Neuroscience, 14*(2), 530–547. https://doi.org/10.3758/s13415-014-0259-x

Fromer, R., Lin, H., Dean Wolf, C. K., Inzlicht, M., & Shenhav, A. (2021). Expectations of reward and efficacy guide cognitive control allocation. *Nature Communications, 12*(1), 1030. https://doi.org/10.1038/s41467-021-21315-z

Gattig, A., & Hendrickx, L. (2007). Judgmental discounting and environmental risk perception: Dimensional similarities, domain differences, and implications for sustainability. *Journal of Social Issues, 63*(1), 21–39. https://doi.org/10.1111/j.1540-4560.2007.00494.x

Genevsky, A., & Knutson, B. (2015). Neural affective mechanisms predict market-level microlending. *Psychological Science, 26*(9), 1411–1422. https://doi.org/10.1177/0956797615588467

Genevsky, A., Västfjäll, D., Slovic, P., & Knutson, B. (2013). Neural underpinnings of the identifiable victim effect: Affect shifts preferences for giving. *Journal of Neuroscience, 33*(43), 17188–17196. https://doi.org/10.1523/Jneurosci.2348-13.2013

Gifford, R. (2011). The dragons of inaction psychological barriers that limit climate change mitigation and adaptation. *American Psychologist, 66*(4), 290–302. https://doi.org/10.1037/a0023566

Glimcher, P. W., & Rustichini, A. (2004). Neuroeconomics: The consilience of brain and decision. *Science, 306*(5695), 447–452.

Graybiel, A. M. (2008). Habits, rituals, and the evaluative brain. *Annual Review of Neuroscience, 31*, 359–387. https://doi.org/10.1146/annurev.neuro.29.051605.112851

Holland, R. W., Verplanken, B., & Van Knippenberg, A. (2002). On the nature of attitude-behavior relations: the strong guide, the weak follow. *European Journal of Social Psychology, 32*(6), 869–876. https://doi.org/10.1002/ejsp.135

Hu, J., Konovalov, A., & Ruff, C. C. (2023). A unified neural account of contextual and individual differences in altruism. *eLife, 12*, e80667. https://doi.org/10.7554/eLife.80667

Hughes, D. M., Yates, M. J., Morton, E. E., & Smillie, L. D. (2015). Asymmetric frontal cortical activity predicts effort expenditure for reward. *Social Cognitive and Affective Neuroscience, 10*(7), 1015–1019. https://doi.org/10.1093/scan/nsu149

Hull, C. (1943). *Principles of behavior*. Appleton-Century-Crofts.

Kahneman, D., Ritov, I., Jacowitz, K. E., & Grant, P. (1993). Stated willingness-to-pay for public-goods – A psychological perspective. *Psychological Science, 4*(5), 310–315. https://doi.org/10.1111/j.1467-9280.1993.tb00570.x

Khaw, M. W., Grab, D. A., Livermore, M. A., Vossler, C. A., & Glimcher, P. W. (2015). The measurement of subjective value and its relation to contingent valuation and environmental public goods. *PLoS One, 10*(7), e0132842. https://doi.org/10.1371/journal.pone.0132842

Knutson, B., & Genevsky, A. (2018). Neuroforecasting aggregate choice. *Current Directions in Psychological Science, 27*(2), 110–115. https://doi.org/10.1177/0963721417737877

Knutson, B., & Greer, S. M. (2008). Anticipatory affect: neural correlates and consequences for choice. *Philosophical Transactions of the Royal Society B-Biological Sciences, 363*(1511), 3771–3786. https://doi.org/10.1098/rstb.2008.0155

Kobayashi, S., & Schultz, W. (2008). Influence of reward delays on responses of dopamine neurons. *Journal of Neuroscience, 28*(31), 7837–7846. https://doi.org/10.1523/Jneurosci.1600-08.2008

Kollmuss, A., & Agyeman, J. (2002). Mind the gap: Why do people act environmentally and what are the barriers to pro-environmental behavior? *Environmental Education Research, 8*(3), 239–260. https://doi.org/10.1080/13504620220145401

Kool, W., & Botvinick, M. (2018). Mental labour. *Nature Human Behaviour, 2*(12), 899–908. https://doi.org/10.1038/s41562-018-0401-9

Kormos, C., & Gifford, R. (2014). The validity of self-report measures of proenvironmental behavior: A meta-analytic review. *Journal of Environmental Psychology, 40*, 359–371. https://doi.org/10.1016/j.jenvp.2014.09.003

Krebs, R. M., Boehler, C. N., Egner, T., & Woldorff, M. G. (2011). The neural underpinnings of how reward associations can both guide and misguide attention. *The Journal of Neuroscience, 31*(26), 9752–9759.

Krebs, R. M., Boehler, C. N., Roberts, K. C., Song, A. W., & Woldorff, M. G. (2012). The involvement of the dopaminergic midbrain and cortico-striatal-thalamic circuits in the integration of reward prospect and attentional task demands. *Cerebral Cortex, 22*(3), 607–615.

Krebs, R. M., Boehler, C. N., Appelbaum, L. G., & Woldorff, M. G. (2013). Reward associations reduce behavioral interference by changing the temporal dynamics of conflict processing. *PLoS One, 8*(1), e53894. https://doi.org/10.1371/journal.pone.0053894

Krebs, R. M., Prevel, A., Hall, J. M., & Hoofs, V. (2023). Think green: Investing cognitive effort for a pro-environmental cause. *Journal of Environmental Psychology, 85*, 101946. https://doi.org/10.1016/j.jenvp.2022.101946

Kühn, S., Strelow, E., & Gallinat, J. (2016). Multiple "buy buttons" in the brain: Forecasting chocolate sales at point-of-sale based on functional brain activation using fMRI. *NeuroImage, 136*, 122–128. https://doi.org/10.1016/j.neuroimage.2016.05.021

Labrakakis, C. (2023). The role of the insular cortex in pain. *International Journal of Molecular Sciences, 24*(6), 5736. https://doi.org/10.3390/ijms24065736

Laine, P., Nenonen, J., Chappel, S., & Knuutila, J. (2019). Zero helium boiloff MEG technology. In S. Supek & C. Aine (Eds.), *Magnetoencephalography* (pp. 1249–1259). Springer.

Lange, F. (2022). Behavioral paradigms for studying pro-environmental behavior: A systematic review. *Behavior Research Methods*. https://doi.org/10.3758/s13428-022-01825-4

Lange, F. (2023). Beyond self-reports: A call for more behavior in environmental psychology. *Journal of Environmental Psychology, 86*, 101965.

Lange, F., & Brick, C. (2021). Changing pro-environmental behavior: Evidence from (Un)successful intervention studies. *Sustainability, 13*(14), 7748. https://doi.org/10.3390/su13147748

Lange, F., & Dewitte, S. (2019). Measuring pro-environmental behavior: Review and recommendations. *Journal of Environmental Psychology, 63*, 92–100. https://doi.org/10.1016/j.jenvp.2019.04.009

Lange, F., & Dewitte, S. (2022). The work for environmental protection task: A consequential web-based procedure for studying pro-environmental behavior. *Behavior Research Methods, 54*(1), 133–145. https://doi.org/10.3758/s13428-021-01617-2

Lange, F., Steinke, A., & Dewitte, S. (2018). The pro-environmental behavior task: A laboratory measure of actual pro-environmental behavior. *Journal of Environmental Psychology, 56*, 46–54. https://doi.org/10.1016/j.jenvp.2018.02.007

Lee, E. J., Kwon, G., Shin, H. J., Yang, S., Lee, S., & Suh, M. (2014). The spell of green: Can frontal EEG activations identify green consumers? *Journal of Business Ethics, 122*(3), 511–521. https://doi.org/10.1007/s10551-013-1775-2

Leeuwis, N., van Bommel, T., & Alimardani, M. (2022). A framework for application of consumer neuroscience in pro-environmental behavior change interventions. *Frontiers in Human Neuroscience, 16*, 886600. https://doi.org/10.3389/fnhum.2022.886600

Linder, N. S., Uhl, G., Fliessbach, K., Trautner, P., Eiger, C. E., & Weber, B. (2010). Organic labeling influences food valuation and choice. *NeuroImage, 53*(1), 215–220. https://doi.org/10.1016/j.neuroimage.2010.05.077

Liu, D. F., Liu, S., Liu, X. M., Zhang, C., Li, A. S. K., Jin, C. G., et al. (2018). Interactive brain activity: Review and progress on EEG-based hyperscanning in social interactions. *Frontiers in Psychology, 9*, 1862. https://doi.org/10.3389/fpsyg.2018.01862

Lockwood, P. L., Hamonet, M., Zhang, S. H., Ratnavel, A., Salmony, F. U., Husain, M., & Apps, M. A. J. (2017). Prosocial apathy for helping others when effort is required. *Nature Human Behaviour, 1*(7), 0131. https://doi.org/10.1038/s41562-017-0131

Lockwood, P. L., Wittmann, M. K., Nili, H., Matsumoto-Ryan, M., Abdurahman, A., Cutler, J., et al. (2022). Distinct neural representations for prosocial and self-benefiting effort. *Current Biology, 32*(19), 4172. https://doi.org/10.1016/j.cub.2022.08.010

Mailath, G. J., & Postlewaite, A. (2003). The social context of economic decisions. *Journal of the European Economic Association, 1*(2–3), 354–362. https://doi.org/10.1162/154247603322390991

Massin, O. (2017). Towards a definition of efforts. *Motivation Science, 3*(3), 230–259.

Molnar-Szakacs, I., & Uddin, L. Q. (2022). Anterior insula as a gatekeeper of executive control. *Neuroscience and Biobehavioral Reviews, 139*, 104736. https://doi.org/10.1016/j.neubiorev.2022.104736

Munuera, J., & Burguière, E. (2022). Can we tackle climate change by behavioral hacking of the dopaminergic system? *Frontiers in Behavioral Neuroscience, 16*, 996955. https://doi.org/10.3389/fnbeh.2022.996955

Nielsen, K. S., Clayton, S., Stern, P. C., Dietz, T., Capstick, S., & Whitmarsh, L. (2021). How psychology can help limit climate change. *American Psychologist, 76*(1), 130–144. https://doi.org/10.1037/amp0000624

Palidis, D. J., & Gribble, P. L. (2020). EEG correlates of physical effort and reward processing during reinforcement learning. *Journal of Neurophysiology, 124*(2), 610–622. https://doi.org/10.1152/jn.00370.2020

Paller, K. A., & Wagner, A. D. (2002). Observing the transformation of experience into memory. *Trends in Cognitive Sciences, 6*(2), 93–102.

Palminteri, S., Justo, D., Jauffret, C., Pavlicek, B., Dauta, A., Delmaire, C., et al. (2012). Critical roles for anterior insula and dorsal striatum in punishment-based avoidance learning. *Neuron, 76*(5), 998–1009. https://doi.org/10.1016/j.neuron.2012.10.017

Parro, C., Dixon, M. L., & Christoff, K. (2018). The neural basis of motivational influences on cognitive control. *Human Brain Mapping, 39*(12), 5097–5111. https://doi.org/10.1002/hbm.24348

Pieters, R. G. M., & Verplanken, B. (1995). Intention behavior consistency – Effects of consideration set size, involvement and need for cognition. *European Journal of Social Psychology, 25*(5), 531–543. https://doi.org/10.1002/ejsp.2420250505

Plassmann, H., O'Doherty, J., & Rangel, A. (2007). Orbitofrontal cortex encodes willingness to pay in everyday economic transactions. *Journal of Neuroscience, 27*(37), 9984–9988. https://doi.org/10.1523/Jneurosci.2131-07.2007

Platt, M. L., & Huettel, S. A. (2008). Risky business: the neuroeconomics of decision making under uncertainty. *Nature Neuroscience, 11*(4), 398–403. https://doi.org/10.1038/nn2062

Pornpattananangkul, N., & Nusslock, R. (2016). Willing to wait: Elevated reward-processing EEG activity associated with a greater preference for larger-but-delayed rewards. *Neuropsychologia, 91*, 141–162. https://doi.org/10.1016/j.neuropsychologia.2016.07.037

Preuschoff, K., Quartz, S. R., & Bossaerts, P. (2008). Human insula activation reflects risk prediction errors as well as risk. *Journal of Neuroscience, 28*(11), 2745–2752. https://doi.org/10.1523/Jneurosci.4286-07.2008

Rae, C. (2023). Why the climate crisis and ecological emergency concern us, and what we can do. *Psychologist, 36*, 22–25.

Rosales, E. G., Baumgartner, T., & Knoch, D. (2022). Interindividual differences in intergenerational sustainable behavior are associated with cortical thickness of the dorsomedial and dorsolateral prefrontal cortex. *NeuroImage, 264*, 119664. https://doi.org/10.1016/j.neuroimage.2022.119664

Samanez-Larkin, G. R., & Knutson, B. (2015). Decision making in the ageing brain: Changes in affective and motivational circuits. *Nature Reviews Neuroscience, 16*(5), 278–289. https://doi.org/10.1038/nrn3917

Sawe, N. (2019). Adapting neuroeconomics for environmental and energy policy. *Behavioural Public Policy, 3*(1), 17–36.

Sawe, N., & Chawla, K. (2021). Environmental neuroeconomics: how neuroscience can inform our understanding of human responses to climate change. *Current Opinion in Behavioral Sciences, 42*, 147–154. https://doi.org/10.1016/j.cobeha.2021.08.002

Sawe, N., & Knutson, B. (2015). Neural valuation of environmental resources. *NeuroImage, 122*, 87–95. https://doi.org/10.1016/j.neuroimage.2015.08.010

Sawe, N., Srirangarajan, T., Sahoo, A., Tang, G. S., & Knutson, B. (2022). Neural responses clarify how ecolabels promote sustainable purchases. *NeuroImage, 263*, 119668. https://doi.org/10.1016/j.neuroimage.2022.119668

Scheres, A., de Water, E., & Mies, G. W. (2013). The neural correlates of temporal reward discounting. *Wiley Interdisciplinary Reviews-Cognitive Science, 4*(5), 523–545. https://doi.org/10.1002/wcs.1246

Schevernels, H., Krebs, R. M., Santens, P., Woldorff, M. G., & Boehler, C. N. (2014). Task preparation processes related to reward prediction precede those related to task-difficulty expectation. *NeuroImage, 84*, 639–647. https://doi.org/10.1016/j.neuroimage.2013.09.039

Shenhav, A., Botvinick, M. M., & Cohen, J. D. (2013). The expected value of control: An integrative theory of anterior cingulate cortex function. *Neuron, 79*(2), 217–240. https://doi.org/10.1016/j.neuron.2013.07.007

Shenhav, A., Fahey, M. P., & Grahek, I. (2021). Decomposing the motivation to exert mental effort. *Current Directions in Psychological Science, 30*(4), 307–314. https://doi.org/10.1177/09637214211009510

Spurlock, C., Belal, S., Fujita, K., & Sawe, N. (2020). *Enabling behavior through personal commitment statements: why do they work?* Lawrence Berkeley National Laboratory.

Treadway, M. T., Buckholtz, J. W., Schwartzman, A. N., Lambert, W. E., & Zald, D. H. (2009). Worth the 'EEfRT'? The effort expenditure for rewards task as an objective measure of motivation and anhedonia. *PLoS One, 4*(8), e6598. https://doi.org/10.1371/journal.pone.0006598

Verguts, T., Vassena, E., & Silvetti, M. (2015). Adaptive effort investment in cognitive and physical tasks: a neurocomputational model. *Frontiers in Behavioral Neuroscience, 9*, 57. https://doi.org/10.3389/fnbeh.2015.0005

Verplanken, B., & Whitmarsh, L. (2021). Habit and climate change. *Current Opinion in Behavioral Sciences, 42*, 42–46. https://doi.org/10.1016/j.cobeha.2021.02.020

Vlasceanu, M., Doell, K. C., et al. (2024). Addressing climate change with behavioral science: A global intervention tournament in 63 Countries. *Science Advances, 10*(6), eadj5778.

Wang, S. S., & van den Berg, B. (2021). Neuroscience and climate change: How brain recordings can help us understand human responses to climate change. *Current Opinion in Psychology, 42*, 126–132. https://doi.org/10.1016/j.copsyc.2021.06.023

Westbrook, A., & Braver, T. S. (2015). Cognitive effort: A neuroeconomic approach. *Cognitive, Affective, & Behavioral Neuroscience, 15*(2), 395–415. https://doi.org/10.3758/s13415-015-0334-y

Westbrook, A., & Braver, T. S. (2016). Dopamine does double duty in motivating cognitive effort. *Neuron, 89*(4), 695–710. https://doi.org/10.1016/j.neuron.2015.12.029

Westbrook, A., Lamichhane, B., & Braver, T. (2019). The subjective value of cognitive effort is encoded by a domain-general valuation network. *Journal of Neuroscience, 39*(20), 3934–3947. https://doi.org/10.1523/Jneurosci.3071-18.2019

Part IV
Neuroscience of the Built Environment/ Architecture

Aesthetics of Architecture: Emerging Insights from the Brain

Oshin Vartanian

Abstract It has long been recognized that design features of built environments can impact how we feel, think, and act. Recently, however, data from brain imaging studies have begun to add to our understanding of the underlying neurological and psychological mechanisms that may underlie the aesthetic preferences that arise in relation to those design features. In this chapter, I will review the brain imaging work that has been conducted in the emerging domain of the cognitive neuroscience of architecture, focusing on the associations between specific design features (e.g., curvature, ceiling height, perceived enclosure) and variations in brain function, as well as aesthetic preferences. The evidence to date suggests that aesthetic evaluations of architecture largely engage a domain-general set of neural systems that underlie aesthetic preferences. The findings have implications for our understanding of the evolutionary basis of aesthetic preferences for built environments.

Keywords Aesthetics · Architecture · Brain · Design · Neuroscience

It has been noted that more than half of the world's population lives in urban areas (Ohly et al., 2016). Furthermore, among those who do, approximately 90% of their time is spent indoors (Klepeis et al., 2001; Ott, 1989). Because we spend such a large proportion of our time within built environments, it is becoming increasingly important to understand the effects that such environments can have on a host of outcome measures relevant to their dwellers. Perhaps not surprisingly, much research to date has focused on the beneficial effects that natural versus human-made environments might have on health and well-being. For example, in accordance with the *biophilia hypothesis*, it has been argued that people have a genetically rooted need to seek natural environments, based on our species' evolution in biological rather than manufactured environments (Wilson, 1984). As such, it stands to reason that spending time in nature would have salubrious effects on humans, by placing them in environments that have a closer resemblance with their evolutionary

---

O. Vartanian (✉)
Department of Psychology, University of Toronto, Toronto, ON, Canada

S. Kühn (ed.), *Environmental Neuroscience*,
https://doi.org/10.1007/978-3-031-64699-7_8

past. Indeed, this is the general underlying premise of Stress Reduction Theory (Ulrich et al., 1991) and Attention Restoration Theory (Kaplan & Kaplan, 1989), according to which natural environments serve to reduce stress and replenish cognitive resources thus leading to improved health and well-being (see Gaekwad et al., 2023; Shuda et al., 2020; Stevenson et al., 2018).

However, aside from their putative effects on health and well-being, built environments can also exert *aesthetic* effects on their dwellers. Aesthetics can be defined as the study of how and why sensory stimuli acquire hedonic value (Skov & Nadal, 2020)—where hedonic value represents the core experience of pleasure and/or displeasure by the organism (Becker et al., 2019). In the context of built environments, this can be understood to mean the ways in which features of architecture and design are associated with states of pleasure and displeasure experienced by their inhabitants (➔ chapters "The Built Environment and the Brain: Review of Emerging Methods to Investigate the Impact of Viewing Architectural Design" by Tawil, and "Applications of Biometrics in Architectural and Environmental Design" by Lavdas and Sussman). Phenomenologically, we are deeply familiar with such effects. It is easy to recall moments when we have experienced satisfaction or dissatisfaction when interacting with the interior or exterior of buildings, following which we ponder features of their design that may have led to our subjective experiences. It is precisely those features that form the focus of the present chapter. Specifically, I will discuss emerging insights from the psychology and neuroscience of architecture that shed light on how features of built environments drive aesthetic evaluations (Coburn et al., 2017).

However, before reviewing this body of work, it is necessary to touch on two ideas that have exerted great influence on research in this area. The first explores the causes and/or bases for our preferences for specific environments. The touchstone theory in this regard is geologist Appleton's (1975/1996) "habitat theory", according to which the aesthetic satisfaction one derives from contemplating a natural scene is proportional to the extent to which its physical features signal environmental conditions favourable or unfavourable to survival and/or reproduction (see also Kellert & Wilson, 1993; Nasar, 1988; Sagan & Druyan, 1992). Indeed, he noted that "Habitat theory postulates that aesthetic pleasure in landscape derives from the observer experiencing an environment favorable to the satisfaction of his biological needs", and that "The point at which we always seem to run against a brick wall is in understanding more precisely how the actual ingredients of landscape operate on the aesthetic sense" (Appleton, 1975/1996). The idea here is that from an evolutionary perspective, one can find the roots of our environmental preferences in impulses that regulate biologically more fundamental behaviours such as survival and reproduction. Furthermore, within this context, one can think of the "actual ingredients" as those design features within the environments that represent a biologically relevant signal to the brain as the basis for computing aesthetic preferences.

Although habitat theory was originally proposed to account for our preferences for natural scenes, Hildebrand (1999) subsequently extended it to the built environment. He argued that Appleton's principle can be applied to understanding the basis of our preferences for interiors, such that we prefer built spaces that facilitate survival and/or reproduction (see also Kaplan, 1987). Indeed, recent psychometric work has demonstrated that there is great overlap between factors that underlie our preferences for natural and built environments (Coburn et al., 2020; Vartanian et al., 2021), suggesting that our preferences could be informed by some of the same "actual ingredients" in both contexts.

The second idea influencing research in this area involves our larger theoretical understanding of how aesthetic experiences emerge, in the context of which aesthetic experiences in relation to architecture represent one example. Chatterjee and Vartanian (2014, 2016) conducted a critical review of the literature on the psychology and neuroscience of aesthetics, which led to their proposal for the *aesthetic triad*. According to this model, aesthetic experiences emerge from the interaction among three large-scale neural systems in the brain: emotion-reward, sensory-motor, and knowledge-meaning (see also Kirsch et al., 2016; Li & Zhang, 2020). Regions within the emotion-reward system are involved in the computation of reward and our affective responses in relation to various stimuli. One of the major findings emerging from this literature has come to be known as the "common currency" hypothesis, according to which the orbitofrontal and ventral striatal neuronal activity underlies the valuation of rewarding stimuli irrespective of the modality giving rise to the rewarding experience (Levy & Glimcher, 2012). The common currency hypothesis is based on the remarkable observation that visual (Kawabata & Zeki, 2004), musical (Blood et al., 1999), moral (Tsukiura & Cabeza, 2011) and even mathematical (Zeki et al., 2014) beauty judgements all correlate with activity in the medial orbitofrontal cortex—a structure associated strongly with reward. This is not to say that modality-specific neural systems do not exist, but rather that a core set of reward regions appears to be attuned to how rewarding objects are across modalities. In turn, regions within the sensory-motor system are sensitive to the sensorial and perceptual qualities of objects. Indeed, all standard models of aesthetic experience acknowledge that the processing of perceptual features of objects occurs early (and sometimes automatically) in the information-processing stream and represents a core aspect of aesthetic processing (Chatterjee, 2003; Leder et al., 2004). In addition, there is growing appreciation that the motor neuron system also plays an important role in aesthetic processing, especially by supporting embodied mechanisms that underlie action simulations, emotions and body sensations (Freedberg & Gallese, 2007). Finally, the knowledge-meaning system reflects the influence of top-down contextual, societal, cultural and experiential factors that are known to influence aesthetic experiences, including formal training, and physical setting and attitudes, among others. Insofar as architecture is concerned, the overarching assumption is that aesthetic experiences in this domain will emerge from the interaction of the same neural systems that support aesthetic experiences in other domains (Coburn et al., 2017). Motivated by these two overarching considerations, I will now review research on the effects of various architectural design features on aesthetic judgements of built environment. I will close by highlighting some outstanding questions and challenges that will likely shape the next phase of research on this topic.

Curvature

Historically, there has been long-standing interest on the effect of curvature on attitudes and preferences (see Hogarth, 1753). Within experimental psychology, there is research dating back almost a century ago that focused on the "affective value" of lines and line drawings and found that people tend to find curved lines serene and graceful, whereas angular lines were perceived as robust and vigorous (Hevner, 1936; Lundholm, 1921; Poffenberger & Barrows, 1924). An important feature of this early research was recognition that this preference could be rooted in our affective responses to curvature. For example, Gordon (1909) stated that "curves are in general *felt* to be more beautiful than straight lines. They are more graceful and pliable, and avoid the harshness of some straight lines" [italics added]. It has since been shown that preference for curvature is a reliable effect across a large set of stimulus categories including everyday objects, abstract shapes, paintings and faces, among others (for review see Gómez-Puerto et al., 2016). The generality of this effect has been reinforced by its observation in infants (Faantz & Miranda, 1975), great apes (Munar et al., 2015) and cultures (Gómez-Puerto et al., 2018). Indeed, a recent comprehensive meta-analysis of this literature reported a medium effect size (Hedges' *g* of 0.39) associated with the effect, suggesting that it is a relatively robust phenomenon (Chuquichambi et al., 2022). Evidence in support of the *causal* impact of curvature on preference was provided recently when Farzanfar et al. (2023) demonstrated that preference for scenes can be increased or decreased experimentally by manipulating their contour properties.

Of particular relevance for the present purposes, it is important to ask whether a similar preference has been found in relation to the perception of built spaces more specifically. In that regard, a number of studies have accumulated to show that preference for curvature is likely also present for interior design and architecture. For example, Dazkir and Read (2012) presented participants with images of interior spaces that were either curved or angular in design and found that the former elicited higher levels of pleasure and low-intensity pleasant emotions such as feelings of relaxation and calm. Participants were also more likely to opt to approach the curved rather than the angular interior spaces. Consistent with Gordon's (1909) hunch, these results were in line with the idea that preference for curvature might be mediated by our affective responses. Next, van Oel and van den Berkhof (2013) presented participants with images of airport passenger areas that varied in many dimensions, including curvature, and found that curvature was the strongest predictor of preference, such that airport passenger areas with curved roofs and layouts were preferred over their angular counterparts.

Building on this literature, Vartanian et al. (2013) conducted a functional magnetic resonance imaging (fMRI) study to examine not only preference for curvature in the context of architectural interiors but also its neural correlates. Examining the neural correlates for aesthetic phenomena is useful to the extent that it can shed light on the psychological mechanisms that underlie the behavioural effects (see Bar & Neta, 2007). Inside the fMRI scanner, participants were presented with two-dimensional (2D) images that varied in curvature (curved vs. angular) under two different task conditions: In the "beauty judgement" condition they were instructed to make choices between binary "beautiful" versus "not beautiful" options. In turn, in the "approach-avoidance" condition, they were instructed to make choices between binary "enter" versus "exit" options. The authors tested the hypotheses that participants would be more likely to find the curved interiors beautiful in the beauty judgement condition, and more likely to opt to enter them in the approach-avoidance condition. Their first prediction was confirmed, such that participants were more likely to find the curved interiors beautiful in the beauty judgement condition. Furthermore, viewing curved compared to angular interiors in the beauty judgement condition was associated with greater activation in the anterior cingulate cortex (ACC). ACC is a core region within the brain's emotion network (Barrett et al., 2007; Barrett & Wager, 2006), and its activation in relation to curved interiors is consistent with the idea that our preference in relation to this design feature is likely driven in part by affective processes. This inference was reinforced further when, upon leaving the fMRI scanner, all participants were instructed to view the images again and to rate each image on beauty and pleasantness. Pleasantness represents the valence dimension of the affect circumplex, the other dimension being intensity. The results demonstrated that approximately 60% of the variance in beauty ratings was explained by pleasantness ratings, offering additional support for the idea that evaluations of beauty are in part based on affective processes. In contrast, the results did not support the authors' second prediction because participants were no more likely to opt to enter curved interiors compared to angular interiors in the approach-avoidance condition. However, the fMRI data demonstrated that there was greater activation in the visual cortex when participants were exposed to curved compared to angular interiors in the approach-avoidance condition, suggesting that the brain is sensitive to variation in this design feature in that context.

An important limitation of Vartanian et al. (2013) was that all the images of architectural spaces were presented in a 2D format, which in turn limited the ecological validity of the study. Furthermore, replication of neuroscientific studies is necessary to establish the reliability of reported findings. In an important electrophysiological study on this topic geared towards understanding the impact of naturalistic three-dimensional (3D) architectural forms on preference and brain activity, Banaei et al. (2017) equipped their participants with mobile electroencephalogram (EEG) (➔ chapter "EEG for the Study of Environmental Neuroscience" by Grassini) units as they walked naturally through different interior forms in virtual reality (VR) (➔ chapter "Extended Reality in Environmental Neuroscience Research" by Li et al.). Behaviourally, their results revealed a preference for spaces with curved geometry. Furthermore, sensitivity to curved design was localized in the ACC—the same structure found to be sensitive to curvature in the beauty-judgement condition in Vartanian et al. (2013). The fact that these two studies using different study designs, stimuli, and neuroimaging technologies nevertheless converged on the ACC suggests that this region is likely sensitive to the effect of curvature when the task involves aesthetic judgement in the domain of architecture. Additional research has served to reinforce the idea that people prefer curved design features in architecture. For example, Shemesh et al. (2017) who also used a VR paradigm were able to demonstrate that participants have a tendency to prefer curved spaces. In turn, focusing on architectural façades rather than interior spaces, Ruta et al. (2019) demonstrated that curved rather than angular façades were the most preferred, and that this held true for most participants.

This pattern of findings might give the impression that preference for curvature in the domain of architecture is a universal phenomenon. However, as Chuquichambi et al. (2022) have shown in their meta-analysis on preference for curvature in the visual domain more broadly, this effect is moderated by several factors such as stimulus type, expertise, task and affordances. For example, Vartanian et al. (2019) presented a subset of the stimuli from Vartanian et al. (2013) to a group of working architects and designers as well as a control group with no formal training in architecture and design recruited from a university, under the same two task conditions (i.e. beauty judgements and approach-avoidance decisions), but this time outside of the MRI scanner. At the outset, it was not clear whether the experts in architecture and design would exhibit a stronger or weaker preference for curvature compared to the control group. On the one hand, earlier data had shown that people who score higher on openness to experience exhibit a stronger preference for curved over angular abstract shapes (Cotter et al., 2017). To the extent that openness to experience denotes greater creative and aesthetic sensitivity, one would expect working architects and designers to exhibit a similar pattern. However, it is also true that experts in the visual arts have been shown to be affected less by low-level visual features and more by compositional and historical features of artworks (e.g., Parsons, 1987), meaning that perhaps they would be affected less by variation in curvature. The results of the study showed that when the task involved beauty judegments, then the experts in architecture and design were more likely than naïve controls to find the curved spaces beautiful. In contrast, when the task involved approach-avoidance decisions, then experts in architecture and design were less likely than naïve controls to opt to enter the spaces. These results demonstrated that the effect of expertise as measured by formal training interacts with task demands to influence choices. Also focusing on individual differences, Palumbo et al. (2020) administered a slightly modified version of the two conditions in Vartanian et al. (2013) using a subset of the same stimuli to neurologically healthy controls, individuals with autism spectrum disorder and quasi-experts in the form of university students enrolled in design courses. Because the curvature effect has typically been explained in terms of affective and perceptual processes (for review see Gómez-Puerto et al., 2016), and it is well established that individuals with autism spectrum disorder exhibit altered patterns of affective and perceptual processing, then the expectation was that preference for curvature would not emerge in this population. Indeed, the results demonstrated that preference for curved interiors was reduced in individuals with autism spectrum disorder. Surprisingly, it was also found that quasi-experts in fact preferred angular over curved interiors, which could be attributed to the specifics of their formal training in design at the university. Taken together, these studies suggest that preference for curvature in the domain of architecture can vary greatly based on task conditions and individual differences.

As noted earlier, an important limitation to consider when assessing the validity of the studies to date is that the majority were conducted in relation to viewing 2D images, which does not mimic the way in which humans interact with built spaces in everyday life. At least in part to address this shortcoming, Tawil et al. (2021) conducted a study in which participants viewed four well-matched indoor living rooms under a free-exploration photorealistic VR paradigm. Of the 33 outcome measures explored, none exhibited a preference for curved spaces, with two exhibiting a preference for angular spaces. One possible explanation for why preference for curvature is reduced when using architectural stimuli that are photorealistic was provided by a re-analysis of Chuquichambi et al.'s (2022) meta-analytic data by Djebarra and Kalantari (2023). They excluded from their analysis studies that had used photographs or 2D images, focusing instead only on studies that had presented participants with VR or real objects. The idea behind this approach was to focus on studies that offered greater levels of *affordances*—understood as possibilities for interaction (Gibson, 1986). Their results demonstrated that when this condition was met, then there was a negative relationship between curvature preference and an object's affordances. Referencing research on embodiment, they argued that when a person has the opportunity to consider a greater repertoire of possible actions, as would be the case in more ecologically valid and realistic settings, factors other than the geometric features of objects (e.g. curvature) will form the basis for their choices. In the domain of architecture, this would mean that organismic needs likely play a large role in our choices as we consider (i.e. mentally simulate) context-specific behaviours. I will return to this point later in this chapter when considering the relative weighting of evolutionary versus individual-differences factors in determining aesthetic preferences for built environment.

Ceiling Height

Perhaps the strongest evidence in support of people's preference for a geometric feature is to examine market data. For example, according to the U.S. National Building Code, the standard ceiling height is eight feet or 2.44 m (Rybczynski, 2009), although people tend to prefer ceilings that are about two feet (0.61 m) higher than this standard. The cost of adding a single foot to the height of a ceiling is not trivial, as it is estimated to be about $20,000 for a 4000-square-foot house in the United States. Nevertheless, this adjustment is frequently made because it is well known that some people are willing to pay more for living spaces with taller than standard ceilings. Why is that? There could be a number of reasons for this. From a purely aesthetic perspective, there are data to show that people exhibit a preference for rooms with higher ceiling. Baird et al. (1978) conducted an early study on this topic by demonstrating a single-peak function that related ceiling height to preference, such that preference increased monotonically from 6 feet (1.83 m) to a peak at 10 feet (3.04 m), decreasing thereafter. One of the many strengths of this study was that data were obtained both when participants examined model rooms, as well as when they stood under adjustable ceilings for a more realistic experience of ceiling height. The results remained the same. However, ceiling height also been shown to have an impact on how we think, which might in turn impact how we feel about spaces with higher ceilings. For example, Meyers-Levy and Zhu (2007) placed participants in rooms with ceilings that could be adjusted between 8 and 10 feet. In Experiment 1, participants were instructed to rate the degree to which six different items reflected their current body state, ranging from items reflecting freedom-related body states (e.g. a sense of being free, unrestricted, and open) to those that reflected confinement-related body states (e.g. a sense of being encumbered, inhibited, and confined). Then in a second task they were presented with anagrams that were related semantically to the concept of freedom (e.g. liberated, unlimited, and emancipated) or to the concept of confinement (e.g., bound, restrained, and restricted). Compared to participants in rooms with 8-foot ceilings, participants in rooms with 10-foot ceilings reported a greater level of freedom-related states than confinement-related states and exhibited faster reaction times to freedom-related anagrams and slower reaction times to confinement-related anagrams. These results are consistent with the idea that ceiling height can prime (i.e. make salient) the concept of freedom over the concept of constraint. To the extent that thinking freely is associated with positive affect, there could also be downstream effects on our emotions. In Experiment 2, the researchers instructed their participants to complete a categorization task in which they identified as many dimensions as they could from a list of disparate items within a broad category (i.e. different sports). They were also asked to categorize the items further into subgroups and to provide descriptive labels for those subgroups. Their results demonstrated that compared to participants in rooms with 8-foot ceilings, participants in rooms with 10-foot ceilings identified more dimensions shared by a number of rather dissimilar items, exhibited a greater degree of abstraction in the dimensions they identified and sorted these items into more inclusive sub-groups per dimension. Consistent with the results of Experiment 1, these results demonstrated that people's thinking is less constrained in rooms with higher ceilings, as they are more likely to engage in overinclusive thinking in rooms with 10-foot than 8-foot ceilings.

To examine whether ceiling height might influence brain function, Vartanian et al. (2015) revisited data from Vartanian et al. (2013). Specifically, in that study, aside from categorizing each stimulus on curvature, the stimuli had also been categorized in terms of ceiling height and perceived enclosure. Focusing on behavioural data, the results demonstrated that compared to interiors with low ceilings, participants were more likely to judge interiors with high ceilings as beautiful but no more likely to opt to enter them. In turn, the fMRI data demonstrated that in the beauty-judgement condition, viewing interiors with higher ceilings was correlated with greater activation in the precuneus and middle frontal gyrus—two structures involved in visuospatial exploration and attention in the dorsal stream (a pathway dedicated to the spatial localization of objects in space). This suggests that our preference for rooms with higher ceilings might be based in part on their ability to promote visual exploration of spaces.

Perceived Enclosure

There has been longstanding interest in the impact that the ability to see has on our aesthetic preferences for scenes and landscapes. For example, drawing on evolutionary principles, Appleton (1975/1996) argued that people tend to prefer open compared to enclosed spaces because such environments facilitate scanning and seeing of objects in the visual scene, a feature that has likely been favourable to survival and reproduction (e.g. seeing predators and prey). More recently, Stamps (2005) has argued that the degree of movement through a space is more accurately described as permeability, which in turn has visual and locomotive variants. Within this framework, perceived enclosure can be defined as the perceived (i.e. visual) degree of movement through space (see Stamps, 2005; Stamps & Krishnan, 2004).

From a biological perspective, there is growing evidence to suggest that perceived enclosure has effects on both the autonomic and central nervous systems. With respect to the former, Fich et al. (2014) conducted an experiment in which they administered a virtual version of the Trier Social Stress Test to participants while also measuring saliva cortisol and heart rate variability. This test is a stress induction paradigm in the course of which the participants perform a series of stressful tasks (e.g. giving a presentation) in front of a committee. The experiment had two conditions: The open condition was characterized by large openings that offered a virtual potential for escape, whereas the enclosed condition did not offer this potential for escape. They found that compared to participants in the open condition, participants in the enclosed condition responded with more pronounced cortisol reactivity to stress induction and continued to show higher levels throughout recovery. This demonstrates that perceived enclosure has an impact on the body's stress response. These results converge very nicely with those of Vartanian et al.'s (2015) fMRI study, who found that when making beauty judgements, participants were more likely to find a space beautiful in the open than enclosed condition, and that when making approach-avoidance decisions, they were more likely to opt to exit the space in the enclosed than open condition. In turn, the fMRI data demonstrated that in the beauty-judgement condition viewing open interiors was correlated with greater activation in left middle temporal gyrus and right superior temporal gyrus—two regions involved in perceived visual motion. In turn, in the approach-avoidance condition viewing enclosed interiors revealed greater activation in a large cluster spanning both hemispheres in the anterior midcingulate cortex region of the cingulate gyrus—a region within the cingulate gyrus with direct projections from the amygdala. This suggests that a reduction in perceived visual and locomotive permeability characteristic of enclosed spaces might elicit an emotional reaction that accompanies exit decisions.

Dimensions of Preference for Interior Spaces

Thus far, the focus on this chapter has been on the relationship between various geometric features and aesthetic preference. However, another way to approach this same problem space is to ask what the psychological dimensions are that account for preference for interior spaces. Coburn et al. (2020) tackled that problem in the form of three experiments. In Experiment 1, they instructed their participants to rate different subsets of images from Vartanian et al. (2013) on 16 aesthetic scales (e.g. beauty, complexity, interest, etc.). Then, using Psychometric Network Analysis (PNA) and Principal Components Analysis (PCA) they identified three components that explained 90% of the variance in ratings: Coherence (i.e. ease with which one organizes and comprehends a scene), Fascination (i.e. a scene's informational richness and generated interest) and Hominess (i.e. extent to which a scene reflects a personal space). Importantly, whereas Coherence and Fascination had been shown to be dimensions that predict preference for landscapes (Kaplan, 1987; Kaplan & Kaplan, 1989), Hominess emerged as a new dimension related to architectural interiors. The 3-dimensional PCA results obtained in Experiment 1 were shown to be robust in Experiment 2 when they were replicated using a factor analysis in an independent sample. Next, in Experiment 3, the authors reanalysed fMRI data from Vartanian et al. (2013) in relation to Coherence, Fascination and Hominess ratings associated with each image. Parametric analyses demonstrated that Fascination covaried with neural activity in the right lingual gyrus across both beauty judgements and approach-avoidance decisions, whereas Coherence covaried with neural activity in the left inferior occipital gyrus only when participants judged beauty, and Hominess covaried with neural activity in the left cuneus only when they made approach-avoidance decisions. Furthermore, these neural activations did not covary in relation to global image properties including self-similarity and complexity scores. In a follow-up study, Vartanian et al. (2021) tested the hypothesis that the contributions of Coherence, Fascination and Hominess to preference might vary based on individual differences. Specifically, they reanalysed data from Palumbo et al. (2020) in which participants with autism spectrum disorder, neurotypical controls and design students had rated images of interiors on liking and approach-avoidance decisions. For design students, only Coherence drove choices, whereas in participants with autism spectrum disorder and neurotypical controls Hominess and Fascination also contributed, respectively. The authors reasoned that for design students Coherence is paramount because it references the structural organization of spaces and is informed by formal training. In turn, for persons with autism spectrum disorder familiarity, physical proximity and difficulty in mental simulation are relevant, which explains the increased weighting of Hominess in shaping preferences. Finally, in neurotypical controls, interest in visual exploration can explain Fascination's role in shaping preferences. Since then, this three-factor structure has also been shown to explain variations in preference due to expertise (Weinberger et al., 2021). Put together, the results from this series of studies suggest that this 3D model can account for individual differences in preference for interior architecture across populations and that it can form the basis as a model for the psychology and neuroscience of spaces (Chatterjee et al., 2021). As such, it can address some of the theoretical shortcomings that have been highlighted in previous research on the neurophysiological bases of emotion in relation to architectural design (Bower et al., 2019).

Causes and Models

In science, what we ultimately seek are explanations for phenomena of interest. With respect to the psychology and neuroscience of architecture, and taking a leaf out of Appleton's (1975/1996) work with respect to landscapes, I have argued that the proximal causes for our choices can be understood in terms of distal, evolutionary causes (see Hildebrand, 1999). Specifically, in the context of "habitat theory", Appleton argued that we tend to prefer landscapes that promoted survival and reproduction among our evolutionary ancestors. This viewpoint suggests that our brains have evolved to represent choices that are evolutionarily informed. The evidence appears to suggest that the same principle applies to indoor environments. In other words, there is no reason to suggest that there is a discontinuity between factors that determine preferences for landscapes versus those that determine preferences for indoor environments. Instead, it appears that factors that determine preferences for indoor environments (i.e. Coherence, Fascination *and* Hominess) simply add to rather than replace factors that determine preferences for landscapes (i.e. Coherence and Fascination) (Coburn et al., 2020). This does not mean that our choices and preferences are invariant to contextual variation (see Djebarra & Kalantari, 2023). Rather, it is likely that factors that have influenced our environmental choices from an evolutionary perspective continue to exert an influence on our preferences today.

Furthermore, our aesthetic responses in relation to features of built environments can be understood within the context of the aesthetic triad (Chatterjee & Vartanian, 2014, 2016), which is meant to provide a domain-general account of the emergence of aesthetic experiences. Indeed, the effects of curvature, ceiling height and perceived enclosure on aesthetic preferences can be understood by making reference to emotion-reward, sensory-motor and knowledge-meaning systems. This is not to say that gaps in knowledge do not exist. For example, although we have begun to isolate specific regions in the brain that are sensitive to variation in specific geometric features such as curvature (Yue et al., 2020), we still do not have a complete neurobiological account of how such perceptual information is combined with reward information to compute preferences that are experienced consciously (Skov, 2019). Similarly, explaining individual differences remains a challenge. For example, it has been shown that people agree less on what is aesthetically pleasing when judging human-made artifacts (artwork, architecture) compared to natural objects (faces, scenes) (Vessel et al., 2018). There are recent data to suggest that both genetic and environmental variations underlie the observed individual differences in aesthetic preference for visual stimuli (Bignardi et al., 2024), but additional work is needed to understand their developmental emergence across the lifespan. Finally, despite the focus of this specific chapter, it is important to remember that our interactions with architecture include but are not limited to aesthetic experiences (Djebbara et al., 2022), a complete explanation of which requires theoretical integration across focus areas. Despite these challenges, there is reason to be optimistic about the prospect of arriving at a better understanding of the neurobiology of aesthetic preferences for architecture and design in the future.

References

Appleton, J. (1975/1996). *The experience of landscape*. John Wiley and Sons.

Baird, J. C., Cassidy, B., & Kurr, J. (1978). Room preference as a function of architectural features and user activities. *Journal of Applied Psychology, 63*, 719–727.

Banaei, M., Hatami, J., Yazdanfar, A., & Gramann, K. (2017). Walking through architectural spaces: The impact of interior forms on human brain dynamics. *Frontiers in Human Neuroscience, 11*, Article 477. https://doi.org/10.3389/fnhum.2017.00477

Bar, M., & Neta, M. (2007). Visual elements of subjective preference modulate amygdala activation. *Neuropsychologia, 45*, 2191–2200. https://doi.org/10.1016/j.neuropsychologia.2007.03.008

Barrett, L. F., & Wager, T. (2006). The structure of emotion: Evidence from the neuroimaging of emotion. *Current Directions in Psychological Science, 15*, 79–85.

Barrett, L. F., Mesquita, B., Ochsner, K. N., & Gross, J. J. (2007). The experience of emotion. *Annual Review of Psychology, 58*, 373–403.

Becker, S., Bräscher, A.-K., Bannister, S., Bensafi, M., Calma-Birling, D., Chan, R. C. K., Eerola, T., Ellingsen, D.-M., Ferdenzi, C., Hanson, J. L., Joffily, M., Lidhar, N. K., Lowe, L. J., Martin, L. J., Musser, E. D., Noll-Hussong, M., Olino, T. M., Lobo, R. P., & Wang, Y. (2019). The role of hedonics in the human Affectome. *Neuroscience and Biobehavioral Reviews, 102*, 221–241. https://doi.org/10.1016/j.neubiorev.2019.05.003

Bignardi, G., Smit, D. J. A., Vessel, E. A., Trupp, M. D., Ticini, L. F., Fisher, S. E., & Polderman, T. J. C. (2024). Genetic effects on variability in visual aesthetic evaluations are partially shared across visual domains. *Communications Biology, 7*, 55. https://doi.org/10.1038/s42003-023-05710-4

Blood, A. J., Zatorre, R. J., Bermudez, P., & Evans, A. C. (1999). Emotional responses to pleasant and unpleasant music correlate with activity in paralimbic brain regions. *Nature Neuroscience, 2*, 382387. https://doi.org/10.1038/7299

Bower, I., Tucker, R., & Enticott, P. G. (2019). Impact of built environment design on emotion measured via neurophysiological correlates and subjective indicators: A systematic review. *Journal of Environmental Psychology, 66*, 101344. https://doi.org/10.1016/j.jenvp.2019.101344

Chatterjee, A. (2003). Prospects for a cognitive neuroscience of visual aesthetics. *Bulletin of Psychology and the Arts, 4*, 55–60.

Chatterjee, A., & Vartanian, O. (2014). Neuroaesthetics. *Trends in Cognitive Sciences, 18*, 370–375.

Chatterjee, A., & Vartanian, O. (2016). Neuroscience of aesthetics. *Annals of the New York Academy of Sciences, 1369*, 172–194. https://doi.org/10.1111/nyas.13035

Chatterjee, A., Coburn, A., & Weinberger, A. (2021). The neuroaesthetics of architectural spaces. *Cognitive Processing, 22*(Suppl 1), 115–120. https://doi.org/10.1007/s10339-021-01043-4

Chuquichambi, E. G., Vartanian, O., Skov, M., Corradi, G. B., Nadal, M., Silvia, P. J., & Munar, E. (2022). How universal is preference for visual curvature? A systematic review and meta-analysis. *Annals of the New York Academy of Sciences, 1518*, 151–165. https://doi.org/10.1111/nyas.14919

Coburn, A., Vartanian, O., & Chatterjee, A. (2017). Buildings, beauty, and the brain: A neuroscience of architectural experience. *Journal of Cognitive Neuroscience, 29*, 1521–1531. https://doi.org/10.1162/jocn_a_01146

Coburn, A., Vartanian, O., Kenett, Y. N., Nadal, M., Hartung, F., Hayn-Leichsenring, G., Navarrete, G., González-Mora, J. L., & Chatterjee, A. (2020). Psychological and neural responses to architectural interiors. *Cortex, 126*, 217–241. https://doi.org/10.1016/j.cortex.2020.01.009

Cotter, K. N., Silvia, P. J., Bertamini, M., Palumbo, L., & Vartanian, O. (2017). Curve appeal: Exploring individual differences in preference for curved versus angular objects. *Perception*. Advance online publication. https://doi.org/10.1177/2041669517693023

Dazkir, S. S., & Read, M. A. (2012). Furniture forms and their influence on our emotional responses toward interior environments. *Environment and Behavior, 44*, 722–734. https://doi.org/10.1177/0013916511402063

Djebbara, Z., & Kalantari, S. (2023). Affordances and curvature preference: The case of real objects and spaces. *Annals of the New York Academy of Sciences, 1527*, 14–19. https://doi.org/10.1111/nyas.15038

Djebbara, Z., Jensen, O. B., Parada, F. J., & Gramann, K. (2022). Neuroscience and architecture: Modulating behavior through sensorimotor responses to the built environment. *Neuroscience and Biobehavioral Reviews, 138*, 104715. https://doi.org/10.1016/j.neubiorev.2022.104715

Fantz, R. L., & Miranda, S. B. (1975). Newborn infant attention to form of contour. *Child Development*, 224–228.

Farzanfar, D., Walther, B., & D. (2023). Changing what you like: Modifying contour properties shifts aesthetic valuations of scenes. *Psychological Science, 34*, 1101–1120. https://doi.org/10.1177/09567976231190546

Fich, L. B., Jönsson, P., Kirkegaard, P. H., Wallergård, M., Garde, A. H., & Hansen, Å. (2014). Can architectural design alter the physiological reaction to psychosocial stress? A virtual TSST experiment. *Physiology & Behavior, 135*, 91–97. https://doi.org/10.1016/j.physbeh.2014.05.034

Freedberg, D., & Gallese, V. (2007). Motion, emotion and empathy in esthetic experience. *Trends in Cognitive Science, 11*, 197–203.

Gaekwad, J. S., Sal Moslehian, A., & Roös, P. B. (2023). A meta-analysis of physiological stress responses to natural environments: Biophilia and Stress Recovery Theory perspectives. *Journal of Environmental Psychology, 90*, 1–17. https://doi.org/10.1016/j.jenvp.2023.102085

Gibson, J. (1986). *The ecological approach to visual perception*. Psychology Press.

Gómez-Puerto, G., Munar, E., & Nadal, M. (2016). Preference for curva- ture: A historical and conceptual framework. *Frontiers in Human Neuroscience, 9*, Article 712. https://doi.org/10.3389/fnhum.2015.00712

Gómez-Puerto, G., Rosselló, J., Corradi, G., Acedo-Carmona, C., Munar, E., & Nadal, M. (2018). Preference for curved contours across cultures. *Psychology of Aesthetics, Creativity, and the Arts, 12*, 432–439.

Gordon, K. (1909). *Esthetics*. Henry Holt. https://doi.org/10.1037/10824-000

Hevner, K. (1936). Experimental studies of the elements of expression in music. *The American Journal of Psychology, 48*, 246–268.

Hildebrand, G. (1999). *Origins of architectural pleasure*. University of California Press.

Hogarth, W. (1753). *The analysis of beauty*. John Reeves.

Kaplan, S. (1987). Aesthetics, affect, and cognition. *Environment and Behavior, 19*, 3–32.

Kaplan, R., & Kaplan, S. (1989). *The experience of nature: A psychological perspective*. Cambridge University Press.

Kawabata, H., & Zeki, S. (2004). Neural correlates of beauty. *Journal of Neurophysiology, 91*, 1699–1705. https://doi.org/10.1152/jn.00696.2003

Kellert, S. R., & Wilson, E. O. (1993). *The biophilia hypothesis*. Island Press.

Kirsch, L. P., Urgesi, C., & Cross, E. S. (2016). Shaping and reshaping the aesthetic brain: Emerging perspectives on the neurobiology of embodied aesthetics. *Neuroscience and Biobehavioral Reviews, 62*, 56–68.

Klepeis, N. E., Nelson, W. C., Ott, W. R., Robinson, J. P., Tsang, A. M., Switzer, P., Behar, J. V., Hern, S. C., & Engelmann, W. H. (2001). The National Human Activity Pattern Survey (NHAPS): A resource for assessing exposure to environmental pollutants. *Journal of Exposure Analysis and Environmental Epidemiology, 11*, 231–252.

Leder, H., Belke, B., Oeberst, A., & Augustin, D. (2004). A model of aesthetic appreciation and aesthetic judgments. *British Journal of Psychology, 95*, 489–508.

Levy, D. J., & Glimcher, P. W. (2012). The root of all value: a neural common currency for choice. *Current Opinion in Neurobiology, 22*, 1027–1038.

Li, R., & Zhang, J. (2020). Review of computational neuroaesthetics: Bridging the gap between neuroaesthetics and computer science. *Brain Informatics, 7*, 16. https://doi.org/10.1186/s40708-020-00118-w

Lundholm, H. (1921). The affective tone of lines: Experimental researches. *Psychological Review, 28*, 43.

Meyers-Levy, J., & Zhu, R. (2007). The influence of ceiling height: The effect of priming on the type of processing that people use. *Journal of Consumer Research, 34*, 174–186. https://doi.org/10.1086/519146

Munar, E., Gómez-Puerto, G., Call, J., & Nadal, M. (2015). Common visual preference for curved contours in humans and great apes. *PLoS One, 10*, e0141106. https://doi.org/10.1371/journal.pone.0141106

Nasar, J. L. (1988). *Environmental aesthetics: Theory, research, and applications*. Cambridge University Press.

Ohly, H., White, M. P., Wheeler, B. W., Bethel, A., Ukoumunne, O. C., Nikolaou, V., & Garside, R. (2016). Attention Restoration Theory: A systematic review of the attention restoration potential of exposure to natural environments. *Journal of Toxicology and Environmental Health, Part B, 19*, 305–343.

Ott, W. R. (1989). Human activity patterns: A review of the literature for estimating time spent indoors, outdoors, and in transit. In *Proceedings of the Research Planning Conference on Human Activity Patterns*, EPA National Exposure Research Laboratory, EPA/600/4-89/004: Las Vegas.

Palumbo, L., Rampone, G., Bertamini, M., Sinico, M., Clarke, E., & Vartanian, O. (2020). Visual preference for abstract curvature and for interior spaces: Beyond undergraduate student samples. *Psychology of Aesthetics, Creativity, and the Arts*. https://doi.org/10.1037/aca0000359

Parsons, M. J. (1987). *How we understand art: A cognitive developmental account of aesthetic experience*. Cambridge University Press.

Poffenberger, A. T., & Barrows, B. (1924). The feeling value of lines. *Journal of Applied Psychology, 8*, 187.

Ruta, N., Mastandrea, S., Penacchio, O., Lamaddalena, S., & Bove, G. (2019). A comparison between preference judgments of curvature and sharpness in architectural façades. *Architectural Science Review, 62*, 171–181. https://doi.org/10.1080/00038628.2018.1558393

Rybczynski, W. (2009, May 20). *How high: Are high ceilings a sign of wretched architectural excess or just good taste*?. In Slate. Retrieved from http://www.slate.com/articles/arts/architecture/2009/05/how_high.html

Sagan, C., & Druyan, A. (1992). *Shadows of forgotten ancestors*. Ballantine.

Shemesh, A., Talmon, R., Karp, O., Amir, I., Bar, M., & Grobman, Y. J. (2017). Affective response to architecture–investigating human reaction to spaces with different geometry. *Architectural Science Review, 60*(2), 116–125.

Shuda, Q., Bougoulias, M. E., & Kass, R. (2020). Effect of nature exposure on perceived and physiologic stress: A systematic review. *Complementary Therapies in Medicine, 53*, 102514. https://doi.org/10.1016/j.ctim.2020.102514

Skov, M. (2019). Aesthetic appreciation: The view from neuroimaging. *Empirical Studies of the Arts, 37*, 220–248. https://doi.org/10.1177/0276237419839257

Skov, M., & Nadal, M. (2020). A farewell to art: Aesthetics as a topic in psychology and neuroscience. *Perspectives on Psychological Science, 15*, 630–642. https://doi.org/10.1177/1745691619897963

Stamps, A. E. (2005). Visual permeability, locomotive permeability, safety, and enclosure. *Environment and Behavior, 37*, 587–619.

Stamps, A. E., & Krishnan, V. V. (2004). Perceived enclosure of space, angle above observer, and distance to boundary. *Perceptual and Motor Skills, 99*, 1187–1192.

Stevenson, M. P., Schilhab, T., & Bentsen, P. (2018). Attention restoration theory II: A systematic review to clarify attention processes affected by exposure to natural environments. *Journal of Toxicology and Environmental Health, Part B, 21*, 227–268. https://doi.org/10.1080/10937404.2018.1505571

Tawil, N., Sztuka, I. M., Pohlmann, K., Sudimac, S., & Kühn, S. (2021). The living space: Psychological well-being and mental health in response to interiors presented in virtual reality. *International Journal of Environmental Research and Public Health, 18*, 12510.

Tsukiura, T., & Cabeza, R. (2011). Shared brain activity for aesthetic and moral judgments: Implications for the Beauty-is-Good stereotype. *Social Cognitive and Affective Neuroscience, 6*, 138–148. https://doi.org/10.1093/scan/nsq025

Ulrich, R. S., Simons, R. F., Losito, B. D., Fiorito, E., Miles, M. A., & Zelson, M. (1991). Stress recovery during exposure to natural and urban environments. *Journal of Environmental Psychology, 11*, 201–230. https://linkinghub.elsevier.com/retrieve/pii/S0272494405801847, https://doi.org/10.1016/S0272-4944(05)80184-7

van Oel, C. J., & van den Berkhof, F. W. (2013). Consumer preferences in the design of airport passenger areas. *Journal of Environmental Psychology, 36*, 280–290. https://doi.org/10.1016/j.jenvp.2013.08.005

Vartanian, O., Navarrete, G., Chatterjee, A., Fich, L., Leder, H., Modroño, C., Nadal, M., Rostrup, N., & Skov, M. (2013). Impact of contour on aesthetic judgments and approach-avoidance decisions in architecture. *Proceedings of the National Academy of Sciences of the United States of America, 110*, 10446–10453. https://doi.org/10.1073/pnas.1301227110

Vartanian, O., Navarrete, G., Chatterjee, A., Fich, L. B., Gonzalez-Mora, J. L., Leder, H., Modroño, C., Nadal, M., Rostrup, N., & Skov, M. (2015). Architectural design and the brain: Effects of ceiling height and perceived enclosure on beauty judgments and approach-avoidance decisions. *Journal of Environmental Psychology, 41*, 10–18. https://doi.org/10.1016/j.jenvp.2014.11.006

Vartanian, O., Navarrete, G., Chatterjee, A., Fich, L. B., Leder, H., Modroño, C., Rostrup, N., Skov, M., Corradi, G., & Nadal, M. (2019). Preference for curvilinear contour in interior architectural spaces: Evidence from experts and nonexperts. *Psychology of Aesthetics, Creativity, and the Arts, 13*, 110–116. https://doi.org/10.1037/aca0000150

Vartanian, O., Navarrete, G., Palumbo, L., & Chatterjee, A. (2021). Individual differences in preference for architectural interiors. *Journal of Environmental Psychology, 77*, 101668. https://doi.org/10.1016/j.jenvp.2021.101668

Vessel, E. A., Maurer, N., Denker, A. H., & Starr, G. G. (2018). Stronger shared taste for natural aesthetic domains than for artifacts of human culture. *Cognition, 179*, 121–131. https://doi.org/10.1016/j.cognition.2018.06.009

Weinberger, A. B., Christensen, A. P., Coburn, A., & Chatterjee, A. (2021). Psychological responses to buildings and natural landscapes. *Journal of Environmental Psychology, 77*, 101676. https://doi.org/10.1016/j.jenvp.2021.101676

Wilson, E. (1984). *Biophilia: The human bond with other species*. Harvard University Press.

Yue, X., Robert, S., & Ungerleider, L. G. (2020). Curvature processing in human visual cortical areas. *NeuroImage, 222*, 117295. https://doi.org/10.1016/j.neuroimage.2020.11729

Zeki, S., Romaya, J. P., Benincasa, D. M. T., & Atiyah, M. F. (2014). The experience of mathematical beauty and its neural correlates. *Frontiers in Human Neuroscience, 8*. https://doi.org/10.3389/fnhum.2014.00068

Skov, M., & Nadal, M. (2020). A farewell to art: Aesthetics as a topic in psychology and neuroscience. *Perspectives on Psychological Science, 15*, 630–642. https://doi.org/[illegible]

[illegible]

[illegible]

The Built Environment and the Brain: Review of Emerging Methods to Investigate the Impact of Viewing Architectural Design

Nour Tawil and Simone Kühn

Abstract The symbiotic relationship between individuals and their surrounding built environment plays a crucial role in shaping human experiences, from the intimacy of indoor spaces to the vastness of the outdoors, influencing thoughts, emotions, and actions. Recently, research into how architecture and design affect psychological and neurophysiological responses has gained momentum, especially as urbanization and mental health concerns rise. Technological advancements, like neuroimaging and virtual reality, have ushered in a new understanding of how the human brain engages with the built environment. However, despite this growing interest, a comprehensive exploration of how the built environment affects neuronal mechanisms remains relatively uncharted. This chapter seeks to address this gap by exploring methodologies at the intersection of architecture and neuroscience. Examining the impact of viewing interior, building, and urban design on the neurophysiological mechanisms underlying cognitive, affective, and behavioural responses, this review offers insights for researchers and professionals in architecture and urban planning to enhance methodologies and create environments positively impacting human well-being. The chapter concludes by addressing methodological considerations and challenges inherent in conducting neuroscientific research on architectural settings, highlighting potential directions for future research.

Keywords Built environment · Neuroimaging · Virtual reality · Brain responses · Architectural design · Review

---

N. Tawil (✉)
Max Planck Institute for Human Development, Center for Environmental Neuroscience, Berlin, Germany
e-mail: tawil@mpib-berlin.mpg.de

S. Kühn
Max Planck Institute for Human Development, Center for Environmental Neuroscience, Berlin, Germany

Department of Psychiatry and Psychotherapy, University Medical Center Hamburg-Eppendorf, Hamburg, Germany

S. Kühn (ed.), *Environmental Neuroscience*,
https://doi.org/10.1007/978-3-031-64699-7_9

Introduction

The intricate interplay between individuals and the spaces they inhabit intertwines humanity with the constructed world. As we navigate our daily routines, various architectural environments continuously envelope us (Evans & McCoy, 1998). Whether within the sanctuary of our homes, the productivity hubs of our workplaces, the rejuvenating havens of our leisure spots, or the vibrant neighbourhoods and cities we call home, built environments leave an indelible mark on our daily lives shaping our thoughts and emotions (Robinson & Pallasmaa, 2015) and potentially affecting our mental health and well-being (Wohlwill, 1976; Evans & McCoy, 1998; Gifford, 2002; Evans, 2003; Staats et al., 2003; Elliot and Maier, 2014; Coburn et al., 2020). Complexly woven into these spaces, human life takes form in multifaceted ways. It is an experience that goes beyond mere aesthetics and functionality (Eberhard, 2009), extending its influence deep into our brains and bodies, as famously articulated by Sir Winston Churchill: *"We shape our buildings and afterwards they shape us"* (Churchill, 1943) (➔ chapters "Aesthetics of Architecture: Emerging Insights from the Brain" by Vartanian, and "Applications of Biometrics in Architectural and Environmental Design" by Lavdas and Sussman).

Over the years, the exploration of how the built environment impacts psychological and neurophysiological responses has evolved into a thriving field of inquiry (Coburn et al., 2017; Lindal & Hartig, 2013; Stamps, 2000). This interest has broadened beyond a niche, engaging researchers, architects, urban design practitioners, policymakers, and various community stakeholders seeking to understand the diverse dimensions of the human experience within built settings (Eberhard, 2009; Karakas & Yildiz, 2020). This imperative gains significance amidst unprecedented rates of urbanization and a growing global incidence of mental health disorders (Krabbendam et al., 2021; Peen et al., 2010). Key questions emerge, prompting exploration: What aspects of architectural design trigger affective responses, alleviate stress, or induce anxiety? How do architectural elements contribute to overall well-being, from influencing emotions and cognitive performance to promoting mental health? Can this knowledge be harnessed to craft spaces that enrich the quality of life for everyone?

Despite the considerable time spent in and around buildings, a surprising gap exists in linking variations in architectural features to psychological states. This gap may be attributed to methodological and disciplinary incongruences between architecture and psychology (Edelstein & Macagno, 2012; Vartanian et al., 2013). Historically, architectural research is based on philosophical constructs and subjective interpretations, while traditional psychological research relies on observation and subjective measures. Neuroscience was proposed to bridge this gap, suggesting the potential of the discipline in explaining how humans perceive and navigate built spaces and how these spaces, in turn, affect cognition, mood, and behaviour (Gepshtein & Snider, 2019; Sternberg & Wilson, 2006). Recent advancements in presentation methods (e.g. virtual reality) and neuroimaging techniques have fuelled significant developments in neuroscientific research investigating architecture, enabling real-time measurement of brain activity and providing insights into the neural processes underlying the responses to different designs and configurations (Higuera-Trujillo et al., 2021; Karakas & Yildiz, 2020; Rad et al., 2023).

Virtual Reality (VR), a computer-generated simulation, immerses users in schematic or realistic, three-dimensional worlds, simulating physical presence in real or fictitious indoor or outdoor settings on a "real" scale (➔ chapter "Extended Reality in Environmental Neuroscience Research" by Li et al.). This technology serves as a potent tool for conducting controlled experiments, allowing for the manipulation of specific aspects of the environment while maintaining other variables constant. Although debates persist about VR's ecological validity for simulating physical environments (Kalantari et al., 2021; Sanchez-Vives & Slater, 2005), research has demonstrated significant neural overlaps between VR and real-world experiences (Kalantari et al., 2021). Cost-efficiency, precise tracking of participant behaviours (Vilar et al., 2014), and the ability to present a wide range of stimuli with high realism enhance the generalizability and applicability of results. There are typically two types of virtual environments (VEs). The Cave Automatic Virtual Environment (CAVE) includes projector screens which create a room space consisting of three walls and a floor. Ultimately, another type includes virtual display of an enclosed, designed space where participants enter by wearing a head-mounted device (HMD). While CAVE provides a room-sized immersive environment, HMD offers a portable and personal VR experience.

Neuroimaging techniques, including electroencephalography (EEG), functional magnetic resonance imaging (fMRI), and near-infrared spectroscopy (NIRS), offer valuable perspectives into brain processes. EEG, known for its excellent temporal resolution (Gevins et al., 1999) and portability, records electrical activity through scalp electrodes (➔ chapter "EEG for the Study of Environmental Neuroscience" by Grassini). Table 1 depicts the major brainwaves detected by EEG and how they are interpreted (Azzazy et al., 2021). Commercial-grade systems (e.g. Emotiv EPOC) complement research-grade devices (Debener et al., 2012; Duvinage et al., 2013). Although a peer-reviewed framework is needed for field-measured EEG with commercial-grade devices (Mavros et al., 2016), these systems have been widely utilized as they offer relatively easy to use brain–computer interfaces to classify raw data into mental states (e.g. Affective Suite; short-term and long-term excitement, frustration, engagement, and meditation). Another method, fMRI, visualizes brain

Table 1 Major brainwaves detected by EEG and their common interpretation

Wave		Range (Hz)	Mental state indication
Alpha	α	7–13	Disengagement, relaxed, unwind, disentangled
Beta	β	13–30	Engagement, focused, computing
Delta	δ	0.5–4	Sleeping, unconscious
Theta	θ	4–7	Lower range: Unconscious, creative inspiration, deep meditation Upper range: Frustration, dissatisfaction
Gamma	γ	>30	Attention, cognition, perception

Azzazy et al. (2021)

activity by detecting changes in blood flow and oxygenation levels, so-called blood-oxygenation-level-dependent (BOLD) signal, which reflects neuronal activity (Logothetis, 2003), offering insights into functional brain organization (➔ chapter "Mapping the Human Brain with Computational Anatomy" by Draganski et al.). Finally, NIRS measures changes in blood oxygenation and volume in the brain's cortical regions (Villringer et al., 1993), providing moderate spatial resolution and tolerance to motion artefacts. A specialized form of NIRS used for functional brain imaging is functional NIRS (fNIRS) (see also chapter "Environmental Neurorehabilitation: Harnessing the power of the environment in rehabilitation following neurological conditions" by Norwood et al.). For the purpose of simplification, we are using the term NIRS as an umbrella term for studies that used either NIRS or fNIRS.

Often, neuroimaging methods are accompanied by other physiological measurements like electrodermal activity (EDA) or galvanic skin response (GSR), heart rate (HR) and heart rate variability (HRV), blood volume pulse (BVP), photoplethysmography (PPG), eye-movements, etc. While physiological measures provide information about, for example, the autonomic nervous system's (ANS) responses, neuroimaging methods go a step further by directly revealing the neural mechanisms at play. For instance, a study using sensors and VR demonstrated EEG's superiority over GSR and PPG in classifying architectural experiences (Zou & Ergan, 2021). Although EEG research mainly focused on frontal areas, the study also emphasized the importance of frontal and occipital asymmetry as well as the effectiveness of both frontal and back channels. In our review, we acknowledge the nascent stage of the field, covering methods with limited channels. However, we report results from physiological measures only when complementing neuroimaging methods to limit the complexity of the chapter.

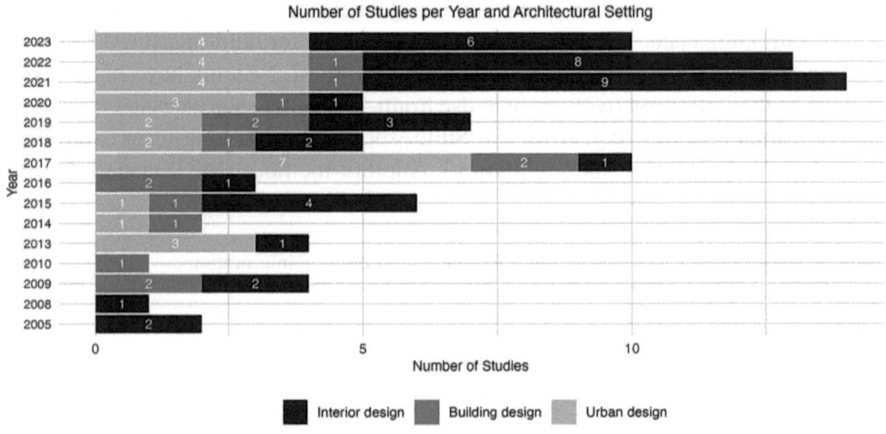

Fig. 1 Distribution of publications per year and architectural setting

Despite the surge of interest in this area, highlighted in the increasing number of publications in recent years (Fig. 1), a systematic exploration of the neurophysiological responses underlying the experience of architectural spaces remains a relatively uncharted territory. This chapter outlines methodologies developed to investigate how we respond to our built surroundings. These methods are at the intersection of neuroscience, environmental psychology, and architecture, and underscore the importance of bridging the gap between the physical and mental realms. In our exploration, we acknowledge the multisensory nature of architecture (Pallasmaa, 2014; Zumthor, 2006). However, we have deliberately focused on the visual system to streamline the review. The results section has three main subsections representing research conducted on interior design, building design, and urban design, following a recently developed framework for investigating architectural design (Azzazy et al., 2021). Subdivisions are classified by mode of presentation (images and videos, virtual environments, and naturalistic environments) on the first level and neuroimaging methods (EEG, fMRI, NIRS) on the second. Studies are grouped by themes. We conclude by discussing methodological considerations and challenges inherent in conducting neuroimaging studies in architectural settings, highlighting potential directions for future investigation.

Methods and Overview

The studies were identified through a comprehensive search across various sources. A thorough exploration of English peer-reviewed papers was conducted using electronic databases from publishers such as PubMed, ScienceDirect, Nature, Springer, Taylor & Francis, and Wiley, along with repositories like Google Scholar. The initial search terms encompassed topics related to architecture, the built environment, indoor space, urban environment, neuroscience, neuroimaging, neuroarchitecture, brain, and activity, as well as specific techniques like EEG, fMRI, and NIRS. Additionally, we reviewed the bibliographic references of the retrieved works. The saturation point was considered achieved when redundancy was observed in most references.

The search was conducted between June and August 2023. The selection of included articles was based on relevance and suitability. A total of 87 research papers, spanning the years 2005–2023, were chosen for analysis. Figure 1 provides an overview of the trends in publication years and the volume of publications according to the different architectural settings (interior, building, and urban design).

Figures 2 and 3 illustrate the distribution of publications according to presentation mode (Fig. 2) and neuroimaging method (Fig. 3).

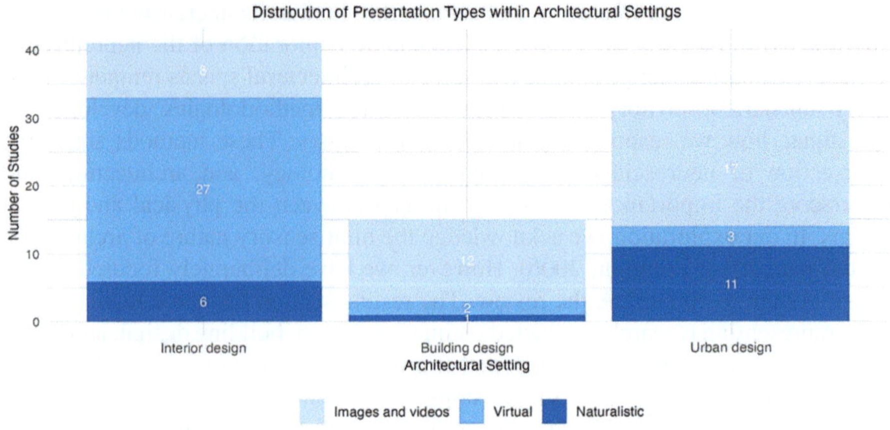

Fig. 2 Distribution of presentation modes per architectural setting

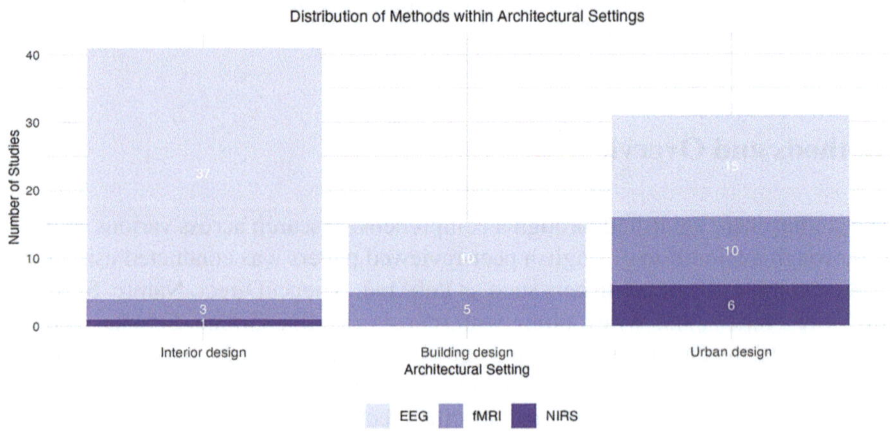

Fig. 3 Distribution of neuroimaging methods per architectural setting

Results

Interior Design

Interior design is primarily concerned with the internal spaces within buildings. Beyond the multi-sensory aspect of interior design, on the visual level, it involves the arrangement and selection of elements within indoor spaces, including geometry, spatial configuration, colour schemes, materials, lighting, furniture, and vegetation, but also other aspects such as functionality and sustainability.

Images and Videos

Researchers often employ static visual stimuli like photographs, drawings, and 3D representations to explore how the brain responds to interior design. Despite an acknowledgement of the role of active exploration in the experience of architecture (Nasar, 1994), the examination of dynamic stimuli, such as videos, remains limited in comparison.

EEG

Evidence from EEG studies illustrates distinct neural processing of different representations of architectural environments. Comparing real and computer-generated bedroom images with different levels of details ($N = 13$), a subjective preference for real and detailed over simpler images was observed (Grima Murcia et al., 2019). Real images activated Brodman 31 brain area (also known as the retrosplenial area) associated with episodic memory retrieval, while images without details engaged areas near Brodman 18 region (also known as the occipital lobe) related to attention to colour, shape, and visuospatial information processing. At later processing stages, both real and images with details activated regions in the superior temporal gyrus and parahippocampal gyrus, associated with reasoning, recognition memory, memory recall, retrieval, and the integration of visual elements into perceptual wholes. Simpler images, such as sketches, activated the medial frontal rotation area, primarily linked to internal mental calculation, memory encoding, and recognition. These differences were attributed to image familiarity and the processing fluency theory (Reber et al., 2004; Winkielman et al., 2006), underlying the significance of realistic computer models in understanding the effects of the built environment.

Shifting focus to occupants' cognitive function and mental health, 36 participants were presented with images and videos of two building interiors, a conventional building and a sustainable building designed according to three sustainable parameters: lighting, view, and spatial arrangement (Hu et al., 2021). In line with the cognitive load theory (Choi et al., 2014), participants exhibited heightened visual system engagement and adjusted their attentional focus and control processing within the sustainable compared to the conventional environment, indicating that sustainable built environments may facilitate a transition towards a more actively engaged mental state akin to mindfulness and emphasizing a greater awareness of the immediate surroundings over internal mental processes. This was evident through the increased theta/delta activities and greater involvement of visual systems and corresponding decreases in frontal activity in the sustainable environment. Although the study identified several parameters as beneficial for cognitive function, making inferences about individual effects remains difficult.

In the exploration of specific design parameters, research underscored the impact of spatial layout and window size on human responses. One study focused on perceived density, defined as the subjective judgment of the number of people fitting in an area (Zur et al., 2023). Thirty-one participants viewed videos depicting equally sized trapezoid-shaped apartments, where the wide wall location (with respect to entrance or edge) and window size were varied. Lower perceived density was observed when the wider wall was at the edge versus entrance. Wider windows were perceived as less dense, especially when the larger wall was at the edge. Positive affect ratings favoured broader windows, especially in the edge condition. The study used two EEG measures associated with movement planning: event-related desynchronization modulation (mu-rhythm) and theta power. Event-related desynchronization refers to a decrease in specific brainwave frequency, and in the case of mu-rhythm, this often occurs when an individual is engaged in motor planning or execution. In this context, variations in theta power may indicate cognitive processes related to planning and coordinating movements. Stronger mu-rhythm in the right primary sensory cortex occurred when apartments became narrow with a wide window or wide with a narrow window. Theta oscillations increased in the left primary sensory cortex for wide window conditions. It was concluded that density perception relates to movement planning, suggesting that design parameters such as an expanding space can modulate early spatial attention.

Building on Ulrich's seminal work, which highlighted the positive effects of natural environments on various psychological aspects (Ulrich, 1983; Ulrich et al., 1991), research explored the impact of including natural elements on neurophysiological responses. When examining a workplace setting ($N = 38$), exposure to natural views or indoor plants reduced anxiety levels, while their absence increased tension and anxiety (Chang & Chen, 2005). Using one electrode placed on the medial prefrontal cortex, and given that alpha activity is present during periods of rest while beta activity is observed when an individual is awake and attentive, scenes with nature views and indoor plants had the most significant impact on EEG-b values, while the presence of a window, irrespective of the view, had higher EEG-a effects. Nature views also resulted in the lowest BVP readings. Self-reported anxiety measures were in line with physiological data, further showing that window views had stronger effects than indoor plants. Importantly, it seems that the density of indoor vegetation can play a role in well-being, impacting perceived restorativeness, stress reduction, and selective visual attention (Rhee et al., 2023). Passively viewing of images depicting indoor living environments with varying vegetation densities ($N = 41$) revealed psychophysiological stress reduction as indoor vegetation increased, peaking at a density range of 13%–24% before plateauing. This was evident through the perceived restorativeness scale scores, as well as the increased relative alpha power spectral density (PSD) values in the analysis of all channels, antero-frontal region, parietal region, and occipital region. However, beta relative PSD in the analysis of all channels and the antero-frontal region showed a continuous increase as the indoor vegetation density increased, indicating heightened selective visual attention.

In sum, EEG studies investigating interior design using images and videos highlight neural distinctions in processing different architectural representations and configurations. Real images activated Brodman 31 area linked to episodic memory, while simpler, less detailed images engaged the Brodman 18 area associated with visual processing (Grima Murcia et al., 2019). Both detailed and real images later activated the superior temporal gyrus and parahippocampal gyrus, involved in memory and visual integration, whereas simpler images activated the medial frontal rotation area related to mental calculation and memory encoding. Sustainable compared to conventional built environments showed heightened visual engagement, increased theta/delta activities, and decreased frontal activity, suggesting a potential transition to a more actively engaged mental state (Hu et al., 2021). Perceived density in trapezoid-shaped apartments, influenced by large wall location and window size, seems to affect brain activations related to movement planning, with stronger mu-rhythm observed in incongruent designs and increased theta oscillations in conditions with wider windows (Zur et al., 2023). Research also underscores the positive impact of natural elements on neurophysiological responses, emphasizing that window views and indoor plants significantly impacted beta values implicating higher attention, while windows, regardless of the view, had higher effects on alpha values, indicating rest (Chang & Chen, 2005). The findings highlight a more positive effect of windows views than indoor plants, and a higher effect of nature than cityscape views. Finally, the density of indoor vegetation correlated with well-being, revealing stress reduction at optimal density ranges of 13%–24%, as indicated by subjective responses as well as increases in relative alpha relative PSD values, while beta PSD values continuously increased with vegetation density, indicating enhanced selective visual attention (Rhee et al., 2023).

fMRI

One of the first fMRI studies to examine the neural responses to interior design stimuli (N = 18) concern an investigation into the effects of contours on aesthetic judgements and approach-avoidance decisions (Vartanian et al., 2013). Curvilinear spaces were more likely to be rated as beautiful compared with rectilinear ones, in line with a long history of evidence within different types of stimulus domains (Bar & Neta, 2007; Dazkir & Read, 2012; Leder & Carbon, 2005; Lundholm, 1921; Silvia & Barona, 2009). Beauty ratings of curvilinear spaces correlated with increased activity in the anterior cingulate cortex (ACC), which underlies emotional salience monitoring (Taylor et al., 2008) and is part of the core circuit for aesthetic processing (Brown et al., 2011). In contrast with a previous fMRI study on objects (Bar & Neta, 2007), no amygdala activity was observed for rectilinear spaces. Importantly, curvilinear spaces did not affect decisions to approach or avoid spaces, only engaging the visual cortex, particularly the lingual and the calcarine gyri, and not, as predicted, areas involved in planning motor movement.

A subsequent study used the same data to examine how ceiling height (high vs. low) and enclosure (open vs. enclosed) influence neural responses (Vartanian et al., 2015). Spaces with high ceilings were found more beautiful, consistent with earlier findings (Baird et al., 1978), but ceiling height did not influence approach or avoidance decisions. Open spaces were also considered more beautiful, while enclosed spaces made participants more likely to respond with "exit" decisions. Neurologically, beauty ratings showed activation in the left precuneus and left middle gyrus, related to visuospatial processing (Kravitz et al., 2011), for high ceilings, but no regions related to affect and pleasure. Open spaces activated the left middle temporal gyrus and right superior temporal gyrus, associated with processing of abstract representations derived from visual motion information. In the enter–exit run, ceiling height did not trigger activations, but enclosure did, specifically in both hemispheres' anterior midcingulate cortex (aMCC) which receives direct input from the amygdala, suggesting a potential fear response according to the authors.

In a recent reanalysis of the data, the focus was on exploring correlations between psychological dimensions (coherence, fascination, and hominess) and neural activation patterns. The aim was to investigate whether these correlations exhibited variations depending on whether participants were assessing beauty or making approach-avoidance decisions (Coburn et al., 2020). Regardless of the task, the level of fascination correlated with activity in the right lingual gyrus. Coherence showed a correlation with activity in the left inferior occipital gyrus, but only when participants were judging beauty. Hominess correlated with activity in the left cuneus, but only when participants were making approach-avoidance decisions. Neural activations did not correlate with global image properties such as self-similarity and complexity scores. It was concluded that the valuation of architectural processing in the visual cortices varies by dimension, as well as by task in the case of coherence and hominess dimensions, highlighting that these neural signatures reflect top-down influences on visual valuation rather than bottom-up psychophysical properties originating from the visual images themselves.

To sum up, fMRI studies investigating neural responses to images representing interior design stimuli revealed that curvilinear spaces were more likely to be perceived as beautiful, correlating with increased activity in the ACC, indicating a pleasant and rewarding effect of curves (Vartanian et al., 2013). Subsequently, examining the effects of ceiling height and enclosure showed that high ceilings were considered more beautiful, activating the left precuneus and left middle gyrus related to visuospatial processing, while open spaces, also found more beautiful, activated the left middle temporal gyrus and right superior temporal gyrus, associated with abstract representations processing, and enclosed areas triggered aMCC activation suggesting a potential fear response (Vartanian et al., 2015). A recent reanalysis explored correlations between psychological dimensions and neural activation patterns in the visual processing areas, revealing that the level of fascination correlated with activity in the right lingual gyrus regardless of the task, coherence correlated with the left inferior occipital gyrus during beauty judgments and hominess correlated with the left cuneus during approach-avoidance decisions (Coburn et al., 2020).

Virtual Environments

Research has also employed virtual environments to investigate the impact of interior design on human emotions, cognition and spatial navigation, unveiling various aspects of how design interacts with the human brain. We note that until present, the most extensive portion of research has been conducted on virtual interiors using EEG.

EEG

Before delving into details of EEG studies that tested virtual interiors, we mention that research has found consistency in the cognitive and neurophysiological responses to virtual compared to real-world environments (Kalantari et al., 2021). In the context of an educational facility, a study observed similar accuracy on cognitive tests in both real and virtual classrooms ($N = 25$). No significant differences were observed in eye blink rate and heart rate, while head acceleration and GSR variance were lower in the virtual classroom. Generally, EEG frequency bandpower did not significantly differ, and analysis of event-related potentials (ERPs) indicated strong similarity between the real and virtual classrooms, with the exception of executive functioning in a colour mismatch task.

Research into the neurophysiological responses to virtual interior environments underscores the significant role of the sensorimotor system in the perception of architectural spaces (Djebbara et al., 2019; Vecchiato et al., 2015a, b), supporting the embodiment theories (Clark, 1999; Freedberg & Gallese, 2007). In one study, EEG responses of 12 participants were recorded in a CAVE setup during the evaluation of residential bedroom interior designs (Vecchiato et al., 2015a). Positive correlations were observed between ratings of novelty, pleasantness, and comfort, with a negative correlation between familiarity and novelty. EEG measures revealed that interiors rated as pleasant, novel, and comfortable led to a desynchronization of the mu-rhythm in the left sensorimotor areas, indicating involvement of embodied simulation mechanisms. Additionally, more pleasant and less familiar interiors increased activations of the left frontal areas, particularly in the theta and alpha frequency bands, along with the involvement of regions associated with spatial navigation, indicating engagement of motivational factors regulating approach (Coan & Allen, 2003). An increase in comfort levels was associated with heightened theta frontal midline activity, potentially linked to attention and positive emotional experience (Aftanas & Golocheikine, 2001). It was suggested that the cerebral processes underlying the appreciation of architecture involve diverse mechanisms regulating bodily, emotional, and cognitive responses. Another publication of the same data aimed to test whether differences in virtual interior designs trigger distinct neural circuits related to the mechanisms of embodiment (Vecchiato et al., 2015b). Higher reported presence was associated with frontal-midline theta activation, indicating the engagement of sensorimotor integration mechanisms. Similar patterns were observed in familiar and comfortable VEs, with pleasant VEs exhibiting increased theta power within visuomotor circuits and activating the alpha band in regions associated with visuospatial exploration and processing of categorical spatial relations. The de-synchronization of the mu-rhythm characterized the perception of pleasant and comfortable VEs, highlighting the involvement of left motor areas and mechanisms related to embodiment. Generally, these findings indicate that EEG correlates of architectural perception involve cerebral circuits associated with sensorimotor integration, spatial navigation, and embodiment. A more recent investigation explored the influence of environmental features on early sensory processing in a go/nogo task involving virtual transitions of different passability levels (Djebbara et al., 2019). The study aimed to contribute to the debate on the relationship between cognition, movement, and environment. There were significant differences in subjective ratings and approach times based on door width in the Go but not NoGo trials. Participants ($N = 20$) perceived narrow doors as more dominant but less arousing and valence-inducing than mid and wide doors. Moreover, they approached wide doors faster than mid and narrow ones. This suggests that environmental affordances impact behaviour. EEG results demonstrated distinct evoked potentials between impassable doors with limited affordances and passable doors, particularly at frontocentral and occipital sites. These differences were most notable between 50 and 200 milliseconds after the initial view of the environment, indicating that affordances impact early cortical potentials. Motor-related brain activity revealed that the narrow door elicited the highest amplitude for anterior positivity and posterior negativity, likely due to the body not fitting well, creating a prediction error during the transition. The findings suggest that early sensory brain activity, even before actual movement, was influenced by affordances, and that movement through transitions was preceded by a motor-related negative component shaped by these affordances. In conclusion, the study proposes that an environment's potential actions significantly impact perception, supporting enactivism as a holistic approach to studying cognition (Jelić et al., 2016).

In the realm of wayfinding research, investigations into spatial navigation within virtual environments emphasized the significant impact of architectural cues on spatial orientation and navigation (Edelstein et al., 2008; Kalantari et al., 2022; Zhu et al., 2022). Utilizing a CAVE system, brain responses to ambiguous and non-ambiguous VEs were examined to test the effectiveness of cues to visual and spatial relationships (Edelstein et al., 2008). Navigation from a first person perspective engaged a network of cortical areas, including occipital, parietal, occipitotemporal, and frontal areas. Differences in brain activity dynamics related to the ambiguity of the starting position were observed, with stronger synchronization in the theta band and stronger desynchronization in the lower alpha band in the parietal and occipitotemporal cortex. This indicates the engagement of the parietal cortex in processing visuospatial information, while both parietal and occipitotemporal regions were involved in direction change and future path planning. Disorientation was linked to increased alpha desynchronization, suggesting heightened attentional demands for processing less informative visuospatial information. Moreover, studies in healthcare settings demonstrated that signage and interior design elements could impact neurophysiological responses and orientation behaviours during wayfinding tasks (Kalantari et al., 2022). Environments enhanced with colours, architectural elements, and distinct graphics on signs (condition C) showed improvements in some orientation behaviours and greater neural activations ($N = 63$). While findings did not translate into improved wayfinding times or decreased self-reported stress, fatigue, or confusion, EEG data indicated increased neural processing in the occipital cortex when viewing a sign and deciding on the next move in conditions C and B (enhanced colours) compared to baseline. Moreover, there was a significant desynchronization indicating higher cognitive engagement when looking at signs in conditions B and C. Analysis also revealed an oscillatory pattern in theta-band desynchronization after seeing a sign, with significant differences between conditions C and B at 1s and 2.5s post-event, each lasting about 1s. Participants likely integrated a broader array of environmental information and exhibited different cognitive and behavioural responses due to variations in visual processing between design conditions. The findings generally suggest that enhancing environmental affordances in hospital design, such as architectural features and distinct sign patterns, improved navigational cognition and orientation more effectively than colour enhancement alone. Additionally, other studies have found that ceiling height can also influence wayfinding behaviours. When immersing 343 participants into virtual mazes with varying ceiling heights and objects, EEG measures collected through a commercial-grade device revealed distinct brain activity patterns related to wayfinding and object mapping experiences (Erkan, 2018). The height of the space affected directional behaviour both cognitively and behaviourally. High-ceiling mazes were exited faster and easier, with better object-mapping and recall rates. In low-ceiling space, individuals experiencing prolonged difficulties in finding their way exhibited anxiety-related beta activity due to specific brain regions being overworked. Interestingly, age, gender and level of education were found to also impact wayfinding behaviours.

Other approaches focused on how architectural elements and parameters interacts with affective responses. Some researchers employed pairs of VEs manipulated to represent two extreme ends of bipolar scales corresponding to positive and negative human experiences. For instance, relaxing versus stressful virtual educational spaces, created by changing the level of luminance, natural daylight, colour of surfaces, openness of spaces, and presence or absence of visual cues, were found to actually influence stress and anxiety (Ergan et al., 2019). Thirty-three participants were asked to perform simple tasks in these VEs, revealing 40% higher EEG oscillations in theta, alpha, and beta frequency bands across all 14 channels in the positive compared to the negative VE. GSR data showed a significant increase in the number of peaks and peak amplitude in both environments compared to baseline, particularly in the negative environment, indicating heightened stress and physiological arousal. HRV decreased in both VEs compared to baseline, with a greater decrease in the negative VE, although differences were not statistically significant. Overall, analyses across sensors indicated a higher emotional response in the negative VE with combined results showing that the positive VE was not stressful. Moreover, participants reported higher pleasantness, focus, and relaxation in the positive VE. While the study demonstrated differential responses to the two VEs, it assessed the combined effects of individual elements, preventing the identification of the specific contribution of each factor to the observed effects. In fact, most of the available research addressed architectural design parameters, manipulated separately or in combinations, in a more controlled way. Factors that define a given space were suggested to impact a perceiver's brain responses. This encompasses geometry and spatial configuration including form, scale, proportions (e.g. width, ceiling height), windows, and interior layout, in addition to other characteristics such as colour and lighting, and the inclusion of natural elements.

One fundamental aspects of architectural design explored in studies is the form of interior spaces. For instance, research showed that walking through living spaces with different forms displayed through a CAVE can trigger rapid emotional responses in the ACC ($N = 15$), with curved shapes leading to stronger activations and more positive affective ratings (Banaei et al., 2017). The observed changes in ACC theta band activity suggested that they were not solely attributable to the emotional effects of form features but also influenced by the cognitive processing of architectural elements. However, other form parameters also significantly impact emotional and cognitive processes, with factors like protrusion, curvature, proportion, and scale having pronounced emotional effects in 112 designers and non-designers (Shemesh et al., 2021, 2022). Among the effects observed, large spaces were linked to higher dwell times, liking scores, larger pupillary diameters, and fewer and shorter fixations, indicating positive emotional responses (Shemesh et al., 2021). Physiological and emotional responses also varied with space proportion, showing stronger reactions in extreme conditions. For instance, narrower spaces caused a negative emotional response, as exhibited in the shorter dwell times, lower liking score, as well as increased pupil dimensions, larger fixation rates, longer fixation durations, and a tendency for higher beta ratio (assessed via a commercial-grade device). In terms of protrusion and curvature, zero curvature and asymmetry increased pupil dimensions indicating interest as protrusion increased and curvature decreased. High curvature, asymmetrical VEs were not positively perceived. Notably, regardless of protrusion and symmetry, curvature produced high maximal GSR measurements indicating strong aesthetic emotional responses. Medium curvature with maximal protrusion resulted in the largest pupil dimensions, dwell time, and liking scores. Unique impacts were observed for complete square and complete sphere shapes on GSR, independent of scale, in addition to higher pupil dimensions and a higher beta ratio for sharp rectangle spaces. Another publication relating to the same data (Shemesh et al., 2022) further demonstrated a link between these four geometrical properties and human emotions with similar conclusions. Specifically, larger, simple, and symmetrical spaces evoked more positive responses, while smaller symmetric spaces elicited negative emotions. Curved, large-scale spaces were linked to more positive emotional reactions compared to large square spaces. There was an inverse relationship between proportions and emotional responsivity. A more protruded and asymmetric space, along with sharper non-curved features, elicited a more positive emotional response. In another research, the individual effects of varying scales were explored using a CAVE setup with 66 participants (Bower et al., 2022b). While scale did not influence autonomic indicators or self-reported emotional states, an increased range in skin conductance response (SCR) and HRV was noted across different conditions. Enlarged scale led to increased EEG power within the beta bandwidth and elevated frontal midline low-gamma and high-gamma power, in addition to increased left lateralization in the gamma bandwidth. This suggests that such changes may be a physiological marker of the scale impact. Surprisingly, no significant effects of scale were observed in frontal midline power or lateralization in the theta or alpha bandwidths, associated with positive emotional response (Ekman & Davidson, 1993). While emotional processing may not directly involve scale, it could potentially impact high-frequency oscillatory processes like working memory and decision-making (Spitzer & Haegens, 2017), although this interpretation remains speculative. Functional connectivity analysis revealed that enlarging or reducing scale enhanced theta connectivity in the left temporoparietal and right frontal region, potentially indicating that scale changes may have activated the visuospatial attentional network while reducing scale increased high-gamma connectivity in the right frontoparietal region (Bower et al., 2023).

As part of the overall composition of a given space, the role of ceiling height, layout proportions, windows, and interior layout have also been scrutinized (Cruz-Garza et al., 2022; Kalantari et al., 2023; Kim et al., 2021; Kong et al., 2022; Llinares Millán et al., 2021). An investigation into a virtual postpartum care unit found that ceiling height, aspect ratio, and window ratio impacted users' relaxation-arousal responses in 33 female participants (Kim et al., 2021). A decrease in the poststimulation ratio of alpha to beta (RAB) waves was observed, indicating arousal, with noted variations across the different architectural combinations. Specific lobes of the brain, particularly the right frontal and left parietal lobes, exhibited significant differences in relaxation-arousal responses, potentially linked to visuospatial and categorical space processing. The study concluded with recommended design standards for these three parameters. Other researchers evaluated the sole effect of room width on attention and memory ($N = 90$) within virtual educational settings (Llinares Millán et al., 2021). Cognitive performance declined as room width increased. The HRV-LFHF metric, related to the balance of sympathetic over parasympathetic activity and attentional control, indicated greater sympathetic system activation in the room with the narrower width. EEG measures collected via a commercial-grade device revealed significant effects of width. The EEG-C3-Beta metric, which higher values suggest increased thinking and focus, showed lower attention with increased width. Similarly, for the EEG-CZ-Highbeta metric, which higher values denote increased cognitive performance and alertness, a lower cognitive performance was observed as width increases. The study found correlations between cognitive and neurophysiological measures, concluding that increased width could lead to reduced arousal thereby negatively impacting attention and memory performance. Another educational setting was explored with 23 participants, finding that characteristics of brain activity during cognitive tasks are influenced by room width as well as window location (Cruz-Garza et al., 2022). Although cognitive scores did not reveal statistical differences between the different configurations, consistent differences in EEG components emerged, more prominently for cognitive tasks that required short-term memory encoding. Particularly, distinctive EEG features were observed in the theta and alpha frequency bands across bilateral occipital, parietal, and frontal brain regions. Connectivity analysis supported these findings, indicating alterations in information transfer from centro-parietal to frontal electrodes across different classroom conditions. Focusing on windows, the influence of window size, aperture designs, and sky type on subjective and physiological responses was investigated ($N = 41$) in virtual workplace environments (Kong et al., 2022). Irregular designs were associated with higher interest, excitement, and complexity, while regular designs, such as curtain walls, with calmness, satisfaction with exterior views, and brightness. However, neither EEG nor electrocardiogram data were affected by aperture design. Although subjective assessments of window sizes did not exhibit significant differences, larger windows were correlated with enhanced beta variations and powers in the prefrontal right lobe areas. As for sky type, clear skies led to more positive subjective responses compared to overcast skies, which were associated with greater mean theta rhythm levels in the left and right prefrontal lobe areas. When investigating interior layouts in retail design, 45 participants were exposed to three virtual grocery stores with increasing number of aisles, finding that a store layout can influence product recall and cognitive workload (Kalantari et al., 2023). Contrary to expectations, no significant correlation was found between the number of aisles and perceived visual complexity, possibly due to the minimal contrast in layout designs. Greater perceived complexity was associated with reduced perceived attractiveness. There was a marginally significant correlation indicating that simpler layouts were associated with better product recall. EEG data revealed minimal differences during initial visual inspection, and significant differences during the subsequent product memorization and recall. In particular, the layout with the least number of aisles was associated with lower alpha power (desynchronization) in the ACC region during both product memorization and recall, indicating increased attention, cognitive engagement, and decision-making (Delaux et al., 2021). Furthermore, the retrosplenial cortex (RSC) exhibited stronger activations in the alpha band in the simpler layout, noting that alpha desynchronization in RSC is typically observed during complex visual memory task and spatial navigation (Delaux et al., 2021). Findings, therefore, confirmed that the lower product recall scores in the more complex environments (with more aisles) were due to a higher cognitive load and reduced processing fluency, highlighting that simpler store layouts might be more effective in terms of consumer responses.

Apart from the above-discussed design parameters, colour, another significant architectural feature, was found to play a crucial role in influencing emotional and cognitive responses in VEs (Bower et al., 2022a; Llinares et al., 2021; Sadeghi Habibabad et al., 2022). One study examined the influence of cold and warm colour hues in virtual classrooms on cognitive attention and memory function of 160 university students (Llinares et al., 2021). Spaces with cold-hues were associated with better error rates in the attention task and higher memory performance. They also increased sympathetic activity and reduced parasympathetic activity, indicating heightened arousal. These spaces also exhibited higher brain activations, particularly in beta and high-beta bands (measured with a commercial-grade device), which are associated with attention and cognitive performance, indicating that cold-hues contribute to achieving higher levels of attention, cognitive performance, and working memory. The study showed significant correlations between psychological and physiological metrics. Focusing on the use of blue as ambient wall colour within a CAVE ($N = 18$), changes in indicators of emotions were observed (Bower et al., 2022a). Blue colour led to an increased range in respiration and SCR, along with an insignificant drop in mean values of HRV. It also increased PSD across all electrodes for theta, alpha, and beta brainwave frequencies, as well as alpha frontal midline power and frontal hemispheric lateralization, indicating a more positive emotional response (Davidson, 1992). However, there was no relationship between subjective and autonomic responses. Functional connectivity analyses revealed heightened theta connectivity in the frontal region when blue was present in the environment (Bower et al., 2023). In comparison with resting state, blue colour increased activity in the low-gamma range, clustered in the right occipital region, suggesting that colour can affect cognitive processes and mental health.

Moving to artificial lighting, a study ($N = 25$) investigated brain responses to artificial lighting in virtual office spaces using 17 conditions varying in illumination (low to high) and correlated colour temperature (CCT) (warm to cool) using a commercial-grade EEG device (Mostafavi et al., 2023). Results revealed an association between illumination levels and CCT with particular EEG band features for both lighting variables. Higher illumination levels affected activity in the frontal and parietal brain regions, which are involved in cognitive processing and spatial perception, respectively. Activity in these regions was correlated in a linear manner across different EEG bands. For instance, greater illumination was linked to increased α/θ and β/θ values at central and parietal electrodes, and decreased α/β and α/γ values at the prefrontal electrodes. Notably, there was a consistent positive linear association between higher beta power compared to theta power and higher illumination in parietal regions, along with a reduction of theta power in frontal regions associated with higher illumination. In terms of colour temperature, higher CCT values (bluer light) positively correlated with increased power across all primary EEG bands in the frontal and left temporal brain region, particularly in the theta, alpha, beta, and gamma bands. Additionally, CCT was negatively associated with the α/γ ratio, signifying that bluer light was linked to lower α/γ ratios. Findings, therefore, indicated common dynamics in brain responses to both higher illumination levels and higher "blue" CCT, such as reductions in the α/γ ratio in frontal brain regions.

As found with other methods, the integration of natural elements has been shown to have positive effects on emotional responses and the underlying brain activity using commercial-grade EEG devices (Jung et al., 2023; Yeom et al., 2021). The effectiveness of (small and large) green walls in relieving stress indoors was examined in 27 participants (Yeom et al., 2021). Self-reported stress measures indicated that both small and large walls reduced anxiety levels although the large wall had a slightly lesser effect. EDA measures showed that the small wall lowered participants' arousal and anxiety levels, while the large wall heightened them. Analysis of the relative alpha power revealed that the small wall had a higher relaxing effect both visually and spatially, evidenced by increased activations in the parietal lobe due to spatial awareness and in the occipital lobe due to visual stimuli. Moreover, the small wall alleviated participants' mental stress and mental fatigue (calculated from relative band power), while the large wall exacerbated them. Findings suggested that the small green wall had the most significant relaxing effect, while the large green wall increased stress levels. Another study investigated how the incorporation of biophilic elements in hospital patient rooms affects emotional and brain responses in 63 participants (Jung et al., 2023). Biophilic designs (with plant wall and/or digital wall) led to an increase in low-frequency brainwave activity associated with relaxation, while decreasing high-frequency activity linked to heightened arousal, indicating a tension-reducing effect. The frontal regions were found to consistently play a significant role in classifying different design conditions, and alpha power in the frontal regions was crucial for distinguishing between the rooms. Additionally, self-reported assessments indicated that biophilic design elicited positive emotional changes, with plant walls contributing to a reduction in negative emotions and a combination of plant walls and digital elements enhancing overall emotional well-being in healthcare environments.

In sum, an extensive body of literature explored the impact of virtual interior environments on the human brain. Notably, research emphasized the consistency in cognitive and neurophysiological responses across real-world and VEs, particularly classrooms (Kalantari et al., 2021). Following the embodied simulation theory, studies have found that, indistinctly of the style, architectural perception is associated with diverse processes integrating emotional, motor, and cognitive mechanisms (Vecchiato et al., 2015a). Further analyses highlighted the role of motor and cognitive processes in evaluating architectural spaces for better interior design (Vecchiato et al., 2015b). Recent evidence underscores that environmental affordances influenced early sensory brain activity before actual movement, and that movements were preceded by a motor-related negative component, also depending on the affordances, indicating that perception is affected by an environment's potential actions (Djebbara et al., 2019). In wayfinding, brain responses to ambiguous and non-ambiguous VEs revealed that navigation activated a network of areas associated with spatial orientation (Edelstein et al., 2008). Stronger theta synchronization and stronger lower alpha desynchronization were observed in the parietal and occipito-temporal regions, suggesting a role of these regions in utilizing visuospatial information, processing direction changes and planning of paths, while disorientation was associated with increased alpha desynchronization indicative of heightened attentional demands. In healthcare settings, enhanced environmental affordances, such as distinct signage patterns and interior design features, led to improved orientation behaviours and greater neural activations (Kalantari et al., 2022). Specifically, there was an increased activity in the occipital cortex, higher cognitive engagement, and an oscillatory pattern in theta-band desynchronization shortly after seeing a sign. Moreover, high ceilings in virtual mazes were associated with distinct brain activity patterns, influencing directional behaviour and object mapping experiences, while also noting an impact of interindividual differences (Erkan, 2018). Other lines of research manipulated VEs to represent extreme ends of bipolar scales, for example, stressful/relaxing, finding higher EEG oscillations in theta, alpha, and beta frequency bands in the positive compared with negative VE, indicating greater activity associated with relaxation and pleasant experiences (Ergan et al., 2019). However, most of the available research focused on specific parameters and elements tested separately or in combinations using factorial designs. Walking through spaces with different forms triggered rapid emotional responses in the ACC when perceiving curved shapes, emphasizing the role of architectural form in shaping emotions (Banaei et al., 2017). In addition to curvature, protrusion, proportion, and scale were found to have pronounced effects on emotional and neurophysiological responses (Shemesh et al., 2021, 2022). For instance, there was a tendency for higher beta ratio in narrow spaces as well as sharp rectangular spaces (Shemesh et al., 2021). Exclusively testing scale, increased size resulted in amplified EEG power in the beta frequency, raised frontal midline power in low-gamma and high-gamma bandwidths, and heightened leftward lateralization in the gamma frequency, suggesting these may be physiological markers of scale effects (Bower et al., 2022b). Additionally, scale changes enhanced theta connectivity in the left temporoparietal and right frontal regions, indicating activation of the visuospatial attentional network and reduced scale increased high-gamma connectivity in the right frontoparietal region (Bower et al., 2023). As for other design parameters, virtual care units with varying ceiling height, aspect ratio, and window ratio influenced RAB waves indicating arousal, with significant changes in the right frontal and left parietal lobes linked to visuospatial and categorical space processing (Kim et al., 2021). Manipulating solely room width revealed positive effects of narrower rooms on attention and memory, with wider rooms leading to declines in cognitive performance and reductions in beta and high-beta frequencies (Llinares Millán et al., 2021). Other research showed that room width and window location influenced brain activity during cognitive tasks that required short-term memory encoding, with distinct EEG features in the theta and alpha frequency bands across bilateral occipital, parietal, and frontal brain regions, and connectivity analysis indicating changes in information transfer from centro-parietal to frontal electrodes (Cruz-Garza et al., 2022). In terms of window composition, larger windows correlated with enhanced beta variations and powers in the prefrontal right lobe, and clear (compared to overcast) skies with decreased mean theta rhythm levels in both the left and right prefrontal lobes, while aperture design did not affect brain activations (Kong et al., 2022). Concerning interior shelving layouts, simpler grocery shops layouts with fewer aisles were associated with better product recall and lower cognitive load (Kalantari et al., 2023). This was indicated by lower alpha power in the ACC during product memorization and recall, reflecting increased attention, cognitive engagement and decision-making, as well as stronger activations in the alpha band in the RSC, which suggests enhanced visual memory and spatial navigation, indicating the effectiveness of simpler layouts in terms of consumer responses (Kalantari et al., 2023). Moving to colours, cold compared to warm-hued virtual classrooms were linked to better attention and memory performance, as well as increased activations in beta and high-beta bands, indicating enhanced cognitive performance and working memory (Llinares et al., 2021). The use of blue as ambient wall colour increased PSD across theta, alpha, and beta brainwave frequencies, increased alpha frontal midline power, and promoted frontal hemispheric lateralization, indicating a positive emotional response (Bower et al., 2022a). Moreover, it heightened theta connectivity in the frontal region and increased activity in the low-gamma range, clustered in the right occipital region, suggesting an impact on cognitive processes and mental health (Bower et al., 2023). As for artificial lighting, higher illuminance levels and higher CCT modulated specific brain activations (Mostafavi et al., 2023). Higher illumination increased activity in the frontal and parietal regions, while higher CCT (bluer light) increased power in the frontal and left temporal regions, with both conditions showing common dynamics in brain responses. Finally, the inclusion of natural elements has been shown to have effects on emotional responses. A small green wall significantly reduced stress and anxiety levels by increasing parietal and occipital lobe activations, while the large green wall had lesser or even adverse effects on stress levels (Yeom et al., 2021). In healthcare settings, plant and/or digital walls evoked heightened low-frequency brainwave activity associated with relaxation and lowered high-frequency activity linked to arousal, indicating a tension-reducing effect, with the frontal regions playing a key role in distinguishing design conditions and alpha power in these regions (Jung et al., 2023).

fMRI

To date, fMRI studies that used VR to investigate environmental stimuli have been limited to the domain of spatial navigation (e.g., Iaria et al., 2003; Spiers & Maguire, 2007). Combining fMRI and VR presents a potent approach in neuroscientific investigations of the built environment, allowing detailed exploration of brain activity during immersive experiences. This synergy enhances ecological validity, providing realistic simulations for studying human cognition and behaviour, and offers high spatial resolution for mapping specific neural responses. This integrated method holds promise for advancing our understanding of the intricate connections between the brain and complex, real-world environments.

Naturalistic Environments

Studies have extended their scope to investigate the impact of interior design in real-life scenarios, utilizing EEG and NIRS as primary methods. These studies highlighted that some architectural parameters and configurations have psychological and neurophysiological effects on individuals.

EEG

In terms of interior layout, a study investigated how bed orientation affects sleep quality (Hekmatmanesh et al., 2019). Polysomnography, a comprehensive test used in sleep medicine including EEG measurements, was conducted on 21 participants during two consecutive 1.5-hour naps over two days in rooms with different bed orientations (east-west and north-south). The results showed that participants had a longer total sleep time in the north-south orientation compared to the other orientation. Moreover, there was a noticeable increase in both rapid eye movement (REM) latency and slow-wave sleep (SWS) latency in the north-south direction, with statistical significance seen only for REM latency. Additionally, the count of arousals and awakenings during sleep significantly increased in the east-west orientation. In terms of EEG data, changes in delta, theta, and alpha frequencies were observed based on bed orientation. The alpha band was the only one displaying a significant energy ratio across the entire signal energy. In the north-south condition, increases in the average energy of delta, theta, and alpha bands were observed. The study concluded that sleeping in the north-south direction might be more beneficial than east-west and that sleep EEG signals can be sensitive to Earth's electromagnetic field.

Initial explorations of real-life environments inspected the impact of colours on brain activity, mood, and performance (Küller et al., 2009). Exposed to colourful (red and patterned wall with black and white check-board floor) and grey full-scale rooms for a 3-hour period each, 12 participants found both rooms slightly unpleasant, closed, and original, with the colourful room higher in complexity while the grey room higher in coherence. Participants' self-reported emotional state revealed decreased activation over time with no room colour differences, but they felt more extraverted with higher emotional control in the grey room, while orientation (attention) and evaluation emotions remained unchanged. Neurophysiological data indicated that the colourful compared to the grey room elicited lower alpha rhythm suggesting higher arousal as well as lower heart rate suggesting cardiac deceleration in environments eliciting cortical arousal. The heart rate reduction was particularly notable in introverted participants, slowing by up to 10%, indicating that this group of people is more susceptible to environmental stimulation. A comparison between a red and a blue room in a subsequent experiment during a 2-hour exposure period ($N = 24$) found that the red room was perceived as more closed with higher affection, while the blue room was considered more potent and slightly higher in social status. Both rooms were rated similarly in terms of pleasantness, complexity, unity, affection, and originality. In terms of emotional responses, participants exhibited increased wakefulness and slight boredom, with no significant differences between the red and blue conditions. Most participants wrote longer stories in the red room although the difference was not statistically significant. EEG and EKG recordings revealed that delta waves, associated with sleepiness, were more abundant in the blue room. Moreover, alpha rhythm tended to be higher in the blue room, especially in the second hour, over the left hemisphere, while beta and theta frequencies showed no significant differences. Heart rate consistently remained lower in the red room. These findings underscore the profound impact of colour on physiological responses and emotional states in real environments, suggesting that the colourful room was the most arousing, followed by the red room, with the blue and grey rooms being the least arousing.

In terms of lighting, a study investigated the effects of direct and indirect lighting on affective and neurophysiological responses of 28 participants in real-scale residential living rooms (Shin et al., 2015). The presence of direct and indirect lighting combined led to a stronger sense of coolness and pleasantness. Theta oscillations were more prominently observed in the right fronto-temporal and left temporo-parietal areas (F4, F8, T4, TP7) when both types of lighting were present. Notably, there was a significant positive correlation between participants' perception of coolness and the theta power in the right frontal region. This suggests that the combination of direct and indirect lighting enhanced feelings of pleasantness, influencing subjective perceptions of the space. It was concluded that a combination of direct/indirect lighting influenced participants' subjective perception, with enhanced pleasantness. The study suggested using theta oscillatory activity as a biological marker for emotional status in lighting environments.

When it comes to natural elements, research aimed to identify the optimal index of greenness for interior spaces in terms of psychophysiological responses and subjective preference by exposing 103 participants to conditions with different greenness indexes (Choi et al., 2016). The 50% greenness index (i.e. amount of greenery in the room) was predominantly subjectively preferred. While no significant differences in HRV values were observed, male participants exhibited increased variability in cardiac normal-to-normal (NN) intervals and greater autonomic activity compared to female participants. EEG data recorded on frontopolar and occipital lobes revealed no significant differences, except that female participants had a higher mean amplitude at the left occipital electrode compared to male participants. The authors concluded that even a small amount of greenery may evoke a relaxing effect in a limited space.

To summarize, in the exploration of real-life environments, EEG research found that architectural parameters such as interior layout, colour, lighting, and natural elements can affect neuronal mechanisms. The orientation of beds in a north-south direction was found to improve sleep quality by increasing total sleep time, increasing REM latency, and altering EEG frequency, particularly enhancing delta, theta, and alpha band energies, suggesting sensitivity to Earth's electromagnetic field and potential advantages for sleep quality in such orientations compared to east-west orientations (Hekmatmanesh et al., 2019). Comparing colourful and grey rooms, lower alpha rhythm was found in colourful environments, indicating higher arousal (Küller et al., 2009). Moreover, comparing blue and red rooms, delta and alpha rhythms were more abundant in the blue room. Lighting studies revealed that the combination of direct and indirect lighting in residential living spaces enhanced feelings of coolness and pleasantness, with increased theta oscillations in the right fronto-temporal and left temporo-parietal areas, and a significant positive correlation between perceptions of coolness and theta power in the right frontal region, suggesting that theta activity could serve as a biological marker for emotional responses to lighting environments (Shin et al., 2015). Transitioning to natural elements, greenery in interior spaces was found to be optimal at a 50% greenness index, with EEG data revealing no significant differences (Choi et al., 2016).

NIRS

In order to examine the effects of visual surroundings in the daily living environments on neurophysiological responses, a study exposed 15 participants to two actual-size rooms with different interior designs (Tsunetsugu et al., 2005). There were no significant differences in subjective responses. Parasympathetic activity was dominant in the standard room, while sympathetic activity was dominant in the designed room indicating a state of arousal. Diastolic blood pressure went down in the standard room supporting a calm state. The regional cerebral blood flow (rCBF) on the left frontal area significantly increased in both rooms suggesting interest. To conclude, the standard room with common interiors caused a calm state, which was interpreted to be due to familiarity, while the designed room caused an active state.

Building Design

Building design encompasses the overall planning and conceptualization of entire structures. It includes the design of exteriors, facades, and the integration of architectural features that contribute to the building's overall appearance and functionality.

Images

EEG

Investigating brain responses in an unconventional virtual environment, 21 participants navigated through videos depicting three types of design representations: abstract, semi-designed, and fully designed (Hakak et al., 2016). Fully designed environments were perceived as easier to navigate but less creative, while semi-designed spaces were liked the most. Creativity correlated with liking, but ease of navigation did not correlate with creativity or liking except in abstract designs. EEG data highlighted distinct brain activity patterns in the comparisons of fully-designed with abstract and semi-designed with abstract environments, particularly in beta2 and beta3 bands, which are typically linked to sensorimotor functions such as movement planning, preparation, and execution (Pavlidou et al., 2014), as well as movement observation and imagination (Schnitzler et al., 1997). Abstract (compared to designed) environments led to activations in the motor network, especially the precentral gyrus, suggesting that sensorimotor areas are involved in the aesthetic appreciation of abstract art, and may reflect an embodied simulation of the environment (Gallese, 2005; Gallese and Sinigaglia, 2011). The study also observed prefrontal cortex (PFC) activation, including ACC, dorsolateral PFC (dLPFC), and superior prefrontal gyrus. The ACC activation in fully-designed versus abstract environments implies an increased engagement of high level cognitive functions such as attention (Bush et al., 2000), error detection and conflict monitoring (Weissman et al., 2005), while activation of the dLPFC in abstract environments might reflect conflict-induced behavioural adjustment (Mansouri et al., 2007) due to the unique challenges and rule violations in such environments. It was concluded that navigating an abstract virtual environment requires more precentral efforts, supporting the idea that beginning the initial design stages with an abstract environment free from phsycial rules and constraints may enhance divergeant thinking and boost creativity during the architectural idea generation phase.

Architectural decorum, a prevalent principle in architectural theory from the Renaissance to the onset of modernism, denotes the suitability of a building's design. This concept, intricately tied to ornamentation, played a crucial role in determining how a building conveyed its position within the broader civic and social hierarchy (Kohane & Hill, 2001). Several EEG studies have explored how decorum and rank influence neural responses and perception (Mecklinger et al., 2014; Oppenheim et al., 2009; Oppenheim et al., 2010). These studies involved ERPs recorded in response to greyscale drawings of fictitious buildings that differ in their ranking according to the Western architectural decorum, ranging from high- to low ranking, while participants categorized images of objects and buildings. Ornaments designating a building as high-ranking are thought to make these buildings more prominent and therefore more memorable. If this was true, high-ranking architectural designs should be more familiar, at least to Western individuals. The first study investigated this claim in 13 participants (Oppenheim et al., 2009). Subjects accurately classified objects and buildings with no significant differences in response times. Moreover, ERP responses between 300 and 600 ms post-stimulus, recorded over frontal lobes, could reliably distinguish between the two classes of architectural stimuli without participants' awareness of these categories. Specifically, high-ranking buildings produced N400-potentials with significantly lower amplitudes compared to low-ranking buildings. Findings were interpreted as evidence that high-ranking buildings evoke a stronger sense of familiarity. After this suggestion that high-ranking buildings may leave traces of their cultural prominence in the perceiver's memory facilitating visual-semantic processing (Oppenheim et al., 2009), another investigation was conducted to probe this assumption and examine whether the hippocampus contributes to the processing of architectural ranking (Oppenheim et al., 2010). Testing patients suffering from medial temporal lobe epilepsy (TLE) with/out hippocampal sclerosis and healthy controls ($N = 44$), the study found that early negative potentials (between 200 and 400 ms) effectively distinguished between high- and low-ranking buildings in both groups. However, late positive potentials (LPPs) (between 400 and 600 ms) exhibited higher amplitudes in response to high-ranking buildings but only in healthy subjects and TLE patients without hippocampal sclerosis. These results suggest that the distinction between high- and low-ranking buildings involves two key processes: early visual object selection and late post-model selection, underscoring the critical contribution of the hippocampus to the latter phase of visual object categorization. A later cross-cultural study explored the influence of cultural expertise on the perception of high- and low-ranking buildings, involving 17 German and 18 Chinese participants (Mecklinger et al., 2014). Classification performance was not affected by cultural expertise, and response times were faster for high-ranking buildings. German participants had a smaller N350 response to high-ranking buildings compared to low-ranking buildings at frontal and central electrode sites, while Chinese participants showed similar N350 responses to both building types, indicating that the initial phase of recognizing visual objects, specifically, selecting object models was made easier for German participants when it comes to high-ranking buildings. However, as the object identification process advanced, incorporating information from semantic and episodic long-term memory, there was no observable distinction between the two cultures, as both groups exhibited an enhanced late positive components response for high-ranking buildings, with this effect extending posteriorly for Chinese participants. It was concluded that the ability to identify architectural hierarchy is influenced by culturally specific expertise in Western-style architecture, particularly during the early stages of processing. Adopting a similar approach, and comparing coloured photographs of noted-architect-designed and ordinary architectures ($N = 18$), master architects' works evoked a reduced P2 amplitude and an increased N2 amplitude, suggesting a positive emotional response and the perception of novelty in the aesthetic evaluation of architecture (Ma et al., 2015). These findings introduce P2 and N2 as discerning neural markers for assessing the delight and novelty aspects expressed by architectural designs.

More recently, contour shapes, a topic well-researched in various disciplines including studies of interior environments, have also been addressed in building design. A study explored the psychological and physiological responses to facades with different window shapes in nine participants (Naghibi Rad et al., 2019). Rectangular, square, circular, and semi-circular arch shapes were perceived as pleasant, whereas triangular and triangular arch shapes were rated as unpleasant. Analysis of ERP components (N1, P2, and P3) revealed higher brain activity for unpleasant window shapes compared to pleasant ones, particularly in frontal (P2 component) and posterio-occipital channels (N1 through P3 peak). The N1 component has been associated with attention to non-spatial object characteristics such as form and colour (Hillyard and Anllo-Vento, 1998), the peak latency of P3 component to stimulus evaluation timing and task processing demands, varying with differences in cognitive capability (Polich, 2007), and the P2 component to feature detection, stimulus processing (Luck et al., 1994), selective attention (Federmeier and Kutas, 2002), and other early sensory stages (Breznitz, 2008). Moreover, there were larger ERP amplitudes in the right hemisphere in response to all stimuli, albeit unpleasant stimuli generated larger ERP amplitudes. P2 amplitudes were more pronounced between frontal and central lobes for unpleasant than pleasant windows. The study, therefore, demonstrated distinguished categorization of pleasant and unpleasant shapes with more significant ERP modulations in the right hemisphere for unpleasant as opposed to pleasant windows, indicating the lateralization of positive emotions towards the left hemisphere and negative emotions towards the right hemisphere.

Building upon the established positive impact of natural scenes, research addressed the influence of viewing height and amount of greenery in visual scenes on the power of alpha oscillations in the frontal cortex using a commercial-grade EEG device (Olszewska-Guizzo et al., 2018b). The study ($N = 33$) revealed that the existence of greenery visible from various floor levels can impact the brainwave patterns, specifically frontal alpha and temporal beta. The results imply that increased greenery increases activation in brainwaves associated with positive emotional states, motivation, and visual attention, with the effects appearing to be more pronounced on higher compared to lower floors.

In summary, EEG studies exploring images and videos of outdoor environments have elucidated many aspects of the relationship between building design and neural responses. Comparing videos of environments with varying abstraction degrees, distinct brain responses were observed while navigating abstract as compared with semi- or fully-designed environments (Hakak et al., 2016). Abstract environments activated the precentral gyrus in the motor network, likely reflecting embodied simulation of the presented environment. In terms of facades, high- compared to low-ranking buildings were found to elicit reduced N400 potentials, indicating a sense of familiarity (Oppenheim et al., 2009). Subsequently, it was shown that high-ranking buildings elicited distinct early (200-400 ms) and late (400-600 ms) ERPs, with a highlighted role of the hippocampus in late-stage visual object categorization (Oppenheim et al., 2010). Further research revealed that cultural expertise influenced early brain responses to high-ranking buildings, with German compared to Chinese participants showing a reduced N350 response (Mecklinger et al., 2014). However, participants from both cultures exhibited similar late positive components, indicating that cultural expertise affects early visual object recognition but not later semantic and episodic processing. Using a similar paradigm to test noted-architect-designed and ordinary architectures, the former elicited a diminished P2 amplitude and an elevated N2 amplitude indicating favourable emotional responses and novelty perception, thus identifying P2 and N2 as neural markers for assessing delight and novelty in architectural design (Ma et al., 2015). Moreover, window shapes on building facades seem to impact psychological and neurophysiological responses, with unpleasant (triangular) shapes generating higher brain activity particularly in frontal and posterio-occipital regions, and larger ERP amplitudes in the right hemisphere, indicating that positive emotions are lateralized to the left hemisphere and negative emotions to the right (Naghibi Rad et al., 2019). Finally, greater visibility of greenery from various floor levels was associated with increased frontal alpha and temporal beta oscillations, enhancing positive emotional states, motivation, and visual attention (Olszewska-Guizzo et al., 2018b).

fMRI

One of the earliest and most common uses of facades in fMRI studies was to investigate visual perception and face/object processing (Kanwisher et al., 1997, Haxby et al., 2001, Keizer et al., 2008, Summerfield et al., 2006, Veuilleumier et al., 2001). Researchers used images of various objects, faces, and houses as stimuli to study how different brain regions respond to specific visual features. These studies helped elucidate the neural networks involved in recognizing different types of visual information in the ventral visual cortex. However, faces and buildings were later used to investigate whether expertise in architecture influences neural activity in regions associated with perception, memory, or reward (Kirk et al., 2009). Both the 11 experts and 13 non-experts exhibited similar behaviour when making aesthetic judgments. However, they displayed differing patterns of neural activation in the bilateral medial orbitofrontal cortex (mOFC) and subcallosal cingulate gyrus (SCG) during aesthetic judgment. On the other hand, unlike the mOFC and SCG, the nucleus accumbens (NAcc) showed a typical response in both subject groups and to both stimuli types, suggesting that these brain areas play distinct roles in reward processing. Interestingly, expertise-related effects were observed in regions associated with memory, such as the hippocampus and precuneus, when participants provided categorical responses, irrespective of their aesthetic ratings. This implies that experts might integrate current input with prior knowledge, using this integration to inform their aesthetic judgments. Overall, findings suggest that expertise influenced both cognitive processing and responses in brain areas related to reward.

In a more recent exploration of the neural representation of architectural styles, 22 architecture and psychology students viewed images of various categories (buildings with different styles, famous architects' building designs, entry-level scene categories, and faces of non-famous men) while performing a one-back task (Choo et al., 2017). Each participant underwent functional localization of several visually active regions of interest (ROIs), including the para-hippocampal place area (PPA), occipital place area (OPA), retrosplenial cortex (RSC), lateral occipital complex (LOC), and fusiform face area (FFA). Findings indicated that specific patterns of neural activity linked to different architectural styles were found in high-level visual brain regions, excluding the primary visual cortex (V1), suggesting complex visual processing beyond the basic features computed in V1. Remarkably, the FFA, typically associated with preferential processing of faces (Kanwisher et al., 1997), was involved in representing architectural styles, alongside several scene-selective regions. Hierarchical clustering of error patterns also indicated that the involvement of the FFA was significantly greater in the neural encoding of architectural styles compared to entry-level scene categories. The study concluded that the FFA plays a crucial role in the detailed neural representation of scenes at a subordinate level, specifically buildings' architectural styles. Notably, there were no observable differences in neural coding for architecture between experts and non-experts within the visual cortex.

While the above-mentioned studies focused on facades, other research used images of exteriors and interiors to study the effects of building design. Healthcare settings were explored to determine whether exposure to sharp versus curved contours triggered specific amygdala activation patterns (Pati et al., 2016), extending prior research on everyday objects (Bar & Neta, 2007). Thirty-six participants were presented with greyscale images varying in contours (sharp, curved, balanced) and content (healthcare objects, interiors, exteriors, landscapes). Responses on the rating task were in line with previous findings, demonstrating a preference for curved over edgy stimuli. However, fMRI data revealed distinct activation patterns based on image content. Sharp conditions in object and landscape categories led to increased bilateral amygdala activation, indicating a potential fear response. Interestingly, curved conditions in exterior and interior categories also showed significantly increased bilateral amygdala activation, indicating a unique precognitive response to contours and content in healthcare settings.

The brain correlates underlying internal states elicited by architecture designed for contemplative experiences were investigated in another study which tested 12 experienced male architects (Bermudez et al., 2017). Viewing images of contemplative compared to ordinary buildings induced attentive, receptive, and absorbing experiences while reducing internal dialogue. These experiences correlated with reduced PFC involvement, which is commonly found to be activated in internally-induced meditative states (Baron Short et al., 2010). This suggests that higher cognitive and executive functions, although important, are not the main focus in contemplative experiences triggered by architecture. Additionally, activations were observed in the occipital lobe, precentral gyrus, and inferior parietal lobule, regions responsible for visual processing, motor control, and sensory information interpretation. The engagement of sensory-motor and premotor regions highlights the importance of embodiment and sensory integration in these experiences. Conversely, ordinary buildings engaged executive control and attention areas such as the medial PFC, along with fundamental motor regions, as evidenced by activations in the left and right cerebellum. Findings suggest that observing buildings designed for contemplation consistently triggered unique experiential and neural patterns compared to everyday buildings. The depth of these externally induced states inversely correlated with the engagement of the default mode network (DMN), which is typically active during self-referential thinking and mind-wandering. This implies that contemplative architecture may induce a meditative state by reducing the need for self-regulation and executive attention, thereby enhancing aesthetic and emotional engagement.

Educational settings were the focus of a more recent study which explored whether pleasant and unpleasant places can stimulate brain regions engaged in emotions in 32 architecture students (Khaleghimoghaddam et al., 2022). Findings revealed distinct patterns in brain activations associated with pleasant and unpleasant places. Pleasant places stimulated emotion-related brain regions more than unpleasant ones. Notably, men exhibited more activations in response to both pleasant and unpleasant places compared to women, although the sample included more men than women. An activation of the left and right amygdala was observed for unpleasant places, indicating negative emotions. Additionally, the study revealed increased blood flow in the frontal lobe, occipital lobe, fusiform, and middle occipital lobe during architectural experiences, correlating with changes in affect. Moreover, a notable rise in brain activity occurred during the induction of pleasant moods in various brain regions, including the parietal lobe, precuneus, cerebellum, lingual gyrus, medial frontal, anterior cingulate, angular gyrus, and middle occipital cortex. Conversely, when observing unpleasant places, increased activity was observed in regions such as the right and left cerebrum, cingulate, middle temporal, left hippocampus, thalamus, insula, and left hippocampus. The authors concluded that architectural settings may impact emotional, perceptual, cognitive, aesthetic judgment, and decision-making processes, emphasizing the importance of attractive elements in architectural design for creating pleasant environments.

To summarize, several fMRI studies examined images of buildings, providing insights into how building design influences neural responses. Earlier investigations revealed that expertise in architecture impacts neural response, modulating activity in reward-related regions such as the mOFC and SCG, while the NAcc responded similarly in both groups, suggesting different roles of these regions in reward processing (Kirk et al., 2009). Moreover, there were expertise effects in memory-related areas such as the hippocampus and precuneus when processing categorical responses, regardless of aesthetic ratings. Exploring architectural styles, it was found that high-level visual brain regions, including the FFA, were involved in representing architectural styles, indicating complex visual processing beyond basic features and highlighting the FFA's crucial role in encoding detailed architectural scenes, although no effect of expertise was observed in the visual cortex (Choo et al., 2017). In the context of healthcare environments, sharp contours increased bilateral amygdala activation in object and landscape categories, suggesting a potential fear response, while curved contours in exterior and exterior categories also showed increased amygdala activation, indicating a precognitive response to contours and content in healthcare settings (Pati et al., 2016). Contemplative compared to ordinary buildings were found to down-regulate the PFC while up-regulating sensorimotor regions, including the premotor, visual areas (occipital lobe), precentral gyrus, and inferior parietal lobule, along with a disruption of the DMN, potentially inducing a meditative state by reducing the need for self-regulation and executive attention (Bermudez et al., 2017). Evaluating architectural experience using positive and negative educational settings, pleasant places activated emotion-related brain regions more than unpleasant ones, with unpleasant places specifically activating the amygdala (Khaleghimoghaddam et al., 2022). Findings highlight the impact of architectural settings on emotions, perception, cognitive processes, aesthetic judgment, and decision-making.

Virtual Environments

EEG

One study explored the neurocognitive effects of "the architecture of vertigo", that is, glass floor platforms, at varying heights from the ground (Erkan, 2021). A total of 110 participants, exposed to scenes of glass floors, showed increased affect scores on glass floors that are 45 m and higher, particularly higher "depression" and "tension" scores at 75 m, with women experiencing greater stress and anxiety than men. GSR measures indicated heightened skin conductivity as glass floor height increased. Changes in HRV were also associated with height, with noticeable peaks after 45 m. Participants at heights between 15 and 30 m had wider focal points compared to other heights, while eye movement speed increased systematically at 45, 60, and 75 m. The effects of alpha waves as indicators of physiological stress were observed in the prefrontal and occipital lobes specifically at the height of 75 m. The study concluded that glass floor built to 15–30 m height is appropriate, while neurocognitive and physiological problems seem to appear at 45 m height and above reducing the perception of the urban textures below.

Other research investigated the neural processes associated with landmark usage and recognition during urban navigation tasks, finding that building shape and height can influence the visual attraction of landmarks (Rounds et al., 2020). Twenty-one participants teleported (while seated) along virtual city routes with salient and non-salient target buildings varying in height, footprint, and facade pattern, while EEG theta band event-related synchronization over posterior areas was evaluated. Recall accuracy, gaze duration, and neural signatures of spatial awareness (posterior lateral EEG theta power) were found to be heightened when gazing at salient buildings, although progression in theta power change was non-linear. There were significant differences in neural activation recorded at left superior posterior regions for buildings with a twist-facade design and nature-containing buildings, but the latter exhibited unexpected results, with a decrease in theta activation instead of the expected increase. Tests for other design features did not yield significant differences, sometimes deviating from gaze duration and recall data, indicating a complex relationship between architectural features and neural responses.

In sum, two EEG studies investigated the neurophysiological responses to virtual building design. The exploration of the neurocognitive effects of glass platforms at varying heights revealed alpha wave changes in the prefrontal and occipital lobes at 75 m, indicating physiological stress (Erkan, 2021). An investigation of landmark usage during navigation revealed enhanced visual attention and memory recall for highly salient buildings, non-linear progression of theta activation in the left posterior superior region related to wayfinding, and complex neural responses to different architectural features, which sometimes deviated from gaze duration and recall data (Rounds et al., 2020).

Naturalistic Environments

EEG

One study explored the psychological and physiological relaxation effects of viewing green versus concrete facades in a sample of 25 Chinese female students (Elsadek et al., 2019). Exposure to the green facade led to higher subjective ratings, along with a significant decrease in negative mood scores and an increase in positive mood. EEG data (collected via a commercial-grade device) revealed increased relative alpha wave power in the occipital lobe when viewing the green facade, indicating enhanced emotional responses, attention, and visual perception. In the prefrontal lobe, relative alpha power was lower when viewing the concrete facade, indicating lower activity in regions that control overall brain function. Physiological data indicated increased parasympathetic activity and decreased sympathetic activity during green facade exposure, as well as a decrease in SCR. It was concluded that green facades enhance relaxation compared with concrete walls.

Urban Design

Urban design operates at a larger scale, involving the planning and organization of entire urban areas or communities. It considers the layout of streets, public spaces, parks, transportation systems, and infrastructure within a city or town. To note that most of the neuroscientific research on urban design often focused on comparing urban and natural environments, frequently overlooking the potential beneficial effects of specific urban elements.

Images and Videos

EEG

One study examined various urban environments (urban green, urban outdoor, urban indoor) with different social density levels (crowded vs. uncrowded) and reported the impact of active walking versus standing on subjective experiences and neurophysiological responses of 42 participants (Mavros et al., 2022). Consistent with previous reports, green spaces were perceived as calmer and more positive, and reduced attentional demands, while outdoor spaces were perceived more positively than indoor ones. Crowded spaces elicited higher arousal, more negative valence, and increased cognitive and attentional demands. Interestingly, EEG data of walking participants revealed that these environments elicited higher frontal alpha asymmetry, indicating positive emotional responses. Overall, EEG data produced a comparable signal-to-noise ratio between walking and standing, and skin-conductance also measured effectively psychophysiological responses. It was concluded that emotional reactions to visual stimuli can be efficiently assessed using mobile EEG and EDA in mobile settings, revealing intricate interactions between walking, social density, and direct and indirect affective responses.

Comparing the restorative effects of natural and urban scenes, a commercial-grade EEG device was used to record the neurophysiological responses of 20 participants (Roe et al., 2013). Ratings showed that landscapes compared with urban scenes were evaluated more positively. EEG data analysis revealed two key components: arousal, which was linked to urban scenes, and interest, which was associated with landscape scenes. Latent class analysis aimed to identify clusters in the data, with a two-cluster model showing the best fit. The clusters were distinguished by image scene and three emotional EEG parameters: excitement, long-term excitement, and meditation. Landscape scenes were associated with higher meditation levels and lower arousal (excitement), while urban scenes were linked to higher arousal. These findings underscore the restorative effects of natural environments, particularly the inclusion of green spaces in urban settings. More recently, a study (N = 32) tested the attention restoration theory (ART) (Kaplan 1985; Kaplan & Kaplan, 1898) and explored the brain's engagement in processing of 830 images of natural versus urban scenery (Grassini et al., 2019). Subjectively, images of natural scenes were found more relaxing than those of urban ones. The cognitive task did not reveal significant differences. However, ERP analysis demonstrated more pronounced P1 wave and more weakened N1, EPN, and P300 waves elicited by images of natural compared with urban scenes. Moreover, natural scenery images induced a more pronounced LPP compared to urban images. In terms of oscillatory activity, natural scenes resulted in a decrease in power, particularly in delta, theta, high alpha, and beta waves and led to an increase in power in low alpha waves. It was concluded that the cognitive and attentional processing demands are lower when perceiving natural environments in contrast to urban ones. Another study tested the physiological dynamics of the interaction with videos of natural, built and neutral settings in 24 participants using subjective, cognitive, and neurophysiological measures (Grassini et al., 2022). Relaxation ratings were the highest for the natural videos, although ratings did not differ between neutral and urban environments. There were no significant differences detecting nature-induced restoration in cognitive performance or EDA responses. EEG results showed that the low alpha spectrum over occipital-temporal and central sensors differed among the three conditions. Videos featuring natural environments increased early alpha waves compared to neutral videos, particularly over occipital-temporal areas, and in comparison to urban videos, particularly over central-frontal areas. The comparison of neutral and urban scenes did not produce any significant differences, suggesting that early alpha effects were driven by the nature and not the urban condition. The results indicate that the exposure to natural environments generates detectable and consistent brain activity associated with restorative processes.

But do landscapes differ in their positive effects? An investigation (N = 32) examined potential differences in brain activity patterns between the processing of landscapes identified as contemplative (CL) and non-contemplative landscapes (NCL) using a commercial-grade EEG device (Olszewska-Guizzo et al., 2018a). No significant differences were observed in frontal alpha asymmetry or frontal beta power between CL and NCL. However, participants showed higher beta power in the right temporal brain regions when viewing CL compared to NCL, indicating that CL captured more visual attention, supporting the ART (Kaplan & Kaplan, 1989) concept of attention-switching systems and suggesting a stress reduction mechanism.

In sum, EEG research used images and videos to examine urban and natural environments. Investigating the impact of different urban environments and social densities during active walking and standing revealed that EEG and EDA effectively captured psychophysiological responses, indicating their suitability in ambulatory settings investigations (Mavros et al., 2022). Focusing on urban versus natural environments, research demonstrated the restorative effects of natural landscapes, with EEG data showing that natural scenes were associated with lower arousal and higher meditation levels, while urban scenes were linked to higher arousal (Roe et al., 2013). In a more recent study, ERPs indicated that natural scenes required less attentional effort when compared with urban scenes, while spectral analysis revealed variations in brain activity and cognitive demand between the two types of environments, concluding that natural environments recruit less cognitive and attentional demands (Grassini et al., 2019). When testing dynamic stimuli, nature scenes produced higher alpha waves, particularly over the central brain, suggesting that such experiences generate measurable brain activity linked to restorative processes (Grassini et al., 2022). Finally, CLs compared to NCLs were found to increase beta power in the right temporal brain regions, indicating enhanced visual attention and potential stress reduction (Olszewska-Guizzo et al., 2018a).

fMRI

Urban scenes were also compared with natural ones in several fMRI investigations. Examining the effects of restorative environments, 24 participants viewed photos with different restorative potential (Martínez-Soto et al., 2013). The high restorative potential (HRP) photos depicted natural environments, while the low restorative potential (LRP) ones showed built environments without nature. When observing HRP environments, specific brain regions, including the middle frontal gyrus, middle and inferior temporal gyrus, insula, inferior parietal lobe, and cuneus, were notably activated. In contrast, viewing LRP settings led to dominant activation in the superior frontal gyrus, precuneus, parahippocampal gyrus, and posterior cingulate. Consistent with the ART (Kaplan, 1995), the authors concluded that HRP environments triggered involuntary, bottom-up attention-related brain regions, while LRP environments engaged areas associated with top-down, directed attention. Another study ($N = 31$) compared the restorative value of four types of environments: urban, mountain, forest, and water scenes (Tang et al., 2017). Urban landscapes were perceived as less restorative than natural landscapes, with water and mountain scenes rated as the most attention-restorative, followed by forests and urban environments. Analysis of brain activity showed that the left and right cuneus, primarily associated with basic visual processing, were particularly involved in differentiating between urban and mountain landscapes. There were no specific brain regions identified for the contrast between urban and forest landscapes. When comparing urban and water landscapes, activations were observed in the left and right cuneus (related to visual processing), as well as the right cingulate gyrus and left precuneus influencing attention focus (Leech & Sharp, 2014). Findings demonstrated that viewing urban as opposed to natural landscapes, particularly water and mountain scenes, increased activations in the visual and attentional brain areas, suggesting that observing natural landscapes might help the attention system to rest. Others attempted to discriminate between brain activation patterns associated with emotional responses (comfortability) to human residences in 13 participants (Kim & Jeong, 2014). An expressed comfort was observed when viewing rural and natural residential environments, contrasting with discomfort felt when observing urban residential scenes. While exposed to comfortable landscapes, notable activations occurred specifically in the calcarine gyrus, associated with visual processing. Conversely, exposure to the uncomfortable scenery resulted in substantial activation in the ventrolateral prefrontal cortex and ACC, associated with cognitive control, as well as the medial prefrontal cortex, hippocampus, parahippocampal gyrus, amygdala, and insula, all of which play roles in emotion regulation and expression. While the above-mentioned studies examined brain function, a more recent investigation assessed differences in whole-brain functional connectivity states while 24 participants passively viewed images representing natural and built environments, equally rated on scenicness (Kühn et al., 2021). No discernible distinctions were found in self-reported measures of perceived stress, rumination, valence, arousal, or dominance. Functional connectivity was notably greater when participants viewed images of natural landscapes compared to built environments. This contrast was evident in circuits involving dorsal attention network (DAN) and ventral attention network (VAN), DAN and DMN, and DMN and somatomotor connections. Additionally, reduced functional connectivity during exposure to natural environments was associated with a higher number of years spent growing up in major cities.

Focusing on green areas within urban environments, and in order to reveal how green content influences brain activity, 44 participants were exposed to images with varying green-space densities (Chang et al., 2021). Viewing green urban landscapes elicited activations in a network of regions including superior parietal lobule (SPL) and superior temporal gyrus (STG), associated with spatial processing, posterior cingulate cortex (PCC), SPL, and medial frontal gyrus (mFG), involved in spatial and executive attention, as well as the cuneus, related to early sensory encoding. While cuneus activity reflects visual differences, the engagement of attention-related regions like the mFG supports the idea that natural landscapes aid attention restoration by reducing attentional demands. Importantly, responses in the ventral posterior cingulate cortex (vPCC) and cuneus varied with green-space densities after accounting for low-level features, while only responses of the vPCC correlated with self-reported stress measures. The authors suggest a role of the vPCC in regulating stress responses, potentially through connections to the prefrontal and hippocampal regions, which project towards the neuroendocrine system. Given the PCC's implication in many neurological diseases and disorders, the findings highlight the therapeutic potential of natural environments for neurological health and urban planning. Another study examined the neural basis of perceiving and appreciating natural landscapes and landscape gardens (Zhang et al., 2019). During scanning, 16 participants performed a scene judgement task with the original photographs, and a colour preference judgement task in lower definition versions of the photographs (baseline condition). Subtracting baseline activations, landscape gardens elicited increased activation in various brain regions, including bilateral middle temporal gyri, right superior temporal gyrus, left superior frontal gyrus, right middle cingulate, left medial orbitofrontal cortex (mOFC), and left precuneus, indicating a rewarding perception. The perception and appreciation of these gardens likely involved neural areas commonly associated with the processing and enjoyment of various visual stimuli, reflecting a coordinated engagement of visual perceptual processing, cognitive processing, and emotional reward, which supports the framework of the aesthetic triad linking artistic and architectural experiences to sensorimotor, knowledge-meaning, and emotion-valuation systems (Coburn et al., 2017). This pattern extended to the appreciation of natural landscapes, which activated the left rolandic operculum, the right ACC expanding to the right mOFC, and the left precuneus. Additionally, conjunction analysis between landscape gardens and natural landscapes highlighted shared brain regions involved in the perception and appreciation of both types of environments, emphasizing attentional orienting region, ventral visual system, and circuits associated with appreciation and reward (precuneus and mOFC). However, contrasting landscape gardens and natural landscapes revealed increased activations in the inferior occipital lobe, the left SPL, the right fusiform gyrus, the right hippocampus, the right cuneus, and supplementary motor area when viewing landscape garden photos. These regions, which seem to play a key role in the appreciation of landscape gardens, suggest involvement in various processing layers, including global configuration perception, architectural style recognition, rewarding emotional experiences, and embodied motivation to approach the structure. Whereas most fMRI research has been focused on static stimuli, a study ($N = 22$) explored the neural mechanisms underlying aesthetic judgement of static versus dynamic landscapes (Zhao et al., 2020). Brain regions such as occipital lobe, frontal lobe, supplementary motor area, cingulate cortex, and insula were activated in the aesthetic judgement of both static and dynamic stimuli. However, dynamic landscapes evoked stronger activations in the middle temporal gyrus (MT/V5) and hippocampus, suggesting greater beauty in dynamic stimuli, as evidenced by the activation of visual processing and emotion-related regions.

In sum, studies on restorative environments found that natural (HRP) and urban (LRP) scenes activated different neural pathways, as the former evoked involuntary, bottom-up attention-related brain regions, while the latter involved areas associated with top-down, directed attention, in line with the ART (Martínez-Soto et al., 2013). Another study found that the differentiation between urban and natural scenes involved the cuneus, related to visual processing, as well as the precuneus and right cingulate gyrus, linked with attention, suggesting that urban environments recruit more attentional demands (Tang et al., 2017). Viewing rural and natural residential environments, perceived as comfortable, activated the calcarine gyrus associated with visual processing, while urban scenes, perceived as uncomfortable, activated the ventrolateral prefrontal cortex and ACC related to cognitive control, along with regions involved in emotion regulation such as the medial prefrontal cortex, hippocampus, parahippocampal gyrus, amygdala, and insula (Kim & Jeong, 2014). Whole-brain functional connectivity studies showed greater connectivity during exposure to natural compared to urban landscapes, implicating DAN and VAN, DAN and DMN, and DMN and somatomotor connections, with reduced connectivity during exposure to natural conditions linked to more years spent growing up in major cities (Kühn et al., 2021). When focusing on natural environments, viewing green urban landscapes with varying green density elicited changes in the vPCC activity, correlating with stress-related responses and influencing attentional and executive brain regions (Chang et al., 2021). This suggests that the vPCC regulates nature dose-dependent stress changes, potentially through connections to the prefrontal and hippocampal region, highlighting the therapeutic potential of natural environments. Examining the neural basis of perceiving natural landscapes and landscape gardens revealed that both activated brain regions associated with perceptual, cognitive, and reward processes, while distinct neural patterns emerged in appreciating landscape gardens, encompassing the inferior occipital lobe, the left SPL, the right fusiform gyrus, the right cuneus, and the right hippocampus (Zhang et al., 2019). In contrast with static landscape stimuli, which activated occipital and frontal lobes, supplementary motor area, cingulate cortex, and insula, dynamic landscapes evoked stronger activations and engaged regions related to visual processing and emotion, such as the middle temporal gyrus and hippocampus, suggesting greater aesthetic appeal (Zhao et al., 2020).

NIRS

To measure the effect of different built urban environments on PFC activation, specifically oxyhemoglobin (HbO) and deoxyhemoglobin (HbR), a pilot study used fNIRS and exposed five participants to videos of urban environments from a pedestrian's perspective (Brunner et al., 2023). Viewing pleasant walking environments, characterized by varied facades, ground-level use, greenery, low traffic volumes, and the availability of spatial elements, was correlated with a decrease in the hemodynamic response in the right and central areas of the PFC, indicating physiological relaxation. In contrast, observing less pleasant walking environments, characterized by monotonous and closed facades, no windows at ground level, no spatial elements or greenery, and high traffic volume led to a slight increase in the hemodynamic response.

Many studies used NIRS methodologies to compare urban and natural environments. In a study investigating the effects of urban and garden visuals, oxygenated haemoglobin levels (O_2Hb) reflecting higher brain activity were monitored in 18 participants (Lee, 2017). Self-reports revealed significant differences in mood states, perceived naturalness, and comfort, indicating higher negative mood states and anxiety in the city. Gender differences were observed, as men showed varying tension–anxiety scores between the city and garden, while women generally reported lower negative mood states in the city landscape. NIRS data showed increased mean O_2Hb levels in the city and decreased levels in the garden in both left and right PFC, with men showing more pronounced responses, especially in the city environment. However, there were no significant differences in blood pressure and pulse rate between the two landscapes. Another pilot examination ($N = 7$) of the prefrontal cortical activation related to pleasant and unpleasant emotions in response to urban versus garden videos using fNIRS showed a significant increase of HbO over the right PFC (Yu et al., 2017). However, no significant activation was observed at the group level for the contrasts between urban versus idle and garden versus idle (idle scene showing a dark blue progress bar on black background). The authors concluded that garden scenes may offer more pleasant and less stressful effects than urban scenes. Aiming to identify the specific regions involved in the mood-improving effect of viewing images of natural compared to built environments, a study ($N = 30$) focused on the orbitofrontal cortex (OFC) and dorsolateral prefrontal cortex (dlPFC), two brain areas linked to affective processing and control (Yamashita et al., 2021). Mood measures revealed greater feelings of comfortableness and relaxation (but not vigor) after viewing images of nature compared to city. Moreover, fNIRS data indicated a significantly lower concentration of oxyhaemoglobin (oxy-Hb) in the right OFC when viewing images of nature compared to city scenes, while there were no significant differences in the other brain regions analysed. However, heart rate measures showed no significant differences. The study, therefore, confirmed the mood-improving effect of viewing images of nature by reducing the activity of the OFC, suggesting therapeutic potentials. Others used fNIRS to examine the haemodynamic activation patterns of the prefrontal and occipital cortices while 12 participants passively viewed videos of three urban sites (urban park, neighbourhood landscape, and city centre) at two different time points, before the Covid-19 pandemic and after lockdown was over (Olszewska-Guizzo et al., 2021). Results indicated a significant decline in average Oxy-Hb levels over time in response to all visual conditions, with a notable decline in the visual cortex for green spaces (urban parks and neighbourhood landscapes) and a marginal decrease in both the visual cortex and the frontal eye fields for the city centre, which lacks green elements. It was suggested that the lockdown experienced by urban residents may have contributed to reduced brain haemodynamics, potentially increasing the risk of mental health issues and cognitive decline. Additionally, the haemodynamic response observed in the busy city centre scenes was associated with stress and anxiety, contrasting with the absence of such effects in urban green scenes, highlighting the necessity of green spaces to mitigate the adverse neuropsychological impact of urban living post-pandemic.

To recapitulate, NIRS was employed to study the responses to images and videos of outdoor stimuli finding that urban and natural environments elicit distinct brain activations. Exploring diverse urban environments' impact on the PFC, pleasant walking areas correlated with reduced haemodynamic response in the right and central PFC, indicating relaxation, while less pleasant environments caused a slight increase in these responses (Brunner et al., 2023). Comparing urban and natural settings revealed increased O_2Hb levels in the left and right PFC in the city, indicating higher brain activity and negative mood states, with pronounced responses in men, while garden visuals decreased these levels, suggesting a calming effect (Lee, 2017). Research with dynamic stimuli, on the other hand, noted increased HbO over the right PFC for the contrast of urban versus garden videos, suggesting less stressful and more pleasant effects of garden scenes (Yu et al., 2017). Focusing on the OFC and dlPFC, the mood-improving effect of nature images was confirmed by reduced activity in the right OFC, evidenced by a lower concentration of oxy-Hb, while no significant differences were observed in other brain regions (Yamashita et al., 2021). Finally, measuring hemodynamic patterns in response to urban and natural scenes before and after the Covid-19 lockdown, a decline in Oxy-Hb levels over time was observed in the visual cortex for green spaces and marginally in both the visual cortex and frontal eye fields for the city centre, indicating the negative effect of the lockdown (A. Olszewska-Guizzo et al., 2021). Additionally, haemodynamic patterns associated with stress and anxiety appeared in response to the urban scenes but not in scenes with natural elements, implying the necessity of green spaces to counteract the negative neuropsychological effects of urban living post-pandemic.

Virtual Environments

EEG

Developing a data-driven approach for evaluating built environment quality based on real-time emotional responses, eight participants were exposed to a virtual urban environment before and after redesign according to six criteria: building types, natural features, street characteristics, architecture features, front porch, and building height (Hu & Roberts, 2020). The post-design environment received higher scores across all criteria, with architectural features density, and land-use mix being the top-rated attributes. While questionnaire results showed a unanimous preference for the post-design setting, EEG data, collected via a commercial-grade device, revealed diverse emotional reactions, ranging from negative and positive to neutral, towards the two built environments. Specifically, the post-design setting evoked greater engagement, interest, and attention, while the pre-design environment induced increased relaxation and moderate stress levels. Further analysis showed that there was a significant distinction between participants' negative and positive emotional responses to the pre- and post-designs and highlight the complexity of measuring the responses to the built environment. To investigate the effects of the comprehensive visual pattern of urban streets on human emotions, a study exposed 26 participants to virtual scenes with a variety of spatial scales: small (two lanes or less), medium (four to six lanes), and large (six lanes or more) conditions using a commercial-grade EEG device and an Empatica wristband (Zhang et al., 2021). The scenes underwent extensive visual pattern auditing by incorporating global landscape metrics that represent scenes' spatial configuration and colour composition. Building models to analyze the impact of selected environmental features on emotional responses, the study expected consistency between psychological and physiological measures. Findings revealed that the proportion of vegetation positively impacted emotional responses, including subjective comfort and vitality. Furthermore, global visual pattern also affected emotions. More specifically, more diverse and less fragmented scenes were associated with positive emotions, with more sensitivity towards configuration than colour composition. Of the five EEG components measured, only excitement and interest showed significance which indicate physiological arousal and valence of the emotion. The study controlled for gender, finding that women exhibited higher interest and lower excitement compared with men. SCR models showed statistically significant results with male compared to female participants exhibiting higher levels of SCRs and SCR amplitude, and an increased proportion of roads positively affecting SCR amplitude, whereas HR models were not statistically significant. Unlike expected, physiological measures were uncorrelated with each other and showed limited correlation with subjective measures, highlighting the need for incorporating qualitative alongside quantitative analyses.

Spatial navigation research also addressed wayfinding and brain responses in virtual environments. Investigating the influence of landmarks on maze-like virtual navigation, 35 participants engaged in navigational tasks in landmark-enriched (LM+) or landmark-devoid (LM-) conditions (Sharma et al., 2017). The LM+ condition resulted in better and more efficient performance, accompanied by increased left hemisphere brain activity. Theta frequency band event-related desynchronization/synchronization indicated activation in the left posterior inferior and superior regions in the LM+ condition. This suggests object-location binding and spatial transformation between representations in this condition. Information flow analysis between brain regions showed that the LM− condition required higher cognitive demands, as evidenced by more active channels. The study highlighted the role of the left-hemispheric region, particularly the parietal cortex, in integrating sensory cues and memory to encode landmark-related contextual information during maze navigation.

In sum, three EEG studies explored virtual urban environments, shedding light on the various effects of design parameters on the human brain. Some researchers developed a data-driven approach for evaluating built environments based on real-time emotional responses and comparing two scenarios before and after redesign, showing diverse emotional responses in EEG data despite a clear subjective preference for the post- compared with pre-design scenario, and highlighting the complexity of measuring physiological responses to the built environment (Hu & Roberts, 2020). Testing the impact of visual patterns of urban streets on emotions, the proportion of vegetation, as well as the fragmentation and diversity of scenes were found to positively affect emotional responses, with inconsistencies observed among the physiological and subjective measures unlike expected, indicating the need for both qualitative and quantitative approaches (Zhang et al., 2021). Finally, spatial navigation research found that landmarks facilitated spatial navigation, with

EEG analyses showing increased left-hemispheric activation, particularly in the parietal cortex, along with specific theta band patterns suggesting enhanced object-location binding and spatial transformation processes in LM+ environments (Sharma et al., 2017).

Naturalistic Environments

EEG

When studying naturalistic urban environments, studies have mainly used commercial-grade, mobile EEG devices to uncover how the brain responds to urban and natural environments. In a study where 10 participants navigated between various city buildings choosing from multiple routes, key moments of their journey, such as moments of decision-making, were correlated with recordings of specific brain waves (Karandinou & Turner, 2017). Critical way-finding decisions were marked by short, intense beta activity peaks, supporting link between beta waves and active decision-making. Major route decisions showed more intense beta activity, while minor decisions showed less intensity. Interactions with people led to increased beta activity, often accompanied by transient rises in theta/alpha range activities, suggesting cognitive and memory performance (Klimesch, 1999). When crossing streets, beta activity decreased at red lights, suggesting a lower level of cognitive engagement, whereas it remained stable at crossings without stop-lights, reflecting the need for heightened attentiveness. Familiar streets generally showed lower beta levels, suggesting reduced cognitive load, while new or unexpected views sparked curiosity increasing beta levels and indicating heightened cognitive functions. In summary, the research underscored the complexity of analysing outdoor urban spaces and the potential for interdisciplinary collaboration to inform design practices.

Comparing urban and natural environments, a pioneering study investigated the emotional impact of real-life urban experiences during walking (Aspinall et al., 2015). Twelve participants took a 25-min walk through an urban shopping street, a green space, and a street in a busy commercial district. The study observed a decrease in frustration, engagement, and arousal, along with an increase in meditation upon entering the green space area, as well as heightened engagement upon exiting it. In line with the restorative theories (Kaplan & Kaplan, 1989; Ulrich et al., 1991), the findings suggest that urban green spaces, in particular, have the potential to reduce stress and promote a meditative state. Subsequently, researchers examined how different urban environments influence mental activity (Al-barrak et al., 2017). Forty female participants visited a café, a supermarket, and a garden for five minutes. Statistical analysis and pattern recognition methodologies were used to automatically recognize the influence of the different places and the associated reported mental states recorded on the frontal lobe (using Neuroplace application). Results revealed that the meditation component showed higher levels in the garden and café compared with the supermarket although the difference between the café and the garden was not statistically significant. Moreover, alpha band waves were observed in these places, while beta band, associated with mental activity, were observed in the supermarket. The findings underline how where we are within our built environment can influence how we feel.

Further research targeted older adults ($N = 90$) and their response to changing urban environment during a 15-min walk from/to urban busy streets (Neale et al., 2017). Engagement levels were higher in the urban green space than in the urban busy and quiet spaces. Conversely, levels of excitement were higher in the urban busy environment. These effects were consistent regardless of the sequence in which participants experienced these environments. The urban green space seemed to exert a restorative effect on this group of older adults as well, further supporting the idea that specific settings can positively impact cognitive and emotional outcomes. A subset of 43 participants were included in another investigation which explored how a 15-min walk from an urban busy street to an urban green street, or vice-versa, affects the moods of older adults adding qualitative methods (Tilley et al., 2017). The study found that participants underwent fluctuations in excitement, engagement, and frustration levels during transitions from a busy urban environment to an urban green space, reflected in both EEG and self-reported data, however with some discrepancies potentially due to the undefined terms in the software. Participants preferred the urban green space over the busy urban area, with factors like flowers and natural elements contributing to its calming effect. The busy urban area was associated with negative feelings due to litter and the need for increased vigilance to avoid obstacles, resulting in a higher cognitive load and potentially more negative mood states. Social interactions with others in these environments also influenced mood, with encounters leading to more positive feelings. Familiarity with the route played a role, with previous positive experiences leading to better moods during the walk. The study emphasized the importance of clearly defining terms in mood measurement tools and recommended integrating quantitative EEG data with qualitative interviews to fully grasp participants' experiences.

More recently, a study investigated the neurocognitive status of 210 depressed and healthy young adults during a 20-min walk in typical or green urban environments (Erkan, 2023). Both groups experienced a reduction in blood pressure following the walk in the urban green environment, while the depressed group exhibited a notable rise in diastolic blood pressure, particularly before and after walking in the typical urban environment. This suggests that the depressed group did not benefit from the walking activity in the typical urban setting. Eye-tracking data showed that the average total gaze time in urban green environments was longer than in the typical one suggesting longer dwelling times. EEG results revealed increased high alpha average values of both groups after the walk in the urban green environment, indicating relaxing effects. Conversely, walking in the typical urban environment was characterized by lower alpha wave activity, suggesting that participants were under stress. On the other hand, high beta activity of the healthy group showed higher activity, and therefore more attention during the walk in the green compared with typical urban environment, while participants with high depression scores showed high beta averages in both environments. Lower alpha power was detected in the left temporal regions and higher beta power in the right temporal regions of both groups. The study concluded that green urban fabric has a positive effect on both groups. The depressed group showed much more positive results in the urban green than in the typical urban environment. This highlights the important role of green spaces in influencing visual and cognitive responses, suggesting that abundant green areas within typical urban environments have relaxation and healing effects.

As with other presentation types, research with naturalistic environments indicated that not all green areas share the same contemplative character. Twenty-two participants passively viewed scenes from a park, a neighbourhood green space, and a busy urban street (as control condition), selected based on contemplative landscape scores (Olszewska-Guizzo et al., 2020). The study measured frontal alpha asymmetry (FAA), linked to approach-related motivation and positive emotions, revealing stronger FAA in parks than the control condition, suggesting a more positive mood. Additionally, there was a trend suggesting that even within the same location, FAA varied depending on the characteristic of a particular scene. The study did not observe improved self-reported mood between before and after the exposures in both green urban sites but decreased mood in the busy urban street site. These findings underline the dynamics between urban green spaces, cognitive states, and emotions.

In summary, EEG research examined the effects of walking in naturalistic urban settings, revealing distinct patterns in response to different environments. Exploring neural dynamics of walking in urban areas, beta activity, which suggests higher cognitive engagement, increased during critical decision points and social interactions, with the latter accompanied by transient rises in theta/alpha activities indicating cognitive and memory performance (Karandinou & Turner, 2017). Additionally, there were lower beta levels in familiar streets and at red lights. Comparing urban and natural environments, decreased frustration, engagement, and arousal, along with increased meditation, were observed in green spaces, aligning with the restorative theories, while heightened engagement was noted upon exiting these areas (Aspinall et al., 2015). Others demonstrated how different outdoor built environments induce varying levels of mental activity, showing higher meditation levels and alpha waves in the garden and café, and increased beta waves, indicating mental activity, in the supermarket, emphasizing the influence of physical surroundings on neural states (Al-barrak et al., 2017). Studies involving older adults underscored the restorative effects of urban green spaces on cognitive and emotional outcomes (Neale et al., 2017; Tilley et al., 2017). Older adults exhibited higher engagement levels and restorative effects in urban green spaces, while urban busy environments elicited higher excitement levels, regardless of the sequence of exposure (Neale et al., 2017). Similarly, incorporating qualitative measures in a subset of participants, another study found that transitions from urban busy streets to green spaces caused fluctuations in excitement, engagement, and frustration in older adults, with interview data reflecting these changes and further showing that urban green spaces were generally preferred due to their calming elements and lower cognitive load (Tilley et al., 2017). The study emphasized the need for combining EEG with qualitative data. Research on depressed and healthy young adults revealed the positive effects of green environments on physiological responses of both groups, showing increased relaxing alpha waves, while typical urban settings induced stress and recruited more attentional mechanisms as evidenced by the lower alpha activity and higher beta activity (Erkan, 2023). Lower alpha power was found in the left temporal regions and higher beta power in the right temporal regions for both groups and depressed participants seemed to yield much more positive data, especially in the urban green environment, highlighting the importance of incorporating green areas within typical urban environments. In the examination of natural scenes, it was found that the exposure to green spaces led to higher frontal asymmetry, indicative of positive emotions and approach-related motivation (Olszewska-Guizzo et al., 2020). Parks, in particular, promoted positive moods, emphasizing the nuanced effects of different green environments on neural states.

fMRI

On the urban scale, an innovative approach was used by a study which examined the association between neighbourhood street network characteristics and brain activity during stress processing (Dimitrov-Discher et al., 2023). The Space Syntax methodology (Hillier & Hanson, 1984) was employed to analyze the address coordinates of 42 participants. This analysis aimed to assess street network characteristics, including global integration, local integration, connectivity, and normalized angular choice (NACH) at two scales: the neighbourhood (within 1500 m from address) and the point address (the closest street to address). Results showed that higher levels of global and local integration (better integrated streets) were associated with reduced brain activity during stress provocation. This effect was observed in brain regions related to detecting salient stimuli and threats, including dorsal anterior cingulate cortex, insula, and thalamus. However, there was no such association found for point address configurations, connectivity, or NACH. The findings highlight the utility of Space Syntax in analysing urban street networks in neuro-urbanistic research and suggest the importance of optimizing street layouts for better urban mental health outcomes.

So far, only two fMRI intervention studies investigated how exposure to naturalistic urban and natural environments influences the human brain (Bratman et al., 2015; Sudimac et al., 2022). In the first study, 38 participants took a 90-min walk in a natural or urban environment, finding that walking in a natural setting decreased self-reported rumination (Bratman et al., 2015). Moreover, there was a decrease in neural activity in the subgenual prefrontal cortex (sgPFC), associated with self-focused withdrawal linked to rumination in both depressed and healthy individuals, indicating that nature experience can improve mental well-being. Focusing on stress-related brain mechanisms, 63 participants went for a one-hour walk in a natural or urban environment (Sudimac et al., 2022). Using a fearful faces task and a social stress task, it was found that amygdala activation decreased after the nature walk while it remained stable after the walk in the urban environment. These findings indicate that taking a stroll in natural environments can positively impact stress-related brain areas, serving as a preventive measure against mental stress and potentially related illnesses. Further analysis revealed that these effects were observed only in women but not in men (Sudimac & Kühn, 2022).

To recap, three fMRI studies addressed the effects of naturalistic urban design on the human brain. Using Space Syntax methodology, it was found that street network characteristics, particularly global and local integration, were associated with decreased activity during stress provocation, especially in areas related to detecting salient stimuli and threat, such as the dorsal anterior cingulate cortex, insula, and thalamus (Dimitrov-Discher et al., 2023). Moreover, two intervention studies found that walking in a natural compared to urban environment reduces rumination and activity in the sgPFC (Bratman et al., 2015), as well as amygdala activity during induced social stress (Sudimac et al., 2022). However, effects on amygdala activity were only observed in women (Sudimac & Kühn, 2022).

NIRS

In order to explore the impact of different types of urban environments on women's emotional health, a study sent 10 female participants to complete 12 road sections by foot (Shaoming et al., 2023). fNIRS data indicated a significant difference in HbO_2 concentration level between participants walking in urban environment compared to those walking in natural environment, with lower HbO_2 concentration levels observed in participants walking in the natural environment. Overall, female participants walking in a positive sequence (from urban to natural environments) exhibited a lower mean HbO_2 concentration level, whereas women walking in a negative sequence (from natural to urban environments) showed a higher HbO_2 concentration level. In terms of EDA, most participants showed a significant emotional response to the nodes where environment type changed. Moreover, women walking in the positive sequence showed significant fluctuations in EDA signals, while this was not observed in the group walking in the negative sequence. There were significant differences in the "sense of relaxation" and "sense of interest" perceived by participants in different types of environments, with emotional scores higher in the natural environments. The study suggested women-friendly planning and design strategies, such as enhancing visibility and creating appropriate transitional spaces, to guide the development of healthy cities that prioritize key populations.

Discussion

The present chapter aims to systematically outline the current approaches and methods for studying how the brain responds to architectural design. The studies are categorized into three main groups, interior design, building design, and urban design, with the objective of offering useful insights into the diverse neurophysiological effects underlying cognitive, affective, and behavioural responses associated with viewing architectural parameters and features in different environmental contexts. These findings hold significance for researchers, designers, policymakers, and community stakeholders engaged in understanding and shaping architectural spaces. They contribute to the academic discourse on the intersection of neuroscience and architecture and have practical implications for refining methodologies, integrating neuroscientific considerations in design, and creating environments that positively influence human emotion and cognition, potentially contributing to overall well-being. Generally, the review serves as a comprehensive resource facilitating interdisciplinary collaboration and aiding informed design decision-making regarding the impact of architectural design on human experiences.

Whereas there has been extensive research on interior design across several functional settings, research on building and urban design is relatively underexplored. Specifically, investigations on interior design covered a variety of functional spaces such as residential, workplace, educational, and healthcare settings. These explorations examined the responses to several key design parameters, finding that geometry and spatial configuration (including forms, scale, proportions, width, ceiling height, aspect ratio, windows, and interior layout), colour, artificial lighting, and the inclusion of natural elements can have psychological and neurophysiological effects. On the other hand, research on building design addressed some functional spaces albeit infrequently (educational, healthcare, and contemplative settings) and investigated parameters such as architectural ranking, style, saliency, glass platforms, window shapes, viewing height and green exposure, and green facades. Conversely, studies exploring urban design have predominantly focused on comparing urban and natural settings. Despite the well-established positive impacts of natural environments on human psychology and neurophysiology, there exists a crucial gap in understanding the effects of urban environments. Beyond the often-proposed detrimental effects, there is a need to explore the influence of urban spaces to inform the creation of improved neighbourhoods and cities (Ellard, 2020). Acknowledging the increased complexity associated with the exploration of building and urban design, collaborative efforts between researchers and professionals in architecture and urban design can significantly contribute to innovative testing paradigms.

In categorizing the studies based on presentation modes, three primary groups were identified such as images and videos, virtual environments, and naturalistic environments. The widespread use of images in investigating architectural stimuli has been notable, but the application of videos has been relatively restricted. It is important to acknowledge that relying on static stimuli has its limitations, particularly in neglecting the impact of human movement on the experience of the environment (Nasar, 1994). Virtual environments have emerged as the most extensively explored presentation mode, particularly in the context of interior design investigations. In contrast, there has been limited research dedicated to understanding naturalistic environments within the scope of this research line. This emphasis on virtual environments suggests a prevailing focus on controlled and simulated settings over real-world, ecological contexts.

In terms of neuroimaging techniques, EEG studies, particularly those exploring virtual interior environments as mentioned earlier, have seen a notable increase. In contrast, a significant gap persists in fMRI research, especially in relation to interior design, with only three identified studies addressing this area, using the same data. Furthermore, the use of VR in fMRI investigations has been restricted to navigation research, missing its potential for enhancing ecological validity. Future studies may want to develop MRI-VR paradigms that can deliver realistic simulations and provide high spatial resolution to map specific neural responses. Notably, despite substantial technological advancements, NIRS methodologies remain underutilized, presenting an opportunity for researchers to explore the evolving capabilities of this technology in studying neural responses to architectural stimuli, especially in naturalistic settings.

When it comes to EEG research in naturalistic urban environments, the collective body of EEG research underlines the complex relationship between real-life built environments and human brain activity. However, this area of study is not without its challenges. Mostly, there are challenges associated with these types of outdoor measurements of online brain signal, specifically those using commercial-grade devices to study urban design. The uniqueness of each participant's input poses a significant challenge. Individuals vary in their cognitive processes, emotional responses, and overall brain activity, making it difficult to establish a uniform baseline for interpreting results. Moreover, variations in the timing of exposure to environments add another layer of complexity. Different individuals may encounter urban spaces at different times, influenced by factors such as daily routines, weather conditions, or personal preferences. This temporal variability can impact the observed brain activity, making it challenging to isolate the specific effects of urban design on neural responses. On the other hand, the use of EEG for studying urban design introduces potential issues related to movement artifacts, particularly with commercial-grade devices, complicating the interpretation of neural data.

Overall, this field of research encounters many challenges. Previous works have identified two methodological constraints in current architectural neuroimaging research (Wang et al., 2022). The first pertains to the narrow emphasis on the aesthetic aspect of architectural experience, often neglecting other aspects like ergonomics, affordances, or functionality (Higuera-Trujillo et al., 2021). Architecture, by definition, endeavours to address diverse human needs, and although it is recognized as one of the "Fine arts", the aesthetic dimension represents just one facet of its cognitive–emotional effects. The second limitation relates to practical constraints imposed by conventional experimental tools and methods, restricting participants' mobility and inhibiting naturalistic interaction with architectural environments (Higuera-Trujillo et al., 2021). Recent studies, especially those conducted in VR, have attempted to cover other cognitive aspects, allowing, in some limited cases, a naturalistic movement through virtual environments.

Moreover, it is noteworthy that the majority, if not all, of the effects discussed primarily pertain to short-term exposures, emphasizing the necessity for longitudinal studies to comprehensively understand the enduring and cumulative influences of architectural and environmental factors. In this context, one way is to employ geographic information system (GIS) methods and Space Syntax in conjunction with large cohort studies to extract and associate environmental characteristics with brain measures collected over multiple time points. For instance, a recent study on 156,075 participants from the U.K. Biobank explored the complex relationship between urban environments and psychiatric symptoms (Xu et al., 2023), identifying three urban environmental profiles correlating with specific psychiatric symptom groups. Positive correlations were observed between an affective symptom group and an environmental profile characterized by social deprivation, air pollution, street network, and urban land-use density. Protective factors, such as greenness and generous destination accessibility, were negatively correlated with an anxiety symptom group.

Another notable limitation in the field is that most of the available research was conducted on WEIRD populations, overlooking the challenges posed by cultural and other interindividual differences. These differences, stemming from factors including but not limited to culture, personal preferences, expertise (Vartanian et al., 2019), personality (Banaei et al., 2020), neurodiversity (Vartanian et al., 2021), but also age differences, introduce complexity and variability in individuals' reactions to architectural elements. In order to advance the development of inclusive and effective strategies, future research efforts should strive to explore and understand these variations more comprehensively using sufficient sample sizes.

Finally, it is essential to note that much of the existing research in the intersection of neuroscience and architecture relied on relationships that are correlational. While these correlations provide valuable insights into the relationships between architectural elements and neurophysiological responses, there is a recognized need to progress towards establishing causality. This requires more rigorous experimental designs and intervention studies. Achieving causality will provide more nuanced insights into how specific architectural features causally impact human cognitive, affective, and behavioural responses, contributing to evidence-based design principles for human well-being.

Conclusion

In conclusion, this chapter explored the neurophysiological responses underlying the cognitive, affective, and behavioural responses associated with viewing architectural designs, categorizing them into interior, building, and urban contexts. Offering extensive insights, it benefits researchers, designers, policymakers, and community stakeholders shaping architectural environments. While interior design features are well explored in various settings, building and urban design research remains relatively underexplored. The chapter emphasizes the need to understand the effects of urban environments, calling for collaborative efforts in neuroscience, architecture and urban design. Presentation modes are categorized into images/videos, virtual environments, and naturalistic environments, with virtual environments extensively studied in interior design. Neuroimaging techniques, particularly EEG in virtual environments, show progress, but fMRI research is limited, presenting an opportunity for exploration. The potential of MRI-VR approaches to enhance ecological validity and provide high spatial resolution for mapping neural responses is noted. NIRS methodologies remain underutilized, despite their major usefulness particularly in naturalistic settings. Challenges include a narrow aesthetic focus, practical constraints, and a focus on short-term exposure studies, emphasizing the necessity for longitudinal research. GIS and Space Syntax combined with large cohort studies offer a promising approach. Another limitation is the reliance on WEIRD populations, urging comprehensive exploration of cultural and interindividual differences. The chapter concludes by highlighting the need to progress from associations to causality, necessitating rigorous experimental designs and intervention studies for evidence-based design principles contributing to human well-being. Overall, interdisciplinary collaboration is crucial for enriching the understanding of how the built environment shapes human psychology and neurophysiology, addressing complexities in architectural experiences.

Acknowledgements During the work on her dissertation, N.T. was a pre-doctoral fellow of the International Max Planck Research School on the Life Course (https://www.imprs-life.mpg.de/) of the Max Planck Institute for Human Development, Berlin, Germany.

References

Aftanas, L., & Golocheikine, S. (2001). Human anterior and frontal midline theta and lower alpha reflect emotionally positive state and internalized attention: High-resolution EEG investigation of meditation. *Neuroscience Letters, 310*, 57–60. https://doi.org/10.1016/S0304-3940(01)02094-8

Al-barrak, L., Kanjo, E., & Younis, E. M. G. (2017). NeuroPlace: Categorizing urban places according to mental states. *PLoS One, 12*(9), e0183890. https://doi.org/10.1371/journal.pone.0183890

Armony, J. L., Driver, J., & Dolan, R. J. (2001). Effects of attention and emotion on face processing in the human brain: An event-related fMRI study. *Neuron, 30*(3), 829–841. https://doi.org/10.1016/s0896-6273(01)00328-2

Aspinall, P., Mavros, P., Coyne, R., & Roe, J. (2015). The urban brain: Analysing outdoor physical activity with mobile EEG. *British Journal of Sports Medicine, 49*(4), 272–276. https://doi.org/10.1136/bjsports-2012-091877

Azzazy, S., Ghaffarianhoseini, A., Ghaffarian Hoseini, A., Naismith, N., & Doborjeh, Z. (2021). A critical review on the impact of built environment on users' measured brain activity. *Architectural Science Review, 64*(4), 319–335. https://doi.org/10.1080/00038628.2020.1749980

Baird, J. C., Cassidy, B., & Kurr, J. (1978). Room preference as a function of architectural features and user activities. *Journal of Applied Psychology, 63*(6), 719–727. https://doi.org/10.1037/0021-9010.63.6.719

Banaei, M., Hatami, J., Yazdanfar, A., & Gramann, K. (2017). Walking through architectural spaces: The impact of interior forms on human brain dynamics. *Frontiers in Human Neuroscience, 11*, 477. https://doi.org/10.3389/fnhum.2017.00477

Banaei, M., Ahmadi, A., Gramann, K., & Hatami, J. (2020). Emotional evaluation of architectural interior forms based on personality differences using virtual reality. *Frontiers of Architectural Research, 9*(1), Article 1. https://doi.org/10.1016/j.foar.2019.07.005

Bar, M., & Neta, M. (2007). Visual elements of subjective preference modulate amygdala activation. *Neuropsychologia, 45*(10) Article 10, 2191. https://doi.org/10.1016/j.neuropsychologia.2007.03.008

Baron Short E, Kose S, Mu Q, Borckardt J, Newberg A, George MS, Kozel FA (2010). Regional brain activation during meditation shows time and practice effects: An exploratory FMRI study. Evidence-based Complementary and Alternative Medicine 7(1):121-7. 10.1093/ecam/nem163. Epub 2007 Dec 27. PMID: 18955268; PMCID: PMC2816391.

Bermudez, J., Krizaj, D., Lipschitz, D. L., Bueler, C. E., Rogowska, J., Yurgelun-Todd, D., & Nakamura, Y. (2017). Externally-induced meditative states: An exploratory fMRI study of architects' responses to contemplative architecture. *Frontiers of Architectural Research, 6*(2), 123–136. https://doi.org/10.1016/j.foar.2017.02.002

Bower, I. S., Clark, G. M., Tucker, R., Hill, A. T., Lum, J. A. G., Mortimer, M. A., & Enticott, P. G. (2022a). Built environment color modulates autonomic and EEG indices of emotional response. *Psychophysiology, 59*(12), e14121. https://doi.org/10.1111/psyp.14121

Bower, I. S., Clark, G. M., Tucker, R., Hill, A. T., Lum, J. A. G., Mortimer, M. A., & Enticott, P. G. (2022b). Enlarged interior built environment scale modulates high-frequency EEG oscillations. *eNeuro, 9*(5), ENEURO.0104. https://doi.org/10.1523/ENEURO.0104-22.2022

Bower, I. S., Hill, A. T., & Enticott, P. G. (2023). Functional brain connectivity during exposure to the scale and color of interior built environments. *Human Brain Mapping, 44*(2), 447–457. https://doi.org/10.1002/hbm.26061

Baron Short, E., Kose, S., Mu, Q., Borckardt, J., Newberg, A., George, M. S., & Kozel, F. A. (2010). Regional brain activation during meditation shows time and practice effects: An exploratory FMRI study. *Evidence-based Complementary and Alternative Medicine, 7*(1), 121–127. https://doi.org/10.1093/ecam/nem163. Epub 2007 Dec 27. PMID: 18955268; PMCID: PMC2816391.

Breznitz, Z. (2008). *Brain research in language*. Springer Science+Business Media, LLC.

Brown, S., Gao, X., Tisdelle, L., Eickhoff, S. B., & Liotti, M. (2011). Naturalizing aesthetics: Brain areas for aesthetic appraisal across sensory modalities. *NeuroImage, 58*(1), 250–258. https://doi.org/10.1016/j.neuroimage.2011.06.012

Brunner, L. M., Hillnhütter, H., Aalto, P. O., Steinert, M., & Dybvik, H. (2023). *Measuring the impact of walking environments on brain activation: Results from an fNIRS pilot study*. REAL CORP. https://ntnuopen.ntnu.no/ntnu-xmlui/handle/11250/3097335

Bush, G., Luu, P., & Posner, M. I. (2000). Cognitive and emotional influences in anterior cingulate cortex. *Trends in Cognitive Sciences, 4*(6), 215–222.

Chang, C.-Y., & Chen, P.-K. (2005). Human response to window views and indoor plants in the workplace. *HortScience, 40*(5), 1354–1359. https://doi.org/10.21273/HORTSCI.40.5.1354

Chang, D. H. F., Jiang, B., Wong, N. H. L., Wong, J. J., Webster, C., & Lee, T. M. C. (2021). The human posterior cingulate and the stress-response benefits of viewing green urban landscapes. *NeuroImage, 226*, 117555. https://doi.org/10.1016/j.neuroimage.2020.117555

Choi, H.-H., van Merriënboer, J. J. G., & Paas, F. (2014). Effects of the physical environment on cognitive load and learning: Towards a new model of cognitive load. *Educational Psychology Review, 26*(2), 225–244. https://doi.org/10.1007/s10648-014-9262-6

Choi, J.-Y., Park, S.-A., Jung, S.-J., Lee, J.-Y., Son, K.-C., An, Y.-J., & Lee, S.-W. (2016). Physiological and psychological responses of humans to the index of greenness of an interior space. *Complementary Therapies in Medicine, 28*, 37–43. https://doi.org/10.1016/j.ctim.2016.08.002

Choo, H., Nasar, J. L., Nikrahei, B., & Walther, D. B. (2017). Neural codes of seeing architectural styles. *Scientific Reports, 7*(1), 40201. https://doi.org/10.1038/srep40201

Churchill, W. (1943, October 28). House of commons rebuilding. Hansard, House of Commons. Retrieved from https://api.parliament.uk/historic-hansard/commons/1943/oct/28/house-of-commons-rebuilding

Clark, A. (1999). An embodied cognitive science? *Trends in Cognitive Sciences, 3*(9), 345–351. https://doi.org/10.1016/s1364-6613(99)01361-3

Coan, J. A., & Allen, J. J. B. (2003). Frontal EEG asymmetry and the behavioral activation and inhibition systems. *Psychophysiology, 40*, 106–114. https://doi.org/10.1111/1469-8986.00011

Coburn, A., Vartanian, O., & Chatterjee, A. (2017). Buildings, beauty, and the brain: A neuroscience of architectural experience. *Journal of Cognitive Neuroscience, 29*(9), 1521–1531. https://doi.org/ezproxy.mpibberlin.mpg.de/10.1162/jocn_a_01146

Coburn, A., Vartanian, O., Kenett, Y. N., Nadal, M., Hartung, F., Hayn-Leichsenring, G., Navarrete, G., González-Mora, J. L., & Chatterjee, A. (2020). Psychological and neural responses to architectural interiors. *Cortex, 126*, 217–241. https://doi.org/10.1016/j.cortex.2020.01.009

Cruz-Garza, J. G., Darfler, M., Rounds, J. D., Gao, E., & Kalantari, S. (2022). EEG-based investigation of the impact of room size and window placement on cognitive performance. *Journal of Building Engineering, 53*, 104540. https://doi.org/10.1016/j.jobe.2022.104540

Davidson, R. J. (1992). Anterior cerebral asymmetry and the nature of emotion. *Brain and Cognition, 20*(1), 125–151. https://doi.org/10.1016/0278-2626(92)90065-T

Dazkir, S. S., & Read, M. A. (2012). Furniture forms and their influence on our emotional responses toward interior environments. *Environment and Behavior, 44*(5), Article 5. https://doi.org/10.1177/0013916511402063

Debener, S., Minow, F., Emkes, R., Gandras, K., & de Vos, M. (2012). How about taking a low-cost, small, and wireless EEG for a walk? *Psychophysiology, 49*(11), 1617–1621. https://doi.org/10.1111/j.1469-8986.2012.01471.x

Delaux, A., de Saint Aubert, J.-B., Ramanoël, S., Bécu, M., Gehrke, L., Klug, M., Chavarriaga, R., Sahel, J.-A., Gramann, K., & Arleo, A. (2021). Mobile brain/body imaging of landmark-based navigation with high-density EEG. *European Journal of Neuroscience, 54*(12), 8256–8282. https://doi.org/10.1111/ejn.15190

Dimitrov-Discher, A., Gu, L., Pandit, L., Veer, I. M., Walter, H., Adli, M., & Knöll, M. (2023). Stress and streets: How the network structure of streets is associated with stress-related brain activation. *Journal of Environmental Psychology, 91*, 102142. https://doi.org/10.1016/j.jenvp.2023.102142

Djebbara, Z., Fich, L. B., Petrini, L., & Gramann, K. (2019). Sensorimotor brain dynamics reflect architectural affordances. *Proceedings of the National Academy of Sciences, 116*(29), 14769–14778. https://doi.org/10.1073/pnas.1900648116

Duvinage, M., Castermans, T., Petieau, M., Hoellinger, T., Cheron, G., & Dutoit, T. (2013). Performance of the Emotiv Epoc headset for P300-based applications. *Biomedical Engineering Online, 12*, 56. https://doi.org/10.1186/1475-925X-12-56

Eberhard, J. P. (2009). Applying neuroscience to architecture. *Neuron, 62*(6), Article 6. https://doi.org/10.1016/j.neuron.2009.06.001

Edelstein, E. A., & Macagno, E. (2012). Form follows function: Bridging neuroscience and architecture. In S. T. Rassia & P. M. Pardalos (Eds.), *Sustainable environmental design in architecture: Impacts on health* (pp. 27–41). Springer. https://doi.org/10.1007/978-1-4419-0745-5_3

Edelstein, E., Gramann, K., Schulze, J., Shamlo, N. B., Erp, E. V., Vankov, A., Makeig, S., Wolszon, L., & Macagno, E. (2008). Neural responses during navigation in the virtual aided design laboratory: Brain dynamics of orientation in architecturally ambiguous space. https://www.semanticscholar.org/paper/Neural-Responses-during-Navigation-in-the-Virtual-%3A-Edelstein-Gramann/1797b007d48fcd171ccc80c8b53968ce5c07c036

Ekman, P., & Davidson, R. J. (1993). Voluntary smiling changes regional brain activity. *Psychological Science, 4*(5), 342–345. https://doi.org/10.1111/j.1467-9280.1993.tb00576.x

Ellard, C. (2020). Neuroscience, wellbeing, and urban design: Our universal attraction to vitality. *Psychological Research on Urban Society, 3*(1), 6. https://doi.org/10.7454/proust.v3i1.81

Elsadek, M., Liu, B., & Lian, Z. (2019). Green façades: Their contribution to stress recovery and well-being in high-density cities. *Urban Forestry & Urban Greening, 46*, 126446. https://doi.org/10.1016/j.ufug.2019.126446

Ergan, S., Radwan, A., Zou, Z., Tseng, H., & Han, X. (2019). Quantifying human experience in architectural spaces with integrated virtual reality and body sensor networks. *Journal of Computing in Civil Engineering, 33*(2), 04018062. https://doi.org/10.1061/(ASCE)CP.1943-5487.0000812

Erkan, İ. (2018). Examining wayfinding behaviours in architectural spaces using brain imaging with electroencephalography (EEG). *Architectural Science Review, 61*(6), 410–428. https://doi.org/10.1080/00038628.2018.1523129

Erkan, İ. (2021). A neuro-cognitive investigation of the impact of glass floors on people. *Architectural Science Review, 64*(4), 336–345. https://doi.org/10.1080/00038628.2020.1858574

Erkan, I. (2023). A neuro-cognitive perspective on urban behavior of people with different moods. *Open House International, 48*(4), 822–839. https://doi.org/10.1108/OHI-10-2022-0252

Elliot, A. J., & Maier, M. A. (2014). Color psychology: Effects of perceiving color on psychological functioning in humans. *Annual Review of Psychology, 65*, 95–120. https://doi.org/10.1146/annurev-psych-010213-115035

Evans, G. W., & McCoy, J. M. (1998). When buildings don't work: The role of architecture in human health. *Journal of Environmental Psychology, 18*(1), Article 1. https://doi.org/10.1006/jevp.1998.0089

Evans, G. W. (2003). The built environment and mental health. *Journal of Urban Health: Bulletin of the New York Academy of Medicine, 80*, 536–555. https://doi.org/10.1093/jurban/jtg063

Federmeier, K. D., & Kutas, M. (2002). Picture the difference: Electrophysiological investigations of picture processing in the two cerebral hemispheres. *Neuropsychologia, 40*, 730–747. https://doi.org/10.1016/s0028-3932(01)00193-2

Freedberg, D., & Gallese, V. (2007). Motion, emotion and empathy in esthetic experience. *Trends in Cognitive Sciences, 11*(5), 197–203. https://doi.org/10.1016/j.tics.2007.02.003

Gallese, V. (2005). Embodied simulation: From neurons to phenomenal experience. *Phenomenology and the Cognitive Sciences, 4*(1), 23–48.

Gallese, V., & Sinigaglia, C. (2011). What is so special about embodied simulation? *Trends in Cognitive Sciences, 15*(11), 512–519.

Gepshtein, S., & Snider, J. (2019). Neuroscience for architecture: The evolving science of perceptual meaning. *Proceedings of the National Academy of Sciences, 116*(29), 14404–14406. https://doi.org/10.1073/pnas.1908868116

Gevins, A., Smith, M. E., McEvoy, L. K., Leong, H., & Le, J. (1999). Electroencephalographic imaging of higher brain function. *Philosophical Transactions of the Royal Society of London. Series B, Biological Sciences, 354*(1387), 1125–1133. https://doi.org/10.1098/rstb.1999.0468

Grassini, S., Revonsuo, A., Castellotti, S., Petrizzo, I., Benedetti, V., & Koivisto, M. (2019). Processing of natural scenery is associated with lower attentional and cognitive load compared with urban ones. *Journal of Environmental Psychology, 62*, 1–11. https://doi.org/10.1016/j.jenvp.2019.01.007

Grassini, S., Segurini, G. V., & Koivisto, M. (2022). Watching nature videos promotes physiological restoration: Evidence from the modulation of alpha waves in electroencephalography. *Frontiers in Psychology, 13*. https://www.frontiersin.org/articles/10.3389/fpsyg.2022.871143

Grima Murcia, M. D., Ortiz, M. J., López-Gordo, M. A., Ferrández, J. M., Sánchez Ferrer, F., & Fernández, E. (2019). Neural representation of different 3D architectural images: An EEG study. *Integrated Computer-Aided Engineering, 26*(2), 197–205. https://doi.org/10.3233/ICA-180591

Gifford, R. (2002). Making a difference: Some ways environmental psychology has improved the world. In R. B. Bechtel & A. Churchman (Eds.), *Handbook of environmental psychology* (pp. 323–334). Wiley.

Hakak, A. M., Bhattacharya, J., Biloria, N., de Kleijn, R., & Shah-Mohammadi, F. (2016). Navigating abstract virtual environment: An eeg study. *Cognitive Neurodynamics, 10*(6), 471–480. https://doi.org/10.1007/s11571-016-9395-z

Haxby, J. V., Gobbini, M. I., Furey, M. L., Ishai, A., Schouten, J. L., & Pietrini, P. (2001). Distributed and overlapping representations of faces and objects in ventral temporal cortex, 293(5539), 2425–2430. https://doi.org/10.1126/science.1063736

Hekmatmanesh, A., Banaei, M., Haghighi, K. S., & Najafi, A. (2019). Bedroom design orientation and sleep electroencephalography signals. *Acta Medica International, 6*(1), 33. https://doi.org/10.4103/ami.ami_60_18

Higuera-Trujillo, J. L., Llinares, C., & Macagno, E. (2021). The cognitive-emotional design and study of architectural space: A scoping review of Neuroarchitecture and its precursor approaches. *Sensors, 21*(6), Article 6. https://doi.org/10.3390/s21062193

Hillier, B., & Hanson, J. (1984). *The social logic of space*. Cambridge University Press. https://doi.org/10.1017/CBO9780511597237

Hillyard, S. A., & Anllo-Vento, L. (1998). Event-related brain potentials in the study of visual selective attention. *Proceedings of the National Academy of Sciences of the United States of America, 95*, 781–787. https://doi.org/10.1073/pnas.95.3.781

Hu, M., & Roberts, J. (2020). Built environment evaluation in virtual reality environments—A cognitive neuroscience approach. *Urban Science, 4*(4), Article 4. https://doi.org/10.3390/urbansci4040048

Hu, M., Simon, M., Fix, S., Vivino, A. A., & Bernat, E. (2021). Exploring a sustainable building's impact on occupant mental health and cognitive function in a virtual environment. *Scientific Reports, 11*(1), 5644. https://doi.org/10.1038/s41598-021-85210-9

Iaria, G., Petrides, M., Dagher, A., Pike, B., & Bohbot, V. D. (2003). Cognitive strategies dependent on the hippocampus and caudate nucleus in human navigation: Variability and change with practice. *Journal of Neuroscience, 23*, 5945–5952.

Jelić, A., Tieri, G., De Matteis, F., Babiloni, F., & Vecchiato, G. (2016). The enactive approach to architectural experience: A neurophysiological perspective on embodiment, motivation, and affordances. *Frontiers in Psychology, 7*. https://www.frontiersin.org/articles/10.3389/fpsyg.2016.00481

Jung, D., Kim, D. I., & Kim, N. (2023). Bringing nature into hospital architecture: Machine learning-based EEG analysis of the biophilia effect in virtual reality. *Journal of Environmental Psychology, 89*, 102033. https://doi.org/10.1016/j.jenvp.2023.102033

Kalantari, S., Rounds, J. D., Kan, J., Tripathi, V., & Cruz-Garza, J. G. (2021). Comparing physiological responses during cognitive tests in virtual environments vs. in identical real-world environments. *Scientific Reports, 11*(1), Article 1. https://doi.org/10.1038/s41598-021-89297-y

Kalantari, S., Tripathi, V., Kan, J., Rounds, J. D., Mostafavi, A., Snell, R., & Cruz-Garza, J. G. (2022). Evaluating the impacts of color, graphics, and architectural features on wayfinding in healthcare settings using EEG data and virtual response testing. *Journal of Environmental Psychology, 79*, 101744. https://doi.org/10.1016/j.jenvp.2021.101744

Kalantari, S., Cruz-Garza, J., Xu, T. B., Mostafavi, A., & Gao, E. (2023). Store layout design and consumer response: A behavioural and EEG study. *Building Research & Information, 51*(8), 897–914. https://doi.org/10.1080/09613218.2023.2201415

Kanwisher, N., McDermott, J., & Chun, M. M. (1997). The fusiform face area: A module in human extrastriate cortex specialized for face perception. *The Journal of Neuroscience: The Official Journal of the Society for Neuroscience, 17*(11), 4302–4311. https://doi.org/10.1523/JNEUROSCI.17-11-04302.1997

Kaplan, S. (1995). The restorative benefits of nature: Toward an integrative framework. *Journal of Environmental Psychology, 15*(3), 169. https://doi.org/10.1016/0272-4944(95)90001-2

Kaplan, R., & Kaplan, S. (1989). *The experience of nature: A psychological perspective* (p. xii, 340). Cambridge University Press.

Karakas, T., & Yildiz, D. (2020). Exploring the influence of the built environment on human experience through a neuroscience approach: A systematic review. *Frontiers of Architectural Research, 9*(1), 236–247. https://doi.org/10.1016/j.foar.2019.10.005

Karandinou, A., & Turner, L. (2017). Architecture and neuroscience; what can the EEG recording of brain activity reveal about a walk through everyday spaces? *International Journal of Parallel, Emergent and Distributed Systems, 32*(sup1), S54–S65. https://doi.org/10.1080/17445760.2017.1390089

Keizer, A. W., Nieuwenhuis, S., Colzato, L. S., Teeuwisse, W., Rombouts, S. A., & Hommel, B. (2008). When moving faces activate the house area: An fMRI study of object-file retrieval. *Behavioral and Brain Functions: BBF*, 4, 50. https://doi.org/10.1186/1744-9081-4-50

Klimesch, W. (1999). EEG alpha and theta oscillations reflect cognitive and memory performance: A review and analysis. *Brain Research Reviews, 29*(2–3), 169–195.

Khaleghimoghaddam, N., Bala, H. A., Özmen, G., & Öztürk, Ş. (2022). Neuroscience and architecture: What does the brain tell to an emotional experience of architecture via a functional MR study? *Frontiers of Architectural Research, 11*(5), 877–890. https://doi.org/10.1016/j.foar.2022.02.007

Kim, G.-W., & Jeong, G.-W. (2014). Brain activation patterns associated with the human comfortability of residential environments: 3.0-T functional MRI. *Neuro Report, 25*(12), 915. https://doi.org/10.1097/WNR.0000000000000205

Kim, S., Park, H., & Choo, S. (2021). Effects of changes to architectural elements on human relaxation-arousal responses: Based on VR and EEG. *International Journal of Environmental Research and Public Health, 18*(8), Article 8. https://doi.org/10.3390/ijerph18084305

Kirk, U., Skov, M., Christensen, M. S., & Nygaard, N. (2009). Brain correlates of aesthetic expertise: A parametric fMRI study. *Brain and Cognition, 69*(2), 306–315. https://doi.org/10.1016/j.bandc.2008.08.004

Kohane, P., & Hill, M. (2001). The eclipse of a commonplace idea: Decorum in architectural theory. *Arq: Architectural Research Quarterly, 5*(1), 63–77. https://doi.org/10.1017/S1359135501001105

Kong, Z., Hou, K., Wang, Z., Chen, F., Li, Y., Liu, X., & Liu, C. (2022). Subjective and physiological responses towards interior natural Lightscape: Influences of aperture design, window size and sky condition. *Buildings, 12*(10), Article 10. https://doi.org/10.3390/buildings12101612

Krabbendam, L., van Vugt, M., Conus, P., Söderström, O., Empson, L. A., van Os, J., & Fett, A.-K. J. (2021). Understanding urbanicity: How interdisciplinary methods help to unravel the effects of the city on mental health. *Psychological Medicine, 51*(7), 1099–1110. https://doi.org/10.1017/S0033291720000355

Kravitz, D., Saleem, K., Baker, C., et al. (2011). A new neural framework for visuospatial processing. *Nature Reviews Neuroscience, 12*, 217–230. https://doi.org/10.1038/nrn3008

Kühn, S., Forlim, C. G., Lender, A., Wirtz, J., & Gallinat, J. (2021). Brain functional connectivity differs when viewing pictures from natural and built environments using fMRI resting state analysis. *Scientific Reports, 11*(1), Article 1. https://doi.org/10.1038/s41598-021-83246-5

Küller, R., Mikellides, B., & Janssens, J. (2009). Color, arousal, and performance—A comparison of three experiments. *Color Research & Application, 34*(2), 141–152. https://doi.org/10.1002/col.20476

Leder, H., & Carbon, C.-C. (2005). Dimensions in appreciation of car interior design. *Applied Cognitive Psychology, 19*(5), Article 5. https://doi.org/10.1002/acp.1088

Lee, J. (2017). Experimental study on the health benefits of garden landscape. *International Journal of Environmental Research and Public Health, 14*(7), Article 7. https://doi.org/10.3390/ijerph14070829

Leech, R., & Sharp, D. J. (2014). The role of the posterior cingulate cortex in cognition and disease. *Brain, 137*(Pt 1), 12–32. https://doi.org/10.1093/brain/awt162.

Lindal, P. J., & Hartig, T. (2013). Architectural variation, building height, and the restorative quality of urban residential streetscapes. *Journal of Environmental Psychology, 33*, 26–36. https://doi.org/10.1016/j.jenvp.2012.09.003

Llinares Millán, C., Higuera-Trujillo, J. L., Montañana i Aviñó, A., Torres, J., & Sentieri, C. (2021). The influence of classroom width on attention and memory: Virtual-reality-based task performance and neurophysiological effects. *Building Research & Information, 49*(7), 813–826. https://doi.org/10.1080/09613218.2021.1899798

Llinares, C., Higuera-Trujillo, J. L., & Serra, J. (2021). Cold and warm coloured classrooms. Effects on students' attention and memory measured through psychological and neurophysiological responses. *Building and Environment, 196*, 107726. https://doi.org/10.1016/j.buildenv.2021.107726

Logothetis, N. K. (2003). The underpinnings of the BOLD functional magnetic resonance imaging signal. *The Journal of Neuroscience: The Official Journal of the Society for Neuroscience, 23*(10), 3963–3971. https://doi.org/10.1523/JNEUROSCI.23-10-03963.2003

Luck, S. J., Hillyard, S. A., Mouloua, M., Woldorff, M. G., Clark, V. P., & Hawkins, H. L. (1994). Effects of spatial cuing on luminance detectability: Psychophysical and electrophysiological evidence for early selection. *Journal of Experimental Psychology: Human Perception and Performance, 20*, 887–904. https://doi.org/10.1037/0096-1523.20.4.887

Lundholm, H. (1921). The affective tone of lines: Experimental researches. *Psychological Review, 28*(1), Article 1. https://doi.org/10.1037/h0072647

Ma, Q., Hu, L., & Wang, X. (2015). Emotion and novelty processing in an implicit aesthetic experience of architectures: Evidence from an event-related potential study. *Neuro Report, 26*(5), 279. https://doi.org/10.1097/WNR.0000000000000344

Mansouri, F. A., Buckley, M. J., & Tanaka, K. (2007). Mnemonic function of the dorsolateral prefrontal cortex in conflictinduced behavioral adjustment. *Science, 318*(5852), 987–990.

Martínez-Soto, J., Gonzales-Santos, L., Pasaye, E., & Barrios, F. A. (2013). Exploration of neural correlates of restorative environment exposure through functional magnetic resonance. *Intelligent Buildings International, 5*(sup1), 10–28. https://doi.org/10.1080/17508975.2013.807765

Mavros, P., Austwick, M. Z., & Smith, A. H. (2016). Geo-EEG: Towards the use of EEG in the study of urban behaviour. *Applied Spatial Analysis and Policy, 9*(2), 191–212. https://doi.org/10.1007/s12061-015-9181-z

Mavros, P., Wälti, J., Nazemi, M., Ong, C. H., & Hölscher, C. (2022). A mobile EEG study on the psychophysiological effects of walking and crowding in indoor and outdoor urban environments. *Scientific Reports, 12*(1), Article 1. https://doi.org/10.1038/s41598-022-20649-y

Mecklinger, A., Kriukova, O., Mühlmann, H., & Grunwald, T. (2014). Cross-cultural differences in processing of architectural ranking: Evidence from an event-related potential study. *Cognitive Neuroscience, 5*(1), 45–53. https://doi.org/10.1080/17588928.2013.869740

Mostafavi, A., Cruz-Garza, J., & Kalantari, S. (2023). Enhancing lighting design through the investigation of illuminance and correlated color Temperature's effects on brain activity: An EEG-VR approach. *Journal of Building Engineering, 75*, 106776. https://doi.org/10.1016/j.jobe.2023.106776.

Naghibi Rad, P., Shahroudi, A. A., Shabani, H., Ajami, S., & Lashgari, R. (2019). Encoding pleasant and unpleasant expression of the architectural window shapes: An ERP study. *Frontiers in Behavioral Neuroscience, 13*. https://www.frontiersin.org/articles/10.3389/fnbeh.2019.00186

Nasar, J. L. (1994). Urban design aesthetics: The evaluative qualities of building exteriors. *Environment and Behavior, 26*(3), Article 3. https://doi.org/10.1177/001391659402600305

Neale, C., Aspinall, P., Roe, J., Tilley, S., Mavros, P., Cinderby, S., Coyne, R., Thin, N., Bennett, G., & Thompson, C. W. (2017). The aging urban brain: Analyzing outdoor physical activity using the emotiv affectiv suite in older people. *Journal of Urban Health, 94*(6), 869–880. https://doi.org/10.1007/s11524-017-0191-9

Olszewska-Guizzo, A. A., Paiva, T. O., & Barbosa, F. (2018a). Effects of 3D contemplative landscape videos on brain activity in a passive exposure EEG experiment. *Frontiers in Psychiatry, 9*. https://www.frontiersin.org/articles/10.3389/fpsyt.2018.00317

Olszewska-Guizzo, A., Escoffier, N., Chan, J., & Puay Yok, T. (2018b). Window view and the brain: Effects of floor level and green cover on the alpha and beta rhythms in a passive exposure EEG experiment. *International Journal of Environmental Research and Public Health, 15*(11), Article 11. https://doi.org/10.3390/ijerph15112358

Olszewska-Guizzo, A., Sia, A., Fogel, A., & Ho, R. (2020). Can exposure to certain urban green spaces trigger frontal alpha asymmetry in the brain?—Preliminary findings from a passive task

EEG study. *International Journal of Environmental Research and Public Health, 17*(2), Article 2. https://doi.org/10.3390/ijerph17020394

Olszewska-Guizzo, A., Mukoyama, A., Naganawa, S., Dan, I., Husain, S. F., Ho, C. S., & Ho, R. (2021). Hemodynamic response to three types of urban spaces before and after lockdown during the COVID-19 pandemic. *International Journal of Environmental Research and Public Health, 18*(11), Article 11. https://doi.org/10.3390/ijerph18116118

Oppenheim, I., Mühlmann, H., Blechinger, G., Mothersill, I. W., Hilfiker, P., Jokeit, H., Kurthen, M., Krämer, G., & Grunwald, T. (2009). Brain electrical responses to high- and low-ranking buildings. *Clinical EEG and Neuroscience, 40*(3), 157–161. https://doi.org/10.1177/155005940904000307

Oppenheim, I., Vannucci, M., Mühlmann, H., Gabriel, R., Jokeit, H., Kurthen, M., Krämer, G., & Grunwald, T. (2010). Hippocampal contributions to the processing of architectural ranking. *NeuroImage, 50*(2), 742–752. https://doi.org/10.1016/j.neuroimage.2009.12.078

Pallasmaa, J. (2014). Space, place and atmosphere. Emotion and peripherical perception in architectural experience. *Lebenswelt. Aesthetics and Philosophy of Experience, 4*, Article 4. https://doi.org/10.13130/2240-9599/4202

Pati, D., O'Boyle, M., Hou, J., Nanda, U., & Ghamari, H. (2016). Can hospital form trigger fear response? *HERD, 9*(3), 162–175. https://doi.org/10.1177/1937586715624210

Pavlidou, A., Schnitzler, A., & Lange, J. (2014). Beta oscillations and their functional role in movement perception. *Translational Neuroscience, 5*(4), 286–292.

Peen, J., Schoevers, R. A., Beekman, A. T., & Dekker, J. (2010). The current status of urban-rural differences in psychiatric disorders. *Acta Psychiatrica Scandinavica, 121*(2), 84–93. https://doi.org/10.1111/j.1600-0447.2009.01438.x

Polich, J. (2007). Updating P300: An integrative theory of P3a and P3b. *Clinical Neurophysiology, 118*, 2128–2148. https://doi.org/10.1016/j.clinph.2007.04.019

Rad, P. N., Behzadi, F., Yazdanfar, S. A., Ghamari, H., Zabeh, E., & Lashgari, R. (2023). Exploring methodological approaches of experimental studies in the field of neuroarchitecture: A systematic review. *HERD: Health Environments Research & Design Journal, 16*(2), 284–309. https://doi.org/10.1177/19375867221133135

Reber, R., Schwarz, N., & Winkielman, P. (2004). Processing fluency and aesthetic pleasure: Is beauty in the Perceiver's processing experience? *Personality and Social Psychology Review, 8*(4), 364–382. https://doi.org/10.1207/s15327957pspr0804_3

Rhee, J. H., Schermer, B., & Cha, S. H. (2023). Effects of indoor vegetation density on human well-being for a healthy built environment. *Developments in the Built Environment, 14*, 100172. https://doi.org/10.1016/j.dibe.2023.100172

Robinson, S., & Pallasmaa, J. (2015). *Mind in architecture: Neuroscience, embodiment, and the future of design*. MIT Press.

Roe, J. J., Aspinall, P. A., Mavros, P., & Coyne, R. (2013). Engaging the brain: The impact of natural versus urban scenes using novel EEG methods in an experimental setting. *Environmental Sciences, 1*, 93–104. https://doi.org/10.12988/es.2013.3109

Rounds, J. D., Cruz-Garza, J. G., & Kalantari, S. (2020). Using posterior EEG theta band to assess the effects of architectural designs on landmark recognition in an urban setting. *Frontiers in Human Neuroscience, 14*. https://www.frontiersin.org/articles/10.3389/fnhum.2020.584385

Sadeghi Habibabad, A., MahdiNejad, J. E., Azemati, H., & Matracchi, P. (2022). Using neurology sciences to investigate the color component and its effect on promoting the sense of spirituality in the interior space of the Vakil Mosque of Shiraz (Using quantitative electroencephalography wave recording). *Journal of Religion and Health, 61*(3), 2398–2415. https://doi.org/10.1007/s10943-019-00937-0

Sanchez-Vives, M. V., & Slater, M. (2005). From presence to consciousness through virtual reality. *Nature Reviews Neuroscience, 6*(4), Article 4. https://doi.org/10.1038/nrn1651

Schnitzler, A., Salenius, S., Salmelin, R., Jousmäki, V., & Hari, R. (1997). Involvement of primary motor cortex in motor imagery: A neuromagnetic study. *NeuroImage, 6*(3), 201–208.

Shaoming, Z., Yuan, Y., & Linting, W. (2023). Impacts of urban environment on women's emotional health and planning improving strategies: An empirical study of Guangzhou based on neuroscience experiments. https://www.proquest.com/docview/2806566170.

Sharma, G., Kaushal, Y., Chandra, S., Singh, V., Mittal, A. P., & Dutt, V. (2017). Influence of landmarks on wayfinding and brain connectivity in immersive virtual reality environment. *Frontiers in Psychology, 8*. https://www.frontiersin.org/articles/10.3389/fpsyg.2017.01220

Shemesh, A., Leisman, G., Bar, M., & Grobman, Y. J. (2021). A neurocognitive study of the emotional impact of geometrical criteria of architectural space. *Architectural Science Review, 64*(4), Article 4. https://doi.org/10.1080/00038628.2021.1940827

Shemesh, A., Leisman, G., Bar, M., & Grobman, Y. J. (2022). The emotional influence of different geometries in virtual spaces: A neurocognitive examination. *Journal of Environmental Psychology, 81*, 101802. https://doi.org/10.1016/j.jenvp.2022.101802

Shin, Y.-B., Woo, S.-H., Kim, D.-H., Kim, J., Kim, J.-J., & Park, J. Y. (2015). The effect on emotions and brain activity by the direct/indirect lighting in the residential environment. *Neuroscience Letters, 584*, 28–32. https://doi.org/10.1016/j.neulet.2014.09.046

Silvia, P. J., & Barona, C. M. (2009). Do people prefer curved objects? Angularity, expertise, and aesthetic preference. *Empirical Studies of the Arts, 27*(1), Article 1. https://doi.org/10.2190/EM.27.1.b

Spiers, H. J., & Maguire, E. A. (2007). A navigational guidance system in the human brain. *Hippocampus, 17*, 618–626.

Spitzer, B., & Haegens, S. (2017). Beyond the status quo: A role for beta oscillations in endogenous content (re)activation. *eNeuro, 4*(4), ENEURO.0170. https://doi.org/10.1523/ENEURO.0170-17.2017

Staats, H., Kieviet, A., & Hartig, T. (2003). Where to recover from attentional fatigue: An expectancy-value analysis of environmental preference. *Journal of Environmental Psychology, 23*, 147–157. https://doi.org/10.1016/S0272-4944(02)00112-3

Stamps, A. (2000). Psychology and the aesthetics of the built environment. Boston: Kluwer Academic Publishers. https://doi.org/10.1007/978-1-4757-6326-3

Sternberg, E. M., & Wilson, M. A. (2006). Neuroscience and architecture: Seeking common ground. *Cell, 127*(2), 239–242. https://doi.org/10.1016/j.cell.2006.10.012

Sudimac, S., & Kühn, S. (2022). A one-hour walk in nature reduces amygdala activity in women, but not in men. *Frontiers in Psychology, 13*. https://www.frontiersin.org/articles/10.3389/fpsyg.2022.931905

Summerfield, C., Egner, T., Mangels, J., & Hirsch, J. (2006). Mistaking a house for a face: Neural correlates of misperception in healthy humans. *Cerebral Cortex (New York, N.Y.: 1991), 16*(4), 500–508. https://doi.org/10.1093/cercor/bhi129

Sudimac, S., Sale, V., & Kühn, S. (2022). *How nature nurtures: Amygdala activity decreases as the result of a one-hour walk in nature*. PsyArXiv. https://doi.org/10.31234/osf.io/tucy7

Tang, I.-C., Tsai, Y.-P., Lin, Y.-J., Chen, J.-H., Hsieh, C.-H., Hung, S.-H., Sullivan, W. C., Tang, H.-F., & Chang, C.-Y. (2017). Using functional magnetic resonance imaging (fMRI) to analyze brain region activity when viewing landscapes. *Landscape and Urban Planning, 162*, 137–144. https://doi.org/10.1016/j.landurbplan.2017.02.007

Taylor, K. S., Seminowicz, D. A., & Davis, K. D. (2008). Two systems of resting state connectivity between the insula and cingulate cortex. *Human Brain Mapping, 30*(9), 2731–2745. https://doi.org/10.1002/hbm.20705

Tilley, S., Neale, C., Patuano, A., & Cinderby, S. (2017). Older people's experiences of mobility and mood in an urban environment: A mixed methods approach using electroencephalography (EEG) and interviews. *International Journal of Environmental Research and Public Health, 14*(2), Article 2. https://doi.org/10.3390/ijerph14020151

Tsunetsugu, Y., Miyazaki, Y., & Sato, H. (2005). Visual effects of interior design in actual-size living rooms on physiological responses. *Building and Environment, 40*(10), 1341–1346. https://doi.org/10.1016/j.buildenv.2004.11.026

Ulrich, R. S. (1983). Aesthetic and affective response to natural environment. In I. Altman & J. F. Wohlwill (Eds.), *Behavior and the natural environment* (pp. 85–125). Springer US. https://doi.org/10.1007/978-1-4613-3539-9_4

Ulrich, R. S., Simons, R. F., Losito, B. D., Fiorito, E., Miles, M. A., & Zelson, M. (1991). Stress recovery during exposure to natural and urban environments. *Journal of Environmental Psychology, 11*(3), Article 3. https://doi.org/10.1016/S0272-4944(05)80184-7

Vartanian, O., Navarrete, G., Chatterjee, A., Fich, L. B., Leder, H., Modrono, C., Nadal, M., Rostrup, N., & Skov, M. (2013). Impact of contour on aesthetic judgments and approach-avoidance decisions in architecture. *Proceedings of the National Academy of Sciences, 110*(Supplement_2), 10446–10453. https://doi.org/10.1073/pnas.1301227110

Vartanian, O., Navarrete, G., Chatterjee, A., Fich, L. B., Gonzalez-Mora, J. L., Leder, H., Modroño, C., Nadal, M., Rostrup, N., & Skov, M. (2015). Architectural design and the brain: Effects of ceiling height and perceived enclosure on beauty judgments and approach-avoidance decisions. *Journal of Environmental Psychology, 41*, 10–18. https://doi.org/10.1016/j.jenvp.2014.11.006

Vartanian, O., Navarrete, G., Chatterjee, A., Fich, L. B., Leder, H., Modroño, C., Rostrup, N., Skov, M., Corradi, G., & Nadal, M. (2019). Preference for curvilinear contour in interior architectural spaces: Evidence from experts and nonexperts. *Psychology of Aesthetics, Creativity, and the Arts, 13*(1), Article 1. https://doi.org/10.1037/aca0000150

Vartanian, O., Navarrete, G., Palumbo, L., & Chatterjee, A. (2021). Individual differences in preference for architectural interiors. *Journal of Environmental Psychology, 77*, 101668. https://doi.org/10.1016/j.jenvp.2021.101668

Vecchiato, G., Jelic, A., Tieri, G., Maglione, A. G., De Matteis, F., & Babiloni, F. (2015a). Neurophysiological correlates of embodiment and motivational factors during the perception of virtual architectural environments. *Cognitive Processing, 16*(S1), 425–429. https://doi.org/10.1007/s10339-015-0725-6

Vecchiato, G., Tieri, G., Jelic, A., De Matteis, F., Maglione, A. G., & Babiloni, F. (2015b). Electroencephalographic correlates of sensorimotor integration and embodiment during the appreciation of virtual architectural environments. *Frontiers in Psychology, 6*. https://doi.org/10.3389/fpsyg.2015.01944

Vilar, E., Rebelo, F., & Noriega, P. (2014). Indoor human wayfinding performance using vertical and horizontal signage in virtual reality. *Human Factors and Ergonomics in Manufacturing & Service Industries, 24*(6), 601–615. https://doi.org/10.1002/hfm.20503

Villringer, A., Planck, J., Hock, C., Schleinkofer, L., & Dirnagl, U. (1993). Near infrared spectroscopy (NIRS): A new tool to study hemodynamic changes during activation of brain function in human adults. *Neuroscience Letters, 154*(1–2), 101–104. https://doi.org/10.1016/0304-3940(93)90181-j

Wang, S., Sanches de Oliveira, G., Djebbara, Z., & Gramann, K. (2022). The embodiment of architectural experience: A methodological perspective on neuro-architecture. *Frontiers in Human Neuroscience, 16*. https://www.frontiersin.org/articles/10.3389/fnhum.2022.833528

Weissman, D. H., Gopalakrishnan, A., Hazlett, C., & Woldorff, M. (2005). Dorsal anterior cingulate cortex resolves conflict from distracting stimuli by boosting attention toward relevant events. *Cerebral Cortex, 15*(2), 229–237.

Winkielman, P., Halberstadt, J., Fazendeiro, T., & Catty, S. (2006). Prototypes are attractive because they are easy on the mind. *Psychological Science, 17*(9), 799–806. https://doi.org/10.1111/j.1467-9280.2006.01785.x

Wohlwill, J. F. (1976). Environmental aesthetics: The environment as a source of affect. In I. Altman & J. F. Wohlwill (Eds.), *Human behavior and environment: Advances in theory and research* (Vol. 1, pp. 37–86). Springer. https://doi.org/10.1007/978-1-4684-2550-5_2

Xu, J., Liu, N., Polemiti, E., Garcia-Mondragon, L., Tang, J., Liu, X., Lett, T., Yu, L., Nöthen, M. M., Feng, J., Yu, C., Marquand, A., Schumann, G., & The Environmental Consortium. (2023). Effects of urban living environments on mental health in adults. *Nature Medicine, 29*(6), 1456–1467. https://doi.org/10.1038/s41591-023-02365-w

Yamashita, R., Chen, C., Matsubara, T., Hagiwara, K., Inamura, M., Aga, K., Hirotsu, M., Seki, T., Takao, A., Nakagawa, E., Kobayashi, A., Fujii, Y., Hirata, K., Ikei, H., Miyazaki, Y., & Nakagawa, S. (2021). The mood-improving effect of viewing images of nature and its neural substrate. *International Journal of Environmental Research and Public Health, 18*(10), Article 10. https://doi.org/10.3390/ijerph18105500

Yeom, S., Kim, H., & Hong, T. (2021). Psychological and physiological effects of a green wall on occupants: A cross-over study in virtual reality. *Building and Environment, 204*, 108134. https://doi.org/10.1016/j.buildenv.2021.108134

Yu, J., Ang, K. K., Ho, S. H., Sia, A., & Ho, R. (2017). *Prefrontal cortical activation while viewing urban and garden scenes: A pilot fNIRS study* (pp. 2546–2549). 2017 39th Annual International Conference of the IEEE Engineering in Medicine and Biology Society (EMBC). https://doi.org/10.1109/EMBC.2017.8037376

Zhang, W., He, X., Liu, S., Li, T., Li, J., Tang, X., & Lai, S. (2019). Neural correlates of appreciating natural landscape and landscape garden: Evidence from an fMRI study. *Brain and Behavior, 9*(7), Article 7. https://doi.org/10.1002/brb3.1335

Zhang, Z., Zhuo, K., Wei, W., Li, F., Yin, J., & Xu, L. (2021). Emotional responses to the visual patterns of urban streets: Evidence from physiological and subjective indicators. *International Journal of Environmental Research and Public Health, 18*(18), 9677. https://doi.org/10.3390/ijerph18189677

Zhao, X., Wang, J., Li, J., Luo, G., Li, T., Chatterjee, A., Zhang, W., & He, X. (2020). The neural mechanism of aesthetic judgments of dynamic landscapes: An fMRI study. *Scientific Reports, 10*(1), Article 1. https://doi.org/10.1038/s41598-020-77658-y

Zhu, B., Cruz-Garza, J. G., Yang, Q., Shoaran, M., & Kalantari, S. (2022). Identifying uncertainty states during wayfinding in indoor environments: An EEG classification study. *Advanced Engineering Informatics, 54*, 101718. https://doi.org/10.1016/j.aei.2022.101718

Zou, Z., & Ergan, S. (2021). Evaluating the effectiveness of biometric sensors and their signal features for classifying human experience in virtual environments. *Advanced Engineering Informatics, 49*, 101358. https://doi.org/10.1016/j.aei.2021.101358

Zumthor, P. (2006). Atmospheres: Architectural environments. Surrounding objects. In *Atmospheres*. Birkhäuser. https://www.degruyter.com/document/isbn/9783764374952/html

Zur, N., Shamay-Tsoory, S., Sterkin, A., & Fisher-Gewirtzman, D. (2023). Perceived density and positive affect ratings of studio apartments: An EEG study. Architectural Science Review, 0(0), 1–11. doi:https://doi.org/10.1080/00038628.2023.2224284.

Applications of Biometrics in Architectural and Environmental Design

Alexandros A. Lavdas and Ann Sussman

Abstract Biometrics is the collection, measurement, and analysis of both physical and behavioral attributes in individuals. While the evaluation of physical traits is commonly employed for authentication purposes, the investigation of quantifiable behavioral characteristics is utilized to observe and measure how individuals react to their environment. Depending on the focus of a study, there may also be a gathering of valence data to be correlated with the physical response data. Biometric measurements can serve as a solid foundation for architects and designers to make data-driven decisions throughout the design process. In cases involving existing structures, biometric data can be applied for post-occupancy analysis, in conjunction with standard methods.

Keywords Biometrics · Eye-tracking · AI · Electrodermal activity · EEG · fMRI · Virtual reality

Introduction

"The purpose of thinking is to let the ideas die instead of us dying", as famously said by English mathematician and philosopher Alfred North Whitehead. Indeed, the mind functions as a dynamic predictor, engaging in perception, comprehension, and action. While in many species behavioral variations are selected at the individual level, the development of critical thought in humans has enabled us to create

A. A. Lavdas (✉)
Eurac Research, Institute for Biomedicine, Affiliated Institute of the University of Lübeck, Bolzano, Italy

Department of Psychology, Webster University, Athens Campus, Athens, Greece

The Human Architecture and Planning Institute, Inc., Concord, MA, USA
e-mail: alexandros.lavdas@eurac.edu

A. Sussman
The Human Architecture and Planning Institute, Inc., Concord, MA, USA

S. Kühn (ed.), *Environmental Neuroscience*,
https://doi.org/10.1007/978-3-031-64699-7_10

"avatars" that can probe the spaces of different possibilities and then die or survive, leading us to safer actions with our physical bodies. Our prefrontal cortex has been the brain locus of this profound evolutionary transformation which can be considered as the basis of abstract thinking and even for the eventual development of the concept of the value of the individual and civilization in general (Kolk & Rakic, 2022).

Viewed from an evolutionary and survival perspective, the primary purpose of the brain is to regulate behavior—as Sherrington had put it, "Life's aim is an act, not a thought" (Zeman, 2007). From this standpoint, the brain's various functions, including those considered as belonging to the cognitive sphere, such as attention, reward, memory, executive function, etc., exist to impact our actions, and even brain regions without an obvious motor function can have significant effects on motor behavior. This line of thinking has led Fine and Hayden to propose that the whole prefrontal cortex can be considered as part of the premotor cortex (Fine & Hayden, 2021).

The promotor cortex is defined by its anatomical and functional relationship with the motor cortex (Wise, 1985; Wise et al., 1997). It is not directly responsible for movement execution, like the motor cortex (Georgopoulos et al., 1986; Graziano et al., 2002; Omrani et al., 2016), but plays a role that is more supervisory, and hierarchically earlier than the motor cortex (Coallier et al., 2015; Dekleva et al., 2018), by potentiating certain motor plans and depotentiating others (Mirabella et al., 2011; Pastor-Bernier & Cisek, 2011).

There is a growing consensus that the prefrontal cortex (PFC) plays a role by holding in working memory information, such as instructions or rules, necessary for directing attention and effectively accomplishing goal-oriented tasks. These representations are believed to provide a framework for the execution of processes carried out by other brain systems responsible for task performance. There are obviously conflicting demands in maintaining and appropriately updating these contextual representations: on the one hand, PFC representations must persist throughout task execution to ensure that attention and behavior remain focused on the goal, without being disrupted by internal or external distractions. On the other hand, PFC representations should be amenable to updating when task requirements or objectives shift, a mechanism in which dopamine signaling plays a role (Cohen & Servan-Schreiber, 1992; D'Ardenne et al., 2012; Higo et al., 2011; Miller & Cohen, 2001; Zanto et al., 2011). So, the PFC seems to have an integrator function.

Groundbreaking experimental work conducted by Zeki and colleagues using fMRI has demonstrated that the experience of beauty in paintings, faces, and across different domains correlates with relative activation of a subregion of the prefrontal cortex, the mOFC (medial orbitofrontal cortex). Increased mOFC activation has also been linked to non-sensory experiences of beauty, including those stemming from mathematics and even sorrow (Ishizu & Zeki, 2011, 2013, 2014, 2017; Zeki et al., 2014). Conversely, the experience of ugliness has been associated with heightened activity in the motor cortex and the amygdala. As the amygdala is involved in the processing of fearful and threatening stimuli, its increased activity is in keeping with the perception of ugliness as something threatening. The observation of an additional positive motor cortical response to ugly stimuli, but not beautiful ones, is intriguing and aligns with the concept of the negativity bias, favoring responses to threats over attractive stimuli, essential for survival (Ohman et al., 2001; Vaish et al., 2008).

While individual brain responses to specific stimuli may be viewed as subjective, the correlation between these responses and the experience of beauty, as well as the connection between this experience and certain structural features of stimuli, supports the idea of a direct, objective causal sequence involving brain reactions, stimulus structure, and subjective experience. As Zeki et al. have recently pointed out (Zeki et al., 2022), the two lines of evidence—the ones focusing on the role of the mOFC in the experience of beauty, and those focusing on its role in decision-making—never seem to converge in the literature. And yet, it is most likely not coincidental that the mOFC plays a role in these various functions, but rather, it reflects their interconnected nature: Beauty, love, and desire, in the broadest sense, are pleasurable and rewarding experiences, which we seek in life almost continuously. To quote Zeki, "the brain is a superb measuring device, and that it continually executes measurements, be they measurements of light intensity or the degree of hate or desire." (Zeki, 2009).

This information puts the idea of biometrics in Architecture and Design more firmly in a practical context: counterintuitively for the naïve observer, by delving into the function of our subjective experience, we can create more objective evaluations of architectural forms, compared to the rather simplistic ad hoc concepts that have been widely used by architects and designers for the last century. Our interaction with architectural spaces involves a blend of sensory experiences that occur concurrently (Pallasmaa, 2014)—and the mOFC can be considered as a site for their integration, as discussed above and later.

However, in the realm of architectural design, there has been a notable absence of comprehensive analysis or detailed observation regarding the impact of the built environment on measurable human responses (➔ chapters "Aesthetics of Architecture: Emerging Insights from the Brain" by Vartanian and "The Built Environment and the Brain: Review of Emerging Methods to Investigate the Impact of Viewing Architectural Design" by Tawil and Kühn). This underscores a distinct gap in our understanding of how the built environment and human behavior correlate. There is a need for further exploration, particularly from a neuropsychological perspective, into the evaluation of both the design-to-construction phase (pre-occupancy) and the construction-to-adaptation phase (post-occupancy). The interdisciplinary field known as neuro-architecture, the exploration of how we perceive and process visual aspects of the built environment and its potential impact on us, is a relatively young field, yet it has gained growing attention in recent years (Albright, 2015; Chatterjee et al., 2021; Coburn et al., 2017; Eberhard, 2009; Higuera Trujillo et al., 2021).

Starting from the classic work of Hubel and Wiesel (Wiesel & Hubel, 1963a, b) the effect of visual input on the visual system has been well documented, and the physiological influence of visual and other environmental enrichment on the brain at the molecular, anatomical and functional level has been extensively studied in animals (for reviews, see (Rosenzweig & Bennett, 1996; Sale et al., 2009; van Praag et al., 2000)). Investigations into human perception and emotion have consistently demonstrated that the built environment profoundly shapes an individual's sensorimotor, emotional, and cognitive responses, and in recent years, many researchers have focused on the potential psychophysiological impact of the visual organization of our surroundings, either natural or artificial (Coburn et al., 2017; Dzhambov et al., 2018; Hartig, 2008; Joye, 2007a). Certain architectural properties, such as room proportions (Cruz-Garza et al., 2021), wall curvature (Shemesh et al., 2021; Vartanian et al., 2015), and window arrangement and size (Cruz-Garza et al., 2021; Ergan et al., 2019; Fich et al., 2014) have been found to correlate with stress responses.

Also, "naturalness," a term used to identify "structural" morphological properties of our environment, has attracted particular attention; for review, see (Bowler et al., 2010). Starting with the famous finding of improved recovery from surgery (Ulrich, 1984), natural surroundings have been shown to exert positive effects (sometimes by mere image viewing) on mood (Barton & Pretty, 2010), stress reduction (Valtchanov et al., 2010), concentration and working memory improvement (Berman et al., 2008, 2012; Berto, 2005; Bratman et al., 2015; Kaplan, 1995), self-perceived health (Kardan et al., 2015), self-esteem (Barton & Pretty, 2010; Pretty et al., 2007) and occurrence of criminal behavior (Kuo & Sullivan, 2001a, b).

The positive effects of nature can be viewed in the light of "the "Biophilia" hypothesis, which posits that humans possess an inherent attraction to the natural environment (Frumkin, 2001). They also suggest that exposure to such environments can influence both mental and physical well-being. Therefore, it is essential to explore the "structural" visual elements that likely mediate these effects, a task that holds significance in both fundamental scientific research and practical applications. These structural features are not limited to natural settings but can also be present in artificial environments, and are not limited to the addition of greenery (Zhong et al., 2022). They may result from deliberate efforts to incorporate "biophilic" design principles or, more intriguingly, exist within pre-modern architecture, where such characteristics are integral to the design or emerge naturally from the choice of materials and construction methods. Some scholars have proposed that exposure to this type of visual organization, even in artificial settings, can yield similar positive effects (Joye, 2007a; Salingaros, 2003, 2012; Taylor, 2006). The presence of this structural order in traditional architecture may actually mirror the structural and functional patterns of our own nervous system (Goldberger, 1996). Our evolution as part of the natural world appears to have profoundly shaped our connection to natural forms. The work of the present authors and their collaborators aligns with these views, and has, so far, mainly used eye-tracking and virtual eye-tracking technologies. These will be presented in a somewhat more detailed manner in the paragraphs that follow, while other methods will be presented more briefly.

Biometrics

Biometrics encompasses the study of analyzing and quantifying human physical and behavioral traits. Analysis of the former usually serves purposes of authentication, while analysis of the latter is commonly utilized for observing how individuals respond to environmental stimuli, in other words, to obtain the measurable human responses mentioned above. Depending on the methodology employed, measurements of valence and arousal can also be recorded to gather and analyze emotional, cognitive, and behavioral information (Mauss & Robinson, 2009). Biometric measurements primarily yield quantitative data, offering architects a foundation for data-driven decision-making throughout the design process. In cases involving existing structures, biometric data can also be harnessed for post-occupancy analysis.

Eye Tracking and AI-Simulated Eye-Tracking

Eye-tracking experiments, based on tracing the movements of the eyes, can offer valuable insights into how we perceive visual information. Eye tracking can be performed either using eye-tracking glasses or by having a camera record the movements of the eyes, in front of a calibrated monitor. The latter solution can be applied in a lab setting or a home setting, using a webcam, while glasses can also be used outdoors. Eye-tracking glasses utilize near-infrared light directed towards the pupil, causing detectable reflections in both the pupil and the cornea. These reflections—the vector between the cornea and the pupil—are tracked by an infrared camera. In the case of eye-tracking glasses, near-infrared cameras are placed around the lenses of the glasses, either on the frame or directly embedded in the lenses. These glasses enable researchers to follow the natural gaze and attention of participants in any environment. Understanding how participants behave in truly natural settings requires moving out of the lab. Eye-tracking glasses are entirely non-invasive, allowing for unprecedented insight into how people see the world. Real-world eye tracking data can be collected using world-leading, research-grade glasses from various companies with gaze, fixation, and saccade data collected and classified.

In addition, nowadays companies offer artificial intelligence (AI) applications created using a substantial amount of eye-tracking experimental data, that can predict initial viewer responses to images (for example, *Visual Attention Software (VAS)* from 3M (2021), Eyequant (2023), Attention Insight (2023), Neurons (2023) and Expoze (2023). Through scans, they produce maps showing the likelihood of fixation points and estimate the sequence of fixation points, effectively unveiling aspects of the subconscious processing of visual stimuli with exceptional accuracy. Initially, such software was developed for applications like product design, advertising, signage, and understanding unconscious user attraction. We believe that it has the potential to become a widely adopted tool for assessing and enhancing architectural and environmental design. There is a number eye-tracking simulation software options available for commercial use, which predict how users will perceive a design in the initial 3–5 s of viewing. This initial period informs us about visual processing that has taken place upon first exposure to a scene.

This technology is particularly well-suited for conducting direct comparative assessments of various structures, allowing for the evaluation of quantitative and semi-quantitative parameters. Researchers, including the current authors, have conducted eye-tracking studies involving images of architecture and constructed environments in general. These studies have employed both human volunteers for actual eye-tracking experiments (Lisińska-Kuśnierz & Krupa, 2020; Rosas et al., 2023; Suárez, 2020; Sussman & Ward, 2019; Zou & Ergan, 2018) and also 3M's artificial intelligence–based *Visual Attention Software (VAS)* (3M, 2021; Hollander et al., 2021; Lavdas & Salingaros, 2022; Lavdas et al., 2021; Salingaros & Sussman, 2020). Such studies trace the sequence and duration of the observed target, indicating when and where the observer's eyes shift from one location to another, including pre-selected Areas of Interest (AOI) (Duchowski, 2017), specific parts of the observed target or design that are of particular interest to the researchers. Tracking eye movements aids human–computer interaction (HCI) researchers in understanding visual and screen-based data processing, user behaviors, and factors that can affect the usability of system interfaces. The recorded eye-tracking data can subsequently be utilized to enhance design and interface usability. In recent years, numerous studies and experiments in the architecture, engineering, and construction (AEC) sector have explored design evaluation through eye tracking (Afrooz et al., 2014; Hollander et al., 2018; Lu & Pesarakli, 2023; Zou & Ergan, 2018).

Early gaze is guided by pre-attentive processes in the brain: neurons in the early stages of perception are specialized to detect fundamental visual attributes, such as differences in brightness and contrasting colors. As processing continues in the visual cortex, they identify characteristics like orientation, direction, and speed of movement (as reviewed by (Itti & Koch, 2001)). These visual features are computed in parallel, creating an early "saliency map" (Tollner et al., 2011). The information from this initial pre-attentive processing, lasting about 200–250 milliseconds, guides the early deployment of selective attention and, potentially, action.

Our visual system has evolved mechanisms to choose relevant or salient information to determine appropriate actions and control their execution. This selection process begins as soon as an image reaches the retina, with computation of low-level visual features starting at this level and continuing in the lateral geniculate nucleus of the thalamus and early visual cortical areas. About 10% of the retinal output follows another pathway, leading to the phylogenetically more primitive system of the superior colliculus and the pulvinar nucleus of the thalamus. This system is responsible for early responses to motion in the peripheral visual field (Banich & Compton, 2018), as well as fear responses triggered directly from the pulvinar to the amygdala, such as when we perceive potential threats, like predators (McFadyen, 2019) -amazingly, primates seem to even have a dedicated circuit for snake recognition in the pulvinar (Van Le et al., 2013).

Given the crucial role of swiftly and appropriately responding to environmental cues for survival, any failure in this early registration mechanism can lead to heightened feelings of potential danger and stress. Essentially, environments that visually deviate from what we evolutionarily consider as “safe” can trigger anxiety. Environments that are challenging to process visually may feel unsafe, as they do not allow a sense of place or “anchoring,” and research has linked stress to surroundings lacking a certain level of organized complexity (Ellard, 2015), an effect related to sensory input deprivation (Merrifield & Danckert, 2014). Visual coherence plays a critical role in efficient information processing. Whether a stimulus leads to a fear response or attraction depends on the specific stimulus. For instance, a building with a coherent design attracts visual attention, while geometric coherence can also induce fear, as seen when viewing a predator. When there are delays in initial information processing, the second evaluation stage is postponed, potentially resulting in stress and a sense of unease. Incoherence in the environment can lead to anxiety because it makes it difficult to assess whether the surroundings are friendly or threatening. Eye tracking by itself does not directly give us any information on emotions, but metrics such as facial expression analysis (see below) alongside visual attention can indicate whether the attachment was positive or negative.

Some of the eye-tracking studies and AI-simulated eye-tracking mentioned above have yielded intriguing findings. One key observation is that the initial gaze, influenced by pre-attentive processing, consistently gravitates toward the presence of people, especially their faces, even within images of architectural or urban scenes. Additionally, these studies demonstrated that the gaze is naturally drawn to specific elements like details, contrasts, and structural aspects that contribute to an overall sense of geometrical coherence (Hollander et al., 2021; Lavdas & Salingaros, 2022; Lavdas et al., 2021; Salingaros & Sussman, 2020).

The term “geometrical coherence” requires clarification here. Christopher Alexander’s pioneering work (Alexander et al., 1977) identified several parameters that foster a unique connection between viewers and their surroundings. Research has shown that exposure to certain fractal visual patterns, both in nature and in architecture, can have measurable physiological effects (Taylor, 2006; Taylor et al., 2011). While the benefits of exposure to natural scenes have been documented extensively (including improved recovery from surgery), similar effects can result from artificial environments that mimic nature’s geometrical qualities (Frumkin, 2001; Joye, 2007b). Notably, these qualities extend beyond fractal properties, as the work of Nikos Salingaros has highlighted (Salingaros, 1995; Salingaros & Mehaffy, 2006). Our processing system is finely attuned to the visual complexity of the natural environment, particularly responding positively to high levels of organized complexity. Fractals are a critical part of this effect, but they do not encompass the entire range of connecting qualities (Salingaros, 2013). The experience of beauty, often overlooked in modern architecture, results from a multiparametric evaluation of a scene and should be considered a necessity, not a luxury (Zeki, 2019).

The present authors have utilized *3M’s VAS* to analyze various images, including buildings in their original states, digitally altered versions, geometric patterns, and sketches. Our analysis with the software has revealed notable disparities between the buildings when compared side by side using scans. Some buildings immediately capture viewers' gaze due to their coherent overall design, while others actively deter, with viewers' gazes directed away from the building's façade. These observations suggest that the underlying visual processes are closely tied to the mathematical coherence or organized complexity present in the design.

The capacity to predict a user's engagement with a building's design, particularly as experienced through its facade, is important both as a study subject and because of potential practical applications.

We have previously (Lavdas & Salingaros, 2022; Lavdas et al., 2021; Salingaros & Sussman, 2020) discussed the differences between designs that create a fragmented gaze heatmap, that falls apart upon zooming in to the image and designs that possess a coherence obvious from the distribution of the gaze heatmap, both initially and at subsequent zooming-in levels (Figs. 1 and 2). This coherent state of unconscious human interest appears to fulfill a profound neurological need. Nevertheless, a nearly evenly distributed visual interest should not be confused with a uniform but empty design, where no point captures the gaze and the scan instead drifts toward the corners or periphery of the image. The apparent uniformity of unconscious visual appeal in a highly coherent structure is remarkable because such designs are far from uniform. Organic designs are highly intricate, rich in contrast

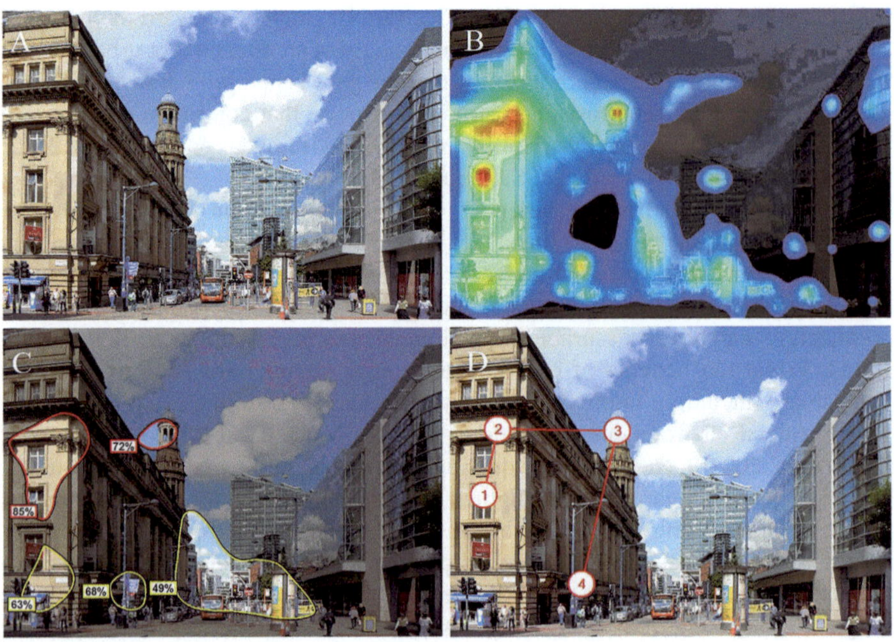

Fig. 1 From Lavdas et al., 2021: Image from the center of Manchester, UK, at the junction of Market and Cross Streets, with the analysis output of the 3M-VAS software. (**a**): Original image. (**b**): Heatmap. (**c**): Simplified version of the heatmap results, showing only areas that are most likely to be seen within the 3–5 s first glance time period. (**d**): Gaze sequence of the four most likely gaze locations, in their most probable viewing order. (Photo: Alexandros Lavdas)

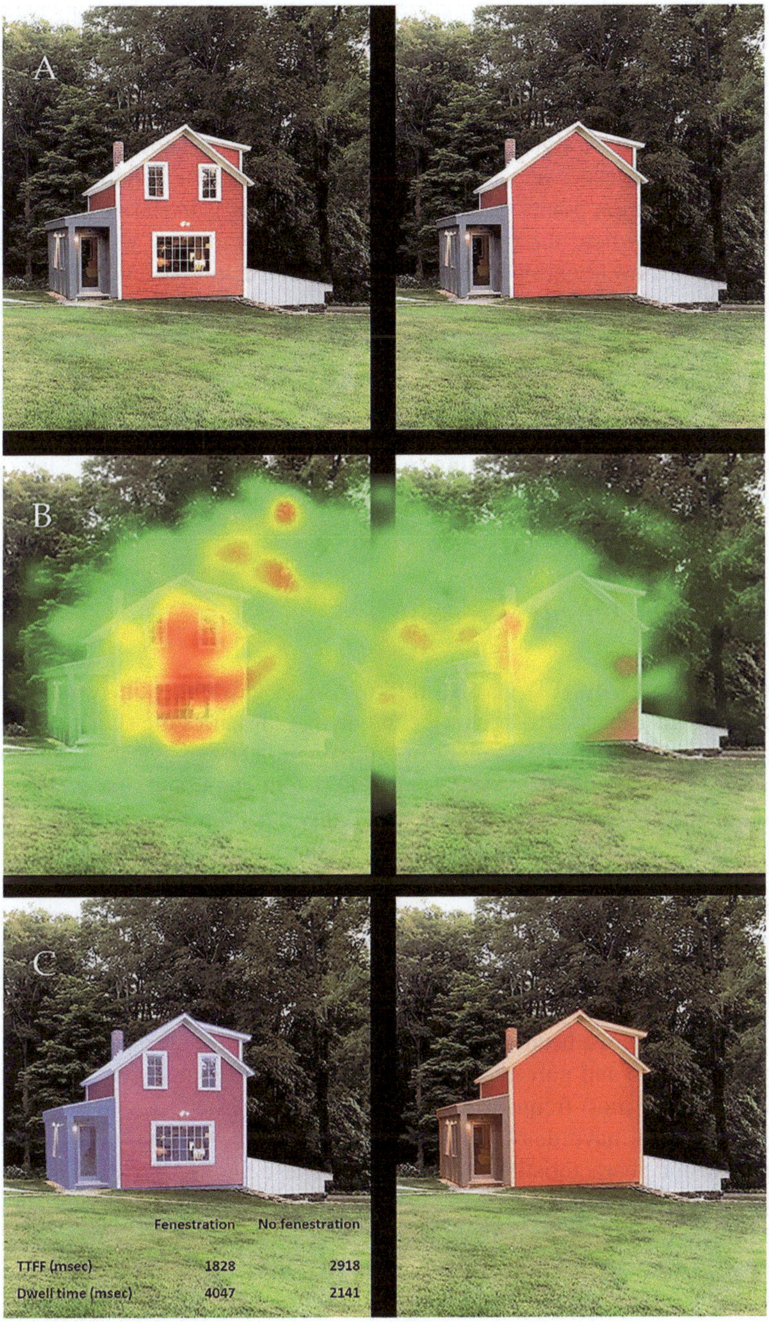

Fig. 2 From Rosas et al., 2023: Eye-tracking results of a paired image of a residence in Dutchess County, NY and a processed version without windows. **(a)** Original and processed images. **(b)** Heatmap of this pairing. **(c)** Tinted areas indicating the Areas of Interest for calculation. The time to first fixation was shorter for the face-like original façade, and the dwell time (indicated by the heatmap) was longer. (Photos: Ann Sussman)

and details. These two distinct situations represent opposite states of the subliminal reality of a viewer's engagement: one is empty and lifeless, showing a significant gap in the scan that the gaze avoids, while the other is displaying coverage that is scalable through iterative zoom levels.

These findings have significant implications for how we experience architecture. Modern buildings, especially those with plain glass facades, tend to be glanced at without clear fixation points on the building itself. This is in stark contrast to traditional buildings, where the presence of organized complexity with nested hierarchies seems to draw attention to the entirety of the structure, in a way that facilitates understanding the relationship between the design's details and the overall structure. For instance, viewers facing older buildings can quickly locate an entrance. Pre-attentive processing appears to attract people to some structures while overlooking others. To create inviting spaces, buildings should incorporate fractal scaling, organized complexity, and repeating symmetries. These elements guide viewers in the right direction, promoting a sense of safety, especially in outdoor environments.

Electrodermal Activity

Electrodermal activity (EDA), also referred to as galvanic skin response (GSR) (Benedek & Kaernbach, 2010), is a technique that measures the skin's conductivity and, through it, can indicate changes in sympathetic nervous system activity which, in turn, reflect emotional states (Shi et al., 2007). Our emotional state adapts in reaction to our surroundings, whether it is fear, danger, happiness, or other emotionally significant stimuli, and this adaptation is accompanied by an elevation in sweat gland activity. EDA/GSR provides a valuable measure of emotional arousal, although obviously it cannot identify the specific emotion being experienced, but only its impact on the periphery (Bakker et al., 2011). Therefore, the identical arousal response could be elicited by positive or negative feelings. A standard GSR setup includes two electrodes, an amplifier and an analog-to-digital signal converter. Human fingers, palms, head, and the soles of our feet have a very high sweat gland density (400/cm^2), and galvanic skin response is usually measured in one of those areas. Fingers are most frequently used, as placing the electrodes there is less intrusive. GSR studies have demonstrated distinct responses to the built environment based on design characteristics (Ellard, 2015). These findings provide evidence that individuals' emotional, cognitive, and physiological states undergo alterations depending on the built environment to which they are exposed.

EDA is frequently used in combination with other techniques to create a more complete understanding of the individual's reaction to the environment. For example, a study conducted by Ergan et al. (2019), where eye tracking and facial electromyography (fEMG, see below) were employed to gather data on participants' visual attention and facial expressions in response to virtual environments (VE), showed that parameters like the presence or absence of natural lighting, the level of luminance in rooms, the color of walls and openness of spaces, and the presence or absence of outside landmarks and of a visible entrance can change the way people perceive the space, and this perception is reflected on the parameters studied (➔ chapter "Extended Reality in Environmental Neuroscience Research" by Li et al.). GSR sensor data showed that skin conductance levels were higher in a negative environment reflecting increased stress, compared to a positive environment. Moreover, heart rate variability (HRV), measured through a photoplethysmogram (PPG, see below), indicated a greater emotional response in the negative space compared to baseline values and those obtained in the positive space. In (Higuera Trujillo et al., 2019) study, stress reduction was monitored using an EDA sensor in different pediatric waiting rooms, first in situ, and then using virtual reality setups, that combined visual features from different waiting rooms, with auditory and olfactory elements, with results suggesting that a combination of environmental satisfaction sources creates an important synergistic effect. In the (Vecchiato et al., 2015) study, on the other hand, while skin conductance levels did not show significant differences, there was a noteworthy increase in heart rate during the observation of virtual environments.

Electrocardiography

The monitoring of the electrical signal emanating from the heart's muscle contractions, revealing its activity and indirectly reflecting blood flow within the heart, is known as electrocardiography (ECG). The heart's rhythm responds to various environmental factors, resulting in changes in its intensity, regularity, and speed. Hence, the ECG, reflecting autonomic nervous system activity, provides a clear and noninvasive means of collecting information about both physiological and emotional conditions. Heart rate variability (HRV) represents a good indicator of autonomic nervous system activity, linked to emotional regulation and cognitive processes (Ferdinando et al., 2014; Lane et al., 2009; Sztajzel, 2004; Williams et al., 2015).

ECG sensors are considered the most suitable source for HRV analysis, but they necessitate the use of intrusive electrodes on the arms, legs, and chest, and of conductive materials (Davila et al., 2017). PPG sensors, integrated into wearable devices like smartwatches and smartphones, are a less intrusive solution, commonly employed to monitor heart activity. To utilize the PPG signal for HRV assessment, accurate measurement of the time interval between consecutive peaks is essential (Alqaraawi et al., 2016). However, HRV alone is not a sufficiently accurate indicator of emotions, given the significant variation in human emotions among individuals and, much like EDA, gives no information about the nature of the stimuli that caused the observed changes. Researchers can synchronize ECG data with data from other sensors to gain a comprehensive understanding of how the heart responds to and is influenced by different stimuli (Fich et al., 2014). The studies mentioned earlier, (Ergan et al., 2019) (Vecchiato et al., 2015) are examples of HRV modulation by differences in presented VEs.

Electroencephalography

Electroencephalography (EEG) records the electrical signals produced by the postsynaptic activity of large neuronal assemblies firing coherently from the higher layers of the cortex, using electrodes placed on the surface of the scalp, and outputs the recording as a series of underlying brain waves (➔ chapter "EEG for the Study of Environmental Neuroscience" by Grassini). This activity can be seen in waves of different frequencies, depending on the location and task at hand. EEG waveforms are categorized based on their frequency range using Greek numerals (Nayak & Anilkumar, 2023). The most commonly examined waveforms include delta (0.5–4 Hz), theta (4–7 Hz), alpha (8–12 Hz), sigma (12–16 Hz), and beta (13–30 Hz). From language and visual processing to executive function, memory encoding, and emotion, EEG data carry information related to concentration and problem-solving (gamma), anxiety, external attention, and relaxation (beta), contemplative, restful, highly relaxed, and passive attention states (alpha), drowsiness, deep relaxation, and inward focus (theta), and sleep and dreaming (delta) (Abhang et al., 2016; Woaswi et al., 2016). Technological advancements have significantly improved EEG recording systems, providing enhanced signal-processing capabilities for innovative applications. The introduction of wireless head-mounted EEG systems has notably improved portability, and made EEG systems useable in mobile electroencephalography, which is ideally suited for studying responses to the built environment (Debener et al., 2012; Kam et al., 2019; Mavros et al., 2016). The number of electrodes used for brain measurement, electrode placement systems, and signal-denoising techniques varies significantly. Electrodes can range from 3 up to 256 depending on the aims of the study, attached to the scalp over the brain areas of interest. Two of the electrodes act as a reference, and the EEG is measured as the voltage difference between each electrode and the references. Traditional wet EEG setups, which involve the application of gel or paste to facilitate conductivity between the skin and the electrode, are time-consuming and require specialized expertise. In contrast, dry electrodes, composed of a single metal facilitating conductivity between the skin and the electrode, eliminate these complexities. Dry electrodes tend to result in higher noise levels (Radüntz, 2018), however, the signal quality of wireless dry EEG systems rivals that of conventional wet EEG setups (Kam et al., 2018).

EEG data processing can be difficult for non-experts, encompassing several steps such as preprocessing, feature extraction, feature selection, and classification. From the perspective of designers, engaging with human biometric responses for design evaluation remains a complex endeavor that necessitates expertise in both neuroscience and psychology. However, newer solutions may provide usability for the non-expert. For example, (Aspinall et al., 2013) demonstrated that a commercial and low-cost EEG technology, (*Emotiv EPOC*), is sensitive to detecting patterns consistent with emotional change in response to different environmental settings, mirroring subjective preferences.

External factors can potentially disrupt biometric measurements, requiring validation from the participants themselves. Consequently, self-reported measurements have been broadly integrated into many studies, employing diverse methodologies and systems. In many instances, the outcomes of self-reported assessments have aligned with the biometric data recorded e.g., an individual expressing relaxation, which was subsequently confirmed by EEG readings, for example (Bower et al., 2019).

EEG data can provide valuable insights into user interactions with their surroundings. For example, the study conducted by (Ergan et al., 2019) showed different responses to different architectural designs, combining EEG, GSR, and EEG data, as already mentioned, and (Vecchiato et al., 2015) demonstrated that the perception of pleasantness in architectural spaces can be objectively validated. Many additional studies have used EEG either for the study of user reactions to architectural/urban designs (Banaei et al., 2017; Bower et al., 2022a, b; Karandinou & Turner, 2017; Mavros et al., 2016; Rounds et al., 2020) or study aspects of Biophilia (Aspinall et al., 2013; Chang & Chen, 2005; Neale et al., 2017; Roe et al., 2013). Of particular interest is the study of (Hägerhäll et al., 2008): when using fractal patterns as stimuli, those having a fractal dimension of approximately 1.3 elicited the most significant changes in the EEG responses of the subjects. These fractals trigger the highest alpha response in the frontal cortex, suggesting that they are particularly restorative and calming, but also the strongest beta response in the parietal region, indicating heightened activation in processing spatial characteristics. This highlights an intriguing interaction between these brain areas in response to mid-range fractals and aligns with the idea emerging from both perception studies and functional magnetic resonance imaging (fMRI) (see below) that some fractals possess a visually distinct quality that gives them a special status in terms of their processing by our perceptual system.

Another important study (Djebbara et al., 2019) demonstrated that the perception of architectural space is influenced by affordance, and the brain's predictive mechanisms can modify spatial perception—a phenomenon not unrelated to the "avatar" concept discussed in the introduction. Using EEG and virtual reality, the study indicates that cortical potentials change based on the affordances for bodily motion provided by the physical setting—in other words, that cognition is intrinsically linked to the potential for action. This suggests that actions play an active role in shaping perception, thereby directly influencing how we perceive our surroundings, and also that navigating through space involves the ongoing construction of an anticipation of a world filled with potential actions. This is something worth to be noted by architects: the progression through an architectural space continually reshapes perception and the overall experiential quality of that space.

Facial Electromyography/Facial Expression Analysis

While recognizing and identifying facial expressions of emotion is an essential interpersonal skill in which individuals typically perform well, there are considerable individual differences in accuracy (Künecke et al., 2014), and a more objective and quantifiable measure is used in quantitative research. Facial electromyography (fEMG) is a psychophysiological method used to detect electrical potentials generated by facial muscles (Read, 2015). EMG activity, measured in microvolts, is linearly related to the amount of muscle contraction as well as the number of contracted muscles. fEMG allows for a better understanding not only of muscle movements and activity themselves but, importantly in the present context, also of their association with certain emotions and behavioral outcomes with which facial expressions correlate (Boxtel, 2010; Dimberg, 1990). The Facial Action Coding System (FACS) is a means of correlating muscle actions during the expression of basic emotions, with those emotions. FACS dissects all observable facial movements into 44 visually distinct and anatomically separate muscle action units (Ekman et al., 1997, 2005). Implementing FACS can range from real-time observations to video-recorded interactions conducted in laboratory conditions (Wolf, 2015), including Automated Facial Coding (AFC) (Valstar et al., 2012).

Existing AFC software can categorize facial movements into emotional or cognitive states (Lewinski et al., 2014). AFC enables the analysis of facial expressions in various contexts, where individuals are observed in their natural environment without the constraints of technical equipment. EMG software can often be integrated with other data analysis tools and software platforms, including EEG, eye tracking, facial expression analysis (see below). EMG devices can be wearable, compact, and lightweight, so they can be used outside a laboratory setting, and have already been implemented, in combination with other techniques, in neuroarchitectural investigations (Balakrishnan & Sundar, 2023; Chang & Chen, 2005).

As an alternative to using fEMG, facial expression analysis can be performed using a camera, even using the participants' computers at home, the same way as eye-tracking at home is performed, using webcams. In fact, data for both eye-tracking and facial data can be collected at the same time, and analyzed either separately or together, with cross-correlation. The company *iMotions* (2023) offers such a solution, with its facial expression analysis module integrating the automated facial coding engines *AFFDEX* by *Affectiva* and *Realeyes*. Using a webcam, one can synchronize expressed facial emotions with stimuli directly in the *iMotions* software, and simply import videos and carry out the relevant analysis.

Tracking facial expressions can, when used in controlled contexts and in collaboration with other biosensors, be a powerful indicator of emotional experiences. Just like the other methods discussed so far, while facial expression data by themselves may not be enough as a means of understanding the responses to the stimulus to which they correlate, they provide insights into its valence and, as part of a synthesis of multiple data streams, into a better understanding of interactions with the environment.

Functional Magnetic Resonance Imaging

Functional Magnetic Resonance Imaging (fMRI), generates images based on detected blood flow which is correlated with neuronal activation (Mier & Mier, 2015). With fMRI, investigators can correlate the brain activity modulated in specific areas, in response to manipulations of independent stimulus variants. Obviously, the strength of the inferences made from the experimental data depends on the functional specificity associated with the investigated region, in other words, the ability to infer a specific function based on the activation of that region (Bub, 2000; Poldrack, 2006). The shortcomings of the technique have to do with the motion restrictions of the subjects and the significantly higher costs in comparison to other techniques, mainly because of equipment maintenance and personnel expenses.

fMRI studies have been used to investigate responses to different scenes, for example, natural vs urban (Kim et al., 2010) or to various parameters of architecture and design, for example, curved vs rectilinear designs (Bar & Neta, 2007; Vartanian et al., 2013) or ceiling height (Vartanian et al., 2014) examining the relation between subject ratings and brain activations. These ratings can be judged by simple tests (such as approach-avoidance tests), or can also be correlated to higher level components: coherence (ease of organizing and comprehending a scene), fascination (informational richness and generated interest), and hominess (personal ease and comfort) (Chatterjee et al., 2021).

Spatial perception, i.e., the appreciation of attributes like height, depth, and geometry, is processed in the occipital, parietal, and temporal lobes (Mallgrave, 2010). Other activations include the thalamus, the parahippocampal region (related to selective responses to places and scenes) and the nucleus accumbens and medial prefrontal cortex (associated with emotional perception, pleasure-related stimuli, and learning). Intriguingly, activations include the Fusiform Face Area (FFA), which has frequently been shown to activate when individuals perceive faces, which seems to be responsible for neural representations of architectural styles (Choo et al., 2017). This tapping into the face recognition neural mechanism to categorize architectural styles is very interesting, given the importance of this mechanism from a very early age on, and its obvious survival value. Infants are instantly attracted to faces, and, amazingly no more than a week after birth they tend to look longer at faces deemed attractive by adults (Slater, et al., 1998), a phenomenon that generalizes across race, sex, and age by the age of 6 months (Langlois, et al., 1991). This is revealing of the early emergence of a capability that is of much higher resolution than simply spotting an abstract facial pattern, and fits well with the adaptation of this same mechanism for stylistic recognition. At the same time, it is challenging to current views of aesthetics as purely socially constructed.

Equally challenging to these views is the activation of the Default Mode Network (DMN) of the brain during the perception of fractals (Fischmeister et al., 2017; Martins et al., 2014). This finding, taken together with findings on the role of fractals in stress reduction (Joye, 2007a, b; Taylor, 2006; Taylor et al., 2011), and data related to specific EEG responses to fractal patterns, as mentioned earlier (Hägerhäll et al., 2008) supports the notion of their privileged status in terms of perceptual fluency.

Interestingly, externally induced meditative states, brought about as a response to exposure to buildings designed for contemplation (churches, etc.) significantly disrupted the DMN activity in architect volunteers, reducing "internal dialogue" (Bermudez et al., 2017) much like what has been found for internally induced meditative states (Brewer et al., 2011; Farb et al., 2007). So, from the early retinal and thalamic responses to specific patterns, mentioned in the introduction, all the way to high-level cortical areas, our responses to architecture and design seem to be deeply rooted in the structural and functional properties of our nervous system.

A number of fMRI studies have focused on the neural correlates of beauty (Chatterjee et al., 2009; Di Dio & Gallese, 2009; Ishai, 2007; Kawabata & Zeki, 2004; Kim et al., 2010; Vartanian & Goel, 2004). The work of Zeki and colleagues mentioned earlier, has managed to correlate activity in a specific part of the prefrontal cortex, the mOFC, with the experience of beauty in paintings, faces, and across different domains, including non-sensory experiences of beauty (Ishizu & Zeki, 2011, 2013, 2014, 2017; Zeki et al., 2014). Conversely, the experience of ugliness across these fields has been associated with heightened activity in the motor cortex and the amygdala, in keeping with the perception of ugliness as something threatening. For the experience of beauty in faces, it has been proposed that it is the collective specific pattern-based activity in face perceptive areas, with activity in the mOFC as a correlate, that constitutes the neural substrate of the experience of facial beauty (Yang et al., 2021). Given that neural representations of architectural styles seem to reside in the FFA, as mentioned earlier (Choo et al., 2017), it would be interesting to investigate the existence of a similar pattern of co-activations that would correlate with the perception of architectural beauty. Furthermore, it is tempting to consider that our ability to appreciate architectural beauty—and, hence, historically, our ability to create it—is a by-product of our exquisite sensitivity to facial features, and of our ability to rank them in a beauty hierarchy. Such a ranking could represent an evolutionary advantage, as a mechanism of selecting the best mate for reproduction. Given the complexity of human cognition, this selection need not only be based on strictly "reproductive" criteria (such as body type and perceived health) but could also be related to other signals of perceived compatibility, detected by the appreciation of facial features.

Using Virtual Reality Environments

Virtual Reality (VR) environments (VE) (➔ chapter "Extended Reality in Environmental Neuroscience Research" by Li et al.) hold significant potential both for neuroscientists and psychologists on the one hand, for basic research purposes, as well as for architects, designers, and stakeholders on the other, in applications ranging from the early design phases to the real estate sector.

The use of VE offers the advantage of conducting all experimental research in a laboratory under controlled conditions which, depending on the technique employed, can be anything from a convenience to an absolute necessity (like in fMRI), without the compromises involved in using 2-dimensional images. For fMRI, VR represents a means of enhancing its "ecological validity" (Reggente et al., 2018).

The aspects of user/environment interaction in VR can be divided into three distinct focal areas: interaction, immersion, and perception/imagination (Diemer et al., 2015; Li et al., 2019). Interaction in VR pertains to the exchange of actions and reactions between users and the virtual scene, in the same manner as it would take place in a natural scene. Immersion refers to users feeling like an integral part of the virtual world depicted in the VR scene. Imagination involves leveraging multidimensional perceptual information provided by VR scenes to evoke the same sensations experienced in the real world. A high pixel density and refresh rate are essential for achieving immersion, interaction, and realism (Weier et al., 2017). The screen resolution of the display and rendering quality of content, including an adequate frame refresh rate, significantly impact the perceived quality of immersive VEs (Zou et al., 2020), consequently affecting the quality of recorded biometric data (Davis et al., 2016), something that applies to both VR and conventional image/video stimuli.

VEs have been used in many studies related to architecture and interior and urban design, for example adding virtual plants (Mostajeran et al., 2023) or manipulating ceiling height, colors, wall curvature, and surface textures (Banaei et al., 2019; Coburn et al., 2020; Elver Boz et al., 2022; Lee et al., 2023; Tawil et al., 2022; Zou et al., 2019). Interestingly, one of the studies focusing on curvature concluded that individuals without design backgrounds tend to favor spaces with curvilinear shapes, whereas designers exhibit a preference for spaces with sharp angles (Shemesh et al., 2016)—a finding that raises, once again, the practical issue of the discord between designers and users. At the same time, there are studies employing VR and Body Sensors, incorporating emotional responses, such as skin conductance, and heart rate, as already mentioned in previous sections, for example (Ergan et al., 2019; Fich et al., 2014; Higuera Trujillo et al., 2019; Vecchiato et al., 2015), while a study by Djebbara and colleagues (Djebbara et al., 2019), as discussed earlier, investigated the effect of affordances on cortical responses.

VE settings can represent an almost limitless repository of potential configurations for research. However, the issue of the incompleteness inherent in virtual experiences presents a potential drawback (Kalantari & Neo, 2020; Valentine, 2023). One significant drawback of VR lies in its inability to replicate gravitational and accelerational sensations, except in specialized setups, which consequently hinders multisensory integration. Undoubtedly, these factors distinguish the VR experience from reality, leading to an argument that, in studies focusing on simple reactions to images, there are scenarios where displaying images on a computer screen might be more suitable than utilizing VR headsets. This is because, in the case of a screen, we perceive what we anticipate, such as a 2-D image on a flat surface, without the perceptual dissonance associated with VR headsets. On the other hand, 2-D representations are inadequate for navigation in 3-D space, and this is where the major advantage of VR lies, provided the stimuli are designed in such a way, that minimize perceptual dissonance.

As we have previously argued (Gorichanaz et al., 2023), what is required is new studies that specifically examine the consequences of altering just one variable, namely, the aspects of the presented geometries, one by one. All other factors, such as exposure duration, characteristics of the headsets, and operating systems, should remain consistent. From an experimental standpoint, this approach appears to be the most suitable way forward.

On the application side, the utilization of VEs as a research tool aligns with a broader trend, emphasizing rigorous evidence-based design (EBD). EBD involves a systematic process that entails the deliberate, explicit, and informed utilization of the most current and relevant evidence derived from research and practice to inform critical decision-making (Hamilton & Watkins, 2008). Modern wearable sensors can detect physiological changes triggered by user's emotions. Such measurements, when they are not accompanied by some measure of valence, cannot by themselves lead us to conclusions about the positive or negative nature of the affect, but can be used as part of broader studies on the user's well-being state (Fernández-Caballero et al., 2016; Marín-Morales et al., 2018). This potential for easily conducting relevant experiments by adapting a virtual environment to influence the users' emotions in a positive direction presents significant opportunities for enhancing human health in real-world environments. We would like to argue that the evidence sought in EBD should not only pertain to some basic, commonly examined, parameters, as we have seen so far (shapes, ceiling heights, colors, etc.) but also to a more fine-grained study of the geometries of structures under investigation, under the prism of what has been discussed in the introductory section of this chapter. At the same time, we are at a stage where VEs should de facto be seen as more than models for the real world; VE, and indeed the Metaverse as a whole, should be seen as an increasingly important part of life, especially in the population of young people. Virtual Architecture should be the subject of research about its impact, both for its potential negative impacts (Valentine, 2023), and as a source of salutogenic effects (Gorichanaz et al., 2023; Lavdas et al., 2023). Negative impacts could stem both from problems with multisensory integration and gravitational perception—what we referred to as perceptual dissonance, which applies to all VR applications—and also by the designs themselves, if they replicate the current prevailing model in the "real" world. At the same time, as we already know about the positive effects of exposure to "virtual nature" (Chan et al., 2021) as well as of the presence of greenery in a virtual urban environment (Hian et al., 2021) and interiors (Mostajeran et al., 2023), further beneficial outcomes can be expected (Fernández-Caballero et al., 2016; Koohsari et al., 2023; Lee et al., 2023; Marín-Morales et al., 2018; Petrigna & Musumeci, 2022), if virtual environments are intentionally crafted to create a therapeutic setting by incorporating more of the right attributes of genuinely biophilic design, along the lines of what has been discussed already.

Conclusions

We are living in a remarkable time, where a plethora of powerful new tools can be used to better understand the mechanisms that drive the human experience of place and, more specifically, to understand how design can impact us in positive and negative ways. This is of profound interest, not only as a research topic but also for its potential applications. We have the opportunity and obligation to use this newfound understanding for the common good.

References

3M. (2021). *3M. Visual attention software*. https://www.3m.com/3M/en_US/visual-attention-software-us/. 3M Corporation. Accessed 16 May 2021

Abhang, P. A., Gawali, B. W., & Mehrotra, S. C. (2016). Chapter 1 – Introduction to emotion, electroencephalography, and speech processing. In P. A. Abhang, B. W. Gawali, & S. C. Mehrotra (Eds.), *Introduction to EEG- and speech-based emotion recognition* (pp. 1–17). Academic. https://doi.org/10.1016/B978-0-12-804490-2.00001-4

Afrooz, A., White, D., & Neuman, M. (2014). Which visual cues are important in way-finding? Measuring the influence of travel mode on visual memory for built environments. *Assistive Technology Research Series, 35*. https://doi.org/10.3233/978-1-61499-403-9-394

Albright, T. D. (2015). Neuroscience for architecture. In S. Robinson & J. Pallasmaa (Eds.), *Mind in architecture: Neuroscience, embodiment, and the future of design* (pp. 197–217). MIT Press.

Alexander, C., Ishikawa, S., Silverstein, M., Jacobson, M., Fiksdahl King, I., & Angel, S. (1977). *A Pattern Language*. Oxford University Press.

Alqaraawi, A., Alwosheel, A., & Alasaad, A. (2016). Heart rate variability estimation in photoplethysmography signals using Bayesian learning approach. *Healthc Technol Lett, 3*(2), 136–142. https://doi.org/10.1049/htl.2016.0006

Aspinall, P., Mavros, P., Coyne, R., & Roe, J. (2013). The urban brain: Analysing outdoor physical activity with mobile EEG. *British Journal of Sports Medicine, 49*, 272. https://doi.org/10.1136/bjsports-2012-091877

Bakker, J., Pechenizkiy, M., & Sidorova, N. (2011, 11 December). What's your current stress level? Detection of stress patterns from GSR sensor data. *2011 IEEE 11th international conference on data mining workshops*.

Balakrishnan, B., & Sundar, S. S. (2023). Capturing affect in architectural visualization – A case for integrating 3-dimensional visualization and psychophysiology. *Communicating Space(s) [24th eCAADe Conference Proceedings / ISBN 0-9541183-5-9] Volos (Greece) 6–9 September 2006, pp. 664–669.*

Banaei, M., Hatami, J., Yazdanfar, S. A. A., & Gramann, K. (2017). Walking through architectural spaces: The impact of interior forms on human brain dynamics. *Frontiers in Human Neuroscience, 11*, 477. https://doi.org/10.3389/fnhum.2017.00477

Banaei, M., Ahmadi, A., Gramann, K., & Hatami, J. (2019). Emotional evaluation of architectural interior forms based on personality differences using virtual reality. *Frontiers of Architectural Research, 9*, 138. https://doi.org/10.1016/j.foar.2019.07.005

Banich, M. T., & Compton, R. J. (2018). *Cognitive neuroscience*. Cambridge University Press.

Bar, M., & Neta, M. (2007). Visual elements of subjective preference modulate amygdala activation. *Neuropsychologia, 45*(10), 2191–2200. https://doi.org/10.1016/j.neuropsychologia.2007.03.008

Barton, J., & Pretty, J. (2010). What is the best dose of nature and green exercise for improving mental health? A multi-study analysis. *Environmental Science & Technology, 44*(10), 3947–3955. https://doi.org/10.1021/es903183r

Benedek, M., & Kaernbach, C. (2010). A continuous measure of phasic electrodermal activity. *Journal of Neuroscience Methods, 190*(1), 80–91. https://doi.org/10.1016/j.jneumeth.2010.04.028

Berman, M. G., Jonides, J., & Kaplan, S. (2008). The cognitive benefits of interacting with nature. *Psychological Science, 19*(12), 1207–1212. https://doi.org/10.1111/j.1467-9280.2008.02225.x

Berman, M. G., Kross, E., Krpan, K. M., Askren, M. K., Burson, A., Deldin, P. J., Kaplan, S., Sherdell, L., Gotlib, I. H., & Jonides, J. (2012). Interacting with nature improves cognition and affect for individuals with depression. *Journal of Affective Disorders, 140*(3), 300–305. https://doi.org/10.1016/j.jad.2012.03.012

Bermudez, J., Krizaj, D., Lipschitz, D. L., Bueler, C. E., Rogowska, J., Yurgelun-Todd, D., & Nakamura, Y. (2017). Externally-induced meditative states: an exploratory fMRI study of architects' responses to contemplative architecture. *Frontiers of Architectural Research, 6*(2), 123–136. https://doi.org/10.1016/j.foar.2017.02.002

Berto, R. (2005). Exposure to restorative environments helps restore attentional capacity. *Journal of Environmental Psychology, 25*(3), 249–259. https://doi.org/10.1016/j.jenvp.2005.07.001

Bower, I., Tucker, R., & Enticott, P. G. (2019). Impact of built environment design on emotion measured via neurophysiological correlates and subjective indicators: A systematic review. *Journal of Environmental Psychology, 66*, 101344. https://doi.org/10.1016/j.jenvp.2019.101344

Bower, I. S., Clark, G. M., Tucker, R., Hill, A. T., Lum, J. A. G., Mortimer, M. A., & Enticott, P. G. (2022a). Built environment color modulates autonomic and EEG indices of emotional response. *Psychophysiology, 59*(12), e14121. https://doi.org/10.1111/psyp.14121

Bower, I. S., Clark, G. M., Tucker, R., Hill, A. T., Lum, J. A. G., Mortimer, M. A., & Enticott, P. G. (2022b). Enlarged interior built environment scale modulates high-frequency EEG oscillations. *eNeuro, 9*(5), ENEURO.0104. https://doi.org/10.1523/ENEURO.0104-22.2022

Bowler, D. E., Buyung-Ali, L. M., Knight, T. M., & Pullin, A. S. (2010). A systematic review of evidence for the added benefits to health of exposure to natural environments. *BMC Public Health, 10*. https://doi.org/10.1186/147-2458-10-456

Boxtel, A. (2010). Facial EMG as a tool for inferring affective states.

Bratman, G. N., Daily, G. C., Levy, B. J., & Gross, J. J. (2015). The benefits of nature experience: Improved affect and cognition. *Landscape and Urban Planning, 138*, 41–50. https://doi.org/10.1016/j.landurbplan.2015.02.005

Brewer, J., Worhunsky, P., Gray, J., Tang, Y.-Y., Weber, J., & Kober, H. (2011). Meditation experience is associated with differences in default mode network activity and connectivity. *Proceedings of the National Academy of Sciences of the United States of America, 108*, 20254–20259. https://doi.org/10.1073/pnas.1112029108

Bub, D. N. (2000). Methodological issues confronting pet and fMRI studies of cognitive function. *Cognitive Neuropsychology, 17*(5), 467–484. https://doi.org/10.1080/026432900410793

Chan, S., Qiu, L., Esposito, G., Mai, K., Tam, K.-P., & Cui, J. (2021). Nature in virtual reality improves mood and reduces stress: Evidence from young adults and senior citizens. *Virtual Reality*. https://doi.org/10.1007/s10055-021-00604-4

Chang, C.-Y., & Chen, P.-K. (2005). Human response to window views and indoor plants in the workplace. *HortScience: a publication of the American Society for Horticultural Science, 40*, 1354–1359. https://doi.org/10.21273/HORTSCI.40.5.1354

Chatterjee, A., Thomas, A., Smith, S., & Aguirre, G. (2009). The neural response to facial attractiveness. *Neuropsychology, 23*, 135–143. https://doi.org/10.1037/a0014430

Chatterjee, A., Coburn, A., & Weinberger, A. (2021). The neuroaesthetics of architectural spaces. *Cognitive Processing, 22*(Suppl 1), 115–120. https://doi.org/10.1007/s10339-021-01043-4

Choo, H., Nasar, J. L., Nikrahei, B., & Walther, D. B. (2017). Neural codes of seeing architectural styles. *Scientific Reports, 7*, 40201. https://doi.org/10.1038/srep40201

Coallier, E., Michelet, T., & Kalaska, J. F. (2015). Dorsal premotor cortex: Neural correlates of reach target decisions based on a color-location matching rule and conflicting sensory evidence. *Journal of Neurophysiology, 113*(10), 3543–3573. https://doi.org/10.1152/jn.00166.2014

Coburn, A., Vartanian, O., & Chatterjee, A. (2017). Buildings, beauty, and the brain: A neuroscience of architectural experience. *Journal of Cognitive Neuroscience, 29*(9), 1521–1531. https://doi.org/10.1162/jocn_a_01146

Coburn, A., Vartanian, O., Kenett, Y., Nadal, M., Hartung, F., Hayn-Leichsenring, G., Navarrete, G., González-Mora, J., & Chatterjee, A. (2020). Psychological and neural responses to architectural interiors. *Cortex, 126*, 217. https://doi.org/10.1016/j.cortex.2020.01.009

Cohen, J. D., & Servan-Schreiber, D. (1992). Context, cortex, and dopamine: A connectionist approach to behavior and biology in schizophrenia. *Psychological Review, 99*(1), 45–77. https://doi.org/10.1037/0033-295x.99.1.45

Cruz-Garza, J., Darfler, M., Rounds, J., Gao, E., & Kalantari, S. (2021). EEG-based investigation of the impact of classroom design on cognitive performance of students.

D'Ardenne, K., Eshel, N., Luka, J., Lenartowicz, A., Nystrom, L. E., & Cohen, J. D. (2012). Role of prefrontal cortex and the midbrain dopamine system in working memory updating. *Proceedings of the National Academy of Sciences of the United States of America, 109*(49), 19900–19909. https://doi.org/10.1073/pnas.1116727109

Davila, M. I., Lewis, G. F., & Porges, S. W. (2017). The PhysioCam: A novel non-contact sensor to measure heart rate variability in clinical and field applications. *Frontiers in Public Health, 5*, 300. https://doi.org/10.3389/fpubh.2017.00300

Davis, P., Creusere, C., & Kroger, J. (2016). The effect of perceptual video quality on EEG power distribution. https://doi.org/10.1109/ICIP.2016.7532793

Debener, S., Minow, F., Emkes, R., Gandras, K., & de Vos, M. (2012). How about taking a low-cost, small, and wireless EEG for a walk? *Psychophysiology, 49*, 1617–1621. https://doi.org/10.1111/j.1469-8986.2012.01471.x

Dekleva, B. M., Kording, K. P., & Miller, L. E. (2018). Single reach plans in dorsal premotor cortex during a two-target task. *Nature Communications, 9*(1), 3556. https://doi.org/10.1038/s41467-018-05959-y

Di Dio, C., & Gallese, V. (2009). Neuroaesthetics: A Review. *Current Opinion in Neurobiology, 19*, 682–687. https://doi.org/10.1016/j.conb.2009.09.001

Diemer, J., Alpers, G., Peperkorn, H., Shiban, Y., & Mühlberger, A. (2015). The impact of perception and presence on emotional reactions: A review of research in virtual reality. *Frontiers in Psychology, 6*. https://doi.org/10.3389/fpsyg.2015.00026

Dimberg, U. (1990). Facial electromyography and emotional reactions. *Psychophysiology, 27*(5), 481–494. https://doi.org/10.1111/j.1469-8986.1990.tb01962.x

Djebbara, Z., Fich, L. B., Petrini, L., & Gramann, K. (2019). Sensorimotor brain dynamics reflect architectural affordances. *Proceedings of the National Academy of Sciences of the United States of America, 116*(29), 14769–14778. https://doi.org/10.1073/pnas.1900648116

Duchowski, A. (2017). Eye tracking methodology. https://doi.org/10.1007/978-3-319-57883-5

Dzhambov, A. M., Markevych, I., Hartig, T., Tilov, B., Arabadzhiev, Z., Stoyanov, D., Gatseva, P., & Dimitrova, D. D. (2018). Multiple pathways link urban green- and bluespace to mental health in young adults. *Environmental Research, 166*, 223–233. https://doi.org/10.1016/j.envres.2018.06.004

Eberhard, J. P. (2009). Applying neuroscience to architecture. *Neuron, 62*(6), 753–756. https://doi.org/10.1016/j.neuron.2009.06.001

Ekman, P., Rosenberg, E.. (1997, 2005). *What the face reveals: Basic and applied studies of spontaneous expression using the Facial Action Coding System (FACS)*. https://doi.org/10.1093/acprof:oso/9780195179644.001.0001.

Ellard, C. (2015). *Places of the heart: The psychogeography of everyday life*. Perseus Books; LLC ed. https://aeon.co/essays/why-boring-streets-make-pedestrians-stressed-and-unhappy

Elver Boz, T., Demirkan, H., & Urgen, B. (2022). Visual perception of the built environment in virtual reality: A systematic characterization of human aesthetic experience in spaces with curved boundaries. *Psychology of Aesthetics, Creativity, and the Arts*. https://doi.org/10.1037/aca0000504

Ergan, S., Radwan, A., Zou, Z., Tseng, H.-a., & Han, X. (2019). Quantifying human experience in architectural spaces with integrated virtual reality and body sensor networks. *Journal of Computing in Civil Engineering, 33*. https://doi.org/10.1061/(ASCE)CP.1943-5487.0000812

Expoze. (2023). Retrieved 1/12/2023 from https://www.expoze.io

Eyequant. (2023). Retrieved 1/12/2023 from https://www.eyequant.com

Farb, N. A. S., Segal, Z. V., Mayberg, H., Bean, J., McKeon, D., Fatima, Z., & Anderson, A. K. (2007). Attending to the present: Mindfulness meditation reveals distinct neural modes of self-reference. *Social Cognitive and Affective Neuroscience, 2*(4), 313–322. https://doi.org/10.1093/scan/nsm030

Ferdinando, H., Ye, L., Seppänen, T., & Alasaarela, E. (2014). Emotion recognition by heart rate variability. *Australian Journal of Basic and Applied Sciences, 8*, 50–55.

Fernández-Caballero, A., Martinez Rodrigo, A., Pastor García, J. M., Castillo, J., Lozano-Monasor, E., López Bonal, M., Zangroniz, R., Latorre, J., & Fernández-Sotos, A. (2016). Smart environment architecture for emotion detection and regulation. *Journal of Biomedical Informatics, 64*, 55. https://doi.org/10.1016/j.jbi.2016.09.015

Fich, L. B., Jönsson, P., Kirkegaard, P. H., Wallergård, M., Garde, A. H., & Hansen, Å. (2014). Can architectural design alter the physiological reaction to psychosocial stress? A virtual TSST experiment. *Physiology & Behavior, 135*, 91–97. https://doi.org/10.1016/j.physbeh.2014.05.034

Fine, J., & Hayden, B. (2021). The whole prefrontal cortex is premotor cortex [Opinion piece]. *Philosophical Transactions of the Royal Society B: Biological Sciences, 377*, 20200524. https://doi.org/10.1098/rstb.2020.0524

Fischmeister, F. P., Martins, M. J. D., Beisteiner, R., & Fitch, W. T. (2017). Self-similarity and recursion as default modes in human cognition. *Cortex, 97*, 183–201. https://doi.org/10.1016/j.cortex.2016.08.016

Frumkin, H. (2001). Beyond toxicity: human health and the natural environment. *American Journal of Preventive Medicine, 20*(3), 234–240. http://www.ncbi.nlm.nih.gov/pubmed/11275453

Georgopoulos, A. P., Schwartz, A. B., & Kettner, R. E. (1986). Neuronal population coding of movement direction. *Science, 233*(4771), 1416–1419. https://doi.org/10.1126/science.3749885

Goldberger, A. L. (1996). Fractals and the birth of Gothic: reflections on the biologic basis of creativity. *Molecular Psychiatry, 1*(2), 99–104. https://www.ncbi.nlm.nih.gov/pubmed/9118332

Gorichanaz, T., Lavdas, A. A., Mehaffy, M., & Salingaros, N. (2023). The impacts of online experience on health and well-being: The overlooked aesthetic dimension. *Virtual Worlds, 2*, 243–266. https://doi.org/10.3390/virtualworlds2030015

Graziano, M. S., Taylor, C. S., & Moore, T. (2002). Complex movements evoked by microstimulation of precentral cortex. *Neuron, 34*(5), 841–851. https://doi.org/10.1016/s0896-6273(02)00698-0

Hägerhäll, C., Laike, T., Taylor, R., Küller, M., Küller, R., & Martin, T. (2008). Investigations of human EEG response to viewing fractal patterns. *Perception, 37*, 1488–1494. https://doi.org/10.1068/p5918

Hamilton, D. K., & Watkins, D. H. (2008). *Evidence-based design for multiple building types*. Wiley.

Hartig, T. (2008). Green space, psychological restoration, and health inequality. *Lancet, 372*(9650), 1614–1615. https://doi.org/10.1016/S0140-6736(08)61669-4

Hian, S., Chan, S., Qiu, L., Esposito, G., & Mai, P. (2021). Vertical greenery buffers against stress: Evidence from psychophysiological responses in virtual reality. *Landscape and Urban Planning, 213*, 104127. https://doi.org/10.1016/j.landurbplan.2021.104127

Higo, T., Mars, R. B., Boorman, E. D., Buch, E. R., & Rushworth, M. F. (2011). Distributed and causal influence of frontal operculum in task control. *Proceedings of the National Academy of Sciences of the United States of America, 108*(10), 4230–4235. https://doi.org/10.1073/pnas.1013361108

Higuera Trujillo, J. L., Llinares, C., Montañana, A., & Rojas, J.-C. (2019). Multisensory stress reduction: A neuro-architecture study of paediatric waiting rooms. *Building Research and Information, 48*, 269. https://doi.org/10.1080/09613218.2019.1612228

Higuera Trujillo, J. L., Llinares, C., & Macagno, E. (2021). The cognitive-emotional design and study of architectural space: A scoping review of neuroarchitecture and its precursor approaches. *Sensors, 21*, 2193. https://doi.org/10.3390/s21062193

Hollander, J., Purdy, A., Wiley, A., Foster, V., Jacob, R., Taylor, H., & Brunyé, T. (2018). Seeing the city: Using eye-tracking technology to explore cognitive responses to the built environment. *Journal of Urbanism: International Research on Placemaking and Urban Sustainability, 12*, 1–16. https://doi.org/10.1080/17549175.2018.1531908

Hollander, J., Sussman, A., Lowitt, P., Angus, N., & Situ, M. (2021). Eye-tracking emulation software: A promising urban design tool. *Architectural Science Review, 64*, 1–11. https://doi.org/10.1080/00038628.2021.1929055

iMotions. (2023). Retrieved 1/12/2023 from https://imotions.com/

Insight, A. (2023). Retrieved 1/12/2023 from https://attentioninsight.com/

Ishai, A. (2007). Sex, beauty and the orbitofrontal cortex. *International Journal of Psychophysiology, 63*, 181–185. https://doi.org/10.1016/j.ijpsycho.2006.03.010

Ishizu, T., & Zeki, S. (2011). Toward a brain-based theory of beauty. *PLoS One, 6*(7), e21852. https://doi.org/10.1371/journal.pone.0021852

Ishizu, T., & Zeki, S. (2013). The brain's specialized systems for aesthetic and perceptual judgment. *European Journal of Neuroscience, 37*(9), 1413–1420. https://doi.org/10.1111/ejn.12135

Ishizu, T., & Zeki, S. (2014). A neurobiological enquiry into the origins of our experience of the sublime and beautiful. *Frontiers in Human Neuroscience, 8*, 891. https://doi.org/10.3389/fnhum.2014.00891

Ishizu, T., & Zeki, S. (2017). The experience of beauty derived from sorrow. *Human Brain Mapping, 38*(8), 4185–4200. https://doi.org/10.1002/hbm.23657

Itti, L., & Koch, C. (2001). Computational modelling of visual attention. *Nature Reviews. Neuroscience, 2*(3), 194–203. https://doi.org/10.1038/35058500

Joye, Y. (2007a). Architectural lessons from environmental psychology: The case of biophilic architecture. *Review of General Psychology, 11*(4), 305–328. https://doi.org/10.1037/1089-2680.11.4.305

Joye, Y. (2007b). Fractal architecture could be good for you. *Nexus Network Journal, 9*(2), 311–320.

Kalantari, S., & Neo, J. R. J. (2020). Virtual environments for design research: Lessons learned from use of fully immersive virtual reality in interior design research. *Journal of Interior Design, 45*, 27. https://doi.org/10.1111/joid.12171

Kam, J., Griffin, S., Shen, A., Patel, S., Hinrichs, H., Heinze, H.-J., Deouell, L., & Knight, R. (2018). Systematic comparison between a wireless EEG system with dry electrodes and a wired EEG system with wet electrodes. *NeuroImage, 184*, 119. https://doi.org/10.1016/j.neuroimage.2018.09.012

Kam, J. W. Y., Griffin, S., Shen, A., Patel, S., Hinrichs, H., Heinze, H. J., Deouell, L. Y., & Knight, R. T. (2019). Systematic comparison between a wireless EEG system with dry electrodes and a wired EEG system with wet electrodes. *NeuroImage, 184*, 119–129. https://doi.org/10.1016/j.neuroimage.2018.09.012

Kaplan, S. (1995). The restorative benefits of nature – Toward an integrative framework. *Journal of Environmental Psychology, 15*(3), 169–182. https://doi.org/10.1016/0272-4944(95)90001-2

Karandinou, A., & Turner, L. (2017). Architecture and neuroscience; what can the EEG recording of brain activity reveal about a walk through everyday spaces? *International Journal of Parallel, Emergent and Distributed Systems, 32*(sup1), S54–S65. https://doi.org/10.1080/17445760.2017.1390089

Kardan, O., Gozdyra, P., Misic, B., Moola, F., Palmer, L. J., Paus, T., & Berman, M. G. (2015). Neighborhood greenspace and health in a large urban center. *Scientific Reports, 5*. https://doi.org/10.1038/srep11610

Kawabata, H., & Zeki, S. (2004). Neural correlates of beauty. *Journal of Neurophysiology, 91*(4), 1699–1705. https://doi.org/10.1152/jn.00696.2003

Kim, T.-H., Jeong, G.-W., Baek, H.-S., Kim, G.-W., Sundaram, T., Kang, H., Lee, S.-W., Kim, H. J., & Song, J. (2010). Human brain activation in response to visual stimulation with rural and urban scenery pictures: A functional magnetic resonance imaging study. *The Science of the Total Environment, 408*, 2600–2607. https://doi.org/10.1016/j.scitotenv.2010.02.025

Kolk, S. M., & Rakic, P. (2022). Development of prefrontal cortex. *Neuropsychopharmacology, 47*(1), 41–57. https://doi.org/10.1038/s41386-021-01137-9

Koohsari, J., McCormack, G., Nakaya, T., Yasunaga, A., Fuller, D., Nagai, Y., & Oka, K. (2023). The metaverse, the built environment, and public health: Opportunities and uncertainties. *Journal of Medical Internet Research, 25*, 1–6. https://doi.org/10.2196/43549

Künecke, J., Hildebrandt, A., Recio, G., Sommer, W., & Wilhelm, O. (2014). Facial EMG responses to emotional expressions are related to emotion perception ability. *PLoS One, 9*(1), e84053. https://doi.org/10.1371/journal.pone.0084053

Kuo, F. E., & Sullivan, W. C. (2001a). Aggression and violence in the inner city – Effects of environment via mental fatigue. *Environment and Behavior, 33*(4), 543–571. https://doi.org/10.1177/00139160121973124

Kuo, F. E., & Sullivan, W. C. (2001b). Environment and crime in the inner city – Does vegetation reduce crime? *Environment and Behavior, 33*(3), 343–367. https://doi.org/10.1177/0013916501333002

Lane, R. D., McRae, K., Reiman, E. M., Chen, K., Ahern, G. L., & Thayer, J. F. (2009). Neural correlates of heart rate variability during emotion. *NeuroImage, 44*(1), 213–222. https://doi.org/10.1016/j.neuroimage.2008.07.056

Langlois, J. H., Ritter, J. M., Roggman, L. A., & Vaughn, L. S. (1991). Facial diversity and infant preferences for attractive faces. *Developmental Psychology, 27*, 79–84.

Lavdas, A. A., & Salingaros, N. A. (2022). Architectural beauty: Developing a measurable and objective scale. *Challenges, 13*(56). https://doi.org/10.3390/challe13020056

Lavdas, A. A., Salingaros, N. A., & Sussman, A. (2021). Visual attention software: A new tool for understanding the "Subliminal" experience of the built environment. *Applied Sciences-Basel, 11*(13), ARTN 6197. https://doi.org/10.3390/app11136197

Lavdas, A. A., Mehaffy, M., & Salingaros, N. (2023). AI, the beauty of places, and the metaverse: beyond "geometrical fundamentalism". https://doi.org/10.1007/s44223-023-00026-z

Lee, K., Park, C.-H., & Kim, J. H. (2023). Examination of user emotions and task performance in indoor space design using mixed-reality. *Buildings, 13*, 1483. https://doi.org/10.3390/buildings13061483

Lewinski, P., Uyl, T., & Butler, C. (2014). Automated facial coding: Validation of basic emotions and FACS AUs in FaceReader. *Journal of Neuroscience, Psychology, and Economics, 7*, 227–236. https://doi.org/10.1037/npe0000028

Li, Y., Huang, J., Feng, T., Hong'an, W., & Guozhong, D. A. I. (2019). Gesture interaction in virtual reality. *Virtual Reality & Intelligent Hardware, 1*, 9. https://doi.org/10.3724/SP.J.2096-5796.2018.0006

Lisińska-Kuśnierz, M., & Krupa, M. (2020). Suitability of eye tracking in assessing the visual perception of architecture-a case study concerning selected projects located in Cologne. *Buildings, 10*(2) https://www.mdpi.com/2075-5309/10/2/20

Lu, Z., & Pesarakli, H. (2023). Seeing is believing: using eye-tracking devices in environmental research. *HERD: Health Environments Research & Design Journal, 16*(1), 15–52. https://doi.org/10.1177/19375867221130806

Mallgrave, H. F. (2010). *The architect's brain: Neuroscience, creativity, and architecture*. Wiley.

Marín-Morales, J., Higuera Trujillo, J. L., Greco, A., Guixeres, J., Llinares, C., Scilingo, E., Alcañiz Raya, M., & Valenza, G. (2018). Affective computing in virtual reality: Emotion recognition from brain and heartbeat dynamics using wearable sensors. *Scientific Reports, 8*, 13657. https://doi.org/10.1038/s41598-018-32063-4

Martins, M. J., Fischmeister, F. P., Puig-Waldmuller, E., Oh, J., Geissler, A., Robinson, S., Fitch, W. T., & Beisteiner, R. (2014). Fractal image perception provides novel insights into hierarchical cognition. *NeuroImage, 96*, 300–308. https://doi.org/10.1016/j.neuroimage.2014.03.064

Mauss, I. B., & Robinson, M. D. (2009). Measures of emotion: A review. *Cognition and Emotion, 23*(2), 209–237. https://doi.org/10.1080/02699930802204677

Mavros, P., Austwick, M., & Hudson-Smith, A. (2016). Geo-EEG: Towards the use of EEG in the study of urban behaviour. *Applied Spatial Analysis and Policy, 9*, 191. https://doi.org/10.1007/s12061-015-9181-z

McFadyen, J. (2019). Investigating the subcortical route to the amygdala across species and in disordered fear responses. *Journal of Experimental Neuroscience, 13*, Artn 1179069519846445. https://doi.org/10.1177/1179069519846445

Merrifield, C., & Danckert, J. (2014). Characterizing the psychophysiological signature of boredom. *Experimental Brain Research, 232*(2), 481–491. https://doi.org/10.1007/s00221-013-3755-2

Mier, W., & Mier, D. (2015). Advantages in functional imaging of the brain. *Frontiers in Human Neuroscience, 9*, 249. https://doi.org/10.3389/fnhum.2015.00249

Miller, E. K., & Cohen, J. D. (2001). An integrative theory of prefrontal cortex function. *Annual Review of Neuroscience, 24*, 167–202. https://doi.org/10.1146/annurev.neuro.24.1.167

Mirabella, G., Pani, P., & Ferraina, S. (2011). Neural correlates of cognitive control of reaching movements in the dorsal premotor cortex of rhesus monkeys. *Journal of Neurophysiology, 106*(3), 1454–1466. https://doi.org/10.1152/jn.00995.2010

Mostajeran, F., Steinicke, F., Reinhart, S., Stuerzlinger, W., Riecke, B., & Kühn, S. (2023). Adding virtual plants leads to higher cognitive performance and psychological well-being in virtual reality. *Scientific Reports, 13*, 8053. https://doi.org/10.1038/s41598-023-34718-3

Nayak, C. S., & Anilkumar, A. C. (2023). EEG normal waveforms. *StatPearls*. https://www.ncbi.nlm.nih.gov/pubmed/30969627

Neale, C., Aspinall, P., Roe, J., Tilley, S., Mavros, P., Cinderby, S., Coyne, R., Thin, N., Bennett, G., & Ward Thompson, C. (2017). The aging urban brain: Analyzing outdoor physical activity using the emotiv affectiv suite in older people. *Journal of Urban Health, 94*, 869. https://doi.org/10.1007/s11524-017-0191-9

Neurons. (2023). www.neuronsinc.com

Ohman, A., Lundqvist, D., & Esteves, F. (2001). The face in the crowd revisited: a threat advantage with schematic stimuli. *Journal of Personality and Social Psychology, 80*(3), 381–396. https://doi.org/10.1037/0022-3514.80.3.381

Omrani, M., Murnaghan, C. D., Pruszynski, J. A., & Scott, S. H. (2016). Distributed task-specific processing of somatosensory feedback for voluntary motor control. *eLife, 5*. https://doi.org/10.7554/eLife.13141

Pallasmaa, J. (2014). Space, place and atmosphere. Emotion and peripherical perception in architectural experience. *Lebenswelt Aesthetics and philosophy of experience, 4*(1), 10.13130/2240-9599/4202.

Pastor-Bernier, A., & Cisek, P. (2011). Neural correlates of biased competition in premotor cortex. *Journal of Neuroscience, 31*(19), 7083–7088. https://doi.org/10.1523/JNEUROSCI.5681-10.2011

Petrigna, L., & Musumeci, G. (2022). The metaverse: A new challenge for the healthcare system: A scoping review. *Journal of Functional Morphology and Kinesiology, 7*, 63. https://doi.org/10.3390/jfmk7030063

Poldrack, R. (2006). Can cognitive processes be inferred from neuroimaging data. *Trends in Cognitive Sciences, 10*, 59–63. https://doi.org/10.1016/j.tics.2005.12.004

Pretty, J., Peacock, J., Hine, R., Sellens, M., South, N., & Griffin, M. (2007). Green exercise in the UK countryside: Effects on health and psychological well-being, and implications for policy and planning. *Journal of Environmental Planning and Management, 50*(2), 211–231. https://doi.org/10.1080/09640560601156466

Radüntz, T. (2018). Signal Quality Evaluation of Emerging EEG Devices. *Frontiers in Physiology, 9*. https://doi.org/10.3389/fphys.2018.00098

Read, G. L. (2015). Facial electromyography (EMG). In *The international encyclopedia of communication research methods* (pp. 1–10). https://doi.org/10.1002/9781118901731.iecrm0100

Reggente, N., Essoe, J., Aghajan, M., & Z., Tavakoli, A., McGuire, J., Suthana, N., & Rissman, J. (2018). Enhancing the ecological validity of fMRI memory research using virtual reality. *Frontiers in Neuroscience, 12*, 408. https://doi.org/10.3389/fnins.2018.00408
Roe, J., Aspinall, P., Mavros, P., & Coyne, R. (2013). Engaging the brain: The impact of natural versus urban scenes using novel EEG methods in an experimental setting. *Environmental Sciences, 1*, 93–104. https://doi.org/10.12988/es.2013.3109
Rosas, H. J., Sussman, A., Sekely, A. C., & Lavdas, A. A. (2023). Using eye tracking to reveal responses to the built environment and its constituents. *Applied Sciences-Basel, 13*(21), ARTN 12071. https://doi.org/10.3390/app132112071
Rosenzweig, M. R., & Bennett, E. L. (1996). Psychobiology of plasticity: effects of training and experience on brain and behavior. *Behavioural Brain Research, 78*(1), 57–65. https://www.ncbi.nlm.nih.gov/pubmed/8793038
Rounds, J., Cruz-Garza, J., & Kalantari, S. (2020). Using posterior EEG theta band to assess the effects of architectural designs on landmark recognition in an urban setting. *Frontiers in Human Neuroscience, 14*. https://doi.org/10.3389/fnhum.2020.584385
Sale, A., Berardi, N., & Maffei, L. (2009). Enrich the environment to empower the brain. *Trends in Neurosciences, 32*(4), 233–239. https://doi.org/10.1016/j.tins.2008.12.004
Salingaros, N. A. (1995). The laws of architecture from a physicist's perspective. *Physics Essays, 8*(4), 638–643.
Salingaros, N. (2003). The sensory value of ornament. *Communication & Cognition, 36*(3), 331–351.
Salingaros, N. (2012). Fractal art and architecture reduce physiological stress. *Journal of Biourbanism, II*(2), 11–28.
Salingaros, N. A. (2013). Unified architectural theory: Form, language, complexity: A companion to Christopher Alexander's. In *The phenomenon of life: The nature of order, book 1* [still image]. Sustasis Foundation.
Salingaros, N. A., & Mehaffy, M. W. (2006). *A theory of architecture*. Umbau-Verlag.
Salingaros, N. A., & Sussman, A. (2020). Biometric pilot-studies reveal the arrangement and shape of windows on a traditional façade to be implicitly "Engaging", whereas contemporary façades are not. *Urban Science, 4*, 26. https://doi.org/10.3390/urbansci4020026
Shemesh, A., Talmon, R., Karp, O., Amir, I., Bar, M., & Grobman, J. (2016). Affective response to architecture—Investigating human reaction to spaces with different geometry. *Architectural Science Review, 60*. https://doi.org/10.1080/00038628.2016.1266597
Shemesh, A., Leisman, G., Bar, M., & Grobman, J. (2021). A neurocognitive study of the emotional impact of geometrical criteria of architectural space. *Architectural Science Review, 64*, 394. https://doi.org/10.1080/00038628.2021.1940827
Shi, Y., Ruiz, N., Taib, R., Choi, E., & Chen, F. (2007). *Galvanic skin response (GSR) as an index of cognitive load*. https://doi.org/10.1145/1240866.1241057
Slater, A., Schulenburg, C. V. D. Brown, E., & Badenoch, M. (1998). Newborn infants prefer attractive faces. *Infant Behavior and Development, 21*, 345–54.
Suárez, L. (2020). Subjective experience and visual attention to a historic building. *Frontiers of Architectural Research, 9*(4), 774–804. https://www.sciencedirect.com/science/article/pii/S2095263520300558?via%3Dihub
Sussman, A., & Ward, J. (2019). Eye-tracking Boston City Hall to better understand human perception and the architectural experience. *New Design Ideas, 3*(1), 53–59.
Sztajzel, J. (2004). Heart rate variability: A noninvasive electrocardiographic method to measure the autonomic nervous system. *Swiss Medical Weekly, 134*, 514–522.
Tawil, N., Ascone, L., & Kühn, S. (2022). The contour effect: Differences in the aesthetic preference and stress response to photo-realistic living environments. *Frontiers in Psychology, 13*. https://doi.org/10.3389/fpsyg.2022.933344
Taylor, R. P. (2006). Reduction of physiological stress using fractal art and architecture. *Leonardo, 39*(3), 245–251.

Taylor, R. P., Spehar, B., Van Donkelaar, P., & Hagerhall, C. M. (2011). Perceptual and physiological responses to Jackson Pollock's fractals. *Frontiers in Human Neuroscience, 5*, 60. https://doi.org/10.3389/fnhum.2011.00060

Tollner, T., Zehetleitner, M., Gramann, K., & Muller, H. J. (2011). Stimulus saliency modulates pre-attentive processing speed in human visual cortex. *PLoS One, 6*(1), e16276. https://doi.org/10.1371/journal.pone.0016276

Ulrich, R. S. (1984). View through a window may influence recovery from surgery. *Science, 224*(4647), 417–419.

Vaish, A., Grossmann, T., & Woodward, A. (2008). Not all emotions are created equal: The negativity bias in social-emotional development. *Psychological Bulletin, 134*(3), 383–403. https://doi.org/10.1037/0033-2909.134.3.383

Valentine, C. (2023). Health implications of virtual architecture: An interdisciplinary exploration of the transferability of findings from neuroarchitecture. *International Journal of Environmental Research and Public Health, 20*(3). https://doi.org/10.3390/ijerph20032735

Valstar, M., Mehu, M., Jiang, B., Pantic, M., & Scherer, K. (2012). Meta-analysis of the first facial expression recognition challenge. *IEEE transactions on systems, man, and cybernetics. Part B, Cybernetics: a publication of the IEEE Systems, Man, and Cybernetics Society, 42*. https://doi.org/10.1109/TSMCB.2012.2200675

Valtchanov, D., Barton, K. R., & Ellard, C. (2010). Restorative effects of virtual nature settings. *Cyberpsychology Behavior and Social Networking, 13*(5), 503–512. https://doi.org/10.1089/cyber.2009.0308

Van Le, Q., Isbell, L. A., Matsumoto, J., Nguyen, M., Hori, E., Maior, R. S., Tomaz, C., Tran, A. H., Ono, T., & Nishijo, H. (2013). Pulvinar neurons reveal neurobiological evidence of past selection for rapid detection of snakes. *Proceedings of the National Academy of Sciences of the United States of America, 110*(47), 19000–19005. https://doi.org/10.1073/pnas.1312648110

van Praag, H., Kempermann, G., & Gage, F. H. (2000). Neural consequences of environmental enrichment. *Nature Reviews. Neuroscience, 1*(3), 191–198. https://doi.org/10.1038/35044558

Vartanian, O., & Goel, V. (2004). Neuroanatomical correlates of aesthetic preference of paintings. *Neuroreport, 15*, 893–897. https://doi.org/10.1097/00001756-200404090-00032

Vartanian, O., Navarrete, G., Chatterjee, A., Fich, L., Leder, H., Modroño, C., Nadal, M., Rostrup, N., & Skov, M. (2013). Impact of contour on aesthetic judgments and approach-avoidance decisions in architecture. *Proceedings of the National Academy of Sciences of the United States of America., 110*, 10446. https://doi.org/10.1073/pnas.1301227110

Vartanian, O., Navarrete, G., Chatterjee, A., Fich, L., González-Mora, J., Leder, H., Modroño, C., Nadal, M., Rostrup, N., & Skov, M. (2014). Architectural design and the brain: Effects of ceiling height and perceived enclosure on beauty judgments and approach-avoidance decisions. *Journal of Environmental Psychology, 41*. https://doi.org/10.1016/j.jenvp.2014.11.006

Vartanian, O., Navarrete, G., Chatterjee, A., Fich, L., Gonzalez-Mora, J., Leder, H., Modroño, C., Nadal, M., Rostrup, N., & Skov, M. (2015). Architectural design and the brain: Effects of ceiling height and perceived enclosure on beauty judgments and approach-avoidance decisions. *Journal of Environmental Psychology, 41*, 10–18. https://doi.org/10.1016/j.jenvp.2014.11.006

Vecchiato, G., Tieri, G., Jelic, A., De Matteis, F., Maglione, A., & Babiloni, F. (2015). Electroencephalographic correlates of sensorimotor integration and embodiment during the appreciation of virtual architectural environments. *Frontiers in Psychology, 6*. https://doi.org/10.3389/fpsyg.2015.01944

Weier, M., Stengel, M., Roth, T., Didyk, P., Eisemann, E., Eisemann, M., Grogorick, S., Hinkenjann, A., Kruijff, E., Magnor, M., Myszkowski, K., & Slusallek, P. (2017). Perception-driven accelerated rendering. *Computer Graphics Forum (Proc. of Eurographics)*, 36. https://doi.org/10.1111/cgf.13150

Wiesel, T. N., & Hubel, D. H. (1963a). Effects of visual deprivation on morphology and physiology of cells in the cats lateral geniculate body. *Journal of Neurophysiology, 26*, 978–993. https://doi.org/10.1152/jn.1963.26.6.978

Wiesel, T. N., & Hubel, D. H. (1963b). Single-cell responses in striate cortex of kittens deprived of vision in one eye. *Journal of Neurophysiology, 26*, 1003–1017. https://doi.org/10.1152/jn.1963.26.6.1003

Williams, D. P., Cash, C., Rankin, C., Bernardi, A., Koenig, J., & Thayer, J. F. (2015). Resting heart rate variability predicts self-reported difficulties in emotion regulation: A focus on different facets of emotion regulation. *Frontiers in Psychology, 6*, 261. https://doi.org/10.3389/fpsyg.2015.00261

Wise, S. P. (1985). The primate premotor cortex: Past, present, and preparatory. *Annual Review of Neuroscience, 8*, 1–19. https://doi.org/10.1146/annurev.ne.08.030185.000245

Wise, S. P., Boussaoud, D., Johnson, P. B., & Caminiti, R. (1997). Premotor and parietal cortex: Corticocortical connectivity and combinatorial computations. *Annual Review of Neuroscience, 20*, 25–42. https://doi.org/10.1146/annurev.neuro.20.1.25

Woaswi, W., Hanif, M., Mohamed, S., Hamzah, N., & Rizman, Z. (2016). Human emotion detection via brain waves study by using electroencephalogram (EEG). *International Journal on Advanced Science, Engineering and Information Technology, 6*, 1005. https://doi.org/10.18517/ijaseit.6.6.1072

Wolf, K. (2015). Measuring facial expression of emotion. *Dialogues in Clinical Neuroscience, 17*(4), 457–462. https://doi.org/10.31887/DCNS.2015.17.4/kwolf

Yang, T., Formuli, A., Paolini, M., & Zeki, S. (2021). The neural determinants of beauty. https://doi.org/10.1101/2021.05.21.444999

Zanto, T. P., Rubens, M. T., Thangavel, A., & Gazzaley, A. (2011). Causal role of the prefrontal cortex in top-down modulation of visual processing and working memory. *Nature Neuroscience, 14*(5), 656–661. https://doi.org/10.1038/nn.2773

Zeki, S. (2009). In praise of subjective truths. *The Journal of Physiology, 587*(Pt 12), 2825–2835. https://doi.org/10.1113/jphysiol.2009.170415

Zeki, S. (2019). Beauty in architecture: Not a luxury – Only a necessity. *Architectural Design, 89*, 14–19. https://doi.org/10.1002/ad.2473

Zeki, S., Romaya, J. P., Benincasa, D. M., & Atiyah, M. F. (2014). The experience of mathematical beauty and its neural correlates. *Frontiers in Human Neuroscience, 8*, 68. https://doi.org/10.3389/fnhum.2014.00068

Zeki, S., Bao, Y., & Pöppel, E. (2022). Inter-disciplinarity versus intra-disciplinarity: A failure in neurobiology. *PsyCh Journal, 11*, 617–618. https://doi.org/10.1002/pchj.607

Zeman, A. (2007). Sherrington's philosophical writings—A 'zest for life'. Brain, 130(Pt 8), 1984–1987. https://doi.org/10.1093/brain/awm075.

Zhong, W., Schröder, T., & Bekkering, J. (2022). Biophilic design in architecture and its contributions to health, well-being, and sustainability: A critical review. *Frontiers of Architectural Research, 11*(1), 114–141. https://doi.org/10.1016/j.foar.2021.07.006

Zou, Z., & Ergan, S. (2018). Where do we look? An eye-tracking study of architectural features in building design. In I. Mutis & T. Hartmann (Eds.), *Advances in informatics and computing in civil and construction engineering*. Springer. https://doi.org/10.1007/978-3-030-00220-6_52

Zou, Z., Yu, X., & Ergan, S. (2019). Integrating biometric sensors, VR, and machine learning to classify EEG signals in alternative architecture designs. *Computing in Civil Engineering, 2019*.

Zou, W., Yang, L., Yang, F., Ma, Z., & Zhao, Q. (2020). The impact of screen resolution of HMD on perceptual quality of immersive videos. https://doi.org/10.1109/ICMEW46912.2020.9105945

The Potential of Biophilic Design and Nature to Improve Health, Creativity and Well-Being

Isabelle A. K. Sjövall and Hugo J. Spiers

Abstract The human brain evolved in adaptive response to natural environments. However, urbanisation has led to the majority of humanity now living in cities, which differ profoundly from the landscape in which our species originated. At the same time, mental health conditions are increasing rapidly worldwide. Studies suggest that city living impacts brain function and structure, activates the stress response to a greater extent, and increases the risk of multiple mental health conditions. Recently, the close-to-nature concepts of biophilia and biophilic design have emerged as a potential approach to counteract these challenges. They may play a pivotal role in achieving the United Nations' Sustainable Development Goals. Here, we explore the potential of biophilic design to balance urban stressors, protect mental health and support sustainable living on a global scale.

Keywords Biophilic Design · Sustainability · Mental Health · Creativity · Wellbeing

Introduction

Mental health is a crucial component of overall health and quality of life (Prince et al., 2007). Mental health conditions are rising worldwide and are now considered leading contributors to global disability, ranging from anxiety and depression to severe mental illnesses such as schizophrenia, and bipolar disorder (Patel et al., 2018; World Health Organization, 2022). Recently, interest in mental health has

---

The original version of the chapter has been revised. A correction to this chapter can be found at https://doi.org/10.1007/978-3-031-64699-7_24

I. A. K. Sjövall (✉) · H. J. Spiers (✉)
Institute of Behavioural Neuroscience, Department of Experimental Psychology, University College London, London, UK
e-mail: isabelle.sjovall.22@ucl.ac.uk; h.spiers@ucl.ac.uk

S. Kühn (ed.), *Environmental Neuroscience*,
https://doi.org/10.1007/978-3-031-64699-7_11

increased due to its vital role in reaching the United Nations' Sustainable Development Goals. For example, the goals aimed at good health and well-being, reduced inequalities and sustainable cities and communities (United Nations, 2023).

Mental health is considerably interconnected with direct and indirect diverse biological, psychological and environmental influences and is significantly associated with overall health and quality of life (Prince et al., 2007; van den Bosch & Meyer-Lindenberg, 2019) (➔ chapter "The CITY & the PSYCHE—A Chronology" by Mehran, Kluge and Adli). Mental health can be defined as the absence of mental disorders, and additionally, the ability to deal with normal stresses of life, acknowledge one's strengths and abilities, productive work and contribute to society (World Health Organization, 2022). Although mental health is multifaceted and complex, with many confounding variables, one fundamental factor impacting behaviour, stress levels and mental health is the surrounding living environment (Clemenson et al., 2015; Evans, 2003). While the world around us has developed at a rapid pace with vast urbanisation and digitalisation, the human brain has not changed significantly in this period. For more than 99 percent of our species' history, humans developed an adaptive response to natural environments, not built or engineered ones (Kellert & Calabrese, 2015). However, now more than 50 percent of the world's population lives in cities, a number expected to reach 70 percent by the year 2050 (Richie & Roser, 2018). It is clear that the urban environments in which the majority of humanity lives today stand in stark contrast to the natural environments in which our brains have evolved (Kühn et al., 2021).

Despite their challenges, cities also offer multiple advantages such as proximity to healthcare, work, education and culture (Hartig et al., 2014). Epidemiological studies have also linked city living to better overall health and longer life expectancy compared to rural inhabitants (Cross et al., 2021; Engemann et al., 2019). At the same time, city dwelling has been associated with increased amygdala activity (Lederbogen et al., 2011; Sudimac et al., 2022), cumulative environmental stressors, and a significantly higher risk of mental health conditions, such as anxiety, depression, psychosis and schizophrenia (Haddad et al., 2015; Lederbogen et al., 2011; Reichert et al., 2020; Sundquist et al., 2004). Additionally, emerging evidence suggests that urban living may alter brain structure and function (Kühn et al., 2017, 2020; Lederbogen et al., 2011). However, the underlying mechanisms remain unclear.

Urban environments tend to activate the human stress response to a greater extent than rural, natural settings (Abbott, 2012; Iglesias-Carrasco et al., 2020), with alterations in the responsiveness of brain regions involved in emotion regulation, such as the amygdala and anterior cingulate cortex (Lederbogen et al., 2011; Sudimac et al., 2022). These findings are concerning, as prolonged stress may influence both the onset and severity of multiple mental health conditions (Marin et al., 2011; Mariotti, 2015). Linking to urban environments, one possible underlying explanation may be the vast and constant amount of stimuli city environments expose the brain and body to. As a result, the cumulative burden of stressors may lead to cognitive and allostatic overload, which in turn may alter the homeostatic balance over time (Abbott, 2012; Isaksson, 2015; Lambert et al., 2015; McEwen, 2004; Ulrich et al., 1991). In sum, there appear to be both advantages and disadvantages of urban exposure on health and wellbeing (Abbott, 2012; Hartig & Kahn, 2016). Exploring both specific mechanisms and collective forces behind risk and resilience factors impacting mental health and wellbeing is therefore pivotal for the development of future urban environments for the global population. Research is critically needed to pinpoint and empirically establish these factors and how they relate to creating a better balance.

The Potential of Biophilic Design

One approach to exploring how cities impact health is attempting to identify which features of the environment might operate as protective factors, helping to lower stress and support mental and physical health. One appealing example is the presence of greenspace and close-to-nature characteristics known as biophilic design (Barbiero & Berto, 2021; Hartig & Kahn, 2016; Kaplan, 1995; Kellert et al., 2011; Kühn et al., 2017; Lindal & Hartig, 2013). Biophilic design has emerged from biophilia, a concept first introduced by Erich Fromm and later popularised by Edward O. Wilson, defined as 'the love of life and of all that is alive' or 'the tendency of humans to focus on and to affiliate with nature' (Fromm, 1964, 1973; Wilson, 1986). Biophilic design involves direct or indirect close-to-nature elements, such as colours, patterns, greenness, motives, lighting, sounds and natural tactile materials, such as for example wood.

In recent years, a growing body of research has acknowledged the important potential of biophilic design as a tool for counteracting stressors and supporting mental health. The findings range from improvements in mood, memory, positive affect and happiness, to cognitive performance, creativity, recovery and well-being (Bolten & Barbiero, 2020; Bratman et al., 2019; Gillis & Gatersleben, 2015; Kellert, 2018; Mostajeran et al., 2023; Stenfors et al., 2019; Ulrich, 1984; Van Hedger et al., 2019; Wijesooriya & Brambilla, 2021). Bringing these findings together, it is tempting to consider future possibilities and positive outcomes that may emerge from making more conscious design choices. Can biophilic design create healthier cities, homes, schools, hospitals and workplaces, helping to better balance risk and resilience factors in future living environments?

Our Approach

In this chapter, we consider the available literature, focusing on how biophilia and biophilic design affect humans psychologically and physiologically linked to mental health and creative cognition. Furthermore, we aim to identify the key features of biophilic design and suggest new ideas for future research to further advance the field. In addition, we propose our vision of how biophilic design qualities can be implemented to shape symbiotic spaces in the real world with the intention of supporting health and sustainable living in future built environments.

Method

A comprehensive search of peer-reviewed papers was completed through PubMed, Google Scholar, ScienceDirect and Cochrane, based on the key terms; 'Biophilia', 'Biophilic Design', 'Mental health', 'Creativity', 'Built Environment'. High-impact journals, including Nature, Science and The Lancet, were independently searched. The included articles were selected due to their relevance, suitability and impact, with careful consideration to match the aims of this chapter. A limited number of books have been included, motivated by their originality and unique contribution to the field. Chat GPT and Gemini have been used to generate ideas, with care taken to consider for confabulations (e.g. no facts listed trusted).

Background and Overview: Biophilia and Biophilic Design

The biophilia hypothesis suggests that humans have an innate connection with nature. Biophilic design can be defined as design elements inspired by nature, typically involving colours, patterns, greenness, sounds, smells, fractals, lighting, motives and natural materials, such as wood (Barbiero & Berto, 2021; Kellert et al., 2011). From a neurobiological and evolutionary perspective, the biophilic design concept is appealing as the brain evolved in adaptive response to natural environments, why a universal preference for close-to-nature characteristics in architecture and design seems logical.

A Dose of Nature

In recent years, there has been increasing interest in biophilia and biophilic design, as a growing number of studies have demonstrated positive psychological and physiological health outcomes associated with nature exposure. Important examples include reduced stress levels; improved immune function, memory, lower blood pressure, slower epigenetic ageing, better sleep quality and quantity, as well as increased performance, creativity, cognition, mood and well-being (Beukeboom et al. 2012; Kim et al., 2023; Li, 2010, Li et al., 2009, 2011; McCoy & Evans, 2002; Ulrich, 1992). These findings are interesting since very few other interventions have the same powerful potential, as nature seems to have multiple domains related to human health outcomes. The next section explores the key elements and characteristics of biophilic design and its various effects on physical and mental health.

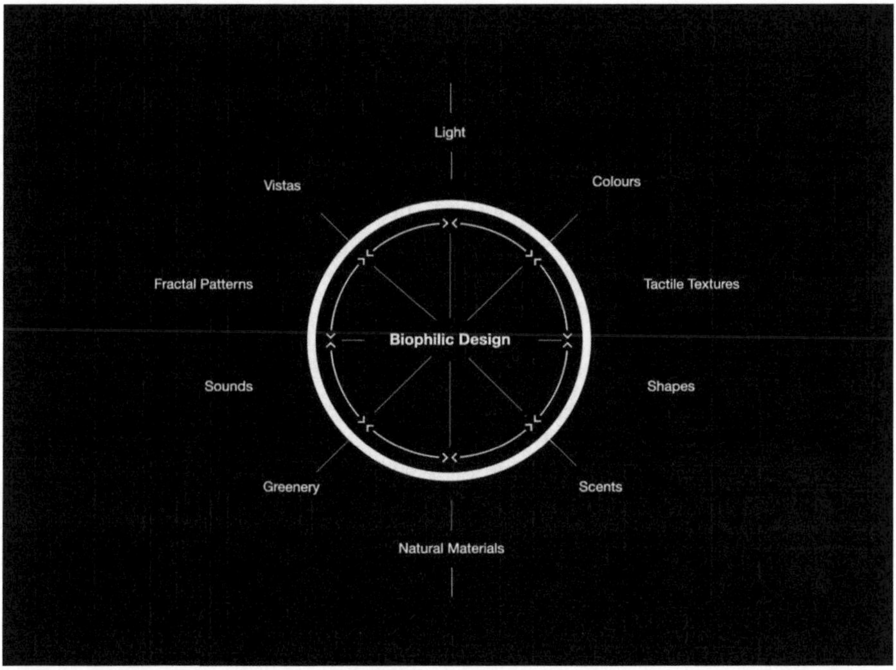

Fig. 1 Visualisation of biophilic design elements. From "The Krebs Cycle of Creativity" by N. Oxman (2016), adapted by Sjövall & Spiers, 2024

Key Elements of Biophilic Design: A Multisensory Approach

Biophilic design is a multifaceted concept that consists of several separate elements. Each element has a close-to-nature look and feel and has been independently linked to positive physical and mental health outcomes. Here, ten core characteristics of biophilic design are presented (Fig. 1).

Natural Materials

Recent developments in the field of ecological sustainability have led to a renewed interest in wooden materials. From a holistic perspective, it is interesting to consider the properties of wood connected to human health, as a growing body of research has found positive effects on mood, cognition and creativity (Beute & De Kort, 2014; Bhatta et al., 2017; McCoy & Evans, 2002; Strobel et al., 2017; Vavrinsky et al., 2019). A clear example is natural materials, such as wood, which provide a multisensory connection with nature (Kellert, 2016). Wooden materials have been found to reduce stress and cortisol levels, lower heart rate and blood pressure and enhance well-being (Beute & de Kort, 2014; Fell, 2010; Kotradyova et al., 2019; Tsunetsugu et al., 2007).

Colours

Colours may play a pivotal role in inducing psychological and physiological states (Elliot & Maier, 2014). Green and blue spaces (e.g. foliage and water) have been found to support cognitive function and mental well-being (White et al., 2021), and Lichtenfeld et al. (2012) proposed that the green colour could facilitate creative performance. Other research by Mehta and Zhu (2009) suggested a link between the colour blue and enhanced creative task performance. These studies suggest a beneficial link between natural nuances (reminiscent of blue skies, water, green plants and vegetation). Some studies have proposed that viewing blue or green may be particularly beneficial for creative performance (Lichtenfeld et al., 2012; Mehta & Zhu, 2009; Küller et al., 2009). Another well-studied colour is red, a poignant colour in biology and nature often associated with arousal and warning signals. For example, in poisonous reptiles and insects (Stevens & Ruxon, 2012), perhaps most importantly, the colour of blood is closely connected to life and death (Elliot & Maier, 2014). Other nuances have not been empirically examined. Although the green/blue/red findings seem well aligned with Kellert's (2008) search into biophilic design, which emphasises that colours often resonate with our evolutionary predisposition towards nature.

Shapes

Organic curvilinear shapes may have originated from an evolutionary mechanism (Bar & Neta, 2006). In recent years, research has explored the cognitive and neural mechanisms supporting preference for natural and organic shapes (Bar and Neta, 2006, 2007; Leder et al., 2011; Vartanian et al., 2013; Bertamini et al., 2016). Vartanian et al. (2013) demonstrated that curvilinear forms activated the medial orbitofrontal cortex, a region involved in emotional processing (Etkin et al., 2011). Moreover, work by Ramachandran and Hirstein (1999) related to aesthetic appreciation suggests that organic forms can induce enhanced pleasure responses, highlighting our biological preference for natural shapes and forms.

Fractal Patterns

Fractals are self-repeating structures found in nature (Taylor, 2021). These patterns are prevalent throughout natural scenery and can be found everywhere in nature and biology, for example on trees, leaves, flowers and coastlines (Taylor et al., 2005). Interestingly, neurons also use fractal geometry to enhance network connectivity (Smith et al., 2021). Research using electroencephalogram (EEG) recording has reported modulation in alpha oscillations in the frontal lobes and the highest beta oscillations in the parietal regions (Hagerhall et al., 2015). These findings indicate a complex interplay between neural networks when experiencing these patterns, and the outcomes have demonstrated both states of relaxation and elevated attention (Hagerhall et al., 2008, 2015). Other studies have found a reduction in physiological stress and enhanced rehabilitation when exposed to fractal patterns in art and architecture (Taylor et al., 2011; Vaz et al., 2020). Fractals could be connected to evolutionary developed patterns aesthetic appreciation. Implementing fractal patterns in architecture and design may therefore have significant potential for future built environments and cities.

Tactile Textures

The sensory experience of touch has a fundamental impact on human physiological, cognitive and emotional responses (Jansson-Boyd, 2011). Perhaps not surprising, as touch provides us with important sensory information and contact with the external world (Barnett, 1972; Gallace, 2012; Gallace & Spence, 2008; Gallace & Spence, 2010). Positive touch responses can stimulate the release of oxytocin, a key chemical for social bonding and relaxation (McGlone et al., 2014; Uvnäs-Moberg et al., 2015; Yu et al., 2022). Studies have shown that tactile touch can lead to significantly lower anxiety levels and increased resilience to stress. The underlying mechanism is suggested to be a reduction in noradrenalin requirements (Henricson et al., 2008; Dagnino-Subiabre, 2022). Gallace and Spence's (2010) research further exemplifies this argument by proposing that varied textures, especially those mimicking nature, can significantly modify mood and overall well-being. Natural tactile materials promote sensory stimulation and provide cognitive, physiological and psychological benefits.

Light

Ranging from natural daylight to artificial lighting sources, light has profound implications for human health and performance. The background is the importance of light for circadian rhythms, as this system orchestrates complicated chemical balances in the brain and the body. Natural daylight has been associated with improved alertness and enhanced cognitive functioning (Choi & Suk, 2016). In contrast, warm light, mimicking sunsets or candlelight, may induce relaxation (Inagaki & Eisenberger, 2013). Exposure timings, colour temperature and strength can influence mood, productivity and overall health (Choi & Suk, 2016; Inagaki & Eisenberger, 2013). Furthermore, Rea et al. (2010) highlight the importance of lighting compatible with circadian rhythms, underscoring its influence on sleep and health. In recent years a growing interest in human-centric lighting (HCL) has emerged. HCL is artificial lighting that aims to provide close-to-nature lighting, tapping into natural circadian rhythms. Properly used, HCL has the potential to improve lighting within the architecture and design process (Houser & Esposito, 2021).

Greenery

Green plants are perhaps the element most associated with biophilic design. For the past few decades, the interest in plants has surged, particularly in office design driven primarily by the wish to boost productivity and performance. Research consistently indicates that plants can improve air quality, reduce stress and improve mood, productivity and cognitive performance (Aydogan & Cerone, 2021; Bringslimark et al., 2007; Hall & Knuth, 2019; Raanaas et al., 2011). Kellert et al. (2011) presented a comprehensive overview of advantages of biophilia in office settings, proposing that the integration of nature may enhance productivity. Later studies support this perspective. One good example is the work by Lei et al. (2021) The study focuses on identifying the optimal greenery dose for workplace well-being, including physiological, psychological and productivity aspects. A repeated-measures experiment was carried out using green coverage ratios of 0%, 0.2%, 5%, 12% and 20%. Health and productivity were measured through questionnaires, tasks, alongside eye tracking to test visual engagement with greenery. The findings indicate that while all levels of greenery improved work performance, only 12% and 20% green coverage ratios significantly enhanced physiological brain activities. Interestingly, a 20% coverage ratio was identified as overwhelming, making the 12% greenery ratio the recommended dose for biophilic design in office environments. This study provides valuable data for biophilic design research and practice, identifying a potential dose–response for enhancing workplace well-being (Lei et al., 2021). Furthermore, Kaplan (1993) underlines the importance of green and natural elements in restorative environments. As restoration is a crucial component for productivity and performance, it is logical that green plants are used as a natural part of architectural design to contribute to reinforcing a healthy work environment. Greenery may also extend to artificial plants, which may be particularly beneficial in environments with no natural light (Beukeboom et al., 2012). On a larger scale, it can be implemented in urban design as green roofs and facades to help improve air quality and keep down rising temperatures linked to global warming.

Sounds

Studies have consistently shown that the sounds of nature such as animals, wind and water, are preferred over anthropogenic sounds such as traffic and industrial noise (Yang & Kang, 2005; Fisher, 1999). The sounds of nature such as diverse birdsong, rustling leaves or a calm water stream can provide positive sensory stimuli to counteract stressing urban sounds such as traffic noise. Research by Alvarsson et al. (2010) found that nature sounds can reduce stress levels, restore cognitive ability and facilitate recovery from sympathetic activation by contributing to relaxation and well-being. Other studies have also highlighted the health advantages of auditory stimuli to better balance city stressors by increasing recovery stress resilience, and demonstrate that birdsongs reduced depressivity, anxiety and paranoia in healthy participants (Gould van Praag et al. 2017; Stobbe et al., 2022). Auditory experiences associated with nature therefore hold a therapeutic promise for architectural design and future urban environments.

Scents

The olfactory dimension and nature's fragrances play a profound role in biophilic design. Incorporating natural scents within the architecture and design associated with nature, can add richness to the multisensory experience and induce a sense of belonging and psychological well-being (Joye & Van den Berg, 2011). Research by Franco et al. (2017) suggests that certain natural scents, such as those from pine trees or fresh soil, may have calming and relaxing effects, as well as reduce anxiety and physiological stress-related biomarkers (Franco et al., 2017). Another example of the positive effects of natural scents is phytoncides, a natural chemical released from trees that have been found to strengthen mood and immune function (Li, 2010; Herz, 2009). Building further on these findings, essential oils may be of benefit. Research has found that olfactory stimulation by orange or rose oil may result in a significant decrease in oxyhemoglobin concentration in the right prefrontal cortex, a region linked to memory for use in speech, and in addition induce positive changes in mood and cognitive function, as well as an increase in a comfortable and relaxed state of mind (Gabrieli et al., 1998; Igarashi et al., 2014). Integrating the olfactory dimension into biophilic design not only amplifies a positive atmosphere but can also support human health and psychological well-being.

Vistas

Robust research shows that visual exposure to nature and landscapes has significant positive outcomes that can lower stress hormones, heart rate and blood pressure, as well as support overall health and well-being (Kaplan & Kaplan, 1989). Furthermore, these views may enhance mood, support attention restoration and activate brain regions associated with higher cognitive functions linked to emotional regulation (Berman et al., 2008). The therapeutic potential of natural views is also accentuated in hospital settings due to its capacity to reduce stress and accelerate recovery (Sternberg, 2009; Ulrich, 1984). A pioneering study by Ulrich (1984) explored recovery records in a hospital environment. Patients who had undergone surgery were assigned to a room either with a window facing a natural setting or a brick wall to examine whether the view might influence recovery. The patients who were assigned to rooms with windows with a natural scene had shorter hospital stays, fewer negative comments in staff's notes, and took less pain medication (Ulrich, 1984). Using nature vistas in built environments may be a fundamental component to enhance health, recovery and well-being. Taken together, these ten key characteristics of biophilic design show great promise and potential to support physiological and mental health. The next section will investigate the stress response linked to the built environment and cities (Fig. 2).

Fig. 2 Example of Biophilic Design. Supertree Grove, Singapore. This provides an example of a rather extreme focus on biophilic design where the location is surrounded by cultivated nature and the structures for walkways and viewing platforms explicitly reproduce patterns in nature—biomimicry. Reproduced under a creative commons licence from Wikipedia

Cities, Stress and the Immune System

From an evolutionary perspective, it has been advantageous for biological organisms such as humans to adapt to the surrounding environment to maintain overall homeostasis and optimise chances of survival. Homeostasis is pivotal for physical and mental health, defined as the process of the internal environment of a living organism to remain steady and balanced (Romero et al., 2009; McEwen & Wingfield, 2010). Prolonged stress can cause homeostatic imbalances and ultimately result in illness (Hoogendoorn et al., 2017). The field of neuroimmunology indicates that chronic or repeated exposure to external or internal stressors may alter homeostatic balance (van den Bosch & Meyer-Lindenberg, 2019). Related to these findings, pioneering research has linked nature exposure to enhanced immune function. The research has shown a decrease in interleukin-6 (IL-6), a key inflammatory biomarker, when regularly exposed to nature. Taken together, these findings demonstrate a robust immune response suggesting that parasympathetic activation contributes to both physical and mental equilibrium (Ulrich & Gilpin, 1999; Ulrich et al., 2010; Li, 2010; Stenfors et al., 2019).

City as a Stressor

Linking to urban environments and city settings, it is critical to contemplate the impact of constant exposure to various stressors, as studies suggest stress as a core component in both the onset and symptom severity of multiple mental health conditions (Abbott, 2012; Adli & Schöndorf, 2020; Sapolsky, 1999; Slavich & Irwin, 2014). The hypothalamic-pituitary-adrenal axis (HPA) is a homeostatic brain-body system playing a fundamental part in psychological well-being and mental health. This system is activated by both internal and external forces, including stressors from the surrounding environment (Sapolsky et al., 2000). Activation of the HPA-axis is initiated in the brain's paraventricular nucleus, leading to the release of corticotropin-releasing hormone. The hormone induces adrenocorticotropic hormone from the pituitary gland into the vascular system and activates the adrenal glands to release the glucocorticoid hormone cortisol. This system acts to regulate daily physiological processes in the body such as respiratory rate, heart rate, glucose levels and immune system response to stress, activating above-normal levels in reaction to fight-or-flight situations (Hoogendoorn et al., 2017; Sapolsky et al., 2000). Long-term exposure to environmental stressors can result in a state of chronic stress and HPA dysregulation, which in turn can stimulate alterations in reward pathways (Adam & Epel, 2007 Gold, 2015; McEwen, 2017). One potential explanation for the elevated stress levels in urban environments may therefore be the exposure to numerous and prolonged stressors, resulting in chronic stress and HPA dysregulation (Valentine, 2023). This accumulative 'cost' is sometimes referred to as allostatic load. Allostatic load indicates the cumulative physiological burden experienced by the body as it constantly adjusts to stressors from the surrounding environment (Sterling & Eyer, 1988). Extensive exposure to stressors can result in allostatic overload, ultimately manifested by disturbances in various physiological systems, including the cardiovascular, metabolic and immune systems (McEwen, 1998). As stress is a well-known mediator in many mental health conditions, this may induce alterations in reward pathways and change serotonergic and noradrenergic receptors in the brain, increasing the risk of mental health conditions (Marin et al., 2011; Mariotti, 2015). To better balance risk and resilience factors could be essential to counteract the mental health challenges that emerge alongside rising urbanisation.

Stress and Inflammation

Emerging evidence suggests that prolonged stress can cause chronic inflammation. Research in recent years has linked inflammatory biomarkers to mental health conditions such as anxiety, depression, psychosis and schizophrenia (Aas et al., 2014; Aruldass et al., 2021; Bullmore, 2018; Fraguas et al., 2019; Pariante, 2017; Yuan et al., 2019). Notably, the same illnesses are more prevalent in urban populations (Sundquist et al., 2004; Peen et al., 2010). These findings align with previous pioneering research by Lederbogen et al. (2011) which found that the amygdala, part of the brain's limbic system associated with initiating acute stress responses, to be more active in city dwellers compared to rural dwellers (Lederbogen et al., 2011). Moreover, individuals who took a forest walk, rather than an urban walk, prior to fMRI scanning showed a lower amygdala response to face stimuli (Sudimac et al., 2022). Therefore, even brief exposure to natural settings may impact brain function in the time period immediately after exposure, supporting the idea that short-term contact with natural settings can have a calming effect on brain regions associated with stress response. Future health policies and urban planning processes should address the complex interplay between the built environment, stressors, inflammatory responses and mental health to aim to better balance these environments with counteracting components such as green space and biophilic design elements.

Neuroanatomical Impacts of Stress: Neuroplasticity and Neurogenesis

Chronic stress has been associated with decreased neuroplasticity and neurogenesis, which are two key processes for good mental health (Uchida et al., 2018). Neuroplasticity refers to the brain's ability to reshape in response to the external world, and neurogenesis is defined as the brain's capacity to form new neurons throughout life (Clemenson et al., 2015; Pittenger & Duman, 2008; Spalding et al., 2013). Robust research in animal models has found that environments rich in natural sensory stimuli can support both neuroplasticity and neurogenesis (Clemenson et al., 2015; van Praag et al., 2000). In humans, it is not clear which environments that can be enriching. However, research in recent years points at nature as one direction in promoting these biological processes (Kühn et al., 2017, 2020). The next section will explore these new studies and in addition, draw conclusions from large epidemiological research linking the environment and mental health.

Green Impact on Mental Health, Brain Structure and Brain Function

Epidemiological research by Engemann and colleagues with over 900,000 participants discovered an association between the amount of green space in one's residential environment during childhood and the risk of developing psychiatric disorders later in life (Engemann et al., 2019). Children growing up in environments with very low levels of green space had an increased prevalence of up to 55% of developing mental health disorders with other risk factors taken into account (Engemann et al.,

2019). These findings are in line with earlier epidemiological studies showing an association between time spent in cities during childhood and adolescence and a greater risk of developing mental health disorders in adulthood (Sundquist et al., 2004). These epidemiological studies indicate a dose-response relationship both in terms of time and the amount of green space.

The association between green spaces and stress resilience is further exemplified in work undertaken by Kühn et al. (2017), which identifies a significant positive relationship between forest coverage and amygdala integrity. Additionally, differences have been observed comparing individuals exclusively raised in urban or rural environments. Using voxel-based morphometry, the results revealed a difference in the left hippocampal formation on group level comparing rural and city exposures. This was positively associated with enhanced cognitive performance in a spatial processing task (Kühn et al., 2020). These results are interesting in the context of urban design and its potential future directions, as proximity to and exposure to green spaces may not only encourage spatial exploration but also build stress resilience and improve mental well-being in city populations.

Spatial Structure of the Environment

The spaces we visit can offer a range of affordances that help meet different needs (Gregorians & Spiers, 2022). It has been argued humans tend to seek out environments that offer both 'prospect' spaces with open views, and 'refuge' spaces for shelter and protection (Appleton, 1984; Dosen & Ostwald, 2016). From an evolutionary perspective, these spatial preferences have optimised chances of survival by supporting threat detection, as well as offering a safe retreat (Heft, 2010). These spatial preferences and constructs may also be activated in contemporary contexts. An example is negative affect when being exposed to high-rise buildings (Mazumder et al., 2022), or the stress one might experience when getting lost in a complex building, as this may tap into evolved neural systems that detect vulnerability to threats in unfamiliar landscapes (Hart & Moore, 1973). These instinctual cognitive states profoundly shape navigation and wayfinding behaviours (Golledge, 1999; Montello, 2005; Ekstrom et al., 2018). Especially how the surrounding environment shapes our ability to navigate (Gregorians & Spiers, 2022). To give one example; growing up inside vs outside cities (particularly griddy cities) has an impact on navigation abilities (Coutrot et al., 2022; Spiers et al., 2023). By testing over 380,000 participants on a virtual navigation test with ecological validity, and asking about the home environment, findings indicate that growing up in griddy cities leads to worse navigation performance, particularly in disorganised environments Coutrot et al., 2018, 2019, 2022). For example, individuals growing up in a city like Buenos Aires in Argentina, which is rather grid-like and well-structured, will on average be worse at navigating compared with someone from Bucharest in Romania, which is less structured (see layouts in Fig. 3). Analysis using agent-based modelling indicated that such cities lead to greater exposure to 90-degree turns and less re-orientation requirements than cities with more organic design or rural settings. Thus, while designing cities with organic layouts may make environments harder to navigate (Yesiltepe et al., 2023), they can also help support better navigation ability in the population.

It is not just city layouts that appear to impact cognitive ability, but what people do when they navigate in such complex urban settings. London licensed taxi drivers spend everyday mentally planning through hundreds of journeys in the 26,000+ streets of London (UK) and their brains appear to be affected by this challenge

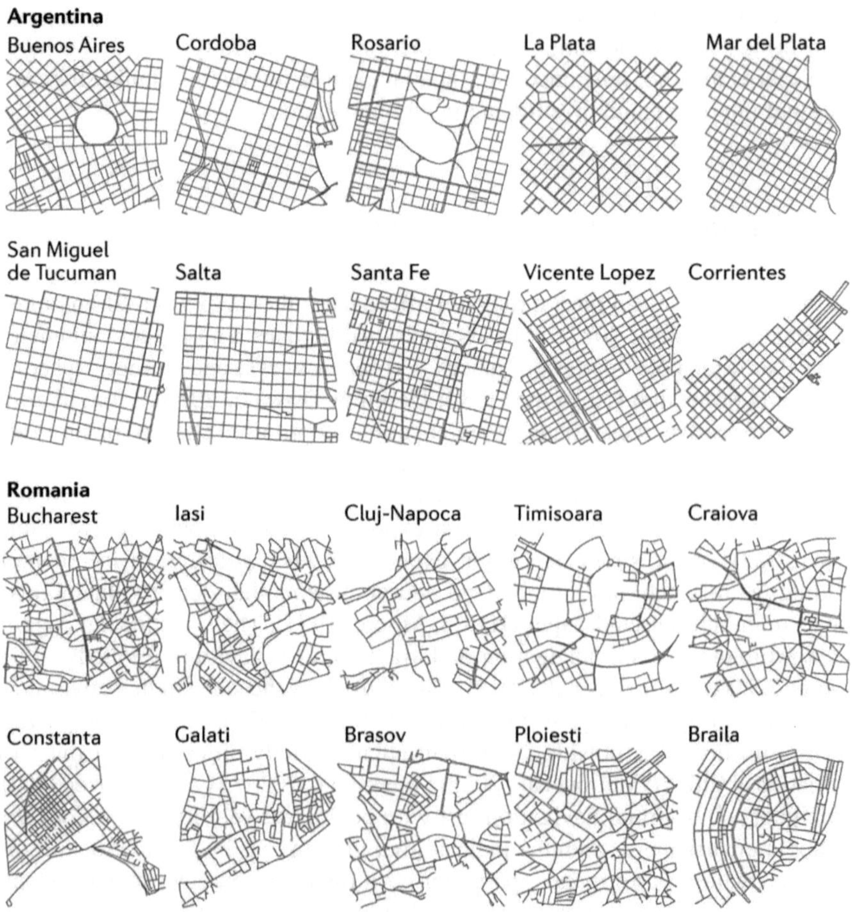

Fig. 3 Difference in the organisational form of cities in two countries, Argentina and Romania. Participants who group up in Argentinian cities (which are grid-like) tend to be worse at navigating compared to those who grew up outside cities. No such effect is visible in Romania or other countries with more organic forms. (Reproduced with Permission from Coutrot et al. (2022)

(Maguire et al., 2000, 2006; Griesbaur et al., 2022). Analysis of MRI data from London taxi drivers has revealed, over repeated studies, that the posterior grey matter of their hippocampus is larger than in non-taxi drivers, including London bus drivers who were exposed to London's streets for a similar period (Maguire et al., 2000, 2006; Griesbauer et al., 2022). Such changes come with expert navigation knowledge, but also slightly reduced performance on some non-spatial memory tests (Maguire et al., 2006). Thus, cities can affect brain structure and cognitive skills in different ways.

The Stress of Being Lost

A key environment where being lost is commonplace is hospital environments. Navigating large hospitals can be a very challenging and stressful experience because of the practical complexity of healthcare spaces and the typical absence of indoor digital map guidance. Additionally, the look-and-feel of the hospital environment is often sterile with few visual variations which makes it even more difficult to navigate. Moreover, poor wayfinding and disorientation are among the first core symptoms of Alzheimer's dementia which is particularly problematic in hospitals (Coughlan et al., 2018). This combination can lead to stress and negative experiences among patients and staff. An appealing example of how green spaces could be used to enhance wayfinding is an experiment by Jiang et al. (2022). By using immersive virtual reality and high-definition virtual models of hospitals, participant's mood, wayfinding performance and spatial experience were examined in hospital settings with or without visible green spaces (see Fig. 4). The results showed better wayfinding performance for the environments with visable green spaces, measured by shorter walking distance, less sign viewing, and fewer stops, more efficient route selection, and less time spent on finding the correct destination. Moreover, the participants also reported enhanced mood states, better spatial experience and higher perceived aesthetics when comparing hospital environments with

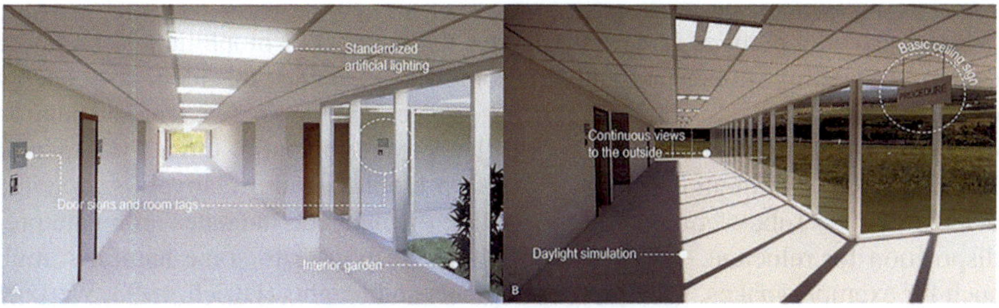

Fig. 4 Use of immersive VR and realistic rendering to explore the impact of environmental design on wayfinding, mood and stress. Left image shows a virtual view of the standard models of a hospital. Right image shows the altered hospital view with a view of outside green space. (Reproduced from Jiang et al. (2022))

visible green spaces to the same space without window views (Jiang et al., 2022). This is another example of how natural elements and green space may help facilitate wayfinding. Because multiple factors were manipulated between the two environments (e.g. signage) it will be useful in future studies to determine the extent to which different elements in the environment play a role.

Awe and Creative Thinking

Building further on this body of research, it is interesting to explore the connection between nature, awe and creativity. Nature exposure has been found to activate the parasympathetic nervous system and stimulate an 'at ease' state, which promotes 'mental equilibrium' and creative thinking (Tversky, 2019). Awe is a relatively new concept in empirical research which can be described as a feeling of connectedness and presence with something greater than oneself, accompanied by wonder or fascination (Shiota et al., 2007). Awe experiences can originate from various stimuli, such as impressive and beautiful architecture, or breath-taking natural views or landscapes (Zhao et al., 2018). Research has also found a link between awe and activation of the default mode network in the brain, a neural network associated with rest, mind-wandering, introspection and daydreaming, a network crucial for cognitive flexibility and creative thinking (Andrews-Hanna, 2012; Beaty et al., 2015; Buckner et al., 2008; Stellar et al., 2018). Relating to these findings, Tversky (2019) suggests that our spatial and navigational thinking, which is often stimulated to a greater extent in nature and natural settings, may positively influence abstract thoughts and creative thinking. Implementing biophilic design and natural elements may contribute to the enhanced interplay between the parasympathetic nervous system, the default mode network, and creativity (Mostarejan et al., 2023).

Biophilia, Biophobia?

While a growing body of research indicates the benefits of biophilia and biophilic design, contrasting studies on biophobia somewhat challenge this idea. The concept of biophobia suggests that humans evolved to react with instinctive fear and caution of certain elements in the natural world that present potential threats (Ohman & Mineka, 2001) Although empirical studies on biophobia are relatively scarce and evidence is less robust compared to biophilia, the research indicates an innate predisposition for reluctant, negative or biophobic responses to some natural stimuli such for example snakes, spiders, certain plants and heights (Ulrich, 1993; Wertz & Wynn, 2014). Studies have explored autonomic nervous system markers for different types of natural stimuli which indicate different responses to geometric figures and flowers, compared to snakes and spiders (Ohman et al., 2001; Ohman 2009; Sabatinelli et al., 2005). From a neurobiological perspective, this is logical, as it is crucial to distinguish risk-relevant and non-risk-relevant stimuli in natural wild environments. Avoiding snakes and other dangerous animals is crucial to enhance chances for survival, while non-risk-relevant natural stimuli may indicate environmental richness, biodiversity and food. To summarise, separating risk-relevant and non-risk-relevant natural stimuli has been, and still is, evolutionary advantageous to optimise human survival.

Conclusion

By 2050, 70 percent of the world's population is estimated to live in urban environments (Richie & Roser, 2018). Cities offer many opportunities such as proximity to work, education and healthcare (Hartig et al., 2014). However, in parallel with rapid urbanisation, mental health conditions are rising (Patel et al., 2018; World Health Organization, 2022). Empirical evidence suggests that exposure to nature can help to counteract these challenges as nature is linked to multiple psychological and physiological pathways, and is associated with mental health, such as restoration, physical activity, and social interaction, as well as immune function, brain development and cognition (Beukeboom et al. 2012; Engemann et al., 2019; Goldy & Piff, 2020; Kaplan & Kaplan, 1989; Kaplan et al., 1989; Kim et al., 2023; Kühn et al., 2017, 2020; Li, 2010, Li et al., 2009, 2011; McCoy & Evans, 2002; Oppezzo & Schwartz, 2014; Ulrich et al., 1991; Ulrich, 1992). As a result of emerging empirical evidence in scientific research, design disciplines outside academia have started to acknowledge the potential of implementing biophilic characteristics in architectural design and built environments (Camargo et al., 2018). Recently, there has been a strong interest among architects, designers, developers and decision-makers to implement and apply biophilic design to the built environment beyond green plants.

Symbiotic Spaces and their Implications for Sustainable Development

The twenty-first century is marked by the escalating urgency to navigate the nexus between human health, ecological protection and economic sustainability. Achieving the Sustainable Development Goals requires innovative solutions and urgent action on a global scale. At the centre of this discussion is the potential of biophilic design as a prospective tool in counteracting these challenges. In the context of sustainable development, the concept of symbiotic spaces stands out as an appealing alternative. The concept brings together natural and man-made elements creating a cycle of mutual benefit. This approach not only nurtures the environment but may also support human health, well-being and economic growth. Thus, there is vast potential creating synergies for both people and the planet, supporting human, ecological and economically sustainable solutions.

Critical Evaluation of the Biophilia Hypothesis: Gaps in Existing Research

While the concept of biophilia has a robust theoretical foundation, the multimodal impact of biophilic design in modern, real-world built environments and potential dose effects needs to be further examined. There may be scenarios or situations where it is less effective and its use needs to be carefully considered (Fitzgerald, 2023. Previous research has mainly originated from fields such as environmental psychology, sociology and anthropology. These studies generally explore how interaction with nature can positively affect human well-being. However, much of this research has relied on self-reported metrics, such as questionnaires and surveys, which are vulnerable to subjective bias.

Furthermore, epidemiological studies have correlated urban living with poorer mental health outcomes. These studies often use registry-based research methods, which provide large sample sizes and strong statistical power; however, they are insufficient for understanding causal relationships between architectural design characteristics and mental health. This limitation indicates a significant opportunity for future research to explore causal links between the built environment and mental health.

A limited number of MRI studies (➔ chapter "Interplay Between the Physical Environment and the Human Brain: A Review of MRI Research" by Beyer et al.) in recent years have identified alterations in stress-responses, and brain structure and function, linking neural changes to greenspace exposure. These studies provide a worthy foundation for further research in the search for causal elements in city environments that may mediate mental health outcomes. Although MRI is still considered the gold standard within neuroscientific research, recent advances in neuroscience offer promising opportunities for complementing objective measures to the existing field of research. For example, novel neurotechnology such as mobile EEG (➔ chapter "EEG for the Study of Environmental Neuroscience", by Grassini) and functional near-infrared spectroscopy (fNIRS) create interesting new opportunities to measure brain activity in real-world environments (e.g. Burgess et al., 2022). When combined with other physiological biomarkers, like smartwatches and body sensors, these technologies and methods have significant potential to add objective qualitative measures to advance the research field. Most studies have primarily focused on the psychological and physiological benefits of exposure to nature and green space. Future research could explore other biophilic aspects of urban environments, such as how architectural design provides access to daylight, encourages movement and fosters social connections.

A clear gap in existing research is the predominant focus on the visual system. A multisensory approach, considering all senses—touch, sight, hearing, smell, and taste—could provide a more comprehensive and realistic perspective that is closely connected to real-world environments. Additional research is also needed to examine social and cross-cultural differences when examining biophilia and close-to-nature experiences, as these may vary. The current body of research has mainly focused on Western populations, often overlooking the role of socio-cultural factors in human-nature experiences and interactions.

Another limitation within biophilic research is the wide definition of biophilia and biophilic design. Clearer definitions and a standardised framework may enable more rigorous research evaluating the theoretical and empirical validity of biophilia and biophilic design. A more nuanced and multifaceted approach could clarify the complex mechanisms behind the potential impact of biophilic design on human cognition, behaviour and mental health, and provide a more holistic perspective.

Vision and Future Directions

More research is needed to gain solid empirical grounding within this field. Applying an interdisciplinary approach integrating insights from multiple scientific disciplines could offer new possibilities for more comprehensive and refined research. Such an approach is essential for a nuanced understanding of the complex mechanisms behind biophilic design's impact on human cognition, behaviour and mental health. Incorporating biophilic design at different scales in future cities, buildings, and interiors by microdosing natural elements could cumulatively provide significant improvements for human health. Implementing biophilic design in future architecture, design, and infrastructure therefore has the potential to create a holistic whole greater than the sum of its parts.

References

Aas, M., Dazzan, P., Mondelli, V., Melle, I., Murray, R. M., & Pariante, C. M. (2014). A systematic review of cognitive function in first-episode psychosis, including a discussion on childhood trauma, stress, and inflammation. *Frontiers in Psychiatry, 4*, 182.

Abbott, A. (2012). Urban decay: Scientists are testing the idea that the stress of modern city life is a breeding ground for psychosis. *Nature, 490*(7419), 162–165.

Adam, T. C., & Epel, E. S. (2007). Stress, eating and the reward system. *Physiology & Behavior, 91*(4), 449–458.

Adli, M., & Schöndorf, J. (2020). Does the city make us ill? The effect of urban stress on emotions, behavior, and mental health. *Bundesgesundheitsblatt-Gesundheitsforschung-Gesundheitsschutz, 63*, 979–986.

Alvarsson, J. J., Wiens, S., & Nilsson, M. E. (2010). Stress recovery during exposure to nature sound and environmental noise. *International Journal of Environmental Research and Public Health, 7*(3), 1036–1046.

Andrews-Hanna, J. R. (2012). The brain's default network and its adaptive role in internal mentation. *The Neuroscientist, 18*(3), 251–270.
Appleton, J. (1984). Prospects and refuges re-visited. *Landscape Journal, 3*(2), 91–103.
Aruldass, A. R., Kitzbichler, M. G., Morgan, S. E., Lim, S., Lynall, M. E., Turner, L., Vertes, P., Wellcome Trust Consortium for Neuroimmunology of Mood Disorders and Alzheimer's Disease (NIMA), Cavanagh, J., Cowen, P., Pariante, C. M., Harrison, N. A., & Bullmore, E. T. (2021). Dysconnectivity of a brain functional network was associated with blood inflammatory markers in depression. *Brain, Behavior, and Immunity, 98*, 299–309. https://doi.org/10.1016/j.bbi.2021.08.226. Epub 2021 Aug 25.
Aydogan, A., & Cerone, R. (2021). Review of the effects of plants on indoor environments. *Indoor and Built Environment, 30*(4), 442–460. https://doi.org/10.1177/1420326X19900213
Bar, M., & Neta, M. (2006). Humans prefer curved visual objects. *Psychological Science, 17*(8), 645–648.
Bar, M., & Neta, M. (2007). Visual elements of subjective preference modulate amygdala activation. *Neuropsychologia, 45*(10), 2191–2200.
Barbiero, G., & Berto, R. (2021). Biophilia as evolutionary adaptation: An onto-and phylogenetic framework for biophilic design. *Frontiers in Psychology, 12*, 700709.
Barnett, K. (1972). A theoretical construct of the concepts of touch as they relate to nursing. *Nursing Research, 21*(2), 102–109.
Beaty, R. E., Benedek, M., Barry Kaufman, S., & Silvia, P. J. (2015). Default and executive network coupling supports creative idea production. *Scientific Reports, 5*(1), 10964.
Berman, M. G., Jonides, J., & Kaplan, S. (2008). The cognitive benefits of interacting with nature. *Psychological Science, 19*(12), 1207–1212.
Bertamini, M., Palumbo, L., Gheorghes, T. N., & Galatsidas, M. (2016). Do observers like curvature or do they dislike angularity? *British Journal of Psychology, 107*(1), 154–178.
Beukeboom, C. J., Langeveld, D., & Tanja-Dijkstra, K. (2012). Stress-reducing effects of real and artificial nature in a hospital waiting room. *The Journal of Alternative and Complementary Medicine, 18*(4), 329–333.
Beute, F., & De Kort, Y. A. W. (2014). Natural resistance: Exposure to nature and self-regulation, mood, and physiology after ego-depletion. *Journal of Environmental Psychology, 40*, 167–178.
Bhatta, S. R., Tiippana, K., Vahtikari, K., Hughes, M., & Kyttä, M. (2017). Sensory and emotional perception of wooden surfaces through fingertip touch. *Frontiers in Psychology, 8*, 367.
Bolten, B., & Barbiero, G. (2020). Biophilic design: How to enhance physical and psychological health and wellbeing in our built environments.
Bratman, G. N., Anderson, C. B., Berman, M. G., Cochran, B., De Vries, S., Flanders, J., et al. (2019). Nature and mental health: An ecosystem service perspective. *Science Advances, 5*(7), eaax0903.
Bringslimark, T., Hartig, T., & Patil, G. G. (2007). Psychological benefits of indoor plants in workplaces: Putting experimental results into context. *HortScience Horts, 42*(3), 581–587. Retrieved Feb 24, 2024, from https://doi.org/10.21273/HORTSCI.42.3.581
Buckner, R. L., Andrews-Hanna, J. R., & Schacter, D. L. (2008). The brain's default network: Anatomy, function, and relevance to disease. *Annals of the New York Academy of Sciences, 1124*(1), 1–38.
Bullmore, E. (2018). Inflamed depression. *The Lancet, 392*(10154), 1189–1190.
Burgess, P. W., Crum, J., Pinti, P., Aichelburg, C., Oliver, D., Lind, F., et al. (2022). Prefrontal cortical activation associated with prospective memory while walking around a real-world street environment. *NeuroImage, 258*, 119392.
Camargo, A., Artus, J., & Spiers, H. J. (2018). *Neuroscience for cities*. Future Cities Catapult.
Choi, K., & Suk, H. J. (2016). Dynamic lighting system for the learning environment: Performance of elementary students. *Optics Express, 24*(10), A907–A916.
Clemenson, G. D., Deng, W., & Gage, F. H. (2015). Environmental enrichment and neurogenesis: From mice to humans. *Current Opinion in Behavioral Sciences, 4*, 56–62. https://doi.org/10.1016/j.cobeha.2015.02.005

Coughlan, G., Laczó, J., Hort, J., Minihane, A. M., & Hornberger, M. (2018). Spatial navigation deficits—Overlooked cognitive marker for preclinical Alzheimer disease? *Nature Reviews Neurology, 14*(8), 496–506.
Coutrot, A., Silva, R., Manley, E., de Cothi, W., Sami, S., Bohbot, V. D., et al. (2018). Global determinants of navigation ability. *Current Biology, 28*(17), 2861–2866.
Coutrot, A., Schmidt, S., Coutrot, L., Pittman, J., Hong, L., Wiener, J. M., et al. (2019). Virtual navigation tested on a mobile app is predictive of real-world wayfinding navigation performance. *PLoS One, 14*(3), e0213272.
Coutrot, A., Manley, E., Goodroe, S., Gahnstrom, C., Filomena, G., Yesiltepe, D., et al. (2022). Entropy of city street networks linked to future spatial navigation ability. *Nature, 604*(7904), 104–110.
Cross, S. H., Califf, R. M., & Warraich, H. J. (2021). Rural-urban disparity in mortality in the US from 1999 to 2019. *JAMA, 325*(22), 2312–2314.
Dagnino-Subiabre, A. (2022). Resilience to stress and social touch. *Current Opinion in Behavioral Sciences, 43*, 75–79.
Dosen, A. S., & Ostwald, M. J. (2016). Evidence for prospect-refuge theory: A meta-analysis of the findings of environmental preference research. *City, Territory and Architecture, 3*(1), 1–14.
Ekstrom, A. D., Spiers, H. J., Bohbot, V. D., & Rosenbaum, R. S. (2018). *Human spatial navigation*. Princeton University Press.
Elliot, A. J., & Maier, M. A. (2014). Color psychology: Effects of perceiving color on psychological functioning in humans. *Annual Review of Psychology, 65*, 95–120.
Engemann, K., Pedersen, C. B., Arge, L., Tsirogiannis, C., Mortensen, P. B., & Svenning, J. C. (2019). Residential green space in childhood is associated with lower risk of psychiatric disorders from adolescence into adulthood. *Proceedings of the National Academy of Sciences, 116*(11), 5188–5193.
Etkin, A., Egner, T., & Kalisch, R. (2011). Emotional processing in anterior cingulate and medial prefrontal cortex. *Trends in Cognitive Sciences, 15*(2), 85–93.
Evans, G. W. (2003). The built environment and mental health. *Journal of Urban Health: Bulletin of the New York Academy of Medicine, 80*(4), 536–555.
Fell, D. R. (2010). *Wood in the human environment: Restorative properties of wood in the built indoorenvironment* (Doctoral dissertation, University of British Columbia).
Fisher, J. A. (1999). The value of natural sounds. *Journal of Aesthetic Education, 33*(3), 26–42.
Fitzgerald, D. (2023). *The living City: Why cities Don't need to be green to be great*. Basic Books.
Fraguas, D., Díaz-Caneja, C. M., Ayora, M., Hernández-Álvarez, F., Rodríguez-Quiroga, A., Recio, S., et al. (2019). Oxidative stress and inflammation in first-episode psychosis: A systematic review and meta-analysis. *Schizophrenia Bulletin, 45*(4), 742–751.
Franco, L. S., Shanahan, D. F., & Fuller, R. A. (2017). A review of the benefits of nature experiences: More than meets the eye. *International Journal of Environmental Research and Public Health, 14*(8), 864.
Fromm, E. (1964). The Heart of Man: Its Genius for Good and Evil. New York: Harper and Row.
Fromm, E. (1973). *The anatomy of human destructiveness*. New York: Fawcett Crest.
Gabrieli, J. D. (1998). Cognitive neuroscience of human memory. *Annual Review of Psychology, 49*(1), 87–115.
Gallace, A. (2012). Living with touch: Understanding tactile interactions. *The Psychologist, 25*(3), 5.
Gallace, A., & Spence, C. (2008). The cognitive and neural correlates of "tactile consciousness": A multisensory perspective. *Consciousness and Cognition, 17*(1), 370–407.
Gallace, A., & Spence, C. (2010). The science of interpersonal touch: An overview. *Neuroscience & Biobehavioral Reviews, 34*(2), 246–259.
Gillis, K., & Gatersleben, B. (2015). A review of psychological literature on the health and wellbeing benefits of biophilic design. *Buildings, 5*(3), 948–963.
Gold, P. W. (2015). The organization of the stress system and its dysregulation in depressive illness. *Molecular Psychiatry, 20*(1), 32–47.

Goldy, S. P., & Piff, P. K. (2020). Toward a social ecology of prosociality: Why, when, and where nature enhances social connection. *Current Opinion in Psychology, 32*, 27–31.

Golledge, R. G. (Ed.). (1999). *Wayfinding behavior: Cognitive mapping and other spatial processes*. JHU Press.

Gould van Praag, C. D., Garfinkel, S. N., Sparasci, O., Mees, A., Philippides, A. O., Ware, M., et al. (2017). Mind-wandering and alterations to default mode network connectivity when listening to naturalistic versus artificial sounds. *Scientific Reports, 7*(1), 45273.

Gregorians, L., & Spiers, H. J. (2022). Affordances for spatial navigation. In *Affordances in everyday life: A multidisciplinary collection of essays* (pp. 99–112).

Griesbauer, E. M., Manley, E., Wiener, J. M., & Spiers, H. J. (2022). London taxi drivers: A review of neurocognitive studies and an exploration of how they build their cognitive map of London. *Hippocampus, 32*(1), 3–20.

Haddad, L., Schäfer, A., Streit, F., Lederbogen, F., Grimm, O., Wüst, S., et al. (2015). Brain structure correlates of urban upbringing, an environmental risk factor for schizophrenia. *Schizophrenia Bulletin, 41*(1), 115–122.

Hagerhall, C. M., Laike, T., Taylor, R. P., Küller, M., Küller, R., & Martin, T. P. (2008). Investigations of human EEG response to viewing fractal patterns. *Perception, 37*(10), 1488–1494.

Hagerhall, C. M., Laike, T., Küller, M., Marcheschi, E., Boydston, C. R., & Taylor, R. P. (2015). Human physiological benefits of viewing nature: EEG responses to exact and statistical fractal patterns. *Nonlinear Dynamics, Psychology, and Life Sciences, 19*(1), 1–12.

Hall, C., & Knuth, M. (2019). An update of the literature supporting the well-being benefits of plants: A review of the emotional and mental health benefits of plants. *Journal of Environmental Horticulture, 37*(1), 30–38.

Hart, R. A., & Moore, G. T. (1973). *The development of spatial cognition: A review*. Aldine Transaction.

Hartig, T., & Kahn, P. H., Jr. (2016). Living in cities, naturally. *Science, 352*(6288), 938–940.

Hartig, T., Mitchell, R., De Vries, S., & Frumkin, H. (2014). Nature and health. *Annual Review of Public Health, 35*, 207–228. https://doi.org/10.1146/ANNUREV-PUBLHEALTH-032013-182443

Heft, H. (2010). Affordances and the perception of landscape. In *Innovative approaches to researching landscape and health* (pp. 9–32).

Henricson, M., Ersson, A., Määttä, S., Segesten, K., & Berglund, A. L. (2008). The outcome of tactile touch on stress parameters in intensive care: A randomized controlled trial. *Complementary Therapies in Clinical Practice, 14*(4), 244–254.

Herz, R. S. (2009). Aromatherapy facts and fictions: A scientific analysis of olfactory effects on mood, physiology and behavior. *International Journal of Neuroscience, 119*(2), 263–290.

Hoogendoorn, C. J., Roy, J. F., & Gonzalez, J. S. (2017). Shared dysregulation of homeostatic brain-body pathways in depression and type 2 diabetes. *Current Diabetes Reports, 17*, 1–10.

Houser, K. W., & Esposito, T. (2021). Human-centric lighting: Foundational considerations and a five-step design process. *Frontiers in Neurology, 12*, 25.

Igarashi, M., Song, C., Ikei, H., Ohira, T., & Miyazaki, Y. (2014). Effect of olfactory stimulation by fresh rose flowers on autonomic nervous activity. *The Journal of Alternative and Complementary Medicine, 20*(9), 727–731.

Iglesias-Carrasco, M., Aich, U., Jennions, M. D., & Head, M. L. (2020). Stress in the city: Meta-analysis indicates no overall evidence for stress in urban vertebrates. *Proceedings of the Royal Society B, 287*(1936), 20201754.

Inagaki, T. K., & Eisenberger, N. I. (2013). Shared neural mechanisms underlying social warmth and physical warmth. *Psychological Science, 24*(11), 2272–2280.

Isaksson, C. (2015). Urbanization, oxidative stress and inflammation: A question of evolving, acclimatizing or coping with urban environmental stress. *Functional Ecology, 29*(7), 913–923.

Jansson-Boyd, C. V. (2011). Touch matters: Exploring the relationship between consumption and tactile interaction. *Social Semiotics, 21*(4), 531–546.

Jiang, S., Allison, D., & Duchowski, A. T. (2022). Hospital greenspaces and the impacts on wayfinding and spatial experience: An explorative experiment through immersive virtual environment (IVE) techniques. *HERD, 15*(3), 206–228. https://doi.org/10.1177/19375867211067539. Epub 2022 Jan 11.

Joye, Y., & Van den Berg, A. (2011). Is love for green in our genes? A critical analysis of evolutionary assumptions in restorative environments research. *Urban Forestry & Urban Greening, 10*(4), 261–268.

Kaplan, R. (1993). The role of nature in the context of the workplace. *Landscape and Urban Planning, 26*(1–4), 193–201.

Kaplan, S. (1995). The restorative benefits of nature: Toward an integrative framework. *Journal of Environmental Psychology, 15*(3), 169–182.

Kaplan, R., & Kaplan, S. (1989). *The experience of nature: A psychological perspective*. Cambridge University Press.

Kaplan, R., Kaplan, S., & Brown, T. (1989). Environmental preference: A comparison of four domains of predictors. *Environment and Behavior, 21*(5), 509–530.

Kellert, S. R. (2008). Dimensions, elements, and attributes of biophilic design. In *Biophilic design: The theory, science, and practice of bringing buildings to life* (pp. 3–19).

Kellert, S. R. (2016). Nature in buildings and health design. In *Green exercise* (pp. 17–25). Routledge.

Kellert, S. R. (2018). *Nature by design: The practice of biophilic design*. Yale University Press.

Kellert, S., & Calabrese, E. (2015). *The practice of biophilic design* (pp. 21–46). Terrapin Bright LLC, 3.

Kellert, S. R., Heerwagen, J., & Mador, M. (2011). *Biophilic design: The theory, science and practice of bringing buildings to life*. John Wiley & Sons.

Kim, K., Joyce, B. T., Nannini, D. R., Zheng, Y., Gordon-Larsen, P., Shikany, J. M., et al. (2023). Inequalities in urban greenness and epigenetic aging: Different associations by race and neighborhood socioeconomic status. *Science. Advances, 9*(26), eadf8140.

Kotradyova, V., Vavrinsky, E., Kalinakova, B., Petro, D., Jansakova, K., Boles, M., & Svobodova, H. (2019). Wood and its impact on humans and environment quality in health care facilities. *International Journal of Environmental Research and Public Health, 16*(18), 3496.

Kühn, S., Düzel, S., Eibich, P., Krekel, C., Wüstemann, H., Kolbe, J., et al. (2017). In search of features that constitute an "enriched environment" in humans: Associations between geographical properties and brain structure. *Scientific Reports, 7*(1), 11920.

Kühn, S., Banaschewski, T., Bokde, A. L., Büchel, C., Quinlan, E. B., Desrivières, S., et al. (2020). Brain structure and habitat: Do the brains of our children tell us where they have been brought up? *NeuroImage, 222*, 117225.

Kühn, S., Forlim, C. G., Lender, A., Wirtz, J., & Gallinat, J. (2021). Brain functional connectivity differs when viewing pictures from natural and built environments using fMRI resting state analysis. *Scientific Reports, 11*(1), 4110.

Küller, R., Mikellides, B., & Janssens, J. (2009). Color, arousal, and performance—A comparison of three experiments. *Color Research & Application: Endorsed by Inter-Society Color Council, The Colour Group (Great Britain), Canadian Society for Color, Color Science Association of Japan, Dutch Society for the Study of Color, The Swedish Colour Centre Foundation, Colour Society of Australia, Centre Français de la Couleur, 34*(2), 141–152.

Lambert, K. G., Nelson, R. J., Jovanovic, T., & Cerdá, M. (2015). Brains in the city: Neurobiological effects of urbanization. *Neuroscience & Biobehavioral Reviews, 58*, 107–122.

Leder, H., Tinio, P. P., & Bar, M. (2011). Emotional valence modulates the preference for curved objects. *Perception, 40*(6), 649–655.

Lederbogen, F., Kirsch, P., Haddad, L., Streit, F., Tost, H., Schuch, P., et al. (2011). City living and urban upbringing affect neural social stress processing in humans. *Nature, 474*(7352), 498–501.

Lei, Q., Yuan, C., & Lau, S. S. Y. (2021). A quantitative study for indoor workplace biophilic design to improve health and productivity performance. *Journal of Cleaner Production, 324*, 129168.

Li, Q. (2010). Effect of forest bathing trips on human immune function. *Environmental Health and Preventive Medicine, 15*, 9–17.
Li, Q., Kobayashi, M., Wakayama, Y., Inagaki, H., Katsumata, M., Hirata, Y., et al. (2009). Effect of phytoncide from trees on human natural killer cell function. *International Journal of Immunopathology and Pharmacology, 22*(4), 951–959.
Li, Q., Otsuka, T., Kobayashi, M., Wakayama, Y., Inagaki, H., Katsumata, M., et al. (2011). Acute effects of walking in forest environments on cardiovascular and metabolic parameters. *European Journal of Applied Physiology, 111*, 2845–2853.
Lichtenfeld, S., Elliot, A. J., Maier, M. A., & Pekrun, R. (2012). Fertile green: Green facilitates creative performance. *Personality and Social Psychology Bulletin, 38*(6), 784–797.
Lindal, P. J., & Hartig, T. (2013). Architectural variation, building height, and the restorative quality of urban residential streetscapes. *Journal of Environmental Psychology, 33*, 26–36.
Maguire, E. A., Gadian, D. G., Johnsrude, I. S., Good, C. D., Ashburner, J., Frackowiak, R. S., & Frith, C. D. (2000). Navigation-related structural change in the hippocampi of taxi drivers. *Proceedings of the National Academy of Sciences, 97*(8), 4398–4403.
Maguire, E. A., Woollett, K., & Spiers, H. J. (2006). London taxi drivers and bus drivers: A structural MRI and neuropsychological analysis. *Hippocampus, 16*(12), 1091–1101.
Marin, M. F., Lord, C., Andrews, J., Juster, R. P., Sindi, S., Arsenault-Lapierre, G., et al. (2011). Chronic stress, cognitive functioning and mental health. *Neurobiology of Learning and Memory, 96*(4), 583–595.
Mariotti, A. (2015). The effects of chronic stress on health: New insights into the molecular mechanisms of brain–body communication. *Future Science OA, 1*(3).
Mazumder, R., Spiers, H. J., & Ellard, C. G. (2022). Exposure to high-rise buildings negatively influences affect: Evidence from real world and 360-degree video. *Cities & Health, 6*(6), 1081–1093.
McCoy, J. M., & Evans, G. W. (2002). The potential role of the physical environment in fostering creativity. *Creativity Research Journal, 14*(3–4), 409–426.
McEwen, B. S. (1998). Stress, adaptation, and disease: Allostasis and allostatic load. *Annals of the New York Academy of Sciences, 840*(1), 33–44.
McEwen, B. S. (2004). Protection and damage from acute and chronic stress: Allostasis and allostatic overload and relevance to the pathophysiology of psychiatric disorders. *Annals of the New York Academy of Sciences, 1032*(1), 1–7.
McEwen, B. S. (2017). Neurobiological and systemic effects of chronic stress. *Chronic Stress, 1*, 2470547017692328.
McEwen, B. S., & Wingfield, J. C. (2010). What's in a name? Integrating homeostasis, allostasis and stress. *Hormones and Behavior, 57*(2), 105.
McGlone, F., Wessberg, J., & Olausson, H. (2014). Discriminative and affective touch: Sensing and feeling. *Neuron, 82*(4), 737–755.
Mehta, R., & Zhu, R. (2009). Blue or red? Exploring the effect of color on cognitive task performances. *Science, 323*(5918), 1226–1229.
Montello, D. R. (2005). *Navigation*. Cambridge University Press.
Mostajeran, F., Fischer, M., Steinicke, F., & Kühn, S. (2023). Effects of exposure to immersive computer-generated virtual nature and control environments on affect and cognition. *Scientific Reports, 13*(1), 220.
Ohman, A. (2009). Of snakes and faces: An evolutionary perspective on the psychology of fear. *Scandinavian Journal of Psychology, 50*(6), 543–552.
Ohman, A., & Mineka, S. (2001). Fears, phobias, and preparedness: Toward an evolved module of fear and fear learning. *Psychological Review, 108*(3), 483.
Ohman, A., Flykt, A., & Esteves, F. (2001). Emotion drives attention: Detecting the snake in the grass. *Journal of Experimental Psychology: General, 130*(3), 466.
Oppezzo, M., & Schwartz, D. L. (2014). Give your ideas some legs: The positive effect of walking on creative thinking. *Journal of Experimental Psychology. Learning, Memory, and Cognition, 40*(4), 1142–1152. https://doi.org/10.1037/a0036577. Epub 2014 Apr 21.

Pariante, C. M. (2017). Why are depressed patients inflamed? A reflection on 20 years of research on depression, glucocorticoid resistance and inflammation. *European Neuropsychopharmacology, 27*(6), 554–559.

Patel, V., Saxena, S., Lund, C., Thornicroft, G., Baingana, F., Bolton, P., et al. (2018). The Lancet Commission on global mental health and sustainable development. *The Lancet, 392*(10157), 1553–1598.

Peen, J., Schoevers, R. A., Beekman, A. T., & Dekker, J. (2010). The current status of urban-rural differences in psychiatric disorders. *Acta Psychiatrica Scandinavica, 121*(2), 84–93.

Pittenger, C., & Duman, R. S. (2008). Stress, depression, and neuroplasticity: A convergence of mechanisms. *Neuropsychopharmacology, 33*(1), 88–109.

Prince, M., Patel, V., Saxena, S., Maj, M., Maselko, J., Phillips, M. R., & Rahman, A. (2007). No health without mental health. *The Lancet, 370*(9590), 859–877.

Raanaas, R. K., Evensen, K. H., Rich, D., Sjøstrøm, G., & Patil, G. (2011). Benefits of indoor plants on attention capacity in an office setting. *Journal of Environmental Psychology, 31*(1), 99–105.

Ramachandran, V. S., & Hirstein, W. (1999). The science of art: A neurological theory of aesthetic experience. *Journal of Consciousness Studies, 6*(6–7), 15–51.

Rea, M. S., Figueiro, M. G., Bierman, A., & Bullough, J. D. (2010). Circadian light. *Journal of Circadian Rhythms, 8*(1), 1–10.

Reichert, M., Braun, U., Lautenbach, S., Zipf, A., Ebner-Priemer, U., Tost, H., & Meyer-Lindenberg, A. (2020). Studying the impact of built environments on human mental health in everyday life: Methodological developments, state-of-the-art and technological frontiers. *Current Opinion in Psychology, 32*, 158–164.

Richie, H., & Roser, M. (2018). *Urbanization*. Our World in Data.

Romero, L. M., Dickens, M. J., & Cyr, N. E. (2009). The reactive scope model—A new model integrating homeostasis, allostasis, and stress. *Hormones and Behavior, 55*(3), 375–389.

Sabatinelli, D., Bradley, M. M., Fitzsimmons, J. R., & Lang, P. J. (2005). Parallel amygdala and inferotemporal activation reflect emotional intensity and fear relevance. *NeuroImage, 24*(4), 1265–1270.

Sapolsky, R. M. (1999). Glucocorticoids, stress, and their adverse neurological effects: Relevance to aging. *Experimental Gerontology, 34*(6), 721–732.

Sapolsky, R. M., Romero, L. M., & Munck, A. U. (2000). How do glucocorticoids influence stress responses? Integrating permissive, suppressive, stimulatory, and preparative actions. *Endocrine Reviews, 21*(1), 55–89.

Shiota, M. N., Keltner, D., & Mossman, A. (2007). The nature of awe: Elicitors, appraisals, and effects on self-concept. *Cognition and Emotion, 21*(5), 944–963.

Slavich, G. M., & Irwin, M. R. (2014). From stress to inflammation and major depressive disorder: A social signal transduction theory of depression. *Psychological Bulletin, 140*(3), 774.

Smith, J. H., Rowland, C., Harland, B., Moslehi, S., Montgomery, R. D., Schobert, K., et al. (2021). How neurons exploit fractal geometry to optimize their network connectivity. *Scientific Reports, 11*(1), 2332.

Spalding, K. L., Bergmann, O., Alkass, K., Bernard, S., Salehpour, M., Huttner, H. B., et al. (2013). Dynamics of hippocampal neurogenesis in adult humans. *Cell, 153*(6), 1219–1227.

Spiers, H. J., Coutrot, A., & Hornberger, M. (2023). Explaining world-wide variation in navigation ability from millions of people: Citizen science project sea hero quest. *Topics in Cognitive Science, 15*(1), 120–138.

Stellar, J. E., Gordon, A., Anderson, C. L., Piff, P. K., McNeil, G. D., & Keltner, D. (2018). Awe and humility. *Journal of Personality and Social Psychology, 114*(2), 258.

Stenfors, C. U., Van Hedger, S. C., Schertz, K. E., Meyer, F. A., Smith, K. E., Norman, G. J., et al. (2019). Positive effects of nature on cognitive performance across multiple experiments: Test order but not affect modulates the cognitive effects. *Frontiers in Psychology, 10*, 1413.

Sterling, P., & Eyer, J. (1988). *Handbook of life stress, cognition and health*. Edited by S. Fisher and J. Reason Copyright 1988 John Wiley &. Sons. Handbook of Life Stress, Cognition and Health.

Sternberg, E. M. (2009). *Healing spaces: The science of place and well-being*. Harvard University Press.

Stevens, M., & Ruxton, G. D. (2012). Linking the evolution and form of warning coloration in nature. *Proceedings of the Royal Society B: Biological Sciences, 279*(1728), 417–426.

Stobbe, E., Sundermann, J., Ascone, L., & Kühn, S. (2022). Birdsongs alleviate anxiety and paranoia in healthy participants. *Scientific Reports, 12*(1), 16414. https://doi.org/10.1038/s41598-022-20841-0. PMID: 36229489; PMCID: PMC9561536.

Strobel, K., Nyrud, A. Q., & Bysheim, K. (2017). Interior wood use: Linking user perceptions to physical properties. *Scandinavian Journal of Forest Research, 32*(8), 798–806.

Sudimac, S., Sale, V., & Kühn, S. (2022). How nature nurtures: Amygdala activity decreases as the result of a one-hour walk in nature. *Molecular Psychiatry, 27*, 4446–4452. https://doi.org/10.1038/s41380-022-01720-6

Sundquist, K., Frank, G., & Sundquist, J. A. N. (2004). Urbanisation and incidence of psychosis and depression: Follow-up study of 4.4 million women and men in Sweden. *The British Journal of Psychiatry, 184*(4), 293–298.

Taylor, R. P. (2021). The potential of biophilic fractal designs to promote health and performance: A review of experiments and applications. *Sustainability, 13*(2), 823.

Taylor, R., Newell, B., Spehar, B., & Clifford, C. (2005). *Fractals: A resonance between art and nature* (pp. 53–63). Springer.

Taylor, R. P., Spehar, B., Van Donkelaar, P., & Hagerhall, C. M. (2011). Perceptual and physiological responses to Jackson Pollock's fractals. *Frontiers in Human Neuroscience, 5*, 60.

Tsunetsugu, Y., Miyazaki, Y., & Sato, H. (2007). Physiological effects in humans induced by the visual stimulation of room interiors with different wood quantities. *Journal of Wood Science, 53*, 11–16.

Tversky, B. (2019). *Mind in motion: How action shapes thought*. Hachette UK.

Uchida, S., Yamagata, H., Seki, T., & Watanabe, Y. (2018). Epigenetic mechanisms of major depression: Targeting neuronal plasticity. *Psychiatry and Clinical Neurosciences, 72*(4), 212–227.

Ulrich, R. S. (1984). View through a window may influence recovery from surgery. *Science, 224*(4647), 420–421.

Ulrich, R. S. (1992). How design impacts wellness. *The Healthcare Forum Journal, 35*(5), 20–25.

Ulrich, R. S. (1993). Biophilia, biophobia, and natural landscapes. *The Biophilia Hypothesis, 7*, 73–137.

Ulrich, R., & Gilpin, L. (1999). Healing arts. *Nursing, 4*(3), 128–133.

Ulrich, R. S., Simons, R. F., Losito, B. D., Fiorito, E., Miles, M. A., & Zelson, M. (1991). Stress recovery during exposure to natural and urban environments. *Journal of Environmental Psychology, 11*(3), 201–230.

Ulrich, R. S., Berry, L. L., Quan, X., & Parish, J. T. (2010). A conceptual framework for the domain of evidence-based design. *HERD: Health Environments Research & Design Journal, 4*(1), 95–114.

United Nations. (2023, December). *The 17 Goals*. https://sdgs.un.org/goals

Uvnas-Moberg, K., Handlin, L., & Petersson, M. (2015). Self-soothing behaviors with particular reference to oxytocin release induced by non-noxious sensory stimulation. *Frontiers in Psychology, 5*, 1529.

van den Bosch, M., & Meyer-Lindenberg, A. (2019). Environmental exposures and depression: Biological mechanisms and epidemiological evidence. *Annual Review of Public Health, 40*, 239–259.

Van Hedger, S. C., Nusbaum, H. C., Clohisy, L., Jaeggi, S. M., Buschkuehl, M., & Berman, M. G. (2019). Of cricket chirps and car horns: The effect of nature sounds on cognitive performance. *Psychonomic Bulletin & Review, 26*, 522–530.

Van Praag, H., Kempermann, G., & Gage, F. H. (2000). Neural consequences of enviromental enrichment. *Nature Reviews Neuroscience, 1*(3), 191–198.

Valentine, C. (2023). Architectural allostatic overloading: exploring a connection between architectural form and allostatic overloading. *International Journal of Environmental Research and Public Health, 20*(9), 5637.

Vartanian, O., Navarrete, G., Chatterjee, A., Fich, L. B., Leder, H., Modroño, C., et al. (2013). Impact of contour on aesthetic judgments and approach-avoidance decisions in architecture. *Proceedings of the National Academy of Sciences, 110*(supplement_2), 10446–10453.

Vavrinsky, E., Kotradyova, V., Svobodova, H., Kopani, M., Donoval, M., Sedlak, S., et al. (2019). Advanced wireless sensors used to monitor the impact of environment design on human physiology. *Advances in Electrical and Electronic Engineering, 17*(3), 320–329.

Vaz, J. R., Knarr, B. A., & Stergiou, N. (2020). Gait complexity is acutely restored in older adults when walking to a fractal-like visual stimulus. *Human Movement Science, 74*, 102677.

Wertz, A. E., & Wynn, K. (2014). Thyme to touch: infants possess strategies that protect them from dangers posed by plants. *Cognition., 130*(1), 44–49. https://doi.org/10.1016/j.cognition.2013.09.002. Epub 2013 Oct 24. PMID: 24161794; PMCID: PMC3896079.

White, M. P., Elliott, L. R., Grellier, J., Economou, T., Bell, S., Bratman, G. N., et al. (2021). Associations between green/blue spaces and mental health across 18 countries. *Scientific Reports, 11*(1), 8903.

Wijesooriya, N., & Brambilla, A. (2021). Bridging biophilic design and environmentally sustainable design: A critical review. *Journal of Cleaner Production, 283*, 124591.

Wilson, E. O. (1986). *Biophilia*. Harvard University Press.

World Health Organization. (June 2022). Mental Health. http://www.who.int/features/qa/84/en/ https://www.who.int/news-room/fact-sheets/detail/mental-health-strengthening-our-response

Yang, W., & Kang, J. (2005). Soundscape and sound preferences in urban squares: A case study in Sheffield. *Journal of Urban Design, 10*(1), 61–80.

Yesiltepe, D., Velasco, P. F., Coutrot, A., Torun, A. O., Wiener, J. M., Holscher, C., et al. (2023). Entropy and a sub-group of geometric measures of paths predict the navigability of an environment. *Cognition, 236*, 105443.

Yu, H., Miao, W., Ji, E., Huang, S., Jin, S., Zhu, X., et al. (2022). Social touch-like tactile stimulation activates a tachykinin 1-oxytocin pathway to promote social interactions. *Neuron, 110*(6), 1051–1067.

Yuan, N., Chen, Y., Xia, Y., Dai, J., & Liu, C. (2019). Inflammation-related biomarkers in major psychiatric disorders: A cross-disorder assessment of reproducibility and specificity in 43 meta-analyses. *Translational Psychiatry, 9*(1), 233.

Zhao, H., Zhang, H., Xu, Y., Lu, J., & He, W. (2018). Relation between awe and environmentalism: The role of social dominance orientation. *Frontiers in Psychology, 9*, 2367.

Part V
Focus on Special Populations

Environmental Exposures and Child Neurodevelopment

Carmen Peuters, Jordi Sunyer, Anne-Claire Binter, and Payam Dadvand

Abstract Environmental exposures such as traffic-related air and noise pollution or surrounding greenness may impact child neurodevelopment through different pathways. Essential processes of neurodevelopment that take place during pregnancy and the first years of life make the brain particularly vulnerable to environmental hazards during these life stages. In this chapter, we present mechanisms that may explain the impact of air pollution, greenspace, and noise on cognitive and behavioral development, and elucidate the scientific evidence on the association between these urban environmental exposures and neurological outcomes in early life. Overall, results demonstrate that early life exposure to air pollution, noise, and lack of greenspace, detrimentally affect cognitive functioning, academic achievement, behavior, or other neurodevelopmental outcomes. Although evidence on the matter is growing, future research is needed to, among others, clarify current inconsistencies in results, determine windows of greatest susceptibility, and study the interrelationship between exposure impacts. Additional challenges concern the specificity and accuracy of exposure assessment, as this will allow for a more precise understanding of the exposure properties accountable for the health effects. Further investigation into early-life environmental exposures and their impact on neurodevelopment is crucial for shaping effective preventive measures.

Keywords Air pollution · Urban green space · Noise pollution · Neurodevelopment · Cognitive functioning · Prenatal life · Infancy · Childhood · Adolescence · Environmental exposures

C. Peuters (✉) · J. Sunyer · A.-C. Binter · P. Dadvand
ISGlobal, Barcelona, Spain

Universitat Pompeu Fabra (UPF), Barcelona, Spain

Spanish Consortium for Research on Epidemiology and Public Health (CIBERESP), Barcelona, Spain
e-mail: carmen.peuters@isglobal.org

S. Kühn (ed.), *Environmental Neuroscience*,
https://doi.org/10.1007/978-3-031-64699-7_12

Introduction

In this chapter, we provide a general description of the evidence on the impact of urban environmental exposures on neurodevelopment and cognition in early life. The chapter covers exposure to air pollution, greenspace, and noise, during prenatal life, childhood, and adolescence. Other environmental exposures, such as chemicals and heat, are not addressed in this chapter, but are not of less importance for child development. The summary of the evidence presented here, is mainly based on previous reviews on early life exposures and child health (Foraster et al., 2023; Sunyer & Dadvand, 2019; Zare Sakhvidi et al., 2022, 2023; Terzakis et al., 2022), complemented with evidence from relevant original manuscripts. Before diving into the pathways and evidence on the link between these environmental exposures and specific neurodevelopmental outcomes, we first elucidate the particular importance of early life for environmental impact on brain development.

Brain Vulnerability in Early Life

The human brain is highly vulnerable in early life mainly due to two reasons. First, given their small body size, an embryo or fetus has relatively higher exposure than the mother. Similar is true for infants or children compared to adults, for which adverse neurological consequences may be observed at lower exposure levels than in adults. Second, from the first weeks of embryonic life until late adolescence is when the brain develops. Exposures to environmental influences during this timeframe have the potential to interfere with normal brain development. The developing brain is therefore often more sensitive to environmental exposures than the adult brain. Consequently, an exposure that is harmless later in life may, however, impact health in early life. Optimal brain maturation requires the sequential completion of multiple complex processes and the different regions of the brain develop at distinct (gestational) ages (Sanes et al., 2011) (see Fig. 1). Critical windows of susceptibility

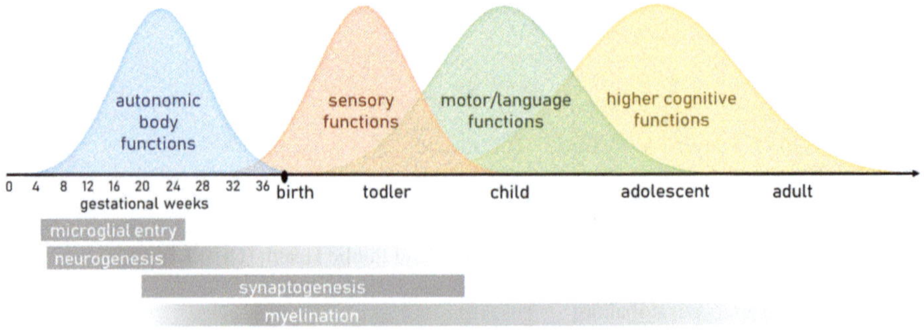

Fig. 1 Neurodevelopment in early life. Peak developmental phases of neuropsychological domains (above timeline) and brain developmental processes (below timeline)

can be different for different environmental factors and for different health outcomes, depending on the underlying mechanisms.

Pregnancy is a particularly vulnerable period for neurodevelopment. The placenta that helps protect the health of the prenatal has limitations when it comes to protection against maternal environmental hazards. Biomarkers of air pollution exposure, such as cotinine (i.e., a nicotine-derived metabolite) and adducts of polycyclic aromatic hydrocarbon (PAH), have been found in cord blood, indicating that tobacco smoke and air pollutants can effectively traverse the placental barrier (Peterson et al., 2015; Bongaerts et al., 2022). In addition to direct exposure to harmful substances, environmental influences can also harm the fetus by affecting properties of the placenta. Exposures during pregnancy may cause alterations in plasma viscosity, systemic low-grade inflammation, hormonal disruption, or epigenetic changes in the mother that may impair the placental functioning (Hettfleisch et al., 2017; Grevendonk et al., 2016). Impaired placental functioning can potentially result in adverse health outcomes such as neurological abnormalities in the fetus and altered brain function and brain connectivity in the newborn (Bronson & Bale, 2016). Relatively short periods of reduced oxygen flow to the fetus, for example, can cause neuronal loss and cerebral white matter damage (Rees & Harding, 2004). As the prenatal life concerns the development of the brainstem, the entry of microglial cells into the brain, and the formation of new neurons that form the basis for later maturation of different brain regions (i.e., neurogenesis); the specific effects of environmental hazards during pregnancy depend on the particular class of neuron or developmental process that is affected.

The human brain continues to develop through infancy and childhood. Up until about 3 to 4 years of age, synaptogenesis is a neurodevelopmental process of synapse formation that plays a vital role in learning, memory formation, and adaptation (Tierney & Nelson III, 2009). This is followed by synaptic pruning, a process of removing the unused synapses, which occurs from early childhood to puberty. Given that during synaptic pruning the pathways that are activated by environmental stimuli are strengthened while those that remain inactive are eliminated, this process is highly dependent on experience. The peak of synaptogenesis and synaptic pruning occurs at different periods depending on the region of the brain. Certain environmental factors may therefore exert a more intense and direct influence on the development of the child in these postnatal years than during pregnancy or the first months of life. Passive smoking during childhood, for example, is associated with a thinner retinal nerve fiber layer (i.e., fewer synapses between ganglion cells) with potential long-term harms for neurodevelopment, independent from maternal smoking during pregnancy (Li et al., 2021). Another vital process for optimal neurodevelopment is myelination. This consists of insulating the axons of the neurons with a fatty myelin membrane, facilitating faster transmission of electrical impulses and hence optimal functioning of the nervous system. The myelination process starts around week 20 of gestation with the base of the brain, and peaks in the first 2 years of age with as main brain regions the sensory and motor areas (Tierney & Nelson III, 2009; Kinney & Volpe, 2018). Hence, the period spanning from mid-pregnancy to 2 years of age is the most vulnerable timeframe for environmental factors that affect the myelination process. Myelination however continues until late adolescence in regions of the brain concerning higher cognitive abilities, such as the prefrontal cortex (Tierney & Nelson III, 2009). Another event during adolescence is the activation of the hypothalamus for producing the gonadotropin-releasing hormone, initiating puberty. Environmental hazards during this period, especially those affecting the neuroendocrine system, could alter metabolic, developmental, and behavioral processes.

The effects of age-specific environmental exposures on neurodevelopment, and in particular on brain alterations, are investigated too scarcely to draw firm conclusions about the critical exposure period for different environmental factors. It is important to better understand the mechanisms underlying the environmental effects in order to develop more efficient preventive measures. Regardless of the incomplete understanding of the mechanisms and windows of greatest susceptibility, there is substantive evidence demonstrating an association between early life exposure to environmental hazards and adverse health outcomes. These outcomes range from cardiovascular and neurological abnormalities to impaired cognitive functioning and behavioral disorders, with longitudinal research showing that neurodevelopmental problems acquired in early life track into adulthood (Sonuga-Barke et al., 2017). In what follows, a summary is provided of the literature linking child neurodevelopment with air pollution, greenspace, and noise.

Air Pollution

Pathways

Air pollution is considered one of the main environmental hazards. Due to urbanization and population growth, an increasing number of people are exposed to higher concentrations of air pollutants, posing serious health risks. Exposure to air pollution during pregnancy could affect brain development by impairing the placental function and decreasing transplacental oxygen and nutrient transport (Clemente et al., 2016; Gruzieva et al., 2017; Carvalho et al., 2016). Other potential mechanisms by which pre- and postnatal air pollution could affect the child's brain, are systemic inflammation and oxidative stress (i.e., excess production of reactive oxygen species (ROS) relative to antioxidant defense) (Grevendonk et al., 2016; Clemente et al., 2016; Saenen et al., 2016; Mozzoni et al., 2022; Costa et al., 2019). In vivo studies in rats demonstrated how air pollution damages the mitochondrial energy metabolism homeostasis, increasing the production of ROS and premature apoptosis in affected tissues (Yan et al., 2015). Preliminary evidence from neuroimaging studies indicates that myelination is likely one of the main brain development processes susceptible to harm from particulate pollution (Sunyer & Dadvand, 2019). There may also be involvement of other processes, such as alterations in the basal ganglia (Kim et al., 2014).

Neurodevelopment

Air pollution has been reported to harm cognitive functioning across all the early life stages, from prenatal development to childhood and adolescence. Most significant effects are found for psychomotor development. Multiple studies investigating exposure to air pollutants including nitrogen dioxide (NO_2) and particles with a diameter of 10 micrometers or less (PM_{10}), found negative associations with psychomotor development. The delay in neurodevelopment was most prominent at 6 months compared to 12 or 24 months of age (Kim et al., 2014) and effects were most significant for NO_2 (Guxens et al., 2014). Poorer motor performance in children has also been shown to be associated with higher copper exposure (Pujol et al., 2016). Other executive functions linked with air pollution levels are working memory and attention. These have been extensively studied in a Spanish cohort in relation to the traffic-related pollutants NO_2, elemental carbon (EC), black carbon (BC), particles with a diameter of 2.5 micrometers or less ($PM_{2.5}$) and ultrafine particles (UFPs or $PM_{0.1}$). Children attending schools with higher levels of EC, NO_2, and UFPs both indoors and outdoors, had substantially smaller growth in working memory and attention (Sunyer et al., 2015). A negative association with working memory was also found for $PM_{2.5}$ exposure at ages 5 and 6 years, with boys showing much higher vulnerability (Rivas et al., 2019); and for children exposed to $PM_{2.5}$ and BC while commuting to school by foot (Alvarez-Pedrerol et al., 2017). Exposure to traffic-related pollutants from the commutes by foot was not associated with attention. Daily ambient levels of NO_2 and EC, however, were negatively associated with daily fluctuations in attention, independently of longer-term exposure (Sunyer et al., 2017). Prenatal $PM_{2.5}$ exposure was associated with decreased executive functioning (Holm et al., 2023). For visual and episodic memory, as well as for language development, findings are inconsistent (Guxens et al., 2014; Suades-González et al., 2015; Castagna et al., 2022; Iglesias-Vázquez et al., 2022).

Furthermore, there is evidence of a link between air pollution and intelligence. Both pre- and postnatal exposure to polycyclic aromatic hydrocarbons (PAHs) were found to be associated with lower global IQ (Suades-González et al., 2015). Moreover, children with intellectual disabilities were found to be significantly more likely to live in areas with high levels of PM, NO_2, and CO (Emerson et al., 2019). Studies investigating intellectual development in babies in relation to prenatal exposure to air pollution did not find effects (Guxens et al., 2014; Iglesias-Vázquez et al., 2022). Regarding academic achievement, evidence is suggestive of a negative association with air pollution, although the quality of the evidence is generally low and the existing body of literature is small. In a systematic review (Stenson et al., 2021), various studies reported higher PM exposure to be related to lower average school grades and lower math scores. For exposure to ozone (O_3), findings are also indicative of a negative impact on math test scores, though the evidence is stronger for $PM_{2.5}$. $PM_{2.5}$ was moreover found to be related to poorer scores on reading, though positive associations between PM and reading have also been reported. A reduction in adolescent academic achievement was also found for short-term exposure to

NO_2. Another study, not included in the systematic review, reported that traffic-related exposure surrounding primary school mediated some of the associations between green space around the school and performance in reading, grammar, and numeracy (Claesen et al., 2021). Most reported studies concerned outdoor exposure, but also indoor air pollutant levels are significantly associated with standardized test scores (Stafford, 2015). Study designs have a wide variation in the timing (short vs. long term) and location (school versus home) of exposure assessment, but the literature suggests that cumulative exposure to air pollution forms the greatest risk for impaired cognitive function (Stenson et al., 2021).

Several studies have investigated the relationship between air pollution and child behavior. For example, a cohort study found higher levels of traffic-related air pollutants to be associated with higher total difficulty scores, in particular for indoor EC, outdoor EC, and outdoor NO_2 (Forns et al., 2016). When considering subgroups of behavior, relationships were strongest for hyperactivity/inattention, peer relationship problems, and prosocial behavior. In the same study, concentrations of BC at children's home addresses were not associated with behavioral problems. NO_2 was also found to be associated with increased internalizing and externalizing problems in a cohort in the mid-South United States, while PM_{10} and road proximity were not (Loftus et al., 2020). Socioeconomic adversity and African American race seemed to increase susceptibility to the detrimental air pollution effects. A study based on six birth cohorts confirmed the importance of air pollution for attention problems and aggressive behaviors, both during pregnancy and childhood (Maitre et al., 2021). The study identified prenatal residential car traffic density on the nearest road and postnatal indoor air $PM_{2.5}$ among the most important exposures associated with increased externalizing symptoms. Next, several systematic reviews reported fairly consistent evidence for an association between pre- and postnatal air pollution exposure and autism spectrum disorder (ASD), in particular for $PM_{2.5}$ but also for NO_2 (Foraster et al., 2023). ASD is a neurodevelopmental disorder manifesting in peculiar social interactions, restrictive or repetitive conduct, and different ways of learning, moving, or paying attention. Various epidemiological studies show moderate to strong associations between pre-and postnatal exposure to air pollution and increased risk of ASD, and controlled animal studies mostly confirm the effect of air pollution on behavioral, biochemical, and morphological aspects also found in persons with ASD (Costa et al., 2019). However, some studies did not find an association between exposure to PM, NO_2, or NO, and autistic traits (Guxens et al., 2016; Gong et al., 2014). Finally, a large body of research has investigated the relation between air pollution exposure and attention deficit hyperactivity disorder (ADHD). Most studies (either cohort or twin studies) did not find an association between estimated traffic-related air pollutants and ADHD symptoms (Forns et al., 2016, 2018; Gong et al., 2014). Only indoor EC was related to ADHD symptomatology, but higher EC related to fewer ADHD symptoms. Other studies did find an increased risk for ADHD for living near pollutant areas (agricultural areas, streets, motorway or dual carriageway, industrial estate) (Saez et al., 2018), prenatal levels of PAHs (Perera et al., 2014), or postnatal exposure to EC (Newman et al., 2013). The exposome study by Maitre and colleagues (Maitre et al., 2021) also indicated that air pollution, including prenatal residential car traffic density on the nearest road and post-natal second-hand smoking, are among the most important exposures associated with an increased ADHD index.

Greenspace

Pathways

Greenness (i.e., level of vegetation) and green spaces (i.e., areas covered by vegetation) are increasingly being associated with multiple aspects of health. Greenspace (used in this chapter to refer to both greenness and green spaces) has been suggested to be linked to neurodevelopment through different pathways. One line of thought concerns the restorative effects of greenspace. Sensory input (e.g., viewing, smelling) of green and other natural environments is theorized to evoke positive emotions that can lower the stress response, hence helping the individual to restore from psychophysiological stress (i.e., the stress reduction theory, SRT) (Ulrich et al., 1991; Hedblom et al., 2019). Greenspace potentially also enhances the recovery from attention fatigue, although the evidence is less clear. The attention restoration theory (ART) posits that natural environments come with "soft" fascinating objects, such as moving clouds in the sky, which attract an individual's attention in an effortless way, thereby alleviating mental fatigue and restoring attention (Ohly et al., 2016). Increased learning opportunities are another potential pathway explaining the link between greenspace and neurodevelopment. Play within natural environments may instill creativity and promote motor skills, as well as enhance the development of a sense of mastery and control (Fjørtoft, 2004; Richardson et al., 2017). In addition, proximity to green spaces may also indirectly affect the brain by promoting healthy behavior (i.e., enhancing physical activity and facilitating social interaction, subsequently supporting cognitive development) and gut microbial diversity (through enriched microbial input from the natural environment) (Markevych et al., 2017). Increased greenspace also mitigates the exposure to urban-related environmental stressors by reducing air pollution, noise, and heat. The evidence is not clear yet about the extent to which each of these mechanisms applies to each of the behavioral outcomes described below.

Neurodevelopment

Overall, the evidence is confirmative of a positive association between greenspace and cognitive functioning in children. For example, a higher amount of greenness within and surrounding schools, as well as during commuting to school, was associated with improved development of working memory over a 1-year timespan

(Dadvand et al., 2015). The authors did not find this association for residential surrounding greenness, but a cross-sectional study did find a significant association between neighborhood greenspace and child spatial working memory (Flouri et al., 2019). Although inconsistencies exist (Luque-García et al., 2022), a recent review also reported substantial evidence for a link of greenspace with attention (Vella-Brodrick & Gilowska, 2022). Especially tree-cover (vegetation higher than 3 m) seems to be associated with improved attention (Bijnens et al., 2022). Evidence also exists on exposure to natural environments and improved motor development (Kabisch et al., 2019; Liao et al., 2019). Moreover, brain imaging studies showed positive associations between lifelong exposure to greenspace and gray and white matter volume related to cognitive functioning (Dadvand et al., 2018).

Related to academic achievement, there are mixed findings on its relationship with greenspace. Different associations have been found for different academic domains. A review of the impact of greenspace within or surrounding schools on academic performance reported that most of the studies showed non-significant associations (Browning & Rigolon, 2019). In particular associations with writing test scores were non-significant. Evidence for a positive association with math and reading test scores was found for trees in the schoolyard and for greenness and green land cover up to 2000 m around the school, but not for agriculture or grass cover within or around the school. Different types of greenspace at 3000 m or more around the school was not associated or was negatively associated with academic performance, with the exception of trees (for which significant positive associations were reported also at more than 3000 m). Window views to greenspace could be especially beneficial to academic performance, but more research is needed (Browning & Rigolon, 2019). Moreover, most of the evidence comes from observational designs at school-level, for which the observed associations between academic outcomes and school-related greenspace might actually be attributed to other school characteristics such as socio-economic status and urbanization. Other studies have, although inconsistently, associated exposure to greenspace with increased intelligence quotient (IQ) (Foraster et al., 2023).

Somewhat more consistent results are found for a beneficial association between exposure to greenspace and child behavior. A recent systematic review on the topic reported that most studies find a beneficial (but not always significant) effect of exposure to greenspace for total behavioral difficulties, externalizing problems, and prosocial behaviors (Zare Sakhvidi et al., 2022). For ADHD symptoms and diagnosis, most reported findings are also suggestive of a protective role of greenspace. For conduct problems, findings are mixed, with half of the studies suggestive of a beneficial association but the other half suggestive of increased conduct problems in case of more exposure to greenspace. For example, a cross-sectional study reported that higher neighborhood greenspace exposure was associated with slightly increased odds of violence in American male youths (Bushover et al., 2020). Regarding ASD, there is some preliminary evidence of the protective role of pre- and postnatal exposure to greenspace (Pagalan et al., 2022; Wu & Jackson, 2017).

Although an increasing number of studies has investigated greenspace exposure in relation to neurodevelopment, the type of greenness has been considered very scarcely and requires deeper investigation. More research is also needed to investigate potential differences in susceptibility of different populations. There are indications that children with autism, for example, do not benefit the same from surrounding greenspace as children without autism (Larson et al., 2018). Additional research is moreover needed to clarify potential windows of susceptibility, as very few studies have considered exposure during pregnancy and early years of age.

Noise

Pathways

Millions of people are exposed to excessive environmental noise levels, with transport being the main source. Because of its harmful effects on physiological and psychological health outcomes, noise pollution is considered a major environmental health concern, and the second most significant cause of ill health in Western Europe (European Environment Agency, 2020). Different pathways have been hypothesized between exposure to noise and impaired neurodevelopment. A straightforward, direct, consequence of noise is that it can injure the auditory system, causing, for example, hearing loss or tinnitus. Traumatized children's hearing can impair cognitive development due to loss of understanding. However, environmental noise exposure usually does not reach harmful levels for causing hearing injury (Stansfeld & Clark, 2015). The impact of environmental noise is more likely to be explained by indirect consequences of noise exposure. Sleep disturbance is a first important non-auditory mediator of the relationship between environmental noise and cognitive development. Noise has been found to cause fragmented sleep and decreased total sleep time in children, which in its turns is linked to reduced daytime alertness and poor concentration (Basner & McGuire, 2018). This may detrimentally affect cognitive development especially in school and study settings. Noise can interfere with learning not only via reduced sleep, but also by interrupting learning activities directly (e.g., a teacher stops talking when a train passes), or by causing annoyance and distraction which may result in mind-wandering during classes (Massonnié et al., 2022). Another main pathway for the link between noise and neurodevelopment concerns the human stress response to prevailing noise exposure (Shukla & Tandel, 2024). Exposure to noise, possibly through noise annoyance, is thought to elicit a stress reaction in the body (Gong et al., 2022; Balk et al., 2023). A study investigating aircraft noise exposure in primary schools indeed revealed a significant relationship with stress symptoms (Haines et al., 2001). Psychophysiological stress reactions affect cardiovascular and neuroendocrine aspects (e.g., cortisol, glucocorticoids, metabolites), and may affect body and brain function (Balk et al., 2023). Findings from animal studies confirm the neurological basis of noise effects, with most evidence for affected areas of the central nervous system and modifications related to neurotransmitters and hormones, such as noradrenergic neurotransmission and adrenocortical activation (Arcangeli et al., 2023). Few studies have evaluated the effects of pre-natal exposure to noise, but it is hypothesized that a sustained stress reaction caused by noise exposure may impair placental functioning and harm child neurodevelopment. Individual differences are observed in the effects that noise may exert. An individual's genetic and acquired physiological reaction to noise may differ depending on personal characteristics. Some individuals are more sensitive for noise, meaning that they—consciously or not—appraise more types of noise or lower degrees of noise sources as a threat, increasing their reactivity to noise (Shepherd et al., 2015). Also coping strategies (e.g., tuning out unwanted noise events) and age group (children are more sensitive to noise stimuli than adults, probably related to the development of coping strategies) may differ between individuals. In addition, specific noise characteristics such as the type of noise exposures and the duration and temporal variations of the noise exposure, may mediate the noise–health relationship (Terzakis et al., 2022).

Neurodevelopment

Scientific evidence on the relationship between noise pollution and human neurodevelopment is still scarce. Especially with regard to exposure during pre-natal and early life, few studies have been conducted. Initial studies on noise exposure in pregnant women suggest an association with pre-eclampsia, a condition that may affect placental functioning and hence neurodevelopment of the infant (Gómez-Roig et al., 2021). Recent animal studies indicate impaired neurodevelopment during embryonic life following noise exposure, which can be expected to be similar in humans (Kesar, 2014). A growing number of studies has however examined noise exposure in childhood, documenting its impact on various aspects of child cognitive functioning.

There is sufficient evidence for detrimental effects of noise on cognitive tasks, including memory and language comprehension. Long-term memory (including recognition memory and recall memory) was found to be more affected than working memory (Stansfeld & Clark, 2015). In a cohort study in Switzerland, for example, road traffic noise at home was found to significantly reduce figural memory (Tangermann et al., 2023). Regarding language comprehension, noise was found to significantly impact speech comprehension, speech intelligibility, and reading (Stansfeld & Clark, 2015; Balk et al., 2023). A meta-analysis from three studies demonstrated that reading comprehension and language abilities were higher in quiet classrooms compared to noisier classrooms (Thompson et al., 2022). Most consistent results for the association between noise exposure and reading were found for children aged between 7 and 11 years (Stansfeld & Clark, 2015). There is furthermore sufficient evidence to assume a relationship between exposure to noise and worsened academic performance. A review reported sufficient evidence for a detrimental impact of noise exposure on standardized academic achievement tests (Stansfeld & Clark, 2015). Results are more consistent for noise at school compared to noise at home (Terzakis et al., 2022). For noise exposure in educational buildings, significant associations have been reported with language, mathematics, and science tasks. For noise at dwellings, significant associations were found with language and mathematics tasks, although these results were not consistent over different noise measures. One study investigating academic achievement in schools with relatively low noise levels did not find an association with road traffic noise exposure, suggesting that harmful effects only occur at high noise levels (Xie et al., 2011).

Less evidence is available for an association between noise exposure and sustained attention. Sustained attention was found to be significantly associated with exposure to aircraft noise, but not road traffic noise (Terzakis et al., 2022). In an exposome study investigating multiple environmental exposures, noise did not appear among the most important predictors for attentional functioning (Julvez et al., 2021). For executive functioning in general, there is moderate quality evidence against an association with aircraft noise (Thompson et al., 2022). No human studies so far investigated the relationship between noise exposure and motor development. Based on animal studies and the known effects of the human stress reaction, however, it is hypothesized that exposure to noise may impair motor coordination and locomotor activity (Hahad et al., 2022).

Regarding child behavior, there are only a few studies investigating this outcome in relation to noise pollution. There is, however, growing evidence for an association of noise with increased hyperactivity symptoms. In multiple studies, an association was found between increased scores on the hyperactivity and inattention subscale measured by the SDQ, and noise from aircrafts and road traffic, both in dwellings and in educational buildings (Terzakis et al., 2022). No significant associations were found between road traffic noise exposure and externalizing or internalizing behavioral problems. Also for conduct problems, most studies reported non-significant associations with noise (Terzakis et al., 2022). The above studies all used the SDQ as measure of child behavior, and additional research is needed to investigate other aspects of behavior that might be a consequence of neurodevelopmental changes caused by environmental noise. In children with ADHD, for example, auditory processing issues related to sensitivity to sounds are reported, which suggests an impact of noise on ADHD symptoms (Balk et al., 2023). A longitudinal study following children from 11 to 27 years of age, however, did not observe consistent associations between noise exposure and ADHD severity (Li et al., 2023). Neither for ASD did this study find evidence for a negative impact of noise.

Exposure to noise in relation to child neurodevelopment is increasingly being studied in recent years, but noise exposure is often measured poorly, and more good quality research is required to allow for precise risk estimation.

Future Research Directions

The existing literature linking environmental exposures with child neurodevelopment highlights the importance to keep broadening our understanding of this relationship, to improve the prevention of adverse health outcomes. A general limitation of the current evidence lays in the measurement of environmental exposures. Most studies have considered air pollution, greenspace, and noise in only one environment, mainly the home or school of the child. Recently, an effort is being made to capture exposures in all environments where mother or child spends time. A combination of GPS tracking (e.g., via the smartphone) with spatial modeling of environmental exposures allows to obtain real-time personal exposure data. Moreover, portable devices capable of measuring air pollution and noise (e.g., attached to a backpack) are providing promising avenues to more accurately capture the levels of environmental pollution to which an individual is exposed. The last decade, big advances are being made also in the study of the exposome, taking a holistic approach by jointly examining the impact of multiple personal and environmental exposures. This is important for considering contextual and environmental factors such as the social environment, stress, or physical activity, which may confound or interact with the effect of air pollution, greenspace, or noise (Sunyer & Dadvand, 2019). The study of the exposome may moreover help disentangle effects from environmental exposures that derive from a similar source, such as traffic-related air pollution and traffic noise. Another emerging field is the use of biomarkers of environmental hazard exposure. Biomarkers of exposure, examined from biological samples such as the blood or urine, can reveal the real effect of an environmental exposure in the body. Biomarkers can moreover help identify biological responses and pathways of how an exposure leads to adverse health effects, referred to as "early pathway perturbations" (Foraster et al., 2023). It could, for example, aid in the understanding of the role of the placenta in the causal relationship of environmental exposures on impaired brain development. This type of biomarker evidence may reinforce arguments to improve implementation of preventive measures. A challenge related to biomonitoring of environmental exposures is the difficulty to capture non-persistent matter (i.e., matter that is eliminated from the body in a few hours or days), and the interpretation of biomarker data for which the physiological characteristics remain unexplained. Furthermore, there are advances for the measurement of neurodevelopment with regard to brain imaging techniques that allow for more precise investigation of alterations in brain development (e.g., magnetic resonance imaging (MRI) to assess myelination (Filimonova et al., 2023) (➔ chapter "Mapping the Human Brain with Computational Anatomy" by Draganski, Joost and Kherif), and this both post- and pre-natal. For example, transabdominal and transvaginal neurosonography make it possible to measure the maturation of specific cortical fissures and sulci in the fetal brain (Volpe et al., 2021), and recent developments in neonatal MRI allow for imaging while the infant is asleep and have improved the efficiency of neuroimaging in neonates (Hughes et al., 2017). Future research is needed applying these techniques to investigate the neurological impact of environmental exposures during the fetal and early life, as there is currently a lack of studies during this early period while the prenatal and first years of life are the most vulnerable stages of brain development with the largest preventive opportunities. There is also a need for more studies in adolescence. Evidence over the entire period of child development is key to discover the windows of greatest sensitivity for effects of environmental exposures on neurodevelopment. In addition, the available studies on environmental exposures and child neurodevelopment are generally performed in high-income countries, and there is a need for studies in low- and middle-income countries. The more quality research is available to understand the link between air pollution, greenspace, and noise with child neurodevelopment, the better we can translate this knowledge into preventive public health interventions.

Acknowledgments We acknowledge support from the grant CEX2023-0001290-S funded by MCIN/AEI/10.13039/501100011033, and support from the Generalitat de Catalunya through the CERCA Program.

References

Alvarez-Pedrerol, M., Rivas, I., López-Vicente, M., et al. (2017). Impact of commuting exposure to traffic-related air pollution on cognitive development in children walking to school. *Environmental Pollution, 231*, 837–844. https://doi.org/10.1016/j.envpol.2017.08.075

Arcangeli, G., Lulli, L. G., Traversini, V., et al. (2023). Neurobehavioral alterations from noise exposure in animals: A systematic review. *International Journal of Environmental Research and Public Health, 20*(1). https://doi.org/10.3390/ijerph20010591

Balk, S. J., Bochner, R. E., Ramdhanie, M. A., & Reilly, B. K. (2023). Preventing excessive noise exposure in infants, children, and adolescents. *Pediatrics, 152*(5). https://doi.org/10.1542/peds.2023-063752

Basner, M., & McGuire, S. (2018). WHO environmental noise guidelines for the european region: A systematic review on environmental noise and effects on sleep. *International Journal of Environmental Research and Public Health, 15*(3). https://doi.org/10.3390/ijerph15030519

Bijnens, E. M., Vos, S., Verheyen, V. V., et al. (2022). Higher surrounding green space is associated with better attention in Flemish adolescents. *Environment International, 159*, 107016. https://doi.org/10.1016/j.envint.2021.107016

Bongaerts, E., Lecante, L. L., Bové, H., et al. (2022). Articles maternal exposure to ambient black carbon particles and their presence in maternal and fetal circulation and organs: An analysis of two independent population-based observational studies. *Lancet Planet Health, 6*, e804–e811. www.thelancet.com/

Bronson, S. L., & Bale, T. L. (2016). The placenta as a mediator of stress effects on neurodevelopmental reprogramming. *Neuropsychopharmacology, 41*(1), 207–218. https://doi.org/10.1038/npp.2015.231

Browning, M. H. E. M., & Rigolon, A. (2019). School green space and its impact on academic performance: A systematic literature review. *International Journal of Environmental Research and Public Health, 16*(3). https://doi.org/10.3390/ijerph16030429

Bushover, B., Miller, E., Bair-Merritt, M., Abebe, K., & Culyba, A. (2020). Physical environment and violence perpetration among male youth in Pittsburgh: A spatial analysis. *Injury Prevention, 26*(6), 588–592. https://doi.org/10.1136/injuryprev-2019-043356

Carvalho, M. A., Bernardes, L. S., Hettfleisch, K., et al. (2016). Associations of maternal personal exposure to air pollution on fetal weight and fetoplacental Doppler: A prospective cohort study. *Reproductive Toxicology, 62*, 9–17. https://doi.org/10.1016/j.reprotox.2016.04.013

Castagna, A., Mascheroni, E., Fustinoni, S., & Montirosso, R. (2022). Air pollution and neurodevelopmental skills in preschool- and school-aged children: A systematic review. *Neuroscience and Biobehavioral Reviews, 136*(March), 104623. https://doi.org/10.1016/j.neubiorev.2022.104623

Claesen, J. L. A., Wheeler, A. J., Klabbers, G., et al. (2021). Associations of traffic-related air pollution and greenery with academic outcomes among primary schoolchildren. *Environmental Research, 199*(March), 111325. https://doi.org/10.1016/j.envres.2021.111325

Clemente, D. B. P., Casas, M., Vilahur, N., et al. (2016). Prenatal ambient air pollution, placental mitochondrial DNA content, and birth weight in the INMA (Spain) and ENVIRONAGE (Belgium) birth Cohorts. *Environmental Health Perspectives, 124*(5), 659–665. https://doi.org/10.1289/ehp.1408981

Costa, L. G., Cole, T. B., Dao, K., Chang, Y. C., & Garrick, J. M. (2019). Developmental impact of air pollution on brain function. *Neurochemistry International, 131*, 104580. https://doi.org/10.1016/j.neuint.2019.104580

Dadvand, P., Nieuwenhuijsen, M. J., Esnaola, M., et al. (2015). Green spaces and cognitive development in primary schoolchildren. *Proceedings of the National Academy of Sciences of the United States of America, 112*(26), 7937–7942. https://doi.org/10.1073/pnas.1503402112

Dadvand, P., Pujol, J., Maciá, D., et al. (2018). The association between lifelong greenspace exposure and 3-dimensional brain magnetic resonance imaging in Barcelona schoolchildren. *Environmental Health Perspectives, 126*(2), 027012. https://doi.org/10.1289/EHP1876

Emerson, E., Robertson, J., Hatton, C., & Baines, S. (2019). Risk of exposure to air pollution among British children with and without intellectual disabilities. *Journal of Intellectual Disability Research, 63*(2), 161–167. https://doi.org/10.1111/jir.12561

European Environment Agency. (2020). *Environmental noise in Europe – 2020: EEA report*. European Environment Agency.

Filimonova, E., Amelina, E., Sazonova, A., Zaitsev, B., & Rzaev, J. (2023). Assessment of normal myelination in infants and young children using the T1w/T2w mapping technique. *Frontiers in Neuroscience, 17*, 1102691. https://doi.org/10.3389/fnins.2023.1102691

Fjørtoft, I. (2004). Landscape as playscape: The effects of Natural Environments on Children's Play and Motor Development. *Children, Youth and Environments, 14*(2), 21–44. https://doi.org/10.7721/chilyoutenvi.14.2.0021

Flouri, E., Papachristou, E., & Midouhas, E. (2019). The role of neighbourhood greenspace in children's spatial working memory. *British Journal of Educational Psychology, 89*(2), 359–373. https://doi.org/10.1111/bjep.12243

Foraster, M., Casas, M., Vrijheid, M., & Sunyer, J. (2023). Early life environmental exposures and Children's health. In M. L. Boulton & R. B. Wallace (Eds.), *Maxcy-Rosenau-last public health & preventive medicine*. McGraw Hill.

Forns, J., Dadvand, P., Foraster, M., et al. (2016). Traffic-related air pollution, noise at school, and behavioral problems in Barcelona schoolchildren: A cross-sectional study. *Environmental Health Perspectives, 124*(4), 529–535. https://doi.org/10.1289/ehp.1409449

Forns, J., Sunyer, J., Garcia-Esteban, R., et al. (2018). Air pollution exposure during pregnancy and symptoms of attention deficit and hyperactivity disorder in children in Europe. *Epidemiology, 29*(5), 618–622. https://doi.org/10.1097/EDE.0000000000000874

Gómez-Roig, M. D., Pascal, R., Cahuana, M. J., et al. (2021). Environmental exposure during pregnancy: Influence on prenatal development and early life: A comprehensive review. *Fetal Diagnosis and Therapy, 48*(4), 245–257. https://doi.org/10.1159/000514884

Gong, T., Almqvist, C., Bölte, S., et al. (2014). Exposure to air pollution from traffic and neurodevelopmental disorders in swedish twins. *Twin Research and Human Genetics, 17*(6), 553–562. https://doi.org/10.1017/thg.2014.58

Gong, X., Fenech, B., Blackmore, C., et al. (2022). Association between noise annoyance and mental health outcomes: A systematic review and meta-analysis. *International Journal of Environmental Research and Public Health, 19*(5). https://doi.org/10.3390/ijerph19052696

Grevendonk, L., Janssen, B. G., Vanpoucke, C., et al. (2016). Mitochondrial oxidative DNA damage and exposure to particulate air pollution in mother-newborn pairs. *Environmental Health, 15*(1), 10. https://doi.org/10.1186/s12940-016-0095-2

Gruzieva, O., Xu, C. J., Breton, C. V., et al. (2017). Epigenome-wide meta-analysis of methylation in children related to prenatal NO2 air pollution exposure. *Environmental Health Perspectives, 125*(1), 104–110. https://doi.org/10.1289/EHP36

Guxens, M., Garcia-Esteban, R., Giorgis-Allemand, L., et al. (2014). Air pollution during pregnancy and childhood cognitive and psychomotor development: Six european birth cohorts. *Epidemiology, 25*(5), 636–647. https://doi.org/10.1097/EDE.0000000000000133

Guxens, M., Ghassabian, A., Gong, T., et al. (2016). Air pollution exposure during pregnancy and childhood autistic traits in four European population-based cohort studies: The ESCAPE project. *Environmental Health Perspectives, 124*(1), 133–140. https://doi.org/10.1289/ehp.1408483

Hahad, O., Bayo Jimenez, M. T., Kuntic, M., et al. (2022). Cerebral consequences of environmental noise exposure. *Environment International, 165*, 107306. https://doi.org/10.1016/J.ENVINT.2022.107306

Haines, M. M., Stansfeld, S. A., Soames Job, R., Berglund, B., & Head, J. (2001). A follow-up study of effects of chronic aircraft noise exposure on child stress responses and cognition. *International Journal of Epidemiology, 30*, 839–845.

Hedblom, M., Gunnarsson, B., Iravani, B., et al. (2019). Reduction of physiological stress by urban green space in a multisensory virtual experiment. *Scientific Reports, 9*(1), 10113. https://doi.org/10.1038/s41598-019-46099-7

Hettfleisch, K., Bernardes, L. S., Carvalho, M. A., et al. (2017). Short-term exposure to urban air pollution and influences on placental vascularization indexes. *Environmental Health Perspectives, 125*(4), 753–759. https://doi.org/10.1289/EHP300

Holm, S. M., Balmes, J. R., Gunier, R. B., Kogut, K., Harley, K. G., & Eskenazi, B. (2023). Cognitive development and prenatal air pollution exposure in the CHAMACOS cohort. *Environmental Health Perspectives, 131*(3), 1–11. https://doi.org/10.1289/EHP10812

Hughes, E. J., Winchman, T., Padormo, F., et al. (2017). A dedicated neonatal brain imaging system. *Magnetic Resonance in Medicine, 78*(2), 794–804. https://doi.org/10.1002/mrm.26462

Iglesias-Vázquez, L., Binter, A. C., Canals, J., et al. (2022). Maternal exposure to air pollution during pregnancy and child's cognitive, language, and motor function: ECLIPSES study. *Environmental Research, 212*(April), 113501. https://doi.org/10.1016/j.envres.2022.113501

Julvez, J., López-Vicente, M., Warembourg, C., et al. (2021). Early life multiple exposures and child cognitive function: A multi-centric birth cohort study in six European countries. *Environmental Pollution, 284*, 117404. https://doi.org/10.1016/J.ENVPOL.2021.117404

Kabisch, N., Alonso, L., Dadvand, P., & van den Bosch, M. (2019). Urban natural environments and motor development in early life. *Environmental Research, 179*, 108774. https://doi.org/10.1016/j.envres.2019.108774

Kesar, A. (2014). Effect of prenatal chronic noise exposure on the growth and development of body and brain of chick embryo. *International Journal of Applied & Basic Medical Research, 4*(1), 3. https://doi.org/10.4103/2229-516x.125666

Kim, E., Park, H., Hong, Y. C., et al. (2014). Prenatal exposure to PM10 and NO2 and children's neurodevelopment from birth to 24 months of age: Mothers and Children's Environmental Health (MOCEH) study. *Science of the Total Environment, 481*(1), 439–445. https://doi.org/10.1016/j.scitotenv.2014.01.107

Kinney, H. C., & Volpe, J. J. (2018). Myelination events. In J. J. Volpe, T. E. Inder, B. T. Darras, et al. (Eds.), *Volpe's neurology of the newborn* (6th ed., pp. 176–188). Elsevier. https://doi.org/10.1016/B978-0-323-42876-7.00008-9. Get rights and content.

Larson, L. R., Barger, B., Ogletree, S., et al. (2018). Gray space and green space proximity associated with higher anxiety in youth with autism. *Health & Place, 53*, 94–102. https://doi.org/10.1016/j.healthplace.2018.07.006

Li, J., Yuan, N., Chu, W. K., et al. (2021). Exposure to secondhand smoke in children is associated with a thinner retinal nerve fiber layer: The Hong Kong Children Eye Study. *American Journal of Ophthalmology, 223*, 91–99. https://doi.org/10.1016/j.ajo.2020.10.016

Li, Y., Xie, T., Cardoso Melo, R. D., et al. (2023). Longitudinal effects of environmental noise and air pollution exposure on autism spectrum disorder and attention-deficit/hyperactivity disorder during adolescence and early adulthood: The TRAILS study. *Environmental Research, 227*, 115704. https://doi.org/10.1016/j.envres.2023.115704

Liao, J., Zhang, B., Xia, W., et al. (2019). Residential exposure to green space and early childhood neurodevelopment. *Environment International, 128*, 70–76. https://doi.org/10.1016/j.envint.2019.03.070

Loftus, C. T., Ni, Y., Szpiro, A. A., et al. (2020). Exposure to ambient air pollution and early childhood behavior: A longitudinal cohort study. *Environmental Research, 183*, 109075. https://doi.org/10.1016/j.envres.2019.109075

Luque-García, L., Corrales, A., Lertxundi, A., Díaz, S., & Ibarluzea, J. (2022). Does exposure to greenness improve children's neuropsychological development and mental health? A Navigation Guide systematic review of observational evidence for associations. *Environmental Research, 206*, 112599. https://doi.org/10.1016/j.envres.2021.112599

Maitre, L., Julvez, J., López-Vicente, M., et al. (2021). Early-life environmental exposure determinants of child behavior in Europe: A longitudinal, population-based study. *Environment International, 153*, 106523. https://doi.org/10.1016/j.envint.2021.106523

Markevych, I., Schoierer, J., Hartig, T., et al. (2017). Exploring pathways linking greenspace to health: Theoretical and methodological guidance. *Environmental Research, 158*, 301–317. https://doi.org/10.1016/j.envres.2017.06.028

Massonnié, J., Frasseto, P., Mareschal, D., & Kirkham, N. Z. (2022). Learning in Noisy classrooms: Children's reports of annoyance and distraction from noise are associated with individual differences in mind-wandering and switching skills. *Environment and Behavior, 54*(1), 58–88. https://doi.org/10.1177/0013916520950277

Mozzoni, P., Iodice, S., Persico, N., et al. (2022). Maternal air pollution exposure during the first trimester of pregnancy and markers of inflammation and endothelial dysfunction. *Environmental Research, 212*, 113216. https://doi.org/10.1016/j.envres.2022.113216

Newman, N. C., Ryan, P., Le Masters, G., et al. (2013). Traffic-related air pollution exposure in the first year of life and behavioral scores at 7 years of age. *Environmental Health Perspectives, 121*(6), 731–736. https://doi.org/10.1289/ehp.1205555

Ohly, H., White, M. P., Wheeler, B. W., et al. (2016). Attention restoration theory: A systematic review of the attention restoration potential of exposure to natural environments. *Journal of Toxicology and Environmental Health. Part B, Critical Reviews, 19*(7), 305–343. https://doi.org/10.1080/10937404.2016.1196155

Pagalan, L., Oberlander, T. F., Hanley, G. E., et al. (2022). The association between prenatal greenspace exposure and Autism spectrum disorder, and the potentially mediating role of air pollution reduction: A population-based birth cohort study. *Environment International, 167*, 107445. https://doi.org/10.1016/j.envint.2022.107445

Perera, F. P., Chang, H. W., Tang, D., et al. (2014). Early-life exposure to polycyclic aromatic hydrocarbons and ADHD behavior problems. *PLoS One, 9*(11), e111670. https://doi.org/10.1371/journal.pone.0111670

Peterson, B. S., Rauh, V. A., Bansal, R., et al. (2015). Effects of prenatal exposure to air pollutants (polycyclic aromatic hydrocarbons) on the development of brain white matter, cognition, and behavior in later childhood. *JAMA Psychiatry, 72*(6), 531–540. https://doi.org/10.1001/jamapsychiatry.2015.57

Pujol, J., Fenoll, R., Macià, D., et al. (2016). Airborne copper exposure in school environments associated with poorer motor performance and altered basal ganglia. *Brain and Behavior: A Cognitive Neuroscience Perspective, 6*(6), e00467. https://doi.org/10.1002/brb3.467

Rees, S., & Harding, R. (2004). Brain development during fetal life: Influences of the intra-uterine environment. *Neuroscience Letters, 361*(1–3), 111–114. https://doi.org/10.1016/j.neulet.2004.02.002

Richardson, E. A., Pearce, J., Shortt, N. K., & Mitchell, R. (2017). The role of public and private natural space in children's social, emotional and behavioural development in Scotland: A longitudinal study. *Environmental Research, 158*, 729–736. https://doi.org/10.1016/j.envres.2017.07.038

Rivas, I., Basagaña, X., Cirach, M., et al. (2019). Association between early life exposure to air pollution and working memory and attention. *Environmental Health Perspectives, 127*(5), 57002. https://doi.org/10.1289/EHP3169

Saenen, N. D., Vrijens, K., Janssen, B. G., et al. (2016). Placental Nitrosative stress and exposure to ambient air pollution during gestation: A population study. *American Journal of Epidemiology, 184*(6), 442–449. https://doi.org/10.1093/aje/kww007

Saez, M., Barceló, M. A., Farrerons, M., & López-Casasnovas, G. (2018). The association between exposure to environmental factors and the occurrence of attention-deficit/hyperactivity disorder (ADHD). A population-based retrospective cohort study. *Environmental Research, 166*, 205–214. https://doi.org/10.1016/j.envres.2018.05.009

Sanes, D. H., Reh, T. A., & Harris, W. A. (2011). *Development of the nervous system* (3rd ed.). Elsevier. https://doi.org/10.1016/C2009-0-02132-1

Shepherd, D., Heinonen-Guzejev, M., Hautus, M. J., & Heikkilä, K. (2015). Elucidating the relationship between noise sensitivity and personality. *Noise & Health, 17*(76), 165–171. https://doi.org/10.4103/1463-1741.155850

Shukla, A., & Tandel, B. N. (2024). Association of road traffic noise exposure and school childrens' cognition: A structural equation model approach. *Environmental Research, 240*, 117388. https://doi.org/10.1016/j.envres.2023.117388

Sonuga-Barke, E. J. S., Kennedy, M., Kumsta, R., et al. (2017). Child-to-adult neurodevelopmental and mental health trajectories after early life deprivation: The young adult follow-up of the longitudinal English and Romanian Adoptees study. *The Lancet, 389*(10078), 1539–1548. https://doi.org/10.1016/S0140-6736(17)30045-4

Stafford, T. M. (2015). Indoor air quality and academic performance. *Journal of Environmental Economics and Management, 70*, 34–50. https://doi.org/10.1016/j.jeem.2014.11.002

Stansfeld, S., & Clark, C. (2015). Health effects of noise exposure in children. *Current Environmental Health Reports, 2*(2), 171–178. https://doi.org/10.1007/S40572-015-0044-1/TABLES/1

Stenson, C., Wheeler, A. J., Carver, A., et al. (2021). The impact of traffic-related air pollution on child and adolescent academic performance: A systematic review. *Environment International, 155*, 106696. https://doi.org/10.1016/j.envint.2021.106696

Suades-González, E., Gascon, M., Guxens, M., & Sunyer, J. (2015). Air pollution and neuropsychological development: A review of the latest evidence. *Endocrinology (United States), 156*(10), 3473–3482. https://doi.org/10.1210/en.2015-1403

Sunyer, J., & Dadvand, P. (2019). Pre-natal brain development as a target for urban air pollution. *Basic & Clinical Pharmacology & Toxicology, 125*(S3), 81–88. https://doi.org/10.1111/bcpt.13226

Sunyer, J., Esnaola, M., Alvarez-Pedrerol, M., et al. (2015). Association between traffic-related air pollution in schools and cognitive development in primary school children: A prospective cohort study. *PLoS Medicine, 12*(3), e1001792. https://doi.org/10.1371/journal.pmed.1001792

Sunyer, J., Suades-González, E., García-Esteban, R., et al. (2017). Traffic-related air pollution and attention in primary school children. *Epidemiology, 28*(2), 181–189. https://doi.org/10.1097/EDE.0000000000000603

Tangermann, L., Vienneau, D., Saucy, A., et al. (2023). The association of road traffic noise with cognition in adolescents: A cohort study in Switzerland. *Environmental Research, 218*, 115031. https://doi.org/10.1016/j.envres.2022.115031

Terzakis, M. E., Dohmen, M., van Kamp, I., & Hornikx, M. (2022). Noise indicators relating to non-auditory health effects in children—A systematic literature review. *International Journal of Environmental Research and Public Health, 19*(23). https://doi.org/10.3390/ijerph192315633

Thompson, R., Smith, R. B., Bou Karim, Y., et al. (2022). Noise pollution and human cognition: An updated systematic review and meta-analysis of recent evidence. *Environment International, 158*, 106905. https://doi.org/10.1016/J.ENVINT.2021.106905

Tierney, A. L., & Nelson, C. A., III. (2009). Brain development and the role of experience in the early years. *Zero Three, 30*(2), 9.

Ulrich, R. S., Simonst, R. F., Lositot, B. D., Fioritot, E., Milest, M. A., & Zelsont, M. (1991). Stress recovery during exposure to natural and urban environments. *Journal of Environmental Psychology, 11*(3), 201–230. https://doi.org/10.1016/S0272-4944(05)80184-7
Vella-Brodrick, D. A., & Gilowska, K. (2022). Effects of nature (greenspace) on cognitive functioning in school children and adolescents: A systematic review. *Educational Psychology Review, 34*(3), 1217–1254. https://doi.org/10.1007/s10648-022-09658-5
Volpe, N., Dall'Asta, A., Di Pasquo, E., Frusca, T., & Ghi, T. (2021). First-trimester fetal neurosonography: Technique and diagnostic potential. *Ultrasound in Obstetrics and Gynecology, 57*(2), 204–214. https://doi.org/10.1002/uog.23149
Wu, J., & Jackson, L. (2017). Inverse relationship between urban green space and childhood autism in California elementary school districts. *Environment International, 107*, 140–146. https://doi.org/10.1016/j.envint.2017.07.010
Xie, H., Kang, J., & Tompsett, R. (2011). The impacts of environmental noise on the academic achievements of secondary school students in Greater London. *Applied Acoustics, 72*(8), 551–555. https://doi.org/10.1016/j.apacoust.2010.10.013
Yan, W., Ji, X., Shi, J., Li, G., & Sang, N. (2015). Acute nitrogen dioxide inhalation induces mitochondrial dysfunction in rat brain. *Environmental Research, 138*, 416–424. https://doi.org/10.1016/j.envres.2015.02.022
Zare Sakhvidi, M. J., Knobel, P., Bauwelinck, M., et al. (2022). Greenspace exposure and children behavior: A systematic review. *Science of the Total Environment, 824*, 824. https://doi.org/10.1016/j.scitotenv.2022.153608
Zare Sakhvidi, M. J., Mehrparvar, A. H., Zare Sakhvidi, F., & Dadvand, P. (2023). Greenspace and health, wellbeing, physical activity, and development in children and adolescents: An overview of the systematic reviews. *Current Opinion in Environmental Science & Health, 32*, 32. https://doi.org/10.1016/j.coesh.2023.100445

Neighbourhood Urban Environments and Cognitive Health in Ageing Populations

Maria V. Soloveva, Govinda Poudel, Anthony Barnett, and Ester Cerin

Abstract Population ageing and urbanisation are two global demographic trends with significant public health implications, including increased rates of cognitive decline and neurogenerative disorders, as well as diminished quality of life. It is known that the neighbourhood urban environment impacts on lifestyle behaviours and physical health in older adults; however, little is known about how key features of the urban environment support cognitive health in late life. The main objective of this chapter is to critically review the current evidence on how physical features of the neighbourhood urban environment impact cognitive health in ageing populations. In particular, we apply a working ecological model of cognitive health that postulates that urban design and its by-products affect cognitive health via behavioural and biological pathways. In accordance with the model, the evidence provided in this chapter demonstrates that neighbourhood physical environments have the potential to promote cognitively healthy ageing and, thus, the creation of urban physical environments that preserve cognitive health is considered a feasible large-scale and sustainable solution. We further highlight the importance to recognise the complexity and diversity of urban design features and their multidimensional interactions with cognitive health in older adults. High-quality empirical studies that adopt ecological models of cognitive health are needed to inform the creation of age-friendly urban environments.

Keywords Urban environments · Cognitive health · Older adults · Neurodegenerative disorders · Cognitive decline · Ecological models · Urban design · Mechanisms

---

M. V. Soloveva · G. Poudel · A. Barnett · E. Cerin (✉)
Behaviour, Environment and Cognition Research Program, Mary MacKillop Institute for Health Research, Australian Catholic University, Melbourne, VIC, Australia
e-mail: ester.cerin@acu.edu.au

S. Kühn (ed.), *Environmental Neuroscience*,
https://doi.org/10.1007/978-3-031-64699-7_13

Overview

Urbanisation and population ageing are two major global demographic trends with significant environmental and social impacts. While in the 1950s, less than a third of the world's population lived in cities (United Nations, 2018), urban dwellers now make up more than half of the global population. This proportion is projected to increase to two-thirds by 2050 (World Bank, 2020). In 2017, the global number of people aged ≥60 years approached 1 billion. This was more than twice the number recorded in the 1980s and less than half the number expected in 2050 (United Nations, 2017). Moreover, the majority of older people live in cities and the ratio of urban vs rural older dwellers can only increase as a result of current global urbanisation trends (United Nations, 2014). These demographic drifts call for the planning and development of urban environments that promote the well-being of older populations and allow them to age in place, that is, continue living independently in their communities as long as possible.

Cities are sources of innovation and economic development that provide better access to healthcare and education (Moore et al., 2003). Through their design, cities can promote an active lifestyle and facilitate social interactions (Giles-Corti et al., 2016) that benefit human health. However, if not carefully planned, urban growth may have unintended adverse effects on health by exacerbating air pollution, noise, environmental degradation and social inequalities (Moore et al., 2003). It is, thus, imperative to implement evidence-based, health-oriented urban planning policies aimed at the creation of healthy cities for all (Lowe et al., 2022). In order to do this, we first need to understand how key features of the urban environment influence physical, mental and cognitive health.

In line with the overarching theme of this book, the main objective of this chapter is to examine the current evidence on how physical characteristics of the neighbourhood urban environment impact cognitive health in late life. We first provide an overview of a proposed ecological model of the effects of urban neighbourhoods on cognitive health in ageing populations followed by a synthesis of empirical evidence. At the end of the chapter, we discuss the limitations and practical implications of current research in this field and outline future research directions.

An Ecological Model of the Effects of Urban Neighbourhood Physical Environments on Cognitive Health in Ageing Populations

Geographical life space, defined as the area within which a person participates in activities during their daily life, typically shrinks with age due to retirement and the onset of chronic health conditions (Barnes et al., 2007). For this reason, older urban dwellers spend most of their time in their residential neighbourhood (Glass & Balfour, 2003). Understanding how to create neighbourhood environments that promote cognitive health and independence in this age group is, therefore, of utmost societal importance. Figure 1 shows a working ecological model of physical features of the neighbourhood environment deemed to impact older adults' cognitive health via behavioural and biological pathways.

Here, neighbourhood is defined as the immediate geographical area surrounding one's place of residence (Swisher, 2023). It may be objectively operationalised as an administrative division (e.g. census area), an area surrounded by physical features such as rivers, roads and railways, or an area within a certain distance from a residential address. Alternatively, neighbourhoods may be defined based on residents' subjective delineations of the areas they consider to be their neighbourhoods (Cerin et al., 2017a, b). Urban neighbourhoods are characterised by a set of interlinked environmental features depicted in Fig. 1. *Density* usually refers to the number of dwellers or dwellings per unit area of a neighbourhood (Ewing & Cervero, 2010) and represents the backbone of urbanisation. Increases in density may potentially result in beneficial as well as detrimental effects on health because they impact all key features of urban neighbourhoods shown in Fig. 1. For example, densification is accompanied by increases in human activities and motorised traffic, which are major sources of *air pollution*, *noise* and *artificial night lighting* known to be harmful to health (Cerin, 2019; Cerin et al., 2020a, b). At the same time, densification increases the demand for *public transport*, *pedestrian/cycling infrastructure* and

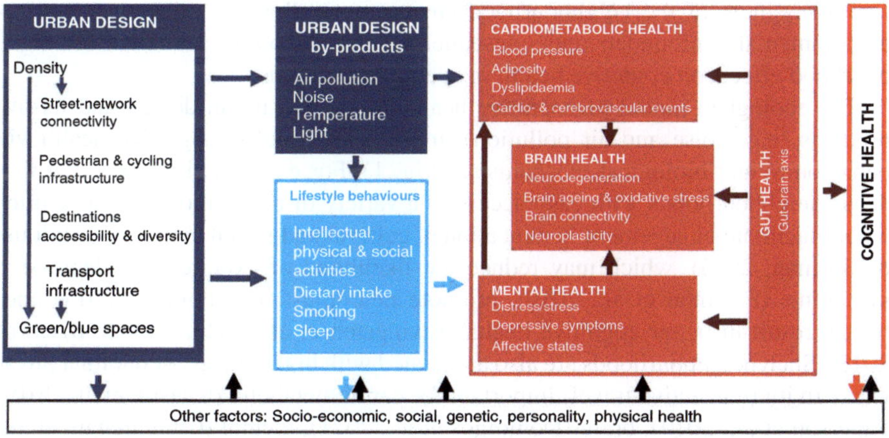

Fig. 1 Conceptual ecological model of the effects of urban design and its by-products on cognitive health in ageing populations. Urban design affects cardiometabolic health, brain health, mental and gut health and, thus, cognitive health, *indirectly* by influencing transport infrastructure (e.g. cycling or motorised transport) and lifestyle behaviours and, more *directly*, by exposing older adults to potentially harmful environmental factors (urban design by-products) and/or limiting exposures to green and blue natural environments. The arrow from the urban design box to the red box through urban design by-products and lifestyle behaviours indicates that urban environments affect all health indicators in the red box. Green and blue spaces are referred to as geographical areas characterised by various physical features (e.g. size, types of vegetation), excluding non-physical features, such as active ingredients (e.g. terpenes)

access to diverse destinations and services which, in turn, provide opportunities for health-enhancing *physical, intellectual and social activities* (Cerin et al., 2020a). Furthermore, the expansion of human settlements and densification may result in a significant loss of natural habitat, *green public open spaces,* and even *blue spaces* (i.e. water bodies such as rivers, lakes and the sea) through land reclamation. While low-density neighbourhoods with a very high proportion of green and blue spaces may not represent the optimal environment for a physically and socially active lifestyle, high-density neighbourhoods with no access to nature may be even worse. In fact, *green* and *blue spaces* have been linked to a variety of positive health outcomes (de Keijzer et al., 2020; Gascon et al., 2017). Yet, what mix of built and natural environment features would make a neighbourhood optimal for older adults' cognitive health, remains unclear. One of the reasons for this uncertainty is the relative lack of studies with comprehensive environmental assessments of all key attributes of the neighbourhood environment (Cerin et al., 2020a). Numerous studies have examined the effects of a single set of neighbourhood environmental features, such as ambient air pollution (Killin et al., 2016; Peters et al., 2019; Power et al., 2016), access to greenspace (de Keijzer et al., 2016) and walkability components on cognitive health (Besser et al., 2017; Chen et al., 2022a). However, only a handful of studies have considered all these sets of features and their causal interrelationships in estimating the effects of the neighbourhood environment on cognitive health (Cerin et al., 2021; 2022, 2023; Soloveva et al., 2023). One of the main motivations for writing this chapter is to propose a model that advocates for a more comprehensive assessment of the neighbourhood environment allowing adjustment for key environmental confounders and, consequently, a more robust estimation of neighbourhood effects on cognitive health in ageing populations.

The ecological model of cognitive health posits that urban design and its by-products (e.g. noise and air pollution) impact older adults' cognitive health via behavioural and biological mechanisms (Fig. 1). For example, it is expected that dense neighbourhoods with easy access to a variety of destinations and amenities would facilitate engagement in educational, cultural and social activities (Alidoust & Bosman, 2015), which may reduce the likelihood of experiencing depressive symptoms (Forsman et al., 2010), promote neuroplasticity (Phillips, 2017) and, hence, result in better cognitive health (Livingston et al., 2020; Park & Bischof, 2013). Such neighbourhoods are also associated with higher levels of habitual physical activity (e.g. active travel) important for cardiometabolic (Chastin et al., 2019; Patterson et al., 2019), brain (Domingos et al., 2021), mental (Cunningham et al., 2020) and gut health (Ramos et al., 2022). Access and exposure to nature may regulate sleep (Shin et al., 2020), reduce stress (de Keijzer et al., 2020) and provide opportunities for physical and social activities (Maas et al., 2009). Healthy sleep patterns and stress levels are essential for cardiometabolic (Dejenie et al., 2022), brain (Kent & Mistlberger, 2017) and mental health (Bao et al., 2017) (➔ chapter "The CITY & the PSYCHE—A Chronology" by Mehran, Kluge and Adli). On the other hand, dense, trafficked neighbourhoods may expose residents to higher levels of air pollution and noise, and the heat island effect arising from the lack of greenery. These, in turn, may adversely affect sleep (Smith et al., 2022), deter residents from being physically active outdoors (An et al., 2019), and increase blood pressure and systemic inflammation (Münzel et al., 2021). All these factors are linked to worse cognitive health outcomes (Livingston et al., 2020; Qin et al., 2023; Walker et al., 2023) as well as worse cardiometabolic (Chastin et al., 2019; Münzel et al., 2021), mental (Bao et al., 2017), brain (Kent & Mistlberger, 2017) and gut health (Ramos et al., 2022; Tsereteli et al., 2022).

Another hallmark of the working model of urban environment effects on cognitive health is its interactionist nature, which is common to ecological models of health (Stokols, 1996). Specifically, we hypothesise that the impact of the neighbourhood physical environment (urban design, nature and by-products of densification) on lifestyle behaviours may depend on the socio-economic and social characteristics of the neighbourhood as well as individual-level characteristics, such as personality, socio-economic status, demographics, social support and physical health. For example, whether neighbourhood recreational facilities result in higher levels of exercise may depend on a person's physical functionality (Cerin et al., 2020c). Furthermore, the impacts of lifestyle behaviours and the direct effects of environmental exposures on cognitive health and related biological mechanisms may be contingent on genetic factors and other health conditions. In fact, for example, apolipoprotein E genotype has been found to moderate the harmful effects of exposure to air pollution on cognitive health, with APOE ε4 carriers showing the strongest effects (Cacciottolo et al., 2017). APOE ε4 carriers also appear to benefit more than their counterparts from being physically active as they show lower amyloid burden in the brain if they accumulate higher levels of physical activity, while this has not been found in APOE ε4 non-carriers (Brown et al., 2013a, b).

In the remainder of this chapter (sections "Urban Environments and Cognitive Health: Epidemiological Evidence" to "Mechanisms Linking Urban Environments with Cognitive Health"), we examine the current evidence on the various components and tenets of the proposed model of the effects of urban neighbourhood physical environments on older adults' cognitive health. We start with an overview of research on neighbourhood physical environmental correlates of neurodegenerative diseases and cognitive function, including moderating factors, and then report on studies that examined the various mechanisms linking the neighbourhood environment with cognitive health depicted in Fig. 1—namely lifestyle behaviours and indicators of cardiometabolic, gut, mental and brain health.

Urban Environments and Cognitive Health: Epidemiological Evidence

Natural Environment and Neurodegenerative Disorders

An emerging body of evidence has started to acknowledge that exposure to green (Ho et al., 2020; Li et al., 2023; Rodriguez-Loureiro et al., 2022; Zijlema et al., 2019) and blue (Crouse Dan et al., 2018; Kasdagli et al., 2023; Zijlema et al., 2019) spaces can reduce the risk of mortality attributed to neurodegeneration. Epidemiological studies demonstrated that older adults living in neighbourhoods with greater amount of greenery were less likely to develop a neurodegenerative disorder, including Alzheimer's disease (Aitken et al., 2021; Jin et al., 2021; Klompmaker et al., 2022; Wu & Jackson, 2021; Zhu et al., 2023), vascular and all-cause dementia (Hu et al., 2023; Yuchi et al., 2020), dementia (Aitken et al., 2021; Astell-Burt et al., 2020; Paul et al., 2020; Slawsky et al., 2022), mild cognitive impairment (Godina et al., 2023) and Parkinson's disease (Jung et al., 2022; Klompmaker et al., 2022; Yu et al., 2021; Yuchi et al., 2020; Zhu et al., 2023). Compared with the cognitive health benefits of green spaces, evidence with respect to blue spaces and the incidence of neurodegenerative disorders is extremely scarce, with two studies showing negative (Klompmaker et al., 2022; Wu & Jackson, 2021) and one study showing positive (Zijlema et al., 2019) associations.

Most consistent evidence of a negative association between greenspace and neurodegenerative diseases comes from longitudinal studies. However, a number of shortcomings remain (Hirsch et al., 2022; Klompmaker et al., 2022). In particular, studies have used various exposure measures, such as the percentage of total greenspace (Astell-Burt et al., 2020; Godina et al., 2023; Hu et al., 2023), tree canopy coverage (Astell-Burt et al., 2020), normalised difference vegetation index (NDVI) (Jung et al., 2022; Paul et al., 2020; Yuchi et al., 2020; Zhu et al., 2023), green space diversity (Godina et al., 2023) as well as composite measures of residential greenspace (Slawsky et al., 2022). Older adults are more vulnerable than younger groups to the adverse effects of environmental hazards, such as extreme heat and outdoor air pollution that are typically pronounced in urban environments (Qiu et al., 2023). Certain green landscape types, such as tree canopies, vegetated streets or woodland may be more relevant for reducing the risk of neurodegenerative disorders in ageing populations because they provide shade, and temperature regulation (Kabisch et al., 2017) and may lower concentrations of airborne air pollutants, which, in turn, may encourage older adults to be physically, cognitively and socially active and, hence, healthier (Sun et al., 2023). For example, people aged 75 years and older living in the highest quartile of forest greenspace had reduced odds of cognitive impairment compared to those living in the lowest quartile of forest greenspace (Godina et al., 2023). In contrast, no associations were found between the total percentage of greenspace and the risk of mild cognitive impairment and dementia (Godina et al., 2023). Similarly, there were no associations between the total percentage of greenspace and incident dementia, although older adults living in neighbourhoods with greater tree canopy coverage were less likely to develop dementia (Li et al., 2023). This issue holds true for studies on greenery-cognition associations (please see section "Natural Environment and Cognitive Function").

Another methodological issue of studies on access to nature and neurodegenerative disorders pertains to the definition and assessment of neurodegenerative disorders. If a clinical diagnosis is not made according to the established standards, it can lead to misclassification and measurement error (Chow et al., 2022). For example, Chow et al. (2022) found that hospital and death data have a sensitivity of 20%, which indicates that only one-fifth of individuals with dementia can be identified with this approach. The majority of studies have employed administrative medical data for classification purposes, such as hospital records (Klompmaker et al., 2022; Paul et al., 2020; Yuchi et al., 2020), medication prescriptions (Astell-Burt et al., 2020; Paul et al., 2020; Yuchi et al., 2020), electronic health and medical claims data (Aitken et al., 2021; Hu et al., 2023; Klompmaker et al., 2022; Paul et al., 2020; Zhu et al., 2023), while a few studies have used other types of assessment that are more accurate and in line with the gold standard, such as the Diagnostic and Statistical Manual of Mental Disorders-4 (DSM-4) (Slawsky et al., 2022), or detailed neurological/neuropsychological examinations (Godina et al., 2023). The field should acknowledge the importance of following established protocols in classifying dementia, Alzheimer's disease or any other neurodegenerative disorder and doing so consistently.

Furthermore, it is critical to highlight that the majority of longitudinal studies in this field have considered only individual-level confounders and/or covariates while failing to adjust for environmental confounders (e.g. residential density, land use mix), known to be potentially causally related to both access to nature and cognitive health, as depicted in Fig. 1 (Cerin et al., 2021, 2022, 2023; Soloveva et al., 2023). Thus, to ensure unbiased estimates of associations between the neighbourhood natural environment and neurodegenerative diseases, it is important to account for the key environmental confounders and distinguish between total and direct environmental effects in multiple-exposure regression models, such as in two recent longitudinal studies (Paul et al., 2020; Jung et al., 2022). For example, in a large, population-based cohort study (N = 1,737,460), increased exposure to urban green spaces (measured by the NDVI) within a 250 m neighbourhood buffer was associated with a 3% reduced incidence of dementia in urban residents aged 55–85 years. These estimates of direct effects were obtained from multiple-exposure regression models adjusted for air pollutants (NO_2, O_3, $PM_{2.5}$), population density and neighbourhood socioeconomic status (Paul et al., 2020).

Natural Environment and Cognitive Function

Current cross-sectional evidence with respect to the effects of the natural urban environment on cognitive function in older adults is largely contradictory, with studies showing no (Besser et al., 2020; Chan et al., 2023; Clarke et al., 2012; Crous-Bou et al., 2020; Zijlema et al., 2017), positive (Astell-Burt & Feng, 2020; Besser et al., 2020; Cerin et al., 2021, 2023) and negative associations (Besser et al., 2020; Wu et al., 2015). However, it is important to highlight that longitudinal findings are rather suggestive of the beneficial effect of urban greenery on cognition (Besser et al., 2021b; Cherrie et al., 2018; de Keijzer et al., 2018; Zhang et al., 2022a; Zhu et al., 2019, 2020).

For example, in the Multi-Ethnic Study of Atherosclerosis (MESA) study (Besser et al., 2020), after adjusting for population density and neighbourhood socioeconomic status, residents of New York and St. Paul showed improved global cognition and processing speed with increased percentage of parkland (>6%). This effect was not present among residents living in Los Angeles and Forsyth. Such discrepancy in findings may be due to the lack of adjustment for other environmental confounders, such as street connectivity and other land uses, or for differences in the quality and safety of parklands. A series of cross-sectional studies by Cerin and colleagues examined parkland-cognition associations by applying an ecological model of cognitive health which, as we discussed in section "An Ecological Model of the Effects of Urban Neighbourhood Physical Environments on Cognitive Health in Ageing Populations" recommends the inclusion of key environmental attributes in multiple-exposure models. These studies found that exposure to a larger percentage of parkland in the neighbourhood was related to better verbal memory and processing speed (Cerin et al., 2021, 2022, 2023), and a smaller cognitive age gap (Soloveva et al., 2023).

Longitudinal studies utilised samples with diverse characteristics (e.g. ethnicity/race) and adopted various measures of exposure (e.g. NDVI, % of parkland), and cognitive function (e.g. memory, processing speed, reasoning, general intelligence, global cognition), residential buffers (e.g. 500 meters, 1.5 km) and time intervals between assessments (6 years, 10 years) (Besser et al., 2021a; Cherrie et al., 2018; de Keijzer et al., 2018; Zhu et al., 2019). Despite this, increased exposure to greenspaces was associated with a slower rate of cognitive decline in older adults. To the best of our knowledge, only one longitudinal study did not find any associations between greenspaces (NDVI) and multiple cognitive outcomes in a relatively young sample ($M = 54.7$ years), though regression models included population density as a covariate.

In general, research into the effects of the neighbourhood environment on older adults' cognitive function has been quite limited in terms of a range of cognitive assessments, primarily focusing on global cognition or processing speed. This is problematic because some facets of cognition in addition to processing speed are more susceptible to ageing, such as inhibitory control, episodic and spatial memory, while speech and language remain relatively intact late in life (Caffò et al., 2020; Lopez et al., 2021; Salthouse, 2009). Ageing is also associated with a decline in social cognitive function—older adults exhibit compromised social perception, cognitive empathy and mentalising, known to be important for quality of life and everyday functioning (Arioli et al., 2018), psychological well-being and social networks (Grainger et al., 2023). Importantly, social cognitive skills are believed to be implicated in loneliness (Beadle & de la Vega, 2019), a precursor of cognitive impairment and neurodegeneration (Harrington et al., 2023) and are partially independent of age-related cognitive decline (Grainger et al., 2023). We, thus, recommend investigation of environmental correlates of social cognitive function in older adults in combination with other facets of cognitive function.

Finally, the link between blue spaces and cognitive function has not been adequately studied. While a couple of recent cross-sectional studies did not support total blue spaces-cognition associations in older adults (Cerin et al., 2021; Soloveva et al., 2023), Cerin et al. (2022) showed positive indirect effects of blue spaces on processing speed through waist circumference and glycated haemoglobin, indicating that access to blue spaces may benefit cognitive health via cardiometabolic health. The mechanisms through which blue spaces may impact cognitive function are yet to be understood in future research (please see section "Mechanisms Linking Urban Environments with Cognitive Health" of this chapter).

Built Environment and Cognitive Health

Population Density

Most evidence suggests a non-linear, dose-dependent relationship between population density and the odds of a neurodegenerative disorder (Lorenzo-López et al., 2017; Luo et al., 2021). For example, there was an inverse relationship between urbanisation and dementia in Spain, with higher urbanicity relating to decreased odds of dementia (Lorenzo-López et al., 2017). Importantly, in China, Luo et al. (2021) showed that at low and middle levels of urbanisation (operationalised as the ratio of non-agricultural population to the total population in %), older adults were less likely to have dementia. However, extremely high urbanicity was associated with increased odds of dementia. Other studies focused on the comparison between rural and urban areas (Abner et al., 2016; Helmes & Van Gerven, 2017; Robbins et al., 2019; Weden et al., 2018) and reported mixed results. Likewise, with respect to cognitive function, evidence is mixed. However, curvilinear population density-cognition associations are commonly reported (Cassarino et al., 2020; Cassarino et al., 2018; Cerin et al., 2021; Wörn et al., 2017). In Worn et al.'s study (2017), higher residential density predicted better global cognition, processing speed, problem solving and episodic memory among older adults. However, at the very high end of residential density, the positive relationship with processing speed seemed to reverse, suggesting a curvilinear relationship. This is expected, as older adults who live in overpopulated and/or overcrowded urban environments are more likely to be exposed to increased concentrations of air pollutants as well as to road traffic and noise (Carey et al., 2018; Dzhambov & Dimitrova, 2016). Therefore, while population density may be beneficial for older adults in terms of provision of services and transport infrastructure, living in very densely populated neighbourhoods may be detrimental to their cognitive health due to the high levels of environmental stressors. Future research is needed to identify the optimal level of population density to maintain and promote cognitive health in older adults.

Destination Accessibility and Diversity

Easy access to a variety of local services and destinations is an important determinant of cognitive health in older adults (Besser et al., 2018; Cerin et al., 2020a; Chan et al., 2023; Chen et al., 2022a; Clarke et al., 2012; Clarke et al., 2015; Fangfang et al., 2022; Finlay et al., 2021a, b; Guo et al., 2019; Wu et al., 2015, 2017, 2020).

For example, the availability of healthy food stores and eating/dining places in the neighbourhood has been linked to lower odds of dementia and mild cognitive impairment (Fangfang et al., 2022; Tani et al., 2019) as well as slower cognitive decline (Finlay et al., 2020, 2022; Hyun et al., 2023). Greater densities of institutional resources (community centres, schools, libraries, senior centres, post offices civic organisations) (Clarke et al., 2012; Finlay et al., 2021a, 2022; Guo et al., 2019) and recreational sites (art centres, museums, gyms, walking paths) (Finlay et al., 2021b, 2022; Guo et al., 2019) were associated with better cognitive function and a slower rate of cognitive decline (Clarke et al., 2015; Luo et al., 2019). Some contradictory findings include those by Besser et al. (2018) and Guo et al. (2019) who did not report positive effects of access to places for social engagement (beauty shops/barbers) and walking facilities (restaurants, parks) on cognitive function in older adults. There are several mechanisms that link access to destinations and services to cognitive health (please see section "Mechanisms Linking Urban Environments with Cognitive Health") and to date, there is no clear information on the types of destinations/services that are the most important for cognitive health in ageing cohorts. Thus, we recommend future studies to identify mixes of destinations that are optimal for cognitive health.

Street Connectivity

Neighbourhoods with superior street connectivity are characterised by well-interconnected streets. They consist of small blocks and shorter street segments. Neighbourhoods with such a street layout offer alternative routes to destinations which, in turn, can facilitate walking for utilitarian purposes in older adults (Saelens et al., 2003). No evidence exists with respect to the protective effects of street layout on the risk of a neurodegenerative disorder and only a handful of studies examined street connectivity as a correlate of cognitive health (Besser et al., 2021b; Cerin et al., 2021; Koohsari et al., 2019; Ng et al., 2017; Watts et al., 2015). For example, greater street connectivity was associated with better global cognition and immediate memory recall (Ng et al., 2018) and with improved attention in older adults over a 2-year period (Watts et al., 2015). However, this association was not found in participants with Alzheimer's disease. In contrast, no cross-sectional or longitudinal street connectivity-cognition associations were reported in studies by Cerin et al. (2021) and Besser et al. (2021a), respectively. A street layout can be alternatively operationalised in terms of street integration, defined by the number of turns or choice points an individual must make to travel from one street to another and/or to access a destination (Bafna, 2003). A study found that older adults who lived in urban neighbourhoods with more integrated street layouts (i.e. requiring fewer turns to reach a destination) exhibited diminished attention and verbal memory over the 2-year follow-up period (Watts et al., 2015), and this cognitive decline was more pronounced in participants with mild Alzheimer's disease, suggesting that integrated streets (less complex for navigation purposes) result in worsened cognitive outcomes. More research is needed to disentangle whether simpler or more complex street layouts and street networks are beneficial for older adults' cognitive health.

Transport Infrastructure and Air Pollution

Accessible and well-maintained transport infrastructure is a critical component of age-friendly cities (Chen et al., 2022a; Gaglione et al., 2021; Ozbilen et al., 2022). Due to the deterioration of physical health and driving skills, older adults may rely on public transport to engage in daily activities and/or access services and destinations that are not within walking distance. Of note, the prevalence of walking for transport increases with advancing age (Hasselder et al., 2022). Although transport infrastructure-cognitive health associations are well-reported in the literature (Chen et al., 2022a), evidence is inconclusive.

For example, in Hong Kong, Chan et al. (2023) found that a greater number of public transport terminals was related to better cognitive health in older adults aged 65–80 years. Likewise, the introduction of free bus passes and more bus lines were, respectively, associated with better global cognitive function and memory (Reinhard et al., 2019) as well as with slower cognitive decline over a 4-year period in older adults (Luo et al., 2019). On the other hand, living in urban areas with too many public transport stops may expose older adults to increased air and noise pollution (Wilker et al., 2023), which, in turn, may decrease their willingness to go out and engage in physical, social or cognitive pursuits that are beneficial for cognitive health (James et al., 2015). In a recent review, Andrade and colleagues highlighted that exposure to NO_2, $PM_{2.5}$, O_3 and PM_{10} during physical activity is harmful for older adults' cardiovascular and respiratory health (Andrade et al., 2023), suggesting that it is important to optimise neighbourhood walkability because benefits of transportation walking can be offset by increased exposure to outdoor air pollutants (Cerin et al., 2023; Chen et al., 2017; Cowie et al., 2016).

Furthermore, studies showed that greater road density and proximity to roads and/or highways were associated with worsened cognitive performance in older adults (Kim et al., 2023b; Ranft et al., 2009; Schikowski et al., 2015; Wellenius et al., 2012) and this detrimental effect was most pronounced in individuals living less than 100 meters from the road. In particular, older adults living 50–100 meters from major roadways in China were 26% more likely to exhibit cognitive impairment, as opposed to individuals living further than 300 m from a major roadway (Yao et al., 2022). Similarly, living in close proximity to major roads increased the odds of all-cause dementia (Chen et al., 2017), non-Alzheimer's dementia, Alzheimer's disease and Parkinson's disease (Yu et al., 2021; Yuchi et al., 2020) and these associations were mediated by NO_2 and $PM_{2.5}$ (Chen et al., 2017). In contrast, at an 18-year follow-up, Clarke et al. (2015) found that proximity to public transit stops was associated with slower rates of cognitive decline, whereas older adults with access to public spaces in poor conditions had a faster rate of cognitive decline over time.

It is important to emphasise that proximity to and availability of public transport and infrastructure for walking do not necessarily imply access and usage (Yu et al., 2023). Specifically, the lack of personal safety while walking and high concentrations of air pollutants in the neighbourhood might discourage older people from using public transport or engaging in active transport (Cerin et al., 2023). Pedestrian-friendly features, such as footpath coverage and benches, longer pedestrian countdown timers and safe crosswalks have been identified as determinants of transportation walking (Cerin et al., 2017; Motohiro et al., 2021; Ozbilen et al., 2022) which, in turn, may benefit cognitive health (Cerin et al., 2021). For example, living in a neighbourhood with a high level of sidewalk installation was associated with reduced incidence of dementia in urban areas (Tani et al., 2021), whereas living in hilly environments was related to a higher likelihood of cognitive decline in rural residents (Motohiro et al., 2021). Further, the lack of cycling infrastructure and support for safe cycling makes this travel mode unattractive for older people (for review please see Cui et al., 2017) and emerging evidence has shown that cycling is beneficial for cognitive health in older adults (Kardan et al., 2023; Leyland et al., 2019). Thus, urban planners should prioritise traffic and pedestrian safety to ensure that older people can access good-quality transport infrastructure to enjoy the benefits of walking and other cognitively stimulating activities (Barnett et al., 2017; Ozbilen et al., 2022).

Modifiers of the Urban Environment-Cognitive Health Association

There is evidence to suggest that the associations between aspects of the natural and built urban environment with cognitive health in older adults depend on both individual (Aneshensel et al., 2011; Chan et al., 2023; Zhu et al., 2020) and neighbourhood-level factors (Cerin et al., 2023; Lang et al., 2008; Yao et al., 2022). For a comprehensive review of the modifiers of the neighbourhood-cognitive health associations, please see published reviews by Besser et al. (2017), Besser et al. (2023), Chen et al. (2022a), Ricciardi et al. (2022), Sullivan et al. (2023) and Zagnoli et al. (2022).

With respect to neurodegeneration, older adults aged 60 + years with primary or no formal education consistently showed a stronger beneficial effect of green spaces on all neurodegenerative disease, Alzheimer's disease and unspecified dementia mortality. In addition, the beneficial effect of residential greenness was strongest for all neurodegenerative disease and unspecified dementia mortality (except for Alzheimer's disease) in older adults living in disadvantaged neighbourhoods (Rodriguez-Loureiro et al., 2022). Beneficial neighbourhood-cognition associations might be more prominent in women (Cherrie et al., 2018), lower socioeconomic groups (Aneshensel et al., 2011; de Keijzer et al., 2018; Cherrie et al., 2018), carriers of APOE ε4 (Besser et al., 2019; Franz et al., 2023; Jin et al., 2021; Zhu et al.,

2020), younger age groups (~ 65 years on average) (Chan et al., 2023) and individuals with disabilities (Chen et al., 2022a).

For example, the protective effects of residential greenness on cognitive function differed by APOE ε4 status, with the APOE ε4 carriers not benefiting from exposure to residential greenness, compared with the non-ε4 carriers (Zhu et al., 2020). Further, associations of proximity to main roads, higher land-use mix and more public transport terminals as well as greater greenness with cognitive health were stronger in older adults aged 65–80 years, in comparison to adults over 80+ years (Chan et al., 2023; Yao et al., 2022; Zhu et al., 2020). Interestingly, urban environments might exert both positive and negative effects on cognitive health depending upon the racial and/or ethnic group (Besser et al., 2018; Clarke et al., 2012; Sullivan et al., 2023; Thierry et al., 2021), which requires further investigation. A few reported neighbourhood-level modifiers include artificial lighting (Stanhope et al., 2021), area-level socioeconomic status (Ailshire et al., 2017; Cerin et al., 2023; Li et al., 2022), traffic accident density (Cherrie et al., 2019), traffic noise (Tzivian et al., 2016) as well as outdoor air pollution (Cerin et al., 2023) and temperature (Delclòs-Alió et al., 2020). Specifically, Cerin et al. (2023) showed that the positive association between a built environment complexity index (similar to a walkability index) and memory was stronger in older adults living in neighbourhoods with low levels of NO_2, suggesting that high concentrations of air pollutants may attenuate the positive effects of walkable environments; the authors also found that *both* NO_2 and area-level socioeconomic status moderated the relationship between the natural environment index and verbal memory, with higher levels of NO_2 attenuating positive effects of natural environments on verbal memory only in older adults living in disadvantaged neighbourhoods. To date, epidemiological evidence has mostly focused on the moderating effects of individual-level socioeconomic factors and there is a pressing need to investigate the moderating roles of individual characteristics and various neighbourhood-level factors conjointly and independently. This will allow us to better understand complex neighbourhood-cognitive health relationships in order to create urban environments that support the cognitive health of all its residents.

Mechanisms Linking Urban Environments with Cognitive Health

Overview

A substantial number of lifestyle behaviours, health conditions and biological processes have been identified as potential causes of cognitive decline and neurodegenerative diseases in midlife to late life (Livingstone et al., 2020; Zhang et al., 2022c). For the purpose of this chapter, we focus on those lifestyle behaviours, health conditions and biological processes that may be plausibly influenced by urban design directly or indirectly (through behaviours or other health conditions and/or biological processes) and, hence, act as mediators of the effects of the neighbourhood physical environment on older adults' cognitive health as proposed in our working model (Fig. 1). We, therefore, do not discuss some important determinants of cognitive impairment and dementia, such as hearing loss and traumatic brain injury (Livingston et al., 2020; Zhang et al., 2022c), for which there is no evidence that they are affected by the neighbourhood physical environment. For each set of mechanisms linking the neighbourhood environment with cognitive health, we provide a short overview of the evidence of their importance for cognitive health, their relationships with urban design features and findings from studies that formally examined their role as mediators of the effects of the neighbourhood environment on cognitive health.

Lifestyle Behaviours

Several lifestyle behaviours have been identified as important modifiable risk factors for cognitive decline and dementias. These, as depicted in Fig. 1, include participation in physical, social and intellectual activities, dietary behaviours, alcohol intake, sleep and smoking (Institute of Medicine, 2015; Livingston et al., 2020; Zhang et al., 2022c). Compared to pharmacotherapies, interventions targeting modifiable lifestyle factors appear to be a cheaper, safer and, currently, more effective alternative for preventing and delaying the onset of cognitive decline and dementia (Institute of Medicine, 2015). As neighbourhood environments can potentially impact entire populations for a sustained amount of time, it is paramount to examine how they shape lifestyle behaviours relevant to cognitive health.

Physical, Social and Intellectual Activities

Regular engagement in physical (e.g. exercise), social (e.g. socialising) and intellectual (e.g. reading, learning a language) activities across life and/or later in life is deemed to positively contribute to cognitive health (Institute of Medicine, 2015), although the effects have been sometimes inconsistent (Livingston et al., 2020; Zhang et al., 2022c) and may be in part due to reverse causality (prodromal dementia) (Livingston et al., 2020). The inconsistent findings could be in part due to lifestyle activities being classified into mutually exclusive categories. It could be argued that no type of activity is purely physical, social or intellectual. For example, walking may be a predominantly physical (solitary walking on a treadmill) or a predominantly social and intellectual activity (navigating unfamiliar environments with friends while walking for recreation). Lifestyle activities varying in social, physical and cognitive aspects likely differ in their effects on cognitive function, with those involving multiple components potentially being more beneficial (Kattenstroth et al., 2013). In general, regular participation in physical activity, especially aerobic exercise (Livingston et al., 2020), has been linked to better cognition and a lower risk of all-cause dementia in mid-to-late life (Zhang et al., 2022c). Several biological mechanisms are thought to underlie these findings, including beneficial effects on cardiovascular risk factors, glucose uptake, neurotrophic and growth factors, inflammation processes (Brown et al., 2013a) and gut microbiota (Cataldi et al., 2022). Health-enhancing levels of physical activity can be accrued by participating in recreational, household and transport-related activities. Among these, recreational (e.g. walking, sports or dancing) and transport-related activities (walking and cycling to go to/from places) are those mostly affected by urban design. Evidence from systematic reviews on the neighbourhood environment and physical activity suggest that older adults living in dense, destination-rich, walkable neighbourhoods are more physically active than those who do not (Cerin et al., 2019). Also, having good access to public transport, public open spaces, recreational facilities and greenspaces is associated with higher levels of total, transport-related and recreational physical activity. Neighbourhoods with a developed pedestrian and cycling infrastructure appear to support both domains of physical activity, while street connectivity is mainly positively related to walking for transport (Cerin et al., 2019).

Only a few studies have explicitly examined physical activity as a mediator of the associations between the neighbourhood environment and older adults' cognitive health. Support was found for the mediating role of total physical activity in the positive association between institutional resources (e.g. community centres, libraries) and global cognition (Clarke et al., 2012); walking for transport as a mediator of the positive associations of street connectivity, population density, commercial land and ambient air pollution with memory and processing speed (Cerin et al., 2021); and total walking as a mediator of the association of perceived but not objectively-assessed neighbourhood walkability and global cognition (Kim et al., 2023a). Predominantly intellectual (e.g. learning new skills, problem-solving, artistic pursuits) and social activities (e.g. social outings) are deemed to benefit cognitive health by promoting neurogenesis and synaptic plasticity (Rothman & Mattson, 2013). Positive social interactions are thought to impact cognitive health by reducing the risk of depressive symptoms (Pemberton & Fuller Tyszkiewicz, 2016; Pérez et al., 2020). Participation in a variety of intellectual and social activities, including cognitively demanding occupations in midlife as well as late life has been linked with maintaining cognition and slower cognitive decline (Livingston et al., 2020). While there is a dearth of studies on the neighbourhood environment and participation in intellectual activities, green public open space, good pedestrian infrastructure and access to destinations appear to facilitate social interaction and participation among older people (Clarke et al., 2019; Schmidt et al., 2019). To our knowledge, only one study examined participation in social activities as a mediator of urban design-cognitive health associations. While they did not report a significant mediating effect, the findings of the study were suggestive of social activities being responsible for a positive link between neighbourhood institutional resources, physical activity and cognition (Clarke et al., 2012).

Dietary Intake (Including Alcohol Consumption)

Adherence to the Mediterranean and Nordic diets—high in vegetables, whole grains, fruits, olive or canola oils, fish, and low in saturated fats and red meat—tend to be associated with better cognitive health (Keenan et al., 2020; Livingston et al., 2020), although the findings are mixed (Livingston et al., 2020). The positive effects of such diets on cognition are thought to be prevalently due to them benefiting cardiometabolic health (Akbaraly et al., 2019). There is some evidence that the availability of certain food outlets in the neighbourhood may impact healthy eating. For example, a systematic review reported that access to supermarkets and availability of fruit and vegetables in shops was associated with higher fruit and vegetables intake and lower fat intake (Giskes et al., 2011). Another study reported the proportion, rather than the absolute number, of healthy food outlets to be predictive of fruit and vegetable intake (Clary et al., 2016). No studies have examined the mediating effect of a healthy diet on the associations between the neighbourhood built environment and cognitive health. However, adherence to a Mediterranean diet was identified as a mediator of a positive relationship between healthy food outlets in the neighbourhood and obesity, a midlife risk factor for cognitive decline and dementia (Chen et al., 2020).

Excessive alcohol intake is also a dietary risk factor for cognitive decline and dementia (Livingston et al., 2020; Zhang et al., 2022c). Studies have reported drinking more than 12–21 units of alcohol per week as being detrimental to cognition (Livingston et al., 2020) possibly due to higher rates of brain atrophy (Sullivan et al., 2018; Topiwala et al., 2017) and adverse effects on cardiometabolic health (e.g. elevated blood pressure and glucose) (White et al., 2023). While a substantial number of studies have reported positive associations of neighbourhood alcohol outlet density with both alcohol consumption and alcohol-related harm (Shortt et al., 2018), no studies, thus far, have examined alcohol consumption as a mediator of the potential effects of neighbourhood alcohol outlets on cognitive health.

Smoking

Cigarette smoking is associated with a higher risk of dementia and cognitive decline (Livingston et al., 2020). Smoking cessation reduces the risk of dementia even if it occurs in late life (Choi et al., 2018). This is not surprising given that cigarette smoking is one of the top causes of stroke (Wolf et al., 1988) and cardiometabolic disorders (Cheng et al., 2023), which are key risk factors for dementia (Livingston et al., 2020). Cigarette smoking has also been implicated in gut microbiota dysbiosis and intestinal dysfunction (ill gut health), which can lead to systemic inflammation and, ultimately, cardiometabolic disorders (Cheng et al., 2023).

The current literature suggests that the neighbourhood built and natural environment influences smoking behaviours (Martin et al., 2020). Neighbourhoods with a higher density of tobacco retailers and those with substantial non-residential land coverage, such as commercial and industrial land, tend to have higher tobacco consumption among residents (Brooks et al., 2021; Vallarta-Robledo et al., 2022). Tobacco retailers, obviously, promote smoking and expose residents to marketing of tobacco products, which may impact smoking adoption and maintenance. Urbanised areas with diverse land use typically have higher density of food and alcohol outlets allowing smoking in a social context. Additionally, they are more densely populated and, hence, may increase stress which, in turn, may lead to smoking (Cui et al., 2012). Stress may also be the mechanism responsible for a negative association between access to green spaces and smoking observed in a recent study (Martin et al., 2020). As for dietary intake, although empirical evidence supports links between the built environment, smoking and cognitive health, no studies have formally examined the mediating effects of smoking in the associations between aspects of the neighbourhood physical environment and cognitive health in any age group.

Sleep

Sleep disturbances, such as insomnia, poor sleep quality and short sleep duration, are risk factors for all-cause dementia and Alzheimer's disease (Livingston et al., 2020; Zhang et al., 2022c). Also, objectively assessed sleep parameters, including lower restlessness, higher sleep efficiency, shorter sleep onset latency and a higher percentage of N2 and REM (rapid eye movement) sleep have been related to better cognitive function in healthy older adults (Qin et al., 2023). Sleep disturbances are deemed to affect cognitive health by increasing β-amyloid depositions and Tau protein levels in the brain, reducing glymphatic clearance, triggering low-grade inflammation and exerting adverse impacts on cardiometabolic health (Livingston et al., 2020).

As sleep appears to be favourably influenced by higher levels of physical activity (Kredlow et al., 2015), low or no alcohol consumption (Britton et al., 2020), a healthy diet (e.g. Mediterranean diet) (Scoditti et al., 2022) and abstaining from smoking (Amiri & Behnezhad, 2020), aspects of the neighbourhood environment related to these lifestyle behaviours are also potentially relevant to sleep. In fact, neighbourhood walkability and access to greenspace have been associated with better sleep outcomes (Kim et al., 2022). In contrast, exposure to urban design by-products, including noise, ambient air pollution and night lighting has been linked to worse sleep quality (Kim et al., 2022). Although there is evidence of a connection between the neighbourhood environment and sleep as well as between sleep and older adults' cognitive health, no studies have to date investigated sleep as a mediator of the associations between the neighbourhood physical environment and cognitive health in this demographic.

Brain Mechanisms

Lifelong exposure to different elements of the urban environment, such as greenspace, urbanicity, air pollution and noise, can alter the trajectory of brain development well into adulthood and ageing (Besser, 2021; Cerin et al., 2017b; Dadvand et al., 2018; Hunt et al., 2021; Pini et al., 2023; Xu et al., 2023). Emerging studies suggest that these brain changes can be quantified using advanced multimodal brain imaging techniques, including Magnetic Resonance Imaging (MRI) (➔ chapter "Mapping the Human Brain with Computational Anatomy" by Draganski, Joost and Kherif) and Electroencephalography (EEG) (➔ chapter "EEG for the Study of Environmental Neuroscience" by Grassini) (Besser, 2021; Hunt et al., 2021; Pini et al., 2023). More specifically, structural MRI (sMRI) has been used to delineate macrostructural anatomical changes, while diffusion-weighted MRI (DWI) has been used to characterise the brain's microstructure. DWI has been found to be particularly useful for measuring self diffusivity of water molecules within brain tissues, which can be used as a marker of microstructural degeneration. Functional MRI (fMRI and EEG) have been found to be suitable for measuring brain activity changes associated with exposure to different environmental elements (Tang et al., 2017; Tost et al., 2019). This section provides insights into structural and functional brain changes associated with exposure to the natural and built environment. Given that there are very few brain imaging studies that have specifically focused on environment-brain nexus in older adults, this section also uses research on brain changes associated with life-long exposure to urban environment in younger adults. This can help delineate potential brain structures and neural pathways that may be implicated in cognitive and mental health changes associated with living in an urban environment in older adults.

Structural and Functional Brain Changes Associated with Exposure to the Natural Environment

Changes in brain structure and microstructure associated with life-long exposure to the natural environment can be measured using sMRI and DWI-based brain imaging technique (Xu et al., 2023). SMRI can provide a good spatial resolution (~1 mm) image of the brain, which is used to segment various cortical and sub-cortical brain structures and measure changes in their volume. A number of structural MRI studies have delineated various brain structures and showed that the amount of green space around the home is associated with the brain's anatomy and structural integrity (Besser et al., 2021c; Kühn et al., 2017). For example, in a large epidemiological study using UK BioBank data ($N = 156{,}075$), decreased exposure to greenspace was linked to reduced gray matter volume in the inferior frontal cortex, amygdala and cerebellar cortex (Xu et al., 2023). More greenspace in the neighbourhood has also been associated with decreased ventricular sizes (Besser et al., 2021c) and increased gray matter in brain structures, such as the anterior cingulate cortex, and left and right prefrontal and motor cortices (Dadvand et al., 2018), and increased white matter in the right prefrontal cortex, the left premotor cortex, and both cerebellar hemispheres (Dadvand et al., 2018). Notably, not just brain volume, but other MRI-based measures of structural integrity, including diffusivity and magnetisation transfer, showed positive relationships with urban greenspace exposure (Kühn et al., 2017). Taken together, these findings provide some insight into brain structures linking urban green space with brain health. The amygdala, along with the prefrontal and cingulate cortices plays a critical role in top-down regulation of emotional response in the brain (McGlashan et al., 2021; Kühn et al., 2017). Individuals suffering from chronic anxiety and stress often show disruption in this neural pathway (Xu et al., 2023). Thus, structural integrity of the amygdala-prefrontal pathway due to lifelong exposure to urban greenness may reduce chronic stress and alleviate the risk of neuroinflammation (Jeanneteau, 2023), leading to better overall brain and, thus, cognitive health in older adults.

Measurement of brain activity using fMRI or EEG has provided another window into the acute brain changes that occur following exposures to different elements of the urban environment. Although most of these studies are conducted in healthy adults, the pattern of brain activity provides spatio-temporal maps of brain function, which can be useful to understand the neuroplastic changes occurring in the brain during life-long exposure. fMRI measures brain activity using blood-oxygen-level-dependent (BOLD) signal and has a high spatial resolution. However, it has limited temporal resolution and is not suitable for obtaining brain activity in natural settings. EEG provides a direct measure of cortical electrical activity with a high temporal resolution and is sensitive to changes in various mental states. EEG can also be portable and is often used to monitor brain activity in the natural environment, such as in nature strips and parks (Mavros et al., 2016). EEG studies have enabled researchers to measure brain activity in real-life, natural settings (Aspinall et al., 2015; Lin et al., 2020; Mavros et al., 2016; Olszewska-Guizzo et al., 2020). EEG brain activity is usually assessed in terms of amplitude and frequency of neuronal activity measured by placing electrodes on the scalp. The four primary EEG frequency bands are: delta ($\delta = 0.5$–4 Hz), reflecting deep sleep, and restfulness, theta ($\theta = 4$–7 Hz), indicating meditative state and drowsiness, alpha ($\alpha = 8$–15 Hz), indicating relaxed alertness while eyes closed, and beta ($\beta = 15$–30 Hz), indicating wakefulness, alertness and task performance.

There are no studies that have specifically tested for brain activity in older adults during exposure to natural environment stimuli. However, some insights regarding changes in the brain's functional activity during exposure to a natural environment can be gained from studies in younger adults. In one such study, brain activity was measured during exposure to different types of images of the natural environment (e.g. natural parks, forests, and mountains). The study showed increased activity in the superior and middle frontal gyrus, superior parietal gyrus, precuneus and anterior cingulate cortex regions (Kim et al., 2010). These findings were replicated in another study which also observed additional brain activity in the basal ganglia system of the brain. These regions represent a multi-sensory system and are often associated with increased internalisation and restfulness (e.g. more active during eyes-closed rest) (Poudel et al., 2021). Kunh et al. (2017) also revealed changes in resting or task-related functional connectivity, whereby young adults exhibited higher levels of connectivity when watching natural scenes compared to images with buildings. Increased functional connectivity suggests that there is stronger and more consistent communication between the brain regions during exposures to natural scenes, which may reflect enhanced information processing in the brain. There are no studies that specifically focused on mapping brain activity during exposure to blue spaces and future research is warranted.

EEG has also been used extensively to monitor brain activity associated with the natural environment. In one of the first mobile EEG studies in a natural environment, Aspinall et al. (2015) compared brain activity in different urban settings including green parks, urban shopping areas and crowded urban spaces. They reported that moving from an urban area to a green area leads to decreases in EEG activity associated with frustration, excitement and engagement (i.e. lower beta activity). In another study, Banaei et al. (2015) measured EEG activity while walking along two different types of trails: park and commercial spaces. Consistent with Aspinall et al. (2015), they also found EEG activity was associated with lowered arousal while walking in the park. Notably, the state of lowered arousal faded after 3 minutes of walking in the park, suggesting that a brief walk in nature may be sufficient to get the beneficial effect. More recent studies have utilised computerised tests and virtual reality technologies (➔ chapter "Extended Reality in Environmental Neuroscience Research" by Li et al.) to measure EEG activity (Chiang et al., 2017; Wang et al., 2020a, b). While these studies do not provide the same ambience of a natural environment, they can evoke similar cognitive and emotional states. A major advantage of this approach is the ability to run experiments in controlled settings and test specific hypotheses around the type of natural stimuli used in such settings. For instance, controlled exposure to digital images of the natural environment was associated with an increase in EEG alpha frequency, particularly when the images comprised dense trees and vegetation. Other studies have also confirmed the findings that exposure to greenspace scenes enhances the state of relaxed wakefulness in the brain (Lin et al., 2020; Olszewska-Guizzo et al., 2020), suggesting that lifelong exposure to greenspace may be associated with long-term reduction in stress, which can have a positive impact on overall well-being and, thus, cognitive health.

In line with the positive role of the natural environment on brain health, exposure to blue spaces has also been associated with changes in EEG activity. For example, exposure to the photographs of blue spaces elicited higher alpha activity than built environment photographs, which is consistent with the brain activity pattern associated with green spaces (Ulrich, 1981). Interestingly, the alpha activity associated with blue spaces was lower than that associated with green vegetation (Ulrich, 1981). This was suggested to be likely due to greater attentional processing while viewing blue spaces, which is supported by recent studies (Deng et al., 2020). Another recent study compared brain activity in different frequency bands (alpha and beta) during exposure to different types of water bodies (falling, flowing and static) compared to streetscapes (Luo et al., 2023). The study showed that falling water bodies were associated with the highest level of alpha brainwaves and the lowest level of beta brainwaves compared to other types of water bodies. They also showed that a 3-minute contact with blue spaces was sufficient to promote positive brain activity. Taken together, these studies suggest a positive impact of exposure to blue spaces on brain activity. However, these studies used photographs/digital images of blue spaces. Studies using mobile EEG systems in real-world blue-space settings (e.g. rivers, oceans, and waterfalls) may provide better and practically relevant insights into their effects on brain activity.

Structural and Functional Brain Changes Associated with Exposure to Built Environment

The built environment factors that affect the human brain can roughly be divided into direct and indirect factors. The direct factors include the toxic effects of air pollutants and urban noise on the brain (Cho et al., 2020; Dimitrov-Discher et al., 2022; Hedges et al., 2019; Lubczynska et al., 2021; Pini et al., 2023). The indirect factors can arise due to physical activity, transportation and social interactions facilitated by different features of the built environment (Cerin et al., 2017a, b). An understanding of the changes in brain structure and brain activity associated with exposure to these elements can provide insights into how the brain evolves in a built environmental setting.

Lifelong exposure to air pollution has been linked to neurotoxic effects on the human brain (Cory-Slechta et al., 2023). Neuroimaging studies have identified different brain structures which show degenerative changes in adults living in urban environments with high levels of air pollutants. However, the findings are highly heterogeneous regarding the types of brain changes observed. For example, exposure to high levels of $PM_{2.5}$ for 7 years prior to MRI scans was associated with a significant decline in white matter, but not in gray matter, among older adults (Chen et al., 2015). Notably, the association was independent of geographical region, demographic and clinical characteristics (cardiometabolic health indicators), socioeconomic status and lifestyle factors (Chen et al., 2015). In contrast, another UK-based study showed that adults living in higher exposure to $PM_{2.5}$ had smaller total brain volumes and a greater risk of brain infarcts, but no changes in white matter volume (Wilker et al., 2015). Another study from the United States, which specifically investigated brain atrophy and the presence of microbleeds, reported a protective effect of higher levels of $PM_{2.5}$ on white matter volume (Wilker et al., 2016). In another multi-site study in the United States (Power et al., 2018), living in an environment with higher mean $PM_{2.5}$ and PM_{10} exposures was associated with smaller deep-gray regional brain volumes and smaller frontal lobe volumes. However, there was no association between air pollutants and the number of microbleeds or changes in white matter volume.

More recent large-scale studies from the UK Biobank data have also confirmed these earlier findings (Gale et al., 2020, Furlong et al., 2022), showing that both particulate matter ($PM_{2.5}$, PM_{10}) and nitrogen oxides concentrations were associated with decreased volume of the prefrontal cortex. There is only one fMRI study, which has investigated the changes in brain activity associated with air pollution, which showed that a brief exposure to diluted diesel was associated with weakened resting-state functional connectivity in the default-mode network (Gawryluk et al., 2023). Decreased resting-state connectivity may be a result of reduced information flow, which over time (and from chronic exposure), may lead to lasting neural activity changes with maladaptive consequences. Taken together, these findings provide strong evidence for the neurotoxic effects of air pollutants, manifesting as degenerative changes in gray matter. Such degenerative changes can manifest as cognitive deficits and an increased risk of neurodegenerative diseases like Alzheimer's and Parkinson's diseases. As urbanisation and industrialisation continue to increase air pollution levels globally, understanding and mitigating the neurotoxicity of these pollutants is crucial for safeguarding public health and promoting environmental sustainability.

Other aspects of the built environment (e.g. open public spaces, neighbourhood walkability, street network) can also impact brain structure (Cerin et al., 2017a, b; Crous-Bou et al., 2020; Kim et al., 2010; Mavros et al., 2016; Xu et al., 2023). For example, living in cities that are more walkable appears to have positive impacts on the brain, stimulating the neural pathways that improve cognitive function. A higher level of neighbourhood walkability—defined as a composite measure of dwelling density, street connectivity and land use mix—has been shown to be associated with increased grey matter volume, increased right hippocampal volume and decreased ventricular volume in older adults (Cerin et al., 2017b). The hippocampus plays a pivotal role in memory and cognitive function, as it converts short-term memories into long-term ones by facilitating information transfer from the hippocampal region to the neocortex (Lisman et al., 2018). Additionally, this structure is intricately linked to spatial navigation and cognitive processes like spatial memory, which is essential for our ability to navigate and remember locations in our environment (Lisman et al., 2018). Dysfunction of the hippocampus can lead to profound memory deficits and impairments in various cognitive functions, highlighting its role in our capacity to learn, remember, and make sense of the world around us. Further, Besteher et al. (2017) assessed the association of urban upbringing for the first 15 years of life with cortical brain regions and found that the urbanicity index was associated with reduced cortical thickness in the left dorsolateral prefrontal cortex, bilateral medial prefrontal cortices as well as temporal cortices including the left superior temporal and left parahippocampal cortex. Taken together, these findings suggest that living in highly urban environment is associated with changes in brain structures important for cognitive and emotional control, which may explain increasing rate of mental health issues in city dwellers.

Similar to studies on the effects of the natural environment, mobile EEG studies have also provided novel insights into brain activity during exposure to the built environment (Aspinall et al., 2015; Lin et al., 2020; Mavros et al., 2016; Olszewska-Guizzo et al., 2020). Acute exposures to built environment scenes (e.g. buildings, street networks) have been associated with increased EEG activity in the beta frequency band, known to be a hallmark of increased arousal and attention (Buschman & Miller, 2007). Individuals who walk in an urban environment are more likely to exhibit increased EEG beta activity compared to those walking in green spaces

(Neale et al., 2020). Higher beta activity can occur in response to increased mental activity or due to increased stress or anxiety (Hayashi et al., 2009). While some level of increased beta (associated with mental activity) is necessary for mental acuity, prolonged periods of elevated beta due to chronic stress or anxiety can have detrimental effects on the mental health of individuals living in urban dwellings.

Cardiometabolic Health

Cardiometabolic health is defined by a set of risk factors for diseases of the cardiovascular system, including the heart, blood and blood vessels (Guo et al., 2014). These include indicators of overweight/obesity, elevated blood glucose, elevated blood pressure, dyslipidaemia and historical cardiovascular events (e.g. stroke or heart attack). This cluster of factors has been associated with a higher risk of dementia and cognitive decline (Livingston et al., 2020; Zhang et al., 2022c) possibly through their influence on brain health (here defined in terms of hippocampal atrophy and reduced brain volume) (Sabia et al., 2019). Obesity and persistent hypertension as risk factors for dementia seem to be particularly important in midlife (Livingston et al., 2020), while studies have reported late-life overweight to be protective of dementia (Zhang et al., 2022c). The latter association has been in part attributed to the potential beneficial effects of leptin secreted by the adipose tissue on hippocampal synaptic function (McGregor & Harvey, 2018) and to lower β-amyloid depositions in the brain among overweight/obese older adults as compared to normal weight counterparts (Sun et al., 2020). Elevated blood glucose (e.g. as in type 2 diabetes) in mid-to-late life has been linked to worse cognitive outcomes even when intensively treated (Livingston et al., 2020). With regard to dyslipidaemia, blood lipid-lowering treatment has been associated with a lower risk of Alzheimer's disease in mid-to-late life (Zhang et al., 2022c) and midlife dyslipidaemia—characterised by high LDL cholesterol and triglycerides levels and/or low levels of HDL cholesterol—was found to be associated with an increased risk of cognitive impairment in later life (Wee et al., 2023).

As noted earlier, being physically active, following a healthy diet, sleeping well and being a non-smoker are behaviours beneficial to cardiometabolic health (Akbaraly et al., 2019; Cheng et al., 2023; Cunningham et al., 2020; Wang et al., 2022). It follows that all neighbourhood attributes that affect such behaviours (see section "Lifestyle Behaviours") could also potentially impact cardiometabolic health. In fact, neighbourhoods with good access to parks and greenery tend to be associated with a lower likelihood of overweight/obesity (Rahimi-Ardabili et al., 2021), hypertension (Fan et al., 2022), dyslipidaemia and hyperglycaemia (Rahimi-Ardabili et al., 2021). In contrast, the overall evidence on the effects of blue space on cardiometabolic health is rather scarce and mixed (Geneshka et al., 2021), although a recent study found that mid-aged and older adults living close to water bodies were leaner and had healthier blood lipid and glucose profiles (Cerin et al., 2022). The evidence of dense, well interconnected and destination-rich neighbourhoods having beneficial effects on weight status, blood pressure and blood glucose is relatively consistent (Chandrabose et al., 2019), as is that of air pollution (Gaio et al., 2019; Liu et al., 2019; Wang et al., 2023) and urban noise (D'Souza et al., 2021) being detrimental to cardiometabolic health.

To our knowledge, only one study examined cardiometabolic health indicators as mediators of the associations between aspects of the neighbourhood physical environment and cognitive health in ageing populations. Using data from an Australian national sample of 4141 mid-aged and older adults, Cerin and colleagues reported detrimental indirect effects of population density, land use mix, parkland and commercial land through annual average $PM_{2.5}$ concentrations on memory which were mediated by waist circumference and HDL cholesterol (Cerin et al., 2022). Beneficial indirect effects of population density and street connectivity on memory were mediated by HDL cholesterol, and those of blue space on memory were channelled through waist circumference and HDL cholesterol. Similar indirect effects were observed for processing speed, with the only difference that glycated haemoglobin was also in the pathway between most of the measured built and natural environment attributes and the measure of cognitive function. Apart from unveiling the mediating role of cardiometabolic health indicators in the associations between the neighbourhood physical environment and older adults' cognitive health, this study revealed the existence of other potential, unmeasured mechanisms of neighbourhood influences on cognitive function.

Mental Health

Compromised mental health is a risk factor for developing neurodegenerative diseases, which include but are not limited to dementia (Bougea et al., 2020; Cherbuin et al., 2015), vascular dementia (Becker et al., 2018; Zhao et al., 2022), mild cognitive impairment (Gallacher et al., 2009) and Alzheimer's disease (Lara et al., 2019; Pietrzak et al., 2015; Wilson et al., 2011) as well as for a cognitive decline in older people (Abramovitch et al., 2021; James et al., 2023). Specifically, increased symptoms of stress (Bougea et al., 2020; Stuart & Padgett, 2020), anxiety (Santabárbara et al., 2020), depression (Cherbuin et al., 2015), loneliness (Lara et al., 2019; Ren et al., 2023; Salinas et al., 2022), apathy (Custo et al., 2022; Fitts et al., 2015), nervousness and exhaustion (Sulkava et al., 2022), and diminished life satisfaction (Mank et al., 2022) contribute to the risk of neurodegeneration. For example, mild depressive symptoms were associated with a 60–70% increased risk of dementia and Alzheimer's disease in older adults (Cherbuin et al., 2015), while higher levels of anxiety were associated with a 45% increase in the risk of Alzheimer disease (Santabárbara et al., 2020). In a 10-year longitudinal study, lonely adults <80 years of age (*M*age = 73 years) without APOE e4 alleles had a three-fold greater risk of developing dementia (Salinas et al., 2022) when compared to not lonely adults. In another study, feeling lonely was associated with a ~60% increased risk of all-cause dementia, after accounting for clinical (e.g. depression) and behavioural (e.g. physical activity, social isolation) risk factors (Sutin et al., 2023).

Furthermore, evidence has consistently shown stronger inverse mental health–cognitive function associations for executive function versus non-executive function-related outcomes in older adults (Abramovitch et al., 2021; John et al., 2019, 2021). In particular, higher depressive and anxiety symptoms were associated with worsened episodic and verbal memory, learning (Gallacher et al., 2009; Gulpers et al., 2019; John et al., 2021; Lindert et al., 2021; Zahodne et al., 2014), processing speed (John et al., 2021) and global cognition (Ahn et al., 2021). Interestingly, an 8-year study using cross-lagged panel network analysis to examine component-to-component mental health-cognition associations in older adults found that depression and anxiety predicted worsened processing speed, episodic and working memory, whereas anxiety was uniquely predictive of poorer attention (Zainal & Newman, 2022), suggesting that mental health symptoms may have a distinct and unique association with cognitive function in older adults. Moreover, older adults who experienced stress and loneliness similarly exhibited worsened cognitive function, such as in explicit memory processes (Boss et al., 2015; Korten et al., 2017; Sandi, 2013) and an accelerated decline in global cognition (Aggarwal et al., 2014). Importantly, and regardless of the profile of mental health symptoms, older people who are subjected to chronic psychological distress are at increased risk of biological 'wear and tear' (McEwen, 1998) that induces neuroinflammatory processes. Neuroinflammatory processes, in turn, trigger an ongoing activation of the hypothalamic pituitary adrenal-axis (Black, 2002), resulting in impaired brain structure and function underlying cognitive performance (Lupien et al., 1998; Lupien et al., 2018) (please see section "Brain Mechanisms").

Emerging evidence has shown that aspects of the neighbourhood natural built and socioeconomic environment are related to mental health (Alcock et al., 2014; Gianfredi et al., 2021; Padeiro et al., 2021). In particular, the presence and availability of green spaces in the neighbourhood was related to lower depression (Gonzales-Inca et al., 2022; Lee & Lee, 2019; Peng et al., 2022; Sarkar et al., 2018; Wang et al., 2020a, b), stress (Astell-Burt & Feng, 2019; John et al., 2023; Lee & Lee, 2019), loneliness (Astell-Burt et al., 2022a, b) as well as enhanced mood (McMahan & Estes, 2015), happiness (Capaldi et al., 2014), healthy ageing score (John et al., 2023), social health (Huang & Lin, 2023) and quality of life (Levasseur et al., 2015; Zhang et al., 2019). On the other hand, older adults were more likely to experience depressive and stress symptoms with increasing distance from their residential home to the nearest park (Reklaitiene et al., 2014; Stigsdotter et al., 2010). Likewise, the feeling of anxiety was 19.3% lower after an urban park walk than after a city area walk (Song et al., 2015). Of note, there is a lack of high-quality epidemiological evidence to support blue space-mental health associations in ageing cohorts (Gascon et al., 2017)—though few studies showed that exposure to coastal blue space (Pasanen et al., 2019; White et al., 2021) and higher freshwater availability (Chen & Yuan, 2020) were associated with better mental health, including reduced depressive symptoms (Dempsey et al., 2018; McDougall et al., 2021),

There is a dearth of knowledge about the ways in which built environments influence mental health in older adults (Barnett et al., 2018) and a large part of the evidence is focused on neighbourhood-depressive symptoms associations, with limited coverage of other mental health symptoms that older adults experience. In a systematic review, Padeiro et al. (2021) identified adequate street layout (e.g. benches), the availability of local services and facilities as well as good public transit to have the strongest associations with better well-being among older adults. Interestingly, Xu et al. (2023) used sparse canonical correlation analyses to link independent urban living environment profiles with mental health symptoms and found that an urban environment profile characterised by high degrees of area-level deprivation, air pollution and traffic, short distances to infrastructural facilities and lack of green space correlated with the frequency of tiredness, unenthusiasm, depressed mood, feelings of being fed-up and loneliness. In contrast, greater density of leisure places, adequate street layout, coast proximity, variation of green spaces and density of mixed urban infrastructure correlated with anxiety symptoms and worry. This suggests that certain attributes of the built environment may be linked to distinct profiles of mental health symptoms in older adults, which warrants future research.

With respect to population density, most studies reported a negative urbanicity-mental health association—residents living in highly urbanised neighbourhoods are more likely to be depressed and lonely (Saarloos et al., 2011; Scharf & de Jong Gierveld, 2008; Sui et al., 2022; Xu et al., 2023) and with mood and anxiety disorders (Motoc et al., 2023; Peen et al., 2010; Walters et al., 2004). Another aspect of the built environment that has been consistently related to mental health is that of access to destinations and services (Besser et al., 2017; Lu et al., 2021; Motoc et al., 2023; Saarloos et al., 2011; Yue et al., 2022); however, effects differ depending on the type of destination and/or facility under examination. For example, residents living in neighbourhoods characterised by good access to retail facilities exhibited a higher risk of incident anxiety and depression (Motoc et al., 2023; Saarloos et al., 2011), whereas at a 4-year follow-up, greater availability of community centres within 500 m from home reduced depressive symptoms in older adults (Lu et al., 2021). On the other hand, access to leisure (e.g. restaurants) and daily living services (e.g. supermarkets, banks) and exercise facilities, underground parking and the presence of elevators within a neighbourhood were positively associated with older adults' mental well-being (Finlay et al. 2020; Yue et al., 2022).

City dwellers are exposed to motorised traffic and the associated noise on a daily basis which, according to Herrmann-Lunecke et al. (2021), induce stress, fear and anger and may prevent older adults from walking (Bornioli et al., 2019). On the other hand, living further from major roads was significantly associated with decreases in depressive and anxiety symptoms among older adults and this was partially mediated by loneliness and air pollution (Pun et al., 2019). This suggests that older adults are likely to walk in their neighbourhood when concentrations of air pollutants are low (Cerin et al., 2023). Access to public transport was found to be associated with reduced loneliness in older adults (*M*age = 81 years) (Hagan, 2020), whereas narrow and deteriorated sidewalks and difficult crossings hindered neighbourhood walking and induced psychological distress (Herrmann-Lunecke et al., 2021).

Lastly, with respect to street connectivity, studies showed both positive (Zhang et al., 2018) and negative associations (Chen et al., 2022b; Guo et al., 2021; Zhang et al., 2022b; Yue et al., 2022), with the authors suggesting that the structure of streets may have a more limited effect on older adults' mental health as compared to other neighbourhood attributes (Barnett et al., 2018); although this needs to be tested in future studies. Moreover, there is a need to examine the curvilinear nature of the urban environment-mental health associations (e.g. street intersection connectivity, Guo et al., 2021; green spaces, Pope et al., 2018, Zagnoli et al., 2022), as in the case of neighbourhood-cognitive health associations (Cerin et al., 2021; Worn et al., 2017).

Despite the evidence showing mental health–cognition and neighbourhood–mental health associations, there is a dearth of studies on the mediating effects of mental health symptoms in the environment-cognitive health associations in older adults. However, such findings need to be interpreted with caution, as people who experience mental health symptoms may be more likely to report that their neighbourhood has problems (Echeverría et al., 2008)—individuals with mild symptoms of depression and anxiety are prone to negative thinking (Chevalier et al., 2023) and, thus, adopting objective measures to assess neighbourhood-mental health-cognition associations is critical in this field of research. In a sample of women (M_{age} = 61.2 years), the green space–cognition association was mediated by *depression* where depression explained 3.95% of the association between green space and psychomotor speed and attention and 6.30% of the association between green space and overall cognition (Jimenez et al., 2022). Further, *stress* mediated the relationship between neighbourhood blue spaces and self-reported health of older people in China (Chen & Yuan, 2020). Taken together, these findings point to the relevance of natural spaces in maintaining and improving mental health states and, thus, reducing the risk of cognitive decline and developing a neurodegenerative disease. Future research is needed to identify the aspects of mental health that mediate the associations between the natural and built neighbourhood environments and cognitive health (including cognitive function and the risk of neurodegeneration) to serve as potential targets for urban planning policies and public health interventions. Further, it is important to examine the mediating effects of commonly reported mental health symptoms in older adults, such as depression, anxiety, stress and loneliness, in relation to distinct attributes of urban environments.

Limitations and Future Research

Although we outlined limitations of neighbourhood-cognitive health studies in ageing populations in sections "An Ecological Model of the Effects of Urban Neighbourhood Physical Environments on Cognitive Health in Ageing Populations", "Urban Environments and Cognitive Health: Epidemiological Evidence", and "Mechanisms Linking Urban Environments with Cognitive Health", there are three major areas of concern we would want to highlight in this section. First, most studies adopted a simplistic theoretical framework focused on a limited number of aspects of the urban environment rather than examining key features of urban design and its by-products (e.g. urban noise and air pollution) and explicitly acknowledging that these affect cognitive health through multiple behavioural (lifestyle behaviours) and biological mechanisms (cardiometabolic, brain and mental health). Given the multilevel and interactionist nature of neighbourhood-cognition associations, we recommend future studies to consider all key individual-level (e.g. age, APOE ε4) and neighbourhood-level factors (e.g. population density, NO_2). Second, there is a dearth of evidence on mechanisms linking urban environments with older adults' cognitive health as proposed in the ecological model and, thus, future studies should investigate the mediating effects of lifestyle behaviours (e.g. smoking, diet, physical activity, high-quality sleep), cardiometabolic (weight status, blood pressure, blood glucose) and brain health (altered structure and function of the brain regions implicated in mental and cognitive function) indicators as well as of gut microbiota and mental health (loneliness, depression, stress, anxiety). Another issue across all studies is the variability of research strategy and design, sample characteristics, neighbourhood definitions, criteria to assess and classify neurodegenerative disorders, measurement tools to assess cognition as well as varied operationalisation of urban design features, making it difficult to compare and synthesise findings. Therefore, high-quality empirical studies that adopt rigorous and standardised methodologies are needed. Other suggestions for future studies are the inclusion of diverse neighbourhood features (e.g. landscape types) and their quality, settings (e.g. different cultures and geographical locations), racial/ethnic groups and domain-specific cognitive- (e.g. inhibitory processes) and social-cognitive (e.g. empathy, mentalising) functions to gain a more nuanced understanding of the effects of the urban environments on cognitive health in city dwellers. Lastly, more research is needed into the associations of population density, street connectivity and blue spaces with cognitive health in older adults.

Practical Implications and Conclusions

Sustainable development goals (SDG) call for creation of sustainable, inclusive, safe and resilient cities (SDG 11), as well as for promotion of healthy lives and well-being across all ages (SDG 3) (United Nations, 2019), with a particular focus on the most vulnerable populations—older people. From the evidence provided in this chapter, and in line with the United Nations goals, neighbourhood physical environments have the potential to promote cognitively healthy ageing. In particular, we found that urban design can both positively and negatively affect older people's cognitive health; however, empirical evidence is limited, lacks rigorous and standardised research methods and there is a need for further research.

The most consistent finding across studies is that neighbourhoods with access to natural environments have been associated with a reduced risk of most neurodegenerative disorders and with maintained/improved cognitive function in older adults, suggesting that green spaces in cities may represent a promising target for urban planners and policymakers to facilitate healthy ageing. This chapter has discussed the importance of applying ecological models (Cerin, 2019) to examine the associations between aspects of urban design and cognitive health in older adults. However, this theoretical model has not been explicitly tested in most of the studies and, without this nuanced knowledge, it might be problematic to inform large-scale, public health policies and community programs aimed at creating age-friendly cities. Importantly, and as demonstrated in this chapter, there is a need to recognise the complexity and diversity of urban design features and their multidimensional interactions with cognitive health in older adults, which might explain some of the mixed findings regarding some of the examined environmental attributes.

For example, at least every second person over the age of 60 years will experience some degree of loneliness and depression (Cherbuin et al., 2015; Landeiro et al., 2017); however, from an urban planning perspective, it might be possible to reduce these negative mental health states, well-known precursors of cognitive impairment, by building and increasing the density of institutional resources (community and senior centres) and recreational sites (parks, walking paths, galleries). Older adults might prioritise destinations and facilities that enable them to socialise regularly and, thus, the presence of such destinations in the neighbourhood has the potential to reduce the risk of neurodegenerative disorders. Similarly, low concentrations of outdoor air pollutants and traffic and pedestrian safety encourage older adults to engage in recreational walking, which is protective against cognitive decline and subsequent risk of a neurodegenerative disorder. Thus, cities must be suited to the needs of older people.

Furthermore, efforts should be made by the research community to better understand the brain and other biological mechanisms that underpin neighbourhood–cognitive health associations in older adults. The reason is that, as can be seen from this chapter, it is possible to detect structural and functional changes early in the ageing process and prior to the manifestation of behavioural or cognitive impairments. Thus, early detection of brain abnormalities can be of relevance, especially from the interventional point of view. If it is known which brain regions or neural networks are affected by certain features of the neighbourhood physical environment, then it is possible to deliver targeted cognitive training interventions (e.g. van Balkom et al., 2020). There are still many unknowns in the field and more high-quality evidence is needed in order to inform the creation of age- and cognition-friendly urban environments.

References

Abner, E. L., Jicha, G. A., Christian, W. J., & Schreurs, B. G. (2016). Rural-urban differences in Alzheimer's disease and related disorders diagnostic prevalence in Kentucky and West Virginia. *The Journal of Rural Health, 32*(3), 314–320. https://doi.org/10.1111/jrh.12155

Abramovitch, A., Short, T., & Schweiger, A. (2021). The C Factor: Cognitive dysfunction as a transdiagnostic dimension in psychopathology. *Clinical Psychology Review, 86*, 102007. https://doi.org/10.1016/j.cpr.2021.102007

Aggarwal, N. T., Wilson, R. S., Beck, T. L., Rajan, K. B., Mendes de Leon, C. F., Evans, D. A., & Everson-Rose, S. A. (2014). Perceived stress and change in cognitive function among adults 65 years and older. *Psychosomatic Medicine, 76*(1), 80–85. https://doi.org/10.1097/psy.0000000000000016

Ahn, S., Mathiason, M. A., & Yu, F. (2021). Longitudinal cognitive profiles by anxiety and depressive symptoms in American older adults with subjective cognitive decline. *Journal of Nursing Scholarship, 53*(6), 698–708. https://doi.org/10.1111/jnu.12692

Ailshire, J., Karraker, A., & Clarke, P. (2017). Neighborhood social stressors, fine particulate matter air pollution, and cognitive function among older U.S. adults. *Social Science & Medicine, 172*, 56–63. https://doi.org/10.1016/j.socscimed.2016.11.019

Aitken, W. W., Lombard, J., Wang, K., Toro, M., Byrne, M., Nardi, M. I., Kardys, J., Parrish, A., Dong, C., Szapocznik, J., Rundek, T., & Brown, S. C. (2021). Relationship of neighborhood greenness to Alzheimer's disease and non-Alzheimer's dementia among 249,405 U.S. *Medicare Beneficiaries. Journal of Alzheimer's Disease: JAD, 81*(2), 597–606. https://doi.org/10.3233/jad-201179

Akbaraly, T. N., Singh-Manoux, A., Dugravot, A., Brunner, E. J., Kivimäki, M., & Sabia, S. (2019). Association of midlife diet with subsequent risk for dementia. *JAMA, 321*(10), 957–968. https://doi.org/10.1001/jama.2019.1432

Alcock, I., White, M. P., Wheeler, B. W., Fleming, L. E., & Depledge, M. H. (2014). Longitudinal effects on mental health of moving to greener and less green urban areas. *Environmental Science & Technology, 48*(2), 1247–1255. https://doi.org/10.1021/es403688w

Alidoust, S., & Bosman, C. (2015). Planning for an ageing population: Links between social health, neighbourhood environment and the elderly. *Australian Planner, 52*(3), 177–186. https://doi.org/10.1080/07293682.2015.1034145

Amiri, S., & Behnezhad, S. (2020). Smoking and risk of sleep-related issues: A systematic review and meta-analysis of prospective studies. *Canadian Journal of Public Health, 111*(5), 775–786. https://doi.org/10.17269/s41997-020-00308-3

An, R., Shen, J., Ying, B., Tainio, M., Andersen, Z. J., & de Nazelle, A. (2019). Impact of ambient air pollution on physical activity and sedentary behavior in China: A systematic review. *Environmental Research, 176*, 108545. https://doi.org/10.1016/j.envres.2019.108545

Andrade, A., D'Oliveira, A., De Souza, L. C., Bastos, A., Dominski, F. H., Stabile, L., & Buonanno, G. (2023). Effects of air pollution on the health of older adults during physical activities: Mapping review. *International Journal of Environmental Research and Public Health, 20*(4). https://doi.org/10.3390/ijerph20043506

Aneshensel, C. S., Ko, M. J., Chodosh, J., & Wight, R. G. (2011). The urban neighborhood and cognitive functioning in late middle age. *Journal of Health and Social Behavior, 52*(2), 163–179. https://doi.org/10.1177/0022146510393974

Arioli, M., Crespi, C., & Canessa, N. (2018). Social cognition through the lens of cognitive and clinical neuroscience. *BioMed Research International*, 4283427. https://doi.org/10.1155/2018/4283427

Aspinall, P., Mavros, P., Coyne, R., & Roe, J. (2015). The urban brain: Analysing outdoor physical activity with mobile EEG. *British Journal of Sports Medicine, 49*(4), 272–276. https://doi.org/10.1136/bjsports-2012-091877

Astell-Burt, T., & Feng, X. (2019). Association of urban green space with mental health and general health among adults in Australia. *JAMA Network Open, 2*(7), e198209–e198209. https://doi.org/10.1001/jamanetworkopen.2019.8209

Astell-Burt, T., & Feng, X. (2020). Greener neighbourhoods, better memory? A longitudinal study. *Health & Place, 65*, 102393. https://doi.org/10.1016/j.healthplace.2020.102393

Astell-Burt, T., Navakatikyan, M. A., & Feng, X. (2020). Urban green space, tree canopy and 11-year risk of dementia in a cohort of 109,688 Australians. *Environment International, 145*, 106102. https://doi.org/10.1016/j.envint.2020.106102

Astell-Burt, T., Hartig, T., Eckermann, S., Nieuwenhuijsen, M., McMunn, A., Frumkin, H., & Feng, X. (2022a). More green, less lonely? A longitudinal cohort study. *International Journal of Epidemiology, 51*(1), 99–110. https://doi.org/10.1093/ije/dyab089

Astell-Burt, T., Hartig, T., Putra, I. G. N. E., Walsan, R., Dendup, T., & Feng, X. (2022b). Green space and loneliness: A systematic review with theoretical and methodological guidance for future research. *Science of the Total Environment, 847*, 157521. https://doi.org/10.1016/j.scitotenv.2022.157521

Bafna, S. (2003). Space syntax: A brief introduction to its logic and analytical techniques. *Environment and Behavior, 35*(1), 17–29. https://doi.org/10.1177/0013916502238863

Banaei, M., Yazdanfar, A., Nooreddin, M., & Yoonessi, A. (2015). Enhancing urban trails design quality by using electroencephalography device. *Procedia—Social and Behavioral Sciences, 201*, 386–396. https://doi.org/10.1016/j.sbspro.2015.08.191

Bao, Y.-P., Han, Y., Ma, J., Wang, R.-J., Shi, L., Wang, T.-Y., He, J., Yue, J.-L., Shi, J., Tang, X.-D., & Lu, L. (2017). Cooccurrence and bidirectional prediction of sleep disturbances and depression in older adults: Meta-analysis and systematic review. *Neuroscience & Biobehavioral Reviews, 75*, 257–273. https://doi.org/10.1016/j.neubiorev.2017.01.032

Barnes, L. L., Wilson, R. S., Bienias, J. L., Mendes de Leon, C. F., Kim, H.-J. N., Buchman, A. S., & Bennett, D. A. (2007). Correlates of life space in a volunteer cohort of older adults. *Experimental Aging Research, 33*(1), 77–93. https://doi.org/10.1080/03610730601006420

Barnett, D. W., Barnett, A., Nathan, A., Van Cauwenberg, J., Cerin, E., & on behalf of the Council on, E., & Physical Activity – Older Adults working, G. (2017). Built environmental correlates of older adults' total physical activity and walking: A systematic review and meta-analysis. *International Journal of Behavioral Nutrition and Physical Activity, 14*(1), 103. https://doi.org/10.1186/s12966-017-0558-z

Barnett, A., Zhang, C. J. P., Johnston, J. M., & Cerin, E. (2018). Relationships between the neighborhood environment and depression in older adults: A systematic review and meta-analysis. *International Psychogeriatrics, 30*(8), 1153–1176. https://doi.org/10.1017/S104161021700271X

Beadle, J. N., & de la Vega, C. E. (2019). Impact of aging on empathy: Review of psychological and neural mechanisms [Review]. *Frontiers in Psychiatry, 10*. https://www.frontiersin.org/articles/10.3389/fpsyt.2019.00331

Becker, E., Orellana Rios, C. L., Lahmann, C., Rücker, G., Bauer, J., & Boeker, M. (2018). Anxiety as a risk factor of Alzheimer's disease and vascular dementia. *The British Journal of Psychiatry, 213*(5), 654–660. https://doi.org/10.1192/bjp.2018.173

Besser, L. (2021). Outdoor green space exposure and brain health measures related to Alzheimer's disease: A rapid review. *BMJ Open, 11*(5), e043456. https://doi.org/10.1136/bmjopen-2020-043456

Besser, L. M., McDonald, N. C., Song, Y., Kukull, W. A., & Rodriguez, D. A. (2017). Neighborhood environment and cognition in older adults: A systematic review. *American Journal of Preventive Medicine, 53*(2), 241–251. https://doi.org/10.1016/j.amepre.2017.02.013

Besser, L. M., Rodriguez, D. A., McDonald, N., Kukull, W. A., Fitzpatrick, A. L., Rapp, S. R., & Seeman, T. (2018). Neighborhood built environment and cognition in non-demented older adults: The Multi-Ethnic Study of Atherosclerosis. *Social Science & Medicine, 200*, 27–35. https://doi.org/10.1016/j.socscimed.2018.01.007

Besser, L., Galvin, J. E., Rodriguez, D., Seeman, T., Kukull, W., Rapp, S. R., & Smith, J. (2019). Associations between neighborhood built environment and cognition vary by apolipoprotein E genotype: Multi-Ethnic Study of Atherosclerosis. *Health & Place, 60*, 102188. https://doi.org/10.1016/j.healthplace.2019.102188

Besser, L. M., Hirsch, J., Galvin, J. E., Renne, J., Park, J., Evenson, K. R., Kaufman, J. D., & Fitzpatrick, A. L. (2020). Associations between neighborhood park space and cognition in older adults vary by US location: The Multi-Ethnic Study of Atherosclerosis. *Health & Place, 66*, 102459. https://doi.org/10.1016/j.healthplace.2020.102459

Besser, L. M., Chang, L.-C., Hirsch, J. A., Rodriguez, D. A., Renne, J., Rapp, S. R., Fitzpatrick, A. L., Heckbert, S. R., Kaufman, J. D., & Hughes, T. M. (2021a). Longitudinal associations between the neighborhood built environment and cognition in US older adults: The multi-ethnic study of atherosclerosis. *International Journal of Environmental Research and Public Health, 18*(15).

Besser, L. M., Chang, L. C., Evenson, K. R., Hirsch, J. A., Michael, Y. L., Galvin, J. E., Rapp, S. R., Fitzpatrick, A. L., Heckbert, S. R., Kaufman, J. D., & Hughes, T. M. (2021b). Associations between neighborhood park access and longitudinal change in cognition in older adults: The multi-ethnic study of atherosclerosis. *Journal of Alzheimer's disease: JAD, 82*(1), 221–233. https://doi.org/10.3233/jad-210370

Besser, L. M., Lovasi, G. S., Michael, Y. L., Garg, P., Hirsch, J. A., Siscovick, D., Hurvitz, P., Biggs, M. L., Galvin, J. E., Bartz, T. M., & Longstreth, W. T. (2021c). Associations between neighborhood greenspace and brain imaging measures in non-demented older adults: The Cardiovascular Health Study. *Social Psychiatry and Psychiatric Epidemiology, 56*(9), 1575–1585. https://doi.org/10.1007/s00127-020-02000-w

Besser, L. M., Jimenez, M. P., Reimer, C. J., Meyer, O. L., Mitsova, D., George, K. M., Adkins-Jackson, P. B., & Galvin, J. E. (2023). Diversity of studies on neighborhood greenspace and brain health by racialized/ethnic group and geographic region: A rapid review. *International Journal of Environmental Research and Public Health, 20*(9). https://doi.org/10.3390/ijerph20095666

Besteher, B., Gaser, C., Spalthoff, R., & Nenadic, I. (2017). Associations between urban upbringing and cortical thickness and gyrification. *Journal of Psychiatric Research, 95*, 114–120. https://doi.org/10.1016/j.jpsychires.2017.08.012

Black, P. H. (2002). Stress and the inflammatory response: A review of neurogenic inflammation. *Brain, Behavior, and Immunity, 16*(6), 622–653. https://doi.org/10.1016/S0889-1591(02)00021-1

Bornioli, A., Parkhurst, G., & Morgan, P. L. (2019). Affective experiences of built environments and the promotion of urban walking. *Transportation Research Part A: Policy and Practice, 123*, 200–215. https://doi.org/10.1016/j.tra.2018.12.006

Boss, L., Kang, D.-H., & Branson, S. (2015). Loneliness and cognitive function in the older adult: A systematic review. *International Psychogeriatrics, 27*(4), 541–553. https://doi.org/10.1017/S1041610214002749

Bougea, A., Anagnostouli, M., Angelopoulou, E., Spanou, I., & Chrousos, G. (2020). Psychosocial and trauma-related stress and risk of dementia: A meta-analytic systematic review of longitudinal studies. *Journal of Geriatric Psychiatry and Neurology, 35*(1), 24–37. https://doi.org/10.1177/0891988720973759

Britton, A., Fat, L. N., & Neligan, A. (2020). The association between alcohol consumption and sleep disorders among older people in the general population. *Scientific Reports, 10*(1), 5275. https://doi.org/10.1038/s41598-020-62227-0

Brooks, M. M., Siegel, S. D., & Curriero, F. C. (2021). Characterizing the spatial relationship between smoking status and tobacco retail exposure: Implications for policy development and evaluation. *Health & Place, 68*, 102530. https://doi.org/10.1016/j.healthplace.2021.102530

Brown, B. M., Peiffer, J. J., & Martins, R. N. (2013a). Multiple effects of physical activity on molecular and cognitive signs of brain aging: Can exercise slow neurodegeneration and delay Alzheimer's disease? *Molecular Psychiatry, 18*(8), 864–874. https://doi.org/10.1038/mp.2012.162

Brown, B. M., Peiffer, J. J., Taddei, K., Lui, J. K., Laws, S. M., Gupta, V. B., Taddei, T., Ward, V. K., Rodrigues, M. A., Burnham, S., Rainey-Smith, S. R., Villemagne, V. L., Bush, A., Ellis, K. A., Masters, C. L., Ames, D., Macaulay, S. L., Szoeke, C., Rowe, C. C., et al. (2013b). Physical activity and amyloid-β plasma and brain levels: Results from the Australian imaging, biomarkers and lifestyle study of ageing. *Molecular Psychiatry, 18*(8), 875–881. https://doi.org/10.1038/mp.2012.107

Buschman, T. J., & Miller, E. K. (2007). Top-down versus bottom-up control of attention in the prefrontal and posterior parietal cortices. *Science, 315*(5820), 1860–1862. https://doi.org/10.1126/science.1138071

Cacciottolo, M., Wang, X., Driscoll, I., Woodward, N., Saffari, A., Reyes, J., Serre, M. L., Vizuete, W., Sioutas, C., Morgan, T. E., Gatz, M., Chui, H. C., Shumaker, S. A., Resnick, S. M., Espeland, M. A., Finch, C. E., & Chen, J. C. (2017). Particulate air pollutants, APOE alleles and their contributions to cognitive impairment in older women and to amyloidogenesis in experimental models. *Translational Psychiatry, 7*(1), e1022–e1022. https://doi.org/10.1038/tp.2016.280

Caffò, A. O., Lopez, A., Spano, G., Stasolla, F., Serino, S., Cipresso, P., Riva, G., & Bosco, A. (2020). The differential effect of normal and pathological aging on egocentric and allocentric spatial memory in navigational and reaching space. *Neurological Sciences, 41*(7), 1741–1749. https://doi.org/10.1007/s10072-020-04261-4

Capaldi, C. A., Dopko, R. L., & Zelenski, J. M. (2014). The relationship between nature connectedness and happiness: A meta-analysis [Original Research]. *Frontiers in Psychology, 5*. https://www.frontiersin.org/articles/10.3389/fpsyg.2014.00976

Carey, I. M., Anderson, H. R., Atkinson, R. W., Beevers, S. D., Cook, D. G., Strachan, D. P., Dajnak, D., Gulliver, J., & Kelly, F. J. (2018). Are noise and air pollution related to the incidence of dementia? A cohort study in London, England. *BMJ Open, 8*(9), e022404. https://doi.org/10.1136/bmjopen-2018-022404

Cassarino, M., O'Sullivan, V., Kenny, R. A., & Setti, A. (2018). Disabilities moderate the association between neighbourhood urbanity and cognitive health: Results from the Irish longitudinal study on ageing. *Disability and Health Journal, 11*(3), 359–366. https://doi.org/10.1016/j.dhjo.2017.12.002

Cassarino, M., Bantry-White, E., & Setti, A. (2020). Neighbourhood environment and cognitive vulnerability—A survey investigation of variations across the lifespan and urbanity levels. *Sustainability, 12*(19). https://doi.org/10.3390/su12197951

Cataldi, S., Poli, L., Şahin, F. N., Patti, A., Santacroce, L., Bianco, A., Greco, G., Ghinassi, B., Di Baldassarre, A., & Fischetti, F. (2022). The effects of physical activity on the gut microbiota and the gut–brain Axis in preclinical and human models: A narrative review. *Nutrients, 14*(16). https://doi.org/10.3390/nu14163293

Cerin, E. (2019). Building the evidence for an ecological model of cognitive health. *Health & Place, 60*, 102206. https://doi.org/10.1016/j.healthplace.2019.102206

Cerin, E., Nathan, A., van Cauwenberg, J., Barnett, D. W., Barnett, A., on behalf of the Council on, E, & Physical Activity – Older Adults working, G. (2017a). The neighbourhood physical environment and active travel in older adults: A systematic review and meta-analysis. *International Journal of Behavioral Nutrition and Physical Activity, 14*(1), 15. https://doi.org/10.1186/s12966-017-0471-5

Cerin, E., Rainey-Smith, S. R., Ames, D., Lautenschlager, N. T., Macaulay, S. L., Fowler, C., Robertson, J. S., Rowe, C. C., Maruff, P., Martins, R. N., Masters, C. L., & Ellis, K. A. (2017b). Associations of neighborhood environment with brain imaging outcomes in the Australian Imaging, Biomarkers and Lifestyle cohort. *Alzheimer's & Dementia Journal, 13*(4), 388–398. https://doi.org/10.1016/j.jalz.2016.06.2364

Cerin, E., Nathan, A., Van Cauwenberg, J., & Barnett, A. (2019). Neighbourhood built environment and older adults' physical activity. In A.P Lane (Ed.), *Urban environments and healthy ageing: A global perspective* (pp. 69–88). Routledge.

Cerin, E., Barnett, A., Chaix, B., Nieuwenhuijsen, M. J., Caeyenberghs, K., Jalaludin, B., Sugiyama, T., Sallis, J. F., Lautenschlager, N. T., Ni, M. Y., Poudel, G., Donaire-Gonzalez, D., Tham, R., Wheeler, A. J., Knibbs, L. D., Tian, L., Chan, Y.-K., Dunstan, D. W., Carver, A., & Anstey, K. J. (2020a). International Mind, Activities and Urban Places (iMAP) study: Methods of a cohort study on environmental and lifestyle influences on brain and cognitive health. *BMJ Open, 10*(3), e036607. https://doi.org/10.1136/bmjopen-2019-036607

Cerin, E., Barnett, A., Zhang, C. J. P., Lai, P. C., Sit, C. H. P., & Lee, R. S. Y. (2020b). How urban densification shapes walking behaviours in older community dwellers: A cross-sectional analysis of potential pathways of influence. *International Journal of Health Geographics, 19*(1), 14. https://doi.org/10.1186/s12942-020-00210-8

Cerin, E., Van Dyck, D., Zhang, C. J. P., Van Cauwenberg, J., Lai, P.-C., & Barnett, A. (2020c). Urban environments and objectively-assessed physical activity and sedentary time in older Belgian and Chinese community dwellers: Potential pathways of influence and the moderating role of physical function. *International Journal of Behavioral Nutrition and Physical Activity, 17*(1), 73. https://doi.org/10.1186/s12966-020-00979-8

Cerin, E., Barnett, A., Shaw, J. E., Martino, E., Knibbs, L. D., Tham, R., Wheeler, A. J., & Anstey, K. J. (2021). From urban neighbourhood environments to cognitive health: A cross-sectional analysis of the role of physical activity and sedentary behaviours. *BMC Public Health, 21*(1), 2320. https://doi.org/10.1186/s12889-021-12375-3

Cerin, E., Barnett, A., Shaw, J. E., Martino, E., Knibbs, L. D., Tham, R., Wheeler, A. J., & Anstey, K. J. (2022). Urban Neighbourhood environments, cardiometabolic health and cognitive function: A national cross-sectional study of middle-aged and older adults in Australia. *Toxics, 10*(1). https://doi.org/10.3390/toxics10010023

Cerin, E., Barnett, A., Wu, Y.-T., Martino, E., Shaw, J. E., Knibbs, L. D., Poudel, G., Jalaludin, B., & Anstey, K. J. (2023). Do neighbourhood traffic-related air pollution and socio-economic status moderate the associations of the neighbourhood physical environment with cognitive function? Findings from the AusDiab study. *Science of the Total Environment, 858*, 160028. https://doi.org/10.1016/j.scitotenv.2022.160028

Chan, O. F., Liu, Y., Guo, Y., Lu, S., Chui, C. H. K., Ho, H. C., Song, Y., Cheng, W., Chiu, R. L. H., Webster, C., & Lum, T. Y. S. (2023). Neighborhood built environments and cognition in later life. *Aging & Mental Health, 27*(3), 466–474. https://doi.org/10.1080/13607863.2022.2046697

Chandrabose, M., Rachele, J. N., Gunn, L., Kavanagh, A., Owen, N., Turrell, G., Giles-Corti, B., & Sugiyama, T. (2019). Built environment and cardio-metabolic health: Systematic review and meta-analysis of longitudinal studies. *Obesity Reviews, 20*(1), 41–54. https://doi.org/10.1111/obr.12759

Chastin, S. F. M., Craemer, M. D., Cocker, K. D., Powell, L., Cauwenberg, J. V., Dall, P., Hamer, M., & Stamatakis, E. (2019). How does light-intensity physical activity associate with adult cardiometabolic health and mortality? Systematic review with meta-analysis of experimental and observational studies. *British Journal of Sports Medicine, 53*(6), 370. https://doi.org/10.1136/bjsports-2017-097563

Chen, Y., & Yuan, Y. (2020). The neighborhood effect of exposure to blue space on elderly individuals' mental health: A case study in Guangzhou, China. *Health & Place, 63*(102), 348. https://doi.org/10.1016/j.healthplace.2020.102348

Chen, J. C., Wang, X., Wellenius, G. A., Serre, M. L., Driscoll, I., Casanova, R., McArdle, J. J., Manson, J. E., Chui, H. C., & Espeland, M. A. (2015). Ambient air pollution and neurotoxicity on brain structure: Evidence from women's health initiative memory study. *Annals of Neurology, 78*(3), 466–476. https://doi.org/10.1002/ana.24460

Chen, H., Kwong, J. C., Copes, R., Tu, K., Villeneuve, P. J., van Donkelaar, A., Hystad, P., Martin, R. V., Murray, B. J., Jessiman, B., Wilton, A. S., Kopp, A., & Burnett, R. T. (2017). Living near major roads and the incidence of dementia, Parkinson's disease, and multiple sclerosis: A population-based cohort study. *The Lancet, 389*(10070), 718–726. https://doi.org/10.1016/S0140-6736(16)32399-6

Chen, M., Howard, V., Harrington, K. F., Creger, T., Judd, S. E., & Fontaine, K. R. (2020). Does adherence to mediterranean diet mediate the association between food environment and obesity among non-hispanic black and white older US Adults? A path analysis. *American Journal of Health Promotion, 34*(6), 652–658. https://doi.org/10.1177/0890117120905240

Chen, X., Lee, C., & Huang, H. (2022a). Neighborhood built environment associated with cognition and dementia risk among older adults: A systematic literature review. *Social Science & Medicine, 292*(114), 560. https://doi.org/10.1016/j.socscimed.2021.114560

Chen, Y. R., Hanazato, M., Koga, C., Ide, K., & Kondo, K. (2022b). The association between street connectivity and depression among older Japanese adults: The JAGES longitudinal study. *Scientific Reports, 12*(1), 13533. https://doi.org/10.1038/s41598-022-17650-w

Cheng, T.-Y., Chang, C.-C., Luo, C.-S., Chen, K.-Y., Yeh, Y.-K., Zheng, J.-Q., & Wu, S.-M. (2023). Targeting lung–gut axis for regulating pollution particle–mediated inflammation and metabolic disorders. *Cells, 12*(6). https://doi.org/10.3390/cells12060901

Cherbuin, N., Kim, S., Anstey, K., & J.. (2015). Dementia risk estimates associated with measures of depression: A systematic review and meta-analysis. *BMJ Open, 5*(12), e008853. https://doi.org/10.1136/bmjopen-2015-008853

Cherrie, M. P. C., Shortt, N. K., Mitchell, R. J., Taylor, A. M., Redmond, P., Thompson, C. W., Starr, J. M., Deary, I. J., & Pearce, J. R. (2018). Green space and cognitive ageing: A retrospective life course analysis in the Lothian Birth Cohort 1936. *Social Science & Medicine, 196*, 56–65. https://doi.org/10.1016/j.socscimed.2017.10.038

Cherrie, M. P. C., Shortt, N. K., Ward Thompson, C., Deary, I. J., & Pearce, J. R. (2019). Association between the activity space exposure to parks in childhood and adolescence and cognitive aging in later life. *International Journal of Environmental Research and Public Health, 16*(4). https://doi.org/10.3390/ijerph16040632

Chevalier, V., Simard, V., & Achim, J. (2023). Meta-analyses of the associations of mentalization and proxy variables with anxiety and internalizing problems. *Journal of Anxiety Disorders, 95*(102), 694. https://doi.org/10.1016/j.janxdis.2023.102694

Chiang, Y. C., Li, D. Y., & Jane, H. A. (2017). Wild or tended nature? The effects of landscape location and vegetation density on physiological and psychological responses. *Landscape and Urban Planning, 167*, 72–83. https://doi.org/10.1016/j.landurbplan.2017.06.001

Cho, J., Noh, Y., Kim, S. Y., Sohn, J., Noh, J., Kim, W., Cho, S. K., Seo, H., Seo, G., Lee, S. K., Seo, S., Koh, S. B., Oh, S. S., Kim, H. J., Seo, S. W., Shin, D. S., Kim, N., Kim, H. H., Lee, J. I., & Kim, C. (2020). Long-term ambient air pollution exposures and brain imaging markers in Korean adults: The Environmental Pollution-Induced Neurological EFfects (EPINEF) study. *Environmental Health Perspectives, 128*(11). https://doi.org/10.1289/EHP7133

Choi, D., Choi, S., & Park, S. M. (2018). Effect of smoking cessation on the risk of dementia: A longitudinal study. *Annals of Clinical and Translational Neurology, 5*(10), 1192–1199. https://doi.org/10.1002/acn3.633

Chow, E. P. F., Hsu, B., Waite, L. M., Blyth, F. M., Handelsman, D. J., Le Couteur, D. G., Naganathan, V., & Stanaway, F. F. (2022). Diagnostic accuracy of linked administrative data for dementia diagnosis in community-dwelling older men in Australia. *BMC Geriatrics, 22*(1), 858. https://doi.org/10.1186/s12877-022-03579-2

Clarke, P. J., Ailshire, J. A., House, J. S., Morenoff, J. D., King, K., Melendez, R., & Langa, K. M. (2012). Cognitive function in the community setting: The neighbourhood as a source of 'cognitive reserve'? *Journal of Epidemiology and Community Health, 66*(8), 730. https://doi.org/10.1136/jech.2010.128116

Clarke, P. J., Weuve, J., Barnes, L., Evans, D. A., & Mendes de Leon, C. F. (2015). Cognitive decline and the neighborhood environment. *Annals of Epidemiology, 25*(11), 849–854. https://doi.org/10.1016/j.annepidem.2015.07.001

Clarke, P., Twardzik, E., Meade, M. A., Peterson, M. D., & Tate, D. (2019). Social participation among adults aging with long-term physical disability: The role of socioenvironmental factors. *Journal of Aging and Health, 31*(10_suppl), 145S–168S. https://doi.org/10.1177/0898264318822238

Clary, C., Lewis, D. J., Flint, E., Smith, N. R., Kestens, Y., & Cummins, S. (2016). The local food environment and fruit and vegetable intake: A geographically weighted regression approach in the ORiEL study. *American Journal of Epidemiology, 184*(11), 837–846. https://doi.org/10.1093/aje/kww073

Cory-Slechta, D. A., & Sobolewski, M. (2023). Neurotoxic effects of air pollution: An urgent public health concern. *Nature Reviews Neuroscience, 24*(3), 129–130. https://doi.org/10.1038/s41583-022-00672-8

Cowie, C. T., Ding, D., Rolfe, M. I., Mayne, D. J., Jalaludin, B., Bauman, A., & Morgan, G. G. (2016). Neighbourhood walkability, road density and socio-economic status in Sydney. *Australia. Environmental Health, 15*, 58. https://doi.org/10.1186/s12940-016-0135-y

Crous-Bou, M., Gascon, M., Gispert, J. D., Cirach, M., Sánchez-Benavides, G., Falcon, C., Arenaza-Urquijo, E. M., Gotsens, X., Fauria, K., Sunyer, J., Nieuwenhuijsen, M. J., & Luis Molinuevo, J. (2020). Impact of urban environmental exposures on cognitive performance and brain structure of healthy individuals at risk for Alzheimer's dementia. *Environment International, 138*(105), 546. https://doi.org/10.1016/j.envint.2020.105546

Crouse Dan, L., Balram, A., Hystad, P., Pinault, L., van den Bosch, M., Chen, H., Rainham, D., Thomson Errol, M., Close Christopher, H., van Donkelaar, A., Martin Randall, V., Ménard, R., Robichaud, A., & Villeneuve Paul, J. (2018). Associations between living near water and risk of mortality among urban Canadians. *Environmental Health Perspectives, 126*(7), 077008. https://doi.org/10.1289/EHP3397

Cui, X., Rockett, I. R. H., Yang, T., & Cao, R. (2012). Work stress, life stress, and smoking among rural–urban migrant workers in China. *BMC Public Health, 12*(1), 979. https://doi.org/10.1186/1471-2458-12-979

Cui, J., Loo, B. P. Y., & Lin, D. (2017). Travel behaviour and mobility needs of older adults in an ageing and car-dependent society. *International Journal of Urban Sciences, 21*(2), 109–128. https://doi.org/10.1080/12265934.2016.1262785

Cunningham, C., O'Sullivan, R., Caserotti, P., & Tully, M. A. (2020). Consequences of physical inactivity in older adults: A systematic review of reviews and meta-analyses. *Scandinavian Journal of Medicine & Science in Sports, 30*(5), 816–827. https://doi.org/10.1111/sms.13616

D'Souza, J., Weuve, J., Brook, R. D., Evans, D. A., Kaufman, J. D., & Adar, S. D. (2021). Long-term exposures to urban noise and blood pressure levels and control among older adults. *Hypertension, 78*(6), 1801–1808. https://doi.org/10.1161/HYPERTENSIONAHA.121.17708

Dadvand, P., Pujol, J., Macia, D., Martinez-Vilavella, G., Blanco-Hinojo, L., Mortamais, M., Alvarez-Pedrerol, M., Fenoll, R., Esnaola, M., Dalmau-Bueno, A., Lopez-Vicente, M., Basagana, X., Jerrett, M., Nieuwenhuijsen, M. J., & Sunyer, J. (2018). The association between lifelong greenspace exposure and 3-dimensional brain magnetic resonance imaging in Barcelona Schoolchildren. *Environmental Health Perspectives, 126*(2). https://doi.org/10.1289/EHP1876

de Keijzer, C., Gascon, M., Nieuwenhuijsen, M. J., & Dadvand, P. (2016). Long-term green space exposure and cognition across the life course: A systematic review. *Current Environmental Health Reports, 3*(4), 468–477. https://doi.org/10.1007/s40572-016-0116-x

de Keijzer, C., Tonne, C., Basagaña, X., Valentín, A., Singh-Manoux, A., Alonso, J., Antó Josep, M., Nieuwenhuijsen Mark, J., Sunyer, J., & Dadvand, P. (2018). Residential surrounding greenness and cognitive decline: A 10-year follow-up of the Whitehall II Cohort. *Environmental Health Perspectives, 126*(7), 077003. https://doi.org/10.1289/EHP2875

de Keijzer, C., Bauwelinck, M., & Dadvand, P. (2020). Long-term exposure to residential greenspace and healthy ageing: A systematic review. *Current Environmental Health Reports, 7*(1), 65–88. https://doi.org/10.1007/s40572-020-00264-7

Dejenie, T. A., G/Medhin, M. T., Admasu, F. T., Adella, G. A., Enyew, E. F., Kifle, Z. D., Seid, M. A., Mengstie, M. A., & Abebe, E. C. (2022). Impact of objectively-measured sleep duration on cardiometabolic health: A systematic review of recent evidence [Review]. *Frontiers in Endocrinology, 13*. https://www.frontiersin.org/articles/10.3389/fendo.2022.1064969

Delclòs-Alió, X., Marquet, O., Vich, G., Schipperijn, J., Zhang, K., Maciejewska, M., & Miralles-Guasch, C. (2020). Temperature and rain moderate the effect of neighborhood walkability on walking time for seniors in Barcelona. *International Journal of Environmental Research and Public Health, 17*(1). https://doi.org/10.3390/ijerph17010014

Dempsey, S., Devine, M. T., Gillespie, T., Lyons, S., & Nolan, A. (2018). Coastal blue space and depression in older adults. *Health & Place, 54*, 110–117. https://doi.org/10.1016/j.healthplace.2018.09.002

Deng, L., Li, X., Luo, H., Fu, E. K., Ma, J., Sun, L. X., Huang, Z., Cai, S. Z., & Jia, Y. (2020). Empirical study of landscape types, landscape elements and landscape components of the urban park promoting physiological and psychological restoration. *Urban Forestry & Urban Greening, 48*. https://doi.org/10.1016/j.ufug.2019.126488

Dimitrov-Discher, A., Wenzel, J., Kabisch, N., Hemmerling, J., Bunz, M., Schondorf, J., Walter, H., Veer, I. M., & Adli, M. (2022). Residential green space and air pollution are associated with brain activation in a social-stress paradigm. *Scientific Reports, 12*(1). https://doi.org/10.1038/s41598-022-14,659-z

Domingos, C., Pêgo, J. M., & Santos, N. C. (2021). Effects of physical activity on brain function and structure in older adults: A systematic review. *Behavioural Brain Research, 402*(113), 061. https://doi.org/10.1016/j.bbr.2020.113061

Dzhambov, A. M., & Dimitrova, D. D. (2016). Exposures to road traffic, noise, and air pollution as risk factors for type 2 diabetes: A feasibility study in Bulgaria. *Noise Health, 18*(82), 133–142. https://doi.org/10.4103/1463-1741.181996

Echeverría, S., Diez-Roux, A. V., Shea, S., Borrell, L. N., & Jackson, S. (2008). Associations of neighborhood problems and neighborhood social cohesion with mental health and health behaviors: The Multi-Ethnic Study of Atherosclerosis. *Health & Place, 14*(4), 853–865. https://doi.org/10.1016/j.healthplace.2008.01.004

Ewing, R., & Cervero, R. (2010). Travel and the built environment. *Journal of the American Planning Association, 76*(3), 265–294. https://doi.org/10.1080/01944361003766766

Fan, S., Feng, W., Zhou, Z., Guo, Y., Xue, Z., Yuan, J., Wang, Y., Li, C., Zhong, Y., He, W., Dong, G., Yang, B., Yang, Z., & Zhang, Z. (2022). Association between residential greenness and overweight/obesity among rural adults in northwestern China. *Environmental Research, 204*(112), 358. https://doi.org/10.1016/j.envres.2021.112358

Fangfang, H., Xiao, H., Shuai, Z., Qiong, W., Jingya, Z., Guodong, S., & Yan, Z. (2022). Living environment, built environment and cognitive function among older chinese adults: Results from a cross-sectional study. *The Journal of Prevention of Alzheimer's Disease, 9*(1), 126–135. https://doi.org/10.14283/jpad.2021.59

Finlay, J., Esposito, M., Tang, S., Gomez-Lopez, I., Sylvers, D., Judd, S., & Clarke, P. (2020). Fast-food for thought: Retail food environments as resources for cognitive health and well-being among aging Americans? *Health & Place, 64*(102), 379. https://doi.org/10.1016/j.healthplace.2020.102379

Finlay, J., Esposito, M., Li, M., Kobayashi, L. C., Khan, A. M., Gomez-Lopez, I., Melendez, R., Colabianchi, N., Judd, S., & Clarke, P. J. (2021a). Can neighborhood social infrastructure modify cognitive function? A mixed-methods study of urban-dwelling aging Americans. *Journal of Aging and Health, 33*(9), 772–785. https://doi.org/10.1177/08982643211008673

Finlay, J., Yu, W., Clarke, P., Li, M., Judd, S., & Esposito, M. (2021b). Neighborhood cognitive amenities? A mixed-methods study of intellectually-stimulating places and cognitive function among older Americans. *Wellbeing, Space and Society, 2*(100), 040. https://doi.org/10.1016/j.wss.2021.100040

Finlay, J., Esposito, M., Langa, K. M., Judd, S., & Clarke, P. (2022). Cognability: An ecological theory of neighborhoods and cognitive aging. *Social Science & Medicine, 309*(115), 220. https://doi.org/10.1016/j.socscimed.2022.115220

Fitts, W., Weintraub, D., Massimo, L., Chahine, L., Chen-Plotkin, A., Duda, J. E., Hurtig, H. I., Rick, J., Trojanowski, J. Q., & Dahodwala, N. (2015). Caregiver report of apathy predicts dementia in Parkinson's disease. *Parkinsonism & Related Disorders, 21*(8), 992–995. https://doi.org/10.1016/j.parkreldis.2015.06.009

Forsman, A. K., Schierenbeck, I., & Wahlbeck, K. (2010). Psychosocial interventions for the prevention of depression in older adults: Systematic review and meta-analysis. *Journal of Aging and Health, 23*(3), 387–416. https://doi.org/10.1177/0898264310378041

Franz, C. E., Gustavson, D. E., Elman, J. A., Fennema-Notestine, C., Hagler, D. J., Jr., Baraff, A., Tu, X. M., Wu, T. C., De Anda, J., Beck, A., Kaufman, J. D., Whitsel, N., Finch, C. E., Chen, J. C., Lyons, M. J., & Kremen, W. S. (2023). Associations between ambient air pollution and cognitive abilities from midlife to early old age: Modification by APOE genotype. *Journal of Alzheimer's disease: JAD, 93*(1), 193–209. https://doi.org/10.3233/jad-221054

Furlong, M. A., Alexander, G. E., Klimentidis, Y. C., & Raichlen, D. A. (2022). Association of air pollution and physical activity with brain volumes. *Neurology, 98*(4), E416–E426. https://doi.org/10.1212/Wnl.0000000000013031

Gaglione, F., Cottrill, C., & Gargiulo, C. (2021). Urban services, pedestrian networks and behaviors to measure elderly accessibility. *Transportation Research Part D: Transport and Environment, 90*(102), 687. https://doi.org/10.1016/j.trd.2020.102687

Gaio, V., Roquette, R., Dias, C. M., & Nunes, B. (2019). Ambient air pollution and lipid profile: Systematic review and meta-analysis. *Environmental Pollution, 254*(113), 036. https://doi.org/10.1016/j.envpol.2019.113036

Gale, S. D., Erickson, L. D., Anderson, J. E., Brown, B. L., & Hedges, D. W. (2020). Association between exposure to air pollution and prefrontal cortical volume in adults: A cross-sectional study from the UK biobank. *Environmental Research, 185*. https://doi.org/10.1016/j.envres.2020.109365

Gallacher, J., Bayer, A., Fish, M., Pickering, J., Pedro, S., Dunstan, F., Ebrahim, S., & Ben-Shlomo, Y. (2009). Does anxiety affect risk of dementia? Findings from the caerphilly prospective study. *Psychosomatic Medicine, 71*(6). https://doi.org/10.1097/PSY.0b013e3181a6177c

Gascon, M., Zijlema, W., Vert, C., White, M. P., & Nieuwenhuijsen, M. J. (2017). Outdoor blue spaces, human health and well-being: A systematic review of quantitative studies. *International Journal of Hygiene and Environmental Health, 220*(8), 1207–1221. https://doi.org/10.1016/j.ijheh.2017.08.004

Gawryluk, J. R., Palombo, D. J., Curran, J., Parker, A., & Carlsten, C. (2023). Brief diesel exhaust exposure acutely impairs functional brain connectivity in humans: A randomized controlled crossover study. *Environmental Health, 22*(1), 7. https://doi.org/10.1186/s12940-023-00961-4

Geneshka, M., Coventry, P., Cruz, J., & Gilbody, S. (2021). Relationship between Green and Blue spaces with mental and physical health: A systematic review of longitudinal observational studies. *International Journal of Environmental Research and Public Health, 18*(17). https://doi.org/10.3390/ijerph18179010

Gianfredi, V., Buffoli, M., Rebecchi, A., Croci, R., Oradini-Alacreu, A., Stirparo, G., Marino, A., Odone, A., Capolongo, S., & Signorelli, C. (2021). Association between urban greenspace and health: A systematic review of literature. *International Journal of Environmental Research and Public Health, 18*(10). https://doi.org/10.3390/ijerph18105137

Giles-Corti, B., Vernez-Moudon, A., Reis, R., Turrell, G., Dannenberg, A. L., Badland, H., Foster, S., Lowe, M., Sallis, J. F., Stevenson, M., & Owen, N. (2016). City planning and population health: a global challenge. *The Lancet, 388*(10062), 2912–2924. https://doi.org/10.1016/S0140-6736(16)30066-6

Giskes, K., van Lenthe, F., Avendano-Pabon, M., & Brug, J. (2011). A systematic review of environmental factors and obesogenic dietary intakes among adults: Are we getting closer to understanding obesogenic environments? *Obesity Reviews, 12*(5), e95–e106. https://doi.org/10.1111/j.1467-789X.2010.00769.x

Glass, T. A., & Balfour, J. L. (2003). Neighborhoods, aging, and functional limitations. In I. Kawachi & L. F. Berkman (Eds.), *Neighborhoods and health* (pp. 303–334). Oxford University Press. https://doi.org/10.1093/acprof:oso/9780195138382.003.0014

Godina, S. L., Rosso, A. L., Hirsch, J. A., Besser, L. M., Lovasi, G. S., Donovan, G. H., Garg, P. K., Platt, J. M., Fitzpatrick, A. L., Lopez, O. L., Carlson, M. C., & Michael, Y. L. (2023). Neighborhood greenspace and cognition: The cardiovascular health study. *Health & Place, 79*(102), 960. https://doi.org/10.1016/j.healthplace.2022.102960

Gonzales-Inca, C., Pentti, J., Stenholm, S., Suominen, S., Vahtera, J., & Kayhko, N. (2022). Residential greenness and risks of depression: Longitudinal associations with different greenness indicators and spatial scales in a Finnish population cohort. *Health & Place, 74*(102), 760. https://doi.org/10.1016/j.healthplace.2022.102760

Grainger, S. A., Crawford, J. D., Riches, J. C., Kochan, N. A., Chander, R. J., Mather, K. A., Sachdev, P. S., & Henry, J. D. (2023). Aging is associated with multidirectional changes in social cognition: Findings from an adult life-span sample ranging from 18 to 101 years. *The Journals of Gerontology: Series B, 78*(1), 62–72. https://doi.org/10.1093/geronb/gbac110

Greig Custo, M. T., Lang, M. K., Barker, W. W., Gonzalez, J., Vélez-Uribe, I., Arruda, F., Conniff, J., Rodriguez, M. J., Loewenstein, D. A., Duara, R., Adjouadi, M., Curiel, R. E., & Rosselli, M. (2022). The association of depression and apathy with Alzheimer's disease biomarkers in a cross-cultural sample. *Applied Neuropsychology: Adult*, 1–17. https://doi.org/10.1080/23279095.2022.2079414

Gulpers, B. J. A., Oude Voshaar, R. C., van Boxtel, M. P. J., Verhey, F. R. J., & Köhler, S. (2019). Anxiety as a risk factor for cognitive decline: A 12-year follow-up cohort study. *The American Journal of Geriatric Psychiatry, 27*(1), 42–52. https://doi.org/10.1016/j.jagp.2018.09.006

Guo, F., Moellering, D. R., & Garvey, W. T. (2014). The progression of cardiometabolic disease: Validation of a new cardiometabolic disease staging system applicable to obesity. *Obesity, 22*(1), 110–118. https://doi.org/10.1002/oby.20585

Guo, Y., Chan, C. H., Chang, Q., Liu, T., & Yip, P. S. F. (2019). Neighborhood environment and cognitive function in older adults: A multilevel analysis in Hong Kong. *Health & Place, 58*(102), 146. https://doi.org/10.1016/j.healthplace.2019.102146

Guo, Y., Liu, Y., Lu, S., Chan, O. F., Chui, C. H. K., & Lum, T. Y. S. (2021). Objective and perceived built environment, sense of community, and mental wellbeing in older adults in Hong Kong: A multilevel structural equation study. *Landscape and Urban Planning, 209*(104), 058. https://doi.org/10.1016/j.landurbplan.2021.104058

Hagan, R. J. (2020). Getting out of the house: The use of community transport as a third place for rural-dwelling older adults. *Ageing & Society, 40*(11), 2519–2539. https://doi.org/10.1017/S0144686X19000722

Harrington, K. D., Vasan, S., Kang, J. E., Sliwinski, M. J., & Lim, M. H. (2023). Loneliness and cognitive function in older adults without dementia: A systematic review and meta-analysis. *Journal of Alzheimer's disease: JAD, 91*(4), 1243–1259. https://doi.org/10.3233/jad-220832

Hasselder, P., Brüchert, T., Baumgart, S., & Bolte, G. (2022). Destinations fostering older adults' walking for transport: A cross-sectional study from Germany. *BMC Geriatrics, 22*(1), 219. https://doi.org/10.1186/s12877-022-02896-w

Hayashi, T., Okamoto, E., Nishimura, H., Mizuno-Matsumoto, Y., Ishii, R., & Ukai, S. (2009). Beta activities in EEG associated with emotional stress. *International Journal of Intelligent Computing in Medical Sciences & Image Processing, 3*(1), 57–68. https://doi.org/10.1080/1931308X.2009.10644171

Hedges, D. W., Erickson, L. D., Kunzelman, J., Brown, B. L., & Gale, S. D. (2019). Association between exposure to air pollution and hippocampal volume in adults in the UK Biobank. *Neurotoxicology, 74*, 108–120. https://doi.org/10.1016/j.neuro.2019.06.005

Helmes, E., & Van Gerven, P. W. M. (2017). Urban residence and higher education do not protect against cognitive decline in aging and dementia: 10-year follow-up of the Canadian Study of Health and Aging. *Educational Gerontology, 43*(11), 552–560. https://doi.org/10.1080/03601277.2017.1372951

Herrmann-Lunecke, M. G., Mora, R., & Vejares, P. (2021). Perception of the built environment and walking in pericentral neighbourhoods in Santiago, Chile. *Travel Behaviour and Society, 23*, 192–206. https://doi.org/10.1016/j.tbs.2021.01.002

Hirsch, J. A., Michael, Y. L., Moore, K. A., Melly, S., Hughes, T. M., Hayden, K., Luchsinger, J. A., Jimenez, M. P., James, P., Besser, L. M., Sánchez, B., & Diez Roux, A. V. (2022). Longitudinal neighbourhood determinants with cognitive health and dementia disparities: protocol of the Multi-Ethnic Study of Atherosclerosis Neighborhoods and Aging prospective cohort study. *BMJ Open, 12*(11), e066971. https://doi.org/10.1136/bmjopen-2022-066971

Ho, H. C., Fong, K. N. K., Chan, T.-C., & Shi, Y. (2020). The associations between social, built and geophysical environment and age-specific dementia mortality among older adults in a high-density Asian city. *International Journal of Health Geographics, 19*(1), 53. https://doi.org/10.1186/s12942-020-00252-y

Hu, H.-Y., Ma, Y.-H., Deng, Y.-T., Ou, Y.-N., Cheng, W., Feng, J.-F., Tan, L., & Yu, J.-T. (2023). Residential greenness and risk of incident dementia: A prospective study of 375,342 participants. *Environmental Research, 216*(114), 703. https://doi.org/10.1016/j.envres.2022.114703

Huang, W., & Lin, G. (2023). The relationship between urban green space and social health of individuals: A scoping review. *Urban Forestry & Urban Greening, 85*(127), 969. https://doi.org/10.1016/j.ufug.2023.127969

Hunt, J. F. V., Vogt, N. M., Jonaitis, E. M., Buckingham, W. R., Koscik, R. L., Zuelsdorff, M., Clark, L. R., Gleason, C. E., Yu, M., Okonkwo, O., Johnson, S. C., Asthana, S., Bendlin, B. B., & Kind, A. J. H. (2021). Association of neighborhood context, cognitive decline, and cortical change in an unimpaired cohort. *Neurology, 96*(20), e2500–e2512. https://doi.org/10.1212/WNL.0000000000011918

Hyun, J., Katz, M. J., Derby, C. A., Roque, N., Muñoz, E., Sliwinski, M. J., Lovasi, G. S., & Lipton, R. B. (2023). Availability of healthy foods, fruit and vegetable consumption, and cognition among urban older adults. *BMC Geriatrics, 23*(1), 302. https://doi.org/10.1186/s12877-023-04003-z

Institute of Medicine. (2015). *Cognitive aging: Progress in understanding and opportunities for action*. The National Academies Press.

James, P., Hart, J. E., & Laden, F. (2015). Neighborhood walkability and particulate air pollution in a nationwide cohort of women. *Environmental Research, 142*, 703–711. https://doi.org/10.1016/j.envres.2015.09.005

James, K. A., Stromin, J. I., Steenkamp, N., & Combrinck, M. I. (2023). Understanding the relationships between physiological and psychosocial stress, cortisol and cognition [Review]. *Frontiers in Endocrinology, 14*. https://www.frontiersin.org/articles/10.3389/fendo.2023.1085950

Jeanneteau, F. (2023). Stress and the risk of Alzheimer dementia: Can deconstructed engrams be rebuilt? *Journal of Neuroendocrinology., e13235*. https://doi.org/10.1111/jne.13235

Jimenez, M. P., Elliott, E. G., DeVille, N. V., Laden, F., Hart, J. E., Weuve, J., Grodstein, F., & James, P. (2022). Residential green space and cognitive function in a large cohort of middle-aged women. *JAMA Network Open, 5*(4), e229306–e229306. https://doi.org/10.1001/jamanetworkopen.2022.9306

Jin, X., Shu, C., Zeng, Y., Liang, L., & Ji, J. S. (2021). Interaction of greenness and polygenic risk score of Alzheimer's disease on risk of cognitive impairment. *Science of The Total Environment, 796*(148), 767. https://doi.org/10.1016/j.scitotenv.2021.148767

John, A., Patel, U., Rusted, J., Richards, M., & Gaysina, D. (2019). Affective problems and decline in cognitive state in older adults: A systematic review and meta-analysis. *Psychological Medicine, 49*(3), 353–365. https://doi.org/10.1017/S0033291718001137

John, A., Rusted, J., Richards, M., & Gaysina, D. (2021). Bidirectional relation between affective symptoms and cognitive function from middle to late adulthood: A population-based birth cohort study. *Aging & Mental Health, 25*(9), 1642–1648. https://doi.org/10.1080/13607863.2020.1758916

John, E. E., Astell-Burt, T., Yu, P., Brennan-Horley, C., & Feng, X. (2023). Green space type and healthy ageing in place: An Australian longitudinal study. *Urban Forestry & Urban Greening, 84*(127), 903. https://doi.org/10.1016/j.ufug.2023.127903

Jung, J., Park, J. Y., Myung, W., Lee, J.-Y., Ko, H., & Lee, H. (2022). Association between residential greenness and incidence of Parkinson's disease: A population-based Cohort Study in South Korea. *International Journal of Environmental Research and Public Health, 19*(6). https://doi.org/10.3390/ijerph19063491

Kabisch, N., Korn, H., Stadler, J., & Bonn, A. (2017). *Nature-based solutions to climate change adaptation in urban areas—Linkages between science, policy and practice*. Springer Nature.

Kardan, M., Akter, T., Iqbal, M., Tcymbal, A., Messing, S., Gelius, P., & Abu-Omar, K. (2023). Cycling in older adults: A scoping review [Review]. *Frontiers in Sports and Active Living, 5*. https://www.frontiersin.org/articles/10.3389/fspor.2023.1157503

Kasdagli, M.-I., Katsouyanni, K., de Hoogh, K., Zafeiratou, S., Dimakopoulou, K., & Samoli, E. (2023). Associations between exposure to blue spaces and natural and cause-specific mortality in Greece: An ecological study. *International Journal of Hygiene and Environmental Health, 249*(114), 137. https://doi.org/10.1016/j.ijheh.2023.114137

Kattenstroth, J.-C., Kalisch, T., Holt, S., Tegenthoff, M., & Dinse, H. (2013). Six months of dance intervention enhances postural, sensorimotor, and cognitive performance in elderly without affecting cardio-respiratory functions [Original Research]. *Frontiers in Aging Neuroscience, 5*. https://www.frontiersin.org/articles/10.3389/fnagi.2013.00005

Keenan, T. D., Agrón, E., Mares, J. A., Clemons, T. E., van Asten, F., Swaroop, A., Chew, E. Y., for the, A, & Groups, A. R. (2020). Adherence to a Mediterranean diet and cognitive function in the Age-Related Eye Disease Studies 1 & 2. *Alzheimer's & Dementia, 16*(6), 831–842. https://doi.org/10.1002/alz.12077

Kent, B. A., & Mistlberger, R. E. (2017). Sleep and hippocampal neurogenesis: Implications for Alzheimer's disease. *Frontiers in Neuroendocrinology, 45*, 35–52. https://doi.org/10.1016/j.yfrne.2017.02.004

Killin, L. O. J., Starr, J. M., Shiue, I. J., & Russ, T. C. (2016). Environmental risk factors for dementia: A systematic review. *BMC Geriatrics, 16*(1), 175. https://doi.org/10.1186/s12877-016-0342-y

Kim, G. W., Jeong, G. W., Kim, T. H., Baek, H. S., Oh, S. K., Kang, H. K., Lee, S. G., Kim, Y. S., & Song, J. K. (2010). Functional neuroanatomy associated with natural and urban scenic views in the human brain: 3.0T functional MR imaging. *Korean Journal of Radiology, 11*(5), 507–513. https://doi.org/10.3348/kjr.2010.11.5.507

Kim, B., Branas, C. C., Rudolph, K. E., Morrison, C. N., Chaix, B., Troxel, W. M., & Duncan, D. T. (2022). Neighborhoods and sleep health among adults: A systematic review. *Sleep Health, 8*(3), 322–333. https://doi.org/10.1016/j.sleh.2022.03.005

Kim, B., Barrington, W. E., Dobra, A., Rosenberg, D., Hurvitz, P., & Belza, B. (2023a). Mediating role of walking between perceived and objective walkability and cognitive function in older adults. *Health & Place, 79*(102), 943. https://doi.org/10.1016/j.healthplace.2022.102943

Kim, B., Rosenberg Dori, E., Dobra, A., Barrington Wendy, E., Hurvitz Philip, M., & Belza, B. (2023b). Association of perceived neighborhood environments with cognitive function in older adults. *Journal of Gerontological Nursing, 49*(8), 35–41. https://doi.org/10.3928/00989134-20230707-04

Klompmaker, J. O., Laden, F., Browning, M. H. E. M., Dominici, F., Jimenez, M. P., Ogletree, S. S., Rigolon, A., Zanobetti, A., Hart, J. E., & James, P. (2022). Associations of greenness, parks, and blue space with neurodegenerative disease hospitalizations among older US adults. *JAMA Network Open, 5*(12), e2247664. https://doi.org/10.1001/jamanetworkopen.2022.47664

Koohsari, M. J., Nakaya, T., McCormack, G. R., Shibata, A., Ishii, K., Yasunaga, A., & Oka, K. (2019). Cognitive function of elderly persons in Japanese neighborhoods: The role of street layout. *American Journal of Alzheimer's Disease & Other Dementias®, 34*(6), 381–389. https://doi.org/10.1177/1533317519844046

Korten, N. C. M., Comijs, H. C., Penninx, B. W. J. H., & Deeg, D. J. H. (2017). Perceived stress and cognitive function in older adults: Which aspect of perceived stress is important? *International Journal of Geriatric Psychiatry, 32*(4), 439–445. https://doi.org/10.1002/gps.4486

Kredlow, M. A., Capozzoli, M. C., Hearon, B. A., Calkins, A. W., & Otto, M. W. (2015). The effects of physical activity on sleep: A meta-analytic review. *Journal of Behavioral Medicine, 38*(3), 427–449. https://doi.org/10.1007/s10865-015-9617-6

Kühn, S., Düzel, S., Eibich, P., Krekel, C., Kolbe, J., Martensson, J., Goebel, J., Gallinat, J., Wagner, G. G., & Lindenberger, U. (2017). In search of features that constitute an "enriched environment" in humans: Associations between geographical properties and brain structure. *Scientific Reports, 7*, 11920. https://doi.org/10.1038/s41598-017-12046-7

Landeiro, F., Barrows, P., Musson, E. N., Gray, A. M., & Leal, J. (2017). Reducing social isolation and loneliness in older people: A systematic review protocol. *BMJ Open, 7*(5), e013778. https://doi.org/10.1136/bmjopen-2016-013778

Lang, I. A., Llewellyn, D. J., Langa, K. M., Wallace, R. B., Huppert, F. A., & Melzer, D. (2008). Neighborhood deprivation, individual socioeconomic status, and cognitive function in older people: Analyses from the English longitudinal study of ageing. *Journal of the American Geriatrics Society, 56*(2), 191–198. https://doi.org/10.1111/j.1532-5415.2007.01557.x

Lara, E., Martín-María, N., De la Torre-Luque, A., Koyanagi, A., Vancampfort, D., Izquierdo, A., & Miret, M. (2019). Does loneliness contribute to mild cognitive impairment and dementia? A systematic review and meta-analysis of longitudinal studies. *Ageing Research Reviews, 52*, 7–16. https://doi.org/10.1016/j.arr.2019.03.002

Lee, H. J., & Lee, D. K. (2019). Do sociodemographic factors and urban green space affect mental health outcomes among the urban elderly population? *International Journal of Environmental Research and Public Health, 16*(5). https://doi.org/10.3390/ijerph16050789

Levasseur, M., Généreux, M., Bruneau, J.-F., Vanasse, A., Chabot, É., Beaulac, C., & Bédard, M.-M. (2015). Importance of proximity to resources, social support, transportation and neighborhood security for mobility and social participation in older adults: Results from a scoping study. *BMC Public Health, 15*(1), 503. https://doi.org/10.1186/s12889-015-1824-0

Leyland, L.-A., Spencer, B., Beale, N., Jones, T., & van Reekum, C. M. (2019). The effect of cycling on cognitive function and well-being in older adults. *PLoS ONE, 14*(2), e0211779. https://doi.org/10.1371/journal.pone.0211779

Li, J., Dai, Y., Wang, C. C., & Sun, J. (2022). Assessment of environmental demands of age-friendly communities from perspectives of different residential groups: A case of Wuhan, China. *International Journal of Environmental Research and Public Health, 19*(15). https://doi.org/10.3390/ijerph19159120

Li, F., Liu, W., Hu, C., Tang, M., Zhang, Y., Ho, H. C., Peng, S., Li, Z., Wang, Q., Li, X., Xu, B., & Li, F. (2023). Global association of greenness exposure with risk of nervous system disease: A systematic review and meta-analysis. *Science of The Total Environment, 877*(162), 773. https://doi.org/10.1016/j.scitotenv.2023.162773

Lin, W., Chen, Q. B., Jiang, M. Y., Tao, J. Y., Liu, Z. F., Zhang, X. X., Wu, L. J., Xu, S., Kang, Y. S., & Zeng, Q. Y. (2020). Sitting or walking? Analyzing the neural emotional indicators of urban green space behavior with mobile EEG. *Journal of Urban Health-Bulletin of the New York Academy of Medicine, 97*(2), 191–203. https://doi.org/10.1007/s11524-019-00407-8

Lindert, J., Paul, K. C., Lachman, M. E., Ritz, B., & Seeman, T. E. (2021). Depression-, anxiety-, and anger and cognitive functions: Findings from a longitudinal prospective study [Original Research]. *Frontiers in Psychiatry, 12*. https://www.frontiersin.org/articles/10.3389/fpsyt.2021.665742

Lisman, J., Buzsaki, G., Eichenbaum, H., Nadel, L., Ranganath, C., & Redish, A. D. (2018). Viewpoints: How the hippocampus contributes to memory, navigation and cognition. *Nature Neuroscience, 21*(7), 1018–1018. https://doi.org/10.1038/s41593-017-0034-8

Liu, F., Chen, G., Huo, W., Wang, C., Liu, S., Li, N., Mao, S., Hou, Y., Lu, Y., & Xiang, H. (2019). Associations between long-term exposure to ambient air pollution and risk of type 2 diabetes mellitus: A systematic review and meta-analysis. *Environmental Pollution, 252*, 1235–1245. https://doi.org/10.1016/j.envpol.2019.06.033

Livingston, G., Huntley, J., Sommerlad, A., Ames, D., Ballard, C., Banerjee, S., Brayne, C., Burns, A., Cohen-Mansfield, J., Cooper, C., Costafreda, S. G., Dias, A., Fox, N., Gitlin, L. N., Howard, R., Kales, H. C., Kivimäki, M., Larson, E. B., Ogunniyi, A.,. .. Mukadam, N. (2020). Dementia prevention, intervention, and care: 2020 report of the Lancet Commission. *The Lancet, 396*(10248), 413–446. https://doi.org/10.1016/S0140-6736(20)30367-6.

Lopez, A., Germani, A., Tinella, L., Caffò, A. O., Postma, A., & Bosco, A. (2021). The road more travelled: The differential effects of spatial experience in young and elderly participants. *International Journal of Environmental Research and Public Health, 18*(2). https://doi.org/10.3390/ijerph18020709

Lorenzo-López, L., Millán-Calenti, J. C., López-López, R., Diego-Diez, C., Laffon, B., Pásaro, E., Valdiglesias, V., & Maseda, A. (2017). Effects of degree of urbanization and lifetime longest-held occupation on cognitive impairment prevalence in an older Spanish population

[Original Research]. *Frontiers in Psychology, 8*. https://www.frontiersin.org/articles/10.3389/fpsyg.2017.00162

Lowe, M., Adlakha, D., Sallis, J. F., Salvo, D., Cerin, E., Moudon, A. V., Higgs, C., Hinckson, E., Arundel, J., Boeing, G., Liu, S., Mansour, P., Gebel, K., Puig-Ribera, A., Mishra, P. B., Bozovic, T., Carson, J., Dygrýn, J., Florindo, A. A., et al. (2022). City planning policies to support health and sustainability: an international comparison of policy indicators for 25 cities. *The Lancet Global Health, 10*(6), e882–e894. https://doi.org/10.1016/S2214-109X(22)00069-9

Lu, S., Liu, Y., Guo, Y., Ho, H. C., Song, Y., Cheng, W., Chui, C., Chan, O. F., Webster, C., Chiu, R. L. H., & Lum, T. Y. (2021). Neighborhood built environment and late-life depression: A multilevel path analysis in a Chinese Society. *Journal of Gerontology: Series B, 76*(10), 2143–2154. https://doi.org/10.1093/geronb/gbab037

Lubczynska, M. J., Muetzel, R. L., El Marroun, H., Hoek, G., Kooter, I. M., Thomson, E. M., Hillegers, M., Vernooij, M. W., White, T., Tiemeier, H., & Guxens, M. (2021). Air pollution exposure during pregnancy and childhood and brain morphology in preadolescents. *Environmental Research, 198*. https://doi.org/10.1016/j.envres.2020.110446

Luo, Y., Zhang, L., & Pan, X. (2019). Neighborhood environments and cognitive decline among middle-aged and older people in China. *The Journals of Gerontology: Series B, 74*(7), e60–e71. https://doi.org/10.1093/geronb/gbz016

Luo, Y., Zhao, Y., Pang, L., Guo, C., Liang, R., & Zheng, X. (2021). Association between urbanicity and dementia in China: A population-based study. *The Journals of Gerontology: Series B, 76*(8), 1644–1651. https://doi.org/10.1093/geronb/gbaa090

Luo, L., Yu, P., & Jiang, B. (2023). Differentiating mental health promotion effects of various blue spaces: An electroencephalography study. *Journal of Environmental Psychology, 88*. https://doi.org/10.1016/j.jenvp.2023.102010

Lupien, S. J., de Leon, M., de Santi, S., Convit, A., Tarshish, C., Nair, N. P. V., Thakur, M., McEwen, B. S., Hauger, R. L., & Meaney, M. J. (1998). Cortisol levels during human aging predict hippocampal atrophy and memory deficits. *Nature Neuroscience, 1*(1), 69–73. https://doi.org/10.1038/271

Lupien, S. J., Juster, R.-P., Raymond, C., & Marin, M.-F. (2018). The effects of chronic stress on the human brain: From neurotoxicity, to vulnerability, to opportunity. *Frontiers in Neuroendocrinology, 49*, 91–105. https://doi.org/10.1016/j.yfrne.2018.02.001

Maas, J., van Dillen, S. M. E., Verheij, R. A., & Groenewegen, P. P. (2009). Social contacts as a possible mechanism behind the relation between green space and health. *Health & Place, 15*(2), 586–595. https://doi.org/10.1016/j.healthplace.2008.09.006

Mank, A., van Maurik, I. S., van Harten, A. C., Rhodius-Meester, H. F. M., Teunissen, C. E., van Berckel, B. N. M., Berkhof, J., van der Flier, W. M., & Rijnhart, J. J. M. (2022). Life satisfaction across the entire trajectory of Alzheimer's disease: A mediation analysis. *Alzheimer's & Dementia: Diagnosis, Assessment & Disease Monitoring, 14*(1), e12389. https://doi.org/10.1002/dad2.12389

Martin, L., White, M. P., Pahl, S., May, J., & Wheeler, B. W. (2020). Neighbourhood greenspace and smoking prevalence: Results from a nationally representative survey in England. *Social Science & Medicine, 265*(113), 448. https://doi.org/10.1016/j.socscimed.2020.113448

Mavros, P., Austwick, M. Z., & Smith, A. H. (2016). Geo-EEG: Towards the use of EEG in the study of urban behaviour. *Applied Spatial Analysis and Policy, 9*(2), 191–212. https://doi.org/10.1007/s12061-015-9181-z

McDougall, C. W., Hanley, N., Quilliam, R. S., Bartie, P. J., Robertson, T., Griffiths, M., & Oliver, D. M. (2021). Neighbourhood blue space and mental health: A nationwide ecological study of antidepressant medication prescribed to older adults. *Landscape and Urban Planning, 214*(104), 132. https://doi.org/10.1016/j.landurbplan.2021.104132

McEwen, B. S. (1998). Stress, adaptation, and disease: Allostasis and allostatic load. *Annals of the New York Academy of Sciences, 840*(1), 33–44. https://doi.org/10.1111/j.1749-6632.1998.tb09546.x

McGlashan, E. M., Poudel, G. R., Jamadar, S. D., Phillips, A. J. K., & Cain, S. W. (2021). Afraid of the dark: Light acutely suppresses activity in the human amygdala. *Plos One, 16*(6). https://doi.org/10.1371/journal.pone.0252350

McGregor, G., & Harvey, J. (2018). Food for thought: Leptin regulation of hippocampal function and its role in Alzheimer's disease. *Neuropharmacology, 136*, 298–306. https://doi.org/10.1016/j.neuropharm.2017.09.038

McMahan, E. A., & Estes, D. (2015). The effect of contact with natural environments on positive and negative affect: A meta-analysis. *The Journal of Positive Psychology, 10*(6), 507–519. https://doi.org/10.1080/17439760.2014.994224

Moore, M., Gould, P., & Keary, B. S. (2003). Global urbanization and impact on health. *International Journal of Hygiene and Environmental Health, 206*(4), 269–278. https://doi.org/10.1078/1438-4639-00223

Motoc, I., Hoogendijk, E. O., Timmermans, E. J., Deeg, D., Penninx, B., & Huisman, M. (2023). Social and physical neighbourhood characteristics and 10-year incidence of depression and anxiety in older adults: Results from the Longitudinal Aging Study Amsterdam. *Social Science & Medicine, 327*(115), 963. https://doi.org/10.1016/j.socscimed.2023.115963

Motohiro, A., Abe, T., Okuyama, K., Onoda, K., Ito, T., Isomura, M., Nabika, T., & Kumakura, S. (2021). Environmental factors affecting cognitive function among community-dwelling older adults: A longitudinal study. *International Journal of Environmental Research and Public Health, 18*(16).

Münzel, T., Sørensen, M., & Daiber, A. (2021). Transportation noise pollution and cardiovascular disease. *Nature Reviews Cardiology, 18*(9), 619–636. https://doi.org/10.1038/s41569-021-00532-5

Neale, C., Aspinall, P., Roe, J., Tilley, S., Mavros, P., Cinderby, S., Coyne, R., Thin, N., & Thompson, C. W. (2020). The impact of walking in different urban environments on brain activity in older people. *Cities & Health, 4*(1), 94–106. https://doi.org/10.1080/23748834.2019.1619893

Ng, T. P., Nyunt, M. S. Z., Shuvo, F. K., Eng, J. Y., Yap, K. B., Hee, L. M., Chan, S. P., & Scherer, S. (2017). The neighborhood built environment and cognitive function of older persons: Results from the Singapore Longitudinal Ageing Study. *Gerontology, 64*(2), 149–156. https://doi.org/10.1159/000480080

Olszewska-Guizzo, A., Sia, A., Fogel, A., & Ho, R. (2020). Can exposure to certain urban green spaces trigger frontal alpha asymmetry in the brain?-Preliminary findings from a passive task EEG study. *International Journal of Environmental Research and Public Health, 17*(2). https://doi.org/10.3390/ijerph17020394

Ozbilen, B., Akar, G., White, K., Dabelko-Schoeny, H., & Cao, Q. (2022). Analysing the travel behaviour of older adults: What are the determinants of sustainable mobility? *Ageing & Society*, 1–29. https://doi.org/10.1017/S0144686X22001180

Padeiro, M., de São José, J., Amado, C., Sousa, L., Roma Oliveira, C., Esteves, A., & McGarrigle, J. (2021). Neighborhood attributes and well-being among older adults in urban areas: A mixed-methods systematic review. *Research on Aging, 44*(5–6), 351–368. https://doi.org/10.1177/0164027521999980

Park, D. C., & Bischof, G. N. (2013). The aging mind: neuroplasticity in response to cognitive training. *Dialogues in Clinical Neuroscience, 15*(1), 109–119. https://doi.org/10.31887/DCNS.2013.15.1/dpark

Pasanen, T. P., White, M. P., Wheeler, B. W., Garrett, J. K., & Elliott, L. R. (2019). Neighbourhood blue space, health and wellbeing: The mediating role of different types of physical activity. *Environment International, 131*(105), 016. https://doi.org/10.1016/j.envint.2019.105016

Patterson, R., Webb, E., Hone, T., Millett, C., & Laverty, A. A. (2019). Associations of public transportation use with cardiometabolic health: A systematic review and meta-analysis. *American Journal of Epidemiology, 188*(4), 785–795. https://doi.org/10.1093/aje/kwz012

Paul, L. A., Hystad, P., Burnett, R. T., Kwong, J. C., Crouse, D. L., van Donkelaar, A., Tu, K., Lavigne, E., Copes, R., Martin, R. V., & Chen, H. (2020). Urban green space and the risks of dementia and stroke. *Environmental Research, 186*(109), 520. https://doi.org/10.1016/j.envres.2020.109520

Peen, J., Schoevers, R. A., Beekman, A. T., & Dekker, J. (2010). The current status of urban-rural differences in psychiatric disorders. *Acta Psychiatrica Scandinavica, 121*(2), 84–93. https://doi.org/10.1111/j.1600-0447.2009.01438.x

Pemberton, R., & Fuller Tyszkiewicz, M. D. (2016). Factors contributing to depressive mood states in everyday life: A systematic review. *Journal of Affective Disorders, 200*, 103–110. https://doi.org/10.1016/j.jad.2016.04.023

Peng, W., Shi, H., Li, M., Li, X., Liu, T., & Wang, Y. (2022). Association of residential greenness with geriatric depression among the elderly covered by long-term care insurance in Shanghai. *Environmental Science and Pollution Research, 29*(8), 12054–12064. https://doi.org/10.1007/s11356-021-16585-5

Pérez, E., Braën, C., Boyer, G., Mercille, G., Rehany, É., Deslauriers, V., Bilodeau, A., & Potvin, L. (2020). Neighbourhood community life and health: A systematic review of reviews. *Health & Place, 61*(102), 238. https://doi.org/10.1016/j.healthplace.2019.102238

Peters, R., Ee, N., Peters, J., Booth, A., Mudway, I., & Anstey, K. J. (2019). Air pollution and dementia: A systematic review. *Journal of Alzheimer's Disease, 70*, S145–S163. https://doi.org/10.3233/JAD-180631

Phillips, C. (2017). Lifestyle modulators of neuroplasticity: How physical activity, mental engagement, and diet promote cognitive health during aging. *Neural Plasticity, 3*(589), 271. https://doi.org/10.1155/2017/3589271

Pietrzak, R. H., Lim, Y. Y., Neumeister, A., Ames, D., Ellis, K. A., Harrington, K., Lautenschlager, N. T., Restrepo, C., Martins, R. N., Masters, C. L., Villemagne, V. L., Rowe, C. C., Maruff, P., & for the Australian Imaging, B., & Lifestyle Research, G. (2015). Amyloid-β, anxiety, and cognitive decline in preclinical Alzheimer disease: A multicenter, prospective cohort study. *JAMA Psychiatry, 72*(3), 284–291. https://doi.org/10.1001/jamapsychiatry.2014.2476

Pini, L., Salvalaggio, A., Wennberg, A. M., Dimakou, A., Matteoli, M., & Corbetta, M. (2023). The pollutome-connectome axis: A putative mechanism to explain pollution effects on neurodegeneration. *Ageing Research Reviews, 86*(101), 867. https://doi.org/10.1016/j.arr.2023.101867

Pope, D., Tisdall, R., Middleton, J., Verma, A., van Ameijden, E., Birt, C., Macherianakis, A., & Bruce, N. G. (2018). Quality of and access to green space in relation to psychological distress: Results from a population-based cross-sectional study as part of the EURO-URHIS 2 project. *European Journal of Public Health, 28*(1), 35–38. https://doi.org/10.1093/eurpub/ckv094

Poudel, G. R., Hawes, S., Innes, C. R. H., Parsons, N., Drummond, S. P. A., Caeyensberghs, K., & Jones, R. D. (2021). RoWDI: Rolling window detection of sleep intrusions in the awake brain using fMRI. *Journal of Neural Engineering, 18*(5). https://doi.org/10.1088/1741-2552/ac2bb9

Power, M. C., Adar, S. D., Yanosky, J. D., & Weuve, J. (2016). Exposure to air pollution as a potential contributor to cognitive function, cognitive decline, brain imaging, and dementia: A systematic review of epidemiologic research. *NeuroToxicology, 56*, 235–253. https://doi.org/10.1016/j.neuro.2016.06.004

Power, M. C., Lamichhane, A. P., Liao, D. P., Xu, X. H., Jack, C. R., Gottesman, R. F., Mosley, T., Stewart, J. D., Yanosky, J. D., & Whitsel, E. A. (2018). The association of long-term exposure to particulate matter air pollution with brain MRI findings: The ARIC study. *Environmental Health Perspectives, 126*(2). https://doi.org/10.1289/EHP2152

Pun, V. C., Manjourides, J., & Suh, H. H. (2019). Close proximity to roadway and urbanicity associated with mental ill-health in older adults. *Science of The Total Environment, 658*, 854–860. https://doi.org/10.1016/j.scitotenv.2018.12.221

Qin, S., Leong, R. L. F., Ong, J. L., & Chee, M. W. L. (2023). Associations between objectively measured sleep parameters and cognition in healthy older adults: A meta-analysis. *Sleep Medicine Reviews, 67*(101), 734. https://doi.org/10.1016/j.smrv.2022.101734

Qiu, X., Shi, L., Kubzansky, L. D., Wei, Y., Castro, E., Li, H., Weisskopf, M. G., & Schwartz, J. D. (2023). Association of long-term exposure to air pollution with late-life depression in older adults in the US. *JAMA Network Open, 6*(2), e2253668. https://doi.org/10.1001/jamanetworkopen.2022.53668

Rahimi-Ardabili, H., Astell-Burt, T., Nguyen, P.-Y., Zhang, J., Jiang, Y., Dong, G.-H., & Feng, X. (2021). Green space and health in Mainland China: A systematic review. *International Journal of Environmental Research and Public Health, 18*(18).

Ramos, C., Gibson, G. R., Walton, G. E., Magistro, D., Kinnear, W., & Hunter, K. (2022). Systematic review of the effects of exercise and physical activity on the gut microbiome of older adults. *Nutrients, 14*(3). https://doi.org/10.3390/nu14030674

Ranft, U., Schikowski, T., Sugiri, D., Krutmann, J., & Krämer, U. (2009). Long-term exposure to traffic-related particulate matter impairs cognitive function in the elderly. *Environmental Research, 109*(8), 1004–1011. https://doi.org/10.1016/j.envres.2009.08.003

Reinhard, E., Carrino, L., Courtin, E., van Lenthe, F. J., & Avendano, M. (2019). Public transportation use and cognitive function in older age: A Quasiexperimental evaluation of the free bus pass policy in the United Kingdom. *American Journal of Epidemiology, 188*(10), 1774–1783. https://doi.org/10.1093/aje/kwz149

Reklaitiene, R., Grazuleviciene, R., Dedele, A., Virviciute, D., Vensloviene, J., Tamosiunas, A., Baceviciene, M., Luksiene, D., Sapranaviciute-Zabazlajeva, L., Radisauskas, R., Bernotiene, G., Bobak, M., & Nieuwenhuijsen, M. J. (2014). The relationship of green space, depressive symptoms and perceived general health in urban population. *Scandinavian Journal of Public Health, 42*(7), 669–676. https://doi.org/10.1177/1403494814544494

Ren, Y., Savadlou, A., Park, S., Siska, P., Epp, J. R., & Sargin, D. (2023). The impact of loneliness and social isolation on the development of cognitive decline and Alzheimer's Disease. *Frontiers in Neuroendocrinology, 69*(101), 061. https://doi.org/10.1016/j.yfrne.2023.101061

Ricciardi, E., Spano, G., Lopez, A., Tinella, L., Clemente, C., Elia, G., Dadvand, P., Sanesi, G., Bosco, A., & Caffò, A. O. (2022). Long-term exposure to greenspace and cognitive function during the lifespan: A systematic review. *International Journal of Environmental Research and Public Health, 19*(18). https://doi.org/10.3390/ijerph191811700

Robbins, R. N., Scott, T., Joska, J. A., & Gouse, H. (2019). Impact of urbanization on cognitive disorders. *Current Opinion in Psychiatry, 32*(3), 210–217. https://doi.org/10.1097/yco.0000000000000490

Rodriguez-Loureiro, L., Gadeyne, S., Bauwelinck, M., Lefebvre, W., Vanpoucke, C., & Casas, L. (2022). Long-term exposure to residential greenness and neurodegenerative disease mortality among older adults: A 13-year follow-up cohort study. *Environmental Health, 21*(1), 49. https://doi.org/10.1186/s12940-022-00863-x

Rothman, S. M., & Mattson, M. P. (2013). Activity-dependent, stress-responsive BDNF signaling and the quest for optimal brain health and resilience throughout the lifespan. *Neuroscience, 239*, 228–240. https://doi.org/10.1016/j.neuroscience.2012.10.014

Saarloos, D., Alfonso, H., Giles-Corti, B., Middleton, N., & Almeida, O. P. (2011). The built environment and depression in later life: The health in men study. *The American Journal of Geriatric Psychiatry, 19*(5), 461–470. https://doi.org/10.1097/JGP.0b013e3181e9b9bf

Sabia, S., Fayosse, A., Dumurgier, J., Schnitzler, A., Empana, J.-P., Ebmeier, K. P., Dugravot, A., Kivimäki, M., & Singh-Manoux, A. (2019). Association of ideal cardiovascular health at age 50 with incidence of dementia: 25 year follow-up of Whitehall II cohort study. *BMJ, 366*, l4414. https://doi.org/10.1136/bmj.l4414

Saelens, B. E., Sallis, J. F., & Frank, L. D. (2003). Environmental correlates of walking and cycling: Findings from the transportation, urban design, and planning literatures. *Annals of Behavioral Medicine, 25*(2), 80–91. https://doi.org/10.1207/S15324796ABM2502_03

Salinas, J., Beiser, A. S., Samra, J. K., O'Donnell, A., DeCarli, C. S., Gonzales, M. M., Aparicio, H. J., & Seshadri, S. (2022). Association of loneliness with 10-year dementia risk and early markers of vulnerability for neurocognitive decline. *Neurology, 98*(13), e1337. https://doi.org/10.1212/WNL.0000000000200039

Salthouse, T. A. (2009). When does age-related cognitive decline begin? *Neurobiology of Aging, 30*(4), 507–514. https://doi.org/10.1016/j.neurobiolaging.2008.09.023

Sandi, C. (2013). Stress and cognition. *WIREs Cognitive Science, 4*(3), 245–261. https://doi.org/10.1002/wcs.1222

Santabárbara, J., Lipnicki, D. M., Bueno-Notivol, J., Olaya-Guzmán, B., Villagrasa, B., & López-Antón, R. (2020). Updating the evidence for an association between anxiety and risk of Alzheimer's disease: A meta-analysis of prospective cohort studies. *Journal of Affective Disorders, 262*, 397–404. https://doi.org/10.1016/j.jad.2019.11.065

Sarkar, C., Webster, C., & Gallacher, J. (2018). Residential greenness and prevalence of major depressive disorders: A cross-sectional, observational, associational study of 94879 adult UK Biobank participants. *Lancet Planet Health, 2*(4), e162-e173. https://doi.org/10.1016/S2542-5196(18)30051-2.

Scharf, T., & de Jong Gierveld, J. (2008). Loneliness in urban neighbourhoods: An Anglo-Dutch comparison. *European Journal of Ageing, 5*(2), 103. https://doi.org/10.1007/s10433-008-0080-x

Schikowski, T., Vossoughi, M., Vierkötter, A., Schulte, T., Teichert, T., Sugiri, D., Fehsel, K., Tzivian, L., Bae, I.-S., Ranft, U., Hoffmann, B., Probst-Hensch, N., Herder, C., Krämer, U., & Luckhaus, C. (2015). Association of air pollution with cognitive functions and its modification by APOE gene variants in elderly women. *Environmental Research, 142*, 10–16. https://doi.org/10.1016/j.envres.2015.06.009

Schmidt, T., Kerr, J., & Schipperijn, J. (2019). Associations between neighborhood open space features and walking and social interaction in older adults—A mixed methods study. *Geriatrics, 4*(3). https://doi.org/10.3390/geriatrics4030041

Scoditti, E., Tumolo, M. R., & Garbarino, S. (2022). Mediterranean diet on sleep: A health alliance. *Nutrients, 14*(14). https://doi.org/10.3390/nu14142998

Shin, J. C., Parab, K. V., An, R., & Grigsby-Toussaint, D. S. (2020). Greenspace exposure and sleep: A systematic review. *Environmental Research, 182*(109), 081. https://doi.org/10.1016/j.envres.2019.109081

Shortt, N. K., Rind, E., Pearce, J., Mitchell, R., & Curtis, S. (2018). Alcohol risk environments, vulnerability, and social inequalities in alcohol consumption. *Annals of the American Association of Geographers, 108*(5), 1210–1227. https://doi.org/10.1080/24694452.2018.1431105

Slawsky, E. D., Hajat, A., Rhew, I. C., Russette, H., Semmens, E. O., Kaufman, J. D., Leary, C. S., & Fitzpatrick, A. L. (2022). Neighborhood greenspace exposure as a protective factor in dementia risk among U.S. adults 75 years or older: A cohort study. *Environmental Health, 21*(1), 14. https://doi.org/10.1186/s12940-022-00830-6

Smith, M. G., Cordoza, M., & Basner, M. (2022). Environmental noise and effects on sleep: An update to the WHO systematic review and meta-analysis. *Environmental Health Perspectives, 130*(7), 076001. https://doi.org/10.1289/EHP10197

Soloveva, M. V., Poudel, G., Barnett, A., Shaw, J. E., Martino, E., Knibbs, L. D., Anstey, K. J., & Cerin, E. (2023). Characteristics of urban neighbourhood environments and cognitive age in mid-age and older adults. *Health & Place, 83*(103), 077. https://doi.org/10.1016/j.healthplace.2023.103077

Song, C., Ikei, H., Igarashi, M., Takagaki, M., & Miyazaki, Y. (2015). Physiological and psychological effects of a walk in urban parks in Fall. *International Journal of Environmental Research and Public Health, 12*(11), 14216–14228. https://doi.org/10.3390/ijerph121114216

Stanhope, J., Liddicoat, C., & Weinstein, P. (2021). Outdoor artificial light at night: A forgotten factor in green space and health research. *Environmental Research, 197*(111), 012. https://doi.org/10.1016/j.envres.2021.111012

Stigsdotter, U. K., Ekholm, O., Schipperijn, J., Toftager, M., Kamper-Jørgensen, F., & Randrup, T. B. (2010). Health promoting outdoor environments - Associations between green space, and health, health-related quality of life and stress based on a Danish national representative survey. *Scandinavian Journal of Public Health, 38*(4), 411–417. https://doi.org/10.1177/1403494810367468

Stokols, D. (1996). Translating social ecological theory into guidelines for community health promotion. *American Journal of Health Promotion, 10*(4), 282–298. https://doi.org/10.4278/0890-1171-10.4.282

Stuart, K. E., & Padgett, C. (2020). A Systematic review of the association between psychological stress and dementia risk in humans. *Journal of Alzheimer's disease: JAD, 78*(1), 335–352. https://doi.org/10.3233/jad-191096

Sui, Y., Ettema, D., & Helbich, M. (2022). Longitudinal associations between the neighborhood social, natural, and built environment and mental health: A systematic review with meta-analyses. *Health & Place, 77*(102), 893. https://doi.org/10.1016/j.healthplace.2022.102893

Sulkava, S., Haukka, J., Sulkava, R., Laatikainen, T., & Paunio, T. (2022). Association between psychological distress and incident dementia in a population-based cohort in Finland. *JAMA Network Open, 5*(12), e2247115–e2247115. https://doi.org/10.1001/jamanetworkopen.2022.47115

Sullivan, E. V., Zahr, N. M., Sassoon, S. A., Thompson, W. K., Kwon, D., Pohl, K. M., & Pfefferbaum, A. (2018). The role of aging, drug dependence, and hepatitis C comorbidity in alcoholism cortical compromise. *JAMA Psychiatry, 75*(5), 474–483. https://doi.org/10.1001/jamapsychiatry.2018.0021

Sullivan, A., Armendariz, M., & Thierry, A. D. (2023). A scoping review of neighborhoods and cognitive health disparities among US midlife and older adults. *Journal of Aging and Health*, 08982643231185379. https://doi.org/10.1177/08982643231185379

Sun, Z., Wang, Z. T., Sun, F. R., Shen, X. N., Xu, W., Ma, Y. H., Dong, Q., Tan, L., & Yu, J. T. (2020). Late-life obesity is a protective factor for prodromal Alzheimer's disease: A longitudinal study. *Aging (Albany NY), 12*(2), 2005–2017. https://doi.org/10.18632/aging.102738

Sun, Y., Molitor, J., Benmarhnia, T., Avila, C., Chiu, V., Slezak, J., Sacks, D. A., Chen, J. C., Getahun, D., & Wu, J. (2023). Association between urban green space and postpartum depression, and the role of physical activity: A retrospective cohort study in Southern California. *The Lancet Regional Health – Americas, 21*, 100,462. https://doi.org/10.1016/j.lana.2023.100462

Sutin, A. R., Luchetti, M., Aschwanden, D., Zhu, X., Stephan, Y., & Terracciano, A. (2023). Loneliness and risk of all-cause, Alzheimer's, vascular, and frontotemporal dementia: A prospective study of 492,322 individuals over 15 years. *International Psychogeriatrics, 35*(6), 283–292. https://doi.org/10.1017/S1041610222001028

Swisher, R. R. Neighbourhood. *Encyclopedia Britannica*. https://www.britannica.com/topic/neighborhood-sociology. Accessed 31 July 2023.

Tang, I. C., Tsai, Y. P., Lin, Y. J., Chen, J. H., Hsieh, C. H., Hung, S. H., Sullivan, W. C., Tang, H. F., & Chang, C. Y. (2017). Using functional Magnetic Resonance Imaging (fMRI) to analyze brain region activity when viewing landscapes. *Landscape and Urban Planning, 162*, 137–144. https://doi.org/10.1016/j.landurbplan.2017.02.007

Tani, Y., Suzuki, N., Fujiwara, T., Hanazato, M., & Kondo, K. (2019). Neighborhood food environment and dementia incidence: The Japan gerontological evaluation study cohort survey. *American Journal of Preventive Medicine, 56*(3), 383–392. https://doi.org/10.1016/j.amepre.2018.10.028

Tani, Y., Hanazato, M., Fujiwara, T., Suzuki, N., & Kondo, K. (2021). Neighborhood sidewalk environment and incidence of dementia in older Japanese adults: The Japan gerontological evaluation study cohort. *American Journal of Epidemiology, 190*(7), 1270–1280. https://doi.org/10.1093/aje/kwab043

Thierry, A. D., Sherman-Wilkins, K., Armendariz, M., Sullivan, A., & Farmer, H. R. (2021). Perceived neighborhood characteristics and cognitive functioning among diverse older adults: An intersectional approach. *International Journal of Environmental Research and Public Health, 18*(5). https://doi.org/10.3390/ijerph18052661

Topiwala, A., Allan, C. L., Valkanova, V., Zsoldos, E., Filippini, N., Sexton, C., Mahmood, A., Fooks, P., Singh-Manoux, A., Mackay, C. E., Kivimäki, M., & Ebmeier, K. P. (2017). Moderate alcohol consumption as risk factor for adverse brain outcomes and cognitive decline: longitudinal cohort study. *BMJ, 357*, j2353. https://doi.org/10.1136/bmj.j2353

Tost, H., Reichert, M., Braun, U., Reinhard, I., Peters, R., Lautenbach, S., Hoell, A., Schwarz, E., Ebner-Priemer, U., Zipf, A., & Meyer-Lindenberg, A. (2019). Neural correlates of individual differences in affective benefit of real-life urban green space exposure. *Nature Neuroscience, 22*(9), 1389–1393. https://doi.org/10.1038/s41593-019-0451-y

Tsereteli, N., Vallat, R., Fernandez-Tajes, J., Delahanty, L. M., Ordovas, J. M., Drew, D. A., Valdes, A. M., Segata, N., Chan, A. T., Wolf, J., Berry, S. E., Walker, M. P., Spector, T. D., & Franks, P. W. (2022). Impact of insufficient sleep on dysregulated blood glucose control under standardised meal conditions. *Diabetologia, 65*(2), 356–365. https://doi.org/10.1007/s00125-021-05608-y

Tzivian, L., Dlugaj, M., Winkler, A., Hennig, F., Fuks, K., Sugiri, D., Schikowski, T., Jakobs, H., Erbel, R., Jöckel, K.-H., Moebus, S., Hoffmann, B., & Weimar, C. (2016). Long-term air pollution and traffic noise exposures and cognitive function: A cross-sectional analysis of the Heinz Nixdorf Recall study. *Journal of Toxicology and Environmental Health, Part A, 79*(22–23), 1057–1069. https://doi.org/10.1080/15287394.2016.1219570

Ulrich, R. S. (1981). Natural versus urban scenes - Some psychophysiological effects. *Environment and Behavior, 13*(5), 523–556. https://doi.org/10.1177/0013916581135O

United Nations Department of Economic and Social Affairs. 2018 Revision of World Urbanization Prospects. https://population.un.org/wup/publications/Files/WUP2018-Methodology.pdf. Accessed 1 July 2023.

United Nations, Department of Economic and Social Affairs. (2019). *The Sustainable development goals report 2019*. United Nations. https://unstats.un.org/sdgs/report/2019/The-Sustainable-Development-Goals-Report-2019.pdf

United Nations, Department of Economic and Social Affairs, Population Division. (2014). *Urban and rural population by age and sex, 1980–2015*. United Nations. https://esa.un.org/unpd/popdev/urpas/urpas2014.aspx

United Nations, World Population Ageing 2017. (2017). *Highlights (ST/ESA/SER.A/397)*. United Nations. https://www.un.org/en/development/desa/population/publications/pdf/ageing/WPA2017_Highlights.pdf

Vallarta-Robledo, J. R., Marques-Vidal, P., Sandoval, J. L., De Ridder, D., Schaffner, E., Humair, J.-P., Cornuz, J., Probst-Hensch, N., Joost, S., & Guessous, I. (2022). The neighborhood environment and its association with the spatio-temporal footprint of tobacco consumption and changes in smoking-related behaviors in a Swiss urban area. *Health & Place, 76*(102), 845. https://doi.org/10.1016/j.healthplace.2022.102845

van Balkom, T. D., van den Heuve, O. A., Berendse, H. W., van der Werf, Y. D., & Vriend, C. (2020). The effects of cognitive training on brain network activity and connectivity in aging and neurodegenerative diseases: A systematic review. *Neuropsychological Review, 30*(2), 267–286. https://doi.org/10.1007/s11065-020-09440-w

Walker, K. A., Le Page, L. M., Terrando, N., Duggan, M. R., Heneka, M. T., & Bettcher, B. M. (2023). The role of peripheral inflammatory insults in Alzheimer's disease: A review and research roadmap. *Molecular Neurodegeneration, 18*(1), 37. https://doi.org/10.1186/s13024-023-00627-2

Walters, K., Breeze, E., Wilkinson, P., Price, G. M., Bulpitt, C. J., & Fletcher, A. (2004). Local area deprivation and urban–rural differences in anxiety and depression among people older than 75 years in Britain. *American Journal of Public Health, 94*(10), 1768–1774. https://doi.org/10.2105/AJPH.94.10.1768

Wang, Y. Q., Jiang, M. Y., Huang, Y. S., Sheng, Z. Y., Huang, X., Lin, W., Chen, Q. B., Li, X., Luo, Z. H., & Lv, B. Y. (2020a). Physiological and psychological effects of watching videos of different durations showing urban bamboo forests with varied structures. *International Journal of Environmental Research and Public Health, 17*(10). https://doi.org/10.3390/ijerph17103434

Wang, R., Yang, B., Yao, Y., Bloom, M. S., Feng, Z., Yuan, Y., Zhang, J., Liu, P., Wu, W., Lu, Y., Baranyi, G., Wu, R., Liu, Y., & Dong, G. (2020b). Residential greenness, air pollution and psychological well-being among urban residents in Guangzhou, China. *Science of the Total Environment, 711*(134), 843. https://doi.org/10.1016/j.scitotenv.2019.134843

Wang, S., Li, Z., Wang, X., Guo, S., Sun, Y., Li, G., Zhao, C., Yuan, W., Li, M., Li, X., & Ai, S. (2022). Associations between sleep duration and cardiovascular diseases: A meta-review and meta-analysis of observational and Mendelian randomization studies [Systematic Review].

Frontiers in Cardiovascular Medicine, 9. https://www.frontiersin.org/articles/10.3389/fcvm.2022.930000

Wang, Y., Tan, H., Zheng, H., Ma, Z., Zhan, Y., Hu, K., Yang, Z., Yao, Y., & Zhang, Y. (2023). Exposure to air pollution and gains in body weight and waist circumference among middle-aged and older adults. *Science of The Total Environment, 869*(161), 895. https://doi.org/10.1016/j.scitotenv.2023.161895

Watts, A., Ferdous, F., Moore, K. D., & Burns, J. M. (2015). Neighborhood integration and connectivity predict cognitive performance and decline. *Gerontology and Geriatric Medicine, 1*. https://doi.org/10.1177/2333721415599141

Weden, M. M., Shih, R. A., Kabeto, M. U., & Langa, K. M. (2018). Secular trends in dementia and cognitive impairment of US rural and urban older adults. *American Journal of Preventive Medicine, 54*(2), 164–172. https://doi.org/10.1016/j.amepre.2017.10.021

Wee, J., Sukudom, S., Bhat, S., Marklund, M., Peiris, N. J., Hoyos, C. M., Patel, S., Naismith, S. L., Dwivedi, G., & Misra, A. (2023). The relationship between midlife dyslipidemia and lifetime incidence of dementia: A systematic review and meta-analysis of cohort studies. *Alzheimer's & Dementia: Diagnosis, Assessment & Disease Monitoring, 15*(1), e12395. https://doi.org/10.1002/dad2.12395

Wellenius, G. A., Boyle, L. D., Coull, B. A., Milberg, W. P., Gryparis, A., Schwartz, J., Mittleman, M. A., & Lipsitz, L. A. (2012). Residential proximity to nearest major roadway and cognitive function in community-dwelling seniors: Results from the MOBILIZE boston study. *Journal of the American Geriatrics Society, 60*(11), 2075–2080. https://doi.org/10.1111/j.1532-5415.2012.04195.x

White, M. P., Elliot, L. R., Grellier, J., Economou, T., Bell, S., Bratman, G. N., et al. (2021). Associations between green/blue spaces and mental health across 18 countries. *Scientific Reports, 11*, 8903. https://doi.org/10.1038/s41598-021-87675-0

White, A. M., Orosz, A., Powell, P. A., & Koob, G. F. (2023). Alcohol and aging—An area of increasing concern. *Alcohol, 107*, 19–27. https://doi.org/10.1016/j.alcohol.2022.07.005

Wilker, E. H., Preis, S. R., Beiser, A. S., Wolf, P. A., Au, R., Kloog, I., Li, W., Schwartz, J., Koutrakis, P., DeCarli, C., Seshadri, S., & Mittleman, M. A. (2015). Long-term exposure to fine particulate matter, residential proximity to major roads and measures of brain structure. *Stroke, 46*(5), 1161–1166. https://doi.org/10.1161/STROKEAHA.114.008348

Wilker, E. H., Martinez-Ramirez, S., Kloog, I., Schwartz, J., Mostofsky, E., Koutrakis, P., Mittleman, M. A., & Viswanathan, A. (2016). Fine particulate matter, residential proximity to major roads, and markers of small vessel disease in a memory study population. *Journal of Alzheimer's Disease, 53*(4), 1315–1323. https://doi.org/10.3233/JAD-151143

Wilker, E. H., Osman, M., & Weisskopf, M. G. (2023). Ambient air pollution and clinical dementia: systematic review and meta-analysis. *BMJ, 381*, e071620. https://doi.org/10.1136/bmj-2022-071620

Wilson, R. S., Begeny, C. T., Boyle, P. A., Schneider, J. A., & Bennett, D. A. (2011). Vulnerability to stress, anxiety, and development of dementia in old age. *The American Journal of Geriatric Psychiatry, 19*(4), 327–334. https://doi.org/10.1097/JGP.0b013e31820119da

Wolf, P. A., D'Agostino, R. B., Kannel, W. B., Bonita, R., & Belanger, A. J. (1988). Cigarette smoking as a risk factor for stroke: The Framingham study. *JAMA, 259*(7), 1025–1029. https://doi.org/10.1001/jama.1988.03720070025028

World Bank. (2020). Urban Development Overview. https://www.worldbank.org/en/topic/urbandevelopment/overview. Accessed 1 July 2023.

Wörn, J., Ellwardt, L., Aartsen, M., & Huisman, M. (2017). Cognitive functioning among Dutch older adults: Do neighborhood socioeconomic status and urbanity matter? *Social Science & Medicine, 187*, 29–38. https://doi.org/10.1016/j.socscimed.2017.05.052

Wu, J., & Jackson, L. (2021). Greenspace inversely associated with the risk of Alzheimer's disease in the mid-Atlantic United States. *Earth, 2*(1), 140–150.

Wu, Y.-T., Prina, A. M., Jones, A. P., Barnes, L. E., Matthews, F. E., Brayne, C., & on the behalf of the Medical Research Council Cognitive, F., & Ageing, S. (2015). Community environment, cognitive impairment and dementia in later life: results from the Cognitive Function and Ageing Study. *Age and Ageing, 44*(6), 1005–1011. https://doi.org/10.1093/ageing/afv137

Wu, Y.-T., Prina, A. M., Jones, A., Matthews, F. E., & Brayne, C. (2017). The built environment and cognitive disorders: Results from the cognitive function and ageing study II. *American Journal of Preventive Medicine, 53*(1), 25–32. https://doi.org/10.1016/j.amepre.2016.11.020

Wu, Y.-T., Brayne, C., Liu, Z., Huang, Y., Sosa, A. L., Acosta, D., & Prina, M. (2020). Neighbourhood environment and dementia in older people from high-, middle- and low-income countries: Results from two population-based cohort studies. *BMC Public Health, 20*(1), 1330. https://doi.org/10.1186/s12889-020-09435-5

Xu, J., Liu, N., Polemiti, E., Garcia-Mondragon, L., Tang, J., Liu, X., Lett, T., Yu, L., Nöthen, M. M., Feng, J., Yu, C., Marquand, A., Schumann, G., Walter, H., Heinz, A., Ralser, M., Twardziok, S., Vaidya, N., Serin, E., et al. (2023). Effects of urban living environments on mental health in adults. *Nature Medicine, 29*(6), 1456–1467. https://doi.org/10.1038/s41591-023-02365-w

Yao, Y., Xurui, J., Zhang, J., & Zeng, Y. (2022). Living closer to major roads may increase the risk of cognitive decline. In Y. Zeng, J. Lu, X. Lei, & X. Shi (Eds.), *Trends and Determinants of Healthy Aging in China* (pp. 135–149). Springer Nature. https://doi.org/10.1007/978-981-19-4154-2_8

Yu, Z., Wei, F., Zhang, X., Wu, M., Lin, H., Shui, L., Jin, M., Wang, J., Tang, M., & Chen, K. (2021). Air pollution, surrounding green, road proximity and Parkinson's disease: A prospective cohort study. *Environmental Research, 197*(111), 170. https://doi.org/10.1016/j.envres.2021.111170

Yu, W., Esposito, M., Li, M., Clarke, P., Judd, S., & Finlay, J. (2023). Neighborhood 'Disamenities': Local barriers and cognitive function among Black and white aging adults. *BMC Public Health, 23*(1), 197. https://doi.org/10.1186/s12889-023-15026-x

Yuchi, W., Sbihi, H., Davies, H., Tamburic, L., & Brauer, M. (2020). Road proximity, air pollution, noise, green space and neurologic disease incidence: A population-based cohort study. *Environmental Health, 19*(1), 8. https://doi.org/10.1186/s12940-020-0565-4

Yue, Y., Yang, D., Owen, N., & Van Dyck, D. (2022). The built environment and mental health among older adults in Dalian: The mediating role of perceived environmental attributes. *Social Science & Medicine, 311*(115), 333. https://doi.org/10.1016/j.socscimed.2022.115333

Zagnoli, F., Filippini, T., Jimenez, M. P., Wise, L. A., Hatch, E. E., & Vinceti, M. (2022). Is greenness associated with dementia? A systematic review and dose–response meta-analysis. *Current Environmental Health Reports, 9*(4), 574–590. https://doi.org/10.1007/s40572-022-00365-5

Zahodne, L. B., Stern, Y., & Manly, J. J. (2014). Depressive symptoms precede memory decline, but not vice versa, in non-demented older adults. *Journal of the American Geriatrics Society, 62*(1), 130–134. https://doi.org/10.1111/jgs.12600

Zainal, N. H., & Newman, M. G. (2022). Elevated anxious and depressed mood relates to future executive dysfunction in older adults: A longitudinal network analysis of psychopathology and cognitive functioning. *Clinical Psychological Science, 11*(2), 218–238. https://doi.org/10.1177/21677026221114076

Zhang, C. J. P., Barnett, A., Sit, C. H. P., Lai, P. C., Johnston, J. M., Lee, R. S. Y., & Cerin, E. (2018). Cross-sectional associations of objectively assessed neighbourhood attributes with depressive symptoms in older adults of an ultra-dense urban environment: the Hong Kong ALECS study. *BMJ Open, 8*(3), e020480. https://doi.org/10.1136/bmjopen-2017-020480

Zhang, C. J. P., Barnett, A., Johnston, J. M., Lai, P.-C., Lee, R. S. Y., Sit, C. H. P., & Cerin, E. (2019). Objectively-measured neighbourhood attributes as correlates and moderators of quality of life in older adults with different living arrangements: The ALECS cross-sectional study. *International Journal of Environmental Research and Public Health, 16*(5). https://doi.org/10.3390/ijerph16050876

Zhang, L., Luo, Y., Zhang, Y., Pan, X., Zhao, D., & Wang, Q. (2022a). Green Space, Air Pollution, Weather, and Cognitive Function in Middle and Old Age in China [Original Research]. *Frontiers in Public Health, 10*. https://www.frontiersin.org/articles/10.3389/fpubh.2022.871104

Zhang, R., He, X., Liu, Y., Li, M., & Zhou, C. (2022b). The relationship between built environment and mental health of older adults: Mediating effects of perceptions of community cohesion and community safety and the moderating effect of income. *Frontiers in Public Health, 10*(881), 169. https://doi.org/10.3389/fpubh.2022.881169

Zhang, Y.-R., Xu, W., Zhang, W., Wang, H.-F., Ou, Y.-N., Qu, Y., Shen, X.-N., Chen, S.-D., Wu, K.-M., Zhao, Q.-H., Zhang, H.-N., Sun, L., Dong, Q., Tan, L., Feng, L., Zhang, C., Evangelou, E., Smith, A. D., & Yu, J.-T. (2022c). Modifiable risk factors for incident dementia and cognitive impairment: An umbrella review of evidence. *Journal of Affective Disorders, 314*, 160–167. https://doi.org/10.1016/j.jad.2022.07.008

Zhao, Q., Xiang, H., Cai, Y., Meng, S. S., Zhang, Y., & Qiu, P. (2022). Systematic evaluation of the associations between mental disorders and dementia: An umbrella review of systematic reviews and meta-analyses. *Journal of Affective Disorders, 307*, 301–309. https://doi.org/10.1016/j.jad.2022.03.010

Zhu, A., Wu, C., Yan, L. L., Wu, C. D., Bai, C., Shi, X., Zeng, Y., & Ji, J. S. (2019). Association between residential greenness and cognitive function: analysis of the Chinese Longitudinal Healthy Longevity Survey. *BMJ Nutrition, Prevention & Health, 2*(2), 72–79. https://doi.org/10.1136/bmjnph-2019-000030

Zhu, A., Yan, L., Shu, C., Zeng, Y., & Ji, J. S. (2020). APOE ε4 modifies Effect of residential greenness on cognitive function among older adults: A longitudinal analysis in China. *Scientific Reports, 10*(1), 82. https://doi.org/10.1038/s41598-019-57082-7

Zhu, Z., Yang, Z., Yu, L., Xu, L., Wu, Y., Zhang, X., Shen, P., Lin, H., Shui, L., Tang, M., Jin, M., Wang, J., & Chen, K. (2023). Residential greenness, air pollution and incident neurodegenerative disease: A cohort study in China. *Science of The Total Environment, 878*(163), 173. https://doi.org/10.1016/j.scitotenv.2023.163173

Zijlema, W. L., Triguero-Mas, M., Smith, G., Cirach, M., Martinez, D., Dadvand, P., Gascon, M., Jones, M., Gidlow, C., Hurst, G., Masterson, D., Ellis, N., van den Berg, M., Maas, J., van Kamp, I., van den Hazel, P., Kruize, H., Nieuwenhuijsen, M. J., & Julvez, J. (2017). The relationship between natural outdoor environments and cognitive functioning and its mediators. *Environmental Research, 155*, 268–275. https://doi.org/10.1016/j.envres.2017.02.017

Zijlema, W. L., Stasinska, A., Blake, D., Dirgawati, M., Flicker, L., Yeap, B. B., Golledge, J., Hankey, G. J., Nieuwenhuijsen, M., & Heyworth, J. (2019). The longitudinal association between natural outdoor environments and mortality in 9218 older men from Perth, Western Australia. *Environment International, 125*, 430–436. https://doi.org/10.1016/j.envint.2019.01.075

The CITY & the PSYCHE—A Chronology

Nassim Mehran, Ulrike Kluge, and Mazda Adli

Abstract This text takes a chronological approach to examining the literature concerning the impact of urban living on mental health, identifying two distinct generations of knowledge production on this topic: (1) the first generation: from philosophical speculation to empirical studies, and (2) the second generation: from city-centrism to embracing the complexity of context. The contrast between these generations is evident in the conceptualization of health and city, the methodologies utilized to examine health, the influence of global dynamics, the rural-urban ratio, and the technological and methodological progression.

It is essential to consider that (1) there is a potential for bias in the text presented since the discussions drawn in a review depending on the results of the included studies and the authors' expertise and access to information, and (2) the text follows a chronological order, and it has referenced existing literature according to their respective timelines. Consequently, specific discussions and literature may appear outdated by contemporary standards or the current state of the art but were pertinent at their time.

Keywords Urbanisation · Urbanity · Rurality · Mental health · Psychiatry · Psychology · Environment

---

N. Mehran (✉) · M. Adli
Department of Psychiatry and Psychotherapy, Campus Mitte, Charité – Universitätsmedizin Berlin, Berlin, Germany
e-mail: nassim.mehran@charite.de

U. Kluge
Department of Psychiatry and Psychotherapy, Campus Mitte, Charité – Universitätsmedizin Berlin, Berlin, Germany

Berlin Institute for Migration and Integration Research, Humboldt-Universität zu Berlin, Berlin, Germany

S. Kühn (ed.), *Environmental Neuroscience*,
https://doi.org/10.1007/978-3-031-64699-7_14

Introduction

An essential characteristic of contemporary times involves the shift in the size of the population and migration from rural regions to urban hubs. Particularly in the past two centuries, this change has sparked considerable interest in the impact of the urban environment on the human condition.

Urban living is not a phenomenon that emerged recently; instead, it is an age-old human practice that had been considered an unproblematic fragment of human settlements parallel to rural areas until the unprecedented surge in urbanization in the recent century. With more than half of the world's societies living in urban areas, projected to increase to 70% by 2050 (UN, 2022), cities are now the predominant human habitat. As stated by UN-Habitat (2022), *"cities are here to stay, and the future of humanity is undoubtedly urban."* Along with this reality, the report of the World Health Organisation (WHO) indicates an escalation in the burden of common mental disorders (CMDs),[1] with depression holding the top position in the global burden of disability (2017). These two processes prompted urgent demands for urban psychopathology, questioning (whether or not) urban living impacts the mental health of the inhabitants. Consequently, as the main living *"environment"* for humans, cities became the epicenter of various interdisciplinary research on mental health.

This article takes a chronological approach to the literature concerning the role of the urban environment on mental health. It delves into the following questions: *Does urban living contribute to a specific type of mental disorder? When and why did cities become crucial in understanding mental health research? What is the state of the art, and what are the theories, methods, and implications that have been derived from existing evidence regarding the role of the environment in causing psychiatric disorders?* Its objective is to move beyond merely selecting and reviewing the most notable contributions. Rather than arriving at a conclusion or making generalizations, it places emphasis on the multidisciplinary and multisectoral aspects of the subject. This includes its associated theoretical and conceptual frameworks, measurement indices, data analysis procedures, and methodological challenges.

Chronologically, we distinguish between two generations of knowledge production on the urban milieu's role in population mental health. The generational disparity is evident in how health and city are conceptualized, the methodologies applied, the intensity of global dynamics, the rural–urban ratio, and the technological progression.

[1] Common Mental disorders include depression, anxiety disorder (GAD), panic disorder and phobias, social anxiety disorder, obsessive-compulsive disorder (OCD), and post-traumatic stress disorder (PTSD) (National Collaborating Centre for Mental Health (UK), 2011).

The First Generation: From Philosophical Speculation to Empirical Studies

The first generation of knowledge production extends from the Industrial Revolution to the early years of the twenty-first century. The onset of industrialization came with a radical shift in social structures and hierarchies. The transition from agrarian to industrialized urban society brought about significant population growth in cities. Modern city planning emerged as a response to inhumane living conditions in nineteenth-century cities. However, over the last centuries, this connection was severed (Barton, 2009), and planning was still considered "relatively straightforward" (Taylor & Lang, 2004). Characterized by abstract utopian visions, urban design was consistently driven by the imperative of top-down theorizing about the urban condition as a whole—something that could be planned and monitored. In this approach, planners and architects showed indifference to the experiences of inhabitants.

The acceleration of urbanization in tandem with industrial processes brought about major urban planning paradigms characterized by extremely specialized land use. At its core, this approach led to the socio-physical disintegration of everyday living spaces into distanced zones, marked by a radical spatial division with immense functional separation of residential, industrial, commercial, administrative, and recreational areas. This, in turn, has been disrupting the pre-established integral order of territoriality determined by historical and social practices (Cappon, 1975), resulting in social and economic hierarchies in terms of affordability and access to the urban infrastructure and participation in decision-making. Consequently, mobility across the city may pose challenges for the youth, the poor, the unemployed, the elderly, and individuals with disabilities, thereby embodying and exacerbating stress, poverty, and discrimination (Gans, 1967; Hillman & Whalley, 1975). These dramatic changes brought about a tangible impact on the health and well-being of the population, capturing the attention of both health practitioners and urban designers. The pivotal shift in focus for planners and medical practitioners toward urban health occurred during the turn of the twentieth century. Cities then grappled with infectious diseases and epidemics, such as influenza, typhus, and tuberculosis, attributed to population crowding. These health challenges resulted in the loss of millions of lives in cities characterized by inadequate sanitation and squalid living conditions (Rosen, 2015).

Theorizing the City as the "Environment"

In tandem with the rapid pace of urbanization and industrialization, and concurrent with the endeavors of urban sociologists and designers to conceptualize these processes, various social and human-centered scopes, such as "anthropology," "ethology," and "psychology," were emerging or undergoing a metamorphosis, transforming from philosophical speculation into scientific disciplines; psychology evolved into a trajectory of empiricism for a comprehensive understanding of the complexities of human existence. It was the time when mental health disorders began to be defined and classified in medicine. Prior to that, the conditions that might now be considered as schizophrenia were often described differently, grouped under broader terms, or sometimes referred to as various forms of insanity or madness. During industrialization and rapid population increase, the emergence of schizophrenia as a significant disorder in cities prompted practitioners to investigate its underlying causes (Cooper & Sartorius, 1977). Consequently, by considering the "city" as the "environment," the science of the mind began to delve into the psychopathological effects of the urban surroundings on individuals and scrutinize the "environmental factors," exclusively associated with "urban living," which have a direct impact on the prevalence and incidence of mental disorders. This movement in perspective brought cities as a new "living environment" to the center of the inquiry. Consequently, a substantial amount of research is dedicated to examining mental health, either in urban or rural areas or making comparisons between the two. In theory, this knowledge can potentially create opportunities for both primary prevention and early-stage treatment of illnesses.

Rural–Urban Dichotomy

For over a century, the predominant approach to understanding the risks of mental disorders in urban areas has involved widespread speculation comparing the impact of urban and rural environments on mental health. According to the records, the discussion on this subject could be traced back to the 1800s. Okkles et al. (2018) cited several studies, including the work of Hübertz, a Danish physician, in 1839, that demonstrated a tangible higher number of people with "a diseased mind" (Seelenkranken) in the metropolis compared to the countryside. A substantial body of literature contrasting urban life with rural settings has given rise to a rural–urban dichotomy. The adverse consequences of industrialization and urbanization have been depicted and expressed in philosophy, sociology, and belletristic literature by various writers such as Charles Dickens, Karl Marx, Robert Musil, Jean-Jacques Rousseau, and Henry David Thoreau, even preceding the exploration of these issues in epidemiology and empirical research. Thinkers of this period developed a contrasting perception that valued the virtues of pristine and benevolent rural life while associating urban life with decay, inequity, distress, and malevolence—the toxicity of cities against the tranquility of villages (Marsella, 1998; Mueller, 1981; Shorter, 2017).

The German sociologist Georg Simmel, in 1903, laid the foundation for a long-standing tradition by indicting the detrimental impact of big-city materialism and anonymity on human relationships and community solidarity that lead to the intensification of emotional experiences and constant fluctuation of external and internal stimuli. This analysis was incorporated into a socially conservative perspective that viewed the city as a "solvent for the entire society," forming the basis for the extensively debated hypothesis on the psychological climate of cities (Simmel, 1903). This perspective formed the basis for comparative studies on the risks of psychiatric disorders in rural and urban areas.

Breeder Hypothesis

In addressing the reason behind geographical disparities in mental illnesses, two contrasting theories emerged to explain the underlying reasons in sequence: (1) the breeder hypothesis and (2) the drift hypothesis. We will discuss the particularities of these theories in the following discussions.

The early studies were mainly based on the "breeder" or "social causation" hypothesis, which posits that environmental factors either cause mental illness or must be present for a predisposed individual to develop the illness. The theory is founded on propositions that consider urban areas as having disorganized societies with excessive mobility, conflict, isolation, and disruptions in communication processes. These factors within urban settings contribute to the emergence of seclusiveness, a prominent trait in schizophrenia (Dunham, 1965; Faris & Dunham, 1939). Verheij (1996) referred to two interconnected mechanisms—exposure and behavior—that underpin the breeder hypothesis. Exposure includes both physical factors (noise, pollution, adverse housing) as well as social factors (stress, life events, poverty, crime) to which individuals are directly subjected. Behavior is ascribed to an individual's own lifestyle and living practices, including smoking, drinking, and substance abuse, as well as exposure to the behavior of others, which can directly or indirectly subject people to specific types of illnesses. Many of the factors mentioned above are more common in urbanized areas (Freeman, 1978; Maas et al., 2006).

Social Load and Information Load Under- and Overstimulation of Receptive Capacities in Cities

Wing (1989) theorized urban environments as a context that often leads to two contrasting extremes of social-environmental processes, under-stimulating and overstimulating, which can negatively impact the mental health of vulnerable individuals. The prolonged period of social and cultural isolation can be viewed as an understimulating environment, particularly in urban settings, where social disorganization is prevalent and establishing an intimate and sympathetic social life is challenging. This, in turn, exacerbates symptoms such as social withdrawal, passivity, inertia, and a lack of initiative, leading to the development of a "shut-in" or "seclusive" personality in schizophrenic patients (Faris & Dunham, 1939; Jaco, 1954; Wing, 1989). In their 1978 study on the impact of social networks on schizophrenia, Hammer et al. hypothesized that possible common mechanisms underlying schizophrenia could involve adverse alterations in individuals' social networks. They proposed a theoretical framework utilizing network variables that could contribute to analyzing the effects of continuities, disruptions, or distortions in individuals' social networks. In addition, concepts like stress and overstimulation could be measured. This framework is particularly useful for assessing large-scale variables such as social class or ethnicity, which could be challenging to evaluate among individuals, especially across diverse cultural contexts.

On the other hand, several studies stated that urban areas could lead to continual over-vigilance in the nervous system. The overload of receptive capacities is caused by stimuli surpassing an individual's processing capacity (Cappon, 1975; Lipowski, 1975; McCarthy & Saegert, 1978). Wing (1989) emphasized a tendency to break down under conditions of social over-stimulation.

Conceptualized by various authors, the considerable size and rapid pace of the crowd, along with high population density and overcrowding, could lead to "social overload." This involves having no autonomy over undesired or spontaneous social encounters and living environment, acting as a stressor (Bornstein & Bornstein, 1976; Magaziner, 1988; McCarthy & Saegert, 1978). A related concept was discussed by Lipowski (1975), who drew attention to the "information overload" in cities that constitute a form of psycho-social stress facilitated by information technologies, the widespread production of symbols, and man-made alteration of the physical environment. These processes, in turn, place significant demands on the adaptive capacity of humans and "psychological habituation," leading to the development of new behavioral norms, different positive or negative modes of coping strategies, and expediting changes in both culture and value systems. Freeman (1978) hypothesized this psychological model as evidently relevant to schizophrenia, a condition currently believed to entail a perpetual state of over-arousal in the central nervous system. Cooper and Sartorius (1977) presented their speculative hypothesis that discusses how industrialization coupled with urbanization and population growth contribute to a greater prevalence of chronically handicapped schizophrenics in the population. Stressful life events and ongoing crises, more commonly found in urban settings, are among the social-environmental factors that constitute a risk factor for the onset of various psychiatric disorders, including depressive disorders and schizophrenia. In this instance, the lack of a traditional social lifestyle and disrupted social networks could act as a vulnerability factor for individuals in the urban middle-class status (Brown & Birley, 1968; Brown & Prudo, 1981).

Epidemiological Psychiatry of Urban Environment

One of the primary approaches to addressing "mental health" in cities was epidemiological investigation, which compared the prevalence and incidence of mental illness cases between urban and rural areas or across different urban areas and addresses variations in illness rates associated with the varying strengths of environmental factors in distinct populations.

The epidemiological study conducted by Faris and Dunham (1939) could be considered the commencement of modern studies on the relationship between the environment and mental health. The studies of this generation are mainly based on the comparative rates of mental illness in different rural and urban areas as well as the geographical distribution and disparities of occurrence between the geo-economic center and periphery dynamic of the cities. These studies rarely addressed direct cause-and-effect relationships between specific environmental features and particular mental conditions. From the late nineteenth century, more psychologists began to put the breeder theory to the test using various methods across the globe. As we delve further into the twentieth century, an increasing number of studies address this question.

Limitation and Inconsistencies

Several authors, including Dohrenwend and Dohrenwend (1974), Freeman (1978), Mueller (1981), Neff (1983), Harpham and Blue (1995), Verheij (1996), and Marsella (1998) have provided a review of early research on the relationship between psychiatric epidemiology and urbanization. Despite the extensive research and popular literature as well as related conceptual and methodological issues regarding the relationship of urbanization to mental health, reviews of findings reveal significant complexities, contradictions, and inconsistencies. Some studies indicate a higher prevalence of mental disorders in cities, while others do not; it remained unclear whether living in urban areas is mentally healthier than in non-urban environments (Freeman & Alpert, 1986). Aside from the large body of literature, the challenge in arriving at conclusive findings lies in its broad coverage across various disciplines, professions, and sectors. Additionally, the literature has a global scale and is published in numerous and diverse international journals and books (Marsella, 1998). In 1978, Freeman raised an inevitable question: *Does such epidemiological information provide any insights into the etiology of mental illnesses in cities?* In other words: *Does residing in a dilapidated central city area actually "cause" the development of psychiatric disorders?*

In the forthcoming section, our objective is to examine the limitations and inconsistencies mentioned by reviewers that impeded the development of definitive conclusions explicitly asserting that cities make us mentally ill.

Problematizing Rural–Urban Dichotomy

The distinction between urban and rural residence is frequently mentioned as a potential risk factor for depression and other mental disorders; however, epidemiological evidence supporting this relationship remains inconclusive.

Deconstructing Rural-Urban Orthodoxy Following a century of growing empirical studies, the orthodox conceptualization of rurality and urbanity began to unravel; cities were no longer merely spaces of chaos, corruption, and cruelty, and rural life was not exclusively tranquil and harmonious. Consequently, a notable ambivalence toward rural and urban life has surfaced (Marsella, 1998; Shorter, 2017). Numerous studies have indicated that people in rural areas may experience a substantial amount of psychopathology attributed to factors such as poverty, unemployment, agricultural crises, declines in rural manufacturing, and conflicting personal identities. The sociologist Ernest Groves, in 1916, addressed the rural–urban migration as the relief of rural restlessness:

> In the city, the sharp contrast between labor and recreation increases without doubt the appeal of the city to many…So great a contrast between work and play—monotony and freedom—is not possible in the country environment. […] In an age restless and mobile, with family traditions less strong, and transportation exceedingly cheap and inviting, it is hardly strange that so many of the young people are eager to leave the country which they pronounce dead—as it literally is to them—for the lively town or city. (Groves, 1916)

Many commentaries discuss how the expanding scale and proportion of cities surpass humans' stimulation capacities. They highlight an increasing sense of fear, powerlessness, and anger among urban residents. In addition, they address individualization, succumbing to anonymity, indifference, and narrow self-interest. However, despite the backdrop of significant social, economic, and political challenges in urban areas, some writers praise the numerous virtues of urban life that provide residents with a broader and more diverse range of intellectual and cultural stimulation, economic opportunities, improved public utilities and services, transportation facilities, accessibility to healthcare services and personal choices in pursuing various social roles, relations, and moral options, et Cetra (Marsella, 1992).

Breslau et al. (2014) analyzed three consecutive annual samples (2009–2011) of adolescents and adults from the National Survey of Drug Use and Health (NSDUH) to assess the prevalence of major depression and other serious mental illnesses across four urbanicity categories that include (1) large metropolitan areas, (2) small metropolitan areas, (3) semi-rural areas, and (4) rural areas, with and without adjustment for other demographic risk factors. The study concluded the inconclusiveness of epidemiological evidence for urban vs. rural residence as a risk factor for depression and other mental disorders in the US:

> The findings […] challenge the accepted wisdom that major depression and other serious mental illnesses are more common in urban than in rural areas in the US. We found no differences in risk between the most rural areas and the largest metropolitan areas for adults or youth. The findings are particularly notable because we were able to distinguish truly rural areas, i.e., those that are not socially or economically integrated with an urban center, from semi-rural areas, which include small towns and their integrated surrounding areas. […] Contrary to expectations, the prevalence of mental disorders was not higher in the most urban compared with the most rural areas, suggesting that the move to identify mechanistic explanations for risk associated with the urban environment is premature. (Breslau et al., 2014)

Furthermore, another analysis of two US government databases suggested that when assessed through various health or mental pathology indicators, the vast central metropolitan areas consistently appear to be significantly safer and healthier than rural areas. Moreover, the size of the place had an inverse relationship with the suicide rate, cigarette smoking in teenagers, severe mental illness in males, and major depression in female adolescents (Meit et al., 2014).

The absence of substantial evidence for urban/rural disparities in psychiatric morbidity does not necessarily negate the hypothesis of added social causation in cities. Instead, it suggests that the conventional urban/rural dichotomy may not accurately represent it. The rationale behind this suggestion is that each large rural or urban area is too diverse to have a uniform impact on all inhabitants from this perspective (Murphy, 1976).[2]

Relativity of Rurality and Urbanity Concepts One of the primary factors contributing to data inconsistency regarding rural–urban differences in psychiatric morbidity is associated with the relativity of the concepts of "urbanity" and "rurality," lack of rural–urban identification scale, and the diversity of measures and definitions employed for urban-rural distinction across studies, geographies, and social and political systems. The variability in the interpretation of these terms complicates comparisons across studies. In 1990, the geographer Keith Hoggart problematized this dichotomy and addressed the ambiguity inherent in the concepts of rurality and urbanity.

> … there is too much laxity in the treatment of rural areas in empirical analysis. […] The central point is that the undifferentiated use of "rural" in a research context is detrimental to the advancement of social theory. […] The broad category "rural" is obfuscatory, whether the aim is a description or theoretical evaluation, since intra-rural differences can be enormous and rural-urban similarities can be sharp….the main objection to the pragmatic approach is simply that it groups together places in which social processes of a very different kind are in agency and structure, the production of new forms of social life and the reproduction of existing forms in differing unknown quantities (and qualities). […] However, could we circumvent this by including a number of places in our studies and conducting comparative analyses? I do not think so. (Hoggart, 1990)

While the term urbanization is commonly used in various disciplines of thought and practice, there is substantial variation in its interpretation among these groups. Rural–urban metrics exhibit notable variability among social demographers, anthropologists, sociologists, economists, psychologists, psychiatrists, and urban planners. Each of these disciplines measures the city and separates the urban-rural contexts through specific aspects and identification scales, including population size growth, subsequent social dynamic alterations, patterns of income, labor and productivity, urban culture, human adjustment, adaptation, et Cetra (Marsella, 1998).

Furthermore, the tangible manifestation of urban and rural phenomena exhibits significant diversity globally; national and international bodies lack consensus on a shared language regarding what may be termed an urban or rural setting based on their unique national population and administrative, political, infrastructural, and socioeconomic dynamics (Vlahov, 2002). This, in turn, has hindered progress in research and conceptual endeavors. Simultaneously, the definition of associated terms, such as rural, countryside, village, town, city, suburb, semi-urban, peri-urban, inner-city, urban, metropolitan, and megalopolis, is also a subject of debate. This stems from the fact that "urban planning" not only eroded the tangible territorial boundaries between urban and rural areas but also created a continuum of rural–urban life through transitional infrastructural spaces like suburban areas. The social and cultural demarcation of rurality and urbanity becomes increasingly ambiguous

[2] Mentioned in Freeman (1994).

as numerous rural inhabitants have embraced urban values, influenced by their exposure to mass media or infrastructure advancements.

Opting for extremes within these rural–urban categories poses an additional risk of generating spurious results when trying to identify significant differences between rural and urban areas; these extremes may encompass variations in economic disparities, cultural and social diversity, population density, living infrastructures, et Cetra. At times, the disparities in psychological morbidity within a particular city's suburbs and inner areas might be more notable than the distinctions between those areas and a village. Thus, it is crucial to elucidate how the concepts of "urban-ness" and "rural-ness" have been operationalized.

Uneven Global Process of Urbanicity and Urbanisation In conversations about a world undergoing increased urbanization, it is crucial to distinguish between two closely connected dynamics often used interchangeably yet bearing distinct meanings: urbanization and urbanicity. Urbanization is a dynamic process and product that arises from the intricate interplay of population, spatial-geographical, historical, cultural, economic, and political forces, contributing to the socio-environmental formation and resulting in patterns of psychosocial adjustment and adaptation (Marsella, 1998). However, this phenomenon does not unfold uniformly across the globe, varying among regions and countries; a significant portion of population growth in cities is anticipated and taking place in low- and middle-income countries.

Cross-national research can offer insights into the essential characteristics of cities and how urbanization impacts population health. However, in the Western world, the focus shifts towards urbanicity as one of the most frequently utilized environmental features in health research. Throughout its development, the term urbanicity has gained increased prominence in scholarly and academic discussions, particularly in the realms of urban studies, psychology, sociology, and related fields (Häfner, 2002; Neff & Husaini, 1987). The term is employed to describe the characteristics and quality of urban life or the degree to which an area exhibits urban characteristics. Urbanicity is often used to classify areas into different types of settlements, such as cities, towns, suburbs, and rural areas. Therefore, when most discussions focus on standardized studies in high-income countries, caution must be exercised when generalizing the concepts of "urbanicity," "urbanization," and "urbanity" and, consequently, "rurality."

Methodological Heterogeneity: Incoherent Case-Finding Instruments and Methods

There is considerable variability in conceptions of psychiatric disorders and methods of case identification in epidemiology (Mueller, 1981). This significant methodological heterogeneity employed in the existing literature hinders conclusive discussions on the influence of urban environment on the mental health of the population (Marsella, 1998); inconsistency exists in global screening instruments, which have utilized various measures ranging from self-report psychological well-being scales to case definition through structured interviews. Dohrenwend (1975) addressed how the methodology applied significantly influences the rates obtained. Conceptual frameworks are lacking, and there is variation in the quality of research and published commentary. The range of disorders and deviancies linked to urbanization is vast, encompassing psychoses, depression, sociopathy, substance abuse, alcoholism, crime, delinquency, vandalism, family disintegration, and alienation (Marsella, 1998). Prevalence rates may or may not be adjusted for different types of confounders (Neff, 1983).

In his critical review, Mueller (1981) discussed the lack of a discernible trend in psychiatric caseness identified through case-finding instruments like the national surveys, Langner scale, or HOS. While some studies controlled for potential confounding sociodemographic variables, the others seemingly did not. Variations in respondent socioeconomic characteristics beyond urban-rural residence could have influenced studies using Langner scale scores and those involving both urban and rural samples that generally did not control for sociodemographic characteristics when comparing rates across urban-rural settings. Therefore, it is important to exercise caution when comparing scores from various studies, as differences in the administration method and context of the scale could impact the results.

In addition, Carstairs and Kapur (1976) highlighted that cross-cultural comparisons face two significant challenges: the need to establish comparable definitions of psychiatric symptoms across cultures and the absence of suitable techniques for measuring the relevant socio-cultural parameters.

> Because of differences in the data collection procedures and in the criteria for ascertaining psychiatric morbidity, the results in this study cannot be strictly compared with those obtained in other studies carried out in India or abroad. (Carstairs & Kapur, 1976)

One of the primary approaches to delve into "mental health" in cities was to compare the prevalence of mental illness cases between urban and rural areas. However, health facilities are typically more accessible in urban areas compared to rural regions. This increased accessibility may lead to higher service utilization and, consequently, an apparent increase in morbidity due to nosocomial factors. Therefore, surveys relying on service usage will be influenced, to some degree, by variations in the utilization rates of these services rather than disparities in morbidity (Dohrenwend, 1975; Freeman, 1978). In addition, due to the demands of industrialization, more families in urban areas may institutionalize their psychotic relatives than in rural areas. This is because they find it challenging to care for them at home, especially in overcrowded conditions that often result from the urban environment, whereas this might have been feasible in their previous agrarian settings (Freeman, 1994). Consequently, the rural–urban rates could vary due to differences in the availability of psychiatric services or their acceptability to the public (Dohrenwend, 1969).

Moreover, the diversity of societal "norms" across studies and countries, as well as the level of institutionalization and allocation of mental health services, play a role in influencing the count of individuals hospitalized or who use mental healthcare services.

Contextual Heterogeneity: Interplay Between Time, Subject, and Site

These studies are scattered across the globe, examining diverse contexts, different targeted populations, and various mental disorders. Regardless of whether in an urban or rural environment, the occurrence depends on the current equilibrium of risk and protective factors unique to that particular context (place, structure, and time) and subject. It is crucial to acknowledge that the majority of environments exhibit both positive and negative qualities, yet all residents do not uniformly experience these attributes; for certain groups of individuals, the positive aspects may outweigh the negative, whereas, for others, the negative aspects may be more significant (Verheij, 1996). Marsella (1992) discussed that examining urban-rural differences alone does not significantly contribute to advancing our knowledge; thus, future research should concentrate on specific sub-populations. However, the significance of sub-populations becomes evident only when contrasting these populations. Milgram (1970), in his leading work, emphasizes the necessity of connecting the demographic conditions of urban life with the psychological experiences of individuals. Almeida Filho et al. (1995)[3] discussed that the rural–urban distinction is an artificial dichotomy. They argued that the genuine concerns lie in the variations of economic and political power dynamics, specifically emphasizing issues of domination and subordination, often characterized as oppression and exploitation. In their publication, Goldsmith and Blakely (1992) took a critical stance on structures perpetuating poverty and urban distress, emphasizing comparable inequalities within the United States.

Furthermore, The fast pace of urban changes brings about swift and substantial social transformations that could directly influence the rates of psychiatric casualties. Consequently, the incidence and prevalence of mental disorders in a particular urban area are directly influenced by its temporal factors and historical development. Freeman (1994), while addressing two case studies, discusses how the distribution of schizophrenia rates could undergo a radical change over time in a particular area. For instance, in Detroit, between the 1930s and 1950s, the incidence of schizophrenia shifted from being densely concentrated in central areas to becoming dispersed in distribution. In another study in Hamburg, the distribution pattern altered between first admission and readmission of schizophrenic patients in the central regions (Klusmann & Angermeyer, 1987). In addition, it should not be presumed that the complete impacts of urban areas are immediately evident. The effects may unfold over more extended periods of time, with an incubation process spanning several generations. During this period, the environment undergoes continuous changes. The relationship between mental conditions and city living may have evolved over time due to different modes of change, further complicating an already highly complex question (Freeman, 1994).

[3] Mentioned in Marsella (1998).

"Selective Migration": Geographical Drift and Social Residue

As previously stated, two primary interpretations exist regarding the relationship between the urban environment and mental health: the first, discussed above, is the breeder hypothesis that emphasizes the role of immediate environmental factors on mental conditions, including social, urban infrastructural, and natural elements. The drift hypothesis and social residue, however, emerged as the second potential mechanism explaining spatial variations and geographical differences in the health status of the population. Both of these theories present varied manifestations of the social selection hypothesis, highlighting the historical processes and social systems as causal factors for mental health conditions, as well as the circumstances that move, migrate, and/or situate individuals to/within their immediate environment. It is hypothesized that if empirical data validate the geographical drift hypothesis, the urban/rural differences would not persist (Verheij, 1996). While these two concepts of drift and residue present reverse processes, they may operate concurrently at the same time and place.

The key driver of urbanization is often attributed to rural–urban migration and population growth in cities. Nevertheless, additional types of migration and population mobility contribute to the complexities of the context. At times, rural–urban movement is bidirectional—escape. In addition, different social processes that result in forced relocation within a city are happening on an enormous scale.

Social Selection Hypothesis Exploration of the etiological importance of socioeconomic variables is one of the crucial and demanding focal points within the field of the sociology of mental illness. Many comparative studies' exclusion or inclusion criteria are related to the level of economic development as it is one of the principal determinants of health (Verheij, 1996). However, in a study conducted in Detroit in 1965, Dunham emphasized the social selection hypothesis as an explanation for the variations in the occurrence of schizophrenia between two contrasting city areas.

> The evidence provided by this study gives support to any proposition that asserts that type of community and social class is related in any manner to the incidence of schizophrenia. Rather, we find that both community and class distributions of schizophrenia are to be explained by factors indigenous to the social system. (Dunham, 1965)

Through this approach, varying rates of schizophrenia are explained by social processes that tend to position individuals with the illness within social systems. In this case, the momentary social situations are not directly involved in causing the origin of the disease. By asserting that the concentration of cases in the lowest socioeconomic class can be explained by social selection factors and patterns of residential mobility, he raised questions about how objective circumstances within an individual's culture could manifest as distortions in that person's thoughts and behavior (Dunham, 1965).

Drift Hypothesis: From "Socio-economic" Mobility to "Spatial" Mobility The initial theory of generational "social drift" proposed by Goldberg & Morrison (1963) held significant heuristic value; it emphasizes the potential "social mobility" and intra- and inter-generational movement down the socio-economic scale of those affected by psychiatric morbidity that results in a downward social shift and attainment of a lower social class (Goldberg & Morrison, 1963).

Following that, the concept has evolved by incorporating "spatial mobility" and the process of "selective migration," which addresses the migration of individuals with adverse mental health situations to socially deprived areas, where there is low social control, fewer social demands, and greater tolerance towards deviant behavior. Numerous studies have depicted the clustering of schizophrenic patients in economically deprived central areas of cities (Faris & Dunham, 1939; Giggs & Cooper, 1987). While the selected areas typically offer inexpensive, single-person accommodation and opportunities for casual but low-paying employment, the lack of a well-knit social structure and sufficient social support in these areas may, in turn, negatively affect individuals' mental health. Based on this perspective, the cause of the onset of schizophrenia and other mental illnesses is not that those affected reside in environmentally unfavorable areas, but rather individuals with schizophrenia, like other socially marginalized individuals, find it challenging to live in more affluent residential districts. Therefore, residing in a poor environmental area is not considered an etiological factor in itself but is instead linked to the development of the disorder (Freeman, 1994). Social drift has been notably associated with individuals with schizophrenia, but the same tendency is observed among alcoholics, drug addicts, individuals with marked personality disorders, and others with diminished social competence and affiliative bonds (Freeman, 1994). According to Cooper and Sartorius (1977), the significant population growth during the nineteenth century, primarily driven by migration, played a crucial role in the rise of schizophrenia as a prominent disorder in that era. Consequently, instances of severe mental illness, which might have been dispersed in rural communities in the past, became concentrated in such quantities that they posed a significant public health challenge.

In 1988, Henderson discussed that the non-random distribution of a disorder's incidence could be attributed not only to the direct effects of environmental factors but also to the influence of the disorder or its early symptoms on individuals' residential choices or a set of factors preceding both mechanisms (Henderson, 1988). The social processes, such as "deinstitutionalization," have also played a role. In Western countries, there has been a widespread failure to establish sufficient community residences for schizophrenic patients transitioning out of mental hospitals. Consequently, they often end up drifting into the most impoverished areas of cities or are resettled there by public services (Goldman, 1983).

Lewis et al. (1992) further discussed "social drift" as a social selection process—involving the choice between relocating or staying—that leads to a greater concentration of either individuals with illnesses (direct selection) or a spatial clustering of more susceptible individuals (indirect selection) in a particular location. Several authors noted that besides potential environmental influences, variations in psychiatric disorder rates across different geographical zones—e.g., between different urban localities or urban and rural regions—could stem from patterns of "selective migration" (Mueller, 1981; Neff, 1983).

Social Residue Theory On the other hand, the social residue theory highlights a reverse social process that involves the migration of well-off individuals with "better" mental health who tend to move away from socially and environmentally undesirable areas, leaving those with relatively lower competence behind (Freeman & Alpert, 1986). Freeman (1994) pointed out several case studies exploring significant variations in mental disorder incidence. This includes scenarios of shrinking cities that experience declining populations due to the outward migration of affluent families, resulting in the concentration of below-average groups, the "residual population," such as individuals with mental and physical disabilities, single-parent families, and the elderly, who are more vulnerable to mental illness (Strömgren, 1987).

The two concepts of drift and residue elucidate two facades of "selective migration," and both may function concurrently at the same time and place. As stated by Mueller (1981) and Neff (1983), certain studies lack clarity regarding whether the observed outcomes related to mental health arise from environmental causation or selective migration. The absence of comparable data that enables controls for selective migration turns the findings as indicative of environmental causation. In a critical examination of two studies, Verheij (1996) elaborates on the concept, asserting that the drift hypothesis should scrutinize "selective migration" not only by questioning the history of migration between two distinct rural and/or urban areas but also relocation across a particular urban area—including place of uprising and residence in the city. Moreover, the individual's history of illness should be examined alongside the history of spatial movement.

To gain more insight into the geographical drift hypothesis (direct and indirect selection effects), large-scale individual-based longitudinal data involving substantial numbers of individuals are essential; that should include analytical models that measure mobility and the history of illness behavior, along with past and present individual determinants of health. Meeting these requirements is rare if it happens at all. Consequently, evidence concerning drift processes remains limited (Moorin et al., 2006; Verheij, 1996). Verheij (1996) criticized the literature existing until the late twentieth century as predominantly grounded in aggregate data, and even when based on individual data, it often had a cross-sectional design. The scarcity of evidence on selective migration makes it nearly impossible to detect geographical drift and spatial mobility effects; as a result, it has led to uncertainty regarding the differential impacts of urban and rural areas on the mental health of the population.

A shift of focus from solely mental disorder outcomes toward urban exposure, namely the characteristics of the urban environment that influence mental health and disorders, could pave the way for the development of appropriate interventions and preventive measures.

Emergence of Ecological Approach in Psychology: A Multi-dimensional Perspective

The review of studies illustrated that psychiatric disorders do not exhibit unique urban patterns and could affect anyone, anywhere, at any time. However, is there something inherent to urban life that either raises or inhibits the occurrence of these mental conditions? That is a separate question, one that has consistently sparked lively interest (Shorter, 2017). The reviews (Dohrenwend & Dohrenwend, 1974; Freeman, 1978; Harpham & Blue, 1995; Marsella, 1998; Mueller, 1981; Neff, 1983; Verheij, 1996) show that establishing direct cause-and-effect relationships between specific environmental features and forms of psychological malfunctioning with an epidemiological approach is highly improbable.

Understanding how the man-made environment affects the population's health requires a highly intricate analytical framework that includes three broad factors: the objective environment, the subjective experience, and physiological, affective, and behavioral responses that serve as mediators (Kasl, 1977). The objective environment includes the social environment, the physical environment, and access to health and social services. Freeman (1978) noted that while European psychiatry recognized a multifactorial perspective, acknowledging the involvement of environmental aspects, their contemporary understanding could only provide a rough assessment of these factors. Moreover, enabling the examination of both breeder and drift hypotheses together requires collecting population-based longitudinal data. Additionally, it was emphasized that establishing urban mental health as a discipline necessitates leveraging the strengths of various academic fields, such as ecology, epidemiology, and sociology; the hypotheses within this field need validation through multidisciplinary research (Lipowski, 1975). Craik (1973) addresses how disciplinary approaches such as psychiatry, psychology, architecture, and planning could diverge in direction but remain complementary.

In addition to epidemiological efforts focused on determining the locations and prevalence of mental disorders in cities, studies started employing ecological methodologies to discern the environmental factors influencing mental health. By incorporating an ecological approach, several psychological theories have sought answers regarding which environmental factors affect human mental health. The field theory of Kurt Lewin is among the pioneering works addressing human psychology in its relation to immediate, intermediate, and distal environments (1939).

> The field theoretical approach is instrumental in integrating divergent physiological, psychological, and sociological facts on the basis of their interdependence. To explain social behavior it is necessary to represent the structure of the total situation and the distribution of the forces in it. Certain geometries are offered as adequate for representing social fields. (Lewin, 1939)

Barker's (1968) ecological psychology, Kelly's (1971) community psychology, Craik's (1973) environmental psychology, and Urie Bronfenbrenner's (1977) experimental ecology of human development are among other theoretical frameworks that have developed ecological perspectives to understand human psychology in their everyday social, environmental contexts. These disciplines sought to comprehend the multifactorial man-environment relation by employing an interdisciplinary, system-oriented approach.

Bagley et al. (1973) adopted an ecological approach to investigate the clustering of social pathologies within isolated small urban areas characterized by high indices of overcrowding, poverty, and adverse social conditions. The resulting data exhibited intricate patterns, including the heterogeneity of most area populations, the interdependence of geographical units of analysis, and the complexity of applying ecological correlations to individuals. In conclusion, the authors highlighted exploring the life histories, social networks, and social connections of individuals both with and without mental illness in high and low-rate areas as crucial. This aims to unveil causal patterns, such as determining the primary or secondary nature of environmental conditions, and explain the connections of various diagnoses in different age and sex groups through enumeration district analysis.

Freeman (1978) reviewed various studies utilizing an ecological approach that sought to identify correlations between environmental factors and the development of psychiatric disorders. He highlights that while research of this kind is pragmatic, establishing a basis for allocating resources to areas with the most pressing needs may not necessarily yield insights into the etiology of psychiatric disorders and social mechanisms. Ecological studies focusing on individual-environment dynamics could bring attention to associated factors; however, they often could not define a controlled environment, primarily relying on single-variable modeling.

The factors influencing mental health are believed to encompass both environmental (aggregate) and individual characteristics. However, extensive sociological and geographic research often leans toward focusing on either the individual or the environment (Verheij, 1996).

The Second Generation: From City-Centrism to Embracing the Complexity of Context

In the twenty-first century, the question of "mental health" is being increasingly understood in conjunction with transformative forces that simultaneously shape human societies. These forces include rapid urbanization, extensive migration alongside demographic changes, neo-liberalization, digitalization, technological progress, climate-related challenges, and violence. Many of these factors are acknowledged as threats to humanity.

Urban dynamics are undergoing a transformation, with resources and infrastructures increasingly centralizing and consolidating in specific areas, making these regions expensive for both living and property acquisition; this contributes to a widening disparity and increased inequalities in numerous cities. Residents in some urban areas may potentially enjoy advantages such as improved social and cultural infrastructure, accessible healthcare services, recreational spaces, and fair employment opportunities. Simultaneously, other areas in cities are characterized by a bewildering expanse of urban sprawl, dissected by hazardous, densely populated areas, polluted highways, and inadequate green spaces. These areas are filled with monotonous tower blocks interspersed with vacant, impersonal spaces. This environment, coupled with heightened exposure to both licit and illicit substance use, as well as more stressful social conditions, appears to pose a threat to psychological health.

While in the nineteenth and twentieth centuries, urban studies were marked by the continuous production of new concepts and ongoing efforts to theorize urban conditions, twenty-first century discourses on cities are shaped by the constant realignment of urban perspectives amidst recurring, multi-scale, and entangled crises (Madden, 2023). Whether in rural or urban settings, our modern living contexts are affected by intertwined crises, including housing speculations, various forms of violence, social disruptions and conflicts, forced migration, the lack of political legitimacy and accountability, intensifying inequalities, natural resource, and water scarcity, escalating pollution, environmental disasters, health deterioration, and the recurrence of pandemics (Madden, 2023). As Madden (2023) noted, today's cities are manifestations of the "urbanization of crises"—the embodiment of the symptoms of the inherent contradictions within the capitalist accumulation process (Harvey, 1978). In this context, the old orthodox understanding of cities has faded into history, and no equivalent new clarity has emerged; we encounter a conceptual disintegration in our comprehension of cities (Taylor & Lang, 2004), which, in turn, makes it challenging to comprehend cities.

Since the initiation of the concepts of "Healthy Cities" and "Healthy Communities" by the World Health Organization (WHO) in the 1986 Ottawa Charter (1986), there has been a surge of interest in cities' population health within the academic community across diverse disciplines, particularly urban planning. While our direct comprehension of the connection between urban physical health and the environment has led to a predominant focus on interventions for enhancing physical well-being (Xu et al., 2023), the inquiry into the factors influencing mental health in urban environments has remained nebulous and abstract.

The endeavor of this era unfolds through a process initiated by the emergence of divergent yet interconnected inter- and transdisciplinary theoretical and methodological approaches. Of these, each seeks answers to questions about urban mental health and then progresses toward developing an integrative lens to unify these diverse attempts. Hereafter, we will discuss these endeavors:

Toward Inter-and Transdisdiciplinarity

As the urban population grows, the variety of living circumstances in cities has increased beyond imagination. Therefore, the human realm and its psychological condition within a transitional society exemplify the phenomenology of diversity (including identity, lifestyles, and perceptual frameworks), multidimensionality, and complexity. This necessitates comprehension through the amalgamation of knowledge within the scope of an inter- and transdisciplinary approach (Guseltseva, 2020), extending beyond the field of psychiatrists and psychologists. The increasingly intricate urban environment eludes comprehension through single-discipline frameworks and cannot be adequately addressed through modest reforms and interventions alone. In response to this challenge, plural yet interconnected inter- and transdisciplinary research approaches began to emerge, developing complex research frameworks to delve into a multi-layered understanding of how the relationship between humans and their (urban) environment influences mental well-being.

Transdisciplinarity involves movement through and beyond disciplinary boundaries; it signifies the essential fluidity of disciplinary borders and the dissolution of paradigmatic constraints. This entails the absence of a preordained matrix for research thinking, with methods, paths, and methodologies evolving organically throughout the course of the study (Guseltseva, 2020).

Two critical elements in the field have propelled advancements in inter- and transdisciplinary frameworks: (1) the development of social models for psychiatric conditions (including the Social Stress Hypothesis), necessitating transdisciplinary approaches, and (2) the evolution of real-life laboratories that incorporate technological innovations into methodological approaches, enabling complex methodological frameworks in transdisciplinarity.

Development of Social Models for Psychiatric Distress: Necessitating Inter- and Transdisciplinarity The development of the Social Stress Hypothesis, linking sociological factors to psychological stress, could be considered a cornerstone in advancing discussions on the impact of urban environments on mental health. The sociological endeavor expanded the conceptualization and measurement of stressors beyond biomedical and epidemiological models to encompass integrated constellations of stressors comprising social events and strains over time. Theorized by Pearlin (1989; Pearlin et al., 1981), the Social Stress Hypothesis emphasizes the consideration of convergence of social events and chronic strains and discusses how simply correlating events or strains with health outcomes does not adequately capture the nuanced process that frequently involves a dynamic relationship between social events and enduring strains. The Social Stress Hypothesis suggests that individuals' mental health is influenced not only by intrapersonal factors but also by the social environment in which they live, such as discrimination, stigma, loneliness, and social isolation. Further, the theory elucidates the enduring relationship between social inequalities and mental health disparities, emphasizing that individuals with disadvantaged social status are predisposed to greater exposure to stressors and exhibit heightened vulnerability to stress due to their limited psychosocial coping resources. This, in turn, increases their risk of developing mental illness. The coping resources, such as social support, self-esteem, and a sense of autonomy over one's life, are shaped by the complex interplay of both momentary social status and environmental influences, as well as personal developmental processes (Mirowsky & Ross, 2017; Mossakowski, 2014; Thoits, 1995). From the mid-twentieth century to the present day, extensive literature has examined the theory. Factors such as poverty, economic crises, unemployment, limited educational attainment, housing insecurity, climate disasters, social density, loneliness, and others have been recognized as social stressors linked to significantly elevated levels of psychological distress and rates of mental illness (Adli et al., 2017; Cutts et al., 2011; Heinz et al., 2020b; Kawachi & Berkman, 2003; Kessler & McLeod, 1984; Meyer et al., 2008; Murphy, 1976; Rapp et al., 2015).

Evolving Real-Life Laboratories: Enabling Inter- and Transdisciplinarity The influence of human-environmental dynamics on mental health displays significant heterogeneity and dynamic features, manifesting variations in inter- and intra-individuals and fluctuations over time. This underscores the need for methodological approaches to capture dynamic moment-to-moment processes, recognize substantial individual differences in these processes, and ultimately leverage this information to improve and fine-tune intervention strategies (Smith & Juarascio, 2019). Over the last decades, utilizing technological advancements, urban real-life laboratories have been established to develop and implement complex methodological frameworks that bring together assessable data of everyday life. These frameworks enable the collection and analysis of extensive, multi-layered inter- and transdisciplinary data across geographies, contexts, subjectivities, and temporalities on urban psychiatric symptomatology (Dye, 2008; Reichert et al., 2020; Tost et al., 2015). Hence, studies in real-life urban settings have been initiated, combining the leveraging of recent methodological advancements with traditional methods for collecting multi-layered subjective and objective data. Some advanced techniques in this field include (1) Neuroimaging, which demonstrates brain-related factors associated with social and environmental stressors. These brain imaging techniques involve functional near-infrared spectroscopy, electroencephalogram (EEG) (➔ chapter "EEG for the Study of Environmental Neuroscience" by Grassini), functional magnetic resonance imaging (fMRI), and portable magnetic-encephalogram caps, as well as physiological measurements like galvanic skin response (GSR), heart rate variability (HRV), pupilometer, and eye tracking (Xu et al., 2023); (2) virtual reality (VR) and augmented reality (AR) (➔ chapter "Extended Reality in Environmental Neuroscience Research" by Li et al.) setups enable increasingly realistic and ecologically valid simulations of complex urban environments; (3) stationary sensors, volunteered geographic information and remote sensing imagery deliver specific context information, including conditions of traffic and weather, timely resolved air pollution data, high-resolution aerial images of urban green spaces, and land registry data providing demographic information such as population density (➔ chapter "Neuroscience and Geographic Information Systems to Investigate the Impact of Global Warming on Mood Disorders and Brain Plasticity in Urban Areas" by Vieira Ruas et al.). Administrative statistics also play a role in this context (Reichert et al., 2020); and (4) mobile devices such as smartphones and wearables gather intensive longitudinal mental health data (➔ chapter "In the Right Place at the Right Time: Leveraging Geographic Ecological Momentary Assessment to Analyze Contexts and Health" Kestens and Kingsbury), encompassing subjective and objective information on both the context and the individual.

Emergence of Neuro-Spatial Disciplines Brain-Environment Assessments

The relationship between psychiatric disorders and brain structure in response to exposure to built environmental factors, whether in urban or other settings, has been examined and conceptualized by architects, urban designers, neuroscientists, and psychologists across history (Eberhard, 2009a). Integrating advanced technologies like neuroimaging into inter- and transdisciplinary frameworks has facilitated the emergence of neuro-spatial disciplines.

The establishment of the interdisciplinary Academy of Neuroscience for Architecture (ANFA) at the Salk Institute for Biological Sciences in 2002 provided pioneering grounds to explore this hypothesis. Focusing on architecture through the lens of neuroscience led to the emergence of "neuroarchitecture," aiming to design environments that optimize human health and well-being (Dougherty & Arbib, 2013). (➔ chapters "Aesthetics of Architecture: Emerging Insights from the Brain" by Vartanian, "The Built Environment and the Brain: Review of Emerging Methods to Investigate the Impact of Viewing Architectural Design" by Tawil and Kühn, "Applications of Biometrics in Architectural and Environmental Design" by Lavdas and Sussman, "The Potential of Biophilic Design and Nature to Improve Health, Creativity and Well-Being" by Sjövall and Spiers). John Paul Eberhard, the first director of the institute, articulated this theme in his book "Brain Landscape: The Coexistence of Neuroscience and Architecture." The text focuses on how, through design, we could reduce stress and improve cognition in different architectural spaces for different people (Eberhard, 2009b). "Neuromorphic architecture" is another term introduced by Michael A. Arbib, addressing methods to integrate insights gained from studying actual biological brains into the development of computational systems rooted in neuroscience findings. These systems could be implemented in intelligent buildings, offering a biologically grounded perspective to the broader concept that future constructions should function as perceptive, active, and adaptive entities (Arbib, 2012). In 2012, a series of articles was presented at the inaugural ANFA International Conference hosted at the Salk Institute in La Jolla, California. Published in an edited issue titled "The Evolution of Neuroscience for Architecture" in the Journal of Intelligent Buildings International, the articles explore the applications of scientific research in uncovering architectural solutions. Encompassing diverse subjects such as behavior, healthcare, and environments for children, the study utilizes various methodologies, including functional magnetic resonance imaging (fMRI), simulation, and observation. The topics encompass a spectrum of complexity, employing both technical analysis and fundamental approaches to perception and measurement. In so doing, they explored implications for the influence of the built environment on brain development during the formative years of childhood, in adult experiences, and among aging populations (Dougherty & Arbib, 2013).

The study by Lederbogen et al. (2011) can be regarded as one of the pivotal experiments integrating neuroscience into researching urban qualities. This study has significantly contributed to shaping the emerging body of knowledge, with a specific focus on the interaction between urbanicity and mental illness, exploring the influence of city living and urban upbringing on the brain's responsiveness towards social stress processing: healthy individuals exhibit heightened activity in stress-processing limbic brain regions (such as the amygdala and anterior cingulate cortex) when engaging in a social stress task, a response that correlates with the size of the city they currently reside in or were raised in.

In 2016, Adli et al. (2016, 2017) called for an interdisciplinary approach spanning neuroscience, environmental psychology, biology, cognitive sciences, urban disciplines, geography, and social sciences together with practicians to "create better environments that will improve the mental wellbeing of individuals and communities in cities, and strengthen the resilience of high-risk individuals." They called this approach Neurourbanism.[4] This call did not remain isolated. Around the world, there has been a continuous development in the discourse regarding the correlation between mental health and urban areas. Over the years, multiple journals, institutions, networks of experts, and initiatives prominently led by German-speaking countries and the Netherlands, including the Center for Urban Mental Health at the University of Amsterdam,[4] have incorporated the subject.

Due to the stationary nature of the devices used, the initial neuroimaging methods were unable to evaluate neural response dynamics in real-world scenarios. The integration of immersive mobile virtual reality setups and mobile brain/body imaging (MoBI) has emerged as a primary method for overcoming these limitations in studies examining neurophysiological correlates during spatial movements (Banaei et al., 2017; Iggena et al., 2023). Further, recent technological developments, including functional near-infrared spectroscopy (fNIRS) or a portable magnetic encephalogram cap, could potentially address and overcome these limitations (Boto et al., 2018; Reichert et al., 2020; Sun et al., 2018).

In a recent endeavor, Dimitrov-Discher et al. (2022) examined data from functional magnetic resonance imaging of 42 male participants, investigating the impact of green space, air pollution, and noise pollution in the context of social stress. The results indicate that a higher percentage of green space correlated with heightened activation in regions that process emotion-based regulation of stress during a social stress paradigm compared to the control condition (i.e., parietal and insular as well as dorsolateral prefrontal regions depending on applied geographical buffer zones). Conversely, increased air pollution was linked to reduced activation in these same (and extended) brain regions.

Another exploration examined the correlation between street network features and brain activation in a social stress paradigm, integrating the space syntax model and MRI scanning techniques. The study indicated that greater mean neighborhood global and local integration, indicative of more interconnected streets in the network, was associated with less activation during stress provocation in several brain regions relevant to detecting salient stimuli and environmental threats (Dimitrov-Discher et al., 2023). A review of the existing neuroimaging studies in the field can be found here (➔ chapter "Interplay Between the Physical Environment and the Human Brain: A Review of MRI Research" by Beyer et al.).

---

[4] University of Amsterdam (n.d.).

Unfolding Sociocultural Psychology: Contemplating Diversity, Inequalities, and Vulnerabilities

Recent significant migration movements do not solely stem from urbanization processes. Influenced by neoliberal austerity policies exacerbating inequalities, as well as the increasing prevalence of war, violence, and climate crises, individuals are compelled to move in search of decent living conditions. The complex interplay between global and urban patterns of displacement as well as the population mobility across different geographic ranges, language zones, and cultural spheres has resulted in the gradual or sudden arrival of diverse populations into every neighborhood within cities, resulting in hyperdiversity as Bun Chul-Han names those processes (Han, 2022). This, in turn, has introduced intricate and varying forms of inequalities (➔ chapter "Experiential Equity: Reconstructing Cities Through Intersecting Theories" by Mazumder), social reproduction, and psychological vulnerabilities across urban areas, emphasizing the need for developing more socially inclusive research and practice frameworks in psychology and psychiatry. As a result, there has been a rise in inter- and transdisciplinary research spanning psychology, anthropology, and sociology, leading to more nuanced frameworks for exploring individuals' intricate social worlds and the interaction of these social stressors with psychological conditions.[5] These endeavors focus on the diversity of individuals and social groups focussing on (forced) immigrant populations, aiming to unfold structural processes and societal and environmental determinants contributing to increased vulnerability and limited psychosocial coping resources. The work of Rapp et al. (2015) represents early research focusing on individuals with a Turkish migration background, the most significant "minority" in Berlin until 2015. Participants were recruited from 11 neighborhoods within the central district of Berlin's inner city (Berlin-Mitte). For the first time, they addressed the role of neighborhood poverty as a social stressor on psychological distress. Regardless of the person's own economic status, neighborhood poverty significantly contributes to more variation in mental health among individuals; this effect was more pronounced for participants with a so-called migration background compared to those without (Rapp et al., 2015). The influence of local poverty was further examined by Darabi et al. (2023), comparing people with and without migration backgrounds. The result indicated that local poverty heightens susceptibility to both a substantial mental health burden and potential environmental stressors associated with climate change—such as noise and air pollution or a lack of greenspaces.

Particularly, with the arrival of a significant number of refugees in European countries after 2015, there has been a notable increase in studies of this nature, addressing patterns of socio-spatial exclusion, racialization, marginalization, and urban poverty as social stressors, employing various qualitative and quantitative methods.

[5] In 2002, the establishment of the Center for Intercultural Psychiatry & Psychotherapy (ZIPP) at the Department of Psychiatry and Psychotherapy at Charité University Medicine Berlin, Campus Mitte, has been among the pioneering institutions incorporating socio-cultural psychology research and practice.

Research on mental health in the context of migration and refuge has underscored the significance of societal stressors in the domain of urban mental health and highlights the need to construct comprehensive frameworks for comprehending psychological conditions. In their 2023 study on substance (mis)use among refugees in urban and rural areas in Germany, Hertner et al. (2023) emphasized the need for a holistic understanding of refugees' psycho-social circumstances that lead to substance (mis)use. They advocated expanding the focus to encompass the social-ecological context.

Embracing the "Complexity" of the "Context"

Despite extensive research, there is a significant gap in our understanding of how various urban factors interact over time with each other and with individuals to contribute to mental health conditions in cities (Galea, 2011; Galea et al., 2005, 2010; Rydin et al., 2012; Van Der Wal et al., 2021). While various studies indicated the association of urban factors with CMD outcomes, the majority of these findings rely mainly on simple correlational designs, providing univariate explanations for psychological disorders (Oishi & Graham, 2010; Van Der Wal et al., 2021).

> Scientific work is concerned centrally with isolating small pieces of larger systems by way of illustrating how particular associations may help explain the natural world. However, there is no one single explanation that can satisfactorily describe how the urban environment may influence the risk of mood–anxiety disorders. (Galea, 2011)

With cities becoming the major living environment, analyzing studies that explore the variations in the prevalence of psychological disorders in rural and urban areas not only continuously unveiled inconsistencies (Gruebner et al., 2017; Peen et al., 2010) but also emphasized diminishing the significance of such comparisons. The conceptualization of "the urban" or "the rural" as a broad factor is growing more obscure and less aligned with scientific rigor, as it lacks specificity and precision in its characterization. Lately, many authors have emphasized the need to move away from centering the city as an abstract factor in research. Instead, they advocate for reintroducing "the complexity of the context" into studies of mental health, highlighting the trajectories and temporal dynamics between the individual, urban environment, and psychology that contribute to the occurrence and transformation of psychiatric conditions.

The recent theoretical frameworks in psychiatry and psychology strive to offer a more comprehensive and holistic understanding of the intricate dynamics at play in the realm of urban mental health. Rather than oversimplifying the complexities involved, they aim to capture the context of human–environment interactions among biological, psychological, social, and environmental elements. These efforts, however, should be viewed as part of the ongoing struggle to reconcile the psychosocial and the biological aspects of medicine in continuation of the biopsychosocial model (Engel, 1977) influenced by the general system theory (Von Bertalanffy, 1952,

1972) which offers a comprehensive and holistic framework for understanding the determinants of disease. This approach, which considers groups of interconnected events as systems demonstrating functions and properties at a holistic level, enables the identification of similarities across various levels of organization, such as molecules, cells, organs, organisms, individuals, families, societies, or the biosphere (Engel, 1977).

The psychological conditions are influenced by a multitude of dynamic interactions and feedback processes over time (Van Der Wal et al., 2021). However, the temporal dynamics and ongoing processes shaping the momentary context of the individual-environment-psyche relationship are disregarded significantly. Uher and Zwicker (2017) addressed two significant aspects of psychiatry that remain underrated and not reflected in most etiological research.

> The first one is the continuity of pathology over the life course. From cohorts with complete and intensive long-term follow up, it has become clear that most cases of mental illness start in childhood and adolescence. (Uher & Zwicker, 2017)

Further, they emphasized that the course of mental illness varies between individuals and is only loosely related to the diagnostic category. By introducing a "poly-gene-environmental framework" for etiology in psychiatry, they emphasized that exposure to (urban) environmental challenges does not evoke consistent and uniform biological pathways and responses in the brain and psychiatric disorders. Instead, individual differences are observed, and genotypes play a significant role in shaping these variations. In addition to differences in individuals' responses to (urban) environmental factors, a particular person's response might vary over time (Uher & Zwicker, 2017).

As neuro-spatial platforms continue to evolve, the philosophical and theoretical basis of cognitive and neurosciences addressing the interaction of the built environment and the human mind is advancing. The Embodied Embedded Cognition theory (EEC) underscores the dynamic interplay among the brain, body, and environment, explaining the practical emergence of specific behaviors (Newen et al., 2018).

The development of the Vulnerability-Stress Model (Diathesis-Stress Model) (Nielsen et al., 2020) and the Unified Model of Depression (Beck & Bredemeier, 2016) proposes cohesive frameworks for comprehending the intricate interactions between cognitive and biological stress responses. These models shed light on how the influence of stressful life events and (urban) environmental factors play a role in initiating and perpetuating psychological conditions. Further advancement of the "Social Stress Hypothesis" (Pearlin et al., 1981) broadened our understanding of stressors beyond biomedical and epidemiological models. It defines social stress as extending from social isolation, population density, various forms of pollution, and limited access to nature to a poorly built environment and beyond.

The reintroduction of social-ecological psychology could be viewed as another endeavor of this nature. Various waves of psychological movements have emerged, sharing a focus on ecological frameworks. Integrating the macro-environmental concerns advocated by pioneers, such as Lewin (1939), Barker (1968), Bronfenbrenner (1977) and others into mainstream psychology. As a branch of psychology, socioecological psychology examines the impact of natural and social environments (social ecology) on the mind and behavior, along with the reciprocal influence of mind and behavior on these environments. The primary objective of this approach is to shed light on the multifaceted nature of the interplay between individuals and their social ecologies, elucidating how they mutually shape each other (Oishi & Graham, 2010). Socioecological psychology seeks to integrate not only objective socioecological factors, exploring aspects like objective physical, interpersonal, economic, and political environments, but also the psychological processes underlying the link between these factors and human cognition, emotion, and behavior (Oishi, 2014). It addresses three scales of investigation: (a) "association studies" that establish links between specific elements of social ecology and psychology, (b) "process studies" that explore how distinct aspects of social ecology evoke psychological states, leading to the emergence of the target cognition, emotion, or behavior, and (c) "niche construction studies" that elucidate how a particular psychological state may contribute to the establishment and maintenance of a specific form of social ecology (Oishi, 2014; Oishi & Graham, 2010).

Capturing the Complexity of the Context

In addition to the development of comprehensive theoretical models to capture the interplay between the environment, individuals, and psychiatric conditions, there is a concerted effort to establish methodological approaches capable of addressing these complex frameworks. Below, we discuss several recent or ongoing studies aimed at grappling with the multi-layered and intricate nature of the subject.

Complexity Science Perspective Applying a "complexity science perspective" (Mitchell, 2009), Van der Wal (2021) introduced a conceptual framework highlighting that factors and interactions influencing the mental health of urban residents are likely components of a complex system. The implicated factors are integral to dynamic processes, extending from the city to the individual and vice versa, influencing mental health in an urban context. This framework introduced city size, population growth, urbanization, migration, and economic development as meta-factors that influence individual, social, and urban factors. The analytical framework not only takes into account the factors but also considers the temporal aspects of the processes through which these factors can exert their influence. Consequently, the system exhibits dynamics involving feedback loops across various timescales. Thus, its multifaceted behavior displays complex features, including non-linearity, path dependency, and abrupt transitions between states. Additionally, by applying the network theory of mental disorders (Borsboom, 2017), they identified symptom clusters that could be influenced by external factors while also recognizing that symptoms can mutually influence each other. Accounting for this concept is an effort to achieve a more profound understanding of the interplay between the urban environment and mental health.

Defining Multimodal Urban Environmental Profiles In another study, Xu et al. (2023) formulated a unified model that integrates multiple environmental, biological, and behavioral modalities. By defining specific urban environmental profiles correlated with distinct sets of psychiatric symptoms and are mediated by different brain mechanisms, the results outline biological pathways underlying complex, real-life environmental adversity. Their objective is to comprehend which combinations of environmental factors (traffic, air and sound pollution, coastal, water, and green space proximity, etc.) hold the most significance for these psychiatric symptoms and how, within these combinations, each factor contributes to the risk or resilience of mental health symptoms. This model has facilitated the identification of complex living environments that impact specific psychiatric symptom clusters and has revealed the underlying biological mechanisms. Quantifying the impact of various environmental factors on brain and psychiatric symptoms, as well as their interactions in urban settings, could help prioritize and target future public health interventions more effectively.

Call for Integrating Transdisciplinarity and "Community Science"

Promoting Democratic and Equitable Research Approaches for Improving Mental Health in Cities

Addressing the complexity of the context in advancing our critical knowledge of the interplay between environment and individuals' mental health should extend beyond academia. Understanding the psychiatric conditions of individuals within a transitive urban environment characterized by variability, multidimensionality, and complexity necessitates broadening the scientific perspective beyond scientific disciplinary boundaries. This involves creating platforms for democratic knowledge and embracing diverse voices, thus constructing a more nuanced and comprehensive understanding of the world (Bergmann et al., 2021; Guseltseva, 2020).

Developing transdiciplinray research by engaging various social actors and stakeholders, as well as partnering with those most affected by psychiatric conditions, has the potential to revolutionize mental health outcomes. "Citizen science" and "community science" hold promise in the field of mental health and can complement other participatory approaches by enabling active involvement, engagement, and knowledge production with the public (Todowede et al., 2023).

The development of mixed methods, including qualitative and quantitative approaches, as well as the advancement and availability of technology, opened new avenues for involving non-academics in the process of knowledge production process. This could potentially narrow the divide between scientists and civil society in a general sense and might offer opportunities to address unequal access among various social groups locally and globally. One significant methodological approach in this category involves applying mobile devices for mental health purposes, enabling

Ecological Momentary Assessments (EMA) and Ambulatory Assessments (AA) (➔ chapter "In the Right Place at the Right Time: Leveraging Geographic Ecological Momentary Assessment to Analyze Contexts and Health" Kestens and Kingsbury).

Since the latter part of the twentieth century, the advances in portable geo-physiological sensors and microcomputer systems integrated into mobile devices like smartphones and wearables have facilitated the development of mobile health (m-health) and AA. These techniques have facilitated a wide range of assessment and field methods that enable intra- and/or inter-subject-environmental assessments and intervention: exploring real-time and longitudinal subjective experiences, as well as the objective biological, physiological, behavioral, and environmental aspects of humans or nonhuman primates in natural or unconstrained settings, both between and within individuals' everyday lives (Ebner-Priemer & Kubiak, 2010; Fahrenberg et al., 2007; Reichert et al., 2020; Trull & Ebner-Priemer, 2014). These techniques serve two distinct purposes: (1) maintaining and facilitating intervention development, including ecological momentary intervention (EMI), and (2) conducting research to inform our understanding of the etiology of psychiatric conditions, including EMA. Unlike their early application in medicine, these innovative approaches have made late and minimal inroads into the field of mental health. Despite numerous research studies exploring the feasibility of these aspirations both for EMI and EMA, more research is still needed in this area (Balaskas et al., 2021; Heinz et al., 2020a). In the realm of clinical and behavioral psychology, the technique has been applied for interventions or evaluating treatments such as monitoring symptoms' dynamics and trajectories, evaluating treatment effects, and averting recurrence and relapse (Heinz et al., 2020a). Over the last few years, ecological momentary assessment (EMA) has been more commonly employed in assessing the longitudinal and momentary indices to examine the individuals' psychological and emotional state while exposed to everyday living environments and social stressors.

In a recent transdisciplinary initiative, the Research Platform Neurourbanism (RPN) is investigating the psychological, neurocognitive, social, and environmental factors that influence the relationship between urban density and the mental well-being of city residents. Alongside two other major methodological approaches—(1) analyzing and aggregating global health data from existing genetic databases to establish correlations between urban density and mental health factors and (2) conducting controlled experiments using Mobile Brain/Body Technologies and Virtual Reality experiments to isolate density-related factors (infrastructural, social, architectural, et Cetra) on immediate physiological responses associated with different subjective experiences—the study employs an ecological momentary assessment (EMA) with a citizen science approach. This approach aims to longitudinally measure stress exposure and emotional well-being in Berlin through "Your Emotional City,"[6] an app-based EMA combined with a satellite-based global positioning system (GPS). In offering the app in multiple languages, the project emphasizes the significance of integrating knowledge co-production to address the complexity of

[6] https://www.neurourbanism.eu/

mental health in cities through a more equitable and democratic approach. This approach represents another step toward generating transformative knowledge concerning a neurourbanistic perspective on urban planning and governance. Additionally, it calls for developing pathways for translating this knowledge to support action, policy design, and urban planning to equitable access to resources for promoting the city's population's mental health (Tonne et al., 2021).

The chapter aimed to shed light on the researchers' attempt to explore urban living's impact on the human psyche for over a century; the subject has unfolded unevenly across different time periods and geographical locations. In recent decades, the reintroduction of health as a holistic condition encompassing both psychological and physical aspects has elevated the significance of mental health from a marginal position to the forefront of all discussions surrounding 'health.' Particularly after the advent of the COVID-19 pandemic, there has been a growing focus on research and practical initiatives to address this question.

Our understanding of the city's influence on the psyche is intricately connected to our comprehension of health across various scales, from the individual body and psyche to the broader environment. In this context, methodological and pertinent ethical considerations continue to pose challenges and play a crucial role in determining how we aim to attain this understanding.

References

Adli, M., Berger, M., Brakemeier, E.-L., Engel, L., Fingerhut, J., Hehl, R., Heinz, A., Mayer, H., Matussek, T., Mehran, N., Tolaas, S., Walter, H., Weiland, U., & Stollmann, J. (2016). Neurourbanistik—Ein Methodischer Schulterschluss zwischen Stadtplanung Und Neurowissenschaften. *Die Psychiatrie, 13*(02), 70–78. https://doi.org/10.1055/S-0038-1670122

Adli, M., Berger, M., Brakemeier, E.-L., Engel, L., Fingerhut, J., Gomez-Carrillo, A., Hehl, R., Heinz, A., Juergen Mayer, H., Mehran, N., Tolaas, S., Walter, H., Weiland, U., & Stollmann, J. (2017). Neurourbanism: Towards a new discipline. *The Lancet Psychiatry, 4*(3), 183–185. https://doi.org/10.1016/S2215-0366(16)30371-6

Almeida-Filho, N., Mari, J., & Coutinho, E. (1995). Rural–urban migration and mental health in Latin America: The debate continues. In I. T. Harpham & I. Blue (Eds.), *Urbanization and mental health in developing countries* (pp. 41–60). Avebury.

Arbib, M. A. (2012). Brains, machines and buildings: Towards A neuromorphic architecture. *Intelligent Buildings International, 4*(3), 147–1681. https://doi.org/10.1080/17508975.2012.702863

Bagley, C., Jacobson, S., & Palmer, C. (1973). Social structure and the ecological distribution of mental illness, suicide, and delinquency. *Psychological Medicine, 3*(2), 177–187. https://doi.org/10.1017/S0033291700048509

Balaskas, A., Schueller, S. M., Cox, A. L., & Doherty, G. (2021). Ecological momentary interventions for mental health: A scoping review. *PLoS One, 16*(3), E0248152. https://doi.org/10.1371/Journal.Pone.0248152

Banaei, M., Hatami, J., Yazdanfar, A., & Gramann, K. (2017). Walking through architectural spaces: The impact of interior forms on human brain dynamics. *Frontiers in Human Neuroscience, 11*, 477. https://doi.org/10.3389/Fnhum.2017.00477

Barker, R. G. (1968). *Ecological psychology: Concepts and methods for studying the environment of human behavior*. Stanford University Press.

Barton, H. (2009). Land use planning and health and Well-being. *Land Use Policy, 26*, S115–S123. https://doi.org/10.1016/J.Landusepol.2009.09.008

Beck, A. T., & Bredemeier, K. (2016). A unified model of depression: Integrating clinical, cognitive, biological, and evolutionary perspectives. *Clinical Psychological Science, 4*(4), 596–619. https://doi.org/10.1177/2167702616628523

Bergmann, M., Schäpke, N., Marg, O., Stelzer, F., Lang, D. J., Bossert, M., Gantert, M., Häußler, E., Marquardt, E., Piontek, F. M., Potthast, T., Rhodius, R., Rudolph, M., Ruddat, M., Seebacher, A., & Sußmann, N. (2021). Transdisciplinary sustainability research in real-world labs: Success factors and methods for change. *Sustainability Science, 16*(2), 541–564. https://doi.org/10.1007/S11625-020-00886-8

Bornstein, M. H., & Bornstein, H. G. (1976). The pace of life. *Nature, 259*, 557–559.

Borsboom, D. (2017). A network theory of Mental disorders. *World Psychiatry, 16*(1), 5–13. https://doi.org/10.1002/Wps.20375

Boto, E., Holmes, N., Leggett, J., Roberts, G., Shah, V., Meyer, S. S., Muñoz, L. D., Mullinger, K. J., Tierney, T. M., Bestmann, S., Barnes, G. R., Bowtell, R., & Brookes, M. J. (2018). Moving magnetoencephalography towards real-world applications with A wearable system. *Nature, 555*(7698), 657–661. https://doi.org/10.1038/Nature26147

Breslau, J., Marshall, G. N., Pincus, H. A., & Brown, R. A. (2014). Are mental disorders more common in urban than rural areas of the United States? *Journal of Psychiatric Research, 56*, 50–55. https://doi.org/10.1016/J.Jpsychires.2014.05.004

Bronfenbrenner, U. (1977). Toward an experimental ecology of human development. *American Psychologist, 32*(7), 513–531. https://doi.org/10.1037/0003-066X.32.7.513

Brown, G. W., & Birley, J. L. T. (1968). Crises and life changes and the onset of schizophrenia. *Journal of Health and Social Behaviour, 9*(3), 203. https://doi.org/10.2307/2948405

Brown, G. W., & Prudo, R. (1981). Psychiatric disorder in a rural and an urban population: 1. Aetiology of depression. *Psychological Medicine, 11*(3), 581–599. https://doi.org/10.1017/S0033291700052880

Cappon, D. (1975). Designs for improvements in the quality of life in downtown cores. *International Journal of Mental Health, 4*(3), 31–47. https://doi.org/10.1080/00207411.1975.11448691

Carstairs, G. M., & Kapur, R. L. (1976). *The great universe of Kota: Stress, change and mental disorder in an Indian Village*. University of California Press.

Cooper, J. E., & Sartorius, N. (1977). Cultural and temporal variations in schizophrenia: A speculation on the importance of industrialization. *British Journal of Psychiatry, 130*(1), 50–55. https://doi.org/10.1192/Bjp.130.1.50

Craik, K. H. (1973). Environmental psychology. *Annual Review of Psychology, 24*(1), 403–422. https://doi.org/10.1146/Annurev.Ps.24.020173.002155

Cutts, D. B., Meyers, A. F., Black, M. M., Casey, P. H., Chilton, M., Cook, J. T., Geppert, J., Ettinger De Cuba, S., Heeren, T., Coleman, S., Rose-Jacobs, R., & Frank, D. A. (2011). US housing insecurity and the health of very young children. *American Journal of Public Health, 101*(8), 1508–1514. https://doi.org/10.2105/AJPH.2011.300139

Darabi, D., Kluge, U., Penka, S., Mundt, A. P., Schouler-Ocak, M., Butler, J., Liu, S., Heinz, A., & Rapp, M. A. (2023). Environmental stress, minority status, and local poverty: Risk factors for mental health in Berlin's inner city. *European Archives of Psychiatry and Clinical Neuroscience, 273*(5), 1201–1206. https://doi.org/10.1007/S00406-022-01508-3

Dimitrov-Discher, A., Wenzel, J., Kabisch, N., Hemmerling, J., Bunz, M., Schöndorf, J., Walter, H., Veer, I. M., & Adli, M. (2022). Residential green space and air pollution are associated with brain activation in A social-stress paradigm. *Scientific Reports, 12*(1), 10614. https://doi.org/10.1038/S41598-022-14659-Z

Dimitrov-Discher, A., Gu, L., Pandit, L., Veer, I. M., Walter, H., Adli, M., & Knöll, M. (2023). Stress and streets: How the network structure of streets is associated with stress-related brain activation. *Journal of Environmental Psychology, 91*, 102142. https://doi.org/10.1016/J.Jenvp.2023.102142

Dohrenwend, B. P. (1969). Social status, stress and psychological symptoms. *The Milbank Memorial Fund Quarterly, 47*(1), 137. https://doi.org/10.2307/3349184
Dohrenwend, B. P. (1975). Sociocultural and social-psychological factors in the genesis of Mental disorders. *Journal of Health and Social Behavior, 16*(4), 365. https://doi.org/10.2307/2136610
Dohrenwend, B., & Dohrenwend, B. (1974). Psychiatric disorders in urban settings. In G. Caplan (Ed.), *American handbook of psychiatry* (pp. 424–447). Basic Books.
Dougherty, B. O., & Arbib, M. A. (2013). The evolution of neuroscience for architecture: Introducing the special issue. *Intelligent Buildings International, 5*(Supl), 4–9. https://doi.org/10.1080/17508975.2013.818763
Dunham, H. W. (1965). *Community of schizophrenia.* Wayne State University Press.
Dye, C. (2008). Health and urban living. *Science, 319*(5864), 766–769. https://doi.org/10.1126/Science.1150198
Eberhard, J. P. (2009a). Applying neuroscience to architecture. *Neuron, 62*(6), 753–756. https://doi.org/10.1016/J.Neuron.2009.06.001
Eberhard, J. P. (2009b). *Brain landscape: The coexistence of neuroscience and architecture.* Oxford University Press.
Ebner-Priemer, U. W., & Kubiak, T. (2010). The decade of behavior revisited: future prospects for Ambulatory Assessment. *European Journal of Psychological Assessment, 26*(3), 151–153. https://doi.org/10.1027/1015-5759/a000021
Engel, G. L. (1977). The need for A new medical model: A challenge for biomedicine. *Science, 196*(4286), 129–136. https://doi.org/10.1126/Science.847460
Fahrenberg, J., Myrtek, M., Pawlik, K., & Perrez, M. (2007). Ambulatory assessment—Monitoring behavior in daily life settings. *European Journal of Psychological Assessment, 23*(4), 206–213. https://doi.org/10.1027/1015-5759.23.4.206
Faris, R. E. L., & Dunham, H. W. (1939). *Mental disorders in urban areas: An ecological study of schizophrenia and other psychoses.* University of Chicago Press.
Freeman, H. (1978). Mental health and the environment. *The British Journal of Psychiatry, 132*(2), 113–124. https://doi.org/10.1192/Bjp.132.2.113
Freeman, H. (1994). Schizophrenia and City residence. *British Journal of Psychiatry, 164*(S23), 39–50. https://doi.org/10.1192/S0007125000292726
Freeman, H., & Alpert, M. (1986). Prevalence of schizophrenia in an urban population. *The British Journal of Psychiatry, 149*(5), 603–611. https://doi.org/10.1192/Bjp.149.5.603
Galea, S. (2011). The urban brain: New directions in research exploring the relation between cities and mood-anxiety disorders. *Depression and Anxiety, 28*(10), 857–862. https://doi.org/10.1002/Da.20868
Galea, S., Rudenstine, S., & Vlahov, D. (2005). Drug use, misuse, and the urban environment. *Drug and Alcohol Review, 24*(2), 127–136. https://doi.org/10.1080/09595230500102509
Galea, S., Riddle, M., & Kaplan, G. A. (2010). Causal thinking and complex system approaches in epidemiology. *International Journal of Epidemiology, 39*(1), 97–106. https://doi.org/10.1093/ije/dyp296
Gans, H. J. (1967). Planning and city planning for mental health. In H. W. Eldredge (Ed.), *Taming megalopolis* (Vol. 2, pp. 897–916). Doubleday.
Giggs, J. A., & Cooper, J. E. (1987). Ecological structure and the distribution of schizophrenia and affective psychoses in Nottingham. *The British Journal of Psychiatry, 151*(5), 627–633. https://doi.org/10.1192/Bjp.151.5.627
Goldberg, E. M., & Morrison, S. L. (1963). Schizophrenia and social class. *British Journal of Psychiatry, 109*(463), 785–802. https://doi.org/10.1192/Bjp.109.463.785
Goldman, H. H. (1983). International perspectives in deinstitutionalization. *International Journal of Mental Health, 11*(4), 1–168.
Goldsmith, W. W., & Blakely, E. J. (1992). *Separate societies: Poverty and inequality in U.S. cities* (2nd ed.). Temple University Press.
Groves, E. R. (1916). Psychic causes of rural migration. *American Journal of Sociology, 21*(5), 623–627. https://doi.org/10.1086/212545

Gruebner, O., Rapp, M. A., Adli, M., Kluge, U., Galea, S., & Heinz, A. (2017). Cities and mental health. *Deutsches Ärzteblatt International*. https://doi.org/10.3238/Arztebl.2017.0121

Guseltseva, M. (2020). Transdisciplinary approach in modern psychology. *Main Issues of Pedagogy and Psychology, 18*(2), 11–22. https://doi.org/10.24234/Miopap.V18i2.375

Häfner, H. (Ed.). (2002). *Risk and protective factors in Schizophrenia: Towards a conceptual model of the disease process*. Steinkopff. https://doi.org/10.1007/978-3-642-57516-7

Hammer, M., Makiesky-Barrow, S., & Gutwirth, L. (1978). Social networks and schizophrenia*. *Schizophrenia Bulletin, 4*(4), 522–545. https://doi.org/10.1093/Schbul/4.4.522

Han, B.-C. (2022). *Hyperculture: Culture and globalization* (D. Steuer, Trans.). Polity Press.

Harpham, T., & Blue, I. (Eds.). (1995). *Urbanization and mental health in developing countries*. Avebury.

Harvey, D. (1978). The urban process under capitalism: A framework for Analysis. *International Journal of Urban and Regional Research, 2*(1–3), 101–131. https://doi.org/10.1111/J.1468-2427.1978.Tb00738.X

Heinz, A., Kiefer, F., Smolka, M. N., Endrass, T., Beste, C., Beck, A., Liu, S., Genauck, A., Romund, L., Banaschewski, T., Bermpohl, F., Deserno, L., Dolan, R. J., Durstewitz, D., Ebner-Priemer, U., Flor, H., Hansson, A. C., Heim, C., Hermann, D., et al. (2020a). Addiction research consortium: Losing and regaining control over drug intake (recode)—From trajectories to mechanisms and interventions. *Addiction Biology, 25*(2), E12866. https://doi.org/10.1111/Adb.12866

Heinz, A., Zhao, X., & Liu, S. (2020b). Implications of the association of social exclusion with mental health. *JAMA Psychiatry, 77*(2), 113. https://doi.org/10.1001/Jamapsychiatry.2019.3009

Henderson, S. (1988). *An introduction to social psychiatry*. Oxford University Press.

Hertner, L., Stylianopoulos, P., Heinz, A., Kluge, U., Schäfer, I., & Penka, S. (2023). Substance (Mis)Use among refugees as a matter of social ecology: Insights into a multi-site rapid assessment in Germany. *Conflict and Health, 17*(1), 1. https://doi.org/10.1186/S13031-023-00499-9

Hillman, M., & Whalley, A. (1975). Land use and travel. *Built Environment Quarterly; London, 1*(2), 105.

Hoggart, K. (1990). Let's do away with rural. *Journal of Rural Studies, 6*(3), 245–257. https://doi.org/10.1016/0743-0167(90)90079-N

Iggena, D., Jeung, S., Maier, P. M., Ploner, C. J., Gramann, K., & Finke, C. (2023). Multisensory input modulates memory-guided spatial navigation in humans. *Communications Biology, 6*(1), 1167. https://doi.org/10.1038/S42003-023-05522-6

Jaco, E. G. (1954). The social isolation hypothesis and schizophrenia. *American Sociological Review, 19*(5), 567. https://doi.org/10.2307/2087795

Kasl, S. V. (1977). The effect of residentional Environmnet on health and behaviour: A review. In L. E. Hinkle & W. C. Loring (Eds.), *The effect of the man-made environment on health and behavior: A report of the inter-university board of collaborators*. Center for Disease Control, Public Health Service, U.S. Department of Health, Education, and Welfare.

Kawachi, I., & Berkman, L. F. (Eds.). (2003). *Neighborhoods and health* (1st ed.). Oxford University Pressnew York. https://doi.org/10.1093/Acprof:Oso/9780195138382.001.0001

Kelly, J. G. (1971). Qualities for the community psychologist. *American Psychologist, 26*(10), 897–903. https://doi.org/10.1037/H0032231

Kessler, R. C., & Mcleod, J. D. (1984). Sex differences in vulnerability to undesirable life events. *American Sociological Review, 49*(5), 620. https://doi.org/10.2307/2095420

Klusmann, D., & Angermeyer, M. C. (1987). Urban ecology and psychiatric admission rates: Results from a study in the City of Hamburg. In M. C. Angermeyer (Ed.), *From social class to social stress* (pp. 16–45). Springer. https://doi.org/10.1007/978-3-642-52057-0_2

Lederbogen, F., Kirsch, P., Haddad, L., Streit, F., Tost, H., Schuch, P., Wüst, S., Pruessner, J. C., Rietschel, M., Deuschle, M., & Meyer-Lindenberg, A. (2011). City living and urban upbringing affect neural social stress processing in humans. *Nature, 474*(7352), 498–501. https://doi.org/10.1038/Nature10190

Lewin, K. (1939). Field theory and experiment in social psychology: Concepts and methods. *American Journal of Sociology, 44*(6), 868–896.
Lewis, G., David, A., Andréassson, S., & Allebeck, P. (1992). Schizophrenia and city life. *The Lancet, 340*(8812), 137–140. https://doi.org/10.1016/0140-6736(92)93213-7
Lipowski, Z. J. (1975). Sensory and information inputs overload: Behavioral effects. *Comprehensive Psychiatry, 16*(3), 199–221. https://doi.org/10.1016/0010-440X(75)90047-4
Maas, J., Verheij, R. A., Groenewegen, P. P., De Vries, S., & Spreeuwenberg, P. (2006). Green space, urbanity, and health: How strong is the relation? *Journal of Epidemiology & Community Health, 60*(7), 587–592. https://doi.org/10.1136/Jech.2005.043125
Madden, D. (2023). Polycritical city? *City, 27*(3–4), 271–274. https://doi.org/10.1080/13604813.2023.2232682
Magaziner, J. (1988). Living density and psychopathology: A re-examination of the negative model. *Psychological Medicine, 18*(2), 419–431. https://doi.org/10.1017/S0033291700007960
Marsella, A. (1992). *Urbanisation and mental health disorders: An overview of theory and research, and recommendations for interventions and research*. WHO.
Marsella. (1998). Urbanization, mental health, and social deviancy: A review of issues and research. *American Psychologist, 53*(6), 624–634. https://doi.org/10.1037/0003-066X.53.6.624
Mccarthy, D., & Saegert, S. (1978). Residential density, social overload, and social withdrawal. *Human Ecology, 6*(3), 253–272. https://doi.org/10.1007/BF00889026
Meit, M., Knudson, A., Gilbert, T., Tzy-Chyi Yu, A., Tanenbaum, E., Ormson, E., Tenbroeck, S., Alycia, B., Popat, S., & NORC Walsh Center for Rural Health Analysis. (2014). *The 2014 update of the rural-urban chartbook*. The Rural Health Reform Policy Research Center.
Meyer, I. H., Schwartz, S., & Frost, D. M. (2008). Social patterning of stress and coping: Does disadvantaged social statuses confer more stress and fewer coping resources? *Social Science & Medicine, 67*(3), 368–379. https://doi.org/10.1016/J.Socscimed.2008.03.012
Milgram, S. (1970). The experience of living in cities. *Science, 167*(3924), 1461–1468.
Mirowsky, J., & Ross, C. E. (2017). *Social causes of psychological distress* (1st ed.). https://doi.org/10.4324/9781315129464
Mitchell, M. (2009). *Complexity: A guided tour*. Oxford University Press.
Moorin, R. E., Holman, C. D. J., Garfield, C., & Brameld, K. J. (2006). Health related migration: Evidence of reduced "urban-drift". *Health & Place, 12*(2), 131–140. https://doi.org/10.1016/J.healthplace.2004.10.013
Mossakowski, K. N. (2014). Stress and mental illness. In W. C. Cockerham, R. Dingwall, & S. Quah (Eds.), *The Wiley Blackwell encyclopedia of health, illness, behavior, and society* (1st ed., pp. 1–5). Wiley. https://doi.org/10.1002/9781118410868.Wbehibs358
Mueller, D. P. (1981). The current status of urban-rural differences in psychiatric disorder: An emerging trend for depression. *The Journal of Nervous and Mental Disease, 169*(1), 18.
Murphy, J. E. (1976). Social causes: The independent variable. In B. H. Kaplan, R. N. Wilson, & A. H. Leighton (Eds.), *Further explorations in social psychiatry*. Basic Books.
National Collaborating Centre for Mental Health (UK). (2011). Common mental health disorders. In *Common mental health disorders: Identification and pathways to care*. The British Psychological Society & The Royal College of Psychiatrists. https://www.ncbi.nlm.nih.gov/books/nbk92254/
Neff, J. A. (1983). Urbanicity and depression reconsidered: The evidence regarding depressive symptomatology. *The Journal of Nervous and Mental Disease, 171*(9), 546–552. https://doi.org/10.1097/00005053-198309000-00004
Neff, J. A., & Husaini, B. A. (1987). Urbanicity, race, and psychological distress. *Journal of Community Psychology, 15*, 520–536.
Newen, A., Gallagher, S., & De Bruin, L. (2018). 4E cognition: Historical roots, key concepts, and central issues. In A. Newen, L. De Bruin, & S. Gallagher (Eds.), *The Oxford handbook of 4E cognition* (pp. 2–16). Oxford University Press. https://doi.org/10.1093/Oxfordhb/9780198735410.013.1

Nielsen, J. D., Mennies, R. J., & Olino, T. M. (2020). Application of a diathesis-stress model to the interplay of cortical structural development and emerging depression in youth. *Clinical Psychology Review, 82*, 101922. https://doi.org/10.1016/J.Cpr.2020.101922

Oishi, S. (2014). Socioecological psychology. *Annual Review of Psychology, 65*(1), 581–609. https://doi.org/10.1146/Annurev-Psych-030413-152156

Oishi, S., & Graham, J. (2010). Social ecology: Lost and found in psychological science. *Perspectives on Psychological Science, 5*(4), 356–377. https://doi.org/10.1177/1745691610374588

Okkels, N., Kristiansen, C. B., Munk-Jørgensen, P., & Sartorius, N. (2018). Urban mental health: Challenges and perspectives. *Current Opinion in Psychiatry, 31*(3), 258. https://doi.org/10.1097/YCO.0000000000000413

Pearlin, L. I. (1989). The sociological study of stress. *Journal of Health and Social Behavior, 30*(3), 241. https://doi.org/10.2307/2136956

Pearlin, L. I., Menaghan, E. G., Lieberman, M. A., & Mullan, J. T. (1981). The stress process. *Journal of Health and Social Behavior, 22*(4), 337. https://doi.org/10.2307/2136676

Peen, J., Schoevers, R. A., Beekman, A. T., & Dekker, J. (2010). The current status of urban-rural differences in psychiatric disorders. *Acta Psychiatrica Scandinavica, 121*(2), 84–93. https://doi.org/10.1111/J.1600-0447.2009.01438.X

Rapp, M. A., Kluge, U., Penka, S., Vardar, A., Aichberger, M. C., Mundt, A. P., Schouler-Ocak, M., Mösko, M., Butler, J., Meyer-Lindenberg, A., & Heinz, A. (2015). When local poverty is more important than your income: Mental health in minorities in inner cities. *World Psychiatry, 14*(2), 249–250. https://doi.org/10.1002/Wps.20221

Reichert, M., Braun, U., Lautenbach, S., Zipf, A., Ebner-Priemer, U., Tost, H., & Meyer-Lindenberg, A. (2020). Studying the impact of built environments on human mental health in everyday life: Methodological developments, state-of-the-art and technological frontiers. *Current Opinion in Psychology, 32*, 158–164. https://doi.org/10.1016/J.Copsyc.2019.08.026

Rosen, G. (2015). *A history of public health* (Revised Expanded ed.). Johns Hopkins University Press.

Rydin, Y., Bleahu, A., Davies, M., Dávila, J. D., Friel, S., De Grandis, G., Groce, N., Hallal, P. C., Hamilton, I., Howden-Chapman, P., Lai, K.-M., Lim, C., Martins, J., Osrin, D., Ridley, I., Scott, I., Taylor, M., Wilkinson, P., & Wilson, J. (2012). Shaping cities for health: Complexity and the planning of urban environments in the 21st century. *The Lancet, 379*(9831), 2079–2108. https://doi.org/10.1016/S0140-6736(12)60435-8

Shorter, E. (2017). History of urban mental illness. In N. Okkels, C. B. Kristiansen, & P. Munk-Jorgensen (Eds.), *Mental health and illness in the city* (pp. 1–9). Springer. https://doi.org/10.1007/978-981-10-0752-1_18-1

Simmel, G. (1903). *Die Großstädte und das Geistesleben*. Petermann.

Smith, K. E., & Juarascio, A. (2019). From Ecological Momentary Assessment (EMA) to Ecological Momentary Intervention (EMI): Past and future directions for ambulatory assessment and interventions in eating disorders. *Current Psychiatry Reports, 21*(7), 53. https://doi.org/10.1007/S11920-019-1046-8

Strömgren, E. (1987). Changes in the incidence of schizophrenia? *British Journal of Psychiatry, 150*(1), 1–7. https://doi.org/10.1192/Bjp.150.1.1

Sun, P.-P., Tan, F.-L., Zhang, Z., Jiang, Y.-H., Zhao, Y., & Zhu, C.-Z. (2018). Feasibility of functional near-infrared spectroscopy (Fnirs) to investigate the mirror neuron system: An experimental study in a real-life situation. *Frontiers in Human Neuroscience, 12*, 86. https://doi.org/10.3389/Fnhum.2018.00086

Taylor, P. J., & Lang, R. E. (2004). The shock of the new: 100 concepts describing recent urban change. *Environment and Planning A, 36*(6), 951–958. https://doi.org/10.1068/A375

Thoits, P. A. (1995). Stress, coping, and social support processes: Where are we? What next? *Journal of Health and Social Behavior, 35*, 53. https://doi.org/10.2307/2626957

Todowede, O., Lewandowski, F., Kotera, Y., Ashmore, A., Rennick-Egglestone, S., Boyd, D., Moran, S., Ørjasæter, K. B., Repper, J., Robotham, D., Rowe, M., Katsampa, D., & Slade, M. (2023). Best practice guidelines for citizen science in mental health research: Systematic

review and evidence synthesis. *Frontiers in Psychiatry, 14*, 1175311. https://doi.org/10.3389/Fpsyt.2023.1175311
Tonne, C., Adair, L., Adlakha, D., Anguelovski, I., Belesova, K., Berger, M., Brelsford, C., Dadvand, P., Dimitrova, A., Giles-Corti, B., Heinz, A., Mehran, N., Nieuwenhuijsen, M., Pelletier, F., Ranzani, O., Rodenstein, M., Rybski, D., Samavati, S., Satterthwaite, D., et al. (2021). Defining pathways to healthy sustainable urban development. *Environment International, 146*, 106236. https://doi.org/10.1016/J.Envint.2020.106236
Tost, H., Champagne, F. A., & Meyer-Lindenberg, A. (2015). Environmental influence in the brain, human welfare and mental health. *Nature Neuroscience, 18*(10), 1421–1431. https://doi.org/10.1038/Nn.4108
Trull, T. J., & Ebner-Priemer, U. (2014). The role of ambulatory assessment in psychological science. *Current Directions in Psychological Science, 23*(6), 466–470. https://doi.org/10.1177/0963721414550706
Uher, R., & Zwicker, A. (2017). Etiology in psychiatry: Embracing the reality of poly-gene-environmental causation of mental illness. *World Psychiatry, 16*(2), 121–129. https://doi.org/10.1002/Wps.20436
UN. (2022). *World population prospects*. Population Division—United Nations. https://population.un.org/wpp/
UN-HABITAT. (2022). *World cities report 2022, envisaging the future of cities*. https://unhabitat.org/wcr/
University of Amsterdam. (n.d.). *The centre for urban mental health*. https://www.centreforurbanmentalhealth.com/home
Van Der Wal, J. M., Van Borkulo, C. D., Deserno, M. K., Breedvelt, J. J. F., Lees, M., Lokman, J. C., Borsboom, D., Denys, D., Van Holst, R. J., Smidt, M. P., Stronks, K., Lucassen, P. J., Van Weert, J. C. M., Sloot, P. M. A., Bockting, C. L., & Wiers, R. W. (2021). Advancing urban mental health research: From complexity science to actionable targets for intervention. *The Lancet Psychiatry, 8*(11), 991–1000. https://doi.org/10.1016/S2215-0366(21)00047-X
Verheij, R. A. (1996). Explaining urban-rural variations in health: A review of interactions between individual and environment. *Social Science & Medicine, 42*(6), 923–935. https://doi.org/10.1016/0277-9536(95)00190-5
Vlahov, D. (2002). Urbanization, urbanicity, and health. *Journal of Urban Health: Bulletin of the New York Academy of Medicine, 79*(90001), 1S–12S. https://doi.org/10.1093/Jurban/79.Suppl_1.S1
Von Bertalanffy, L. (1952). Problems of life. An evaluation of modern biological thought. *AIBS Bulletin, 2*(5), 14–14. https://doi.org/10.1093/Aibsbulletin/2.5.14-H
Von Bertalanffy, L. (1972). The history and status of general systems theory. *Academy of Management Journal, 15*(4), 407–426. https://doi.org/10.2307/255139
WHO. (1986). *Creating healthy cities*. https://www.who.int/activities/creating-healthy-cities
WHO. (2017). *Depression and other common mental disorders global health estimates*. https://iris.who.int/bitstream/handle/10665/254610/who-msd-mer-2017.2-eng.pdf
Wing, J. K. (1989). The concept of negative symptoms. *British Journal of Psychiatry, 155*(S7), 10–14. https://doi.org/10.1192/S000712500029140X
Xu, J., Liu, N., Polemiti, E., Garcia-Mondragon, L., Tang, J., Liu, X., Lett, T., Yu, L., Nöthen, M. M., Feng, J., Yu, C., Marquand, A., Schumann, G., The Environmental Consortium, Walter, H., Heinz, A., Ralser, M., Twardziok, S., Vaidya, N., et al. (2023). Effects of urban living environments on mental health in adults. *Nature Medicine, 29*(6), 1456–1467. https://doi.org/10.1038/S41591-023-02365-W

Environmental Neurorehabilitation: Harnessing the Power of the Environment in Rehabilitation Following Neurological Conditions

Michael Norwood, Danielle Pretty, Chelsea Marsh, Camila Shirota, Coral Gillet, Heidi Zeeman, and Elizabeth Kendall

Abstract The global need for rehabilitation is steadily increasing due to the greater likelihood of survival following severe neurological injury. However, survival means more people are living with disability, resulting in demand for health and rehabilitation systems, most of which are stretched beyond their capacity. The environment is a constant factor in our lives and can have a significant impact on brain structure and activity. Despite this potentially important impact of the environment on recovery following brain injury, it is usually overlooked in neurorehabilitation. In the early stages of rehabilitation, the environment is designed for clinical efficiency rather than for individual well-being. In the later stages of recovery, the person's environment can become increasingly diminished due to the loss of employment, social connections, and other roles. Environmental neuroscience is well-placed to inform patient-centred approaches to rehabilitation and health care through improved and facilitative environmental design. In this chapter, we propose the purposeful inclusion of the environment in rehabilitation services at all stages of recovery and across all contexts. First, we outline how fundamental concepts such as neuroplasticity and neurogenesis and the environment are interrelated, and the implications for neurorehabilitation. We discuss the role of technology and virtual environments, the built and natural environments, and the importance of brain imaging techniques that can determine environmental interactions. The implications for research, challenges, and future directions are also discussed. Readers will hopefully be convinced that is important to understand how—or rather where—neurorehabilitation is delivered in the future.

Keywords Disability · Rehabilitation · Environment · Neuroplasticity · Brain Injury

M. Norwood · D. Pretty · C. Marsh · C. Shirota · C. Gillet · H. Zeeman · E. Kendall (✉)
The Hopkins Centre, Griffith University, Nathan, QLD, Australia
e-mail: e.kendall@griffith.edu.au

S. Kühn (ed.), *Environmental Neuroscience*,
https://doi.org/10.1007/978-3-031-64699-7_15

Introduction

Neurological conditions are an ongoing global health challenge affecting more people each year (Feigin et al., 2019; Ding et al., 2022). They are a leading cause of death and disability, with approximately ten million deaths worldwide from Parkinson's disease, stroke, and acquired brain injury in 2019 (Ding et al., 2022). Although this is a sizeable number in absolute terms, more people are surviving neurological diseases and living longer with disability. As the rates of people with neurological impairment continue to grow, our struggling health and rehabilitation systems are stretched beyond capacity. Indeed, the global need for rehabilitation is currently unmet, with the World Health Organisation noting that in some low- and middle-income countries, more than 50% of people do not receive the rehabilitation services they require.

To keep pace with the growing demand for neurological rehabilitation and to enable continuous delivery over time, there is an urgent need to improve its capacity (WHO Rehabilitation, 2030). As a result, considerable attention has been given to the ways in which novel approaches and technologies can improve the efficiency and effectiveness of rehabilitation (for example see, Laver et al., 2017; Mehrholz et al., 2020; Möller et al., 2021). However, little thought has been given to how we can capitalise on the environment to deliver, reinforce, and complement positive rehabilitation outcomes both in the hospital and for years post-discharge.

In addition to being linked to the onset of neurological conditions (e.g. through accidents, lifestyle, toxic exposures, or social harms), the environment remains one of the most constant and consistent elements of the lifelong rehabilitation journey. Everything happens in places, and these places can have independent positive or negative impacts on people's health and recovery. The environment also provides a potentially fruitful focus for facilitating long-term adjustment and independence, through modified housing, accessible equipment, and urban design that maximises well-being and social engagement. Further, the environment has profound effects on the brain (e.g. see, Mackes et al., 2022; Miller et al., 2013; Weissman et al., 2020), which may have a significant impact on neurorehabilitation.

In this chapter, we introduce the environment as an important, often overlooked, aspect of neurorehabilitation. We present a case for the deliberate inclusion of the environment in rehabilitation delivery from early in the recovery and treatment processes through to long-term life in the community with neurological disability. Although the effects of the environment on human health and well-being have been thoroughly researched in some domains, this knowledge has rarely been applied to the world of rehabilitation.

The chapter first outlines fundamental concepts such as neuroplasticity and neurogenesis, exploring how they relate to the environment and the implications for neurorehabilitation. Some environmental factors that can contribute to neurorehabilitation are outlined, including the role of technology, and the built and natural environments. The implications for research, measurement challenges, and future directions are explored. By the end of the chapter, the reader will have a basic background on the effects of the environment on the brain, how this can be measured, and how rehabilitation can adopt this knowledge.

Neuroplasticity and Neurogenesis

For many decades, the conventional understanding of neurological damage was that individuals recovered rapidly during the first few years, after which time little change was likely to occur, meaning that any residual disabilities were permanent. However, the notion that damaged brains remain stable over time has now been debunked through sophisticated animal research (Monaco et al., 2013) and studies in human neuroscience (Ng et al., 2008). Practitioners who work in rehabilitation have also described instances of unexpected deterioration in residual skills (Till et al., 2008) many years after discharge from the hospital. Indeed, magnetic resonance imaging (MRI) studies (➔ chapter "Mapping the Human Brain with Computational Anatomy" by Draganski, Joost and Kherif) have shown that brain mass continues to deteriorate many years after the initial injury (Till et al., 2008), presumably because healthy areas of the brain become disconnected by lesions that break the neuronal pathways, causing atrophy in undamaged brain tissue. However, protective impacts on brain health have been observed in ageing and injured brains in response to a range of learning experiences, physical exercise, heightened emotional states, and during cognitive activity (Domingos et al., 2021; Martínez-Molina et al., 2021; Miller et al., 2013). As a result of observations like these, interest has been generated in the potential of neuroplasticity and neurogenesis for rehabilitation.

Neuroplasticity is the brain's ability to adapt through structural and functional adjustments that occur in response to experiences, underscoring its intrinsic dynamism and capacity to constantly reshape its neural pathways and connections (Berlucchi & Buchtel, 2009; Pascual-Leone et al., 2005). There are two primary forms of neuroplasticity; structural and functional. Structural or synaptic plasticity involves physical changes in the brain's structure, mostly through the strength or number of synapses (connections between neurons may be pruned, strengthened, or newly formed). It allows the brain to refine its circuitry based on experience, allowing new skills to be acquired and unused skills to be lost. Functional plasticity, in contrast, pertains to the brain's ability to redistribute its functions in response to changes or damage. If a brain region becomes compromised, neighbouring areas might take on new functions to ensure overall continuity. This phenomenon allows the brain to optimise its functioning, enabling individuals to maintain previous skills, adjust to changes, recover from injuries, and even compensate for deficiencies.

The ability of the adult brain, notably the hippocampus, to produce fresh neurons (neurogenesis) is more contested than neuroplasticity. Eminent figures like Ramón y Cajal challenged the plausibility of this concept, advocating for a more rigid neurological morphology. Even after the first evidence of neurogenesis was presented

(Altman, 1962; Altman & Das, 1965), it was largely dismissed until the entrenched "central dogma" of the fixed adult brain was finally debunked in the 1990s (see Oppenheim, 2019 for a brief description and history). Indeed, adult neurogenesis was, until recently, thought to be impossible.

Neurogenesis occurs when neural stem cells divide and mature into neural progenitor cells, which then mature into neurons (Eriksson et al., 1998; Kumar et al., 2019). Prior to our current understanding of neuroplasticity, it was believed that there was an even more finite window for neurogenesis, namely that the brain ceased to produce new neurons after birth (Kumar et al., 2019; Owji & Shoja, 2020). There is no doubt that neurogenesis is most prominent during critical periods of foetal development, whereas neuroplasticity dominates early childhood. However, we now know that both continue, albeit to a lesser degree, throughout life (Boldrini et al., 2018; Noack et al., 2009), especially in the hippocampus (Gage, 2019).

There is evidence that the processes of neurogenesis and neuroplasticity may be automatically triggered immediately after damage to the brain, as evidenced by a distinct pattern of brain activity, including increases in hippocampal volume (Manivannan et al., 2021; Redell et al., 2020) and increased activity in contra-lesion brain regions (see Alia et al., 2017 for a review; Jones, 1999; Jones & Jefferson, 2011). Although the latter occurs at unpredictable times during recovery (Jones & Pons, 1998; Nicolelis, 1997), these windows of time allow reorganisation and regeneration of the nervous system and are critical to outcomes. They also represent points of opportunity for rehabilitation (i.e. by understanding how to trigger these events, optimise their impact, or capitalise on them as a time to deliver optimal interventions).

Our understanding of the timing and mechanisms of neuroplasticity and neurogenesis continues to evolve, but they have significant implications for fields such as education, mental health, and rehabilitation. These processes underscore the potential for lifelong learning and adaptation, as well as the brain's innate ability to recover, adapt, and, perhaps, repair itself after injuries or illnesses with the right input. However, the nature of the input required to increase the likelihood of neurogenesis and neuroplasticity remains elusive. Evidence suggests that the environment, especially a rich environment, might be critical in creating the right circumstances to promote neurogenesis and neuroplasticity.

Environmental enrichment has been described as a non-invasive external stimulus capable of eliciting internal changes (Ceanga et al., 2021). It entails the deliberate design and development of settings (Forbes et al., 2020) that promote and facilitate active engagement with the environment (McDonald et al., 2018; Nithianantharajah & Hannan, 2006). Environmental enrichment encompasses a range of activities that make the environment more interesting and more novel, including physical activity, social interaction, and cognitive exploration (Ceanga et al., 2021). Optimal enrichment is achieved by incorporating a combination of environmental modalities (Johansson & Ohlsson, 1996), encompassing physical decor, views, lighting, smells, and sounds (Huisman et al., 2012; Kang & Liu, 2022; Olszewska et al., 2018).

The theory of environmental enrichment has long been applied to the care of captive animals, prompting animal keepers to design environments that can stimulate animals and prevent boredom. Rodents and mammals exposed to enriched environments have shown increased hippocampal neurogenesis (Benarroch, 2013; Grońska-Pęski et al., 2021), including dendritic branching, synaptogenesis, and the survival of new neurons (Gage, 2019; Van Praag et al., 2000). It has been hypothesised that these changes may then exert antidepressant effects (Planchez et al., 2020) and impact cognitive function (Hannan, 2014; Marin-Burgin & Schinder, 2012). In a review of 34 studies, environmental enrichment was also reported to have behavioural benefits in rodents exposed to experimental traumatic brain injury (Bondi et al., 2014). In the last few years, further research has demonstrated that non-invasive environmental stimulation is beneficial in attenuating cognitive decline following experimental brain lesions in animals.

Studies focused on humans who have sustained brain injury demonstrate similar differences in behaviour and brain responses between those living in enriched environments and those in control (unmodified) environments. Enriched environments have been found to be associated with increased human hippocampal volume (Clemenson et al., 2015; Weissman et al., 2020; Yu et al., 2018), coma arousal (Megha et al., 2013), and relaxation (Poza et al., 2013). Miller et al. (2013) found a link between neural decline and environmental deprivation that was attenuated over the next 28 months by a relatively short period of environmental stimulation. Control participants whose environment remained unchanged do not show a similar reduction in atrophy (Frasca et al., 2013; Janssen et al., 2010; Miller et al., 2013; Ng et al., 2008).

Neurorehabilitation and the Environment

Rehabilitation for neurological conditions focuses largely on managing impairments and functional difficulties to facilitate the highest possible levels of independence, participation in everyday activities (e.g. school, work, sports, and recreation), and meaningful life roles. Rehabilitation must, therefore, simultaneously address physical, functional, cognitive, psychological, and emotional needs. It requires a multidisciplinary approach involving nurses, medical staff, and allied healthcare professionals, such as occupational therapists, speech therapists, psychologists, physiotherapists, and social workers. Interventions are likely to be more intense during the early inpatient period when maximum benefit can be gained but the need for rehabilitation continues throughout life as new challenges emerge. Once people are discharged from the hospital, the intensity of rehabilitation declines and is dependent on the home and community environments in which people live. As a result, gains achieved through expensive rehabilitation interventions may not be maintained after discharge.

The WHO defines rehabilitation as 'a set of interventions designed to optimise functioning and reduce disability in individuals with health conditions *in interaction with their environment*' (WHO, 2021). The possibility that neuroplasticity and adult neurogenesis are even partially dependent on environmental change makes this observation even more important for how—or rather where—neurorehabilitation is delivered. However, in the early stages of recovery and rehabilitation following an injury or diagnosis, the environment is often maximised for clinical efficiency, usually within a health system or service, rather than for individual well-being. Later in the recovery process, the environment may be modified to maximise functional support, but its use as a rehabilitation tool that can promote physical, psychological, and emotional well-being remains limited.

There is evidence that environments become increasingly depleted following neurological impairment, perhaps compromising the possibility of positive outcomes. For instance, it is not uncommon for people with neurological impairments to experience high levels of unemployment (Doctor et al., 2005), loss of friends (Weddell et al., 1980), exclusion from social activities (Temkin et al., 2009), fatigue (Draper et al., 2007), physical restrictions due to poor mobility, or inadequate resources (Lamontagne et al., 2013). These consequences can, in turn, lead to loneliness, helplessness, and social withdrawal, resulting in even less environmental stimulation (Till et al., 2008). To complicate this picture, the services provided to people with permanent disability typically focus on compensating for residual impairments and delivering daily care rather than providing positive and enriching environments (Kendrick et al., 2012). People with neurological impairment rely heavily on human assistance to perform basic activities and social roles (Lamontagne et al., 2009), which can ironically deplete one's sense of control over one's own environment (Schmidt & Ownsworth, 2022) by continually reinforcing dependence and incompetence. Low levels of self-efficacy deplete people's motivation to try new activities (Dodds, 1989) further reducing the likelihood of environmental stimulation and establishing a negative cycle of environmental deprivation.

Some research has found that conditions of environmental deprivation are intimately related to neurogenesis and neuroplasticity (Buchanan et al., 2013; Gage, 2019). For example, chronic stress and depression can have structural and atrophic impacts on the human hippocampus; dendritic retraction in the hippocampus is characterised by decreased dendritic length and fewer dendritic branches. It has been suggested that hippocampal size could be seen as a marker of depression as it may be smaller during a depressive episode but not in recovery (Nolan et al., 2020). Further, exposure to childhood violence is associated with smaller hippocampal volumes, which may indirectly cause later depression through higher sensitivity to stressful events (Weissman et al., 2020). Indeed, studies also show early childhood deprivation results in substantially smaller brain volumes and poorer behavioural and cognitive outcomes (Mackes et al., 2022). Smaller hippocampal (and amygdala) volumes are also commonly found in brain imaging studies of people with posttraumatic stress disorder (Henigsberg et al., 2019). These alterations in hippocampal structure and function may decrease neurogenesis but are also reversible (Conrad, 2006, 2008; Rosenbrock et al., 2005). For instance, neurogenesis has been found to be affected by the quality of environmental experiences such as learning, stress, exercise, and enriched activities (Van Praag et al., 2000).

Further, research has established the link between a person's experience and long-lasting effects on the brain and behaviour. Essex et al. (2013) investigated the impact of parental stress during a child's infancy and preschool on DNA methylation by mid-adolescence. They found that exposure to maternal stress during infancy and exposure to paternal stress during preschool was predictive of substantial differential DNA methylation in 15-year-olds. The genes in question have a known connection to brain development and behaviour. This was the first study to show that adverse environments during childhood could affect epigenetic profiles related to the brain and behaviour years later, highlighting the interaction between genes, the brain, and the environment.

These studies demonstrate that the environment can shape our brains, both positively and negatively, so should be an important component of any neurorehabilitation program. Although the evidence base still requires clarity, one way of capitalising on the relationship between environment and brain health is through deliberate environmental enrichment.

Environmental Enrichment in Neurorehabilitation

In the context of neurorehabilitation, there are multiple environments that impact on people during their recovery. They encounter a range of hospital and health environments, but the larger portion of their lives is played out in their home living environments within communities, consisting of both built and natural settings. Virtual environments have also been taking an increasing role in society and will continue to grow in the future (➔ chapter "Extended Reality in Environmental Neuroscience Research" by Li et al.). Each section below examines the opportunity provided by these different environments and ways in which they have been used to enrich the human experience to promote the quality of life of people with disability. The future application to neurorehabilitation is then explored.

Hospital Environments

Hospital inpatient environments can be stressful places for patients who have experienced neurotrauma but simultaneously offer the most consistent aspect of treatment. However, there is evidence that hospital environments are not yet optimal for people with neurological conditions. For instance, in one study, brain injury patients commented on how the hospital environment depersonalised them and depleted their sense of self-identity (Colley et al., 2020). Some research has suggested that pleasant health environments are essential to enhance patient engagement (Jahandideh et al., 2018) as this allows the benefit of rehabilitation to be realised. Indeed, stroke patients exposed to an enriched environment have been found to engage in rehabilitation more actively, including social and cognitive pursuits, compared to those receiving rehabilitation in a standard rehabilitation environment (Janssen et al., 2014).

An enriched hospital environment requires equipment and space for therapeutic and recreational activities to alleviate boredom, promote active participation, and promote rehabilitation for all patients (Colley et al., 2020). This change requires an organisational focus on growth and individual strengths, which encourages increased patient freedom and self-determination and a decrease in therapist control (Andrews et al., 2010).

Environmental characteristics such as colour, light, views, shape, and physical features all affect how we experience place and are all relevant to how a person will engage with or experience rehabilitation (Smith & McDonald, 2015). In contrast, the characteristics of noise and space in rehabilitation environments can create stress for patients (Pryor, 2004; Ulrich et al., 1991). However, these features of the environment have also been found to have a direct impact on the brain. For instance, colour has been found to affect cognition, attention, physical activity, and the experience of pain (Chen et al., 2023; Elliot et al., 2007; Wiercioch-Kuzianik et al., 2023; Yang et al., 2023). Noise (Bokyo et al., 2017; Yelden et al., 2015), lighting (Bernhofer et al., 2014; Yelden et al., 2015), and temperature (Linder & Christian, 2012) can all affect sleep, which is vital for people during neurorehabilitation. Woo et al. (2023) found that older adults exposed to odours such as rose, orange, eucalyptus, lemon, peppermint, rosemary, and lavender, for 2 hours a night for 6 months showed increased performance in verbal learning and memory compared to a control condition. Using fMRI, they also found significant changes in mean diffusivity of the left uncinate fasciculus, an area that deteriorates in ageing and Alzheimer's disease, and plays a role in memory and language.

Home and Community Environments

Once people leave the hospital, the home, and community environment must be able to accommodate any residual disability, but also create a positive and healthy environment in which people can thrive. In this regard, Meredith et al. (2022) studied a sample of over 11,000 young people across 10 years and found higher levels of general cognitive ability, executive functioning, and learning/memory performance for those whose community environments were enriched (e.g. healthy families, safe neighbourhoods, and quality schooling environments). Importantly, this impact remained apparent after accounting for the variance associated with socio-demographic factors.

Urban planning for disability justice is an emerging field that will rely on evidence from environmental neuroscience (Stafford, 2022). Apart from denying people with disability access to their physical environment and generating social exclusion, poor urban design can exacerbate the impact of cognitive and physical impairment (e.g. proximity to regular school bells, sirens, traffic noise, nurses stations, and lack of access to resources can all trigger negative behaviours). In contrast, the careful selection of locations for supported accommodation or rehabilitation hospitals could contribute to recovery following discharge and facilitate inclusion in society (Douglas et al., 2023).

Home environments can also reduce care costs and are critical to rehabilitation outcomes in terms of facilitating independence through accessible design and access to assistive technology (Carnemolla & Bridge, 2019). Some simple home modifications can minimise environmental overstimulation for people with cognitive and sensory impairments (e.g. soft closing drawers and cupboards to reduce loud/sudden noises, block out blinds to control light levels at certain times of day, smart home automation to regulate temperature, micro-prompting devices to assist with memory and the execution of daily tasks). The quality of the home environment has also been found to have both direct and indirect effects on the social behaviour of young people with intellectual disabilities (Olivier-Pijpers et al., 2019). However, in a unique study of a smart home environment, Berrezueta-Guzman et al. (2020) demonstrated how the home environment could become an active aspect of neurorehabilitation for young children with ADHD. They established a home that could monitor and prompt children in a timely way during tasks like homework (e.g. through sensors tracking attention and providing reminders). Children with ADHD accepted the robotic intervention, spent more time on task, and performed better, demonstrating the potential of the home environment to act as a therapeutic tool for people with neurological conditions.

Natural Environments

The benefits of nature in health settings have been known for years and as far back as the eighth century, nature was being used purposefully in Islamic psychiatric hospitals, and in the early 1900s accidentally in US asylums (Roe, 2008). The positive benefits of being in and near natural environments have been evidenced throughout a person's life. The impact on health starts before birth resulting in healthier birth weights being associated with access to greenspace (Agay-Shay et al., 2014; Dadvand et al., 2012). Throughout childhood, children choose to play and learn in green settings over indoor and other environments (Wistoft, 2013; Zamani, 2016) and display improved behaviour and engagement in them (Norwood et al., 2021, 2022b). By college age, natural settings are seen as being more restorative than built ones (Hartig & Staats, 2006), and adults with their own children choose to visit green parks over alternative locations (Tilt, 2010). There is even an association between walkable green spaces and a longer length of life (Takano et al., 2002).

There is also considerable justification for the use of nature in the design of adult rehabilitation environments. In the earliest research on nature and medicine, exposure to natural views was found to result in speedier recovery from surgery, reduced use of analgesics (Ulrich, 1983), and greater well-being for inpatients (Raanaas et al., 2012; Verderber & Reuman, 1987). Natural environments are now known to promote physical, psychological, cognitive, and social benefits (Chawla, 2015; Gill, 2014; Kuo et al., 2019; Lakhani et al., 2019; McCormick, 2017; Norwood et al., 2019a, b; Ye et al., 2022). Proximity to a green environment has positive effects on both physical and mental health (e.g. De Vries et al., 2003; Sugiyama et al., 2008) and promotes a feeling of community regardless of how often it is visited (Francis et al., 2012; Sugiyama et al., 2008). Accessible nature promotes physical exercise such as walking, running, and sports (Dadvand et al., 2015a, b, 2016; Ye et al.,

2022). Greenspace impacts air quality, temperature, and noise to affect health benefits (Amoly et al., 2014; Dadvand et al., 2015b; Markevych et al., 2019).

There is also a significant body of research showing how interactions with nature can affect brain structure and brain activity (Norwood et al., 2019b). Brain activity patterns have been found to vary between natural and non-natural settings, with exposure to natural environments inducing low-frequency EEG in various brain locations, including the frontal cortex, associated with relaxation or calmness (Chang & Chen, 2005; Ulrich et al., 1991) (➔ chapter "EEG for the Study of Environmental Neuroscience" by Grassini). Urban scenes have been shown to be associated with fear responses, sadness, effortful cognition, and activation of areas of the brain such as the limbic system, frontal and prefrontal lobes, and cuneus and cingulate gyrus; whereas natural scenes have been shown to elicit feelings of happiness and cognitive restoration with lower activation of the prefrontal cortex (see Norwood et al., 2019b for review).

There is recent evidence that nature can have an immediate, structural impact on the brain. Sudimac and Kühn (2023) showed that even after just a 1-hour walk in a forest, when compared to a busy urban street setting, participants showed significant increases in volume in the subiculum, a part of the hippocampal formation associated with stress inhibition. This change was associated with self-reported decreases in rumination, providing neurological data to support the decades of research on the restorative effects of nature. These findings suggest that not only is environmental enrichment an important consideration in neurorehabilitation, but also the specific inclusion of the natural environment might provide rapidly improved neurological outcomes. More research is needed to establish the link between nature-based brain stimulation and adult neurogenesis.

The benefits of nature have contributed to the growing popularity of 'biophilic' design. Biophilic design is the integration of nature into the design of built environments, with the intention of reproducing the health benefits associated with nature exposure. Some examples of biophilic design are 'green walls' for hospital workers to promote relaxation and positive feelings (Youn et al., 2022). Noting how nature in architecture could reduce stress, increase positive emotions, and improve cognitive performance, Zhong et al. (2022) proposed a framework for biophilic design that prompted architects to consider nature incorporation, inspiration, and interaction. Bringing nature into architecture involves careful planning and maintenance, but Zhong and colleagues recognised that the form and feel of nature could also be mimicked in architectural design. There is also an opportunity for biophilic design to be culturally relevant, especially for First Nations communities who have strong connections to place and country.

The Rise of Virtual Environments

One advanced method that is currently being used to mimic nature inside buildings is virtual or augmented reality. Although there is a distinction between virtual reality, augmented reality, and also different levels of immersion, for the sake of brevity we refer to these environments as VR here. VR relies on computer hardware and software that mediates the interaction between the user and a virtual environment (Gaggioli, 2009). Broadly speaking, VR replaces a person's perceived reality of the external world within a computer-generated 3D environment (Gorman & Gustafsson, 2020). It occurs in real time, typically through a headset and hand controllers, allowing the user to explore and interact with environments that closely approximate reality. Virtual environments have the potential to facilitate rehabilitation goals and could be an invaluable tool in examining the neural correlates of everyday cognition.

VR gives us access to environments that may otherwise be impossible or unsafe when in rehabilitation from brain injury (Farrow & Reid, 2004; Kim et al., 2007, 2010). Due to the complex combinations of visual, motor, sensory, and cognitive impairments that can arise following brain injury tasks such as road crossing may become difficult. VR software has been designed so people with visuospatial neglect can use virtual avatars to practice crossing roads in a safe environment (Kim et al., 2007).

In research settings, VR has been applied to a range of neurological populations (i.e. stroke, Parkinson's, and brain injury) to improve their ability to negotiate obstacles in the environment (Kim et al., 2019; Liao et al., 2015). VR has also been applied to people who cannot move their bodies in space due to paralysis, but the perception of being "in action" has been shown to positively influence cognitive processing (Repetto et al., 2016).

VR allows for remote delivery and monitoring of rehabilitation (Burdea, 2003; Morse et al., 2022; Threapleton et al., 2016). During the COVID-19 pandemic, the call for telehealth in rehabilitation was loud and rapid. It resulted in a feasible, well-accepted, and effective alternative method for providing rehabilitation to neurological patients who could not attend hospitals. It also increased participation and allowed continuity of care from the hospital back into the home environment (Mantovani et al., 2020; Tieri et al., 2018). There is evidence to suggest that the VR environment improves transfer of rehabilitation outcomes by allowing practice to occur within representation of the person's own home.

VR offers the capacity for real-time, objective measures of how each human brain is affected by the environment (Smith & Mcdonald, 2015). By combining neuroimaging with VR, we can observe how the targeted brain areas of each individual respond to stimulation during enrichment programs (Ansado et al., 2021; Teo et al., 2016). This line of research can allow us to personalise therapeutic programs. Virtual environments have already been applied to the assessment and treatment of brain injury in many domains including cognitive, physical, behavioural, and psychological (e.g. Domínguez-Téllez et al., 2020; Norwood et al., 2022a; Zhang et al., 2010). Indeed, virtual reality tasks have been found to be superior to traditional pen-paper assessments in identifying spatial neglect (Knobel et al., 2020; Painter et al., 2023, 2024).

In summary, VR can provide a controlled, accessible, customisable, and potentially engaging environment for rehabilitation (Aida et al., 2018). Despite early identification of this potential, the application of VR in rehabilitation has been limited (Chedid et al., 2013). Inadequate education about its benefits, limited time and finances, and poor access to technology have all been identified as contributing to the lack of uptake (Brouns et al., 2018; McGrath & Astell, 2017; Musselman et al., 2018). However, VR hardware and software are now easily accessible on the consumer market, allowing VR to be increasingly explored in healthcare environments. Within stroke rehabilitation settings, adverse events associated with VR are rare and those that do occur tend to not be serious (Laver et al., 2017). In contrast, traditional rehabilitation programs that are often repetitive and monotonous can be transformed into highly engaging gamified programs in VR (Pietrzak et al., 2014; Thornton et al., 2005). It has the potential to promote engagement in rehabilitation sessions within a multisensory environment (Adlakha et al., 2020; Lohse et al., 2014; Parker et al., 2013; Huygelier et al., 2021). VR may also provide a viable enriched environment that can activate attention and motivation and facilitate neuroplasticity and functional recovery (Danzl et al., 2012; Mantovani et al., 2020; Matamala-Gomez et al., 2021).

Measuring the Impact of Environment

Neuroplasticity can be observed in a variety of ways; structurally in grey matter (measured by increases or decreases in brain volume) and changes to white matter microstructure (using measurements such as fractional anisotropy, radial diffusivity, and axial diffusivity). It can also be evidenced by measuring changes in the organisation and communication between areas of the brain. Finally, neuroplastic changes can be observed by assessing differences in how the brain functions due to an environmental change—for example how the neurophysiological response to a task or event alters before and after exposure to a specific type of environment using excitability, timings of event-related potentials, and amplitude change (Rossinia & Ferreri, 2013). Different techniques can be used, such as electroencephalography (EEG), magnetoencephalography (MEG), and functional magnetic resonance imaging (fMRI)—each with advantages and disadvantages. However, the greatest potential for determining the impact of environment on the brain is functional near-infrared spectroscopy (fNIRS). Recent developments have shown the potential importance of hyperscanning, or the simultaneous measurement of two or more people during interactions. This method shows great promise for understanding the role of the environment in human relationships.

Functional Magnetic Resonance Imaging (fMRI)

The champion of neuroimaging, fMRI measures changes in blood flow to indirectly measure brain activity, relying on the assumption that higher neuronal activity is coupled with higher energy use and thus blood flow. fMRI has high spatiotemporal resolution—although this is dependent on the strength of the scanner (e.g. 1.5 T compared to 4 T) (Logothetis, 2008). Aside from measuring the effects of brain injury, fMRI can be used to understand which areas are critical for certain cognitive processes, what activities result in increased grey matter volume (or less atrophy), what areas of the brain are active at the same time (functional connectivity), and what patterns of activity are the most efficient (using graph theory analysis). Accordingly, fMRI has highlighted important areas in many enriching activities such as music (Gordon et al., 2018), dance (Brown et al., 2006; Calvo-Merino et al., 2005), mindfulness (Doll et al., 2015), viewing natural environments (Kühn et al., 2021; Tang et al., 2017), and art (Cupchik et al., 2009). It has also helped highlight the longitudinal impact of some of these activities (Bar & DeSouza, 2016; Ellis et al., 2013) and the neurorehabilitation potential (Sarkamo et al., 2014; Siponkoski et al., 2020).

However, assessing neuroplastic change associated with the environment via fMRI is difficult because scanners are very large, and measurements are highly sensitive to even minor motion. fMRI is useful in the measurement of environmental exposure that is immobile, such as viewing images or videos, imagining motor activities, or performing activities with only minor movements such as finger tapping or drawing circles with a finger. Some evidence (Manes et al., 2021) has shown that even standard tasks, such as speaking or writing, are fundamentally altered within the fMRI environment (i.e. inside the scanner), making it a cumbersome tool for measuring the impact of the environment. However, in carefully designed studies (Sudimac & Kühn, 2022) where participants undergo fMRI scanning immediately before and after exposure to different environments, differential effects of nature and urban environments have been shown on activity in the subgenual prefrontal cortex and the amygdala (Bratman et al., 2015).

Functional Near Infrared Spectroscopy (fNIRS)

fNIRS is a non-invasive imaging technique that emits near-infrared light into the brain and uses the amount of reflected light to infer brain activity. The underlying premise is that increased neural activity causes increased oxygen consumption and increased blood flow (Scholkmann et al., 2014). Further to this, oxyhaemoglobin (HbO_2) and de-oxy-haemoglobin (HHb) molecules in the blood are absorbed in the 700–900 nm spectrum (Severinghaus, 2007). The amount of near-infrared light that is absorbed is measured and converted into HHb and HbO_2 using the Modified Beer Lambert Law (Delpy et al., 1988). fNIRS, therefore, provides an indirect measure of cortical activity and has been found to be highly correlated with fMRI results (Cui et al., 2011; Steinbrink et al., 2006).

However, unlike fMRI, fNIRS can withstand movement and is transportable away from the laboratory with a wireless backpack. Accordingly, fNIRS can be used to measure the brain response in different in vivo environments (Strangman et al., 2002). Specifically, fNIRS can measure brain activity *during* environmental enrichment activities such as complex multimodal tasks (Ono et al., 2014; Tachibana et al., 2011), and during walking and exercising in many different environments (Jones & Wheat, 2023; McKendrick et al., 2017; Vitorio et al., 2017). The hemodynamic response with increased experience can also be tracked over time, demonstrating the immediate neural response to changing environments.

Currently, fNIRS is not being used to its full potential in neurorehabilitation. Although fNIRS has been used to measure the impact of the environment on other clinical populations such as in gambling disorder (Ochiai et al., 2023), in the elderly (Song et al., 2018), and in clinically depressed adults (Olszewska-Guizzo et al., 2022), it has predominantly been used to assess the neurological impact of mainstream interventions (see Ni et al., 2022; Stilling et al., 2019; Urushidani et al., 2018).

Adoption of fNIRS into neurorehabilitation could allow us to not only understand the benefit of environmental enrichment in rehabilitation, but to streamline those interventions by providing information about the impact of duration and intensity of exposure. Rather than being unimodal (i.e., a specific treatment that is applied at a particular point in time), environmental enrichment is a multimodal program that requires personalised data to unravel its impacts (Pinto et al., 2022). In a similar vein, Paraskevopoulos and Herholz (2013) posed the important question of how structural and functional individual differences in the brain, as might occur through different types of brain damage, alter the effectiveness of complex multimodal interventions, such as environmental enrichment. There is significant variation in how individuals experience enrichment interventions, necessitating personalised rehabilitation plans. Enrichment allows for individual behaviour which results in unique individual patterns of brain plasticity (Kempermann, 2019).

The benefit of applying fNIRS to the evaluation of environmental enrichment programs in neurorehabilitation is clear, particularly given that it offers a non-invasive and portable measurement tool, with the ecological validity of real-time in vivo monitoring of multi-modal environmental stimuli across time and across participants.

Measuring the Complexity of Interactive Environments

One fundamental aspect of many enriched environments is social interaction, but other than self-reports of social engagement, there are few ways of measuring the impact on the brain. Hyperscanning (which can be conducted with other neuroimaging assessment tools) is a breakthrough method for understanding the neural underpinnings of social interactions. The basic premise of hyperscanning is that the neurological impact of being in a person-to-person social interaction will differ from the impact of viewing or thinking about social interactions (Montague et al., 2002) and that this will show in brain activity. Specifically, real-time interactional tasks can create interneural synchrony (INS) (Babiloni & Astolfi, 2014), affecting both parties to the interaction. Hyperscanning aims to collect two or more time-synchronised brain recordings while participants perform interactional tasks (Babiloni & Astolfi, 2014) such as in natural interactions (Nozawa et al., 2019), co-ordinated motor tasks such as dancing (Stone et al., 2019), teamwork tasks (Liu et al., 2021), and other creative adventures such as music improvisation (Lindenberger et al., 2009; Müller et al., 2013).

There are multiple beneficial applications of hyperscanning to environmental enrichment research. For example, researchers have investigated the role of inter-neural synchrony in social interactions in modulating pain and perceptions of social bonding (Tarr et al., 2015), building rapport (Nozawa et al., 2019), improving student learning (Dikker et al., 2017), and increasing self-esteem (Lumsden et al., 2014). However, there are still many under-researched areas. For example, there is a call for hyperscanning to investigate digital compared to face-to-face communication (Balters et al., 2020) and whether one's ability to achieve synchrony can be altered (i.e. inter-brain plasticity) (Sened et al., 2022). Its application to neurorehabilitation is also only in its infancy. For example, although synchrony has been hypothesised to be important in neurorehabilitation (Short et al., 2021) (Basso et al., 2020; Behrends et al., 2012), hyperscanning research in neurorehabilitation, psychopathology, and neurological disease is limited to a few pilot studies in music therapy between parents and children with severe disabilities (Samadani et al., 2021) and stroke rehabilitation (Tucek et al., 2022).

Challenges, Opportunities, Future Directions, and Research Implications

A Reliance on Animal Studies

Oppenheim (2019) highlighted a critical limitation in the existing body of research on environmental enrichment and adult hippocampal neurogenesis, namely the predominant use of highly inbred, genetically homogeneous, laboratory-raised mice and rats housed in controlled environments (Kempermann, 2012). These domesticated animals exhibit behaviours and developmental trajectories that may differ substantially from their wild counterparts, as emphasised by Nottebohm's assertion, "Take nature away and all your insight is in a biological vacuum" (Specter, 2001, p. 53). Moreover, even wild animals are subjected to handling procedures and environmental conditions that can influence their behaviour and brain activity (Oppenheim, 2019). Although these studies have undoubtedly contributed to our understanding of environmental enrichment (EE) and brain plasticity, they may not fully capture the complexity of behaviour and physiological patterns observed in natural environments by the same animals (Oppenheim, 2019). Unsurprisingly, the applicability of research primarily focused on domesticated rodents to human populations remains even more uncertain. The current body of evidence regarding the effects of EE on brain plasticity and neurogenesis lacks controlled human-based trials and is a line of research that remains subject to controversy (see Gage (2019) and Oppenheim (2019)).

Limitations of Current Healthcare Design Principles

The considerable investment in the construction and development of healthcare buildings has a focus on safety, efficient care, and cost-effectiveness, with a goal of meeting clinical and medical requirements. The investment reflects a commitment from government and private providers to integrate advanced health and medical approaches and technology, as well as clinical education and research into their health service. Investment in these areas provides the opportunity for the provision of the best medical care, and innovation in medical practice, to improve overall healthcare outcomes. However, the focus on clinical functionality disregards the holistic nature of healthcare and neglects the effect the healthcare environment has on individual patients.

A design approach that considers user needs and experiences could address this and create rehabilitation environments that can promote a more holistic rehabilitation. However, a person's experience of their physical environment and the impact it may have on their rehabilitation is not considered in healthcare design (Valentin et al., 2011; Halpern, 2014; Tronstad et al., 2021), creating intimidating, depersonalising, and anxiety-provoking spaces (Colley et al., 2020; Horsburgh, 1995; Ulrich et al., 1991). The designed rehabilitation environment should facilitate physical, psychological, and emotional rehabilitation, not obstruct them (Smith & McDonald, 2015).

Environmental neuroscience integrates the environment, brain, and behaviour and is well placed to inform patient-centred approaches to built healthcare design. Knowledge of how the physical environment influences the brain, behaviour, and psychological and emotional well-being has obvious implications for rehabilitation from a physically and emotionally traumatic event such as brain injury. However, currently, environmental neuroscience expertise is not routinely included in healthcare design. By understanding the impact of the physical and sensorial environment such as noise, light, colours, spatial layout, and nature on neurology and human behaviour, healthcare environments can be designed to improve the patient's rehabilitation experience and outcomes. A more comprehensive approach to healthcare environment design, informed by environmental neuroscience, recognises that rehabilitation requires more than medicalised settings for delivery of clinical care. These environments require us to employ empathic approaches to design that reflect how the rehabilitation environment makes users feel. These environments should encourage environmental engagement to facilitate rehabilitation and brain plasticity, taking advantage of the neuroplastic processes occurring at the early stages of rehabilitation from neurotrauma.

Harnessing the Natural Environment

Including greenspace in built and urban design comes with its own challenges that can affect anyone, including those rehabilitating from neurotrauma. It is in cities that natural areas and parks can be at their least effective. Urban greenspaces can be seen as stressful, unsafe, even fearful places (Fullagar & Harrington 2009; Gidlow & Ellis, 2011; Obasanjo, 1998). Sometimes, the design of these spaces promotes the feeling of fear due to poor visual and physical accessibility (Tilt, 2010), and a lack of maintenance or lighting (Groshong et al., 2020). To counter these potential negative effects, increased community engagement, accessible design, and maintenance of urban parks are needed (Groshong et al., 2020).

Challenges also arise in non-urban nature interactions. Our past experiences with an environment, including those from childhood, have a large effect on the 'perceived' current environment and what benefits or otherwise we gain from it (Collado et al., 2016; Kaplan, 1983). Thus, exposure to nature can have differing individualised effects. Despite the established evidence on human–environment interactions, answers to why, how, and for whom have not yet been adequately addressed to enable clinical application. In the context of wilderness therapy, Fernee et al. (2017) labelled this gap in knowledge the need to unpack the "black-box"—wilderness therapy appears to be effective, but under what circumstances is it not effective, for whom does it work best and why and for whom does it not work? These are the questions that drive ethical clinical application.

The recent surge in interest in biophilic design may be more vital than ever in connecting people with nature, as one of the main challenges of the natural environment may be accessing it. By 2050, it is anticipated that nearly 70% of people will live in urban areas (UN-Habitat, 2022). However, inpatient neurotrauma patients are a population that even today cannot typically or easily access natural environments. One reason for the inability to engage with nature results from design and accessibility issues (Lakhani et al., 2019). Where nature is available, negative feelings are reported when accessibility and weather form barriers to use (Hernandez, 2007). Issues such as difficulty in opening doors, sharp turns, narrow paths, and glare from sunshine prevent use in some dementia care homes and see a reversal of the positive effects of accessible green space such as reductions in agitation in patients (Murphy et al., 2010). When gardens are accessible in care homes, we see positive outcomes such as in emotional health through feelings of joy and happiness and a reduction of stress and agitation.

Equity in the Design and Application of Rehabilitation Research

There are several areas of inequity that need to be addressed on societal and individual scales. There are higher rates of brain injury in low-income countries compared to high-income, and within these populations, rehabilitation is more likely to be an unmet need in low- to middle-income countries—in some countries more than half of people who require rehabilitation services do not receive them. Indeed, this chapter started with the statistic that although in absolute terms the number of deaths from neurological disorders has increased, it is actually decreasing in terms of rate of death; however, this is not the case when comparing high-income countries with low- and middle-income countries; low- and middle-income countries are seeing an increase in death and disability from neurological disorders such as stroke where high-income countries are seeing a decrease. The burden of neurological disorders is unequally distributed across the world (Knauss et al., 2019).

The field of environmental neuroscience studies environment, brain, and behaviour, but to address the inequalities listed and provide rehabilitation solutions in those low-middle income countries that perhaps need it the most, research needs to address its WIERD problem. That is, many research participants are largely from Western, Industrialized, Educated, Rich, and Democratic societies (Henrich et al., 2010). Even neuroscience has a WIERD problem. Our understanding of brain, cognition, and behaviour are often spoken of as universal, but they are studied only in a very small percentage of people representing a tiny and unitary fraction of the population, and often these studies are conducted by a similarly narrow representation of the population (Dhamala et al., 2023; Ricard et al., 2023). Research on particular stereotypical profiles does not necessarily apply to people who fall outside that profile (Greene et al., 2022), for example, studies conducted on white males, do not necessarily translate to women or to other races (Jiang et al., 2019; Li et al., 2022; Sanchis-Segura et al., 2020). There is a clear need to conduct context and population-specific research, in collaboration with end-users, to ensure the most effective use of environmental neuroscience in neurotrauma rehabilitation and to help address inequalities in the long-term burdens of neurotrauma.

Addressing Individual Variation

Although the evidence appears to be pointing towards the overall benefits of environmental enrichment for neurorehabilitation, there is some evidence that interventions can trigger variable responses across individuals (Janssen, 2014). This finding raises questions about inter-individual differences and the underlying neural mechanisms that might account for this difference. This presents challenges, but also opportunities in conducting and applying research. Some of the most pressing questions for this area are:

- How and when can environmental enrichment optimise recovery and rehabilitation?
- How much enrichment is optimal for which patients, and at what time (i.e. “dose”)?
- What types of environmental enrichment are best for which rehabilitation outcomes (i.e. recreation, technology, built and natural environments, and virtual environments)?
- Does enrichment need to be actively applied or can passive changes in the environment (e.g. architectural design, biophilic design) trigger neuroplasticity?
- How can we apply new technologies to the enrichment of environments and the measurement of their impact?
- How can we use home environments to capitalise on neuroplasticity beyond inpatient rehabilitation (i.e. home-based therapies and design)?
- Do advances in neurofeedback offer the opportunity to build a self-managed approach to neurorehabilitation?

Predicting the Environments of the Future

What will the environments of the future be like? Advances in technology are poised to transform our relationships with the natural and built environments. As innovation continues to accelerate, a convergence of technological breakthroughs will fundamentally change the nature of the environment and our interactions. Rehabilitation itself is currently being transformed by rapid advances in technology, leading to innovative physical rehabilitation tools and methods that can enrich a person's environment by providing access to information, employment, education, leisure activities, and social events (Mattar et al., 2015).

Virtual reality, robotics, and brain–computer interfaces are beginning to be utilised to enable intensive, engaging, and task-specific training. These technologies can simulate real-life situations, offer immediate feedback, and promote repetitive practice, all of which are essential for neuroplastic changes to occur. The rise of smart devices also offers new avenues for environmental monitoring and management, particularly for people with neurological conditions. Wearable and environmental sensors, instrumented objects, and data analytics can be deployed to track the well-being of humans in their environments (e.g. temperature regulation, skin integrity, and movement) (Elliott et al., 2022; Lee et al., 2022; Orejel Bustos et al., 2023). This real-time data can inform rehabilitation decision-making and prompt swift responses to any issues (Orejel Bustos et al., 2023; Oyibo et al., 2023; Sundaram et al., 2023).

Beyond the home, the concept of smart cities leverages data-driven insights that will enable the design of accessible and inclusive public spaces and transportation systems (Yang et al., 2022). Augmented reality (AR) and navigation apps are enhancing wayfinding, with clear potential benefits for those with cognitive and visuospatial impairments, ensuring independent movement within cities (Feng et al., 2023; Qiu et al., 2023).

Artificial intelligence (AI) and machine learning can now identify patterns in complex environmental data and link this to brain patterns. AI-powered simulations can test various human-environment interfaces, aiding in the design of optimal and facilitative environments. Drones and satellite technology are also revolutionising environmental monitoring, bringing more control over our interactions with the environment (e.g. understanding and managing the impact of climate on our bodies). Brain–computer interfaces (BCIs) have the potential to revolutionise how we interact with the environment in the future, bringing even further control to the individual (Liu et al., 2020; Wang et al., 2013). These interfaces establish a direct communication pathway between the brain and external devices without the involvement of peripheral sensory or motor pathways (Bockbrader et al., 2018).

Architectural design is generally improving with ramps, elevators, tactile cues, and adjustable features integrated into many buildings and natural spaces. Advances in 3D printing technology are enabling custom-designed living spaces that perfectly accommodate human requirements. Innovations like smart home systems and voice-activated controls are transforming houses into spaces that respond intuitively to the needs of people with disability, improving independence, participation, and quality of life (Han et al., 2022; Jamwal et al., 2022; Smith et al., 2023). Nature itself is becoming more accessible and integrated into architectural design. Adaptive equipment, enhanced by technology, allows people with disability to engage in outdoor activities like hiking, kayaking, or even wildlife observation (Lian et al., 2023; Theodorou et al., 2023; Verdonck et al., 2023). Augmented and virtual reality are opening up opportunities for individuals with limited mobility to experience landscapes, rare ecosystems, and natural wonders from the comfort of their homes (Li et al., 2021).

Communication barriers are crumbling, as advances in assistive technologies enable people with speech and hearing impairments to engage seamlessly in social environments. Precision speech recognition software, haptic feedback devices, and innovative communication apps will foster richer interactions and meaningful connections with the environment for a range of people. In virtual spaces, online communities and social platforms are providing avenues for people to interact beyond physical spaces.

This new world will bring a multitude of ethical, privacy, and security challenges for us to address (Buis et al., 2022; Wang et al., 2023). It will also raise professional debate about the extent to which these new environments facilitate autonomy for people with severe disabling conditions who would not otherwise have had any voice in society. We will need sophisticated moral frameworks to manage inherent bias and stereotypes about the nature of life for people who find agency through these new emerging environments (Lubna et al., 2023).

A Case Example: The Attention Atlas—Virtual Reality for Spatial Neglect Assessment

To exemplify some of the points made in this chapter, we provide a case study of a virtual reality-based environment for the assessment of visuospatial neglect, called the Attention Atlas (AA; Norwood et al., 2022a) developed in our Brain Environment Enrichment Lab (The BEEHive). The Attention Atlas came about through discussions between researchers and clinicians about the potential uses of virtual reality in a hospital neurorehabilitation ward. The current pen-paper tests or observation assessment of spatial neglect was identified by clinicians as being inadequate, with the potential of being improved by a 3D virtual immersive environment. In response, the Attention Atlas, a virtual tool to map where a person pays attention in their 3D space was born. The AA is essentially a game, but also an example of how virtual environments can be harnessed to map brain function through its behavioural expression. Players of the game must find a target amongst distractors inside the virtual environment. They progress through levels and receive simple positive feedback when they correctly find the target. The output is based on several metrics including reaction time, accuracy, spatial distributions of responses and head movement, and in some systems eye-tracking.

Clinicians can immediately view results and gain an understanding of any biases in a persons' attention through an illustrated map of responses. If a person misses targets on one side of space, the visual map reflects this, and neglect is implicated. Results suggest this approach may be more sensitive to identifying spatial neglect than current standards (Painter et al., 2023, 2024). The potential of this example also demonstrates the possible impact of environmental neuroscience as an approach to, and tool for, neurorehabilitation. Virtual reality and gaming increased patient engagement with their rehabilitation, with patients wanting to repeat their experience with the AA. The AA also provided a cognitively enriched environment for rehabilitation, and the next step is to examine measures of neuroplasticity and neurogenesis. The virtual reality equipment was inexpensive and portable, providing a distinct advantage over other expensive neuroimaging and environmental simulations. Portability also makes continuity of care possible because patients can continue their rehabilitation at home, with clinicians tracking progress remotely.

The AA also provides an example of potential challenges in any clinical research, especially where technology is involved. Translation and implementation from a research environment into a clinical setting is difficult, with the need for multiple approvals. Technological support and software updates require financial backing to support implementation, and the route for commercialisation of potential products such as this is complex and long.

In summary, the AA is an example of *environmental neurorehabilitation* wherein the cognitive functioning of the brain can be measured through observations of behaviour in responses to a virtual enriched environment. However, it also highlights the challenges of translating environmental neuroscience into neurorehabilitation practice.

Conclusion

Environmental enrichment is increasingly recognised as a significant modulator of neuroplasticity following injury or illness, and it holds promise as an integral component of rehabilitation and recovery programs. The evidence suggests that neurorehabilitation should aim to create an environment that can provide the impetus for neuroplasticity to occur. This involves designing environments that target-specific neural circuits such as the hippocampus, where we see neurogenesis following brain injury. It also means designing environments that capitalise on the brain's plasticity, stimulating repetition yet engagement, curiosity, and novelty. Rehabilitation programs usually incorporate a combination of physical therapy, occupational therapy, speech therapy, social activities, and cognitive training to promote functional gains, but the environment in which these therapies take place also needs consideration. To maintain the benefits of rehabilitation long after discharge, it is important to design environments that emphasise strengths rather than deficits, encourage regular engagement and activity, and minimise barriers while stimulating positive social responses and opportunities.

The hospital, home, and community environments all have potential for enrichment but have different needs, obstacles, and opportunities for change. However, as the environment is the one constant experienced throughout the course of rehabilitation, a focus on its enrichment has the potential to maximise outcomes over time. As McDonald et al. (2018) concluded, the implementation of environmental enrichment as an intervention for human neurorehabilitation faces challenges such as standardising the conditions for delivery, understanding the underlying mechanisms, and determining the optimal "dose" of enrichment. Although there is ample evidence of the role of environment in brain development and decline, and preclinical studies have successfully used environmental enrichment to promote functional recovery, we still lack knowledge about the potential underlying mechanisms of this effect. The application of environmental enrichment to human neurorehabilitation requires clarity about what to do, when, and how to do it as well as a thorough understanding of the ethics and measurement of these effects. In humans, the cost of environmental overstimulation may be greater than the benefit for some people, but this type of clinical detail remains hidden. There is clearly a need to align preclinical and clinical rehabilitation research to fully address gaps in knowledge. Only then will we be able to facilitate the implementation of environmental enrichment in the rehabilitation setting.

References

Adlakha, S., Chhabra, D., & Shukla, P. (2020). Effectiveness of gamification for the rehabilitation of neurodegenerative disorders. *Chaos, Solitons and Fractals: The Interdisciplinary Journal of Nonlinear Science, and Nonequilibrium and Complex Phenomena, 140*. https://doi.org/10.1016/j.chaos.2020.110192

Agay-Shay, K., Peled, A., Valentín, C. A., Peretz, C., Amitai, Y., Linn, S., Friger, M., & Nieuwenhuijsen, M. J. (2014). Green spaces and adverse pregnancy outcomes. *Occupational and Environmental Medicine, 71*(8), 562–562. https://doi.org/10.1136/oemed-2013-101961

Aida, J., Chau, B., & Dunn, J. (2018). Immersive virtual reality in traumatic brain injury rehabilitation: A literature review. *NeuroRehabilitation, 42*(4), 441–448. https://doi.org/10.3233/NRE-172361

Alia, C., Spalletti, C., Lai, S., Panarese, A., Lamola, G., Bertolucci, F., Vallone, F., Di Garbo, A., Chisari, C., Micera, S., & Caleo, M. (2017). Neuroplastic changes following brain ischemia and their contribution to stroke recovery: Novel approaches in neurorehabilitation. *Frontiers in Cellular Neuroscience, 11*. https://doi.org/10.3389/fncel.2017.00076

Altman, J. (1962). Autoradiographic study of degenerative and regenerative proliferation of neuroglia cells with tritiated thymidine. *Experimental Neurology, 5*(4), 302–318. https://doi.org/10.1016/0014-4886(62)90040-7

Altman, J., & Das, G. D. (1965). Autoradiographic and histological evidence of postnatal hippocampal neurogenesis in rats. *Journal of Comparative Neurology, 124*(3), 319–335. https://doi.org/10.1002/cne.901240303

Amoly, E., Dadvand, P., Forns, J., Lopez-Vicente, M., Basagana, X., Julvez, J., Alvarez-Pedrerol, M., Nieuwenhuijsen, M. J., & Sunyer, J. (2014). Green and blue spaces and behavioral development in Barcelona schoolchildren: The BREATHE project. *Environmental Health Perspectives, 122*(12), 1351–1358. https://doi.org/10.1289/ehp.1408215

Andrews, S., Gerhart, K. A., Hosack, K., & Virden, J. S. (2010). Therapeutic recreation in traumatic brain injury rehabilitation. In M. Ashley (Ed.), *Traumatic brain injury:Rehabilitation, treatment, and case* (3rd ed., pp. 799–815). CRC Press.

Ansado, J., Chasen, C., Bouchard, S., & Northoff, G. (2021). How brain imaging provides predictive biomarkers for therapeutic success in the context of virtual reality cognitive training. *Neuroscience and Biobehavioral Reviews, 120*, 583–594. https://doi.org/10.1016/j.neubiorev.2020.05.018

Babiloni, F., & Astolfi, L. (2014). Social neuroscience and hyperscanning techniques: Past, present and future. *Neuroscience and Biobehavioral Reviews, 44*, 76–93. https://doi.org/10.1016/j.neubiorev.2012.07.006

Balters, S., Baker, J. M., Hawthorne, G., & Reiss, A. L. (2020). Capturing human interaction in the virtual age: A perspective on the future of fNIRS Hyperscanning. *Frontiers in Human Neuroscience, 14*, 588494. https://doi.org/10.3389/fnhum.2020.588494

Bar, R., & DeSouza, J. (2016). Tracking plasticity: Effects of long-term rehearsal in expert dancers encoding music to movement. *PLoS One, 11*(1), e0147731. https://doi.org/10.1371/journal.pone.0147731

Basso, J. C., Satyal, M. K., & Rugh, R. (2020). Dance on the brain: Enhancing intra- and inter-brain synchrony. *Frontiers in Human Neuroscience, 14*, 584312. https://doi.org/10.3389/fnhum.2020.584312

Behrends, A., Müller, S., & Dziobek, I. (2012). Moving in and out of synchrony: A concept for a new intervention fostering empathy through interactional movement and dance. *The Arts in Psychotherapy, 39*(2), 107–116. https://doi.org/10.1016/j.aip.2012.02.003

Benarroch, E. E. (2013). Adult neurogenesis in the dentate gyrus: General concepts and potential implications. *Neurology, 81*(16), 1443–1452. https://doi.org/10.1212/WNL.0b013e3182a9a156

Berlucchi, G., & Buchtel, H. A. (2009). Neuronal plasticity: Historical roots and evolution of meaning. *Experimental Brain Research, 192*(3), 307–319. https://doi.org/10.1007/s00221-008-1611-6

Bernhofer, E. I., Higgins, P. A., Daly, B. J., Burant, C. J., & Hornick, T. R. (2014). Hospital lighting and its association with sleep, mood and pain in medical inpatients. *Journal of Advanced Nursing, 70*(5), 1164–1173. https://doi.org/10.1111/jan.12282

Berrezueta-Guzman, J., Pau, I., Martin-Ruiz, M.-L., & Maximo-Bocanegra, N. (2020). Smart-home environment to support homework activities for children. *IEEE Access, 8*. https://doi.org/10.1109/ACCESS.2020.3020734

Bockbrader, M. A., Francisco, G., Lee, R., Olson, J., Solinsky, R., & Boninger, M. L. (2018). Brain computer interfaces in rehabilitation medicine. *PM&R, 10*, 243. https://doi.org/10.1016/j.pmrj.2018.05.028

Boldrini, M., Fulmore, C. A., Tartt, A. N., Simeon, L. R., Pavlova, I., Poposka, V., Rosoklija, G. B., Stankov, A., Arango, V., Dwork, A. J., René, H., & Mann, J. J. (2018). Human hippocampal neurogenesis persists throughout aging. *Cell Stem Cell, 22*(4), 589–599. https://doi.org/10.1016/j.stem.2018.03.015

Bondi, C. O., Klitsch, K. C., Leary, J. B., & Kline, A. E. (2014). Environmental enrichment as a viable neurorehabilitation strategy for experimental traumatic brain injury. *Journal of Neurotrauma, 31*(10), 873–888. https://doi.org/10.1089/neu.2014.3328

Boyko, Y., Jennum, P., Nikolic, M., Holst, R., Oerding, H., & Toft, P. (2017). Sleep in intensive care unit: The role of environment. *Journal of Critical Care, 37*, 99–105. https://doi.org/10.1016/j.jcrc.2016.09.005

Bratman, G. N., Hamilton, J. P., Hahn, K. S., Daily, G. C., & Gross, J. J. (2015). Nature experience reduces rumination and subgenual prefrontal cortex activation. *Proceedings of the National Academy of Sciences, 112*(28), 8567–8572. https://doi.org/10.1073/pnas.1510459112

Brouns, B., Meesters, J. J. L., Wentink, M. M., de Kloet, A. J., Arwert, H. J., Vliet Vlieland, T. P. M., Boyce, L. W., & van Bodegom-Vos, L. (2018). Why the uptake of erehabilitation programs in stroke care is so difficult—A focus group study in The Netherlands. *Implementation Science, 13*(1), 1–11. https://doi.org/10.1186/s13012-018-0827-5

Brown, S., Martinez, M. J., & Parsons, L. M. (2006). The neural basis of human dance. *Cerebral cortex (New York, N.Y. : 1991), 16*(8), 1157–1167. https://doi.org/10.1093/cercor/bhj057

Buchanan, K. L., Grindstaff, J. L., & Pravosudov, V. V. (2013). Condition dependence, developmental plasticity, and cognition: Implications for ecology and evolution. *Trends in Ecology & Evolution, 28*(5), 290–296. https://doi.org/10.1016/j.tree.2013.02.004

Buis, L., Bhadra, P., Salisbury, J., Predel, C., Timmermann, C., Ursin, F., Orzechowski, M., Ropinski, T., & Steger, F. (2022). Conflicting aims and values in the application of smart sensors in geriatric rehabilitation: Ethical analysis. *JMIR mHealth and uHealth, 10*(6). https://doi.org/10.2196/32910

Burdea, G. C. (2003). Virtual rehabilitation--Benefits and challenges. *Methods of Information in Medicine, 42*(5), 519–523.

Calvo-Merino, B., Glaser, D. E., Grèzes, J., Passingham, R. E., & Haggard, P. (2005). Action observation and acquired motor skills: An FMRI study with expert dancers. *Cerebral cortex (New York, N.Y. : 1991), 15*(8), 1243–1249. https://doi.org/10.1093/cercor/bhi007

Carnemolla, P., & Bridge, C. (2019). Housing design and community care: How home modifications reduce care needs of older people and people with disability. *International Journal of Environmental Research and Public Health, 16*, 1951–1963. https://doi.org/10.3390/ijerph16111951

Ceanga, M., Dahab, M., Witte, O. W., & Keiner, S. (2021). Adult neurogenesis and stroke: A tale of two neurogenic niches. *Frontiers in Neuroscience, 15*, 922. https://doi.org/10.3389/fnins.2021.700297

Chang, C.-Y., & Chen, P.-K. (2005). Human response to window views and indoor plants in the workplace. *HortScience, 40*(5), 1354–1359.

Chawla, L. (2015). Benefits of nature contact for children. *Journal of Planning Literature, 30*(4), 433–452. https://doi.org/10.1177/0885412215595441

Chedid, R. J., Dew, A., & Veitch, C. (2013). Barriers to the use of information and communication technology by occupational therapists working in a rural area of new south Wales, Australia. *Australian Occupational Therapy Journal, 60*(3), 197–205. https://doi.org/10.1111/1440-1630.12016

Chen, J., Lehto, X., Lehto, M., & Day, J. (2023). Can colored sidewalk nudge city tourists to walk? An experimental study of the effect of nudges. *Tourism Management, 95*, 104683. https://doi.org/10.1016/j.tourman.2022.104683

Clemenson, G. D., Deng, W., & Gage, F. H. (2015). Environmental enrichment and neurogenesis: From mice to humans. *Current Opinion in Behavioral Sciences, 4*, 56–62. https://doi.org/10.1016/j.cobeha.2015.02.005

Collado, S., Staats, H., & Sorrel, M. A. (2016). Helping out on the land: Effects of children's role in agriculture on reported psychological restoration. *Journal of Environmental Psychology, 45*, 201–209. https://doi.org/10.1016/j.jenvp.2016.01.005

Colley, J., Zeeman, H., & Kendall, E. (2020). How the built environment matters in recovery after neurotrauma: A qualitative examination of first-person experiences across two inpatient settings. *Design for Health, 4*(3), 365–383. https://doi.org/10.1080/24735132.2020.1848975

Conrad, C. D. (2006). What is the functional significance of chronic stress-induced CA3 dendritic retraction within the hippocampus? *Behavioral and Cognitive Neuroscience Reviews, 5*(1), 41–60. https://doi.org/10.1177/1534582306289043

Conrad, C. D. (2008). Chronic stress-induced hippocampal vulnerability: The glucocorticoid vulnerability hypothesis. *Reviews in the Neurosciences, 19*(6), 395–412. https://doi.org/10.1515/revneuro.2008.19.6.395

Cui, X., Bray, S., Bryant, D. M., Glover, G. H., & Reiss, A. L. (2011). A quantitative comparison of NIRS and fMRI across multiple cognitive tasks. *NeuroImage, 54*(4), 2808–2821. https://doi.org/10.1016/j.neuroimage.2010.10.069

Cupchik, G. C., Vartanian, O., Crawley, A., & Mikulis, D. J. (2009). Viewing artworks: Contributions of cognitive control and perceptual facilitation to aesthetic experience. *Brain and Cognition, 70*(1), 84–91. https://doi.org/10.1016/j.bandc.2009.01.003

Dadvand, P., Sunyer, J., Basagana, X., Ballester, F., Lertxundi, A., Fernandez-Somoano, A., Estarlich, M., Garcia-Esteban, R., Mendez, M. A., & Nieuwenhuijsen, M. J. (2012). Surrounding greenness and pregnancy outcomes in four Spanish birth cohorts. *Environmental Health Perspectives, 120*(10), 1481–1487. https://doi.org/10.1289/ehp.1205244

Dadvand, P., Villanueva, C. M., Font-Ribera, L., Martinez, D., Basagaña, X., Belmonte, J., Vrijheid, M., Gražulevičienė, R., Kogevinas, M., & Nieuwenhuijsen, M. J. (2015a). Risks and benefits of green spaces for children: A cross-sectional study of associations with sedentary behavior, obesity, asthma, and allergy. *Environmental Health Perspectives, 122*(12), 1329–1335. https://doi.org/10.1289/ehp.1308038

Dadvand, P., Nieuwenhuijsen, M. J., Esnaola, M., Forns, J., Basagana, X., Alvarez-Pedrerol, M., Rivas, I., Lopez-Vicente, M., De Castro Pascual, M., Su, J., Jerrett, M., Querol, X., & Sunyer, J. (2015b). Green spaces and cognitive development in primary schoolchildren. *Proceedings of the National Academy of Sciences of the United States of America, 112*(26), 7937–7942. https://doi.org/10.1073/pnas.1503402112

Dadvand, P., Bartoll, X., Basagaña, X., Dalmau-Bueno, A., Martinez, D., Ambros, A., Cirach, M., Triguero-Mas, M., Gascon, M., Borrell, C., & Nieuwenhuijsen, M. J. (2016). Green spaces and General Health: Roles of mental health status, social support, and physical activity. *Environment International, 91*, 161–167. https://doi.org/10.1016/j.envint.2016.02.029

Danzl, M. M., Etter, N. M., Andreatta, R. D., & Kitzman, P. H. (2012). Facilitating neurorehabilitation through principles of engagement. *Journal of Allied Health, 41*(1), 35–41.

De Vries, S., Verheij, R. A., Groenewegen, P. P., & Spreeuwenberg, P. (2003). Natural environments - Healthy environments? An exploratory analysis of the relationship between greenspace and health. *Environment and Planning, 35*, 1717–1731. https://doi.org/10.1068/a35111

Delpy, D. T., Cope, M., Van der Zee, P., Arridge, S., Wrayt, S., & Wyatt, J. (1988). Estimation of optical pathlength through tissue from direct time of flight measurement. *Physics in Medicine and Biology, 33*(12), 1433–1442.

Dhamala, E., Yeo, B. T., & Holmes, A. J. (2023). One size does not fit all: Methodological considerations for brain-based predictive modeling in psychiatry. *Biological Psychiatry, 93*(8), 717–728.

Dikker, S., Wan, L., Davidesco, I., Kaggen, L., Oostrik, M., McClintock, J., et al. (2017). Brain-to-brain synchrony tracks real-world dynamic group interactions in the classroom. *Current Biology, 27*(9), 1375–1380. https://doi.org/10.1016/j.cub.2017.04.002

Ding, C., Wu, Y., Chen, X., Chen, Y., Wu, Z., Lin, Z., Kang, D., Fang, W., & Chen, F. (2022). Global, regional, and national burden and attributable risk factors of neurological disorders: The global burden of disease study 1990-2019. *Frontiers in Public Health, 10*, 952161–952161. https://doi.org/10.3389/fpubh.2022.952161

Doctor, J. N., Castro, J., Temkin, N. R., Fraser, R. T., Machamer, J. E., & Dikmen, S. S. (2005). Workers' risk of unemployment after traumatic brain injury: A normed comparison. *Journal of the International Neuropsychological Society : Jins, 11*(6), 747–752.

Dodds, A. G. (1989). Motivation reconsidered: The importance of self-efficacy in rehabilitation. *British Journal of Visual Impairment, 7*(1), 11–15. https://doi.org/10.1177/026461968900700103

Doll, A., Holzel, B. K., Boucard, C. C., Wohlschlager, A. M., & Sorg, C. (2015). Mindfulness is associated with intrinsic functional connectivity between default mode and salience networks. *Frontiers in Human Neuroscience, 9*, 461. https://doi.org/10.3389/fnhum.2015.00461

Domingos, C., Pêgo, J. M., & Santos, N. C. (2021). Effects of physical activity on brain function and structure in older adults: A systematic review. *Behavioural Brain Research, 402*, 113061. https://doi.org/10.1016/j.bbr.2020.113061

Domínguez-Téllez, P., Moral-Muñoz, J. A., Salazar, A., Casado-Fernández, E., & Lucena-Antón, D. (2020). Game-based virtual reality interventions to improve upper limb motor function and quality of life after stroke: Systematic review and meta-analysis. *Games for Health Journal, 9*(1), 1–10. https://doi.org/10.1089/g4h.2019.0043

Douglas, J., Winkler, D., Oliver, S., Liddicoat, S., & D'Cruz, K. (2023). Moving into new housing designed for people with disability: Preliminary evaluation of outcomes. *Disability and Rehabilitation, 45*(8), 1370–1378. https://doi.org/10.1080/09638288.2022.2060343

Draper, K., Ponsford, J., & Michael, S. (2007). Psychosocial and emotional outcomes 10 years following traumatic brain injury. *Journal of Head Trauma Rehabilitation, 22*(5), 278–287. https://doi.org/10.1097/01.HTR.0000290972.63753.a7

Elliot, A. J., Maier, M. A., Moller, A. C., Friedman, R., & Meinhardt, J. (2007). Color and psychological functioning: The effect of red on performance attainment. *Journal of Experimental Psychology: General, 136*(1), 154–168. https://doi.org/10.1037/0096-3445.136.1.154

Elliott, J. E., Tinsley, C. E., Reynolds, C., Olson, R. J., Weymann, K. B., Au-Yeung, W. M., Wilkerson, A., Kaye, J. A., & Lim, M. M. (2022). Tunable White Light for Elders (TWLITE): A protocol demonstrating feasibility and acceptability for deployment, remote data collection, and analysis of a home-based lighting intervention in older adults. *Sensors (Basel, Switzerland), 22*(14), 5372. https://doi.org/10.3390/s22145372

Ellis, R. J., Bruijn, B., Norton, A. C., Winner, E., & Schlaug, G. (2013). Training-mediated leftward asymmetries during music processing: A cross-sectional and longitudinal fMRI analysis. *NeuroImage, 75*, 97–107. https://doi.org/10.1016/j.neuroimage.2013.02.045

Eriksson, P. S., Perfilieva, E., Björk-Eriksson, T., Alborn, A. M., Nordborg, C., Peterson, D. A., & Gage, F. H. (1998). Neurogenesis in the adult human hippocampus. *Nature Medicine, 4*(11), 1313–1317.

Essex, M. J., Thomas Boyce, W., Hertzman, C., Lam, L. L., Armstrong, J. M., Neumann, S. M., & Kobor, M. S. (2013). Epigenetic vestiges of early developmental adversity: Childhood stress exposure and DNA methylation in adolescence. *Child Development, 84*(1), 58–75. https://doi.org/10.1111/j.1467-8624.2011.01641.x

Farrow, S., & Reid, D. (2004). Stroke survivors' perceptions of a leisure-based virtual reality program. *Technology and Disability, 16*(2), 69–81. https://doi.org/10.3233/TAD-2004-16201

Feigin, V. L., Nichols, E., Alam, T., Bannick, M. S., Beghi, E., Blake, N., Culpepper, W. J., Dorsey, E. R., Elbaz, A., Ellenbogen, R. G., Fisher, J. L., Fitzmaurice, C., Giussani, G., Glennie, L., James, S. L., Johnson, C. O., Kassebaum, N. J., Logroscino, G., Marin, B., et al. (2019). Global, regional, and national burden of neurological disorders, 1990-2016: A systematic analysis for the global burden of disease study 2016. *The Lancet. Neurology, 18*(5), 459–480. https://doi.org/10.1016/S1474-4422(18)30499-X

Feng, J., Beheshti, M., Philipson, M., Ramsaywack, Y., Porfiri, M., & Rizzo, J.-R. (2023). Commute booster: A mobile application for first/last mile and middle mile navigation support for people with blindness and low vision. *IEEE Journal of Translational Engineering in Health and Medicine* (p. 99). https://doi.org/10.1109/JTEHM.2023.3293450.

Fernee, C. R., Gabrielsen, L. E., Andersen, A. J. W., & Mesel, T. (2017). Unpacking the black box of wilderness therapy: A realist synthesis. *Qualitative Health Research, 27*(1), 114–129. https://doi.org/10.1177/1049732316655776

Forbes, T. A., Goldstein, E. Z., Dupree, J. L., Jablonska, B., Scafidi, J., Adams, K. L., et al. (2020). Environmental enrichment ameliorates perinatal brain injury and promotes functional white matter recovery. *Nature Communications, 11*(1), 1–17. https://doi.org/10.1038/s41467-020-14762-7

Francis, J., Giles-Corti, B., Wood, L., & Knuiman, M. (2012). Creating sense of community: The role of public space. *Journal of Environmental Psychology, 32*(4), 401–409. https://doi.org/10.1016/j.jenvp.2012.07.002

Frasca, D., Tomaszczyk, J., McFadyen, B. J., & Green, R. E. (2013). Traumatic brain injury and post-acute decline: What role does environmental enrichment play? A scoping review. *Frontiers in Human Neuroscience, 7*, 31–31. https://doi.org/10.3389/fnhum.2013.00031

Fullagar, S. P., & Harrington, M. A. (2009). Negotiating the policy imperative to be healthy: Australian family repertoires of risk, leisure, and healthy lifestyles. *Annals of Leisure Research, 12*(2), 195–215. https://doi.org/10.1080/11745398.2009.9686818

Gage, F. H. (2019). Adult neurogenesis in mammals. *Science, 364*(6443), 827–828. https://doi.org/10.1126/science.aav6885

Gaggioli, A. (2009). *Advanced technologies in rehabilitation : Empowering cognitive, physical, social and communicative skills through virtual reality, robots, wearable systems and brain-computer interfaces* (Studies in health technology and informatics, Vol. 145). IOS Press. Retrieved August 31, 2023.

Gidlow, C. J., & Ellis, N. J. (2011). Neighbourhood green space in deprived urban communities: Issues and barriers to use. *Local Environment, 16*(10), 989–1002. https://doi.org/10.1080/13549839.2011.582861

Gill, T. (2014). The benefits of children's engagement with nature: A systematic literature review. *Children, Youth and Environments, 24*(2), 10–34.

Gordon, C. L., Cobb, P. R., & Balasubramaniam, R. (2018). Recruitment of the motor system during music listening: An ALE meta-analysis of fMRI data. *PLoS One, 13*(11), e0207213. https://doi.org/10.1371/journal.pone.0207213

Gorman, C., & Gustafsson, L. (2020). The use of augmented reality for rehabilitation after stroke: A narrative review. *Disability and Rehabilitation: Assistive Technology*, 1–9. https://doi.org/10.1080/17483107.2020.1791264

Greene, A. S., Shen, X., Noble, S., Horien, C., Hahn, C. A., Arora, J., et al. (2022). Brain–phenotype models fail for individuals who defy sample stereotypes. *Nature, 609*(7925), 109–118.

Grońska-Pęski, M., Gonçalves, J. T., & Hébert, J. M. (2021). Enriched environment promotes adult hippocampal neurogenesis through FGFRs. *Journal of Neuroscience, 41*(13), 2899–2910. https://doi.org/10.1523/JNEUROSCI.2286-20.2021

Groshong, L., Wilhelm Stanis, S. A., Kaczynski, A. T., & Hipp, J. A. (2020). Attitudes about perceived park safety among residents in low-income and high minority Kansas City, Missouri, neighborhoods. *Environment and Behavior, 52*(6), 639–665. https://doi.org/10.1177/0013916518814291

Halpern, N. A. (2014). Innovative designs for the smart ICU. Part 2: The ICU. *Chest, 145*(3), 646–658. https://doi.org/10.1378/chest.13-0004

Han, S. S., White, K., & Cisek, E. (2022). A feasibility study of individuals living at home with alzheimer's disease and related dementias: Utilization of visual mapping assistive technology to enhance quality of life and reduce caregiver burden. *Clinical Interventions in Aging, 17*, 1885–1892. https://doi.org/10.2147/CIA.S387255

Hannan, A. J. (2014). Environmental enrichment and brain repair: Harnessing the therapeutic effects of cognitive stimulation and physical activity to enhance experience-dependent plasticity. *Neuropathology and Applied Neurobiology, 40*(1), 13–25. https://doi.org/10.1111/nan.12102

Hartig, T., & Staats, H. (2006). The need for psychological restoration as a determinant of environmental preferences. *Journal of Environmental Psychology, 26*(3), 215–226. https://doi.org/10.1016/j.jenvp.2006.07.007

Henigsberg, N., Kalember, P., Petrović, Z. K., & Šečić, A. (2019). Neuroimaging research in posttraumatic stress disorder - Focus on amygdala, hippocampus and prefrontal cortex. *Progress in Neuropsychopharmacology & Biological Psychiatry, 90*, 37–42. https://doi.org/10.1016/j.pnpbp.2018.11.003

Henrich, J., Heine, S. J., & Norenzayan, A. (2010). The weirdest people in the world? *Behavioral and Brain Sciences, 33*(2–3), 61–83.

Hernandez, R. O. (2007). Effects of therapeutic gardens in special care units for people with dementia. *Journal of Housing for the Elderly, 21*(1–2), 117–152. https://doi.org/10.1300/J081v21n0107

Horsburgh, C. R. (1995). Healing by design. *The New England Journal of Medicine, 333*(11), 735–740. https://doi.org/10.1056/NEJM199509143331117

Huisman, E. R. C. M., Morales, E., van Hoof, J., & Kort, H. S. M. (2012). Healing environment: A review of the impact of physical environmental factors on users. *Building and Environment, 58*, 70–80. https://doi.org/10.1016/j.buildenv.2012.06.016

Huygelier, H., Mattheus, E., Abeele, V. V., van Ee, R., & Gillebert, C. R. (2021). The use of the term virtual reality in post-stroke rehabilitation: A scoping review and commentary. *Psychologica Belgica, 61*(1), 145–162. https://doi.org/10.5334/pb.1033

Jahandideh, S., Kendall, E., Low-Choy, S., Donald, K., & Jayasinghe, R. (2018). Understanding patient engagement in outpatient cardiac rehabilitation programs through the model of therapeutic engagement. *Examines in Physical Medicine and Rehabilitation: Open Access, 1*(3). https://doi.org/10.31031/EPMR.2018.01.000514

Jamwal, R., Jarman, H. K., Roseingrave, E., Douglas, J., & Winkler, D. (2022). Smart home and communication technology for people with disability: A scoping review. *Disability and Rehabilitation. Assistive Technology, 17*(6), 624–644. https://doi.org/10.1080/17483107.2020.1818138

Janssen, H., Bernhardt, J., Collier, J. M., Sena, E. S., McElduff, P., Attia, J., Pollack, M., Howells, D. W., Nilsson, M., Calford, M. B., & Spratt, N. J. (2010). An enriched environment improves sensorimotor function post-ischemic stroke. *Neurorehabilitation and Neural Repair, 24*(9), 802–813. https://doi.org/10.1177/1545968310372092

Janssen, H., Ada, L., Bernhardt, J., McElduff, P., Pollack, M., Nilsson, M., & Spratt, N. J. (2014). An enriched environment increases activity in stroke patients undergoing rehabilitation in a mixed rehabilitation unit: A pilot non-randomized controlled trial. *Disability and Rehabilitation, 36*(3), 255–262. https://doi.org/10.3109/09638288.2013.788218

Jiang, R., Calhoun, V. D., Cui, Y., Qi, S., Zhuo, C., Li, J., Jung, R., Yang, J., Du, Y., Jiang, T., & Sui, J. (2019). Multimodal data revealed different neurobiological correlates of intelligence between males and females. *Brain Imaging and Behavior, 14*(5), 1979–1993. https://doi.org/10.1007/s11682-019-00146-z

Johansson, B. B., & Ohlsson, A. L. (1996). Environment, social interaction, and physical activity as determinants of functional outcome after cerebral infarction in the rat. *Experimental Neurology, 139*(2), 322–327. https://doi.org/10.1006/exnr.1996.0106

Jones, T. A. (1999). Multiple synapse formation in the motor cortex opposite unilateral sensorimotor cortex lesions in adult rats. *The Journal of Comparative Neurology, 414*(1), 57–66.

Jones, T. A., & Jefferson, S. C. (2011). Reflections of experience-expectant development in repair of the adult damaged brain. *Developmental Psychobiology, 53*(5), 466–475. https://doi.org/10.1002/dev.20557

Jones, E. G., & Pons, T. P. (1998). Thalamic and brainstem contributions to large-scale plasticity of primate somatosensory cortex. *Science, 282*(5391), 1121–1125. https://doi.org/10.1126/science.282.5391.1121

Jones, L., & Wheat, J. (2023). Green and pleasant lands: The affective and cerebral hemodynamic effects of presence in virtual environments during exercise. *Perceptual and Motor Skills, 130*(2), 826–843. https://doi.org/10.1177/00315125221146614

Kang, N., & Liu, C. (2022). Towards landscape visual quality evaluation: Methodologies, technologies, and recommendations. *Ecological Indicators, 142*, 109174. https://doi.org/10.1016/j.ecolind.2022.109174

Kaplan, S. (1983). A model of person-environment compatibility. *Environment and Behavior, 15*(3), 311–332. https://doi.org/10.1177/0013916583153003

Kempermann, G. (2012). New neurons for 'survival of the fittest'. *Nature Reviews. Neuroscience, 13*(10), 727–736. https://doi.org/10.1038/nrn3319

Kempermann, G. (2019). Environmental enrichment, new neurons and the neurobiology of individuality. *Nature Reviews Neuroscience, 20*(4), 235–245. https://doi.org/10.1038/s41583-019-0120-x

Kendrick, D., Silverberg, N. D., Barlow, S., Miller, W. C., & Moffat, J. (2012). Acquired brain injury self-management programme: A pilot study. *Brain Injury, 26*(10), 1243–1249. https://doi.org/10.3109/02699052.2012.672787

Kim, J., Kim, K., Kim, D. Y., Chang, W. H., Park, C.-I., Ohn, S. H., Han, K., Ku, J., Nam, S. W., Kim, I. Y., & Kim, S. I. (2007). Virtual environment training system for rehabilitation of stroke patients with unilateral neglect: Crossing the virtual street, *10*(1), 7–15. https://doi.org/10.1089/cpb.2006.9998

Kim, D. Y., Ku, J., Chang, W. H., Park, T. H., Lim, J. Y., Han, K., Kim, I. Y., & Kim, S. I. (2010). Assessment of post-stroke extrapersonal neglect using a three-dimensional immersive virtual street crossing program. *Acta Neurologica Scandinavica, 121*(3), 171–177. https://doi.org/10.1111/j.1600-0404.2009.01194.x

Kim, A., Schweighofer, N., & Finley, J. M. (2019). Locomotor skill acquisition in virtual reality shows sustained transfer to the real world. *Journal of Neuroengineering and Rehabilitation, 16*(1), 113–113. https://doi.org/10.1186/s12984-019-0584-y

Knauss, S., Stelzle, D., Emmrich, J. V., Korsnes, M. S., Sejvar, J. J., & Winkler, A. S. (2019). An emphasis on neurology in low and middle-income countries. *The Lancet. Neurology, 18*(12), 1078–1079. https://doi.org/10.1016/S1474-4422(19)30392-8

Knobel, S. E. J., Kaufmann, B. C., Gerber, S. M., Cazzoli, D., Müri, R. M., Nyffeler, T., & Nef, T. (2020). Immersive 3d virtual reality cancellation task for visual neglect assessment: A pilot study. *Frontiers in Human Neuroscience, 14*, 180–180. https://doi.org/10.3389/fnhum.2020.00180

Kuhn, S., Forlim, C. G., Lender, A., Wirtz, J., & Gallinat, J. (2021). Brain functional connectivity differs when viewing pictures from natural and built environments using fMRI resting state analysis. *Scientific Reports, 11*(1), 4110. https://doi.org/10.1038/s41598-021-83246-5

Kumar, A., Pareek, V., Faiq, M. A., Ghosh, S. K., & Kumari, C. (2019). Adult neurogenesis in humans: A review of basic concepts, history, current research, and clinical implications. *Innovations in Clinical Neuroscience, 16*(5–6), 30–37.

Kuo, M., Barnes, M., & Jordan, C. (2019). Do experiences with nature promote learning? Converging evidence of a cause-and-effect relationship. *Frontiers in Psychology, 10*, 305.

Lakhani, A., Norwood, M. F., Watling, D., Zeeman, H., & Kendall, E. (2019). Using the natural environment to address the psychosocial impact of neurological disability: A systematic review. *Health & Place, 55*, 188–201.

Lamontagne, M.-E., Ouellet, M.-C., & Simard, J.-F. (2009). A descriptive portrait of human assistance required by individuals with brain injury. *Brain Injury, 23*(7), 693–701. https://doi.org/10.1080/02699050902970760 2

Lamontagne, M.-E., Poncet, F., Careau, E., Sirois, M.-J., & Boucher, N. (2013). Life habits performance of individuals with brain injury in different living environments. *Brain Injury, 27*(2), 135–144. https://doi.org/10.3109/02699052.2012.722253

Laver, K. E., Lange, B., George, S., Deutsch, J. E., Saposnik, G., & Crotty, M. (2017). Virtual reality for stroke rehabilitation. Cochrane database of systematic reviews 2017, *11*, CD008349. https://doi.org/10.1002/14651858.CD008349.pub4

Lee, M., Mishra, R. K., Momin, A., El-Refaei, N., Bagheri, A. B., York, M. K., Kunik, M. E., Derhammer, M., Fatehi, B., Lim, J., Cole, R., Barchard, G., Vaziri, A., & Najafi, B. (2022). Smart-home concept for remote monitoring of instrumental activities of daily living (IADL) in older adults with cognitive impairment: A proof of concept and feasibility study. *Sensors (Basel, Switzerland), 22*(18). https://doi.org/10.3390/s22186745.

Li, H., Zhang, X., Wang, H., Yang, Z., Liu, H., Cao, Y., & Zhang, G. (2021). Access to nature via virtual reality: a mini-review. *Frontiers in Psychology, 12*, 725288–725288. https://doi.org/10.3389/fpsyg.2021.725288

Li, J., Bzdok, D., Chen, J., Tam, A., Ooi, L. Q. R., Holmes, A. J., Ge, T., Patil, K. R., Jabbi, M., Eickhoff, S. B., Yeo, B. T. T., & Genon, S. (2022). Cross-ethnicity/race generalization failure of behavioral prediction from resting-state functional connectivity. *Science Advances, 8*(11), 1812. https://doi.org/10.1126/sciadv.abj1812

Lian, Y., Liu, D.-E., & Ji, W.-Z. (2023). Survey and analysis of the current status of research in the field of outdoor navigation for the blind. *Disability and Rehabilitation. Assistive Technology, 1-19*, 1–19. https://doi.org/10.1080/17483107.2023.2227224

Liao, Y.-Y., Yang, Y.-R., Cheng, S.-J., Wu, Y.-R., Fuh, J.-L., & Wang, R.-Y. (2015). Virtual reality-based training to improve obstacle-crossing performance and dynamic balance in patients with parkinson's disease. *Neurorehabilitation and Neural Repair, 29*(7), 658–667. https://doi.org/10.1177/1545968314562111

Lindenberger, U., Li, S. C., Gruber, W., & Müller, V. (2009). Brains swinging in concert: Cortical phase synchronization while playing guitar. *BMC Neuroscience, 10*. https://doi.org/10.1186/1471-2202-10-22

Linder, L. A., & Christian, B. J. (2012). Nighttime sleep disruptions, the hospital care environment, and symptoms in elementary school-age children with cancer. *Oncology Nursing Forum, 39*(6), 553–561. https://doi.org/10.1188/12.ONF.553-561

Liu, Y., Liu, Y., Tang, J., Yin, E., Hu, D., & Zhou, Z. (2020). A self-paced bci prototype system based on the incorporation of an intelligent environment-understanding approach for rehabilitation hospital environmental control. *Computers in Biology and Medicine, 118*, 103618–103618. https://doi.org/10.1016/j.compbiomed.2020.103618

Liu, T., Duan, L., Dai, R., Pelowski, M., & Zhu, C. (2021). Team-work, team-brain: Exploring synchrony and team interdependence in a nine-person drumming task via multiparticipant hyperscanning and inter-brain network topology with fNIRS. *NeuroImage, 237*, 118147. https://doi.org/10.1016/j.neuroimage.2021.118147

Logothetis, N. K. (2008). What we can do and what we cannot do with fMRI. *Nature, 453*(7197), 869–878. https://doi.org/10.1038/nature06976

Lohse, K. R., Hilderman, C. G. E., Cheung, K. L., Tatla, S., Van der Loos, H. F. M., & Quinn, T. J. (2014). Virtual reality therapy for adults post-stroke: A systematic review and meta-analysis exploring virtual environments and commercial games in therapy. *PLoS One, 9*(3). https://doi.org/10.1371/journal.pone.0093318

Lubna, L. D., Noorain, M., Bhawani, S. C., & Thomas, N. (2023). *Ethical dilemmas and privacy issues in emerging technologies: A review, 23*(1151), 1151–1151. https://doi.org/10.3390/s23031151

Lumsden, J., Miles, L. K., & Macrae, C. N. (2014). Sync or sink? Interpersonal synchrony impacts self-esteem. *Frontiers in Psychology, 5*, 1064. https://doi.org/10.3389/fpsyg.2014.01064

Mackes, N. K., Mehta, M. A., Beyh, A., Nkrumah, R. O., Golm, D., Sarkar, S., ... & Sonuga-Barke, E. J. (2022). A prospective study of the impact of severe childhood deprivation on brain White matter in adult adoptees: Widespread localized reductions in volume but unaffected microstructural organization. *eNeuro, 9*(6). https://doi.org/10.1523/ENEURO.0188-22.2022

Manes, J. L., Herschel, E., Aveni, K., Tjaden, K., Parrish, T., Simuni, T., Corcos, D. M., & Roberts, A. C. (2021). The effects of a simulated fmri environment on voice intensity in individuals with parkinson's disease hypophonia and older healthy adults. *Journal of Communication Disorders, 94*, 106149–106149. https://doi.org/10.1016/j.jcomdis.2021.106149

Manivannan, S., Marei, O., Elalfy, O., & Zaben, M. (2021). Neurogenesis after traumatic brain injury-the complex role of HMGB1 and neuroinflammation. *Neuropharmacology*, 183, 108400. https://doi.org/10.1016/j.neuropharm.2020.108400

Mantovani, E., Zucchella, C., Bottiroli, S., Federico, A., Giugno, R., Sandrini, G., Chiamulera, C., & Tamburin, S. (2020). Telemedicine and virtual reality for cognitive rehabilitation: A roadmap for the covid-19 pandemic. *Frontiers in Neurology, 11*, 926–926. https://doi.org/10.3389/fneur.2020.00926

Marín-Burgin, A., & Schinder, A. F. (2012). Requirement of adult-born neurons for hippocampus-dependent learning. *Behavioural Brain Research, 227*(2), 391–399. https://doi.org/10.1016/j.bbr.2011.07.001

Markevych, I., Feng, X., Astell-Burt, T., Standl, M., Sugiri, D., Schikowski, T., et al. (2019). Residential and school greenspace and academic performance: Evidence from the GINIplus and LISA longitudinal studies of German adolescents. *Environmental Pollution, 245*, 71–76. https://doi.org/10.1016/j.envpol.2018.10.053

Martínez-Molina, N., Siponkoski, S. T., Kuusela, L., Laitinen, S., Holma, M., Ahlfors, M., et al. (2021). Resting-state network plasticity induced by music therapy after traumatic brain injury. *Neural Plasticity, 2021*, 1–18. https://doi.org/10.1155/2021/6682471

Matamala-Gomez, M., Bottiroli, S., Realdon, O., Riva, G., Galvagni, L., Platz, T., Sandrini, G., De Icco, R., & Tassorelli, C. (2021). Telemedicine and virtual reality at time of covid-19 pandemic: An overview for future perspectives in neurorehabilitation. *Frontiers in Neurology, 12*. https://doi.org/10.3389/fneur.2021.646902

Mattar, A. A. G., Hitzig, S. L., & McGillivray, C. F. (2015). A qualitative study on the use of personal information technology by persons with spinal cord injury. *Disability and Rehabilitation, 37*(15), 1362–1371. https://doi.org/10.3109/09638288.2014.963708

McCormick, R. (2017). Does access to Green space impact the mental well-being of children: A systematic review. *Journal of Pediatric Nursing, 37*, 3–7. https://doi.org/10.1016/j.pedn.2017.08.027

McDonald, M. W., Hayward, K. S., Rosbergen, I. C. M., Jeffers, M. S., & Corbett, D. (2018). Is environmental enrichment ready for clinical application in human post-stroke rehabilitation? *Frontiers in Behavioral Neuroscience, 12*, 135. https://doi.org/10.3389/fnbeh.2018.00135

McGrath, C., & Astell, A. (2017). The benefits and barriers to technology acquisition: Understanding the decision-making processes of older adults with age-related vision loss (arvl). *British Journal of Occupational Therapy, 80*(2), 123–131. https://doi.org/10.1177/0308022616667959

McKendrick, R., Mehta, R., Ayaz, H., Scheldrup, M., & Parasuraman, R. (2017). Prefrontal hemodynamics of physical activity and environmental complexity during cognitive work. *Human Factors, 59*(1), 147–162. https://doi.org/10.1177/0018720816675053

Megha, M., Harpreet, S., & Nayeem, Z. (2013). Effect of frequency of multimodal coma stimulation on the consciousness levels of traumatic brain injury comatose patients. *Brain Injury, 27*(5), 570–577. https://doi.org/10.3109/02699052.2013.767937

Mehrholz, J., Thomas, S., Kugler, J., Pohl, M., & Elsner, B. (2020). Electromechanical-assisted training for walking after stroke. *Cochrane Database of Systematic Reviews, 10*, CD006185. https://doi.org/10.1002/14651858.CD006185.pub5

Meredith, W. J., Cardenas-Iniguez, C., Berman, M. G., & Rosenberg, M. D. (2022). Effects of the physical and social environment on youth cognitive performance. *Developmental Psychobiology, 64*(4), 22258. https://doi.org/10.1002/dev.22258

Miller, L. S., Colella, B., Mikulis, D., Maller, J., & Green, R. E. (2013). Environmental enrichment may protect against hippocampal atrophy in the chronic stages of traumatic brain injury. *Frontiers in Human Neuroscience, 7*, 506. https://doi.org/10.3389/fnhum.2013.00506

Möller, J. C., Zutter, D., & Riener, R. (2021). Technology-based neurorehabilitation in Parkinson's disease—A narrative review. *Clinical and Translational Neuroscience, 5*(3), 23. MDPI AG. Retrieved from https://doi.org/10.3390/ctn5030023

Monaco, C. M., Mattiola, V. V., Folweiler, K. A., Tay, J. K., Yelleswarapu, N. K., Curatolo, L. M., Matter, A. M., Cheng, J. P., & Kline, A. E. (2013). Environmental enrichment promotes robust functional and histological benefits in female rats after controlled cortical impact injury. *Experimental Neurology, 247*, 410–418. https://doi.org/10.1016/j.expneurol.2013.01.007

Montague, P. R., Berns, G. S., Cohen, J. D., McClure, S. M., Pagnoni, G., Dhamala, M., et al. (2002). Hyperscanning: Simultaneous fMRI during linked social interactions. *NeuroImage, 16*(4), 1159–1164. https://doi.org/10.1006/nimg.2002.1150

Morse, H., Biggart, L., Pomeroy, V., & Rossit, S. (2022). Exploring perspectives from stroke survivors, carers and clinicians on virtual reality as a precursor to using telerehabilitation for spatial neglect post-stroke. *Neuropsychological Rehabilitation, 32*(5), 707–731. https://doi.org/10.1080/09602011.2020.1819827

Muller, V., Sanger, J., & Lindenberger, U. (2013). Intra- and inter-brain synchronization during musical improvisation on the guitar. *PLoS One, 8*(9), e73852. https://doi.org/10.1371/journal.pone.0073852

Murphy, P. F., Miyazaki, Y., Detweiler, M. B., & Kim, K. Y. (2010). Longitudinal analysis of differential effects on agitation of a therapeutic wander garden for dementia patients based on ambulation ability. *Dementia, 9*(3), 355–373. https://doi.org/10.1177/1471301210375336

Musselman, K. E., Shah, M., & José, Z. (2018). Rehabilitation technologies and interventions for individuals with spinal cord injury: Translational potential of current trends. *Journal of Neuroengineering and Rehabilitation, 15*(1), 1–8. https://doi.org/10.1186/s12984-018-0386-7

Ng, K., Mikulis, D. J., Glazer, J., Kabani, N., Till, C., Greenberg, G., Thompson, A., Lazinski, D., Agid, R., Colella, B., & Green, R. E. (2008). Magnetic resonance imaging evidence of progression of subacute brain atrophy in moderate to severe traumatic brain injury. *Archives of Physical Medicine and Rehabilitation, 89*(12), 44. https://doi.org/10.1016/j.apmr.2008.07.006

Ni, J., Jiang, W., Gong, X., Fan, Y., Qiu, H., Dou, J., et al. (2022). Effect of rTMS intervention on upper limb motor function after stroke: A study based on fNIRS. *Frontiers in Aging Neuroscience, 14*, 1077218. https://doi.org/10.3389/fnagi.2022.1077218

Nicolelis, M. A. (1997). Dynamic and distributed somatosensory representations as the substrate for cortical and subcortical plasticity. *In Seminars in Neuroscience, 9*(1), 24–33. https://doi.org/10.1006/smns.1997.0103

Nithianantharajah, J., & Hannan, A. J. (2006). Enriched environments, experience-dependent plasticity and disorders of the nervous system. *Nature Reviews Neuroscience, 7*(9), 697–709. https://doi.org/10.1038/nrn1970

Noack, H., Lövdén, M., Schmiedek, F., & Lindenberger, U. (2009). Cognitive plasticity in adulthood and old age: Gauging the generality of cognitive intervention effects. *Restorative Neurology and Neuroscience, 27*(5), 435–453. https://doi.org/10.3233/RNN-2009-0496

Nolan, M., Roman, E., Nasa, A., Levins, K. J., O'Hanlon, E., O'Keane, V., & Willian Roddy, D. (2020). Hippocampal and amygdalar volume changes in major depressive disorder: A targeted review and focus on stress. *Chronic Stress, 4*. https://doi.org/10.1177/2470547020944553

Norwood, M. F., Lakhani, A., Fullagar, S., Maujean, A., Downes, M., Byrne, J., et al. (2019a). A narrative and systematic review of the behavioural, cognitive and emotional effects of passive nature exposure on young people: Evidence for prescribing change. *Landscape and Urban Planning, 189*, 71–79.

Norwood, M. F., Lakhani, A., Maujean, A., Zeeman, H., Creux, O., & Kendall, E. (2019b). Brain activity, underlying mood and the environment: A systematic review. *Journal of Environmental Psychology, 65*. https://doi.org/10.1016/j.jenvp.2019.101321

Norwood, M. F., Lakhani, A., & Kendall, E. (2021). Teaching traditional indoor school lessons in nature: The effects on student learning and behaviour. *Landscape and Urban Planning, 206*, 103963.

Norwood, M. F., Painter, D. R., Marsh, C. H., Reid, C., Hine, T., Harvie, D. S., Jones, S., Dungey, K., Chen, B., Libera, M., Gan, L., Bernhardt, J., Kendall, E., & Zeeman, H. (2022a). The attention atlas virtual reality platform maps three-dimensional (3d) attention in unilateral spatial neglect patients: A protocol. *Brain Impairment, 1-20*, 1–20. https://doi.org/10.1017/BrImp.2022.15

Norwood, M. F., Lakhani, A., & Kendall, E. (2022b). 'Almost pooped on—dislike!': Student and teacher reactions to nature-based learning and resulting practical advice for implementing in secondary schools. *Journal of Adventure Education and Outdoor Learning, 22*(4), 298–312.

Nozawa, T., Sakaki, K., Ikeda, S., Jeong, H., Yamazaki, S., dos Santos Kawata, K. H., et al. (2019). Prior physical synchrony enhances rapport and inter-brain synchronization during subsequent educational communication. https://doi.org/10.1101/601385.

Obasanjo, O. O. (1998). *The impact of the physical environment on adolescents in the inner city*. Doctoral Dissertation. University of Michigan.

Ochiai, H., Ikei, H., Jo, H., Ohishi, M., & Miyazaki, Y. (2023). Relaxation effect of nature sound exposure on gambling disorder patients: A crossover study. *Journal of Integrative and Complementary Medicine, 29*(8), 518–525. https://doi.org/10.1089/jicm.2022.0611

Olivier-Pijpers, V. C., Cramm, J. M., & Nieboer, A. P. (2019). Influence of the organizational environment on challenging behaviour in people with intellectual disabilities: Professionals' views. *Journal of Applied Research in Intellectual Disabilities, 32*(3), 610–621. https://doi.org/10.1111/jar.12555

Olszewska, A. A., Marques, P. F., Ryan, R. L., & Barbosa, F. (2018). What makes a landscape contemplative? *Environment and Planning B: Urban Analytics and City Science, 45*(1), 7–25. https://doi.org/10.1177/0265813516660716

Olszewska-Guizzo, A., Fogel, A., Escoffier, N., Sia, A., Nakazawa, K., Kumagai, A., et al. (2022). Therapeutic garden with contemplative features induces desirable changes in mood and brain activity in depressed adults. *Frontiers in Psychiatry, 13*. https://doi.org/10.3389/fpsyt.2022.757056

Ono, Y., Nomoto, Y., Tanaka, S., Sato, K., Shimada, S., Tachibana, A., et al. (2014). Frontotemporal oxyhemoglobin dynamics predict performance accuracy of dance simulation gameplay: Temporal characteristics of top-down and bottom-up cortical activities. *NeuroImage, 85*(Pt 1), 461–470. https://doi.org/10.1016/j.neuroimage.2013.05.071

Oppenheim, R. W. (2019). Adult hippocampal neurogenesis in mammals (and humans): The death of a central dogma in neuroscience and its replacement by a new dogma. *Developmental Neurobiology, 79*(3), 268–280. https://doi.org/10.1002/dneu.22674

Orejel Bustos, A. S., Tramontano, M., Morone, G., Ciancarelli, I., Panza, G., Minnetti, A., Picelli, A., Smania, N., Iosa, M., & Vannozzi, G. (2023). Ambient assisted living systems for falls monitoring at home. *Expert Review of Medical Devices*, 1–8. https://doi.org/10.1080/17434440.2023.2245320

Owji, S., & Shoja, M. M. (2020). The history of discovery of adult neurogenesis. *Clinical Anatomy (New York, N.Y.), 33*(1), 41–55. https://doi.org/10.1002/ca.23447

Oyibo, K., Wang, K., & Morita, P. P. (2023). Using smart home technologies to promote physical activity among the general and aging populations: Scoping review. *Journal of Medical Internet Research, 25*, 41942. https://doi.org/10.2196/41942

Painter, D. R., Norwood, M. F., Marsh, C. H., Hine, T., Harvie, D., Libera, M., Bernhardt, J., Gan, L., & Zeeman, H. (2023). Immersive virtual reality gameplay detects visuospatial atypicality, including unilateral spatial neglect, following brain injury: A pilot study. *Under Review.*

Painter, D. R., Norwood, M. F., Marsh, C. H., Hine, T., Woodman, C., Libera, M., ... & Zeeman, H. (2024). Virtual reality gameplay classification illustrates the multidimensionality of visuospatial neglect. *Brain Communications*, fcae145. https://doi.org/10.1093/braincomms/fcae145

Paraskevopoulos, E., & Herholz, S. (2013). Multisensory integration and neuroplasticity in the human cerebral cortex. *Translational Neuroscience, 4*(3). https://doi.org/10.2478/s13380-013-0134-1

Parker, A. M., Needham, D. M., & Lord, R. K. (2013). Increasing the dose of acute rehabilitation: Is there a benefit? *BMC Medicine, 11*(1). https://doi.org/10.1186/1741-7015-11-199

Pascual-Leone, A., Amedi, A., Fregni, F., & Merabet, L. B. (2005). The plastic human brain cortex. *Annual Review of Neuroscience, 28*, 377–401. https://doi.org/10.1146/annurev.neuro.27.070203.144216

Pietrzak, E., Pullman, S., & McGuire, A. (2014). Using virtual reality and videogames for traumatic brain injury rehabilitation: A structured literature review. *Games for Health: Research, Development, and Clinical Applications, 3*(4), 202–214. https://doi.org/10.1089/g4h.2014.0013

Pinto, J. O., Dores, A. R., Peixoto, B., Geraldo, A., & Barbosa, F. (2022). Systematic review of sensory stimulation programs in the rehabilitation of acquired brain injury. *European Psychologist, 27*(1), 20–40. https://doi.org/10.1027/1016-9040/a000421

Planchez, B., Surget, A., & Belzung, C. (2020). Adult hippocampal neurogenesis and antidepressants effects. *Current Opinion in Pharmacology, 50*, 88–95. https://doi.org/10.1016/j.coph.2019.11.009

Poza, J., Gómez, C., Gutiérrez, M. T., Mendoza, N., & Hornero, R. (2013). Effects of a multisensory environment on brain-injured patients: Assessment of spectral patterns. *Medical Engineering & Physics, 35*(3), 365–375. https://doi.org/10.1016/j.medengphy.2012.06.001

Pryor, J. (2004). What environmental factors irritate people with acquired brain injury? *Disability and Rehabilitation, 26*(16), 974–980. https://doi.org/10.1080/09638280410001702397

Qiu, X., Yang, Z., Yang, J., Wang, Q., & Wang, D. (2023). Impact of ar navigation display methods on wayfinding performance and spatial knowledge acquisition. *International Journal of Human–Computer Interaction*, 1–21. https://doi.org/10.1080/10447318.2023.2169524

Raanaas, R. K., Patil, G. G., & Hartig, T. (2012). Health benefits of a view of nature through the window: A quasi-experimental study of patients in a residential rehabilitation center. *Clinical Rehabilitation, 26*(1), 21–32.

Redell, J. B., Maynard, M. E., Underwood, E. L., Vita, S. M., Dash, P. K., & Kobori, N. (2020). Traumatic brain injury and hippocampal neurogenesis: Functional implications. *Experimental Neurology, 331*, 113372. https://doi.org/10.1016/j.expneurol.2020.113372

Repetto, C., Serino, S., Macedonia, M., & Riva, G. (2016). Virtual reality as an embodied tool to enhance episodic memory in elderly. *Frontiers in Psychology, 7*. https://doi.org/10.3389/fpsyg.2016.01839

Ricard, J. A., Parker, T. C., Dhamala, E., Kwasa, J., Allsop, A., & Holmes, A. J. (2023). Confronting racially exclusionary practices in the acquisition and analyses of neuroimaging data. *Nature Neuroscience, 26*(1), 4–11.

Roe, J. (2008). *The restorative power of natural and built environments*. Heriot-Watt University. Published PhD, School of Built Environment.

Rosenbrock, H., Koros, E., Bloching, A., Podhorna, J., & Borsini, F. (2005). Effect of chronic intermittent restraint stress on hippocampal expression of marker proteins for synaptic plasticity and progenitor cell proliferation in rats. *Brain Research, 1040*(1–2), 55–63. https://doi.org/10.1016/j.brainres.2005.01.065

Rossinia, P. M., & Ferreri, F. (2013). Neurophysiological techniques in the study of the excitability, connectivity, and plasticity of the human brain. *Supplements to Clinical Neurophysiology, 62*, 1–17. https://doi.org/10.1016/b978-0-7020-5307-8.00001-6

Samadani, A., Kim, S., Moon, J., Kang, K., & Chau, T. (2021). Neurophysiological synchrony between children with severe physical disabilities and their parents during music therapy. *Frontiers in Neuroscience, 15*, 531915. https://doi.org/10.3389/fnins.2021.531915

Sanchis-Segura, C., Victoria, I.-G. M., Aguirre, N., Javier, C.-G. Á., & Forn, C. (2020). Effects of different intracranial volume correction methods on univariate sex differences in grey matter volume and multivariate sex prediction. *Scientific Reports, 10*(1). https://doi.org/10.1038/s41598-020-69361-9

Sarkamo, T., Ripolles, P., Vepsalainen, H., Autti, T., Silvennoinen, H. M., Salli, E., et al. (2014). Structural changes induced by daily music listening in the recovering brain after middle cerebral artery stroke: A voxel-based morphometry study. *Frontiers in Human Neuroscience, 8*, 245. https://doi.org/10.3389/fnhum.2014.00245

Schmidt, J., & Ownsworth, T. (2022). Special issue editorial: The self after brain injury. *Neuropsychological Rehabilitation, 32*(8), 1669–1675. https://doi.org/10.1080/09602011.2022.2120504

Scholkmann, F., Kleiser, S., Metz, A. J., Zimmermann, R., Mata Pavia, J., Wolf, U., & Wolf, M. (2014). A review on continuous wave functional near-infrared spectroscopy and imaging instrumentation and methodology. *NeuroImage, 85*(Pt 1), 6–27. https://doi.org/10.1016/j.neuroimage.2013.05.004

Sened, H., Zilcha-Mano, S., & Shamay-Tsoory, S. (2022). Inter-brain plasticity as a biological mechanism of change in psychotherapy: A review and integrative model. *Frontiers in Human Neuroscience, 16*, 955238. https://doi.org/10.3389/fnhum.2022.955238

Severinghaus, J. (2007). Takuo Aoyagi: Discovery of pulse oximetry. *Anaesthesia & Analgesia, 105*(S1–S4). https://doi.org/10.1213/01.ane.0000269514.31660.09

Short, M. R., Hernandez-Pavon, J. C., Jones, A., & Pons, J. L. (2021). EEG hyperscanning in motor rehabilitation: A position paper. *Journal of Neuroengineering and Rehabilitation, 18*(1), 98. https://doi.org/10.1186/s12984-021-00892-6

Siponkoski, S. T., Martinez-Molina, N., Kuusela, L., Laitinen, S., Holma, M., Ahlfors, M., et al. (2020). Music therapy enhances executive functions and prefrontal structural neuroplasticity after traumatic brain injury: Evidence from a randomized controlled trial. *Journal of Neurotrauma, 37*(4), 618–634. https://doi.org/10.1089/neu.2019.6413

Smith, D., & Macdonald, J. (2015). Design as a rehabilitative tool for people with acquired brain injury: Mapping the field. *Journal of Interior Design, 40*(3), 21–38. https://doi.org/10.1111/joid.12048

Smith, E., Sumner, P., Hedge, C., & Powell, G. (2023). Smart-speaker technology and intellectual disabilities: Agency and wellbeing. *Disability and Rehabilitation: Assistive Technology, 18*(4), 432–442. https://doi.org/10.1080/17483107.2020.1864670

Song, C., Ikei, H., Nara, M., Takayama, D., & Miyazaki, Y. (2018). Physiological effects of viewing bonsai in elderly patients undergoing rehabilitation. *International Journal of Environmental Research and Public Health, 15*(12). https://doi.org/10.3390/ijerph15122635

Specter, M. (2001). Rethinking the brain. *The New Yorker*, July 23. Issue:42–53.

Stafford, L. (2022). Planners, we need to talk about ableism. *Planning Theory and Practice, 23*(1), 101–142. https://doi.org/10.1080/14649357.2022.2035545

Steinbrink, J., Villringer, A., Kempf, F., Haux, D., Boden, S., & Obrig, H. (2006). Illuminating the BOLD signal: Combined fMRI-fNIRS studies. *Magnetic Resonance Imaging, 24*(4), 495–505. https://doi.org/10.1016/j.mri.2005.12.034

Stilling, J. M., Duszynski, C. C., Oni, I., Paxman, E., Dunn, J. F., & Debert, C. T. (2019). Using functional near-infrared spectroscopy to study the effect of repetitive transcranial magnetic stimulation in concussion: A two-patient case study. *Frontiers in Neurology, 10*, 476. https://doi.org/10.3389/fneur.2019.00476

Stone, D. B., Tamburro, G., Filho, E., di Fronso, S., Robazza, C., Bertollo, M., & Comani, S. (2019). Hyperscanning of interactive juggling: Expertise influence on source level functional connectivity. *Frontiers in Human Neuroscience, 13*, 321. https://doi.org/10.3389/fnhum.2019.00321

Strangman, G., Boas, D., & Sutton, J. (2002). Non-invasive neuroimaging using near-infrared light.

Sudimac, S., & Kühn, S. (2022). A one-hour walk in nature reduces amygdala activity in women, but not in men. *Frontiers in Psychology, 13*, 931905. https://doi.org/10.3389/fpsyg.2022.931905

Sudimac, S., & Kühn, S. (2023) Can a nature walk change your brain? Investigating Hippocampal brain plasticity after one hour in a forest. *Investigating hippocampal brain plasticity after one hour in a forest.*

Sugiyama, T., Leslie, E., Giles-Corti, B., & Owen, N. (2008). Associations of neighbourhood greenness with physical and mental health: Do walking, social coherence and local social interaction explain the relationships? *Journal of Epidemiology & Community Health, 62*(5). https://doi.org/10.1136/jech.2007.064287

Sundaram, S. A., Chung, C.-S., Gebrosky, B., Brown, J., Grindle, G. G., Deepak, N., Cooper, R., & Cooper, R. A. (2023). Participatory action design and engineering of a manual wheelchair virtual coach including in-home and community usage. *The Journal of Spinal Cord Medicine, 46*(4), 546–559. https://doi.org/10.1080/10790268.2022.2107352

Tachibana, A., Noah, J. A., Bronner, S., Ono, Y., & Onozuka, M. (2011). Parietal and temporal activity during a multimodal dance video game: An fNIRS study. *Neuroscience Letters, 503*(2), 125–130. https://doi.org/10.1016/j.neulet.2011.08.023

Takano, T., Nakamura, K., & Watanabe, M. (2002). Urban residential environments and senior citizens' longevity in megacity areas the imprtance of walkable green spaces. *Journal of Epidemiology and Community Health, 56*(12), 913–918. http://www.jstor.org/stable/25569881

Tang, I. C., Tsai, Y.-P., Lin, Y.-J., Chen, J.-H., Hsieh, C.-H., Hung, S.-H., et al. (2017). Using functional magnetic resonance imaging (fMRI) to analyze brain region activity when viewing landscapes. *Landscape and Urban Planning, 162*, 137–144. https://doi.org/10.1016/j.landurbplan.2017.02.007

Tarr, B., Launay, J., Cohen, E., & Dunbar, R. (2015). Synchrony and exertion during dance independently raise pain threshold and encourage social bonding. *Biology Letters, 11*(10). https://doi.org/10.1098/rsbl.2015.0767

Temkin, N. R., Corrigan, J. D., Dikmen, S. S., & Machamer, J. (2009). Social functioning after traumatic brain injury. *The Journal of Head Trauma Rehabilitation, 24*(6), 460–467. https://doi.org/10.1097/HTR.0b013e3181c13413

Teo, W.-P., Muthalib, M., Yamin, S., Hendy, A. M., Bramstedt, K., Kotsopoulos, E., Perrey, S., & Ayaz, H. (2016). Does a combination of virtual reality, neuromodulation and neuroimaging provide a comprehensive platform for neurorehabilitation?—A narrative review of the literature. *Frontiers in Human Neuroscience, 10*. https://doi.org/10.3389/fnhum.2016.00284

Theodorou, P., Tsiligkos, K., & Meliones, A. (2023). Multi-sensor data fusion solutions for blind and visually impaired: Research and commercial navigation applications for indoor and outdoor spaces. *Sensors (Basel, Switzerland), 23*(12). https://doi.org/10.3390/s23125411

Thornton, M., Marshall, S., McComas, J., Finestone, H., McCormick, A., & Sveistrup, H. (2005). Benefits of activity and virtual reality based balance exercise programmes for adults with traumatic brain injury: Perceptions of participants and their caregivers. *Brain Injury, 19*(12), 989–1000.

Threapleton, K., Drummond, A. E. R., & Standen, P. (2016). Virtual rehabilitation: What are the practical barriers for home-based research? https://doi.org/10.1177/2055207616641302

Tieri, G., Morone, G., Paolucci, S., & Iosa, M. (2018). Virtual reality in cognitive and motor rehabilitation: Facts, fiction and fallacies. *Expert Review of Medical Devices, 15*(2), 107–117. https://doi.org/10.1080/17434440.2018.1425613

Till, C., Colella, B., Verwegen, J., & Green, R. E. (2008). Postrecovery cognitive decline in adults with traumatic brain injury. *Archives of Physical Medicine and Rehabilitation, 89*(12), 34. https://doi.org/10.1016/j.apmr.2008.07.004

Tilt, J. H. (2010). Walking trips to parks: Exploring demographic, environmental factors, and preferences for adults with children in the household. *Preventive Medicine, 50*(Suppl 1), S69–S73. https://doi.org/10.1016/j.ypmed.2009.07.026

Tronstad, O., Flaws, D., Fraser, J. F., & Patterson, S. (2021). Doing time in an Australian ICU; the experience and environment from the perspective of patients and family members. *Australian Critical Care, 34*(3), 254–262. https://doi.org/10.1016/j.aucc.2020.06.006

Tucek, G., Maidhof, C., Vogl, J., Heine, A., Zeppelzauer, M., Steinhoff, N., & Fachner, J. (2022). EEG Hyperscanning and qualitative analysis of moments of interest in music therapy for stroke rehabilitation-a feasibility study. *Brain Sciences, 12*(5). https://doi.org/10.3390/brainsci12050565

Ulrich, R. S. (1983). Aesthetic and affective response to natural environment. In I. Altman & J. F. Wohlwill (Eds.), *Behaviour and the natural environment* (Vol. 6, pp. 85–125). Springer. https://doi.org/10.1007/978-1-4613-3539-9_4

Ulrich, R. S., Simons, R. F., Losito, B. D., Fiorito, E., Miles, M. A., & Zelson, M. (1991). Stress recovery during exposure to natural and urban environments. *Journal of Environmental Psychology, 11*(3), 201–230.

United Nations Human Settlements Programme (UN-Habitat). (2022). *World Cities Report 2022: Envisaging the Future of Cities*. https://unhabitat.org/sites/default/files/2022/06/wcr_2022.pdf

Urushidani, N., Kinoshita, S., Okamoto, T., Tamashiro, H., & Abo, M. (2018). Low-frequency rTMS and intensive occupational therapy improve upper limb motor function and cortical reorganization assessed by functional near-infrared spectroscopy in a subacute stroke patient. *Case Reports in Neurology, 10*(2), 223–231. https://doi.org/10.1159/000492381

Valentin, A., Ferdinande, P., & ESICM Working Group on Quality Improvement. (2011). Recommendations on basic requirements for intensive care units: Structural and organizational aspects. *Intensive Care Medicine, 37*(10), 1575–1587. https://doi.org/10.1007/s00134-011-2300-7

Van Praag, H., Kempermann, G., & Gage, F. H. (2000). Neural consequences of enviromental enrichment. *Nature Reviews Neuroscience, 1*(3), 191–198. https://doi.org/10.1038/35044558

Verderber, S., & Reuman, D. (1987). Windows, views, and health status in hospital therapeutic environments. *Journal of Architectural and Planning Research*, 120–133.

Verdonck, M., Wiles, L., & Broome, K. (2023). Lived experience of using assistive technology for sandy beach based leisure for australian people with mobility limitations. Disability and Rehabilitation. Assistive Technology, 1-11, 1–11. https://doi.org/10.1080/17483107.2023.2217859.

Vitorio, R., Stuart, S., Rochester, L., Alcock, L., & Pantall, A. (2017). fNIRS response during walking - Artefact or cortical activity? A systematic review. *Neuroscience and Biobehavioral Reviews, 83*, 160–172. https://doi.org/10.1016/j.neubiorev.2017.10.002

Wang, P., King, C., Chui, L., Do, A., & Nenadic, Z. (2013). Operation of a brain-computer interface walking simulator for individuals with spinal cord injury. *Journal of Neuroengineering and Rehabilitation, 10*(1), 1–14. https://doi.org/10.1186/1743-0003-10-77

Wang, R. H., Tannou, T., Bier, N., Couture, M., & Aubry, R. (2023). Proactive and ongoing analysis and management of ethical concerns in the development, evaluation, and implementation of smart homes for older adults with frailty. *JMIR Aging, 6*, 41322. https://doi.org/10.2196/41322

Weddell, R., Oddy, M., & Jenkins, D. (1980). Social adjustment after rehabilitation: A two year follow-up of patients with severe head injury. *Psychological Medicine, 10*(2), 257–263.

Weissman, D. G., Lambert, H. K., Rodman, A. M., Peverill, M., Sheridan, M. A., & McLaughlin, K. A. (2020). Reduced hippocampal and amygdala volume as a mechanism underlying stress sensitization to depression following childhood trauma. *Depression and Anxiety, 37*(9), 916–925. https://doi.org/10.1002/da.23062

Wiercioch-Kuzianik, K., Brączyk, J., Bieniek, H., & Bąbel, P. (2023). Red induces hyperalgesia and white induces hypoalgesia regardless of pain modality. *Scientific Reports, 13*(1), 6360. https://doi.org/10.1038/s41598-023-33313-w

Wistoft, K. (2013). The desire to learn as a kind of love: Gardening, cooking, and passion in outdoor education. *Journal of Adventure Education and Outdoor Learning, 13*(2), 125–141.

Woo, C. C., Miranda, B., Sathishkumar, M., Dehkordi-Vakil, F., Yassa, M. A., & Leon, M. (2023). Overnight olfactory enrichment using an odorant diffuser improves memory and modifies the uncinate fasciculus in older adults. *Frontiers in Neuroscience, 17*. https://doi.org/10.3389/fnins.2023.1200448

World Health Organization.(2021). *Rehabilitation*. World Health Organization. https://www.who.int/news-room/fact-sheets/detail/rehabilitation

World Health Organization (WHO). (2030). *Rehabilitation*. https://www.who.int/initiatives/rehabilitation-2030

Yang, H. F., Ling, Y., Kopca, C., Ricord, S., & Wang, Y. (2022). Cooperative traffic signal assistance system for non-motorized users and disabilities empowered by computer vision and edge artificial intelligence. *Transportation Research Part C: Emerging Technologies, 145*, 103896.

Yang, C. C., Tsujimura, S. I., & Yeh, S. L. (2023). Blue-light background impairs visual exogenous attention shift. *Scientific Reports, 13*(1), 3794. https://doi.org/10.1038/s41598-022-24862-7

Ye, T., Yu, P., Wen, B., Yang, Z., Huang, W., Guo, Y., Abramson, M. J., & Li, S. (2022). Greenspace and health outcomes in children and adolescents: A systematic review. *Environmental Pollution*, 120193. https://doi.org/10.1016/j.envpol.2022.120193

Yelden, K., Duport, S., Kempny, A., & Playford, E. D. (2015). A rehabilitation unit at night: Environmental characteristics of patient rooms. *Disability and Rehabilitation, 37*(1), 91–96. https://doi.org/10.3109/09638288.2014.906662

Youn, C. H., Chung, L. B., Kang, M., Kim, S. J., Choi, H. B., & Lee, J. (2022). Effects of green walls on prefrontal cerebral hemodynamics in hospital workers. *Journal of People, Plants, and Environment, 25*(6), 717–728. https://doi.org/10.11628/ksppe.2022.25.6.717

Yu, Q., Daugherty, A. M., Anderson, D. M., Nishimura, M., Brush, D., Hardwick, A., et al. (2018). Socioeconomic status and hippocampal volume in children and young adults. *Developmental Science, 21*(3), e12561. https://doi.org/10.1111/desc.12561

Zamani, Z. (2016). 'The woods is a more free space for children to be creative; their imagination kind of sparks out there': Exploring young children's cognitive play opportunities in natural, manufactured and mixed outdoor preschool zones. *Journal of Adventure Education and Outdoor Learning, 16*(2), 172–189.

Zhang, L., Chi, Y. M., Edelstein, E., Schulze, J., Gramann, K., Cauwenberghs, G., MacAgno, E., & Velasquez, A. (2010). Wireless physiological monitoring and ocular tracking: 3d calibration in a fully-immersive virtual health care environment. In *2010 Annual International Conference of the Ieee Engineering in Medicine and Biology Society, Embc'10, 4464-4467* (pp. 4464–4467). https://doi.org/10.1109/IEMBS.2010.5625969

Zhong, W., Schröder, T., & Bekkering, J. (2022). Biophilic design in architecture and its contributions to health, Well-being, and sustainability: A critical review. *Frontiers of Architectural Research, 11*(1), 114–141. https://doi.org/10.1016/j.foar.2021.07.006

Experiential Equity: Reconstructing Cities Through Intersecting Theories

Robin Mazumder

Abstract Considering the unprecedented rate of global urbanisation, attention must be paid to the implications of city environments on human health. Urban infrastructures, like parks, greenspace, active transportation, and other forms of public space, are seen as health amenities. The concept of distributional equity acknowledges the disparities in access to these health-supporting spaces; however, more granularity is needed in understanding how these spaces themselves are experienced, and the impact of marginalisation and other forms of oppression on the health benefits must be examined. Experiential equity is a framework that can be used to conceptualise and quantify the energetic cost of experiencing unpredictable urban environments, using the free energy principle to unite the energetic costs of the emotional and cognitive burden of marginalisation, and views these costs through an Intersectional Theoretical lens to establish the energetic disparities that exist depending on aspects of identity. The cognitive and emotional costs of uncertainty in urban environments can be understood through the Theory of Constructed Emotion, the model is also informed by environmental prediction and provides a link to the psychophysiology of Racism-Related Vigilance and Stereotype Threat. Applying the Person-Environment-Occupation Model and Affordance Theory, in conjunction, we can conceptualise the environment and examine the ways in which it can constrain or enable engagement with it. Leveraging machine learning to quantify aspects of the built environment and using neuroscientific methodologies like electroencephalography, electrodermal activity, and heart rate variability, we can attempt to metricise the energetic costs of societal and systemic oppression; the consideration of the environment allows potential insights into how spatial variables mediate these experiences and can have implications on road, park, and general urban policy and design.

R. Mazumder (✉)
Future Cities Institute, School of Planning, University of Waterloo, Waterloo, ON, Canada

Urban Realities Lab, Department of Psychology, University of Waterloo, Waterloo, Ontario, Canada
e-mail: robin.mazumder@uwaterloo.ca

S. Kühn (ed.), *Environmental Neuroscience*,
https://doi.org/10.1007/978-3-031-64699-7_16

Keywords Free energy principle · Intersectionality · Urban design · Neurourbanism

Introduction: Preface, Positionality, and the Opportunities in Scientific Blind Spots

I believe that all academic research should open with positionality statements, as our perspectives on the reality we all study undeniably hinges on the intersections of our identities, and our lived experiences. Bourke (2014) states, "positionality represents a space in which objectivism and subjectivism meet...to achieve a pure objectivism is a naïve quest, and we can never truly divorce ourselves from subjectivity" (p. 3). Accordingly, I believe that to deny subjectivity and the power of experience on perspective formation would enable an incomplete science, particularly as it pertains to the focus of my research on cities: where places and people, and the objective, and subjective meet. At the very least, positionality offers insight into where a scientist draws inspiration, or why they research questions they do and draw the corresponding conclusions. Beames et al. (2021) state, "lived experience perspectives can be used to collaboratively identify gaps or problem areas, formulate research questions, interpret empirical findings, comment on empirical gaps, evaluate what implications mean for them and whether they are practically relevant, and help to form a judgment about whether future research on a topic is worthwhile" (p. 2). Positionality and its use originated in qualitative inquiry, but I believe is also pertinent to quantitative research. Environmental Neuroscience, in the granularity and breadth demonstrated in this textbook, includes both qualitative and quantitative methodologies, or a mix. As per my positionality, I identify as a racialised straight cis-gendered male of South Asian Descent. I also was born and raised in an upper middle-class North American context by parents who were at one time poor immigrants—this I believe is relevant to the ideas I present below on cities, and why and how Environmental Neuroscience can inform their design. The intersectional lens of positionality, the consideration of overlapping identities, is also directly related to Intersectionality Theory (Crenshaw, 1989); the theory is both a foundation and framework for the concept of experiential equity, the focus of this chapter. Intersections define our cities in many ways. Racism directly impacts class, race restricting social mobility, enabling social stratification and spectrum of access to healthcare and education necessary for generational health, an aspirational metric of societal well-being, meaning that health equity includes all generations from toddlers to seniors to have the resources to support robust physical and mental health. An example of direct experience, women experience more fear in public as compared to men, and the eye-tracking paths of their environment also massively differ, scanning the periphery to monitor for threats. These are experiences I can never truly understand, and so I establish those limitations from the get go, I do share the same spaces, however. I must also highlight my privilege as a non-Black person of colour; while my skin is dark, I am not Black, and with that comes privileges in how I am treated by society, healthcare, and the judicial system, including the police. Recently, Okeke et al. (2023), through the use of diffusion tensor imaging (DTI) and the measure of fractional anisotropy on Black women who've experienced racism, found evidence of racial discrimination being associated with neural tracts, particularly the corpus callosum, which plays a crucial role in emotional and cognitive dysfunction, and is suggested to have downstream effects on both physical and mental health. I also am a cis-gendered man, so my perspectives on safety in the urban environment are limited to my capacity to empathise with a female-identifying person. Transgender people are four times more likely to experience crime victimisation as compared to a cis-gender person (Flores et al., 2021). This is not to say that I do not experience fear in public. However, I acknowledge I am safer, statistically, if a police officer answers a mental health distress call for me, as opposed to a Black man. In the Canadian context, the disproportionate deaths resulting from police wellness checks have prompted inquiry into the racialisation of treatment (Cooke, 2020). In the realm of cities, the ability is also an important factor, particularly as it pertains to the affordances allowed in the urban environment, an inherently environmental neuroscientific concern. In terms of my lived experience, I live with a compounded disability. I was born with hereditary neuropathy with a liability for pressure palsies (HNPP), a rare degenerative neurological disorder that affects my peripheral nervous system. A heritable mutation in my Peripheral Myelin Protein 22 gene can, unpredictably, yield parts of my body, like my ankles, legs, arms, and fingers, often numb for months, sometimes paralysed permanently, making the city disabling in different intensities based on my fluctuating health status. I often learn of the disability experientially as I discover I cannot unbutton a shirt or whether I have the ankle sensation and strength to walk on an icy sidewalk safely. I also live with bipolar disorder, a mental illness considered treatable but incurable. Both have posed their own unique challenges, inherently influencing how I see the world. Particularly cities. Accessibility issues plague cities internationally, leaving people who use assistive devices disabled by their environments, causing not just momentary stress but also long-term exclusion. Bipolar Disorder is also associated with sensory processing differences, including hypersensitivity (Parker et al., 2018), therefore the stimuli of the city may be more agitating to my brain and body as compared to a "typical" person; my understanding of the limits of sensation and emotion should be understood through a neurodivergent lens that acknowledges the influence of experience on describing and measuring the shared phenomena of feelings. Furthermore, research shows that Bipolar Disorder, which is 80% genetic and 20% environmental, is linked to urbanicity. In their paper, "Bipolar disorder: An evolutionary psychoneuroimmunological approach", evolutionary biologist Markus Rantala et al. (2021) discuss bipolar disorder as a stress-related illness and a modern phenomena, at its core neuroinflammation, which can they indicate can be caused by "many social stressors unique to contemporary urban environments" (p. 33) (--> Chapter Mehran & Adli). Evidently, as per my positionality on the topic of how cities disproportionately contribute to stress and mental illness development, my lived experience and subjective emotional relationship to "the city" and its concept, deeply informs my research and theoretical work on the subject matter. Experiential equity considers the body as an experimental condition for experience itself, in that a core question is on how conditioning, including micro and macro aggressions—social aspects of the environment—influences the experience of the physical contexts of public space. In conducting research and theoretical development that considers this subjectivity, I believe we can tell a more complete and collective story about the diversity of experience. Experiential equity is therefore inherently positional, in that a spatial concept of the experience of difference must acknowledge distance itself, in both its physical and sociological manifestations.

Cities Through the Lens of Environmental Neuroscience

Environmental Neuroscience has been described to be a discipline that considers the external environments in which humans live, the social and physical, in understanding human behaviour, to the extent of understanding the underlying neural correlates; this neurological understanding of the impact of environmental contexts stands to not only expand the field of neuroscience but also can provide practical insights with powerful implications on how our society is structured and governed, at the core of which is the modern city, a confluence of physical structures and social policies. Cities are necessary to study, particularly from the perspective of public health, considering the United Nations (2018) is projecting 68% of the world's population to live in urban centres by 2050—less than three decades away. Cities by no means are simple, however, and should be treated, and studied, like the complex systems that they are.

Therefore, Environmental Neuroscience is well positioned to study these pressing issues and complex intersections of body, space, and place. Within the urban context, public space is an essential element. It is where we come together, in the best-case scenario to connect and celebrate diversity. However, the reality is that cites offer differing (levels of) affordances depending on where people reside, as it pertains to access to good quality public infrastructure, like well-being supporting greenspace or separated cycling infrastructure. This variability of access to health-supporting infrastructure is referred to as an issue of the concept of equity, which will be defined and expanded below, following which I will address its experiential component.

Part 1: Understanding Equity

The World Health Organization (2024) defines equity as:

> the absence of unfair, avoidable, or remediable differences among groups of people, whether those groups are defined socially, economically, demographically, or geographically

or by other dimensions of inequality (e.g. sex, gender, ethnicity, disability, or sexual orientation).

Two things are apparent in this definition of equity (1) equity is a complex construct and therefore achieving it and measuring it is also complex and (2) there is an emphasis on groups marked by differences and aspects of identity, bolstering the necessity for an intersectional approach to assessing and measuring equity.

Social Equity: A Flexible Metric of Fairness

In their paper on social equity, Guy and McCandless (2012) make a point of distinction between equity and equality stating "while equality can be converted into a mathematical measure in which equal parts are identical in size or number, equity is a more flexible measure allowing for equivalency while not demanding exact sameness". This flexible measure that does not demand exact sameness offers neuroscience a valuable lens, and a point of reflection on reductionistic and essentialist thinking, and its possible consequence of erasure. Guy and McCandless go on to state, "social equity draws attention to the human factor in terms of economic fairness and advantage". The human factor mentioned is the foundation of the neuroscientific research methodology that can provide insights into physiology, psychology, and ultimately, health.

Health Equity

Intrinsically linked to social equity is health equity; in other words, health outcomes are demonstrably linked to social forces. The WHO (2024) states "health equity is achieved when everyone can attain their full potential for health and well-being". The definition is as much a description as it is a call to action—aspirational if not ideal, but such is the work of an organisation tasked with achieving global health. Braveman and Gruskin (2003) suggest that the concept of equity is an issue of ethics and closely related to human rights principles. The argument can be made that equity, particularly as it pertains to health, is a collective issue requiring a systems and complexity science approach. Accordingly, a society is not truly healthy until everyone has equal access to healthcare resources, as per the WHO. But also, I would contend, equal experiential access to and of free and safe experiences of health-supporting urban infrastructure such as greenspace and active transportation infrastructure. Under these assumptions, urban health is inherently a spatial matter, which alongside social structures, can be examined in the combined effects on cognition, emotion, and behaviour, and possibly, brain health.

Spatial Equity: Where Theory Hits the Road

Cities are products of social policy; bylaws dictate densification standards, building height, road width, and public art investment. These are just a few ways in which blueprint-determining political documents take on physical manifestation. Österle (2002) states, "equity is widely accepted as an objective in social policy. Explicitly or implicitly social policy definitions at their core often include notions such as equity, justice, equality or fairness". Therefore, a principal motivator for urban planning in modern democracy should be the concept of equity. However, as it stands, it would appear that equity is an ethic and aspiration that requires operationalisation. This aspiration of practicality is a key motivator for this work; I see an ethical and human rights focused Environmental Neuroscience as an important tool with the potential to help level the playing field that is the modern city, as a spatially focused social science.

Examining urban greenspace equity through a spatial lens, Nesbitt et al. (2019) propose distributional equity, which refers to equal spatial distribution of and access to urban greenspace, as well as the complementary concept of recognitional equity, which entails involving marginalised groups in the design of the infrastructure they are being given increased access to.

Experiential Equity

Using the lens of Environmental Neuroscience, and building on the concepts of distributional and recognitional equity, I propose the layer of experiential equity (Mazumder, 2021), which contends that the experience of equitably accessible amenities differs psychologically and physiologically depending on the person experiencing it, specifically depending on their intersectional identity. Legal scholar Kimberle Crenshaw (1989) originally proposed the idea of intersectionality to capture the disparities in pay for Black women, highlighting the intersection of gender and race, and it has grown to capture how oppression and violence to the body can be magnified through the other intersecting aspects of our identity, including sexual orientation, socioeconomic status, ability, and gender orientation (not exhaustive). Cho et al. (2013, p. 795), state:

> what makes an analysis intersectional—whatever terms it deploys, whatever its iteration, whatever its field or discipline—is its adoption of an intersectional way of thinking about the problem of sameness and difference and its relation to power. This framing—conceiving of categories not as distinct but as always permeated by other categories, fluid and changing, always in the process of creating and being created by dynamics of power—emphasises what intersectionality does rather than what intersectionality is.

Intersectional Environmental Neuroscience

Exploring the application of an intersectional lens to quantitative (Bauer et al., 2021), state, "Measurements and quantitative analyses consistent with the central theoretical tenets of intersectionality can contribute to the analysis of health problems in micro and macro socio-structural ways using large-sample data, with the potential to impact the development and implementation of public policy, and ultimately health equity" (p. 9). In their review of work exploring intersectionality more specifically in the context of neuroscience, Duschene and Trujillo (2021) identify three theme areas: (1) how social structures create health inequality in individuals with intersecting social identities, (2) how individuals process intersecting identities, and (3) how epistemology can benefit from intersectionality. Here we can see an opportunity to build on the work on the health impacts of social structures with the addition of a physical lens that considers how the two intersect. Fundamentally, however, these perceptions and experiences of social and physical structures occur through and within the nervous system, for which we have an array of methodologies for observing and assessing. First, we need to establish how the brain is understood in its purpose and function to establish an argument for its response to the layered urban environment.

Part 2: Framing Experience

Free-Energy as a Unifying Lens for Urban Neural Processing Cost

Describing his impactful free energy principle (FEP), neuroscientist Karl Friston proposes a framework, lens, and model of the brain with the potential to unify accounts of its mechanisms and processes, establishing common ground for conceptualisation of brain and behaviour. Friston (2010) states "the free-energy principle says that any self-organising system that is at equilibrium with its environment must minimise its free energy" (p. 127). Here we are presented with a clear and concise model of both environmental engagement and behavioural motivation. The FEP is also of further value to environmental neuroscientists in that its minimalist explanatory model (Fig. 1) involves the environment and sensory processing, and behavioural responses to the sensations, as fundamental aspects to an organism achieving a state of homeostasis through resistance to what Friston refers to as "disorder", making order an inherently environmental factor, with prediction playing a central role in its perception.

Friston goes on to identify surprise as the core variable in the consumption of free energy. Essentially, the capacity of an organism to predict their environment and not be surprised by their experiences and their causes provides it an advantage in that it conserves more energy. This necessity for prediction in the urban

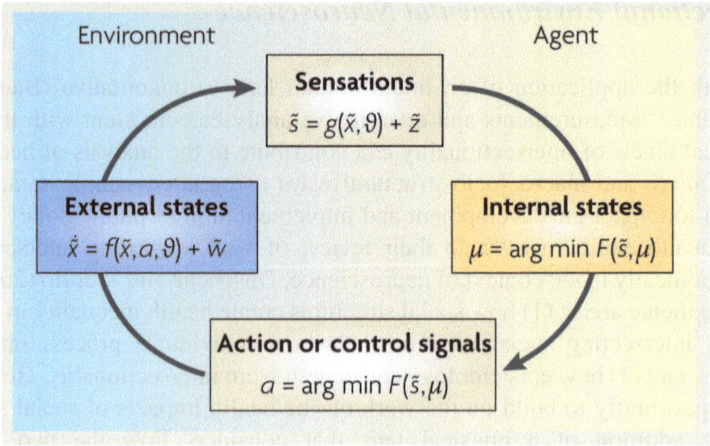

Fig. 1 Model of free energy principle. (Taken from Friston, 2010)

environment is involved in the application of the laws of physics to anticipate the trajectory of moving vehicles to ensure bodily safety, or predicting the movement of bodies on a sidewalk, similarly having implications on bodily safety. In recent years, there has been public discussion on the phenomena and neologism of "manslamming", which describes the experience women have navigating public space and the bodies of men who do not give way. In a Refinery 29 article (Truong, 2017), women who experimented with standing their ground on sidewalks and other public spaces reported not just collisions but also verbal and non-verbal directives to give way to men. The anecdotal take-away, in relation to free energy, is that some women find they have to invest energy into predicting where men are walking and also deviate their paths should they not want to be walked into by men. Is there a disparity in energy consumption and what are the implications on health equity? The anecdote serves as inspiration for other research questions that can be asked to help unpack the smaller yet impactful components, and moments, of urban stress. Grohmann in discussing manslamming (2018) states:

> men determine women's position in, and trajectory through, space in myriads of different ways, and their culturally scripted way of spatial relating is designed to not only establish, but reinforce dominance hierarchies…gender as a social hierarchy, is also and especially spatially produced.

She contends that we create experiences of gender through spatial structures that are meant to establish social hierarchies through forces of dominance. Interestingly, of the few behavioural studies related to manslamming, there is some psychological research on "pedestrian displacement". Rawdon and Willis (1993), in their observational study of London UK found that regardless of gender, "lone white pedestrians tended to displace lone black pedestrians (p. 636)".

They contrasted their findings as opposing those from a previous study done in the United States, where white people moved out of the way of Black people, but also stood further away from them when speaking to them, perhaps allowing the inference that cultural context can be a determinant of proxemics, asserted by Edward T. Hall who coined the term, which refers to the study of how people occupy space in relationship to other bodies and objects. Homeostasis, then, in the urban context, varies on the intersectional identity of the individual who navigates its social and spatial structures. This paper does not aim to propose mathematical or formulaic models, but rather uses the FEP as a theoretical foundation that emphasises the centrality of the urban experience as one that inherently requires prediction, and the emotional and cognitive resources that consume the brain's energy in responding to uncertainty, forced to calculate probability densities for a situation founded on flawed thinking, like the concept of racism. Urban living can involve prediction's role in survival in the realm of bodily harm and avoiding traffic violence, but also, importantly, anticipating police surveillance, and the possibility of death.

Planning for Complexity: Uncertainty and Stress in the City

In their article, Complexity and Uncertainty: Implications for Urban Planning, Moroni and Chiffi (2021) state, "uncertainty is the core feature of complexity". They go on to describe two case studies of where uncertainty plays a role in urban planning decision makers: (a) how decision makers themselves respond to uncertainty themselves in relation to infrastructural decisions, i.e. where to build what the authors refer to as an artefact (i.e. bridge, metro station) and (b) how decision makers anticipate, design for, and ultimately regulate the behaviour of those who interact with artefacts. This complexity of designed and regulated engagement also requires an appreciation for the complexity and variability of experience. Using the array of environmental neuroscientific (Karakas & Yildiz, 2020), we can attempt to quantify these complex experiences of the urban built environment, viewing psychophysiology findings through an intersectional approach to inform health equity practices and policies. Specific methodological examples will be discussed in the section, *Applying Experiential Equity*.

In their paper on uncertainty and disease development (2017), Peters, McEwen and Friston argue that the essence of stress is uncertainty, stating that within the paradigm of the FEP, uncertainty is a measure of entropy, or "expected surprise". This extends contemplation on the psychology of the expectation of threat within the urban context, the associated physiological energetic costs, and the implications on health equity. It could be argued that, ultimately, these biological disparities are of pertinence to survival and genetic fitness, of individuals and groups, positioning cities as passive tools of eugenics. They also indicate that reducing uncertainty requires energy consumption, and therefore from a Darwinian perspective lies advantage (Barrett, 2013), or by another word, privilege. They go on to state, "if the brain cannot reduce uncertainty, a persistent cerebral energy crisis may develop, burdening the individual by 'allostatic load' that contributes to systemic and brain malfunction (Peters et al., 2017, p. 164)". To et al. (2021) identify uncertainty and ambiguity at the core of the emotional experience of racism, and I would argue would apply it to other forms of oppressive treatment, rendering experiences of marginalisation measurable neurological consequences of computing and emotionally processing the improbability of threat.

Where Predictions Become Perceptions: Urban Emotion Embodied as Lived Experience

In reference to the consequences of an allostatic overload, in her paper on the Theory of Constructed Emotion (TOC), Lisa Feldman Barrett (2017) states that the brain's core purpose is to maintain allostasis (homeostasis) meaning ensuring appropriate resources for an organism to grow, survive, and reproduce. Maintaining an optimal state involves predicting and simulating environments based on prior experience, comparing the actual momentary experience to those simulations, and then updating an internal model as necessary. Similar to FEP, this environmental schema updating process is postulated to be motivated by the energetic efficiencies of minimising prediction error. The embodied experience of emotion including negative tone is a result of the interoceptive sensations produced in response to prediction error, and as prediction error decreases, the closer the experiencer approaches a direct experience of reality, or the present moment—to use Barrett's words, "a prediction becomes a perception or an experience" (p. 12). The TOC provides a compatible framework with Friston's, in that both rely on the concepts of prediction and uncertainty for theoretical explanatory power, and also neural resource consumption. Urban stress, then, can be viewed as a process that requires continuous inference formation, both consciously and unconsciously, about the threat(s) present in our environment. Barrett further proposes the concept of an affective niche, stating "our affective niche is shaped by the state of our bodies and our expectations about our bodies. These expectations are informed by our past experiences and our culture". The definition offers attention to the physical environmental aspect of the term while also acknowledging the impact of lived experience and culture, considering both in understanding emotion, and consequently stress.

Double Consciousness, Cognitive Load, and the Free Energy Cost of Being Othered

Using Friston's FEP and Feldman-Barrett's TOC, we can contextualise the concept of double consciousness, or twoness, proposed by Black Sociologist W.E.B Du Bois in 1899, who described the conflictual experience as a:

> sense of always looking at one's self through the eyes of others, of measuring one's soul by the tape of a world that looks on in amused contempt and pity. One ever feels his two-ness – an American, a Negro – two souls, two thoughts, two unreconciled strivings; two warring ideals in one dark body, whose dogged strength alone keeps it from being torn asunder. (Du Bois, 1903, p. 47)

Du Bois' sense-making of double consciousness refers to the cognitive energy required to monitor not only one's own behaviour but also the energy required in the prediction of the behaviour of others who stand to pose a threat, as per the societal norms of his time, which still persist today as demonstrated in societal and systemic racism. This monitoring of the environment, or predicting its disorder, is not relegated to race as a sole factor, but also gender, ability, and other intersections of identity.

This cognitive burden of racism, Double Consciousness, can be understood neurologically using Schmader et al.'s (2008) Integrated Model of Stereotype Threat. They propose stereotype pressures have physiological stress consequences that affect working memory, which they define as "a limited-capacity executive process that coordinates cognition and controls behaviour to achieve performance goals in the presence of exogenous or endogenous information that competes for attention" (Schmader et al., 2008).

Du Bois's Theory of Double Consciousness would suggest racism, and its consequences on self-concept, constitutes the exogenous information (experience of racism) and endogenous information (anticipation of racism) which is competing for cognitive resources, and consuming energy, as per the free energy principle. Furthermore, there is a growing body of literature examining the physiological consequences of experiencing and anticipating oppression. Clark et al. (1999) proposed a Biopsychosocial Model of Racism, identifying the experience of racism as a cause of health disparities. This model builds on the Transactional Model of Coping and Stress originally proposed by Lazarus and Folkman (1984), who state that "the perception of an environmental stimulus as racist results in exaggerated psychological and physiological stress responses that are influenced by constitutional factors, sociodemographic factors, psychological and behavioural factors, and coping responses" (Lazarus & Folkman, 1984. p. 806).

A consequence of this sustained and chronic racial stressor is the prolonged activation of the sympathetic nervous system (Clark et al.,1999). Hicken et al. (2013) found that Black people reported higher levels of vigilance while performing daily activities compared to White and Hispanic people, as well as sleep disruption. Sewell and Jefferson (2016) further extend the impacts of institutional racism on well-being, suggesting that police surveillance is a considerable contributor to racism-related vigilance. This is an incredibly important nuance that must be considered by environmental neuroscientists, particularly those who study urban environments, where racism is most often perpetrated. Applying another lens of the intersectional experience of marginalisation, Meyer (2003) proposes the concept of minority stress, which particularly acknowledges the stress of homophobia.

Hostile Environments and Ambient Stressors

These persistent forms of stress can also be layered upon other urban stressors, like noise pollution and vehicle threat. Campbell (1983) proposes that ambient stressors have 5 main characteristics: they are perceptible, chronic, non-urgent, intractable, and have a negative tone. The word "ambient" is used to highlight the fact that these stressors are often in the background and not at the forefront of our awareness. Ambient stressors can, over the long term, have negative impacts on well-being (Campbell,1981). In discussing perceptibility, Campbell suggests that ambient stressors are typically perceivable, but may not be noticed, but that noticing them may have no bearing on their deleterious effects. Chronicity refers to the extent that the stressor is present; as cited earlier, chronicity of racism has been established as a contributing factor to stress-related diseases. Non-urgency refers to the immediate threat posed by a stressor; the problem with experiences of non-verbal, verbal, physical, and institutional expressions of oppressive violence is the intangibility of threat, as well as its conditioning. The experience is unpredictable for an individual, yet shown to be predicted by geographical location; there are areas of cities where there is more documented racism than others, as demonstrated in the City of Edmonton's Lighthouse Project (City of Edmonton, 2024), which uses cell phone camera technology, artificial intelligence, and geospatial technology to develop insights into where hate symbols and hate crimes occur in cities, to help inform intervention at a community level. Intractability, which is closely related to non-urgency, entails that the individual cannot moderate the experience of the ambient stressor by cognitive reframing—racism, and other forms of oppression, cannot be easily cognitively reframed, and is highly dependent on the individual, their experiences, and their access to resources, such as therapy, that can assist with in the resolution of racial trauma. Lastly, negative tone refers to the affective quality of the ambient stressor which has a negative psychological influence on the experiencer. Extensive research has shown the negative psychological and physiological impacts of the experience of different forms of oppression, like racism, homophobia, and misogyny. The concept of ambient stressors can perhaps help unite physical and social environmental factors that influence urban behaviours.

Urban Design as Behavioural Determinant

How does the physical design of our city influence our affective states and behaviour? Anthony Giddens a prominent sociologist, proposed Structuration Theory (1984), to understand human behaviour, in which there was an agent that continually monitors its environment, and structures which determined the agent's action and autonomy; this theoretical framework aligns well with an environmental neuroscientific perspective, particularly as it pertains to predictive models of human behaviour in relation to its interaction with their environments. Furthermore,

Giddens' delineation between agent and structure easily translates into Friston's FEP model of agent and environment, allowing for a joint lens to view both the cognitive effort and emotional processing as a metric of free energy consumption involved in navigating dangerous environments. While the human brain is primarily concerned with prediction, it could be argued the built environment's most prominent psychological role is behavioural determinism, mediating agency, and free-will. Experiencing the built environment can also result in affective responses, like awe or other forms of affect, as demonstrated by Mazumder et al. (2022), examining the effects of building height, being in the presence of high rise vs low rise buildings, on emotion.

In providing context for the development of Structuration Theory in his book Making Sense of Modernity, written with Christopher Pierson, Giddens indicates that the relationship between individuals and their relationship to society was at the core. Structuration Theory allows us to separate individuals from the environment while providing the theoretical assumption that our lives are shaped by the different environments we live in, by a power that is often correlated with a worldview with a deep appreciation for the power structures that exist in our society. Using Structuration Theory as an explanatory model for city life, we can see how urban design dictates in large part our daily behaviours and experiences which includes the opportunities available to engage in occupations, to borrow a health lens and refer to occupational therapy literature. In their Person-Environment-Occupation Model Law et al. (1996) view the person as a holistic being that considers mind, body, and soul who can assume multiple simultaneous roles. Those roles can be broken down into occupations, defined as "groups of self-directed, functional tasks and activities in which a person engages over the lifespan" (Law et al., 1996, p. 16).

The authors contend that the more overlap there is between a person, considered with all of their unique attributes and characteristics, the occupation—the task or activity they are engaged in—and the environment they perform the occupation in, the greater the possibility the behaviour is health enabling. Law et al. (1996) define the environment broadly, giving "equal importance to the cultural, socio-economic, institutional, physical and social considerations". They also establish assumptions about the environment: (1) the context in which occupational performance occurs (2) it influences and is influenced by human behaviour, (3) it is dynamic and can have constraining or restraining effects on behaviour, (4) the environment is easier to change than the person. This changeability of the environment offers many insights into how environmental design can play a key role in ensuring collective well-being.

Affordance Theory, from perceptual psychologist JJ Gibson, fits well in Structuration Theory's deterministic view of the uses of space and the behaviours contained within it. Affordances are defined as "what the environment offers the animal, what it provides or furnishes, either for good or ill" (Gibson, 1977, p. 69). The concept of affordance also relates to the term occupational possibilities—which Laliberte Rudman (2010) states, "provides a particular lens to examine transactions between structure and agency in the shaping of occupation at individual and collective levels (p. 55)". These enablements can also be understood using the concept of affordances. The human animal can engage with spaces in numerous ways based on the offerings designed into the environment. For example, a playground adds to a park numerous diverse opportunities for children to engage with, explore, and develop skills essential to life, such as motor development and social connection. Conversely, as demonstrated in the advent and practice of hostile architecture, the environment can not only constrain what can occur in spaces but can also through the presence of objects communicate not only what behaviour is allowed, but who is. This is evident in the presence of homeless spikes or intentionally designed benches that interfere with sleeping in public and seeking shelter for the homeless. Furthermore, it needs to be considered that affordances are a subjective inference, in that different people will have different perceptions on how spaces can be used, through their own unique lens, which involves how they have been conditioned by experiences in those spaces. To appeal to the free energy principle and the energy savings of prediction, what can be said about habituating to threat? Unwarranted police surveillance has had documented impacts on how many Black, Indigenous, and other People of Colour (BIPOC), primarily Black people, experience public space (Smith Lee & Robinson, 2019). Tobias and Joseph (2020) refer to the practice of carding, or stop-and-frisk, as a form of psychological abuse). This experience includes the predictive computing burden of having to reconcile the feeling of expecting danger from those to whom we entrust our safety, and our trust in broader society, what Du Bois referred to as twoness or double consciousness. Accordingly, layered upon legislated, permissible, or encouraged activities is the experiential component of the space itself. For example, knowledge of incidents of gendered violence in a local park, the presence of hate symbols in a particular neighbourhood, or a history of experiencing police surveillance may make a person feel unsafe as a baseline in an environment, and thus not able to benefit from the reported benefits of the access to greenspace. The notion of affordances has transcended the disciplinary walls of psychology and is applied widely in contemporary architectural theory and practise given its emphasis on a methodical approach to examine the ways in which an agent can engage with a space. However, the subjectivity, or positionality, involved in the perception of affordances should be considered. Viewing public space through an intersectional lens of affordances provides a practical foundation to build truly multi-use spaces that also ensure the feelings of safety and security of a multicultural city.

A Democracy of Emotions: Measuring Diverse Experiences of Stress in Urban Contexts

Speaking to the feelings of a diverse society, Anthony Giddens (1994) states:

> To the extent that it comes into being, a democracy of the emotions would have major implications for the furtherance of formal public democracy. Individuals who have a good understanding of their own emotional make-up, and who are able to communicate

> effectively on a personal basis, are likely to be well prepared for the wider tasks and responsibilities of citizenship.

Experiential equity aims to better understand emotional make-up—and its complexities—in relation to the physical and social structures that govern the urban experience and establish that equity should not only be a societal ideal but also an actionable goal that can be achieved by breaking down the parts of the experience of oppression on a psychophysiological level. Giddens suggests that a necessary progress in public democracy requires emotional health as it is necessary for hospitable encounters in public space; emotional health is influenced by the imminent, ambient, and aggressive stressors experienced in the city; and has documented downstream effects on both physical and mental health. Feelings of trust are necessary for these hospitable encounters and effective communication; Potts et al. (2019) found that acute stress was a predictor of trust. They also found that trust was predicted by individual differences in physiological reactivity. They suggest that trust is an analogue of risk taking, risk referring to the awareness of the improbability of an outcome, which inherently involves the neural process of uncertainty. Mellinger (1956) examined the influence of trust on interpersonal communication in a large government laboratory and found decreased trust lowers disclosure of authentic attitude about issues and therefore the recipient of the information has an impaired accuracy of perception of the issue being discussed, and therefore, communication is impaired, and impaired communication can result in social tension. Sillars and Parry (1982) suggest that social conflict and its associated stress impairs cognition, particularly the capacity for complex information and perspective taking, a necessary contemplation on the conversation on cities and the role they play on our aspiration for progression as a society.

Applying Experiential Equity

Identifying Environments of Interest

Charles T. Brown is a pracademic and founder of leading consultancy Equitable Cities, which focuses on achieving urban equity with an appreciation for multidimensionality. Brown (2021) proposes the arrested mobility framework (Fig. 2) and states:

> Arrested Mobility is the direct manifestation of racism across four distinct realms of racism, from the personal to interpersonal and institutional and the cultural. These collective realms of racism have resulted in the social construct of race, locally and globally, and have thus led to the intentional and oftentimes deliberate targeting and over policing of Black people. This includes their physical mobility, of which there's a direct relationship, and their social mobility, of which there is an indirect relationship.

Brown goes on to emphasise these race-related forms of restriction, and consequences of oppression can be extended to understand other intersectional

Fig. 2 Framework of arrested mobility. (Brown, 2021)

experiences of marginalisation. The framework of arrested mobility establishes racism and overpolicing as "arresting" different forms of mobility in the city (i.e. walking, using a wheelchair, or cycling), which has implications on direct experience of stress, overall well-being, and, ultimately, mortality. Experiential equity methodology can help develop finer grain understandings of the experience of space, and thus translate research into policy, which Brown also strongly argues for as an action for the framework. In the realm of urban travel safety, Sheller (2004) states automobiles have "become associated with feelings of protection, security, and safety (as emphasised in advertising of the 'family car')" (p. 236). Therefore, automobile is not just a form of transportation but also a physical buffer to the stress and danger posed to an adjacent individual riding a bike on an infrastructureless roadway in an under-invested neighbourhood or the threat exposed to a female transit user in a city with high rates of sexual violence perpetrated in empty subway stations and overcrowded buses, all potential environments and experiences of interest coming with their inherent experimental and methodological challenges.

Given the variability, measuring the urban experience requires a balance of ecological validity in approach; the inherent complexity of the experimental condition can contribute confounding factors affecting the variable of interest, and noise to the signal being used to measure it. Bronfenbrenner (1977, p. 140) described ecological validity as "the extent to which the environment experienced by the subjects in a scientific investigation has the properties it is supposed or assumed to have by the experimenter". Therefore, in addition to real-world research, we can also consider virtual reality, simulated environments, and 360-degree video, presented in head-mount displays as a valid approach to establishing experimental conditions. It has been shown that 360-degree video can offer a level of simulation that parallels real-world environments of urban contexts (Mazumder et al., 2022).

Electroencephalography for Assessing Restoration and Cognitive Load

In both real and virtual reality settings, we can assess states of restoration likelihood, using EEG to provide insights into environmental design. Using videos in a lab-based setting, Grassini et al. (2019) used electroencephalography to compare responses to natural and built environments and found nature exposure marked increases in alpha wave activity, the pattern associated with increased relaxation. They also suggest that their findings indicate higher attentional and cognitive load is necessary in urban environments. Examining the benefits of greenspace exposure from the perspective of task demands, i.e. navigating a busy traffic intersection as compared to sitting on a secluded park bench, we can imagine the respective multitude of the task demands; however, it is greenspace that is seen as having healing qualities through exposure, an assertion that I believe requires nuance. Both Stress Restoration Theory (Ulrich, et al., 1991) and Attention Restoration Theory (Kaplan, 1995) require the restoring environments to not contain threat, as threat disrupts the restorative potential of what true physiological rest provides. Threat, however, can be measured much more directly, using other measures, described below.

Electrodermal Activity as Urban Stress Index

Electrodermal activity, also known as skin conductance, offers another mobile metric of stress and refers to the measure of the rate of sweat gland response. It is generally considered to be a reliable index of sympathetic autonomic nervous system activity and indexes roughly with the concept of stress and could also be seen as a reflection of the free energy consumption required for its neural processing. Wood et al. (2014) found that the amygdala modulated skin conductance responses to threat. A tonic approach is typically used, referred to as skin conductance level, as opposed to skin conductance responses, when assessing the effects of environmental exposure (Dawson et al., 2011), an important consideration as it pertains to examining both the effects of environmental context and the threats they contain using skin conductance.

Heart Rate Variability, the Wearable City, and a Promise of Big Urban Data

While electrodermal activity is a more direct measure of autonomic nervous system activation by virtue of innervation, heart rate, and heart rate variability are exciting psychophysiological candidates to consider in urban health, for their adoption by the general public. In their meta-analysis and review, Kim et al. (2018) found that heart rate variability, defined as the fluctuation of intervals of heartbeats, was a reliable indicator of autonomic nervous system activity and stress. In a study done in the United States over 2019–2020, Dhingra et al. (2023) found that 29% of the population used a wearable device capable of providing behavioural and physiological data. Apple Health and other wearable platforms are expanding in their user adoption and sophistication of technology, as it pertains to providing insights into psychogeography. In their review on using the Apple watch for mental health-related physiological insights, Lui et al. (2022) state, "the results are encouraging regarding the application of the Apple Watch in mental health, particularly as heart rate variability is a key indicator of changes in both physical and emotional states". In their iOS 17 update, Apple (2024) offered a new Mental Wellbeing feature in their Health App which allows a user to track their Emotion: "How you feel right now", and/or their Mood: How you've felt today. Both questionnaires involve a graphical pleasantness slider scale, feeling descriptor prompts (i.e. anxious, sad, ashamed), and life impact exploration (i.e. spirituality, health, community). Both the Apple Watch psychophysiological and the iOS Mental Wellbeing data can be accessed through Apple Research Kit, which through their Core Location function allows for geotagging (Apple, 2018, 2024). These are potential which, when considered through an intersectional lens of the user, could provide a large data set to work with, albeit biased in its population by virtue of the socioeconomic status and access privilege necessary to acquire an iPhone and/or an Apple Watch. But, perhaps as adoption of wearable technology increases, so will its democratisation. Apple, along with other similar platforms, offers a potential big-data solution to the issue of WEIRD samples as the foundations of psychological research and the challenges posed not just in generalisation of findings but also the erasure of diverse experience.

Challenges and Opportunities for Intersectional Machine Learning

Machine learning, with its expansive computing capacity, has the potential to help disentangle the complexities of intersectionality as it pertains to relationship to both experience and environment. Foulds et al. (2020) cites data sparsity—or the suitable sample size that allows for examining diversity—as a challenge in applying an intersectional approach to machine learning concepts and practices. However, they propose the following criteria should be considered in an intersectional approach to machine learning: multiple protected attributes should be considered, all of the intersecting values of the protected attributes should be protected by the definition, we should still also ensure that protection is provided on individual protected attribute values, the definition should protect minority groups, who are often particularly affected by discrimination in society, the definition should ensure that systematic differences between the protected groups, assumed to be due to structural oppression, are rectified, rather than codified. In discussing the implications and opportunities of intersectional artificial intelligence, Ghosh et al. (2021) applied a worst-case scenario approach to intersectional fairness, using a min-max paradigm for metricisation, and state, the "min-max ratio technique allows us to encompass the entire breadth of possible subgroups in a dataset, by considering the worst case scenario in terms of adverse impact" (Ghosh et al., 2021, p. 25). In their approach to capturing unfairness, Molina and Loiseau (2022) state "the most straightforward approach is to estimate the probability mass function and then to compute the unfairness over the estimated distributions" (p. 3). Wang et al. (2022) make the point that applying an intersectional lens to model development while keeping data set and sample size in mind, establishing rationale for labels included bears importance and "recommend evaluating on the most granular intersecting identities available in the dataset, but combining domain knowledge with experiments to understand which are best to include when training models" (p. 346). Accordingly, I would suggest future work in the application of intersectional approach to model development should be informed by existing literature on psychological, physiological, and neurological disparities to establish the case for the inclusion of intersectional data labels, and their associated weighting as it pertains to the research question in particular.

Conclusion

Summarised, experiential equity is a framework to conceptualise and quantify the energetic cost of experiencing unpredictable urban environments, using Friston's free energy principle to unite the energetic costs of the emotional and cognitive burden of marginalisation, and views these costs through an Intersectional Theoretical lens to establish the energetic disparities that exist depending on aspects of identity. Experiential *equity* is also framed as a form of equity to position it as an actionable principle of policy in an attempt to not only highlight disparities but also inform urban design and policy that can perhaps help eliminate them. A concept of cognitive burden proposed by Du Bois of Double Consciousness Theory suggests that the experience of marginalisation requires additional cognitive resources, aligned with the Integrated Model of Stereotype threat, which has demonstrated impacts on cognitive processing. The cognitive and emotional costs of uncertainty in urban environments can be understood through Barrett's Theory of Constructed Emotion, the model also informed by environmental prediction, and provides a link to the psychophysiology of Racism-Related Vigilance, the area of research informing the Biopsychosocial Model of Racism led by the late Black Psychologist Dr. Rodney Clark. These measurable and contextualised cognitive and emotional outcomes can then be viewed through the lens of Structuration Theory to take into account power and privilege, and the physical and social structures that maintain societal inequity. Following the flow of examination of structures, the Person-Environment-Occupation Model and Affordance Theory, in conjunction, help us conceptualise the environment and examine the ways in which it can constrain or enable our engagement with it, including the activities we participate in. These spaces, within the urban context, have been demonstrated to vary in their accessibility, along social and racial lines, leading to concepts like distributional equity and recognitional equity. Leveraging machine learning to quantify aspects of the built environment, and using neuroscientific methodologies like electroencephalography, electrodermal activity, and heart rate variability, we can attempt to metricise the energetic costs of societal and systemic oppression; the consideration of the environment allows potential insights into how spatial variables mediate these experiences and can have implications on road, park, and general urban policy and design.

References

Apple. (2018). *Core location*. Accessible at: https://developer.apple.com/documentation/corelocation

Apple. (2024). *ORKLocationRecorder class reference*. Accessible at: https://researchkit.org/docs/Classes/ORKLocationRecorder.html

Barrett, L. F. (2013). Psychological construction: The Darwinian approach to the science of emotion. *Emotion Review, 5*(4), 379–389.

Barrett, L. F. (2017). The theory of constructed emotion: An active inference account of interoception and categorization. *Social Cognitive and Affective Neuroscience, 12*(1), 1–23.

Bauer, G. R., Churchill, S. M., Mahendran, M., Walwyn, C., Lizotte, D., & Villa-Rueda, A. A. (2021). Intersectionality in quantitative research: A systematic review of its emergence and applications of theory and methods. *SSM-Population Health, 14*, 100798.

Beames, J. R., Kikas, K., O'Gradey-Lee, M., Gale, N., Werner-Seidler, A., Boydell, K. M., & Hudson, J. L. (2021). A new normal: integrating lived experience into scientific data syntheses. *Frontiers in Psychiatry, 12*, 763005.

Bourke, B. (2014). Positionality: Reflecting on the research process. *The Qualitative Report, 19*(33), 1–9.

Braveman, P., & Gruskin, S. (2003). Defining equity in health. *Journal of Epidemiology & Community Health, 57*(4), 254–258.

Bronfenbrenner, U. (1977). Toward an experimental ecology of human development. *American Psychologist, 32*(7), 513–531. https://doi.org/10.1037/0003-066x.32.7.513

Brown, C.T (2021, November 11). *Arrested mobility: Exploring the impacts of over-policing Black mobility in the U.S.* [lecture]. Cities @ Tufts, Tufts University.

Campbell, J. M. (1983). Ambient stressors. *Environment and Behavior, 15*(3), 355–380.

Cho, S., Crenshaw, K. W., & McCall, L. (2013). Toward a field of intersectionality studies: Theory, applications, and praxis. *Signs: Journal of Women in Culture and Society, 38*(4), 785–810.

City of Edmonton. (2024). *Lighthouse: Fighting hate for good*. Accessible at: https://www.edmonton.ca/city_government/initiatives_innovation/lighthouse-fighting-hate-good

Clark, R., Anderson, N. B., Clark, V. R., & Williams, D. R. (1999). Racism as a stressor for African Americans: A biopsychosocial model. *American Psychologist, 54*(10), 805.

Cooke, A. (2020). *Recent deaths prompt questions about police wellness checks*. Canadian Broadcasting corporation. Accessible at: https://www.cbc.ca/news/canada/nova-scotia/police-wellness-checks-deaths-indigenous-black-1.5622320.

Crenshaw, K. (1989). Demarginalizing the intersection of race and sex: A black feminist critique of antidiscrimination doctrine, feminist theory and antiracist politics. *University of Chicago Legal Forum, 1*, 139–167.

Dawson, M. E., Schell, A. M., & Courtney, C. G. (2011). The skin conductance response, anticipation, and decision-making. *Journal of Neuroscience, Psychology, and Economics, 4*(2), 111–116. https://doi.org/10.1037/a0022619

Dhingra, L. S., Aminorroaya, A., Oikonomou, E. K., Nargesi, A. A., Wilson, F. P., Krumholz, H. M., & Khera, R. (2023). Use of wearable devices in individuals with or at risk for cardiovascular disease in the US, 2019 to 2020. *JAMA Network Open, 6*(6), e2316634–e2316634.

Du Bois, W. E. B. (1903). *The souls of Black folk: Essays and sketches*. Fawcett.

Duchesne, A., & Kaiser Trujillo, A. (2021). Reflections on neurofeminism and intersectionality using insights from psychology. *Frontiers in Human Neuroscience, 15*, 684412.

Flores, A. R., Meyer, I. H., Langton, L., & Herman, J. L. (2021). Gender identity disparities in criminal victimization: National Crime Victimization Survey, 2017–2018. *American Journal of Public Health, 111*(4), 726–729.

Ghosh, A., Genuit, L., Reagan, M. (2021). Characterizing intersectional group fairness with worst-case comparisons. Proceedings of 2nd Workshop on Diversity in Artificial Intelligence (AIDBEI), in *Proceedings of Machine Learning Research, 142*, 2–34. https://doi.org/10.48550/arXiv.2101.01673

Foulds, J. R., Islam, R., Keya, K. N., & Pan, S. (2020, April). An intersectional definition of fairness. In *2020 IEEE 36th international conference on data engineering (ICDE)* (pp. 1918–1921). IEEE.

Friston, K. (2010). The free-energy principle: A unified brain theory? *Nature Reviews Neuroscience, 11*(2), 127–138.

Gibson, J. J. (1977). *The theory of affordances* (1(2)) (pp. 67–82).

Giddens, A. (1984). *The constitution of society: Outline of the theory of structuration*. University of California Press.

Giddens, A. (1994). *Beyond left and right: The future of radical politics*. Stanford University Press.

Grassini, S., Revonsuo, A., Castellotti, S., Petrizzo, I., Benedetti, V., & Koivisto, M. (2019). Processing of natural scenery is associated with lower attentional and cognitive load compared with urban ones. *Journal of Environmental Psychology, 62*, 1–11.

Grohmann, S. (2018). Making space for free subjects: Squatting, resistance, and the possibility of ethics. *HAU: Journal of Ethnographic Theory, 8*(3), 506–521.

Guy, M. E., & McCandless, S. A. (2012). Social equity: Its legacy, its promise. *Public Administration Review, 72*(s1), S5–S13.

Hicken, M. T., Lee, H., Ailshire, J., Burgard, S. A., & Williams, D. R. (2013). "Every shut eye, ain't sleep": The role of racism-related vigilance in racial/ethnic disparities in sleep difficulty. *Race and Social Problems, 5*, 100–112.

Kaplan, S. (1995). The restorative benefits of nature: Toward an integrative framework. *Journal of Environmental Psychology, 15*(3), 169–182.

Karakas, T., & Yildiz, D. (2020). Exploring the influence of the built environment on human experience through a neuroscience approach: A systematic review. *Frontiers of Architectural Research, 9*(1), 236–247.

Kim, H. G., Cheon, E. J., Bai, D. S., Lee, Y. H., & Koo, B. H. (2018). Stress and heart rate variability: A meta-analysis and review of the literature. *Psychiatry Investigation, 15*(3), 235.

Law, M., Cooper, B., Strong, S., Stewart, D., Rigby, P., & Letts, L. (1996). The person-environment-occupation model: A transactive approach to occupational performance. *Canadian Journal of Occupational Therapy, 63*(1), 9–23.

Lazarus, R. S., & Folkman, S. (1984). *Stress, appraisal, and coping*. Springer Publishing Company.

Lui, G. Y., Loughnane, D., Polley, C., Jayarathna, T., & Breen, P. P. (2022). The apple watch for monitoring mental health–related physiological symptoms: Literature review. *JMIR Mental Health, 9*(9), e37354.

Mazumder, R. (2021). Experiential equity: An environmental neuroscientific lens for disparities in urban stress. In *Volume 1: Community and society* (pp. 187–196). Bristol University Press.

Mazumder, R., Spiers, H. J., & Ellard, C. G. (2022). Exposure to high-rise buildings negatively influences affect: Evidence from real world and 360-degree video. *Cities & Health, 6*(6), 1081–1093.

Mellinger, G. D. (1956). Interpersonal trust as a factor in communication. *The Journal of Abnormal and Social Psychology, 52*(3), 304.

Meyer, I. H. (2003). Prejudice, social stress, and mental health in lesbian, gay, and bisexual populations: Conceptual issues and research evidence. *Psychological Bulletin, 129*(5), 674.

Molina, M., & Loiseau, P. (2022). Bounding and approximating intersectional fairness through marginal fairness. *Advances in Neural Information Processing Systems, 35*, 16796–16807.

Moroni, S., & Chiffi, D. (2021). Complexity and uncertainty: Implications for urban planning. In *Handbook on cities and complexity* (pp. 319–330). Edward Elgar Publishing.

Nesbitt, L., Meitner, M. J., Girling, C., & Sheppard, S. R. (2019). Urban green equity on the ground: Practice-based models of urban green equity in three multicultural cities. *Urban Forestry & Urban Greening, 44*, 126433.

Okeke, O., Elbasheir, A., Carter, S. E., Powers, A., Mekawi, Y., Gillespie, C. F., et al. (2023). Indirect effects of racial discrimination on health outcomes through prefrontal cortical white matter integrity. *Biological Psychiatry: Cognitive Neuroscience and Neuroimaging, 8*(7), 741–749.

Österle, A. (2002). Evaluating equity in social policy: A framework for comparative analysis. *Evaluation, 8*(1), 46–59.

Parker, G., Paterson, A., Romano, M., & Granville Smith, I. (2018). Suprasensory phenomena in those with a bipolar disorder. *Australasian Psychiatry, 26*(4), 384–387.

Peters, A., McEwen, B. S., & Friston, K. (2017). Uncertainty and stress: Why it causes diseases and how it is mastered by the brain. *Progress in Neurobiology, 156*, 164–188.

Potts, S. R., McCuddy, W. T., Jayan, D., & Porcelli, A. J. (2019). To trust, or not to trust? Individual differences in physiological reactivity predict trust under acute stress. *Psychoneuroendocrinology, 100*, 75–84.

Rantala, M. J., Luoto, S., Borráz-León, J. I., & Krams, I. (2021). Bipolar disorder: An evolutionary psychoneuroimmunological approach. *Neuroscience & Biobehavioral Reviews, 122*, 28–37.

Rawdon, V. A., & Willis, F. N. (1993). Spatial displacement among pedestrians: A cross-cultural replication. *Perceptual and Motor Skills, 77*(2), 635–641.

Rudman, D. L. (2010). Occupational terminology: Occupational possibilities. *Journal of Occupational Science, 17*(1), 55–59.

Schmader, T., Johns, M., & Forbes, C. (2008). An integrated process model of stereotype threat effects on performance. *Psychological Review, 115*(2), 336.

Sewell, A. A., & Jefferson, K. A. (2016). Collateral damage: The health effects of invasive police encounters in New York City. *Journal of Urban Health, 93*(Suppl 1), 42–67.

Sheller, M. (2004). Automotive emotions: Feeling the car. *Theory, Culture & Society, 21*(4–5), 221–242. https://doi.org/10.1177/0263276404046

Sillars, A., & Parry, D. (1982). Stress, cognition, and communication in interpersonal conflicts. *Communication Research, 9*(2), 201–226.

Smith Lee, J. R., & Robinson, M. A. (2019). "That's my number one fear in life. It's the police": Examining young Black men's exposures to trauma and loss resulting from police violence and police killings. *Journal of Black Psychology, 45*(3), 143–184.

To, A., Carey, H., Kaufman, G., & Hammer, J. (2021, May). Reducing uncertainty and offering comfort: Designing technology for coping with interpersonal racism. In *Proceedings of the 2021 CHI conference on human factors in computing systems* (pp. 1–17).

Tobias, H., & Joseph, A. (2020). Sustaining systemic racism through psychological gaslighting: Denials of racial profiling and justifications of carding by police utilng local news media. *Race and Justice, 10*(4), 424–455.

Truong, K. (2017). *We didn't get out of men's way for a week—This is how many injuries we sustained.* Refinery 29. Accessible at: https://www.refinery29.com/en-us/2017/08/167417/manslamming-experiment-personal-story#slide-1

Ulrich, R. S., Simons, R. F., Losito, B. D., Fiorito, E., Miles, M. A., & Zelson, M. (1991). Stress recovery during exposure to natural and urban environments. *Journal of Environmental Psychology, 11*(3), 201–230.

United Nations. (2018). *68% of the world population projected to live in urban areas by 2050, says UN*. Accessible at: https://www.un.org/development/desa/en/news/population/2018-revision-of-world-urbanization-prospects.html

Wang, A., Ramaswamy, V. V., & Russakovsky, O. (2022, June). Towards intersectionality in machine learning: Including more identities, handling underrepresentation, and performing evaluation. In *Proceedings of the 2022 ACM conference on fairness, accountability, and transparency* (pp. 336–349).

Wood, K. H., Ver Hoef, L. W., & Knight, D. C. (2014). The amygdala mediates the emotional modulation of threat-elicited skin conductance response. *Emotion, 14*(4), 693.

World Health Organization. (2024). *Health equity*. Accessible at: https://www.who.int/health-topics/health-equity#tab=tab_1

Part VI
Specific Methodological Approaches

Extended Reality in Environmental Neuroscience Research

Fu Li, Fariba Mostajeran, Frank Steinicke, and Matthew H. E. M. Browning

Abstract Extended reality (XR) technologies provide unique opportunities for environmental neuroscience research. XR can recreate complex environments and stimuli with heightened levels of immersion, presence, and ecological validity compared to traditional lab experiments or field studies alone. By combining the experimental control of lab studies with the authenticity of real-world environments, XR enables researchers to examine neural, cognitive, and behavioral responses to environmental conditions in more naturalistic and systematic ways. This chapter outlines the history, key concepts, and features of XR and explores its applications in research on environmental cognition, emotion, spatial navigation, and more. XR can facilitate investigating environments that may be logistically challenging, hazardous, or impossible in the real world. XR also enables researchers to develop adaptive environments that respond to users' psychological, behavioral, and neural feedback. However, fully realizing the potential of XR requires overcoming challenges regarding the quality of simulations, user experience, accessibility, integrating neuroimaging modalities, and validating its effectiveness compared to conventional methods. Advancements in XR and related computer and environmental science fields may lead to closed-loop brain–computer interaction paradigms offering immersive and personalized experiences for diverse research questions and applications. With thoughtful development focused on human values and well-being, XR's future promises expanded possibilities for understanding and improving the profound ways our brains interact with the surrounding world.

Keywords Extended reality · Virtual reality · Environmental neuroscience · Ecological validity · Cognition

F. Li (✉) · Matthew H. E. M. Browning
Virtual Reality and Nature Lab, Department of Park, Recreation, and Tourism Management, Clemson University, Clemson, SC, USA
e-mail: ful@clemson.edu

F. Mostajeran · F. Steinicke
Human-Computer Interaction Group, Department of Informatics, Universität Hamburg, Hamburg, Germany

S. Kühn (ed.), *Environmental Neuroscience*,
https://doi.org/10.1007/978-3-031-64699-7_17

Introduction

Environmental neuroscience research explores how the human brain, mind, and body interact with the surrounding world. This interdisciplinary field examines how various external stimuli shape our neurological, cognitive, behavioral, and emotional responses (Berman et al., 2019; Golledge et al., 2018). Researchers investigate how nuances of perception, cognition, and emotion relate to physical or social aspects of our environment (Devlin, 2001; Gibson, 1979; Nerb & Spada, 2001).

Traditionally, environmental psychology and neuroscience research has relied on field studies for ecological validity or lab experiments for control. However, these approaches have important limitations. Field research lacks experimental control, whereas classical lab studies oversimplify the complexity of real-world environments (Winkel et al., 2009). Extended reality (XR) technologies, including virtual reality (VR), augmented reality (AR), and mixed reality (MR), offer alternative paradigms. By blending immersive digital environments with systematic investigation, XR can bridge the gap between field and lab research (Smith, 2015).

XR systems recreate aspects of the physical world through interactive three dimensional (3D) simulations. These simulations approximate natural sensorimotor contingencies, allowing users to intuitively perceive and interact with the virtual content (Slater & Wilbur, 1997). XR leverages illusions of presence, plausibility, co-presence, and embodiment to immerse users in these synthetic environments (Maselli & Slater, 2013; Slater et al., 2022). This heightened sense of presence enables environmental neuroscience researchers to simulate complex, ecologically valid stimuli with experimental control (Rizzo & Kim, 2005; Smith, 2015).

This chapter explores some of the potential of using XR in the field of environmental neuroscience. We first outline the history and key concepts underlying XR technologies. Next, we examine various environmental psychology and neuroscience research topics that could benefit from incorporating these tools. Finally, we discuss some of the challenges and ethical considerations that must be addressed to fully realize the potential of XR in this interdisciplinary field.

Simulations of the Real World: Extended Reality and Its History

XR is an umbrella term that describes VR, AR, and MR technologies with differences in the "immersiveness" of the virtual content. The range of reality to immersive virtual environments has been represented in the "Reality - Virtuality Continuum" (Skarbez et al., 2021) (Fig. 1). In short, from VR, AR to MR, the content gets progressively more integrated and "mixed" with real-world reality. VR allows users to feel fully immersed in a computer-generated virtual environment (VE), which typically uses a specialized head-mounted display (HMD) to provide the virtual environment and isolate the user from the real world. In AR, users can

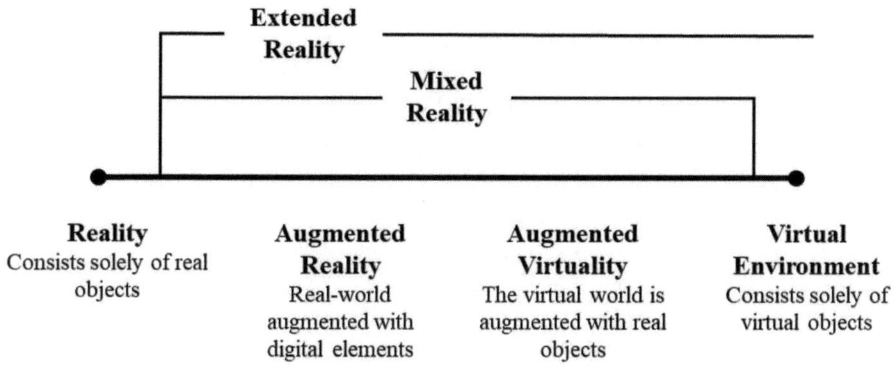

Fig. 1 Reality—virtuality continuum. (Adapted from Milgram and Kishino, 1994)

see the real world (either through cameras or see-through displays) augmented with additional virtual objects (Schmalstieg & Hollerer, 2016). MR denotes the combination of different core elements of VR and AR experiences, providing a more first-person perspective and immersive experience (Speicher et al., 2019). In both AR and MR experiences, the virtual content can adapt and interact with real-world environments and human interactions. Because MR typically provides a first-person perspective experience and needs to constantly adapt the virtual objects to the real environment in an HMD, it requires more sophisticated hardware and software systems to achieve this higher degree of integration (Rokhsaritalemi et al., 2020).

The foundations of XR theory, content creation, and technology have a rich history. Many fundamental principles and concepts can be traced back to fine art and painting. Artists use multiple perspective techniques in their drawings to capture depth and spatial relationships, such as one, two, and three-point perspectives varying in the number of vanishing points (Fig. 2) (Huang & Chang, 2016; Yu, 2023). For instance, panoramas were emphasized in the work by Robert Barker in 1787 to visualize perspective from multiple viewpoints (Otto, 2007; Uricchio, 2011). Barker painted the interior walls of a vast circular dome with a 360-degree view of another location from all angles (Fig. 3). Visitors "immersed" themselves in the painted environment that differed dramatically from the building's location where the paintings were located (Ellis, 2008). This immersive effect can be seen as one of the primitive cases of VR.

Another concept of XR related to fine art is "binocular vision." Binocular vision is achieved in humans by two horizontally separated eyes, allowing depth perception (Blake & Wilson, 2011). In 1838, Sir Charles Wheatstone outlined this concept of binocular vision and developed the first stereoscopes (Bowers, 2001), which took a pair of static images to create perceptions of 3D images with illusions of depth (Fig. 4). A more modern case of a stereoscopic is the View-Master, which generates 3D viewing experiences with thin cardboard disks containing seven pairs of small transparent color photographs on film (Fig. 5) (Gruber, 2014). Stereoscopic displays

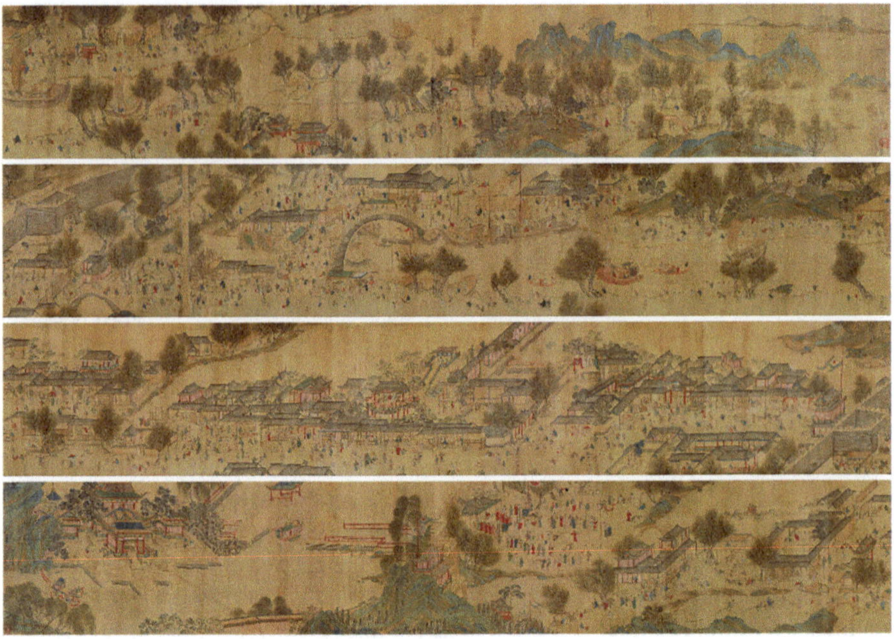

Fig. 2 Chinese painting "Along the River during the Qingming Festival" by Zhang Zeduan, 25.5 cm × 525 cm, 12 AD. An example of using multiple perspective techniques in the drawing. The multiple perspective technique means we can perceive the environment and the people from different angles with more "panoramic" views to tell more stories in one painting. This technique is one of the foundations of panoramas and later XR technology

are still used in VR HMD systems to bring depth perception to digital images, enhancing the feeling of immersion (Brown & Green, 2016).

In the 1950s, the journey of more modern versions of XR began with humble but visionary attempts. Cinematographer Morton Heilig developed the Sensorama in 1955, which was an all-encompassing movie booth that could display movies with immersive views (Jones & Dawkins, 2018). The Sensorama combined 3D visuals, audio, smells, and a vibrating chair (Fig. 6a), but it was also expensive, with little video content, and included additional limitations that prevented it from gaining widespread popularity. Following the Sensorama, engineers at the Philadelphia Storage Battery Company (Philco) sought to incorporate stereoscopic technology into a military tracking system called the Philco Headsight in 1961 (Fig. 6b) (Uruthiralingam & Rea, 2020). Though only partially successful, this project gave rise to what has been considered as the first-ever VR headset. This application further developed into the heads-up-display (HUD) in aircraft (Fig. 6c), which could be considered a use case of AR (Peddie, 2017).

This period in XR history was also when Ivan Sutherland's paper, "The Ultimate Display" proposed the vision of a computer-generated system that closely resembled real life. The Ultimate Display marked a significant step toward what would

Fig. 3 Cross-section of the Panorama building in Leicester Square showing the viewing platforms, 1801. It was designed by Robert Mitchell to exhibit the panoramic paintings of Robert Barker (1737–1806). Staircases led to a number of viewing platforms from which visitors could take in two panoramas, one large and one of smaller dimensions. The large panorama depicted here is a view from the hills around Edinburgh

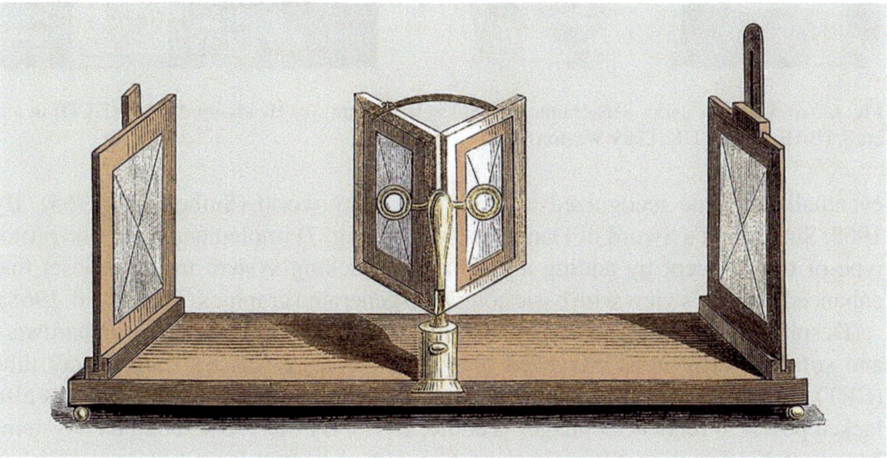

Fig. 4 Illustration of Wheatstone's stereosco.pe, 1838. The device has two mirrors at a 90-degree angle; observers position their eyes close to each side of the mirror join. (Illustraed by Sheila Terry, 2016, https://www.sciencephoto.com/media/702212/view/wheatstone-s-reflective-stereoscope)

Fig. 5 Product image of The View-Master. The View-Master is a stereoscope designed for kids, displaying 3D images on circular reels. Each reel has paired images that create a 3D effect when viewed. It's a popular educational tool, making learning visual and engaging for children. (Mattel and Target)

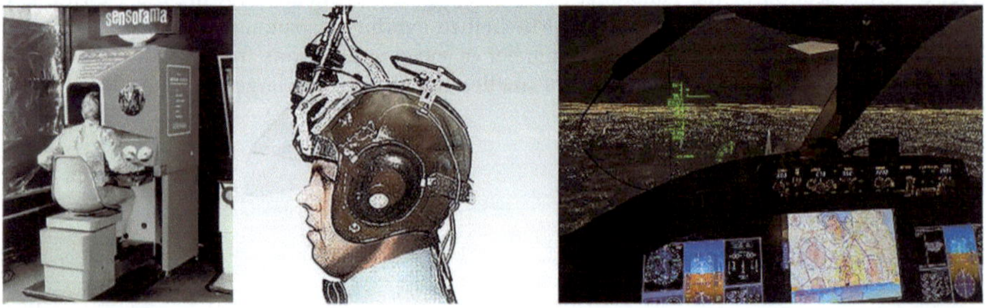

Fig. 6 (**a**) A viewer using Sensorama; (**b**) Philco Headsight; (**c**) Heads-up-display (HUD) in aircraft. (BAE System UK, Gary Weston)

eventually become recognized as a virtual reality world (Sutherland, 1965). By 1968, Sutherland's Sword of Damocles system (Fig. 7) implemented the first prototype of this concept by adding a mechanical tracking system to the headset that enhanced the user's view with basic computer-generated graphics (Sutherland, 1968).

Despite these pioneering efforts, XR devices faced major limitations in hardware and software. The most advanced display technology was the cathode-ray tube (CRT), which could not fit in small, lightweight headsets. Computer systems also lacked power to render 3D images (Peddie, 2017). By the 1980s, computer systems began catching up with XR aspirations. Computer scientist Jaron Lanier established the VR-orientated business VPL Research Inc. and coined the term "virtual reality"around the 1990s (Lanier, 1992). In 1990, the term "augmented reality" was

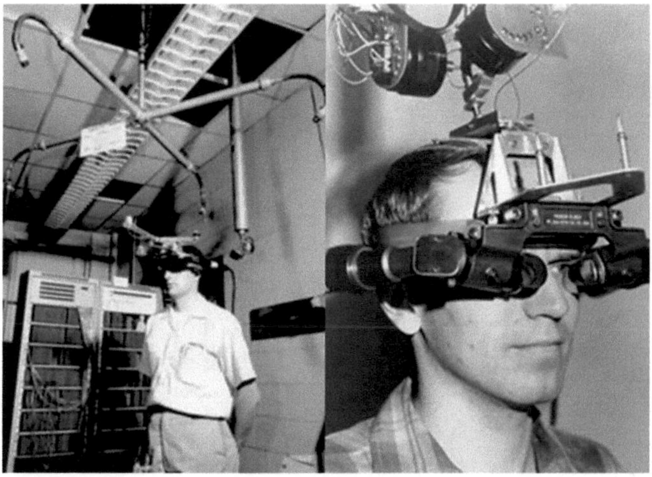

Fig. 7 Sword of Damocles, the first VR Head Mounted Display, 1968. (Sutherland, 1968)

coined by Boeing researcher Tom Caudell (1995). The 1990s saw significant progress for XR in commercial use cases, with the introduction of VR arcade machines and affordable VR headsets like the CyberMaxx and Nintendo Virtual Boy becoming accessible for home use (Greuter & Roberts, 2014) (Fig. 8).

In the early twenty-first century, VR systems remained dependent on cables running to computer graphic cards for processing power and graphical fidelity. In 2010, Palmer Luckey's prototype for the Oculus Rift VR headset marked a turning point in the stand-alone XR experience (Desai et al., 2014). Mobile computing, camera tracking, and a series of open-source projects rekindled the interest in VR among technology megacompanies (Erol-Kantarci & Sukhmani, 2018). In 2014, the HTC Vive was developed, featuring "room-scale" technology via a laser sensing base station and enabling users to move freely in 3D space (Dempsey, 2016). Vive partnered with the trending game-selling platform Steam and game development engine Unity3D to set new standards for VR technology, including rendering, tracking, input systems, and locomotives. AR and MR also garnered attention around this time, with devices like Google Glass (2013) and Microsoft HoloLens (2016). AR also found success in mobile applications, exemplified by popular apps like Pokémon Go (2016), with promising applications in various industries beyond gaming. The OpenXR by Khronos Group, an open-source standard API system launched in 2017 and available in 2019, further addressing the fragmentation of AR/VR/MR development, enables developers to build apps that run on a wide range of devices within their preferred development tool. Today, more specialized or generic manufacturers (e.g., Pico and Apple) have joined the development of consumer devices. Content producers and distributors (e.g., Disney, Netflix, and YouTube) are joining by providing access to 360-degree videos, movies and immersive storytelling. These advancements are further enriching the ecosystem of the XR industry, making it accessible for consumers, industry, and scientists.

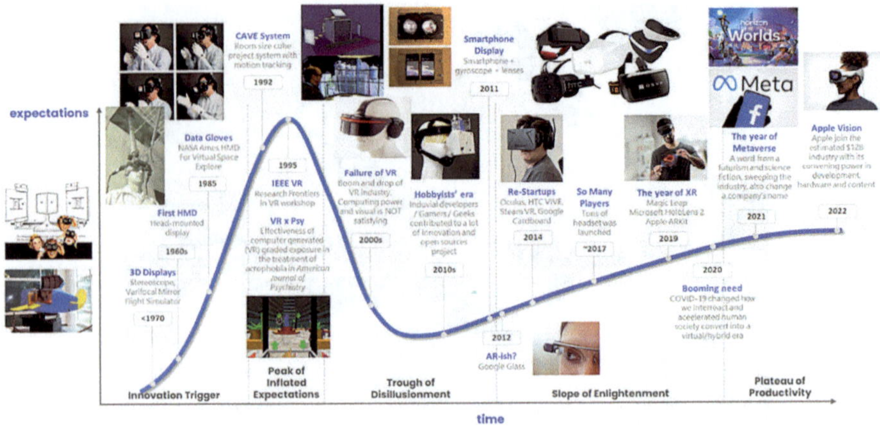

Fig. 8 Developmental timeline of modern XR systems. (From Author)

While XR was developed as a new category of personal computation, it has many applications for environmental neuroscience research. Crafted illusions offer a potent method for immersive stimuli with high degrees of experimental control (Mostajeran et al., 2021). These effects can benefit studies in psychology, behavioral science, and neuroscience in lab settings and real-world environments (Depledge et al., 2011).

Key Features of XR in Environmental Neuroscience Research

XR Systems

The basic components of an XR system for environmental neuroscience research include the display, input, content, and measurement. There are two main types of display technologies in XR: headset-based and screen-based. Headset-based displays are widely used in all the XR types, while screen-based displays are typically used for AR (Fig. 9). Using a headset-based XR system is like wearing eyeglasses, with delicate optical modules projecting images directly on to the user's eyes through head-mounted displays (HMD). Using a screen-based device is like taking a video with a smartphone. These systems overlay virtual objects on the real-world image captured by the devices' camera. The Cave Automatic Virtual Environment (CAVE) is a special type of screen-based system that presents XR content in a room with projections or LED displays on the walls, floor, and ceiling, eliminating the need to wear or hold up a device to interact in VR (Cruz-Neira et al., 1992). A CAVE system can also be equipped with motion-tracking sensors and cameras to track physical objects and real humans within its space, which enables the creation of multi-user and interactive scenarios (Schmidt et al., 2018). Variations of this type

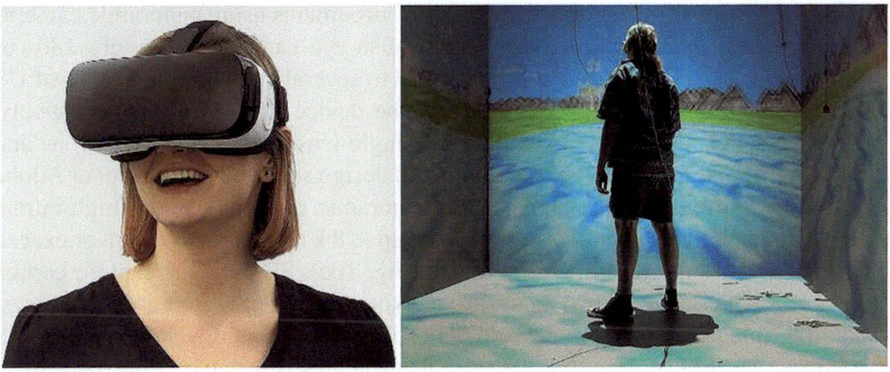

Fig. 9 Examples of headset-based (on the left) and screen-based XR (on the right). (Pikist. WikiCommon)

of XR have also been developed for use in screen-based 3D spatial experiences without 3D glasses, or immersive dark rides at amusement parks (Dobbert, 2006; Zika, 2018).

To create immersive experiences, virtual environments must render each moment's perspective with a millisecond refresh rate that follows the user's eyesight. Most devices locate perspectives using color cameras or light detection and ranging sensors (LiDAR). XR systems calculate the precise movement of devices from the real-world reference points or spatial structures with computational algorithms. AR and CAVE systems use additional cameras that face the users to detect the locations of users' heads. Most displays can be refreshed at 60–120 Hz, rendering an image at a millisecond level.

Just like a keyboard and mouse for computers, input systems like controllers are the basic interaction tools for users of XR. Typical XR systems can detect the motion of the user's head and hands, providing a more natural interaction with the virtual environment and sensing behavioral data. Additional sensors can be integrated into such systems in order to track the user's full body motions as well as fine movements of their facial muscles. Some HMDs have built-in eye-tracking sensors, allowing real-time recording of participants' gazing and eye movements in the environment, providing insights into their attention and responses to various stimuli. Location and teleportation inputs are multifaceted and span cursor-teleporting with controllers to one-direction or omni-direction treadmills, bicycles, or driving simulators. These setups move users in virtual environments, allow the monitoring of movement trajectories and viewing angles, and help researchers understand what objects influence human behavior. Some AR and MR systems utilize more sophisticated techniques, such as a Global Positioning System (GPS), cloud-based street-view references, and artificial intelligence (AI) algorithms, to match users' movements and perspectives in the real world at neighborhood or city scales.

XR visual content largely falls into image/video capturing (also called immersive videos) and 3D modeling categories. The most common tool for creating immersive videos is through capturing real environments using panoramic cameras or 3D cameras. Panoramic cameras utilize fish-eye cameras to capture 180- or 360-degree imagery. They cost a few hundred to several or tens of thousands of US dollars (USD), depending on the grade of the device (Fig. 10a,b). Alternatively, photos can be captured with a smartphone, single-lens reflex camera, or drone and stitched into a panorama post-production with design software like PTGui or Adobe Photoshop. Currently, professional class panoramic cameras capture high-bitrate all-direction monoscopic video streams with up to 8 k resolution to match or exceed the rendering resolution of most headsets (Fig. 10c). 3D cameras capture certain viewpoints with depth of field so users can experience limited stereoscopic effects. Capturing images from the real world is beneficial for creating visual content with the highest level of realism, but they can be difficult to change or edit. Also, since the captured images cannot change with the user's position and viewpoint, they may not provide a fully immersive experience with an interactive and responsive XR system.

Fig. 10 (**a**) Consumer-grade and (**b**) professional-grade panoramic cameras. (**c**) A raw format of a panoramic image. (From Author)

On the other hand, 3D modeling provides a wider range of virtual experiences, from replicating real-world environments to creations based entirely on imagination. Content designers and researchers create 3D objects and environments using a variety of methods, like scanning real-world objects and environments, downloading from 3D content libraries and stores, or modeling from scratch. Scanning real-world objects requires photogrammetry or LiDAR. These technologies are gradually being integrated into consumer-grade devices and software, such as iPhone and iPad's Pro series devices from Apple. Detailed 3D models can be calculated in a few minutes to several hours based on the size of the object and expected resolution. For content creators and researchers with experience in 3D modeling, such as those with architectural, landscape, interior, or gaming design backgrounds, 3D content library and modeling from scratch are reasonable options. Content creators can create objects and environments using 3D software such as 3DMax, Blender, and Sketch Up (Fig. 11).

In order to be displayed on an HMD, visual content created using 3D models or immersive videos needs to be imported into XR-specific engines or software. The Unity3D game engine is one of the most commonly used software for this purpose, with other options including Unreal, Apple ARKit, Google ARCore, and the native XR standard. Most device manufacturers provide a native player application for video and picture playback. Alternatively, users can download third-party players such as Skybox VR Video Player from application stores. Some HMDs such as HTC Vive Pro have to be tethered to a personal computer or a gaming laptop. Other and most recent standalone HMDs such as Meta Quest (Quest 1, Quest 2, Quest 3, and Quest Pro), PICO (PICO 3, PICO 4), and Apple Vision Pro are tether-less and rely on the integrated computing power of the headset. In order to achieve a high-quality experience with high-definition materials and realistic lighting effects, the

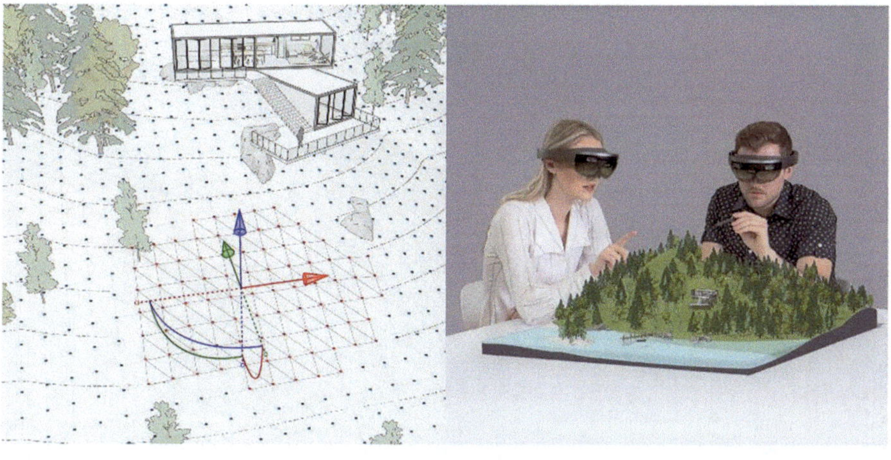

Fig. 11 3D modeling with SketchUp by computer and Mixed Reality. SketchUp is a popular 3D design tool for architects, urban designers, landscape architects, and interior designers. (Trimble SketchUp)

latest hardware is often required. This includes functional central processing units (CPU), graphics processing units (GPU), high speed and large memory (RAM), fast read-write speed solid-state drives (SSD) for tethered HMDs, or more advanced systems on chips (SoC, like Qualcomm Snapdragon XR or Apple Silicone) for standalone HMDs.

Several other emerging technologies can further enhance interactions with virtual environments. Eye-movement-based gazing introduces new types of interaction that can enhance rendering efficiency (Clay et al., 2019). Artificial intelligence-generated content (AIGC) technology can generate, and modify environments and objects in real-time according to user needs, providing more personalized environments and making it easier for researchers to predict user preferences (Fig. 12) (Wu et al., 2022; Yuan et al., 2024). High-quality virtual humans and animals can further enhance the realism of virtual living objects to counteract the uncanny valley effect, which is the fear or discomfort from seeing a medium-quality simulated human (Fang et al., 2021; McMahan et al., 2016). Touch and body-sensation technologies help explore participants' interactions and perceptions within virtual environments, deepening our understanding of the sensory system (Hutson, 2018).

Measurements are also essential for environmental neuroscience research. XR can integrate the collection of many types of psychological and behavioral data, such as questionnaires, movement tracking, eye-tracking, response tests, and cognitive tests. With user experience (UX) design, researchers can easily embed psychological scales in XR without interrupting immersive experiences. Some neurological data (i.e., electroencephalography [EEG] and electromyography [EMG]) and physiological data (i.e., heart rate, skin temperature, and respiration) can also be integrated with XR devices (Baumgartner et al., 2006; Blana et al., 2016; Sun et al., 2023; Wiederhold et al., 2014; Xu et al., 2022; Zhang et al., 2021). The synchronous collection of measures with XR offers insights into participants' emotional and

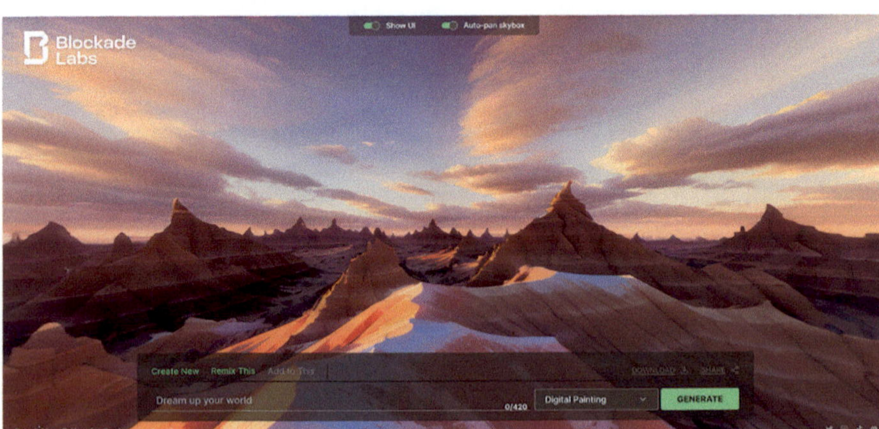

Fig. 12 Skybox AI from Blockade Labs is a AIGC provider specified for generative 360 degree panorama images from text description input. (Blocade Labs)

experiential responses (Spanlang et al., 2014). XR devices fit well into the research paradigm of environmental psychology and environmental neuroscience, in which XR was adapted to provide more immersive stimuli with created illusions.

XR Illusions

An essential feature of XR technologies is their support for sensorimotor contingencies. These refer to the actions that we perform to perceive something. For instance, we move our head and eyes to direct our gaze to see something or we turn around and bend over to see underneath an object (O'Regan & noë, 2001). A typical HMD system with its tracking system provides head tracking, which computes head position to corresponding views in the virtual environment. Therefore, when a user wears an HMD and sees a view of the virtual environment by turning their head, their view of the virtual world changes accordingly as if they were seeing a physical environment. Additional input systems, such as controllers, enable interaction with the virtual environment through touching, grabbing, or manipulating the virtual objects. As a result, an immersive XR system provides users with an omnidirectional view of the virtual environment and enables them to perceive, feel, and interact similar to reality. Such XR systems that approximate perceptions of physical reality can generate various types of illusions for their users. For instance, these devices can give the illusion of being physically located in a virtual environment. This illusion is called place illusion and is also known as the "sense of being there" in the environment depicted by the virtual display (Fig. 13a). This illusion often happens in spite of certain knowledge of the user that this is not the case and they are not located in the virtual environment in reality (Slater et al., 2022).

A second illusion that an immersive XR system can create is called plausibility illusion (Fig. 13b). This happens when users believe that what is perceived in the virtual environment is happening and that events that occur in the virtual world are actually happening in real life (Slater, 2009). This illusion again happens with certain knowledge of the user that the experience is not the case and the events in the virtual world are not happening in reality. For example, in an underwater virtual world, a virtual whale approaching a user may be perceived as real and trigger anxiety, although they know that they are not in an underwater world and the whale is digitally generated. Therefore, place illusion is about how the virtual world is perceived, and plausibility illusion is about what is perceived there. These two illusions are dimensions of a "sense of presence." However, the sense of presence has been extended by recent works to cover other types of XR illusions as well (Slater et al., 2022).

The third illusion is called body ownership (Fig. 13c) and arises when users have a sense of ownership over the virtual body that they have received in the virtual world (Maselli & Slater, 2013). The virtual body can represent various forms and sizes. For instance, it could depict a life-size 3D reconstructed model of the user, a generic humanoid representation or parts of it, or even a body model of a different

Fig. 13 Four types of illusion. (**a**) Richie's Plank Experience video game is an example of experiencing the place illusion in VR where participants feel being physically located on a plank in height. (**b**) The Blu video game is an example of plausibility illusion in VR, where participants believe that a computer-generated whale is actually swimming towards them. (**c**) This is an example of the body ownership illusion, in which participants feel that the virtual body presented in the virtual environment is actually their real body (Jun et al., 2018). (**d**) An example of social and co-presence with virtual humans in VR is depicted in this figure. (Mostajeran et al., 2020)

size or demographics. The user can observe this body in the virtual environment from the first-person perspective by looking down toward themselves or by looking into a virtual mirror. In either case and with the help of real-time body tracking, the virtual body should follow the same movements as the user's actual physical movements. Again, this illusion of believing that the virtual body is their real one happens with the certain knowledge of the users that the depicted digitalized representation of their body in the virtual world is not their real body. Originally, successful replications of the rubber hand illusion (Botvinick & Cohen, 1998) in VR led to the discovery of this type of illusion in virtual environments and was extended to further aspects in XR (IJsselsteijn et al., 2006).

The fourth illusion is called social and co-presence (Fig. 13d). This occurs when users feel that the computer-generated virtual humans or characters are with them in the real environment (Mostajeran et al., 2020). This illusion combines all previous illusions with the social aspect of using XR. For users to feel together in the same space, they should first feel present in the virtual environment (i.e., place illusion). Then they should have a virtual body and have the illusion of its ownership in the virtual environment (i.e., body ownership). Then they should see other users' virtual bodies in the virtual world and perceive their actions and reactions as if they were real. For example, when some events, such as when other virtual humans walk toward or away from a user or when they smile or nod their heads in a conversation are perceived as if they were real, the fourth illusion of social and co-presence has been met.

Such illusion combined with the research paradigm and technology brings many opportunities to environmental neuroscience research. XR can simulate real-world environments with high precision and convenience to the researcher and participant, allowing highly controlled and systematically presented stimuli. In contrast, researching responses in a systematic and highly controlled fashion in field experiments is challenging. Unpredictable and imbalanced conditions between participants limit the level of control available to researchers. Such conditions are quite broad, ranging from the weather to crowding, other visitor behavior, and wildlife interactions. Multiple exposures or environments require substantial amounts of time, travel, and costs for the researcher and participant, limiting the feasibility of generalizable results from larger sample sizes and comprehensive sets of environmental conditions. Some environments of interest may be unaccessible, such as isolated or confined environments (i.e., space travel) (Ascone et al., 2020) or extreme and unsafe (i.e., volcanoes and deep sea settings) (Elor et al., 2021; Tibaldi et al., 2020), yet these environments may be of interest to researchers attempting to promote broad awareness to the impacts of climate change or developing interventions for people required to live in these environments for their employers. Conducting experiments in laboratory studies with less immersive two dimensional (2D) imagery partially solves these constraints; classic examples are early environmental psychology studies with photographs or videos of numerous environments. However, such early studies also have restrictions related to the limited ecological validity of nonimmersive stimuli.

XR can also accelerate the research process. It allows the researcher rapid prototyping. Instead of building and managing a physical environment, researchers can create and modify virtual environment swiftly, reducing time and cost. This could benefit in the research process of finding the key variables or setting, allowing the researcher's freedom for experimentation. Before any main study, XR can be used for pilot testing, training sessions, or to familiarize participants with procedures at a lower cost, ensuring that the data collection phase is efficient. It also can ensure the reproducibility of research. As a major criticism of environmental research, reproducibility allows the research community to test a phenomenon or finding in the way it used to be cannot be achieved. The virtual environment can be shared, and research can be reproduced at different times and locations around the globe. This means research on certain environments is no longer restrained by the geological location. XR can also facilitate participants from individuals across different geographic locations, eliminating the barriers of time, space, climate, or culture. This can significantly expand the pool of participants and the diversity of data collected, further enhancing the validity.

Opportunities of XR in Environmental Neuroscience Research

XR has several potential benefits for environmental neuroscience research. One major opportunity is the capability to recreate complex environments in a controlled manner, allowing researchers to control variables and study psychological or neurological responses between participants and between studies (Kuliga et al., 2015; Spanlang et al., 2014). XR can also facilitate the research of human–environment interactions in settings that might be logistically challenging, hazardous, or even impossible to replicate in field research (Ascone et al., 2020). Additionally, it offers a platform to develop adaptive environments that respond to psychological, behavioral, and neural feedback. This allows the researcher to create a dynamic loop between the brain and the environment (Spanlang et al., 2014). This not only helps in understanding how different environmental stimuli affect the neural processes but also offers potential mitigation or therapeutic applications (Mostajeran et al., 2020). The integration of XR in environmental neuroscience research paves an alternative way for deeper insights into the intricate relationships between the brain, cognition, behavior, and environment (Forlim et al., 2019). Such benefits provide opportunities for various environmental neuroscience research topics to adapt XR. Such wide-ranging topics include environmental cognition, perception, behaviors, and emotion, which develop into questions of aspects like behavior, life-span, education, training, wayfinding, and environmental esthetics.

Environmental Cognition and Behaviors

Environmental cognition and behavior studies aim to understand how individuals perceive and process information about the real-world environment and how this information influences their cognitive process, decision making, and behavior (Golledge et al., 2018; Lynch, 1984). Researchers can use XR to design and facilitate research with different stimuli and scenarios to assess participants' cognitive abilities, memory, attention, and decision making in a more ecologically valid setting. The research question and outcome can be important to various fields of applications, like architecture and interior design, landscape architecture, urban planning and design, human factors, ergonomics, and education, among others.

With XR, researchers can create realistic simulations of complex environments, allowing participants to navigate in buildings (Erkan, 2018), browse virtual stores (Casson & Trimble, 2018; Mostajeran et al., 2019) or drive on simulated roads (Jiang et al., 2021). Brain activities, body behaviors, decisions, and interactions with the environment can be tracked in real time as these simulations are presented to users.

Vehicle driving is a typical and common research scenario (Fig. 14). Many studies have used XR to implement laboratory environments to study the effects of factors including lanes, traffic flow, pedestrians, intersection complexity, vegetation, and other environmental elements on the driver's cognitive load, reaction time, and other behavioral metrics (Deb et al., 2020; Jiang et al., 2021; Lin et al., 2005, 2007; Lorenz et al., 2014; Tran et al., 2021). Researchers can localize which brain regions or neurological responses are involved in simulated driving, further explaining and helping to solve real-world driving problems (Calhoun et al., 2005; Carvalho et al., 2006). Applications extend beyond automobiles to aviation and other forms of

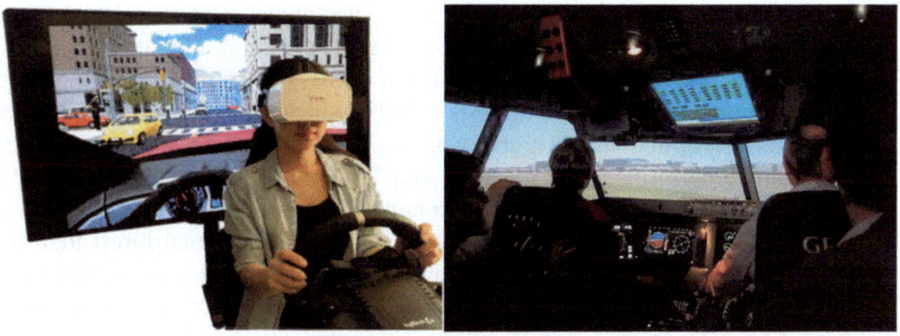

Fig. 14 XR-based car driving simulators and full flight simulators. (Xie et al., 2021)

transportation. For instance, by incorporating a screen-based or headset XR flight simulator with full cockpit motion, researchers can analyze situational awareness, decision making, and behavior of pilots in intense but safe and realistic scenarios (Harrivel et al., 2017; Huttunen et al., 2011; Zhang et al., 2017).

XR can also create other less "intense" scenarios, such as encounters with animals or viewing artwork, for the purpose of studying diverse human behavior (Carter et al., 2020; Mu et al., 2022). And the behavior is not limited to "actual" body behavior. Using XR as an alternative environment for decision making and gaming in an economic setting (Innocenti, 2017) or exploring how people learn and understand languages in different contexts (Tromp et al., 2018).

Recent studies with XR have shown that the type of environment shown to users can influence their attention and their cognitive performances. Comparisons have been made mainly between virtual nature and urban environments (Nukarinen et al., 2022), and, sometimes, with abstract, neutral environments as controls (Mostajeran et al., 2023a). These studies are based on established theories in the field of environmental psychology, which provide explanations for the positive effects of natural environments. Positive effects of nature on attention and cognition could be explained by Attention Restoration Theory (ART) (Kaplan, 1995), which suggests that nature grabs attention in a bottom-up fashion, and, therefore, allows top-down directed-attention abilities to be restored. Thus, it consumes less cognitive resources, allows an interruption of attention-grabbing tasks, and, thereby, elicits attention restoration and recovery from mental fatigue (Hartig et al., 1991; Kaplan & Kaplan, 1989; Tennessen & Cimprich, 1995). Several XR studies have validated this theory. For instance, one study compared exposure to immersive videos of a forest and an urban environment without greenery in VR and reported better performance in a mental arithmetic task after exposure to an immersive forest environment than an immersive urban environment. The authors also reported lower mood disturbances and lower cybersickness as well as a higher sense of presence for the virtual forest environment (Mostajeran et al., 2021). In another study, a computer-generated forest environment was compared to a virtual control environment with a comparable amount of virtual objects, which was an abstract replication of the virtual forest environment (Mostajeran, Fischer, et al., 2023a). The authors measured cognitive performance using the digit span backward test (DSB), which measures short-term working memory, and trail making tests (TMT), which measures executive functions such as task switching and higher working memory load. Their results showed better cognitive performance in the DSB and TMT as a result of exposure to the virtual forest than the virtual control environment. In addition, perceived stress was lower, and positive affect was higher after exposure to the virtual forest than the control environment. Not only virtual nature environments such as forests have been shown beneficial to users but also the presence of natural objects in a virtual environment. For instance, one study compared a virtual office environment with or without virtual plants and observed better cognitive performance with plants compared to a virtual environment without plants (Mostajeran et al., 2023b; Yin et al., 2019). Psychological well-being measures including positive affect and attentive coping were also higher values whereas anger and aggression measurements were lower when plants were present.

In addition to research on cognition and behavioral observation, there are also XR studies focusing on brain activity as an outcome. For instance, one study recorded electroencephalography (EEG) (➔ chapter "EEG for the Study of Environmental Neuroscience" by Grassini) to capture involuntary attention restoration while participants performed an auditory oddball task (Chung et al., 2018). Their results showed that in comparison to immersive videos of fireworks, exposure to immersive videos of natural environments (seaside, grassland, and hilly scenes) improved cognitive functioning and restored involuntary attention. One study used functional near-infrared spectroscopy (fNIRS) to assess the reaction and training outcome for industrial emergencies. By measuring the hemodynamic responses, fNIR provides data on the functional connectivity of participants' cortex, further determining the outcome during repeat training with different levels of stress (Shi et al., 2020). Some other studies use functional magnetic resonance imaging (fMRI) (➔ chapter "Interplay Between the Physical Environment and the Human Brain: A Review of MRI Research" by Beyer et al.) to measure the stress-response benefits of viewing urban green (Chang et al., 2020), or visual attention to environmental elements(Lorentz et al., 2023). Those brain activity measurements, along with self-reporting data, shed light on the process of human cognition and behavioral processes to a certain environment.

Emotion, Stress, and Psychological Disorders

Environmental research can also focus on how settings affect human emotions and how environments can be changed to regulate emotional states (Nerb & Spada, 2001). Although there are divergent theories about the mechanisms that elicit emotional states, this interest has led to the development of different laboratory methods, so-called mood induction procedures (Jallais & Gilet, 2010; Westermann et al.,

1996). XR can be an effective tool to create realistic environments designed to elicit specific emotions, which can be useful in psychological and environmental neuroscience studies.

The mechanisms underlying the perception of emotions are complex. Among these, restoration and stress-reducing effects from natural environments are perhaps the most discussed and empirically supported. In addition to ART, which explains positive effects on attention and cognition, Stress Reduction Theory (SRT) provides an explanation for the positive effects on psychological well-being (Ulrich et al., 1991). SRT suggests that natural environments promote stress recovery by normalizing physiological arousal responses, enhancing positive emotions, and reducing negative or stress-related feelings (Ulrich, 1984; Ulrich et al., 1991). These qualities lead to stronger psychological well-being after exposure to and interaction with natural environments. The Biophilia hypothesis (Kellert & Wilson, 1993; Ulrich, 1993) suggests that evolution may have equipped humankind with an innate connection to nature, and as a result, natural environments tend to show more restorative qualities compared to predominantly artificial and urban environments. Greenery in natural and urban environments such as parks can improve human cognitive and psychological well-being (Berman et al., 2021; Kaplan, 1995; Ulrich, 1981). Such research has migrated to and been confirmed by VR research; as a result, nature exposure in VR is becoming a therapeutic tool for day-to-day emotional regulation (Bohil et al., 2011; Browning et al., 2020b; Jiang et al., 2014; Yin et al., 2018). Such benefits are supported at the neurological level in XR-based experimental studies (Jung et al., 2023; Kühn et al., 2021; Souza & Naves, 2021). While virtual nature cannot fully replace the benefits of being in actual nature, it can still provide a change of environment and a temporary escape from stressors in day-to-day life (Fig. 15) (Beverly et al., 2022; Browning et al., 2020a; Masters et al., 2022).

Another mainstream mechanism for interpreting the perception of emotion is the Emotional Valence and Arousal (Lang et al., 1990). This mechanism has two dimensions: one describes the positive and negative nature of emotional experiences (i.e., positive valence/happy and negative valence /sad) and the other one describes intensity (i.e., low arousal/calm/relaxed to high arousal/excited/stressed), which together form the complex emotional experience. This mechanism is empirically supported in neuroscience (Heilman, 1997; Mneimne et al., 2010) and further extends the evidence that the environment influences complex emotional changes by changing the type of virtual environmental stimulus (Magdin et al., 2021; Presti et al., 2022). The combination of XR and brain activity shows promise in exploring the evidence of how human emotional valence and arousal changed from the environment. For instance, one study used self-reported data and EEG to assess the emotional perception and fear to nightscape in a virtual reality headset, while controlling greenery, architecture, and other human's presence (Kim et al., 2019). Both self-reported fear and alpha wave ratio increased, suggesting that built environment and adult presence contribute to more fear, while nature-dominant nightscape was less fearful. Also, XR is an effective tool for eliciting emotions, such as anxiety, relaxation, fear, and joy, in a laboratory setting, and focusing on the integration of subjective and

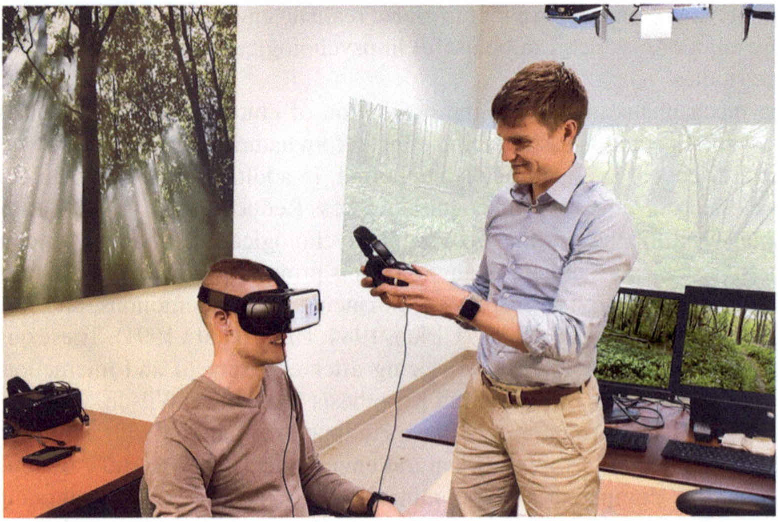

Fig. 15 Virtual nature experience that can provide mental escape for participants. (From author)

physiological outcomes as well as facilitating immersion, could further enhance the ecological validity of this paradigm (Diniz Bernardo et al., 2021).

XR also offers great potential for studying various emotional and psychological challenges by creating controllable environments and simulations. Research has utilized XR to help psychologists examine and diagnose psychological and neurological disorders like autism spectrum disorder (Parsons & Mitchell, 2002) and depression (Bittner et al., 2018). XR has also been demonstrated as an effective treatment approach for several psychological disorders, including phobias, social anxiety disorder, and post-traumatic stress disorder by exposing them to the environment they fear (Cieślik et al., 2020; Rizzo & Koenig, 2017). Through immersive and customizable environments, XR enables exposure and behavior therapies in controlled, supportive settings with lower disengage rate (Baus & Bouchard, 2014; Garcia-Palacios et al., 2007; Mostajeran et al., 2021; Nichols & Patel, 2002). XR-based treatments can help individuals with autism spectrum disorder improve social skills and have shown benefits for other neurological conditions by providing an adaptable yet near to real-life environment (Mesa-Gresa et al., 2018; Parsons & Mitchell, 2002). In summary, XR is a valuable technology for both studying and treating a range of emotional or psychological issues that involve the environment.

Spatial Cognition, Navigation, and Wayfinding

Navigation is a basic skill required by all animals and is one of the basic functions of our nervous system (Devlin, 2001; Geva-Sagiv et al., 2015) (➔ chapter "The Potential of Biophilic Design and Nature to Improve Health, Creativity and

Well-Being" by Sjövall and Spiers). The research on spatial cognition, navigation, and wayfinding majorly focuses on understanding questions like "Why and how do such processes lead individuals' wayfinding abilities and navigation in familiar and unfamiliar environments?"

Spatial cognition involves assessing how individuals mentally represent and organize spatial information about their environment. This can include cognitive maps, verbal descriptions, or mental representations of places (Epstein et al., 2017; Franke & Schweikart, 2017). Wayfinding refers to the process of determining and following a route or path to reach a destination within a physical environment (Baskaya et al., 2012; Lynch, 1960). It involves using spatial cues, landmarks, maps, and other environmental information to make decisions about the best route to take (Andre & Koonce, 1991; Bjerva & Sigurjónsson, 2017; Cadwallader, 1975). Neuroscientists have identified brain regions and their specific cellular structures that perform spatial cognition in mice (Fyhn et al., 2004; O'Keefe & Dostrovsky, 1971) and have further extended this to non-human primates (Hori et al., 2003; Killian et al., 2012), inspiring important insights into how the human nervous system contributes to the recognition of spatial location and orientation in the environment (Park et al., 2018; Watrous et al., 2013).

XR provides unique opportunities to simulate and study spatial environments in controlled and immersive settings, like a city, a building, a park, or a complex indoor space, allowing researchers to investigate human behavior, cognition, and decision-making in a wide range of spatial contexts (Çöltekin et al., 2020; Iftikhar et al., 2020; Jamshidi et al., 2020; Suzer et al., 2018; Zhang et al., 2020). Many past studies on these topics have tested real-world scenarios; however, real-world environments are uncontrollable and precise behavioral data cannot be collected efficiently (Kiefer et al., 2017; Ladouce et al., 2017). Scientists can easily build mazes for rats but can't build complex enough real environments in the lab for humans and monitor brain activity simultaneously. Therefore, XR provides a cost–benefit solution for setting up these scenarios. Meanwhile, peripheral input devices such as omnidirectional treadmills and cockpits provide a various mobility experience for wayfinding studies (Carvalho et al., 2006; Nilsson et al., 2018).

A major research direction is to discover the mechanisms of spatial cognition in humans' brains and compare them with evidence found from other animals. For instance, a study used screen-based VR maze with EEG and NIRs to determine key regions for navigation and spatial processing in the human brain (Kober et al., 2013). They discover that the hemodynamic changes (oxy- and deoxy-Hb) over the whole cortex and EEG theta power at centro-parietal are realted to the spatial task. Other studies using screen-based VR and fMRI have demonstrated the hippocampal and striatal systems, and the presence of grid-like signals in the entorhinal cortex in humans is related to different spatial cognitions (Brown et al., 2012; Doeller et al., 2010; Park et al., 2018). Another research direction is more pragmatically oriented to better optimize the built environment by understanding spatial cognition and decision-making mechanisms. One study use HMD VR and EEG to evaluate the impacts of color, graphics, and architectural features on wayfinding in a complex hospital (Kalantari et al., 2022). Their findings tentatively supported combining

Fig. 16 Virtual shopping mall to test the effect of crowdedness on human wayfinding and locomotion performance. (Li et al., 2019)

color and environmental affordances, which refer to possibilities suggested by a structure, to improve wayfinding cognition compared to designs with just colored signs or neither. Another study using VR to examine the effect of crowdedness on human wayfinding and locomotion in a multilevel virtual shopping mall (Li et al., 2019) (Fig. 16). They found that high levels of crowding did not affect wayfinding strategies or initial route choice, but participants tended to avoid crowds by approaching environmental boundaries in highly crowded situations. Another particular topic that has become popular recently is the effect of AR or MR-based information on wayfinding efficiency (Kim et al., 2015). Those research expands the experimental literature that compares the effects of maps versus mobile GPS navigation on wayfinding efficiency (Kiefer et al., 2017; Krugliak & Clarke, 2022; Stangl et al., 2023). Understanding how individuals navigate and interact with spaces helps designers create environments that support human needs, enhance accessibility, and promote positive user experiences.

Environmental Aesthetics, Preference, and Analysis

Research on environmental aesthetics, preference, and analysis examines how attributes of our surroundings shape human perceptions, emotions, and preferences. This field investigates how varying sensory qualities, like colors, textures, lighting, acoustics, and spatial configurations, impact esthetic judgments and sentiments (Berlyne, 1974; Nasar, 1994). Historically, studying aesthetic responses involved presenting 2D representations of environments, like photos or videos. XR enables embodied exposure to 3D spaces, allowing for more naturalistic and ecologically valid testing. Participants can realistically experience varying scales, walkthroughs, and durations within virtual environments (Birenboim et al., 2021; Kuliga et al., 2015).

For architect and designers, findings from environmental aesthetics research using XR can provide guidance for creating built environments that foster well-being. Understanding how people perceive and evaluate spaces, façades, layouts, materials, lighting, greenery, and other attributes allow professionals to craft designs that are visually pleasing, emotionally uplifting, functionally intuitive, and psychologically restorative (Bruns & Chamberlain, 2019; Mouratidis & Hassan, 2020; Portman et al., 2015; Yin et al., 2018). For example, to study a person's perception of comfort in different light conditions, it is possible to use virtual reality in conjunction with a small number of additional ambient light and heat sources to conduct tests with low cost but high confidence (Chamilothori et al., 2019; Chinazzo et al., 2021). Neuroimaging tools like EEG and functional near-infrared spectroscopy (fNIRS) can reveal how different design elements engage brain regions involved in emotion, cognition, and reward processing (Jung et al., 2023; Kalantari et al., 2022; Mostafavi, 2022). This neural feedback helps identify attributes that generate positive psychological states and lower cognitive load. Urban planners can also gain insight into how varying environments impact walkability, wayfinding, and overall user experience, enabling the creation of more navigable and user friendly cities with XR (Birenboim et al., 2021; Zhang et al., 2021). Discerning reactions to different scales, configurations, and components support shaping spaces that feel aesthetically inspiring yet remain conducive to productivity, creativity, and connection(Bruns & Chamberlain, 2019; Hu & Roberts, 2020). While there is still a lack of systematic findings on how specific or general built environments affect humans at neurological pathways levels since both the built environment and human brain are complex systems, this area of research is now potentially critical to accelerating evidence-based design practice and enriching human's lives and well-being (Ancora et al., 2022; Lambert et al., 2015).

Challenges of Environmental Neuroscience Research

Despite the potentials of XR in research and application, integrating XR into environmental neuroscience presents a myriad of challenges that must be carefully addressed to ensure its success and reliability. One significant obstacle is the quality of the simulations. The current level of realism remains limited by the modeling of environments and rendering capacity of systems. Although the XR industry has made remarkable progress, achieving a truly total immersive "Metaverse" that is indistinguishable from reality remains an elusive goal for the near future.

The issue of user experience is also crucial when using XR in research. Users may experience motion sickness or discomfort in XR headsets (Freiwald et al., 2020), potentially compromising the accuracy of data collected during experiments. The physical weight of the headset and issues related to wearing comfort, such as sweating, still pose barriers to prolonged and comfortable use of XR devices.

Accessibility is another pressing concern with XR. The cost of an XR system can range from hundreds to thousands of dollars, making it a considerable financial barrier for many researchers and users. The current studies predominantly drew samples from WEIRD populations (White, Educated, Industrialized, Rich, and Democratic) (Henrich, 2020), leaving a gap in our understanding of how findings apply to more diverse populations worldwide. In addition, the lack of motivation or physical constraints to deploy research in other contexts may result in limited access to XR research for more remote and underserved populations. This limitation raises concerns about the generalizability of research findings to a broader and more diverse population.

Another set of challenges arises from the nature of neuroscience research paradigms. Comparing the efficacy of XR-based interventions to real world or existing treatments requires careful consideration and investigation. While XR shows promising potential as a tool for therapy and rehabilitation, its effectiveness in various contexts needs further exploration and evidence.

Adapting neuroimaging techniques to XR requires a better assessment of their reliability and limitations. Imaging techniques like EEG (Choi et al., 2023) and fNIRS (Dong et al., 2017; Piper et al., 2014; Shi et al., 2020) have been adapted to XR research with minor modifications (Fig. 17). These noninvasive methods offer the advantage of being relatively portable and well-suited for use in dynamic and immersive XR environments, providing valuable insights into neural responses to virtual stimuli (Park et al., 2018). However, integrating other powerful neuroimaging methods like functional magnetic resonance imaging (fMRI), magnetoencephalography (MEG), or Positron emission tomography (PET) may prove incompatible with XR setups, posing obstacles to integrating these powerful tools into environmental neuroscience research. These techniques often require large and fixed equipment, making them difficult to implement in dynamic XR environments where participants move and interact freely. Also, electromagnetic interference, safety risks, and hazards between the headset and image devices further limit adaptation and application. The studies of VR systems inside MRI scanners use conventional

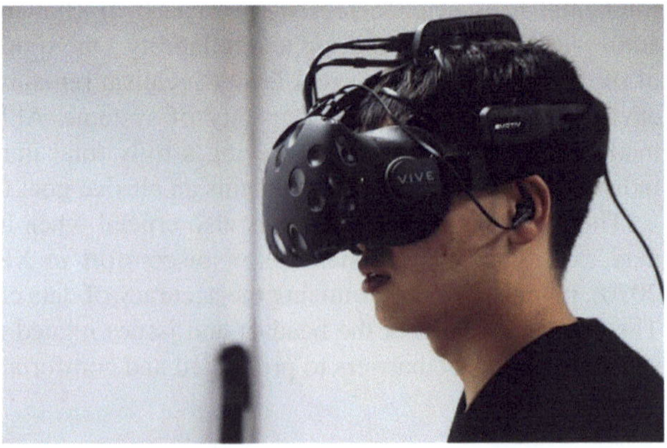

Fig. 17 EEG helmets are relatively easyto adapt to HMDs. (From author)

screens with back-projections via a mirror or compatible goggles with limited fields of view (Forlim et al., 2019; Wiederhold & Wiederhold, 2000). Thus, participants have not benefited from fully immersive VR in these imaging devices.

Addressing these multifaceted challenges is crucial for advancing our understanding of the complex interactions between the human brain and the environment. By overcoming these obstacles, researchers can unlock the full potential of XR in environmental neuroscience research, ultimately leading to better interventions and solutions for various neurological and environmental issues. Collaborative efforts from technology developers, scientists, and policymakers will be essential in driving the field forward and realizing the promising applications of XR in understanding and improving the human brain's responses to the environment.

Conclusion and Vision

Extended reality (XR) holds immense potential for advancing environmental neuroscience research and generating critical insights into the intricate relationship between the human brain, cognition, behavior, and the environment. As outlined in this chapter, XR can recreate complex environments and stimuli with heightened levels of immersion, presence, and ecological validity compared to traditional lab experiments or field studies alone. By combining the experimental control of lab studies with the authenticity of real-world environments, XR enables researchers to examine neural, cognitive, and behavioral responses to environmental conditions in more naturalistic and systematic ways.

However, fully realizing the potential of XR in environmental neuroscience research requires overcoming key challenges regarding the quality of simulations, user experience, accessibility, integrating neuroimaging modalities, and validating its effectiveness compared to conventional methods. While the simulations created using current XR systems may not yet achieve complete verisimilitude with reality, rapid advancements in graphics, computing power, tracking capabilities, and display technologies are steadily enhancing the realism of virtual environments. Developing simulations and experiments collaboratively with domain experts can optimize content accuracy. Research is needed to further address usability constraints posed by discomfort or cybersickness during prolonged XR use. Making XR research more inclusive and accessible worldwide requires reducing costs and physical barriers to use through technological and policy-level efforts. Adapting mobile and less expensive neuroimaging systems appears promising, but interference remains a barrier to implementing XR setups. Carefully designed comparative studies can delineate the advantages, limitations, and validity of XR versus traditional approaches.

As XR technologies continue maturing, they are poised to become a mainstream tool in the environmental neuroscience research toolkit: enhancing ecological validity, facilitating new research paradigms, and enabling studies not feasible previously. With interdisciplinary collaboration and human-centered design, XR systems can be optimized to provide customizable and adaptive environments that dynamically respond to users' neural and behavioral feedback. These advancements pave the way for closed-loop brain–computer interaction paradigms that offer immersive and personalized experiences for myriad research questions and practical applications. Integrating XR with artificial intelligence, multisensory simulations, and telepresence technology can further bridge geographical divides and enhance social presence. Such innovations may transform various facets of society, like revolutionizing clinical treatments, enhancing learning and training, ameliorating workplace environments, and enriching quality of life across the lifespan. Thoughtful development centered on human values is required to ensure that XR's future expands the horizons of possibility for understanding and improving the profound ways our brains interact with the world around us.

References

Ancora, L. A., Blanco-Mora, D. A., Alves, I., Bonifácio, A., Morgado, P., & Miranda, B. (2022). Cities and neuroscience research: A systematic literature review. *Frontiers in Psychiatry, 13*. https://doi.org/10.3389/fpsyt.2022.983352

Andre, A. D., & Koonce, J. M. (1991). Spatial orientation and wayfinding in airport passenger terminals: Implications for environmental design. *Proceedings of the Human Factors Society Annual Meeting, 35*(8), 561–565. https://doi.org/10.1518/107118191786754761

Ascone, L., Ney, K., Mostajeran, F., Steinicke, F., Moritz, S., Gallinat, J., & Kühn, S. (2020). *Virtual reality for individuals with occasional paranoid thoughts*. pp. 1–8.

Baskaya, A., Wilson, C., & Özcan, Y. Z. (2012). Wayfinding in an unfamiliar environment: Different spatial settings of two polyclinics. *Environment and Behavior, 36*(6), 839–867. https://doi.org/10.1177/0013916504265445

Baumgartner, T., Valko, L., Esslen, M., & Jäncke, L. (2006). Neural correlate of spatial presence in an arousing and noninteractive virtual reality: An EEG and psychophysiology study. *Cyberpsychology & Behavior, 9*(1), 30–45. https://doi.org/10.1089/cpb.2006.9.30

Baus, O., & Bouchard, S. (2014). Moving from virtual reality exposure-based therapy to augmented reality exposure-based therapy: A review. *Frontiers in Human Neuroscience, 8*. https://doi.org/10.3389/fnhum.2014.00112

Berlyne, D. E. (1974). *Studies in the new experimental aesthetics: Steps toward an objective psychology of aesthetic appreciation*. Hemisphere Publishing Corporation.

Berman, M. G., Kardan, O., Kotabe, H. P., Nusbaum, H. C., & London, S. E. (2019). The promise of environmental neuroscience. *Nature Human Behaviour, 3*(5), Article 5. https://doi.org/10.1038/s41562-019-0577-7

Berman, M. G., Cardenas-Iniguez, C., & Meidenbauer, K. L. (2021). An environmental neuroscience perspective on the benefits of nature. In A. R. Schutte, J. C. Torquati, & J. R. Stevens (Eds.), *Nature and psychology: Biological, cognitive, developmental, and social pathways to Well-being* (pp. 61–88). Springer International Publishing. https://doi.org/10.1007/978-3-030-69020-5_4

Beverly, E., Hommema, L., Coates, K., Duncan, G., Gable, B., Gutman, T., Love, M., Love, C., Pershing, M., & Stevens, N. (2022). A tranquil virtual reality experience to reduce subjective stress among COVID-19 frontline healthcare workers. *PLoS One, 17*(2), e0262703. https://doi.org/10.1371/journal.pone.0262703

Birenboim, A., Bloom, P. B. N., Levit, H., & Omer, I. (2021). The study of walking, walkability and wellbeing in immersive virtual environments. *International Journal of Environmental Research and Public Health, 18*(2), 1–18. https://doi.org/10.3390/ijerph18020364

Bittner, L., Mostajeran, F., Steinicke, F., Gallinat, J., & Kühn, S. (2018). Evaluation of flowvr: A virtual reality game for improvement of depressive mood. *Biorxiv*, 451245.

Bjerva, T., & Sigurjónsson, T. (2017). Wayfinding by means of maps in real-world settings: A critical review. *Journal of Navigation, 70*(2), 263–275. https://doi.org/10.1017/S0373463316000643

Blake, R., & Wilson, H. (2011). Binocular vision. *Vision Research, 51*(7), 754–770. https://doi.org/10.1016/j.visres.2010.10.009

Blana, D., Kyriacou, T., Lambrecht, J. M., & Chadwick, E. K. (2016). Feasibility of using combined EMG and kinematic signals for prosthesis control: A simulation study using a virtual reality environment. *Journal of Electromyography and Kinesiology, 29*, 21–27. https://doi.org/10.1016/j.jelekin.2015.06.010

Bohil, C. J., Alicea, B., & Biocca, F. A. (2011). Virtual reality in neuroscience research and therapy. *Nature Reviews Neuroscience, 12*(12), 752–762. https://doi.org/10.1038/nrn3122

Botvinick, M., & Cohen, J. (1998). Rubber hands 'feel'touch that eyes see. *Nature, 391*(6669), 756–756. https://doi.org/10.1038/35784

Bowers, B. (2001). *Sir Charles Wheatstone FRS: 1802–1875* (Issue 29). IET.

Brown, A., & Green, T. (2016). Virtual reality: Low-cost tools and resources for the classroom. *TechTrends, 60*, 517–519. https://doi.org/10.1007/s11528-016-0102-z

Brown, T. I., Ross, R. S., Tobyne, S. M., & Stern, C. E. (2012). Cooperative interactions between hippocampal and striatal systems support flexible navigation. *NeuroImage, 60*(2), 1316–1330. https://doi.org/10.1016/j.neuroimage.2012.01.046

Browning, M. H. E. M., Mimnaugh, K. J., van Riper, C. J., Laurent, H. K., & LaValle, S. M. (2020a). Can simulated nature support mental health? Comparing short, single-doses of 360-degree nature videos in virtual reality with the outdoors. *Frontiers in Psychology, 10*(January), 1–14. https://doi.org/10.3389/fpsyg.2019.02667

Browning, M., Suppakittpaisarn, P., Jiang, S., Joseph, A., & Diao, Z. (2020b). Human health assessments of green infrastructure designs using virtual reality. *Landscape Architecture, 27*, 10.14085/j.fjyl.2020.09.0035.15.

Bruns, C. R., & Chamberlain, B. C. (2019). The influence of landmarks and urban form on cognitive maps using virtual reality. *Landscape and Urban Planning, 189*, 296–306. https://doi.org/10.1016/j.landurbplan.2019.05.006

Cadwallader, M. (1975). A behavioral model of consumer spatial decision making. *Economic Geography, 51*(4), 339–349. https://doi.org/10.2307/142918

Calhoun, V. D., Carvalho, K., Astur, R., & Pearlson, G. D. (2005). Using virtual reality to study alcohol intoxication effects on the neural correlates of simulated driving. *Applied Psychophysiology and Biofeedback, 30*(3), 285–306. https://doi.org/10.1007/s10484-005-6384-0

Carter, M., Webber, S., Rawson, S., Smith, W., Purdam, J., & McLeod, E. (2020). Virtual reality in the zoo: A qualitative evaluation of a stereoscopic virtual reality video encounter with little penguins (Eudyptula minor). *Journal of Zoo and Aquarium Research, 8*(4), Article 4. https://doi.org/10.19227/jzar.v8i4.500

Carvalho, K. N., Pearlson, G. D., Astur, R. S., & Calhoun, V. D. (2006). Simulated driving and brain imaging: Combining behavior, brain activity, and virtual reality. *CNS Spectrums, 11*(1), 52–62. https://doi.org/10.1017/S1092852900024214

Casson, A. J., & Trimble, E. V. (2018). Enabling free movement EEG tasks by eye fixation and gyroscope motion correction: EEG effects of color priming in dress shopping. *IEEE Access, 6*, 62975–62987. https://doi.org/10.1109/ACCESS.2018.2877158

Caudell, T. P. (1995). *Introduction to augmented and virtual reality*. 2351, pp. 272–281.

Chamilothori, K., Wienold, J., & Andersen, M. (2019). Adequacy of immersive virtual reality for the perception of Daylit spaces: Comparison of real and virtual environments. *LEUKOS, 15*(2–3), 203–226. https://doi.org/10.1080/15502724.2017.1404918

Chang, D. H. F., Jiang, B., Wong, N. H. L., Wong, J. J., Webster, C., & Lee, T. M. C. (2020). The human posterior cingulate and the stress-response benefits of viewing green urban landscapes. *NeuroImage, 226*, 117555–117555. https://doi.org/10.1016/j.neuroimage.2020.117555

Chinazzo, G., Chamilothori, K., Wienold, J., & Andersen, M. (2021). Temperature–color interaction: Subjective indoor environmental perception and physiological responses in virtual reality. *Human Factors, 63*(3), 474–502. https://doi.org/10.1177/0018720819892383

Choi, J. W., Kwon, H., Choi, J., Kaongoen, N., Hwang, C., Kim, M., Kim, B. H., & Jo, S. (2023). Neural applications using immersive virtual reality: A review on EEG studies. *IEEE Transactions on Neural Systems and Rehabilitation Engineering., 31*, 1645. https://doi.org/10.1109/TNSRE.2023.3254551

Chung, K., Lee, D., & Park, J. Y. (2018). Involuntary attention restoration during exposure to mobile-based 360 virtual nature in healthy adults with different levels of restorative experience: Event-related potential study. *Journal of Medical Internet Research, 20*(11), e11152. https://doi.org/10.2196/11152

Cieślik, B., Mazurek, J., Rutkowski, S., Kiper, P., Turolla, A., & Szczepańska-Gieracha, J. (2020). Virtual reality in psychiatric disorders: A systematic review of reviews. *Complementary Therapies in Medicine, 52*, 102480. https://doi.org/10.1016/j.ctim.2020.102480

Clay, V., König, P., & König, S. U. (2019). Eye tracking in virtual reality. *Journal of Eye Movement Research, 12*(1), 10.16910/jemr.12.1.3.

Çöltekin, A., Lochhead, I., Madden, M., Christophe, S., Devaux, A., Pettit, C., Lock, O., Shukla, S., Herman, L., Stachoň, Z., Kubíček, P., Snopková, D., Bernardes, S., & Hedley, N. (2020). Extended reality in spatial sciences: A review of research challenges and future directions. *ISPRS International Journal of Geo-Information, 9*(7), Article 7. https://doi.org/10.3390/ijgi9070439

Cruz-Neira, C., Sandin, D. J., DeFanti, T. A., Kenyon, R. V., & Hart, J. C. (1992). The CAVE: audio visual experience automatic virtual environment. *Communications of the ACM, 35*(6), 64–73. https://doi.org/10.1145/129888.129892

Deb, S., Carruth, D. W., & Hudson, C. R. (2020). How communicating features can help pedestrian safety in the presence of self-driving vehicles: Virtual reality experiment. *IEEE Transactions on Human-Machine Systems, 50*(2), 176–186. https://doi.org/10.1109/THMS.2019.2960517

Dempsey, P. (2016). The teardown: HTC vive VR headset. *Engineering & Technology, 11*(7–8), 80–81.

Depledge, M. H., Stone, R. J., & Bird, W. (2011). Can natural and virtual environments be used to promote improved human health and wellbeing? *Environmental Science & Technology, 45*(11), 4660–4665. https://doi.org/10.1021/es103907m

Desai, P. R., Desai, P. N., Ajmera, K. D., & Mehta, K. (2014). A review paper on oculus rift-a virtual reality headset. *arXiv Preprint arXiv:1408.1173.*

Devlin, A. S. (2001). *Mind and maze: Spatial cognition and environmental behavior*. Praeger Publishers/Greenwood Publishing Group.

Diniz Bernardo, P., Bains, A., Westwood, S., & Mograbi, D. C. (2021). Mood induction using virtual reality: A systematic review of recent findings. *Journal of Technology in Behavioral Science, 6*(1), 3–24. https://doi.org/10.1007/s41347-020-00152-9

Dobbert, T. (2006). *Matchmoving: The invisible art of camera tracking*. Wiley.

Doeller, C. F., Barry, C., & Burgess, N. (2010). Evidence for grid cells in a human memory network. *Nature, 463*(7281), 657–661. https://doi.org/10.1038/nature08704

Dong, D., Wong, L. K., & Luo, Z. (2017). Assessment of prospective memory using fNIRS in immersive virtual reality environment. *Journal of Behavioral and Brain Science, 7*(06), 247. https://doi.org/10.4236/jbbs.2017.76018

Ellis, M. (2008). 'Spectacles within doors': Panoramas of London in the 1790s. *Romanticism, 14*(2), 133–148. https://doi.org/10.3366/E1354991X0800024X

Elor, A., Thang, T., Hughes, B. P., Crosby, A., Phung, A., Gonzalez, E., Katija, K., Haddock, S. H., Martin, E. J., & Erwin, B. E. (2021). *Catching jellies in immersive virtual reality: A comparative teleoperation study of ROVs in underwater capture tasks*. pp. 1–10.

Epstein, R. A., Patai, E. Z., Julian, J. B., & Spiers, H. J. (2017). The cognitive map in humans: Spatial navigation and beyond. *Nature Neuroscience, 20*(11), 1504–1513. https://doi.org/10.1038/nn.4656

Erkan, İ. (2018). Examining wayfinding behaviours in architectural spaces using brain imaging with electroencephalography (EEG). *Architectural Science Review, 61*(6), 410–428. https://doi.org/10.1080/00038628.2018.1523129

Erol-Kantarci, M., & Sukhmani, S. (2018). *Caching and computing at the edge for mobile augmented reality and virtual reality (AR/VR) in 5G*. pp. 169–177.

Fang, Z., Cai, L., & Wang, G. (2021). *MetaHuman creator the starting point of the metaverse*. pp. 154–157.

Forlim, C. G., Bittner, L., Mostajeran, F., Steinicke, F., Gallinat, J., & Kühn, S. (2019). Stereoscopic rendering via goggles elicits higher functional connectivity during virtual reality gaming. *Frontiers in Human Neuroscience, 13*, 365. https://doi.org/10.3389/fnhum.2019.00365

Franke, C., & Schweikart, J. (2017). Mental representation of landmarks on maps: Investigating cartographic visualization methods with eye tracking technology. *Spatial Cognition & Computation, 17*(1–2), 20–38. https://doi.org/10.1080/13875868.2016.1219912

Freiwald, J. P., Göbel, Y., Mostajeran, F., & Steinicke, F. (2020). The cybersickness susceptibility questionnaire: Predicting virtual reality tolerance. In *Proceedings of mensch und computer 2020* (pp. 115–118).

Fyhn, M., Molden, S., Witter, M. P., Moser, E. I., & Moser, M.-B. (2004). Spatial representation in the entorhinal cortex. *Science, 305*(5688), 1258–1264. https://doi.org/10.1126/science.1099901

Garcia-Palacios, A., Botella, C., Hoffman, H., & Fabregat, S. (2007). Comparing acceptance and refusal rates of virtual reality exposure vs. in vivo exposure by patients with specific phobias. *Cyberpsychology & Behavior, 10*(5), 722–724. https://doi.org/10.1089/cpb.2007.9962

Geva-Sagiv, M., Las, L., Yovel, Y., & Ulanovsky, N. (2015). Spatial cognition in bats and rats: From sensory acquisition to multiscale maps and navigation. *Nature Reviews Neuroscience, 16*(2), Article 2. https://doi.org/10.1038/nrn3888

Gibson, J. J. (1979). *The ecological approach to visual perception*. Houghton Mifflin.

Golledge, R. G., Moore, G. T., Briggs, R., Cadwallader, M. T., Devlin, A. S., George, D. L., Zannaras, G., Kreimer, A., Nigl, A. J., & Fishbein, H. D. (2018). Environmental cognition. In *Environmental design research* (pp. 182–260). Routledge.

Greuter, S., & Roberts, D. J. (2014). *Spacewalk: Movement and interaction in virtual space with commodity hardware*. pp. 1–7.

Gruber, G. J. (2014). *VIEW MASTER: The biography of William B. Gruber*. Hillcrest Publishing Group.

Harrivel, A. R., Stephens, C. L., Milletich, R. J., Heinich, C. M., Last, M. C., Napoli, N. J., Abraham, N., Prinzel, L. J., Motter, M. A., & Pope, A. T. (2017, January 9). Prediction of cognitive states during flight simulation using multimodal psychophysiological sensing. *AIAA Information Systems-AIAA Infotech @ Aerospace*. AIAA Information Systems-AIAA Infotech @ Aerospace, Grapevine, Texas. https://doi.org/10.2514/6.2017-1135

Hartig, T., Mang, M., & Evans, G. W. (1991). Restorative effects of natural environment experiences. *Environment and Behavior, 23*(1), 3–26. https://doi.org/10.1177/0013916591231001

Heilman, K. M. (1997). The neurobiology of emotional experience. *The Journal of Neuropsychiatry and Clinical Neurosciences, 9*(3), 439–448. https://doi.org/10.1176/jnp.9.3.439

Henrich, J. (2020). *The WEIRDest people in the world: How the west became psychologically peculiar and particularly prosperous*. Farrar, Straus and Giroux.

Hori, E., Tabuchi, E., Matsumura, N., Tamura, R., Eifuku, S., Endo, S., Nishijo, H., & Ono, T. (2003). Representation of place by monkey hippocampal neurons in real and virtual translocation. *Hippocampus, 13*(2), 190–196. https://doi.org/10.1002/hipo.10062

Hu, M., & Roberts, J. (2020). Built environment evaluation in virtual reality environments—A cognitive neuroscience approach. *Urban Science, 4*(4), 48. https://doi.org/10.3390/urbansci4040048

Huang, S.-L. (Jully), & Chang, W.-C. (2016). Along the river during the Qingming festival. 天行書苑.
Hutson, M. (2018). Reaching out to touch virtual reality. *Knowable Magazine from Annual Reviews, 9.* https://www.knowablemagazine.org/article/technology/2018/reaching-out-touch-virtual-reality.
Huttunen, K., Keränen, H., Väyrynen, E., Pääkkönen, R., & Leino, T. (2011). Effect of cognitive load on speech prosody in aviation: Evidence from military simulator flights. *Applied Ergonomics, 42*(2), 348–357. https://doi.org/10.1016/j.apergo.2010.08.005
Iftikhar, H., Shah, P., & Luximon, Y. (2020). Human wayfinding behaviour and metrics in complex environments: A systematic literature review. *Architectural Science Review*, 1. https://doi.org/10.1080/00038628.2020.1777386
IJsselsteijn, W. A., de Kort, Y. A. W., & Haans, A. (2006). Is this my hand I see before me? The rubber hand illusion in reality, virtual reality, and mixed reality. *Presence: Teleoperators and Virtual Environments, 15*(4), 455–464. https://doi.org/10.1162/pres.15.4.455
Innocenti, A. (2017). Virtual reality experiments in economics. *Journal of Behavioral and Experimental Economics, 69*, 71–77. https://doi.org/10.1016/j.socec.2017.06.001
Jallais, C., & Gilet, A.-L. (2010). Inducing changes in arousal and valence: Comparison of two mood induction procedures. *Behavior Research Methods, 42*(1), 318–325. https://doi.org/10.3758/BRM.42.1.318
Jamshidi, S., Ensafi, M., & Pati, D. (2020). Wayfinding in interior environments: An integrative review. *Frontiers in Psychology, 11*. https://doi.org/10.3389/fpsyg.2020.549628
Jiang, B., Chang, C. Y., & Sullivan, W. C. (2014). A dose of nature: Tree cover, stress reduction, and gender differences. *Landscape and Urban Planning, 132*, 26–36. https://doi.org/10.1016/j.landurbplan.2014.08.005
Jiang, B., He, J., Chen, J., Larsen, L., & Wang, H. (2021). Perceived Green at speed: A simulated driving experiment raises new questions for attention restoration theory and stress reduction theory. *Environment and Behavior, 53*(3), 296–335. https://doi.org/10.1177/0013916520947111
Jones, S., & Dawkins, S. (2018). The Sensorama revisited: Evaluating the application of multisensory input on the sense of presence in 360-degree immersive film in virtual reality. In T. Jung & M. C. tom Dieck (Eds.), *Augmented reality and virtual reality: Empowering human, place and business* (pp. 183–197). Springer International Publishing. https://doi.org/10.1007/978-3-319-64027-3_13
Jun, J., Jung, M., Kim, S.-Y., & Kim, K. (Kenny). (2018). Full-Body Ownership Illusion Can Change Our Emotion. In *Proceedings of the 2018 CHI conference on human factors in computing systems* (pp. 1–11). https://doi.org/10.1145/3173574.3174175
Jung, D., Kim, D. I., & Kim, N. (2023). Bringing nature into hospital architecture: Machine learning-based EEG analysis of the biophilia effect in virtual reality. *Journal of Environmental Psychology, 89*, 102033. https://doi.org/10.1016/j.jenvp.2023.102033
Kalantari, S., Tripathi, V., Kan, J., Rounds, J. D., Mostafavi, A., Snell, R., & Cruz-Garza, J. G. (2022). Evaluating the impacts of color, graphics, and architectural features on wayfinding in healthcare settings using EEG data and virtual response testing. *Journal of Environmental Psychology, 79*, 101744. https://doi.org/10.1016/j.jenvp.2021.101744
Kaplan, S. (1995). The restorative benefits of nature: Toward an integrative framework. *Journal of Environmental Psychology, 15*(3), 169–182. https://doi.org/10.1016/0272-4944(95)90001-2
Kaplan, R., & Kaplan, S. (1989). *The experience of nature: A psychological perspective*. Cambridge University Press.
Kellert, S. R., & Wilson, E. O. (1993). *The biophilia hypothesis*. Island Press.
Kiefer, P., Giannopoulos, I., Raubal, M., & Duchowski, A. (2017). Eye tracking for spatial research: Cognition, computation, challenges. *Spatial Cognition & Computation, 17*(1–2), 1–19. https://doi.org/10.1080/13875868.2016.1254634
Killian, N. J., Jutras, M. J., & Buffalo, E. A. (2012). A map of visual space in the primate entorhinal cortex. *Nature, 491*(7426), Article 7426. https://doi.org/10.1038/nature11587

Kim, M. J., Wang, X., Han, S., & Wang, Y. (2015). Implementing an augmented reality-enabled wayfinding system through studying user experience and requirements in complex environments. *Visualization in Engineering, 3*(1), 1–12. https://doi.org/10.1186/s40327-015-0026-2

Kim, M., Cheon, S., & Kang, Y. (2019). Use of electroencephalography (EEG) for the analysis of emotional perception and fear to nightscapes. *Sustainability, 11*(1), 233. https://doi.org/10.3390/su11010233

Kober, S. E., Wood, G., & Neuper, C. (2013). Measuring brain activation during spatial navigation in virtual reality: A combined EEG-NIRS study. *International Journal of Computer Research, 20*(1), 1.

Krugliak, A., & Clarke, A. (2022). Towards real-world neuroscience using mobile EEG and augmented reality. *Scientific Reports, 12*(1), Article 1. https://doi.org/10.1038/s41598-022-06296-3

Kühn, S., Forlim, C. G., Lender, A., Wirtz, J., & Gallinat, J. (2021). Brain functional connectivity differs when viewing pictures from natural and built environments using fMRI resting state analysis. *Scientific Reports, 11*(1), Article 1. https://doi.org/10.1038/s41598-021-83246-5

Kuliga, S. F., Thrash, T., Dalton, R. C., & Hölscher, C. (2015). Virtual reality as an empirical research tool—Exploring user experience in a real building and a corresponding virtual model. *Computers, Environment and Urban Systems, 54*(Complete), 363–375. https://doi.org/10.1016/j.compenvurbsys.2015.09.006

Ladouce, S., Donaldson, D. I., Dudchenko, P. A., & Ietswaart, M. (2017). Understanding minds in real-world environments: Toward a mobile cognition approach. *Frontiers in Human Neuroscience, 10*. https://doi.org/10.3389/fnhum.2016.00694

Lambert, K. G., Nelson, R. J., Jovanovic, T., & Cerdá, M. (2015). Brains in the city: Neurobiological effects of urbanization. *Neuroscience & Biobehavioral Reviews, 58*, 107–122. https://doi.org/10.1016/j.neubiorev.2015.04.007

Lang, P. J., Bradley, M. M., & Cuthbert, B. N. (1990). Emotion, attention, and the startle reflex. *Psychological Review, 97*(3), 377. https://doi.org/10.1037/0033-295X.97.3.377

Lanier, J. (1992). Virtual reality: The promise of the future. *Interactive Learning International, 8*(4), 275–279.

Li, H., Thrash, T., Hölscher, C., & Schinazi, V. R. (2019). The effect of crowdedness on human wayfinding and locomotion in a multi-level virtual shopping mall. *Journal of Environmental Psychology, 65*, 101320. https://doi.org/10.1016/j.jenvp.2019.101320

Lin, C.-T., Wu, R.-C., Jung, T.-P., Liang, S.-F., & Huang, T.-Y. (2005). Estimating driving performance based on EEG spectrum analysis. *EURASIP Journal on Advances in Signal Processing, 2005*(19), 521368. https://doi.org/10.1155/ASP.2005.3165

Lin, C.-T., Chung, I.-F., Ko, L.-W., Chen, Y.-C., Liang, S.-F., & Duann, J.-R. (2007). EEG-based assessment of driver cognitive responses in a dynamic virtual-reality driving environment. *IEEE Transactions on Biomedical Engineering, 54*(7), 1349–1352. https://doi.org/10.1109/TBME.2007.891164

Lorentz, L., Schüppen, A., Suchan, B., & Binkofski, F. (2023). Neural correlates of virtual reality-based attention training: An fMRI study. *NeuroImage, 284*, 120454. https://doi.org/10.1016/j.neuroimage.2023.120454

Lorenz, L., Kerschbaum, P., & Schumann, J. (2014). Designing take over scenarios for automated driving: How does augmented reality support the driver to get back into the loop? *Proceedings of the Human Factors and Ergonomics Society Annual Meeting, 58*(1), 1681–1685.

Lynch, K. (1960). *The image of the city* (Vol. 11). MIT press. https://books.google.com/books?hl=en&lr=&id=_phRPWsSpAgC&oi=fnd&pg=PA1&ots=jHHa6i1Fid&sig=NWtElimJabUSben44JRAO-C7EHw

Lynch, K. (1984). Reconsidering the image of the City. In L. Rodwin & R. M. Hollister (Eds.), *Cities of the mind: Images and themes of the city in the social sciences* (pp. 151–161). Springer US. https://doi.org/10.1007/978-1-4757-9697-1_9

Magdin, M., Balogh, Z., Reichel, J., Francisti, J., Koprda, Š., & György, M. (2021). Automatic detection and classification of emotional states in virtual reality and standard environments

(LCD): Comparing valence and arousal of induced emotions. *Virtual Reality, 25*(4), 1029–1041. https://doi.org/10.1007/s10055-021-00506-5

Maselli, A., & Slater, M. (2013). The building blocks of the full body ownership illusion. *Frontiers in Human Neuroscience, 7*. https://doi.org/10.3389/fnhum.2013.00083

Masters, R., Interrante, V., Watts, M., & Ortega, F. (2022). Virtual nature: Investigating the effect of biomass on immersive virtual reality forest bathing applications for stress reduction. *ACM symposium on applied perception, 2022*, 1–10. https://doi.org/10.1145/3548814.3551459

McMahan, R. P., Lai, C., & Pal, S. K. (2016). Interaction fidelity: The uncanny valley of virtual reality interactions. In S. Lackey & R. Shumaker (Eds.), *Virtual, augmented and mixed reality* (pp. 59–70). Springer International Publishing. https://doi.org/10.1007/978-3-319-39907-2_6

Mesa-Gresa, P., Gil-Gómez, H., Lozano-Quilis, J.-A., & Gil-Gómez, J.-A. (2018). Effectiveness of virtual reality for children and adolescents with autism spectrum disorder: An evidence-based systematic review. *Sensors, 18*(8), Article 8. https://doi.org/10.3390/s18082486

Milgram, P., & Kishino, F. (1994). A taxonomy of mixed reality visual displays. IEICE TRANSACTIONS on Information and Systems, 77(12), 1321–1329.

Mneimne, M., Powers, A. S., Walton, K. E., Kosson, D. S., Fonda, S., & Simonetti, J. (2010). Emotional valence and arousal effects on memory and hemispheric asymmetries. *Brain and Cognition, 74*(1), 10–17. https://doi.org/10.1016/j.bandc.2010.05.011

Mostafavi, A. (2022). Architecture, biometrics, and virtual environments triangulation: A research review. *Architectural Science Review, 65*(6), 504–521. https://doi.org/10.1080/00038628.2021.2008300

Mostajeran, F., Kirsten, A., Steinicke, F., Gallinat, J., & Kühn, S. (2019). *Towards gamified alcohol use disorder therapy in virtual reality: A preliminary usability study* (pp. 1471–1476).

Mostajeran, F., Balci, M. B., Steinicke, F., Kühn, S., & Gallinat, J. (2020). The effects of virtual audience size on social anxiety during public speaking. In *2020 IEEE conference on virtual reality and 3D user interfaces (VR)* (pp. 303–312). https://doi.org/10.1109/VR46266.2020.00050

Mostajeran, F., Krzikawski, J., Steinicke, F., & Kühn, S. (2021). Effects of exposure to immersive videos and photo slideshows of forest and urban environments. *Scientific Reports, 11*(1), Article 1. https://doi.org/10.1038/s41598-021-83277-y

Mostajeran, F., Fischer, M., Steinicke, F., & Kühn, S. (2023a). Effects of exposure to immersive computer-generated virtual nature and control environments on affect and cognition. *Scientific Reports, 13*(1), 220. https://doi.org/10.1038/s41598-022-26750-6

Mostajeran, F., Steinicke, F., Reinhart, S., Stuerzlinger, W., Riecke, B. E., & Kühn, S. (2023b). Adding virtual plants leads to higher cognitive performance and psychological well-being in virtual reality. *Scientific Reports, 13*(1), Article 1. https://doi.org/10.1038/s41598-023-34718-3

Mouratidis, K., & Hassan, R. (2020). Contemporary versus traditional styles in architecture and public space: A virtual reality study with 360-degree videos. *Cities, 97*(June 2019), 102499. https://doi.org/10.1016/j.cities.2019.102499

Mu, M., Dohan, M., Goodyear, A., Hill, G., Johns, C., & Mauthe, A. (2022). User attention and behaviour in virtual reality art encounter. *Multimedia Tools and Applications., 83*, 46595. https://doi.org/10.1007/s11042-022-13365-2

Nasar, J. L. (1994). Urban design aesthetics: The evaluative qualities of building exteriors. *Environment and Behavior, 26*(3), 377–401. https://doi.org/10.1177/001391659402600305

Nerb, J., & Spada, H. (2001). Evaluation of environmental problems: A coherence model of cognition and emotion. *Cognition & Emotion, 15*(4), 521–551.

Nichols, S., & Patel, H. (2002). Health and safety implications of virtual reality: A review of empirical evidence. *Applied Ergonomics, 33*(3), 251–271. https://doi.org/10.1016/S0003-6870(02)00020-0

Nilsson, N. C., Serafin, S., Steinicke, F., & Nordahl, R. (2018). Natural walking in virtual reality: A review. *Computers in Entertainment, 16*(2), 8:1–8:22. https://doi.org/10.1145/3180658

Nukarinen, T., Rantala, J., Korpela, K., Browning, M. H. E. M., Istance, H. O., Surakka, V., & Raisamo, R. (2022). Measures and modalities in restorative virtual natural environments:

An integrative narrative review. *Computers in Human Behavior, 126*, 107008. https://doi.org/10.1016/j.chb.2021.107008
O'Keefe, J., & Dostrovsky, J. (1971). The hippocampus as a spatial map: Preliminary evidence from unit activity in the freely-moving rat. *Brain Research., 34*, 171. https://doi.org/10.1016/0006-8993(71)90358-1
O'Regan, J. K., & noë, A. (2001). What it is like to see: A sensorimotor theory of perceptual experience. *Synthese, 129*(1), 79–103. https://doi.org/10.1023/A:1012699224677
Otto, P. (2007). Between the virtual and the actual: Robert Barker's panorama of London and the multiplication of the real in late eighteenth-century London. *Romanticism on the Net, 46*. https://doi.org/10.7202/016130ar
Park, J. L., Dudchenko, P. A., & Donaldson, D. I. (2018). Navigation in real-world environments: New opportunities afforded by advances in mobile brain imaging. *Frontiers in Human Neuroscience, 12*. https://doi.org/10.3389/fnhum.2018.00361
Parsons, S., & Mitchell, P. (2002). The potential of virtual reality in social skills training for people with autistic spectrum disorders. *Journal of Intellectual Disability Research, 46*(5), 430–443. https://doi.org/10.1046/j.1365-2788.2002.00425.x
Peddie, J. (2017). Historical overview. In J. Peddie (Ed.), *Augmented reality: Where we will all live* (pp. 59–86). Springer International Publishing. https://doi.org/10.1007/978-3-319-54502-8_5
Piper, S. K., Krueger, A., Koch, S. P., Mehnert, J., Habermehl, C., Steinbrink, J., Obrig, H., & Schmitz, C. H. (2014). A wearable multi-channel fNIRS system for brain imaging in freely moving subjects. *NeuroImage, 85*, 64–71. https://doi.org/10.1016/j.neuroimage.2013.06.062
Portman, M. E., Natapov, A., & Fisher-Gewirtzman, D. (2015). To go where no man has gone before: Virtual reality in architecture, landscape architecture and environmental planning. *Computers, Environment and Urban Systems, 54*, 376–384. https://doi.org/10.1016/j.compenvurbsys.2015.05.001
Presti, P., Ruzzon, D., Avanzini, P., Caruana, F., Rizzolatti, G., & Vecchiato, G. (2022). Measuring arousal and valence generated by the dynamic experience of architectural forms in virtual environments. *Scientific Reports, 12*(1), Article 1. https://doi.org/10.1038/s41598-022-17689-9
Rizzo, A. "Skip," & Kim, G. J. (2005). A SWOT analysis of the field of virtual reality rehabilitation and therapy. *Presence, 14*(2), 119–146. https://doi.org/10.1162/1054746053967094
Rizzo, A., & Koenig, S. T. (2017). Is clinical virtual reality ready for primetime? *Neuropsychology, 31*(8), 877. https://doi.org/10.1037/neu0000405
Rokhsaritalemi, S., Sadeghi-Niaraki, A., & Choi, S.-M. (2020). A review on mixed reality: Current trends, challenges and prospects. *Applied Sciences, 10*(2), Article 2. https://doi.org/10.3390/app10020636
Schmalstieg, D., & Hollerer, T. (2016). *Augmented reality: Principles and practice*. Addison-Wesley Professional.
Schmidt, S., Steinicke, F., Irlitti, A., & Thomas, B. H. (2018). *Floor-projected guidance cues for collaborative exploration of spatial augmented reality setups* (pp. 279–289).
Shi, Y., Zhu, Y., Mehta, R. K., & Du, J. (2020). A neurophysiological approach to assess training outcome under stress: A virtual reality experiment of industrial shutdown maintenance using Functional Near-Infrared Spectroscopy (fNIRS). *Advanced Engineering Informatics, 46*, 101153. https://doi.org/10.1016/j.aei.2020.101153
Skarbez, R., Smith, M., & Whitton, M. C. (2021). Revisiting Milgram and Kishino's reality-virtuality continuum. *Frontiers in Virtual Reality, 2*, 647997. https://doi.org/10.3389/frvir.2021.647997
Slater, M. (2009). Place illusion and plausibility can lead to realistic behaviour in immersive virtual environments. *Philosophical Transactions of the Royal Society B: Biological Sciences, 364*(1535), 3549–3557. https://doi.org/10.1098/rstb.2009.0138
Slater, M., & Wilbur, S. (1997). A framework for immersive virtual environments (FIVE): Speculations on the role of presence in virtual environments. *Presence: Teleoperators & Virtual Environments, 6*(6), 603–616. https://doi.org/10.1162/pres.1997.6.6.603

Slater, M., Banakou, D., Beacco, A., Gallego, J., Macia-Varela, F., & Oliva, R. (2022). A separate reality: An update on place illusion and plausibility in virtual reality. *Frontiers in Virtual Reality, 3*. https://doi.org/10.3389/frvir.2022.914392

Smith, J. W. (2015). Immersive virtual environment technology to supplement environmental perception, preference and behavior research: A review with applications. *International Journal of Environmental Research and Public Health, 12*(9), 11486–11505. https://doi.org/10.3390/ijerph120911486

Souza, R. H. C. E., & Naves, E. L. M. (2021). Attention detection in virtual environments using EEG signals: A scoping review. *Frontiers in Physiology, 12*. https://doi.org/10.3389/fphys.2021.727840

Spanlang, B., Normand, J.-M., Borland, D., Kilteni, K., Giannopoulos, E., Pomes, A., Gonzalez-Franco, M., Perez-Marcos, D., Arroyo-Palacios, J., Muncunill, X. N., & Slater, M. (2014). How to build an embodiment lab: Achieving body representation illusions in virtual reality. *Frontiers in Robotics and AI*, 1. https://doi.org/10.3389/frobt.2014.00009

Speicher, M., Hall, B. D., & Nebeling, M. (2019). *What is mixed reality?* (pp. 1–15).

Stangl, M., Maoz, S. L., & Suthana, N. (2023). Mobile cognition: Imaging the human brain in the 'real world'. *Nature Reviews Neuroscience, 24*(6), Article 6. https://doi.org/10.1038/s41583-023-00692-y

Sun, Y., Li, F., He, T., Meng, Y., Yin, J., Yim, I. S., Xu, L., & Wu, J. (2023). Physiological and affective responses to green space virtual reality among pregnant women. *Environmental Research, 216*, 114499. https://doi.org/10.1016/j.envres.2022.114499

Sutherland, I. E. (1965). *The ultimate display, 2*, 506–508.

Sutherland, I. E. (1968). A head-mounted three dimensional display. In *Proceedings of the December 9–11, 1968, Fall Joint Computer Conference, Part I on – AFIPS '68 (Fall, Part I)*, p. 757. https://doi.org/10.1145/1476589.1476686.

Suzer, O. K., Olgunturk, N., & Guvenc, D. (2018). The effects of correlated colour temperature on wayfinding: A study in a virtual airport environment. *Displays, 51*, 9–19. Q2/Q3. https://doi.org/10.1016/j.displa.2018.01.003

Tennessen, C. M., & Cimprich, B. (1995). Views to nature: Effects on attention. *Journal of Environmental Psychology, 15*(1), 77–85. https://doi.org/10.1016/0272-4944(95)90016-0

Tibaldi, A., Bonali, F. L., Vitello, F., Delage, E., Nomikou, P., Antoniou, V., Becciani, U., de Vries, B. V. W., Krokos, M., & Whitworth, M. (2020). Real world–based immersive virtual reality for research, teaching and communication in volcanology. *Bulletin of Volcanology, 82*(5), 38. https://doi.org/10.1007/s00445-020-01376-6

Tran, T. T. M., Parker, C., & Tomitsch, M. (2021). A review of virtual reality studies on autonomous vehicle–pedestrian interaction. *IEEE Transactions on Human-Machine Systems, 51*(6), 641–652. https://doi.org/10.1109/THMS.2021.3107517

Tromp, J., Peeters, D., Meyer, A. S., & Hagoort, P. (2018). The combined use of virtual reality and EEG to study language processing in naturalistic environments. *Behavior Research Methods, 50*(2), 862–869. https://doi.org/10.3758/s13428-017-0911-9

Ulrich, R. S. (1981). Natural versus urban scenes: Some psychophysiological effects. *Environment and Behavior, 13*(5), 523–556. https://doi.org/10.1177/0013916581135001

Ulrich, R. S. (1984). View through a window may influence recovery from surgery. *Science, 224*(4647), 420–421. https://doi.org/10.1126/science.6143402

Ulrich, R. S. (1993). Biophilia, biophobia, and natural landscapes. *The Biophilia Hypothesis, 7*, 73–137.

Ulrich, R. S., Simons, R. F., Losito, B. D., Fiorito, E., Miles, M. A., & Zelson, M. S. (1991). Stress recovery during exposure to natural and urban environments. *Journal of Environmental Psychology, 11*, 201–230. https://doi.org/10.1016/S0272-4944(05)80184-7

Uricchio, W. (2011). A 'proper point of view': The panorama and some of its early media iterations. *Early Popular Visual Culture, 9*(3), 225–238. https://doi.org/10.1080/17460654.2011.601165

Uruthiralingam, U., & Rea, P. M. (2020). Augmented and virtual reality in anatomical education–a systematic review. *Biomedical Visualisation, 6*, 89–101. https://doi.org/10.1007/978-3-030-37639-0_5

Watrous, A. J., Lee, D. J., Izadi, A., Gurkoff, G. G., Shahlaie, K., & Ekstrom, A. D. (2013). A comparative study of human and rat hippocampal low-frequency oscillations during spatial navigation. *Hippocampus, 23*(8), 656–661. https://doi.org/10.1002/hipo.22124

Westermann, R., Spies, K., Stahl, G., & Hesse, F. W. (1996). Relative effectiveness and validity of mood induction procedures: A meta-analysis. *European Journal of Social Psychology, 26*(4), 557–580. https://doi.org/10.1002/(SICI)1099-0992(199607)26:4<557::AID-EJSP769>3.0.CO;2-4

Wiederhold, B. K., & Wiederhold, M. D. (2000). Lessons learned from 600 virtual reality sessions. *Cyberpsychology & Behavior, 3*(3), 393–400. https://doi.org/10.1089/10949310050078841

Wiederhold, M. D., Gao, K., & Wiederhold, B. K. (2014). Clinical use of virtual reality distraction system to reduce anxiety and pain in dental procedures. *Cyberpsychology, Behavior, and Social Networking, 17*(6), 359–365.

Winkel, G., Saegert, S., & Evans, G. W. (2009). An ecological perspective on theory, methods, and analysis in environmental psychology: Advances and challenges. *Journal of Environmental Psychology, 29*(3), 318–328. https://doi.org/10.1016/j.jenvp.2009.02.005

Wu, S., Tang, H., Jing, X.-Y., Qian, J., Sebe, N., Yan, Y., & Zhang, Q. (2022). Cross-view panorama image synthesis with progressive attention GANs. *Pattern Recognition, 131*, 108884. https://doi.org/10.1016/j.patcog.2022.108884

Xie, B., Liu, H., Alghofaili, R., Zhang, Y., Jiang, Y., Lobo, F. D., Li, C., Li, W., Huang, H., Akdere, M., Mousas, C., & Yu, L.-F. (2021). A review on virtual reality skill training applications. *Frontiers in Virtual Reality, 2*. https://doi.org/10.3389/frvir.2021.645153

Xu, L., Zhang, Y., Li, F., & Yin, J. (2022). Perceptual difference of urban public spaces between design professionals and 'laypersons': Evidence, health implications and ready-made urban design templates. *Indoor and Built Environment*, 1420326X221116318. https://doi.org/10.1177/1420326X221116318

Yin, J., Zhu, S., MacNaughton, P., Allen, J. G., & Spengler, J. D. (2018). Physiological and cognitive performance of exposure to biophilic indoor environment. *Building and Environment, 132*, 255–262. https://doi.org/10.1016/j.buildenv.2018.01.006

Yin, J., Arfaei, N., MacNaughton, P., Catalano, P. J., Allen, J. G., & Spengler, J. D. (2019). Effects of biophilic interventions in office on stress reaction and cognitive function: A randomized crossover study in virtual reality. *Indoor Air, 29*(6), 1028–1039. https://doi.org/10.1111/ina.12593

Yu, H. (2023). A city in a scroll: Hidden crisis in "along the river during the Qingming festival". In *All about Chinese culture: An illustrated story of China in 10 cultural relics* (pp. 225–275). Springer Nature. https://doi.org/10.1007/978-981-99-3451-5_8

Yuan, S., Li, F., Browning, M. H., Bardhan, M., Zhang, K., McAnirlin, O., ... & Reuben, A. (2024). Leveraging and exercising caution with ChatGPT and other generative artificial intelligence tools in environmental psychology research. *Frontiers in Psychology, 15*, 1295275. https://doi.org/10.3389/fpsyg.2024.1295275

Zhang, S., Zhang, Y., Sun, Y., Thakor, N., & Bezerianos, A. (2017). Graph theoretical analysis of EEG functional network during multi-workload flight simulation experiment in virtual reality environment. In *2017 39th annual international conference of the IEEE engineering in medicine and biology society (EMBC)* (pp. 3957–3960). https://doi.org/10.1109/EMBC.2017.8037722

Zhang, Y., Liu, H., Kang, S.-C., & Al-Hussein, M. (2020). Virtual reality applications for the built environment: Research trends and opportunities. *Automation in Construction, 118*, 103311. https://doi.org/10.1016/j.autcon.2020.103311

Zhang, Z., Zhuo, K., Wei, W., Li, F., Yin, J., & Xu, L. (2021). Emotional responses to the visual patterns of urban streets: Evidence from physiological and subjective indicators. *International Journal of Environmental Research and Public Health, 18*(18), 9677. https://doi.org/10.3390/ijerph18189677

Zika, J. (2018). Dark rides and the evolution of immersive media. *Journal of Themed Experience and Attractions Studies, 1*(1), 54–60.

Yuan, S., Li, F., Browning, M. H., Bardhan, M., Zhang, K., McAnirlin, O., ... & Reuben, A. (2024). Leveraging and exercising caution with ChatGPT and other generative artificial intelligence tools in environmental psychology research. Frontiers in Psychology, 15, 1295275. https://doi.org/10.3389/fpsyg.2024.1295275

Neuroscience and Geographic Information Systems to Investigate the Impact of Global Warming on Mood Disorders and Brain Plasticity in Urban Areas

Marco Vieira Ruas, Romana Paganini, Bogdan Draganski, and Stéphane Joost

Abstract Geographic Information Systems (GIS) are advanced computational systems for acquiring, storing, analyzing, and visualizing spatial data, enabling comprehensive exploration of geographical relationships and patterns across diverse domains. GIS play a crucial role in analyzing the intersection of mood and brain data in a geo-environmental context. Through spatial analysis techniques, GIS

M. Vieira Ruas
Geospatial Molecular Epidemiology group (GEOME), Laboratory for Biological Geochemistry (LGB), Ecole Polytechnique Fédérale de Lausanne (EPFL), Lausanne, Switzerland

Geographic Information Research and Analysis in Population Health (GIRAPH), Geneva University Hospitals, Geneva, Switzerland

R. Paganini
Section of Environmental Science and Engineering (SIE), Ecole Polytechnique Fédérale de Lausanne (EPFL), Lausanne, Switzerland

B. Draganski
Laboratory for Research in Neuroimaging, Department of Clinical Neurosciences, Lausanne University Hospital and University of Lausanne, Lausanne, Switzerland

Neurocentre, Neurology Department, Inselspital - University Hospital Bern, Bern, Switzerland

S. Joost (✉)
Geospatial Molecular Epidemiology group (GEOME), Laboratory for Biological Geochemistry (LGB), Ecole Polytechnique Fédérale de Lausanne (EPFL), Lausanne, Switzerland

Geographic Information Research and Analysis in Population Health (GIRAPH), Geneva University Hospitals, Geneva, Switzerland

Unit of Population Epidemiology, Division of Primary Care Medicine, Geneva University Hospitals, Geneva, Switzerland

La Source School of Nursing, University of Applied Sciences and Arts Western Switzerland (HES-SO), Lausanne, Switzerland
e-mail: stephane.joost@epfl.ch

S. Kühn (ed.), *Environmental Neuroscience*,
https://doi.org/10.1007/978-3-031-64699-7_18

allows researchers to visualize and understand the geographical distribution of mood-related phenomena and their correlation with brain activity patterns.

In this chapter, we show that GIS can offer a powerful framework integrating spatial perspectives for examining the intricate interplay between mood, brain data, and global warming. This interdisciplinary approach enhances our understanding of the psychological impacts of environmental shifts, aiding in the development of targeted interventions to mitigate the mental health consequences of global warming.

On the example of mood disorders, we exemplarily demonstrate how GIS can be implemented to investigate the effects of climate warming on mental health and brain plasticity. We used satellite images to calculate land surface temperature (LST) and LST changes over time in the urban context of the city of Lausanne, Switzerland. These data were compared to mental health and magnetic resonance imaging (MRI) information from a population study in Lausanne.

The distribution of mood disorders exhibited a discernible spatial pattern across Lausanne, indicating that they were not randomly distributed in the geographical space. Spatial clusters with significantly higher rates of mood disorders were associated with significantly higher LST. Conversely, areas with lower LST at residential locations showed significantly better mental health performance.

Keywords Geographic Information Systems (GIS) · Spatial statistics · Spatial dependence · Magnetic Resonance Imaging (MRI) · Urban environment · Climate change · Heat island · Brain plasticity

Introduction

Geographic Information and Neuroscience

In recent years, there has been a growing interest in leveraging GIScience (Goodchild, 1992) and geospatial data to inform public health. The genesis of Geographical Information Systems (GIS) traces back to the 1960s, when the integration of spatial data with computer technology marked a shift in the way geographic information was visualized and analyzed (Chrisman, 2020). This approach has since evolved to integrate hardware, software, and data facilitating a wide range of applications such as environmental conservation (Caniani et al., 2016), urban planning (Zoungrana et al., 2024), agriculture sustainability (Mathenge et al., 2022), and public health surveillance (Di Lorenzo et al., 2023). The utility of GIS extends beyond data visualization; it serves as a powerful analytical tool capable of revealing and examining spatial patterns and relationships (Longley, 2005) that inform decision and policy-making processes (Pedro et al., 2019). In urban settings, an escalating interest has emerged in exploiting georeferenced measurements to study the intricate relationship between social and environmental determinants—i.e., the exposome, brain, and behavioral outcomes. Indeed, the integration of spatial information with brain data has the potential to reveal important insights into neurological disorders and enhances our understanding of how spatial contexts influence brain health. This interdisciplinary approach, sometimes referred to as neuro-geography (Pykett, 2018), seeks to understand how geographic factors influence brain health and cognitive function (Colantonio et al., 2011; León and Burga-León, 2015). Researchers use a variety of geospatial data sources, such as land use, air quality, proximity to green spaces, and socioeconomic factors, to assess their impact on brain health (Park et al., 2021). For example, studies have explored how exposure to urban green spaces can positively affect mental well-being (Jabbar et al. 2022), or how air pollution may be linked to cognitive decline (Delgado-Saborit et al., 2021). GIS tools are employed to map and visualize the spatial distribution of these environmental variables, allowing researchers to correlate them with neuroimaging data and cognitive assessments (Kühn et al. 2017). The integration of GIScience and geodata provides valuable insights into the complex interplay between the brain and the natural or built environment, shedding light on potential risk factors and strategies for promoting well-being and cognitive health. This field holds promise for enhancing our understanding of the brain-behavior-environment relationships and ultimately improving public health initiatives and urban planning strategies. In the current context of climate change, the ability of GIS to integrate brain data and the characteristics of the urban environment is particularly interesting, and this ability can be used to analyze the possible impact of global warming on human health through effects on brain anatomy.

Climate Change and Mental Health

The effects of climate change are multifaceted and manifest in various aspects of our lives. The impact of global warming and the resulting consequences on human health was investigated in various studies. It is likely to affect not only physical but also mental health (Padhy et al., 2015). Previous research has shown evidence connecting mental disorders, specifically anxiety, with extreme weather events (Palinkas & Wong, 2020). However, merely contemplating climate change and its potential consequences in the future can also contribute to mental distress (Palinkas & Wong, 2020).

The trend of rising land surface temperature (LST) has resulted in more prolonged and intense heatwaves. In urban areas, the air temperature is consistently higher compared to rural areas, a disparity that becomes especially pronounced during the summer months due to the heat island effect. During nighttime, urban areas can show temperatures that are 5–7 °C higher than natural surroundings (Wicki et al. 2018; MeteoSwiss, 2023). This poses a significant risk to human well-being (Berry et al., 2010) and health (Hondula et al., 2015). A well-established association was observed between heatwaves and mental disorders (Thompson et al., 2018), and Obradovich et al. (2018) showed that a temperature increase of 1 °C over 5 years was linked to a 2% rise in mental health problems.

A less explored research area is the potential effect of rising temperatures on brain structures. As a result, the extent to which heat islands affect brain structures is not yet understood and it remains difficult to clearly identify the influence of global warming on neurological disorders (Amiri et al., 2021). Still, the evidence available to date indicates that climate change will have a significant impact on the prevalence and management of major neurological diseases in the near future (Louis et al., 2023).

In this chapter, we exemplarily use georeferenced information (place of residence), GIS and spatial statistics to demonstrate how one can investigate relationships between land surface temperature, mental health, and brain plasticity. We focus on population-based data collected in Lausanne, western Switzerland (Firmann et al., 2008; Preisig et al., 2009). The aims of this study are to examine the impact of increased LST on mood disorders, as well as on brain white matter structure characteristics as recorded using magnetic resonance imaging (MRI). We hypothesized that high LST values are associated with low mental health and high anxiety and that relationships exist between global warming—as translated by higher land surface temperature—mood disorders and characteristics of white matter microstructure.

Data

Geographic Information Systems Data

GIS data have distinct characteristics that enable their inclusion in mapping and in spatial data analyses frameworks. It consists primarily in vector and raster data (Fig. 1), two distinct types corresponding to two different ways of storing digital geographic information.

Vector data are represented through points, lines, and polygons, enabling precise delineation of discrete geographical entities. Points mark specific locations, lines embody linear features such as rivers or roads, and polygons outline areas like countries, lakes, or park boundaries. This data format is complemented by the integration of attribute tables. The latter store information about the features represented in vector data, effectively linking non-spatial data to spatial features. Each row in an attribute table corresponds to a specific spatial feature, while each column represents a distinct attribute of that feature, such as name, type, population, or any other descriptive property.

Raster data, on the other hand, consist of a matrix of cells or pixels, with each cell holding a value representing information such as temperature, elevation, or vegetation density. This data type is ideal for modeling continuous spatial phenomena and is characterized by its resolution, determining the level of detail that can be represented. The resolution is defined by the size each pixel represents on the ground. A finer resolution, characterized by smaller pixels, results in a larger

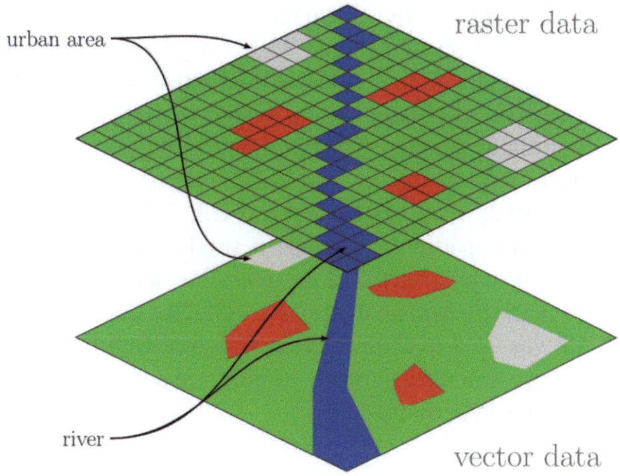

Fig. 1 Representation of superimposed vector and raster data types. (Wegmann, CC BY-SA 3.0, via Wikimedia Commons)

density of values over a given area, thereby enhancing the precision of the data measure and representation.

Vector and raster data are integral to GIS, each serving specific analytical needs. The utilization of both data types relies on the usage of precise coordinates (longitude and latitude) that align the spatial data with a recognized reference system. This alignment allows the accurate mapping and analysis of the data in correspondence with the real world.

In the following data analysis example, the study area is depicted using polygon type data, while the health data are captured within attribute tables linked to the participants' addresses, which are represented as points. Lastly, the LST measurements are presented as a raster grid.

Study Area

The study area consists of the urban area of the city of Lausanne, Switzerland (150,000 inhabitants in 2023). Lausanne is located on the northern shore of Lake Geneva. It is the capital of the canton of Vaud and the administrative center of the Lausanne district. It is the country's fourth-largest city in Switzerland in terms of population, after Zurich, Geneva, and Basel.

Health Data

We utilized health data from the CoLaus-PsyColaus cohort, a longitudinal study in the community dwelling population of Lausanne, Switzerland, initiated in 2003 with 5-yearly follow-ups. At baseline, the cohort involved over 6700 individuals aged 35–75 years old (Firmann et al., 2008; Preisig et al., 2009). The complete list of Lausanne residents aged 35 to 75 years was provided by the city's population register, and a simple, nonstratified random sample of 35% of this population was drawn. The 8121 people who agreed to participate represented 41% of the population initially sampled. The baseline CoLaus-PsyColaus study recruited 6733 participants (3544 women), of whom 5064 agreed to be recontacted for follow-ups (FU1: 2009–2012; FU2: 2014–2017; FU3: 2018–2021). At baseline and subsequent follow-ups, participants made a single visit to the Lausanne University Hospital (CHUV), which included an interview and physical examination. The mood data used in the present analysis were collected during the second follow-up. In addition, MRI was used to acquire brain structure data from subjects during follow-ups two and three. For each subject, we selected a single scan—their earliest one completed during these periods.

Mood Data

For assessment of anxiety levels, we used Spielberger's et al. (1983; Julian, 2011) State Trait Anxiety Inventory (STAI), divided into STAI state and STAI trait subscores. For each of the subsets, the items' scores were summed to obtain a final score ranging between 20 and 80, with higher values corresponding to higher levels of anxiety. We focused on STAI state, which provides an indication of an individual's current anxiety. For the affective components of social, occupational, and psychological well-being we used the Global Assessment of Functioning (GAF) assessing lifetime (GAF-L), worst (GAF-W), and current (GAF-C) affective state (Bell, 1994). Higher GAF scores range from 1 to 100. Higher scores translate into a better overall well-being.

STAI state and GAF scores were adjusted for age, gender, and neighborhood median income (i.e., median revenue of the neighborhood of the place of the subject's residence in 2017) (MicroGIS, 2017).

Brain Data

We analyzed the brain's white matter (WM) microstructure. Specifically, we investigated 31 tracts of interest identified on the basis of four MRI-derived maps in the context of a previous study that investigated the relationship between WM and cardiovascular risk factors (Trofimova et al., 2023). The four maps correspond to measurements of myelin content (MTsat), axonal density (ICVF), free water (ISOVF), and tract volume (number of voxels) (➔ chapter "Mapping the Human Brain with Computational Anatomy" by Draganski, Joost and Kherif). The methodology and analytical procedures used in the present work are based on the approach described in the aforementioned study (Trofimova et al., 2023). The values of the maps for each tract were adjusted for age, gender and total intracranial volume (TIV). For each map and tract, only values within the 3-sigma interval around the mean value were included in the analysis to remove errors in the data.

Environmental Data

We used two distinct types of land surface temperature (LST) measurements to assess its relationship with mood and white matter microstructure. These measurements characterize the place of residence of the participants. First, we computed the LST for the follow-up assessment examination year, and second, the LST delta between two periods: period one—between 2004 and 2012—represents the timeframe before follow-up two; and period two—between 2014 and 2021—including follow-ups two and three.

For each year analyzed, LST was calculated using values from June, July, and August, since these months are typically when the heat island effect[1] is most likely to occur in urban areas.

Land Surface Temperature (LST) Calculation

LST measurements were calculated using Landsat satellite imagery (https://landsat.gsfc.nasa.gov/) from the United States Geological Survey (USGS). We employed the Google Earth Engine platform (GEE; https://earthengine.google.com/) for efficient processing and analysis of the satellite imagery. We used Landsat 5 imagery for the years preceding 2012, and Landsat 8 for the subsequent years. The GEE platform was used to derive the LST data from the satellite images using the Statistical Mono-Window algorithm developed by Ermida et al. (2020). To generate the LST data for each year, we collected all relevant images captured during the defined timeframe and for the months of June, July, and August. On average, there were 14 usable images per year. On the basis of these images, we created a single composite image for each summer by calculating the median value for each pixel across all captures within that year. The use of the median value is robust to outliers, such as those caused by cloud cover. This resulting composite image provided a representative depiction of the average LST during the summer for each year, which was crucial for analyzing the variables of interest in conjunction with the actual LST values of the examination year. In a second step, the median images obtained for each year were further aggregated by calculating the mean value over several

[1] We refer here to the definition of heat island as proposed by the United States Environmental Protection Agency (EPA): "Heat islands are urbanized areas that experience higher temperatures than outlying areas. Structures such as buildings, roads, and other infrastructure absorb and re-emit the sun's heat more than natural landscapes such as forests and water bodies. Urban areas, where these structures are highly concentrated and greenery is limited, become "islands" of higher temperatures relative to outlying areas".

years. This aggregation process was performed separately for the periods one and two defined above. To obtain the delta LST variable, the mean image derived from period one was subtracted from the mean image of period 2, allowing for the examination of temperature change over time.

Methods

As regards mood data, we initially calculated the spatial dependence of the variables of interest (GAF and STAI) to identify significant geographic clusters of high or low mood values. Then, we conducted variance analysis to determine if significant differences exist in LST variables (year of examination and delta) across the different mood cluster classes identified (see the spatial statistics section hereunder).

For brain white matter data, we first calculated a linear regression between the four MRI-derived maps investigated and LST measurements, for each of the 31 tracts of interest. We only considered significant associations for subsequent spatial examination and variance analysis across the identified clusters.

Georeferencing

We utilized georeferencing, to assign the home addresses provided by participants during the follow-ups to specific longitude and latitude coordinates. To achieve this, we employed the Swiss Confederation GeoAdmin API.[2] In Switzerland, the extensive cataloging and geolocation of every street and road facilitate the precise georeferencing of every building. By submitting an address through an API request, we receive a list of attributes, including the coordinates corresponding to that address.

Assigning Exposome Values to Individuals and Buffer Zones

To assign the calculated LST measurements to each corresponding individual, we leveraged the intersection between the participants' georeferenced addresses, depicted as points, and the corresponding LST raster pixel values. Using a GIS sampling tool, we matched the address points with the LST data contained within the 30 × 30 m pixels of their location. We subsequently stored the extracted values from the raster layers in an appropriate column in the attribute table of the address points.

[2]Coordination and Geo-Information Services (COGIS) of the Swiss Confederation, API REST Services—GeoAdmin API 3.0, https://api3.geo.admin.ch/services/sdiservices.html?#search (accessed 9.14.23).

In this instance, we employed a straightforward geographic intersection between points and raster pixels, aiming to obtain LST values near the home address. However, the broader assessment of the environmental exposome typically involves creating buffer zones around home (or other) addresses. This approach enables the calculation of crow-fly median values at various radii surrounding the subjects' home addresses, providing a more comprehensive evaluation of environmental exposure.

The chosen radius length is dictated by the potential impact range of the environmental exposome factor under study. For instance, certain environmental exposome factors, like nighttime traffic noise, necessitate smaller buffer zones—typically no more than 25–50 m—since their effects diminish rapidly with distance from the source. Conversely, vegetation density assessment may require broader ranges, from 25 m for immediate vicinity considerations to up to 5 km (Dimitrov-Discher et al., 2022) for evaluating access to public parks rather than direct visibility or exposure at the residence. Understanding the spatial extent and lag of the effects is crucial in spatial analysis to measure spatial dependence. Accordingly, methodologies have been refined to mathematically ascertain spatial weight matrices (Yu & Fotheringham, 2022).

Furthermore, the choice of buffer zones for assessing environmental exposome factors, especially in studies focusing on accessibility, may shift from simple crow-fly buffers to those that incorporate the specific street network accessible from a participant's home address. This approach enables the measurement of distances based on the actual time it takes for a subject to traverse the path, allowing for the addition of "costs" to the journey. These costs could account for factors such as slope or road obstructions, which impact the distance subjects can cover within a given time frame.

Spatial Statistics

Spatial dependence denotes the tendency for nearby locations to influence each other within a certain distance and share similar features (Goodchild, 1992). Objects or individuals in proximity tend to be more similar than objects or individuals that are farther apart (Tobler, 1970). This principle holds true when examining disease occurrence across a territory, or the geographical distribution of health indicators. Indeed, individuals who share the same environment are exposed to similar social and environmental determinants. Habits and automatisms are easily shared (Marteau et al., 2012) by mimicry (Duffy et al., 2020) within neighborhoods. Spatial autocorrelation is the statistical measure that quantifies this similarity of attributes between sample point locations—i.e., where individuals live. Additionally, the principle that geographical location significantly influences measured values is recognized as spatial dependence.

Several statistics have been developed to measure spatial autocorrelation. The most used is Moran's I (Moran, 1950), which quantifies globally over a given territory the similarity between the attributes of all objects or individuals and their N neighbors, or those located within a given radius. Local versions of this statistic are the local Moran's I (Anselin, 1995) or the Getis-Ord Gi* statistics (Getis & Ord, 1992), capable of quantifying and categorizing locally the similarity or dissimilarity between behaviors. Their calculation amounts to a Z score, with the null hypothesis assuming a random spatial pattern for the analyzed values.

We assessed the spatial dependence of the mood measurements and brain anatomy with the Getis-Ord statistics implemented in Python. Statistical significance was tested using 999 permutations with a significance level of 0.05. Statistically significant positive and negative Z scores are represented respectively by hot spots (HS)—clusters of high values, and cold spots (CS)—clusters of low values. Neutral sampling sites are labeled nonsignificant (NS). On the maps produced, hot spots are shown in red, cold spots in blue and nonsignificant in white. Getis-Ord Gi* was calculated using spatial lags of 200 m, 400 m, 600 m, and 800 m (sensitivity analysis). We selected the spatial lag of 600 m to produce the maps shown on Figs. 3a, 4a and 6a. This specific spatial lag highlights interesting spatial patterns while avoiding excessive spatial aggregation.

The application of spatial statistics to precisely georeferenced individual locations is key as this avoids errors that could be caused by a problem called the modifiable areal unit problem (MAUP; Openshaw, 1984). MAUP—a form of ecological fallacy—is a critical issue in spatial analysis, highlighting how the outcomes of such analyses can be significantly influenced by the scale of aggregation and the delineation of spatial units. This problem manifests in two key ways: through the scale effect, where the granularity of data aggregation (e.g., hectometric grid, city, region, and country) can alter analytical results, potentially masking local variations at broader scales, and the zoning effect, where variations in the configuration of spatial unit boundaries can lead to different interpretations of the same underlying data. MAUP's implications emphasize the necessity for carefully justified choices of scale and zoning to ensure accurate spatial analysis outcomes. One of our example study's strengths lies in its use of fine-scale georeferenced measurements and analysis at an individual level, which surpasses the limitations of studies using predefined units.

Variance Analysis

We calculated group averages of LST and delta LST measurements for the spatial clusters obtained from the Getis-Ord calculation. We assessed homogeneity of LST data across clusters with the Levene's test (Levene, 1960). The Levene's test evaluates the null hypothesis according to which the distinct groups of individuals analyzed are of equal variance. For LST data showing homogeneity across clusters, we calculated a one-way ANOVA (Ståhle and Wold, 1989) and the subsequent Tukey post-hoc test (Tukey, 1949) for multiple pairwise comparisons. In cases of heterogeneous LST measurements, we performed a one-way Kruskal–Wallis (KW) test (Kruskal and Wallis, 1952) and a non-parametric post-hoc Conover test.

In addition, the eta-squared from both the ANOVA and the KW test provides information about the effect size observed. It is the amount of variation explained by the predictor variable. Effect sizes larger than 0.01 indicate a small effect (Coe, 2002). Therefore, any effect size below this threshold is considered negligible.

Linear Regression

We computed ordinary least squares (OLS) linear regression to calculate the associations between the four MRI-derived maps (myelin content, axonal density, free water, and tract volume) and LST of the examination year and LST delta. This analysis was performed for all 31 tracts.

Results

Subjects Included in the Analysis

Only a subset of the total population was selected for our analyses. On the one hand, we focused on the dense urban part of the city, within a perimeter defined by 11 communes (Lausanne, Renens, Paudex, Pully, Belmont-sur-Lausanne, Ecublens, Prilly, Chavannes-près-Renens, Le Mont, Bussigny, and Epalinges). Additionally, a small number of participants were excluded from the analysis due to missing information regarding confounding variables (age, gender, income, and TIV).

As regards WM data, additional participants were excluded when their values fell outside of the three-sigma interval around the mean. These data points were considered to be erroneous data. However, 98.9% of the data were within this interval. We therefore have a different number of individuals for each type of variable. Information on the number of subjects per variable is given in Table 1.

LST and Delta LST

Land surface temperature of the urbanized areas, particularly in the city center of Lausanne and in the western region, exhibits higher values and highlight the urban heat island effect (Fig. 2a). In some places, LST reached up to more than 37 °C. In contrast, the peripheral regions including the northern districts, show substantially lower LST (between 25 and 30 °C). Notably, the shores of the lake and the forest

Table 1 Participant's characteristics for each health variable

	GAF	STAI	Brain
Follow-up	F2	F2	First examination in F2 and F3
Total number of subjects	3647	2470	1582
# of subjects analyzed	2991	2174	1258[a]
% men/women[b]	43.7/56.3	42.8/57.2	47.6/52.4
Mean age (sd) [years][b]	62.9 (10.3)	63.5 (10.1)	63.5 (9.5)

[a]Depending on the tract and map, subjects with outstanding values (outside the 3-sigma interval around the mean) were excluded from the analysis (on average, 98.9% of the values were within the interval)
[b]Referring to the number of subjects analyzed

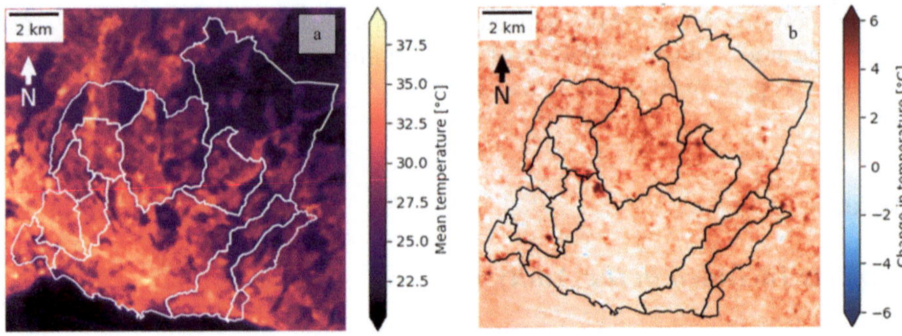

Fig. 2 On the left (**a**), average land surface temperature (LST) calculated for each pixel (30 m × 30 m) between 2014 and 2021 during the months of June, July and August. On the right (**b**), average difference in LST between 2014 and 2021. Delta LST = mean(2014–2021)—mean(2004–2012)

areas, such as the district in the north-east, show the lowest temperatures indicating their cooling effect.

As regards delta LST (Fig. 2b), we can observe a general signal of temperature increase across the area of interest. The magnitude of temperature deltas locally varies, with some areas experiencing even a decrease in dense urban areas (see blue spots). Most pixels show either a slight rise or a more substantial increase of up to 6 °C. Notably, the districts to the north of the urbanized area (center of the image) experience the highest temperature increase.

Mood Variables

Global Assessment of Functioning (GAF)

Variance analysis of LST of the examination year across GAF-Current clusters assessed with spatial lags of 600 m and 800 m showed the most substantial effect sizes. We found that the results across GAF-lifetime and GAF-worst clusters were negligible. Additionally, our findings show that the variance attributable to delta LST was predominantly significant across GAF-current clusters, although results across GAF-lifetime and to a lesser extent GAF-worst were above the 0.01 threshold. We present in Fig. 3a the spatial distribution of GAF-current clusters with a 600 m spatial lag and the associated LST measurements in boxplots (Figs. 3b and 3c).

We identified a primary GAF-Current coldspot (lower current well-being) in the city center, north of the main train station. Conversely, we found GAF-current hotspots (better current well-being) in the peripheric regions. The boxplots in Figs. 3b and 3c illustrate the distribution of LST measurements at the residences of subjects within hotspots, coldspots and neutral geographic space. Both the LST of the examination year and delta LST demonstrated dose-response effects across clusters, albeit in an inverse relationship: average LST measurements of the examination year were higher in GAF-Current coldspots ($p \leq 0.001$), whilst average delta LST measurements were higher in hotspots. Specifically, hotspots recorded high LST deltas (mean = +1 °C), in contrast to the lower LST deltas in coldspots (mean = +0.3 °C). The effect sizes for the corresponding ANOVA analyses, 0.014 and 0.015, were small but non-negligible.

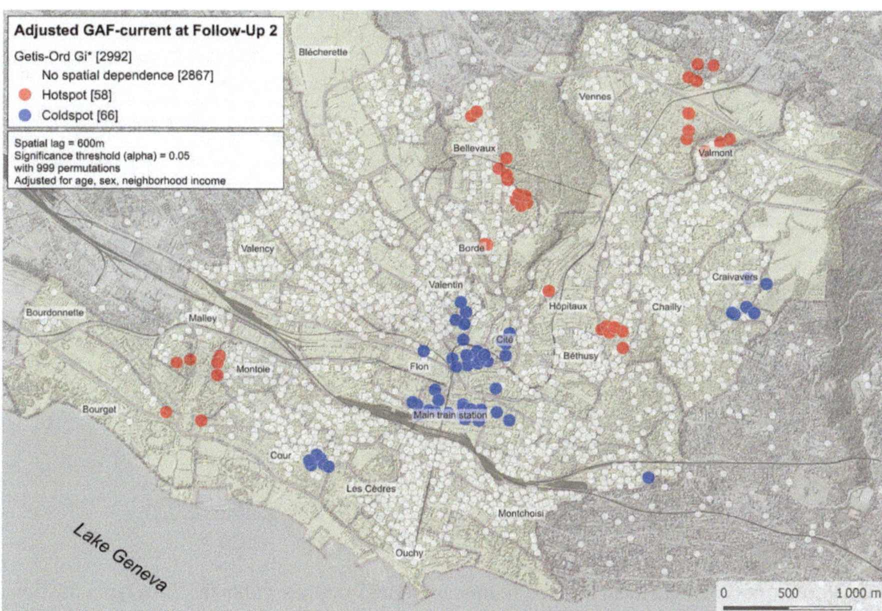

Fig. 3a A dot represents a single subject. Subjects belong either to Global Assessment of Functioning (GAF-Current) hotspots (HS) in red (significant grouping of 58 high GAF-Current values), to GAF-Current coldspots (CS) in blue (significant grouping of 66 low GAF-Current values), or to the neutral geographic space (NS) in white, for a spatial lag of 600 m. GAF-Current was adjusted for age, sex and neighborhood median income. Significance threshold is 0.05 for 999 permutations

Fig. 3b Boxplots of land surface temperature (LST) values within the three types of Global Assessment of Functioning (GAF-Current) clusters shown on Fig. 3a. Low GAF-Current values show low LST, while high GAF-Current show high LST. GAF-Current values where geography's influence is neutral show intermediate LST values. All pairwise comparisons (Tukey HSD) between Getis-Ord classes are significant. Not significant: $p \geq 0.05$; *: $p \leq 0.05$; **: $p \leq 0.01$; ***: $p \leq 0.001$

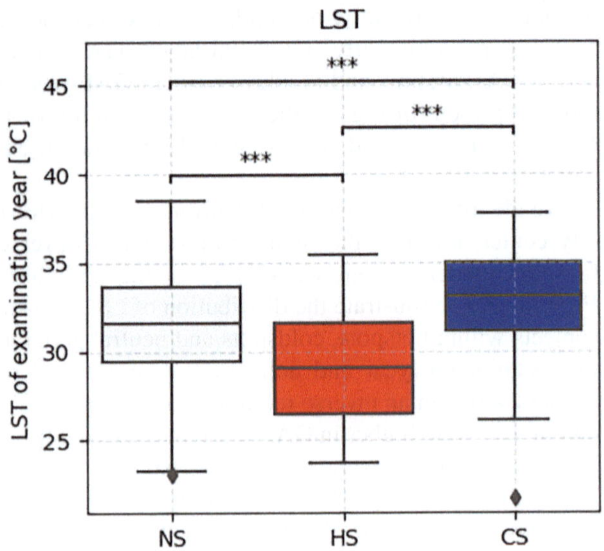

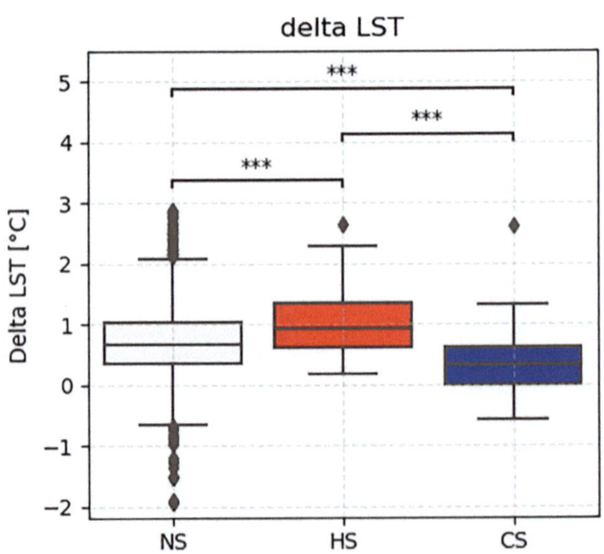

Fig. 3c Boxplots of delta Land Surface Temperature (LST) values within the three types of Global Assessment of Functioning (GAF-Current) clusters shown on Fig. 3a. High GAF-Current values show high delta LST, while low GAF-Current show low delta LST. GAF-Current values where geography's influence is neutral show intermediate delta LST values. All pairwise comparisons (Tukey HSD) between Getis-Ord classes are significant. Not significant: $p \geq 0.05$; *: $p \leq 0.05$; **: $p \leq 0.01$; ***: $p \leq 0.001$

State-Trait Anxiety Inventory (STAI)

Variance analysis revealed that both LST of the examination year and delta LST measurements presented the highest effect size across clusters of STAI state calculated with a 600 m spatial lag. We detected a primary STAI state hotspot (higher anxiety state levels) to the northwest and a secondary one to the east of the city center (Fig. 4a). Moreover, we found a large STAI state coldspot (lower anxiety state levels) in the northern part of the city and smaller ones in the western and southern parts. Subjects residing in the southern city area beneath the railway typically belonged to the non-significant cluster class. LST of the examination year was higher ($p \leq 0.0001$) in anxiety state hotspots (Fig. 4b), whereas delta LST was lower ($p \leq 0.0001$) these hotspots (Fig. 4c). The effect sizes for the corresponding respective KW analyses were 0.028 and 0.021.

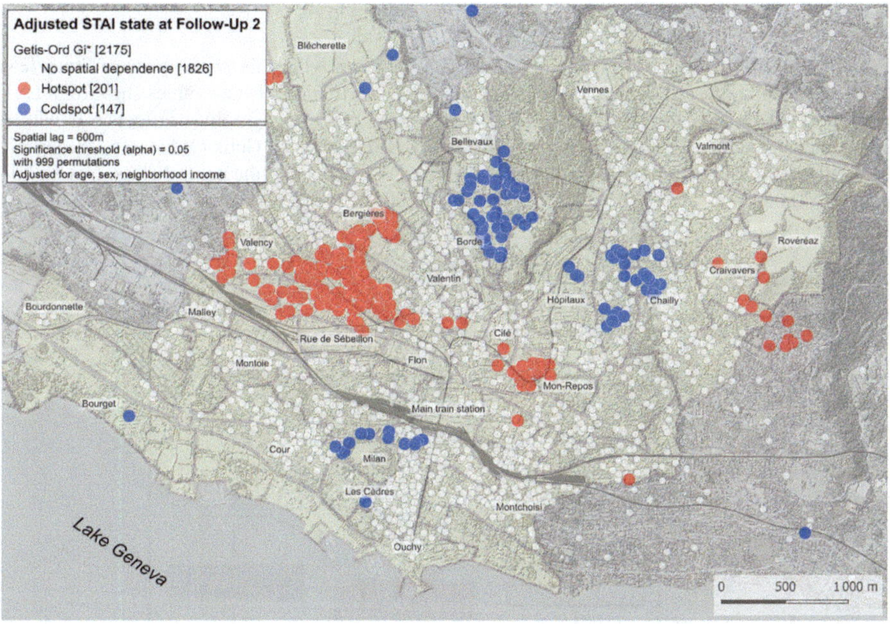

Fig. 4a A dot represents a single subject. Subjects belong either to State Trait Anxiety Inventory (STAI) state hotspots (HS) in red (significant grouping of 201 high STAI state values), to STAI state coldspots (CS) in blue (significant grouping of 147 low STAI state values), or to the neutral geographic space (NS) in white, for a spatial lag of 600 m. STAI state was adjusted for age, sex, and neighborhood median income. Significance threshold is 0.05 for 999 permutations

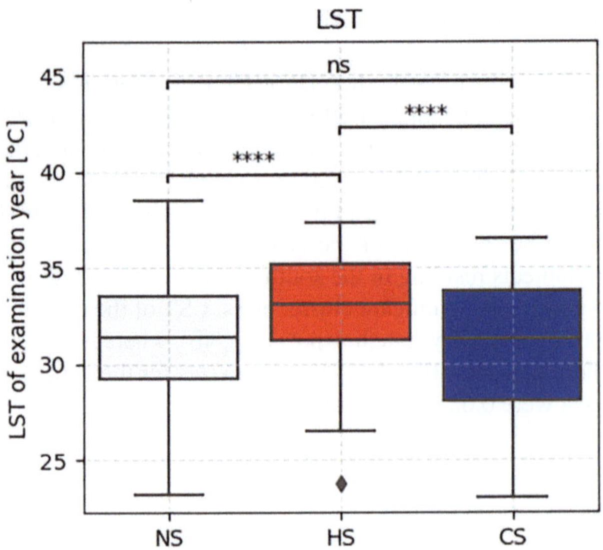

Fig. 4b Boxplots of land surface temperature (LST) values within the three types of State Trait Anxiety Inventory (STAI) state clusters shown on Fig. 4a. High STAI state values show high LST, while low STAI state show low LST. STAI state values where geography's influence is neutral show intermediate LST values. Comparisons (Tukey HSD) between Getis-Ord classes are significant only between hotspots and coldspots and between hotspots and the neutral class. Not significant: $p \geq 0.05$; *: $p \leq 0.05$; **: $p \leq 0.01$; ***: $p \leq 0.001$; ****: $p \leq 0.0001$

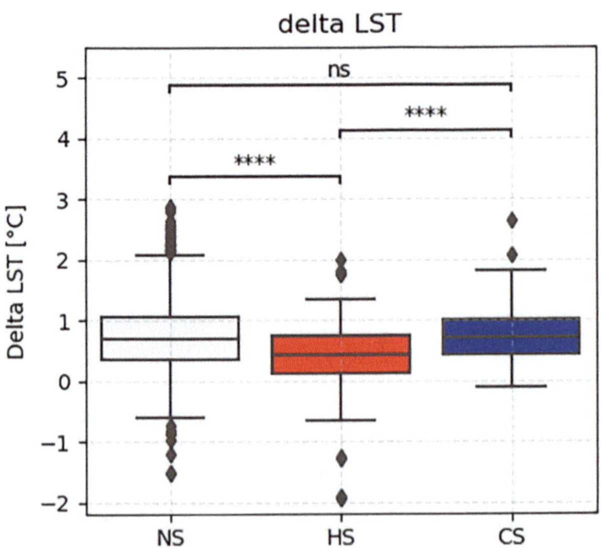

Fig. 4c Boxplots of delta land surface temperature (LST) values within the 3 types of State Trait Anxiety Inventory (STAI) state clusters shown on Fig. 4a. Low STAI state values show low delta LST, while low STAI state show high delta LST. STAI state values where space is neutral show similar delta LST values like in the coldspots. Comparisons (Tukey HSD) between Getis-Ord classes are significant only between hotspots and coldspots and between hotspots and the neutral class. Not significant: $p \geq 0.05$; *: $p \leq 0.05$; **: $p \leq 0.01$; ***: $p \leq 0.001$; ****: $p \leq 0.0001$

Brain's White Matter Microstructure

To represent the strength and significance of the relationships between the four MRI-derived maps (myelin content, axonal density, free water, and tract volume) for the 31 tracts investigated and the LST measurements, we examined the standardized parameter estimates of association, displayed in Fig. 5. LST of the examination year is the explanatory variable in the upper graph, while delta LST is shown in the lower graph.

The myelin content (MT), the axonal density (ICVF), and the tract volume (number of voxels) were not significantly associated with LST. However, we observed a positive correlation between LST and water content (ISOVF) for 20 out of the 31 tracts.

The relationship between delta LST and the tracts revealed only one significant association. Delta LST was weakly but negatively correlated with the number of voxels in the OR left tract. Due to the limited magnitude of this association and to the absence of significance in other parameter estimates, we opted not to conduct further investigation on this specific result.

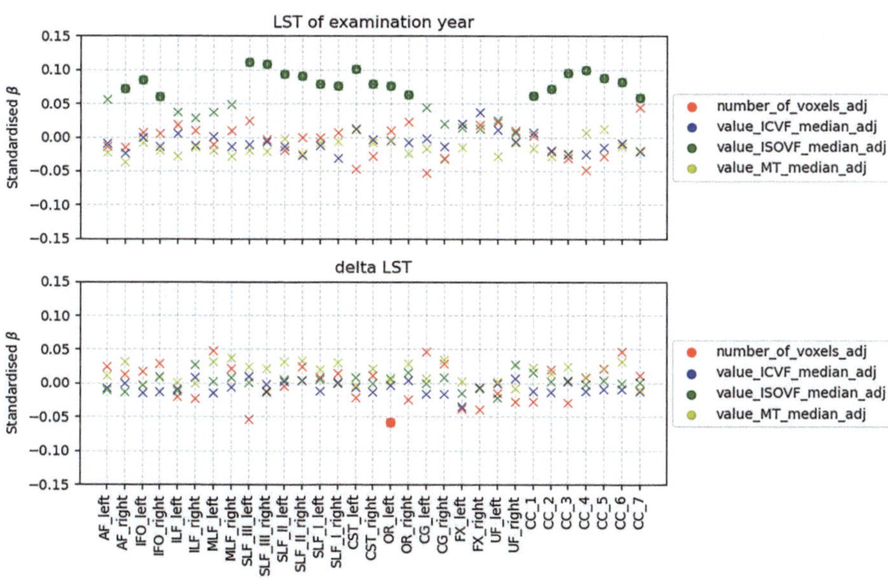

Fig. 5 Relationship between the land surface temperature (LST) (top) and delta LST (bottom) variables and the four MRI-derived maps (myelin content, axonal density, free water, and tract volume) for each of the 31 tracts of interest (on the x-axis). The y-axis shows the standardized parameter estimates. A cross indicates that the parameter estimate for the corresponding map in that tract is not significant, and a dot means that the relationship is significantKey: AF arcuate fasciculus, CG cingulum bundle, CST cortico-spinal tract, FX fornix, IFOF inferior fronto-occipital fasciculus, ILF inferior longitudinal fasciculus, MLF middle longitudinal fasciculus, OR optic radiation, SLF superior longitudinal fasciculus, UF uncinate fasciculus, CC corpus callosum, ICVF intra-cellular volume fraction, ISOVF isotropic volume fraction

The CST left, SLF III left, and CC 4 tracts exhibited the highest significant parameter estimates in the OLS (Fig. 5, top graph). Consequently, we performed a Getis-Ord analysis with a 600 m spatial lag on the free water (ISOVF) value in combination with the LST of the examination year, focused on these selected tracts. By selecting one tract from the association tracts (SLF), the projection tracts (CST), and the corpus callosum (CC), we ensured a representative sample of all tracts of interest.

The obtained three spatial maps show similar spatial patterns. We selected the spatial autocorrelation clusters for free water content in tract SLF III left (largest effect size) to illustrate the results (Fig. 6a). We detected a large coldspot in the northeastern part of the city, whereas hotspots were distributed throughout the remaining urban area.

To assess the differences in variance across Getis-Ord clusters, we conducted a Levene's test, revealing nonhomogeneous variances for the three tracts within the cluster classes. Through the subsequent application of the KW test, we observed effect sizes of 0.023, 0.032, and 0.020 for CST left, SLF III left, and CC 4, respectively. The boxplots in Fig. 6b display the distribution of LST within the spatial clusters of free water content in SLF III left. The primary finding is that LST is significantly higher in hotspots of free water content compared to coldspots. Individuals residing in areas with lower LST equally exhibit lower water content in the CST left and CC 4 tracts.

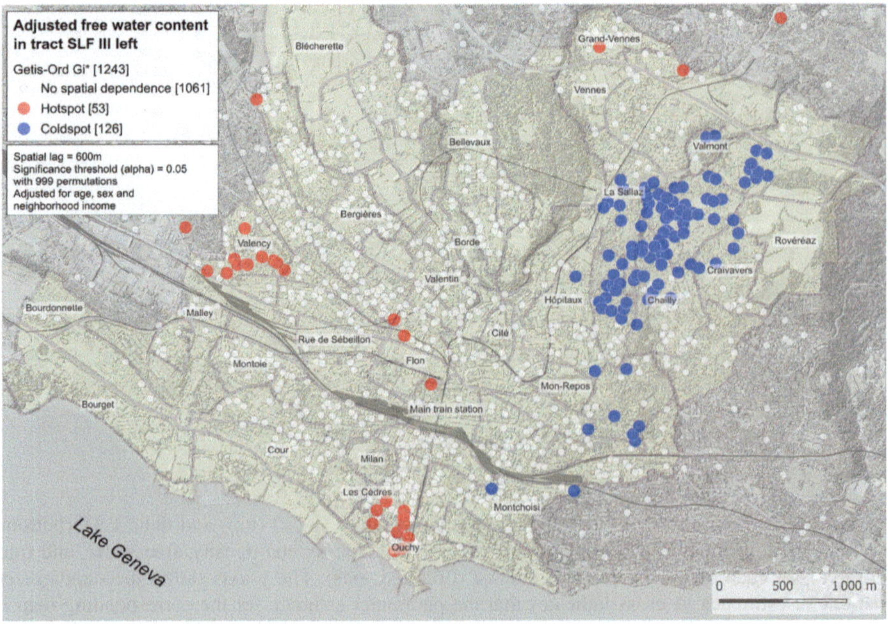

Fig. 6a A dot represents a single subject. Subjects belong either to free water content in tract superior longitudinal fasciculus (SLF) III left hotspots (HS) in red (significant grouping of 53 high water content values), to water content coldspots (CS) in blue (significant grouping of 126 low water content values), or to the neutral geographic space (NS) in white, for a spatial lag of 600 m. Water content in tract SLF III left was adjusted for age, gender, and neighborhood median income. Significance threshold is 0.05 for 999 permutations

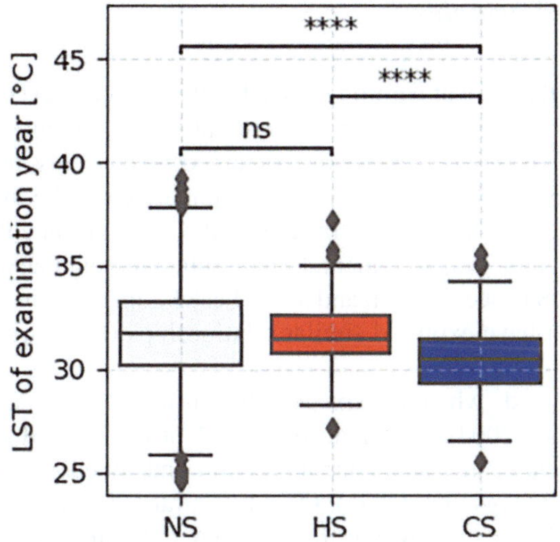

Fig. 6b Boxplots of land surface temperature (LST) values within the three types of clusters of free water content (ISOVF) in tract superior longitudinal fasciculus (SLF) III left shown on Fig. 6a. ISOVF hotspots show higher LST than in coldspots, while their LST mean is not significantly different from the neutral geographic space. Comparisons (Tukey HSD) between Getis-Ord classes are significant only between hotspots and coldspots and between coldspots and the neutral class. Not significant: $p \geq 0.05$; *: $p \leq 0.05$; **: $p \leq 0.01$; ***: $p \leq 0.001$; ****: $p \leq 0.0001$

Discussion

Land Surface Temperature

The geographic distribution of LST values across the urban area of interest (Fig. 1) aligns with our expectations, i.e., densely built-up areas exhibit significantly higher soil temperatures. Conversely, vegetated areas show lower soil temperatures. This is consistent with findings in the existing literature (Gago et al., 2013). However, the average temperature increase over the two periods (delta LST), as illustrated in Fig. 2b, reveals interesting patterns. It was not expected that the city center would experience a lower temperature increase compared to the surrounding areas. This observation highlights the contrasting temperature dynamics between the densely built downtown area where high LST values already existed in 2004, and its more rural surrounding regions. While the urban core maintained consistently high temperatures, the peripheral areas had undergone notable warming over the studied period, mainly because of urban expansion, extensive construction activities of new housings and new industrial zones (Dutta et al. 2019).

Mental Health Variables

We hypothesized that global warming, specifically the occurrence of heat waves in urban areas, may impact well-being and mental health, potentially leading to an increase in mood disorders. The results obtained in fact only partly support this assumption as they point to a temporal distinction affecting the impact of soil temperature on mental health variables. Indeed, both GAF-current coldspots and STAI hotspots, which denote impaired mental functioning and high anxiety, are the areas with the highest LST (see Figs. 3b and 4b). These findings are in concordance with earlier studies that have explored similar relationships (Hansen et al. 2008; Berry et al., 2010).

On the other hand, when we consider the difference (delta) in average LST between the periods 2004–2012 and 2014–2021, it turns out that we observe the opposite phenomenon, i.e., better mental health and moderate anxiety where the delta is the highest (Figs. 3c and 4c). On the contrary, people are worse off where the LST delta is lower (i.e., where the LST has only slightly increased, or even decreased). This finding not only contradicts our initial hypothesis but also previous research (Obradovich et al., 2018). However, this matches specific observations underlined in a systematic review on the impact of climate change on mental health, where Cancioni and colleagues (2020) mention the observation that climate change may have different impacts in different geographical areas, with different timings and with different types of threats to public health. They stress the fact that the effects of climate change can be direct or indirect, and also short-term or long-term. Acute events are likely to act through mechanisms similar to that of traumatic stress, and this is what we observe with the association between punctual high LST and poorer mental health. But we could find no information in the literature regarding a possible mechanism of (mental) adaptation to a gradual increase in delta LST over decades. Finally, to explain why areas with high LST deltas show good mental functioning and low anxiety may be understood as a logical consequence of the results obtained from the LST analysis. As we have observed that areas with high LST already experienced high temperatures 10 years ago, it may be the case that a deterioration in mental health already occurred in the concerned subjects at the end of the previous decade (1995–2004 compared to 2004–2012) and therewith outside the time window of the study.

White Matter Data

The analysis and interpretation of the relationship between WM tracts, mood disorder, and environmental stress like high land surface temperatures in urban areas come within the scope of a largely unexplored field (Amiri et al., 2021). The link between WM tracts and mood disorder is mainly documented by a meta-analysis of 37 different imaging studies of the brain's white matter in patients with depression, bipolar disorder, and anxiety disorders proposing that these diseases share a common pattern of white matter disruption (Jenkins et al., 2016) and recently confirmed by Parsaei et al. (2024) in their review on WM alterations to be associated with social anxiety disorders.

Our statistical investigations mainly revealed a weak but significant association between LST and water content (ISOVF) in several WM tracts. Spatial autocorrelation analysis showed that spatial clusters of lower ISOVF values were found in areas with lower LST. However, the opposite trend—where higher LST would characterize hotspots of high ISOVF values—was not observed, what would have been consistent with the fact that increased free water is associated with chronic bipolar disorders (Tuozzo et al., 2018), and thus confirming the link observed between mental disorders and high LST. Here the low water content cluster is probably rather due to an older population than in other urban districts of the city. Indeed, this coldspot fits the *Chailly*, *Craivavers*, and *Rovéréaz* districts where 19,6% of inhabitants are 65 years of age and more, while the downtown urban districts like "Centre" typically show 10,7% of inhabitants of the same age class. The brain undergoes structural changes with age, and there may be a gradual decrease in water content in both gray and white matter (Giorgio et al. 2010). This goes along with neurological disorders (multiple sclerosis, leukodystrophies, or other demyelinating disorders) leading to alterations in the brain's WM, including changes in water content (Liu et al. 2017). Anyways, all observations and spatial patterns (Fig. 6a) were consistent for all three brain tracts investigated. The interpretation of these findings, however, requires further investigation.

Limitations and Strengths

While extracting LST from satellite imagery provided valuable insights into surface temperature patterns, it is important to recognize that the perceived air temperature by the local population may differ from the LST values obtained from satellite data. Variables like wind speed, humidity, and local microclimates can impact individuals' perception and experience of temperature, influencing human comfort, well-being and mental health differently across various city areas. Ideally, one should consider complementing LST data with air temperature measurements from local weather stations and potentially include surveys on human thermal comfort for a more accurate representation of the population's experienced temperature.

The current study's analysis of delta LST provides a generalized representation of the average temperature increase over two periods of approximately 10 years. A longer-term longitudinal study would permit a more detailed examination of the impact of specific local temperature changes experienced at each subject's residence. Another limitation of our study arises from the delta LST measurements extending through 2021, to align with brain scans conducted during the third follow-up, while the assessment of mood variables was completed during the second follow-up, concluding in 2017.

Conversely, a significant merit of this study lies in analyzing mental health variables and cerebral tracts in a geographic context, leveraging spatial statistics, and quantifying spatial dependence. The geographical dimension facilitates the integration of environmental and neurological information to better comprehend health challenges associated with the accentuation of the climate warming phenomenon, particularly affecting urban populations.

Conclusion

Employing spatial analysis techniques, we investigated the impact of global warming, specifically focusing on urban heat islands, on mood disorders and WM tracts in the human brain. The study narrowed its scope to urban districts constituting the agglomeration of Lausanne, Switzerland, analyzing data from participants in the CoLaus-PsyColaus study. Positive and statistically significant associations were observed between mood disorders, characterized by the GAF and STAI state and land surface temperature (LST). Higher levels of mental well-being were identified in areas with significantly lower LST, while regions with elevated LST exhibited associations with poorer mental health. The analysis of LST increase revealed a significantly higher delta LST for subjects with better mental health scores, whereas individuals with mood disorders experienced a greater temperature rise (only for the GAF). Additionally, the assessment of various WM tracts in the brain unveiled a significant positive correlation between water content (ISOVF) and LST. Lower water content was linked to lower LST, with no significant evidence found for higher water content in areas with high LST.

References

Amiri, M., Peinkhofer, C., Othman, M. H., Vecchi, T. D., Nersesjan, V., & Kondziella, D. (2021). Global warming and neurological practice: Systematic review. *PeerJ, 9*, e11941. https://doi.org/10.7717/peerj.11941

Anselin, L. (1995). Local indicators of spatial association—LISA. *Geographical Analysis, 27*, 93–115. https://doi.org/10.1111/j.1538-4632.1995.tb00338.x

Bell, C. C. (1994). DSM-IV: Diagnostic and statistical manual of mental disorders. *JAMA, 272*(10), 828–829. https://doi.org/10.1001/jama.1994.03520100096046

Berry, H. L., Bowen, K., & Kjellstrom, T. (2010). Climate change and mental health: A causal pathways framework. *International Journal of Public Health, 55*(2), 123–132. https://doi.org/10.1007/s00038-009-0112-0

Caniani, D., Labella, A., Lioi, D. S., Mancini, I. M., & Masi, S. (2016). Habitat ecological integrity and environmental impact assessment of anthropic activities: A GIS-based fuzzy logic model for sites of high biodiversity conservation interest. *Ecological Indicators, 67*, 238–249. https://doi.org/10.1016/j.ecolind.2016.02.038

Chrisman, N. R. (2020). History of geographic information systems. *International Encyclopedia of Human Geography*, 43–47. https://doi.org/10.1016/b978-0-08-102295-5.10555-4

Cianconi, P., Betrò, S., & Janiri, L. (2020). The impact of climate change on mental health: A systematic descriptive review. *Frontiers in Psychiatry, 11*. https://www.frontiersin.org/journals/psychiatry/articles/10.3389/fpsyt.2020.00074

Coe, R. (2002). *It's the effect size, stupid ! What effect size is and why it is important*. British Educational Research Association annual conference, September 12–14, Exeter, United Kingdom.

Colantonio, A., Moldofsky, B., Escobar, M., Vernich, L., Chipman, M., & McLellan, B. (2011). Using geographical information systems mapping to identify areas presenting high risk for traumatic brain injury. *Emerging Themes in Epidemiology, 8*(1), 7. https://doi.org/10.1186/1742-7622-8-7

Delgado-Saborit, J. M., Guercio, V., Gowers, A. M., Shaddick, G., Fox, N. C., & Love, S. (2021). A critical review of the epidemiological evidence of effects of air pollution on dementia, cognitive function and cognitive decline in adult population. *Science of the Total Environment, 757*, 143734. https://doi.org/10.1016/j.scitotenv.2020.143734

Di Lorenzo, A., Mangone, I., Colangeli, P., Cioci, D., Curini, V., Vincifori, G., Mercante, M. T., Di Pasquale, A., Radomski, N., & Iannetti, S. (2023). One health system supporting surveillance during COVID-19 epidemic in Abruzzo region, southern Italy. *One Health, 16*, 100471. https://doi.org/10.1016/j.onehlt.2022.100471

Dimitrov-Discher, A., Wenzel, J., Kabisch, N., Hemmerling, J., Bunz, M., Schöndorf, J., Walter, H., Veer, I. M., & Adli, M. (2022). Residential green space and air pollution are associated with brain activation in a social-stress paradigm. *Scientific Reports, 12*(1), 10614. https://doi.org/10.1038/s41598-022-14659-z

Duffy, K. A., Green, P. A., & Chartrand, T. L. (2020). Mimicry and modeling of health(-risk) behaviors: How others impact our health(-risk) behaviors without our awareness. *Journal of Nonverbal Behavior, 44*(1), 5–40. https://doi.org/10.1007/s10919-019-00318-x

Dutta, D., Rahman, A., Paul, S. K., & Kundu, A. (2019). Changing pattern of urban landscape and its effect on land surface temperature in and around Delhi. *Environmental Monitoring and Assessment, 191*(9), 551. https://doi.org/10.1007/s10661-019-7645-3

Ermida, S. L., Soares, P., Mantas, V., Göttsche, F.-M., & Trigo, I. F. (2020). Google earth engine open-source code for land surface temperature estimation from the Landsat series. *Remote Sensing, 12*(9), 1471. https://doi.org/10.3390/rs12091471

Firmann, M., Mayor, V., Vidal, P. M., Bochud, M., Pécoud, A., Hayoz, D., Paccaud, F., Preisig, M., Song, K. S., Yuan, X., Danoff, T. M., Stirnadel, H. A., Waterworth, D., Mooser, V., Waeber, G., & Vollenweider, P. (2008). The CoLaus study: A population-based study to investigate the epidemiology and genetic determinants of cardiovascular risk factors and metabolic syndrome. *BMC Cardiovascular Disorders, 8*(1), 6. https://doi.org/10.1186/1471-2261-8-6

Gago, E. J., Roldan, J., Pacheco-Torres, R., & Ordóñez, J. (2013). The city and urban heat islands: A review of strategies to mitigate adverse effects. *Renewable and Sustainable Energy Reviews, 25*, 749–758. https://doi.org/10.1016/j.rser.2013.05.057

Getis, A., & Ord, J. K. (1992). The analysis of spatial association by use of distance statistics. *Geographical Analysis, 24*(3), 189–206. https://doi.org/10.1111/j.1538-4632.1992.tb00261.x

Giorgio, A., Santelli, L., Tomassini, V., Bosnell, R., Smith, S., De Stefano, N., & Johansen-Berg, H. (2010). Age-related changes in grey and white matter structure throughout adulthood. *NeuroImage, 51*(3), 943–951. https://doi.org/10.1016/j.neuroimage.2010.03.004

Goodchild, M. F. (1992). Geographical information science. *International Journal of Geographical Information Systems, 6*(1), 31–45. https://doi.org/10.1080/02693799208901893

Hansen, A., Bi, P., Nitschke, M., Ryan, P., Pisaniello, D., & Tucker, G. (2008). The effect of heat waves on mental health in a temperate Australian City. *Environmental Health Perspectives, 116*(10), 1369–1375. https://doi.org/10.1289/ehp.11339

Hondula, D. M., Balling, R. C., Vanos, J. K., & Georgescu, M. (2015). Rising temperatures, human health, and the role of adaptation. *Current Climate Change Reports, 1*(3), 144–154. https://doi.org/10.1007/s40641-015-0016-4

Jabbar, M., Yusoff, M. M., & Shafie, A. (2022). Assessing the role of urban green spaces for human Well-being: A systematic review. *GeoJournal, 87*(5), 4405–4423. https://doi.org/10.1007/s10708-021-10474-7

Jenkins, L. M., Barba, A., Campbell, M., Lamar, M., Shankman, S. A., Leow, A. D., Ajilore, O., & Langenecker, S. A. (2016). Shared white matter alterations across emotional disorders: A voxel-based meta-analysis of fractional anisotropy. *NeuroImage: Clinical, 12*, 1022–1034. https://doi.org/10.1016/j.nicl.2016.09.001

Julian, L. (2011). Measures of anxiety: State-Trait Anxiety Inventory (STAI), Beck Anxiety Inventory (BAI), and Hospital Anxiety and Depression Scale-Anxiety (HADS-A). *Arthritis Care & Research, 63 Suppl 11*, S467. https://doi.org/10.1002/acr.20561

Kruskal, W. H., & Wallis, W. A. (1952). Use of ranks in one-criterion variance analysis. *Journal of the American Statistical Association, 47*(260), 583–621. https://doi.org/10.1080/01621459.1952.10483441

Kühn, S., Düzel, S., Eibich, P., Krekel, C., Wüstemann, H., Kolbe, J., Martensson, J., Goebel, J., Gallinat, J., Wagner, G. G., & Lindenberger, U. (2017). In search of features that constitute an "enriched environment" in humans: Associations between geographical properties and brain structure. *Scientific Reports, 7*(1), Article 1. https://doi.org/10.1038/s41598-017-12046-7

León, F. R., & Burga-León, A. (2015). How geography influences complex cognitive ability. *Intelligence, 50*, 221–227. https://doi.org/10.1016/j.intell.2015.04.011

Levene, H. (1960). Robust tests for equality of variances. In I. Olkin, H. Hotelling, et al. (Eds.), *Contributions to probability and statistics: Essays in honor of Harold Hotelling* (pp. 278–292). Stanford University Press.

Liu, H., Yang, Y., Xia, Y., Zhu, W., Leak, R. K., Wei, Z., Wang, J., & Hu, X. (2017). Aging of cerebral white matter. *Ageing Research Reviews, 34*, 64–76. https://doi.org/10.1016/j.arr.2016.11.006

Longley, P. (2005). *Geographical information systems: Principles techniques applications and management* (2nd ed.). Wiley.

Louis, S., Carlson, A. K., Suresh, A., Rim, J., Mays, M., Ontaneda, D., & Dhawan, A. (2023). Impacts of Climate Change and Air Pollution on Neurologic Health, Disease, and Practice. *Neurology, 100*(10), 474–483. https://doi.org/10.1212/WNL.0000000000201630

Marteau, T. M., Hollands, G. J., & Fletcher, P. C. (2012). Changing human behavior to prevent disease: The importance of targeting automatic processes. *Science (New York, N.Y.), 337*(6101), 1492–1495. https://doi.org/10.1126/science.1226918

Mathenge, M., Sonneveld, B. G. J. S., & Broerse, J. E. W. (2022). Application of GIS in agriculture in promoting evidence-informed decision making for improving agriculture sustainability: A systematic review. *Sustainability, 14*(16), Article 16. https://doi.org/10.3390/su14169974

MeteoSwiss. (2023). *Urban heat—MeteoSwiss*. https://www.meteoswiss.admin.ch/climate/the-climate-of-switzerland/urban-heat.html

MicroGIS. (2017). *Données Statistiques – Swiss Facts*, https://microgis.ch/prestations/donnees-statistiques/

Moran, P. A. P. (1950). Notes on continuous stochastic phenomena. *Biometrika, 37*(1/2), 17–23. https://doi.org/10.2307/2332142

Obradovich, N., Migliorini, R., Paulus, M. P., & Rahwan, I. (2018). Empirical evidence of mental health risks posed by climate change. *Proceedings of the National Academy of Sciences, 115*(43), 10953–10958. https://doi.org/10.1073/pnas.1801528115

Openshaw, S. (1984). Ecological fallacies and the analysis of areal census data. *Environment and Planning A: Economy and Space, 16*(1), 17–31. https://doi.org/10.1068/a160017

Padhy, S. K., Sarkar, S., Panigrahi, M., & Paul, S. (2015). Mental health effects of climate change. *Indian Journal Of Occupational And Environmental Medicine, 19*(1), 3–7. https://doi.org/10.4103/0019-5278.156997

Palinkas, L. A., & Wong, M. (2020). Global climate change and mental health. *Current Opinion in Psychology, 32*, 12–16. https://doi.org/10.1016/j.copsyc.2019.06.023

Park, Y.-S., McMorris, B. J., Pruinelli, L., Song, Y., Kaas, M. J., & Wyman, J. F. (2021). Use of geographic information systems to explore associations between neighborhood attributes and mental health outcomes in adults: A systematic review. *International Journal of Environmental Research and Public Health, 18*(16), Article 16. https://doi.org/10.3390/ijerph18168597

Parsaei, M., Hasehmi, S. M., Seyedmirzaei, H., Cattarinussi, G., Sambataro, F., Brambilla, P., & Delvecchio, G. (2024). Microstructural white matter alterations associated with social anxiety disorders: A systematic review. *Journal of Affective Disorders, 350*, 78–88. https://doi.org/10.1016/j.jad.2024.01.118

Pedro, J., Silva, C., & Pinheiro, M. D. (2019). Integrating GIS spatial dimension into BREEAM communities sustainability assessment to support urban planning policies, Lisbon case study. *Land Use Policy, 83*, 424–434. https://doi.org/10.1016/j.landusepol.2019.02.003

Preisig, M., Waeber, G., Vollenweider, P., Bovet, P., Rothen, S., Vandeleur, C., Guex, P., Middleton, L., Waterworth, D., Mooser, V., Tozzi, F., & Muglia, P. (2009). The PsyCoLaus study: Methodology and characteristics of the sample of a population-based survey on psychiatric disorders and their association with genetic and cardiovascular risk factors. *BMC Psychiatry, 9*, 9. https://doi.org/10.1186/1471-244X-9-9

Pykett, J. (2018). Geography and neuroscience: Critical engagements with geography's "neural turn". *Transactions of the Institute of British Geographers, 43*(2), 154–169. https://doi.org/10.1111/tran.12213

Spielberger, C., Gorsuch, R., Lushene, R., Vagg, P., & Jacobs, G. (1983). *Manual for the state-trait anxiety inventory*. Consulting Psychologists Press.

Ståhle, L., & Wold, S. (1989). Analysis of variance (ANOVA). *Chemometrics and Intelligent Laboratory Systems, 6*, 259–272. https://doi.org/10.1016/0169-7439(89)80095-4

Thompson, R., Hornigold, R., Page, L., & Waite, T. (2018). Associations between high ambient temperatures and heat waves with mental health outcomes: A systematic review. *Public Health, 161*, 171–191. https://doi.org/10.1016/j.puhe.2018.06.008

Tobler, W. R. (1970). A computer movie simulating urban growth in the Detroit region. *Economic Geography, 46*, 234–240. https://doi.org/10.2307/143141

Trofimova, O., Latypova, A., DiDomenicantonio, G., Lutti, A., de Lange, A.-M. G., Kliegel, M., Stringhini, S., Marques-Vidal, P., Vaucher, J., Vollenweider, P., Strippoli, M.-P. F., Preisig, M., Kherif, F., & Draganski, B. (2023). Topography of associations between cardiovascular risk factors and myelin loss in the ageing human brain. *Communications Biology, 6*(1), Art. 1. https://doi.org/10.1038/s42003-023-04741-1

Tukey, J. W. (1949). Comparing individual means in the analysis of variance. *Biometrics, 5*(2), 99–114. https://doi.org/10.2307/3001913

Tuozzo, C., Lyall, A. E., Pasternak, O., James, A. C. D., Crow, T. J., & Kubicki, M. (2018). Patients with chronic bipolar disorder exhibit widespread increases in extracellular free water. *Bipolar Disorders, 20*(6), 523–530. https://doi.org/10.1111/bdi.12588

Wicki, A., Parlow, E., & Feigenwinter, C. (2018). Evaluation and modeling of urban Heat Island intensity in Basel, Switzerland. *Climate, 6*(3), Article 3. https://doi.org/10.3390/cli6030055

Yu, H., & Fotheringham, A. S. (2022). A multiscale measure of spatial dependence based on a discrete Fourier transform. *International Journal of Geographical Information Science, 36*(5), 849–872. https://doi.org/10.1080/13658816.2021.2017440

Zoungrana, M., Andrianisa, H. A., Yonaba, R., Mabia, A. G., Thiam, S., & Bonkian, B. (2024). A GIS-based approach for improving urban sanitation planning and services delivery: A case study from Ouagadougou, Burkina Faso. *Habitat International, 143*, 102993. https://doi.org/10.1016/j.habitatint.2023.102993

In the Right Place at the Right Time: Leveraging Geographic Ecological Momentary Assessment to Analyze Contexts and Health

Yan Kestens and Célia Kingsbury

Abstract A variety of disciplines contribute to our understanding of the interplay between humans and their environments, particularly regarding how environmental exposures shape population health profiles and inequities. Traditionally, research on "neighborhood effects" has relied on relatively crude measures of contextual conditions around one's home, limiting our ability to identify specific mechanisms and actionable targets for intervention. However, the recent proliferation of wearable devices capable of tracking location and making it possible to "at that moment" physiological and behavioral data offers exciting new avenues for investigation. This chapter explores the potential of geographic ecological momentary assessment (GEMA), which combines ecological momentary assessment (EMA) with location tracking and geographic information systems (GIS). GEMA enables the linking of objective environmental data with momentary measures of psychological states, behaviors, and physical health, making it possible to better assess how daily living environments contribute to health outcomes and health inequities. This chapter discusses the theoretical underpinnings of GEMA, its methodological advantages, and its potential to generate evidence for building health-promoting and equitable communities.

Keywords Geographic ecological momentary assessment (GEMA) · Environmental health · Health inequities · Wearable devices

Y. Kestens (✉)
École de santé publique de l'Université de Montréal, Montreal, Quebec, Canada
e-mail: yan.kestens@umontreal.ca

C. Kingsbury
Centre de recherche en santé publique, Montreal, Canada

S. Kühn (ed.), *Environmental Neuroscience*,
https://doi.org/10.1007/978-3-031-64699-7_19

Introduction and Conceptual Underpinnings

Historically, the importance of context on human behavior and health can be seen as resulting from various disciplinary developments. Core to these may lay two fundamental and linked considerations: First, the understanding that the conditions in which individuals are born, live, work, play, and die are strongly influenced by external factors—in other words, that we are partly the result of what surrounds us. Second, variations in health between individuals and populations are tied to differences in environmental conditions. These differences are, to some extent, avoidable, or more specifically, unjust. This implies that by intervening on environments, one can address and minimize disparities or iniquities in health. The first line of thought mostly arises from observational studies and a resulting socioecological understanding of systems that point to explanatory models or representations of how processes linking context—in its broad sense—and health unfold (Golden et al., 2015). This means causal perspectives can help explain current situations, and possibly be used to simulate future states and interventions that could change one's health "life-course" (Laddu et al., 2021; Nathan et al., 2018). The notion of "health determinants" is of course at the core of this causal understanding. The second consideration can be seen as merely a consequence of the realization of the first: if (unequal) contexts are the result of processes that can be acted upon, it seems morally necessary to do something to rectify that (unjust) situation. In other words, purposefully generating or at least, letting health inequalities arise is fundamentally wrong; inequalities are, in that sense, iniquities. The acknowledgment of the socioecological perspective (context matters) and the fundamental moral belief that our societies should be just (justice matters) have underpinned the field of public health, and more specifically health promotion (Peterson et al., 2021). A vast variety of fields have contributed to such intersecting perspectives, and, more broadly, to our understanding of the interplay between context and health.

The notion of context is by itself an interesting one. The expression "putting in context" means explaining the whereabouts of a situation, allowing us to better understand "how and why things are." Contextual elements are by definition linked to time (when?) and space (where?), and may apply to individuals, groups of individuals, or populations (who or with whom?). Of course, both time and space are difficult concepts to grasp. "Now" is never existing, or can be seen as a transient state between the past and the future. In other words, the past determines the now, which is closely linked to the future, although how exactly time should be conceptualized remains open for discussion, or at least, interpretation (Zimmerman, 2007). Similarly, space—and place—does not exist in a vacuum, nor as a container. Here is there and there is here, space being defined by relations (Thrift, 2006). Boundaries are often fuzzy, making it difficult to grasp where "here"—or context—stops. While time and space are core, another dimension of importance is the what - what is there, at that moment?

The role of context has been acknowledged for a long time, but the approach to how we transform it has evolved over time. Roughly 2500 years ago, ancient Greek physiocratic thinkers already identified the role of environmental conditions and behavior on health (Tountas, 2009). However, it is only recently, during the second half of the nineteenth century, that our understanding and conceptualization of contextual determinants of health—and inequities—significantly improved. In 1854, Dr. John Snow's seminal mapping of cholera cases in London brought him to identify a public water pump as the potential source of disease (Tulchinsky, 2018). Since then, a growing number of people—among which many outside the medical profession—started exploring, mapping, and linking how social, environmental, and health conditions relate, underlining the role of contexts on health (Health I of M (US) C for the S of the F of P, 1988). In his book "How the Other Half Lives," journalist and photographer Jacob Riis documented the detrimental environmental and health conditions of immigrants in overcrowded New York tenement houses (Riis, 1890). His work led to important modifications of New York building codes, which imposed the construction of air shafts and larger windows, improving air quality, among others. During that same period, the "Muckraking" writers' movement helped spread the word about how social, legislative, and industry forces impacted both environmental conditions, health and healthcare. Samuel Hopkin Adams' exposé of "The great American Fraud" published in a 1905 issue of Adams (1905) came out just before the first Pure Food and Drugs Act of 1906—which he contributed to draft—was signed. Important documentation of social conditions and social class, as done by Charles Booth's 1900 map of poverty in London—but also from work in sociology such as Durkheim's study of suicide (Durkheim, 2002)—further led to increasing realization of environmental and social impacts on health and health inequities and the potential for targeted interventions.

More recently, the 1986 Ottawa Charter posits the importance of "where" people "learn, work, play and love" and that the "inextricable links between people and their environment constitutes the basis for a socio-ecological approach to health" (World Health Organization, 1986). Leading to this charter, a range of disciplines had indirectly contributed, both conceptually and empirically, to refinements and the development of the socio-ecological framework. This understanding of health acknowledges the multilevel and dynamic system perspective of human and natural processes. In the 1960s and 1970s, works from biology, ecology, or ethology made important contributions to the study of spatio-temporal patterns and behavior of living species (Stuhrmann, 2022) and to the development of system thinking (Van Eetvelde & Aagaard, 2023). Increasingly, subdisciplines bridging traditional silos develop, as a logical response to the necessity of considering multi-disciplinary perspectives to grasp the complex interplay between humans and how they shape and are being shaped by their environment. Fields like environmental neuroscience and psychology, behavioral geography (Hart, 1987), medical geography, or social ecology, to name just a few, are at the crossroads of disciplinary perspectives. Each contributes to that vast objective of understanding how social and environmental conditions shape who and how (un)healthy we are. Among the various disciplinary contributions toward that goal, this chapter focuses on recent developments in our capacity to quantitatively evaluate the impact of daily and even momentary environmental conditions on health outcomes. It specifically explores relatively novel geographic ecological momentary assessment study protocols.

Technological Contributions Toward Geographic Consideration for Exposure Assessment

As is often the case in science, there has been a dialectic relation between theoretical developments and technological advances. This is particularly the case in our conceptualization and ability to objectively measure components of our "subsystem" of interest, which we could define for our purpose, as an individual (population) and its environment(s).

Let's start with technological developments. Increasing computing capacity at the turn of the twenty-first century has opened important possibilities to "digitalize" information, both about humans, their behavior, and the environment. This has led to the development of Geographic Information Systems (GIS) and GIS science, a new field about the science of collecting, storing, and handling spatial—and, less perfectly, spatiotemporal—data (Carré & Hamdani, 2021) (➔ chapter "Neuroscience and Geographic Information Systems to Investigate the Impact of Global Warming on Mood Disorders and Brain Plasticity in Urban Areas" by Vieira Ruas et al.). Through a multilayered representation of our world, GIS has made it possible to link observations about individuals' location, to information characterizing that location. Doing so, it has largely helped develop environmental exposure measures. At the turn of the twenty-first century, an increasing number of observation studies have explored links between such built and social environmental exposure measures and individual or population health characteristics. Such studies were initially mainly conducted at the area level, linking area-level health profiles to environmental characteristics of these areas (Diez Roux, 2001). Because of various issues linked to ecological fallacy and the modifiable area unit problem, both outcome and exposure measures were then increasingly considered at the individual and environmental levels, including through the adoption of multilevel frameworks (Riva et al., 2007). Knowing a person's residential location provided a first spatial anchor for exposure assessment. Various methods—whether through direct spatial correspondence, through buffers to account for spatial proximity, or through density measures providing local estimates of access—were used to evaluate how one's (mostly residential) surroundings could explain one's health profile. Yet, the conclusions that could be drawn from this line of work were partly limited because people move around and are exposed to a variety of environments at different scales, at different times, as they "live, work, play, and love"—also called the exposome (Hahad et al., 2024). At the turn of the 2010, increasing availability of technologies able to monitor or to document individual mobility—mainly through Global Positioning System (GPS) receivers but also through map-based online surveys—contributed to the (re) development of "activity-space"-based studies. Activity spaces are defined as "the subset of all locations within which an individual has direct contact as a result of his or her day-to-day activities" (Golledge & Stimson, 1997; Perchoux et al., 2013). Once more, these advancements benefited from contributions across various fields. Action space, activity space, or potential path area conceptualizations of human behavior linked back to theoretical work from space-time geography mostly developed in the '70s (Hägerstraand, 1970; Lenntorp, 1976). Since then, it has been largely used in the area of transportation studies, focusing on potential or realised daily mobility, destination and mode choice (Golledge & Stimson, 1997). Mobility surveys, mostly conducted for transportation planning purposes, also provided daily mobility and activity space data useful for health studies. Such data were, for example, used to assess multiple daily exposures to food environments (Kestens et al., 2010).

At the turn of the century, portable devices capturing location became reality. The first cellphone with GPS capabilities was produced in 1999. In 2000, the US government removed the "selective availability" on its GPS signal, increasing spatial precision for civilian applications ten-fold. Since then, the cost for GPS receivers decreased dramatically, making their integration in many mobile devices ubiquitous. This contributed to the development of location-based services, in-car navigation, and of course, location capabilities of individual smartphones. Both the existence of dedicated wearable GPS devices and GPS capabilities of smartphones have opened new possibilities for individualized continuous mobility measurement, enabling a better understanding of how the characteristics of our living environments "get under the skin" (Taylor et al., 1997).

With continuous location tracking providing information about where individuals are at any time, linkage with environmental data layers using GIS provides possibilities for localised, continuous or cumulative exposure to different environmental characteristics. While technology has maximized this possibility to measure our daily exposome, former methods of collecting mobility data—using diaries, filled on paper or on handheld devices—existed before. These could be used to document where people went, but also what activities they conducted, whom they met, how they felt, or how spontaneous a chosen activity or location had been (Ramadier et al., 2005). Indeed, location can today be tracked passively—that is, without any active input from the participant beyond his/her consent and her willingness to carry a smartphone or any other type of device that contains a GPS receiver. However, additional information such as the type of activities being conducted, specific health behavior such as diet or tobacco consumption, current emotional states, and social interactions still require active self-report. This data collection is often conducted using electronic diaries or short questionnaires. Several similar terms are used to describe this practice of collecting high-frequency (daily, or several times a day) observational data: Ambulatory Assessment (Conner & Mehl, 2015), Experience Sampling Method (Larson & Csikszentmihalyi, 2014), Ecological Momentary Assessment (Shiffman et al., 2008). Advantages of collecting data during participants' daily lives are numerous, including reducing retrospective bias. A comprehensive assessment of the positives of momentary data collection is described in Doherty et al. (2020). As the authors conclude in their introduction "EMA, by framing questions in the moment of daily life, can therefore produce a very different portrayal of experience—one less subject to the complex set of processes that shape retrospective reports." Originally conducted using paper diaries, today's EMA studies are increasingly conducted using wearable sensors and smartphones, with either specific ambulatory assessment applications, or, increasingly, using web applications (Conner & Mehl, 2015; Christensen et al., 2003; Stone & Shiffman, 1994). Ecological Momentary Assessment that also collects location data and links this location information to environmental context measures is called *Geographical or Geographically Explicit Ecological Momentary Assessment* (GEMA) (Kirchner & Shiffman, 2016).

GEMA for Health Research

We will now review key elements of GEMA study designs, using for illustrative purposes a number of GEMA studies identified in a systematic review conducted in 2023 which gave rise to a STROBE-adapted reporting guideline for GEMA (Kingsbury et al., 2024). From a data collection perspective, numerous considerations pertaining to GEMA studies are identical to EMA studies. Particularly, most of what relates to the study or prompting design, or the instruments participants interact with when providing self-report data can be identical. Yet, because GEMA studies additionally collect location information, new possibilities for data collection itself (prompting) and for (ecological) analysis arise. GEMA studies may explore how noise exposure and perceived noise annoyance relate to psychological stress in multiple geographic contexts and at different temporal scales (Tao et al., 2020) (REF), or if social interaction and surrounding environmental greenness are associated with momentary affective states in walking situations in everyday life (Bollenbach et al., 2022), or if teens' alcohol consumption was related to multiple exposure to alcohol outlets and perceived disorganisation (Byrnes et al., 2017). We will briefly discuss common aspects of EMA and GEMA studies, and further explore specificities of what GEMA studies allow in terms of people–place interaction analysis. Various statistics about EMA studies presented here come from a comprehensive meta-analysis of EMA studies published in 2023 that included some 477 articles (Wrzus & Neubauer, 2023).

Study Design and Prompts

GEMA or EMA studies rely on the principle of collecting data "at that very moment," knowing that self-reports about past events are fraught with retrospective bias (Stone & Shiffman, 1994). With an EMA or GEMA design, one can collect repeated information throughout the day, using various instruments for self-report, sometimes combined with sensors that might provide individual-level physiological or behavioral measures. Additionally, environmental data can be captured, for example using noise or air pollution sensors (Bista et al., 2023). The duration for which such daily assessments are done can vary. Most often, EMA studies collect data for 7 consecutive days, ensuring capture of both week and weekend days. Wrusz et al.'s meta-analysis reports durations ranging from 2 to 180 days. Due to participant burden, the number of days for which data is collected is inversely correlated to the number of prompts or diary records asked from the participant. Break days—that is, days without any data collection—may also be included to lower burden and increase compliance: among the studies reviewed, the median number of prompts per day was 6, but ranged from 2 to 81 (!).

As mentioned, sensor-based data collection is often strictly passive—a smartphone microphone might collect noise data every 5 min, without the participants having to intervene. For self-reports of any sort, however, some type of cue needs to be provided to the participant to let him know when he/she should actively engage in providing information, most often answering a questionnaire, but also possibly taking photos, or recording sound. These calls for action are either based on instructions to the participants or on prompts, which may take the form of participants receiving a push notification or a SMS on their smartphone. Interval-contingent schedules rely on a predetermined programming of prompting times, which may take many forms. For example, prompts might be generated three times per day, one in the morning randomly between 9:00 a.m. and 11:00 a.m., one in the afternoon between 2:00 p.m. and 4:00 p.m., and one in the evening between 7:00 p.m. and 9:00 p.m. Prompting strategies related to situations or events are tied to participants' own behavior or feeling (i.e., through self identification of when a diary record should be filled), or triggered by the automatic detection of thresholds or rule completion. These triggers can be based on physiological, location, or movement measures. In a German study on momentary affective states, walking-triggered e-diaries were used to assess the impact of greenness and social interaction (Bollenbach et al., 2022). Prompts were activated based on combined readings of accelerometry—when movement acceleration exceeded a predetermined threshold—and location movement—when participants' location was identified as nonstationary. Location-based triggers are also often based on geofencing techniques: the prompt is triggered when the GPS receiver detects reaching a predetermined location or crossing a spatial boundary, for example, when one walks into a park, or walks along a waterway. In the Addressing People and Place Microenvironments (APP-Me) weight loss randomized trial testing for the impact of APP-Me messaging conducted in Indianapolis, Indiana, push notifications were sent when the app, using GPS coordinates, detected being at home, at work, or at friends/family home (Clark et al., 2022).

In a Dutch study on happiness, in addition to an interval-contingent schedule, a prompt was sent 10 min after a participant entered an area categorised by "natural" or "built-up area" land-uses, with up to two prompts sent per land use per day. The identification of land-uses was made possible by assessing the main land-use within a 125 m buffer radius of the GPS location, through linkage to a high-resolution topographic map (de Vries et al., 2021). Event-contingent EMA data collection can also be linked to participants' self-identification of behavior or feelings. In a study on tobacco consumption among adult smokers with attention deficit hyperactivity disorder (ADHD) (Mitchell et al., 2014), participants were instructed to provide self-initiated entries about the urge to smoke immediately prior to actually smoking a cigarette. As one can imagine, prompting designs can be varied, using combinations of time and event schedules, depending on the purpose of the study, the disponibility of sensors, or the capacity to link sensors to other types of information such as environmental data layers in near real-time. Comprehensive reviews of EMA design aspects can be found both in Doherty et al. (2020) and in Wrzus and Neubauer (2023).

Location Data

The added value of collecting location information in ambulatory assessments lies in the possibility to better describe participants' momentary context, through either real-time or post-processing linkage with environmental datasets. In many GEMA studies, location data are collected when participants answer an EMA prompt. However, in some cases, it is collected in a continuous manner, providing a more comprehensive mobility profile. This continuous collection allows continuous or cumulative exposure assessments throughout the day. When such mobility data are available, the impact of exposure to environmental features before responding to the prompt can also be evaluated. But capturing—and analyzing–location data does not come without challenges. Most location data are obtained from GPS receivers—either in dedicated GPS trackers or embedded within mobile objects such as smartphones—that acquire signals from the GPS or from similar European Galileo or Russian Glonass satellite constellations, which orbit earth from roughly 20,000 km altitude. However, GPS signals are blocked by hard surfaces, meaning they do generally not reach indoor or underground locations. In other words, GPS tracking is often interrupted when participants are indoors or when traveling in underground transportations such as subways. Location imprecision also arises in denser urban environments. Due to the "bouncing" of GPS signals on buildings, an erroneous estimate of distance to the satellite when the signal reaches the receiver can occur, resulting in an erroneous location estimation on earth (Kerr et al., 2011). Today, most smartphones have assisted GPS functionalities, where the location estimation is "enhanced" with other types of signals for which the source location is known, such as cell towers or Wi-Fi routers. Because of these issues, certain EMA answers have missing location data or may be geographically inaccurate. In a recent review analyzing 20 GEMA studies, eight studies reported the proportion of answered prompts for which GPS data were present, which varied from 29.9% to 100%. In one study, 54% of participants were removed because of missing GPS data (Watkins et al., 2014). While studies that use continuous location tracking may also suffer from periods of missing location data, they can estimate exposure measures for longer periods, including exposures prior to answering an EMA prompt. In a GEMA study conducted in Paris, participants answered electronic diaries three times a day to report momentary depressive symptomatology, while wearing a GPS tracker that continuously collected location data at every second. Exposures during daily mobility paths, computed for a given period before people answered the questionnaire, could be computed (Fancello et al., 2023).

Linking Location Data with Environmental Information

There are many ways to link location data obtained from GEMA studies with environmental information to provide exposure measures and partly depends on the research question. Most of current EMA or GEMA studies are interested in health behaviors (e.g., smoking, alcohol consumption, physical activity, diet, and social behavior), psychological states and affects (e.g., well-being, self-control, stress, mood states), or physical health (e.g., pain), or, more broadly, aim to better understand environmental patterns or inequities in exposures (e.g., air pollution) and health. Environmental correlates, or at least hypothesized environmental determinants of health, are numerous. Environmental health has been mostly looking at the impacts of air pollution or noise, while broader health promotion or social epidemiology considerations have investigated the role of built and social environments on health. Various conceptual underpinnings have posited how such built and social environments might "get under the skin," with some conceptual models identifying various domains and rules of access linked to neighborhood characteristics (Bernard et al., 2007). Other models coming from an eco-social system perspective integrate causal processes linking environments to individual perceptions and behaviors and sometimes biological markers toward health outcomes (Daniel et al., 2008). Most often, such models are limited in their capacity to account for daily mobility and corresponding multiple exposures. Some have tried to integrate daily mobility to better illustrate the interplay between multiple exposures and health and health inequalities (Matthews & Yang, 2013; Sharp et al., 2015). Sharp's work, for example, shows how contextual exposure to disadvantage, residential and non-residential, is independently associated with a higher likelihood of reporting poor or fair health (Sharp et al., 2015). In most of these conceptual models, built and social environmental features one get exposed to are seen as resources (generally positive effect) or externalities (generally negative effect) that shape perceptions, feelings, behaviors, and health. In other words, exposures either "at that moment" or prior to answering a prompt may explain part of the health outcome of interest, whether it relates to physical activity, affect, diet, or craving, to name just a few. A GEMA study on happiness conducted in the Netherlands linked location data when participants rated their level of happiness to the dominant land use within a 125 m radius, but also to noise levels and weather conditions (momentary temperature and rainfall within the past 30 min) (de Vries et al., 2021). Various types of natural environments were positively linked to happiness, especially being exposed to water bodies. Other environments with vegetation, including forests, parks but also agricultural grassland and arable land, were positively associated. Interestingly, happiness was highest along natural coasts and in areas with low-lying natural vegetation. EMA data collection also showed self-perceived peacefulness and fascinatingness of the environment better predicted happiness than did scenic beauty. This aligns with the Attention Restoration Theory (ART), which posits that the restorative capacity of the environment is linked to the non-demanding variety of fascinatingness (soft fascination) (Kaplan, 1995). It is complementary to the stress reduction theory (SRT), which posits that peacefulness or serenity is indicative of the stress-reducing qualities of the environment (Beute & de Kort, 2018; Mennis et al., 2018), features that are linked to both green and blue spaces (Collins et al., 2020).

When looking at how exposure to resources might influence behavior, most GEMA studies assess how much of such resources are found within a close vicinity, either when answering a prompt, or for a given time period prior to that answer, when continuous mobility assessment is available. In a study comparing how perceived vs. objective measures of food environments related to actual food intake, the authors used a 50 m buffer radius to estimate how many food outlets were found around the location of a participant (Elliston et al., 2020). Then they compared how this objective measure fared compared to perception of exposure, and how these measures related to eating, with perceived measures more strongly related to eating behavior.

The study evaluating smoking urges during a quit attempt similarly showed proximity to the closest tobacco outlet and density of tobacco outlets was related to urges, but only when closer (less than one miles) to home. This suggests the impact of a given environmental context might itself vary, depending on its location—here relative to familiar anchor points which may have specific meanings, such as home (Watkins et al., 2014). Another GEMA study on alcohol consumption among teens living in the San Francisco area also computed the number of alcohol outlets within different radius of 50, 100, and 200 m from the prompt's response. Results showed that alcohol consumption was related to the number of outlets found within 100 and 200 m radii (Byrnes et al., 2017).

While GEMA allows testing the relation between self-reports and environmental conditions at the prompt's location, one should be cautious when assuming an immediate "dose–response." For some outcomes, a certain delay between exposure and outcome might be expected. Similarly, when the dose–response relationship involves exceeding a certain threshold, exposure may need to be maintained for a certain period to elicit a given outcome. Hence, GEMA protocols that collect location or mobility data in a continuous manner or at least with repeated location assessments enable testing how various exposure time windows relate to the outcome of interest. This was done in a study conducted in the Paris region in France that used continuous mobility assessment with a wearable GPS tracker and prompted participants regarding depressive symptoms three times a day (Fancello et al., 2023). Environmental exposure measures were calculated within a time window of 2 h prior to reporting depressive symptoms. GPS locations available during that time window were used to establish a measure of 'daily path exposure' using a combination of 50 m street buffers and 25 m line buffers to mimic participants' street viewpoint experience. Within these micro areas, various streetscape environment variables relating to pleasantness, distinctiveness, attractiveness, visual permeability, comfort and safety were derived, using measures of green and blue spaces, presence of landmarks and architectural elements, openness, walkability, noise, traffic calming, and social indicators of population density, income and elderly ratio. Models further distinguished locations (e.g., exposures) when participants were outdoors from indoors. Various environmental elements, including water, commerce, leisure and cultural attractors, and walkable paths, were associated with better momentary mental well-being, and associations were stronger when only outdoor GPS locations were considered.

Of course, the temporal and spatial information included in GPS coordinates also makes it possible to link locations to momentary weather conditions, often available for the closest meteorological station at the hour epoch through public weather data repositories. Such analyses have shown higher temperatures are for example positively linked to happiness (premium when air temperature is above 25 °C in the Netherlands compared to less than 10° (de Vries et al., 2021)) and to positive moods (Bundo et al., 2023), although part of this association was explained by daily sunshine duration), while precipitation was negatively linked to happiness (precipitation within 30 min prior to evaluation).

Modeling GEMA Data

By design, GEMA, as EMA, presents a hierarchical structure, with observations (momentary answered prompts) that can be nested in days, themselves nested in individuals. Momentary variables include the outcome of interest and may also include contextual variables of interest—e.g., environmental exposures, whether objective or self-reported, or complementary contextual indications, such as momentary social interaction. Throughout the day, contextual factors can vary due to people moving around or fluctuations in social interactions. Consequently, momentary correlates of outcomes may vary, with participants with smaller activity spaces or with less daily mobility may show less variance in contexts. At this observation level, the time or period of the day is also often used, potentially revealing temporal within-day variations in outcomes. At the day level, measures generally include day descriptors (specific days of the week or weekdays vs. weekend days) and may also include weather conditions (e.g., average daily temperature and total daily precipitation), although these may also be available at the observation level (e.g., precipitations in the last hour). Finally, individual-level descriptors often concern sociodemographics or health characteristics. Because of the nested structure of EMA data, mixed models or multilevel models are often used, although less often with three levels, and more often with two levels (observations nested in individuals, day-level omitted). Bayesian approaches are well adapted to analyzing EMA data structures (see (Dora et al., 2024) for an online tutorial). Mixed models allow testing either fixed or random effects at each level. In Fancello et al. (2023), tests based on the widely applicable information criterion (WAIC) and on the leave-one-out cross-validation (LOO) indices revealed models performed better with random effects at the day and individual level. They developed Bayesian linear mixed effect models with Monte Carlo Chain using the Stan modeling language and further added a first-order auto-regression function of day-time (within participants) to account for the temporal autocorrelation in the outcome, a mood measure. Because of the longitudinal and nested structure inherent to EMA data, a number of modeling strategies can be applied. Further use of geographic information in GEMA studies may also involve models that account for the spatial autocorrelation of neighbouring observations. A study compared five different analytic strategies adapted for EMA datasets, that is, generalized estimating equations, generalized linear mixed models, Markov models, generalized linear mixed-effects Markov models, and differential eqs. (Shao et al., 2022). Models were developed with an EMA dataset, examining how men-who-have-sex-with-men with mild-to-moderate alcohol use disorder maintain and/or might resist high-risk drinking. This investigation was based on brief treatments that allow for moderation, combining a modified behavioral self-control therapy and medication naltrexone, thought to work by reducing both craving and the pleasure obtained from drinking (Mereish et al., 2018). The authors state that different modeling strategies may be complementary, serving slightly different purposes in the understanding of the complex relationships being explored. While some types of models such as generalized linear mixed models, Markov models, generalized linear mixed-effects Markov models provide indications on the direct relationship of the predictors on some form of the observed response, differential equations quantify "the complex and nonlinear relationship between the predictor variables and the rate of change of the mean of the response variable" (Dora et al., 2024, pp. 22). They conclude that the research question should guide the modeling process and the form of the data used—mentioning that time-scale averages do for example not always dilute the power of EMA, and citing a study of alcohol treatment clinical trials that observed that models testing for main effects of treatments with aggregate data performed as well or even better than when more complicated data structures were used (Hallgren et al., 2016).

The fact that GEMA leverages repeated measures, accounts for localized contexts, and makes it possible to analyze the covariance of various environment-outcome distributions within and between days and between individuals enhances the causality assumptions about environmental "effects," compared to more traditional cross-sectional designs. However, because participants choose—with varying degrees of freedom—their destinations, the directionality of the environment–outcomes association may in some cases remain questionable. Indeed, people's choice of destinations, or in a certain way, choice of "exposures," may be linked to the outcome of interest itself. This issue is known as the selective daily mobility bias (Chaix et al., 2013). For example, a person who feels relaxed may purposefully choose to go for a stroll in a park. The somewhat simultaneous observation of exposure to greenness and relaxed feeling can lead to significant associations in the models, suggesting exposure has had a relaxing effect, whereas in this example, the relaxed feeling has led to being exposed in the first place. In other words, the directionality of an observed exposure–outcome association may be questioned, and various sensitivity tests including using varying time lags between exposures and outcomes may help reinforcing causality assumptions. It is similar to the residential selection bias when looking at residence-based environmental contexts and linking those to certain outcomes (e.g., active living environments and physical activity).

Conclusion

In an accelerating world, where rapidly developing technological capacity is translating into increasing volumes of fine-grained real-time data, geographic ecological momentary assessment (GEMA) methods are poised to become powerful approaches for understanding how the health of humans, animals, and ecosystems relate (Europe WHORO for, 2022). While location tracking is central to evaluating contexts, other sensors embedded in our mobile devices will further provide information on noise and light levels, co-presence of other devices and people, or physiological measures, to name just a few. As for any development strongly dependent on technology, it is crucial to be aware of who participates, and which conclusions are drawn from these studies. However, there is a strong potential to develop strong study protocols that will provide evidence that can help shape environments and societies that promote health and equity.

References

Adams, S. H. (1905). The great American fraud. *Journal of the American Medical Association, XLV*(16), 1171.

Bernard, P., Charafeddine, R., Frohlich, K. L., Daniel, M., Kestens, Y., & Potvin, L. (2007). Health inequalities and place: A theoretical conception of neighbourhood. *Social Science & Medicine, 65*(9), 1839–1852.

Beute, F., & de Kort, Y. A. W. (2018). The natural context of wellbeing: Ecological momentary assessment of the influence of nature and daylight on affect and stress for individuals with depression levels varying from none to clinical. *Health & Place, 49*, 7–18.

Bista, S., Chatzidiakou, L., Jones, R. L., Benmarhnia, T., Postel-Vinay, N., & Chaix, B. (2023). Associations of air pollution mixtures with ambulatory blood pressure: The MobiliSense sensor-based study. *Environmental Research, 227*, 115720.

Bollenbach, L., Schmitz, J., Niermann, C., & Kanning, M. (2022). How do people feel while walking in the city? Using walking-triggered e-diaries to investigate the association of social interaction and environmental greenness during everyday life walking. *Frontiers in Psychology, 13*; [cited 2024 Jan 16]. Available from: https://www.frontiersin.org/articles/10.3389/fpsyg.2022.970336

Bundo, M., Preisig, M., Merikangas, K., Glaus, J., Vaucher, J., Waeber, G., et al. (2023). How ambient temperature affects mood: An ecological momentary assessment study in Switzerland. *Environmental Health, 22*(1), 52.

Byrnes, H. F., Miller, B. A., Morrison, C. N., Wiebe, D. J., Woychik, M., & Wiehe, S. E. (2017). Association of environmental indicators with teen alcohol use and problem behavior: Teens' observations vs. objectively-measured indicators. *Health & Place, 43*, 151–157.

Carré, C., & Hamdani, Y. (2021). Pyramidal framework: Guidance for the next generation of GIS spatial-temporal models. *ISPRS International Journal of Geo-Information, 10*(3), 188.

Chaix, B., Méline, J., Duncan, S., Merrien, C., Karusisi, N., Perchoux, C., et al. (2013). GPS tracking in neighborhood and health studies: A step forward for environmental exposure assessment, a step backward for causal inference? *Health & Place, 21*, 46–51.

Christensen, T. C., Barrett, L. F., Bliss-Moreau, E., Lebo, K., & Kaschub, C. (2003). A practical guide to experience sampling procedures. *Journal of Happiness Studies, 4*, 53–78.

Clark, D. O., Keith, N. R., Ofner, S., Hackett, J., Li, R., Agarwal, N., et al. (2022). Environments and situations as correlates of eating and drinking among women living with obesity and urban poverty. *Obesity Science and Practice, 8*(2), 153–163.

Collins, R. M., Spake, R., Brown, K. A., Ogutu, B. O., Smith, D., & Eigenbrod, F. (2020). A systematic map of research exploring the effect of greenspace on mental health. *Landscape and Urban Planning, 201*, 103823.

Conner, T. S., & Mehl, M. R. (2015). Ambulatory assessment: Methods for studying everyday life. In *Emerging trends in the social and behavioral sciences* (pp. 1–15). Wiley; [cited 2023 Nov 29]. Available from: https://onlinelibrary.wiley.com/doi/abs/10.1002/9781118900772.etrds0010

Daniel, M., Moore, S., & Kestens, Y. (2008). Framing the biosocial pathways underlying associations between place and cardiometabolic disease. *Health & Place, 14*(2), 117–132.

de Vries, S., Nieuwenhuizen, W., Farjon, H., van Hinsberg, A., & Dirkx, J. (2021). In which natural environments are people happiest? Large-scale experience sampling in The Netherlands. *Landscape and Urban Planning, 205*, 103972.

Diez Roux, A. V. (2001). Investigating neighborhood and area effects on health. *American Journal of Public Health, 91*(11), 1783–1789.

Doherty, K., Balaskas, A., & Doherty, G. (2020). The Design of Ecological Momentary Assessment Technologies. *Interacting with Computers, 32*(3), 257–278.

Dora, J., McCabe, C. J., van Lissa, C. J., Witkiewitz, K., & King, K. M. (2024). A tutorial on analyzing ecological momentary assessment data in psychological research with Bayesian (generalized) mixed-effects mode. *Advances in Methods and Practices in Psychological Science*; [cited 2024 Feb 19]. Available from: Preprint at https://osf.io/preprints/psyarxiv/jc938

Durkheim, E. (2002). *Suicide: A study in sociology* (2nd ed., p. 426). Routledge.

Elliston, K. G., Schüz, B., Albion, T., & Ferguson, S. G. (2020). Comparison of geographic information system and subjective assessments of momentary food environments as predictors of food intake: An ecological momentary assessment study. *JMIR mHealth and uHealth, 8*(7), e15948.

Europe WHORO for. (2022). *A health perspective on the role of the environment in One Health*; [cited 2024 Jan 16]. Available from: https://iris.who.int/handle/10665/354574

Fancello, G., Vallée, J., Sueur, C., van Lenthe, F. J., Kestens, Y., Montanari, A., et al. (2023). Micro urban spaces and mental well-being: Measuring the exposure to urban landscapes along daily mobility paths and their effects on momentary depressive symptomatology among older population. *Environment International, 178*, 108095.

Golden, S. D., McLeroy, K. R., Green, L. W., Earp, J. A. L., & Lieberman, L. D. (2015). Upending the social ecological model to guide health promotion efforts toward policy and environmental change. *Health Education & Behavior, 42*(1_suppl), 8S–14S.

Golledge, R. G., & Stimson, R. J. (1997). *Spatial behavior: A geographic perspective*. Guilford Press; [cited 2024 Jan 16]. Available from: https://www.guilford.com/books/Spatial-Behavior/Golledge-Stimson/9781572300507

Hägerstraand, T. (1970). What about people in regional science? *Papers in Regional Science, 24*(1), 7–24.

Hahad, O., Al-Kindi, S., Lelieveld, J., Münzel, T., & Daiber, A. (2024). Supporting and implementing the beneficial parts of the exposome: The environment can be the problem, but it can also be the solution. *International Journal of Hygiene and Environmental Health, 255*, 114290.

Hallgren, K. A., Atkins, D. C., & Witkiewitz, K. (2016). Aggregating and analyzing daily drinking data in clinical trials: A comparison of type I errors, power, and bias. *Journal of Studies on Alcohol and Drugs, 77*(6), 986–991.

Hart, R. A. (1987). Environmental psychology or behavioral geography? Either way it was a good start. *Journal of Environmental Psychology, 7*(4), 321–329.

Health I of M (US) C for the S of the F of P. (1988). A history of the public health system. In *The future of public health*. National Academies Press (US); [cited 2024 Jan 16]. Available from: https://www.ncbi.nlm.nih.gov/books/NBK218224/

Kaplan, S. (1995). The restorative benefits of nature: Toward an integrative framework. *Journal of Environmental Psychology, 15*(3), 169–182.

Kerr, J., Duncan, S., & Schipperjin, J. (2011). Using global positioning Systems in Health Research: A practical approach to data collection and processing. *American Journal of Preventive Medicine, 41*(5), 532–540.

Kestens, Y., Lebel, A., Daniel, M., Thériault, M., & Pampalon, R. (2010). Using experienced activity spaces to measure foodscape exposure. *Health & Place, 16*(6), 1094–1103.

Kingsbury, C., Buzzi, M., Chaix, B., Kanning, M., Kherzi, S., Kiani, B., et al. (2024). STROBE-GEMA: A STROBE extension for reporting of geographically explicit ecological momentary assessment studies. *Social Science & Medicine*. https://doi.org/10.1186/s13690-024-01310-8

Kirchner, T. R., & Shiffman, S. (2016). Spatio-temporal determinants of mental health and well-being: Advances in geographically-explicit ecological momentary assessment (GEMA). *Social Psychiatry and Psychiatric Epidemiology, 51*(9), 1211–1223.

Laddu, D., Paluch, A. E., & LaMonte, M. J. (2021). The role of the built environment in promoting movement and physical activity across the lifespan: Implications for public health. *Progress in Cardiovascular Diseases, 64*, 33–40.

Larson, R., & Csikszentmihalyi, M. (2014). The experience sampling method. In M. Csikszentmihalyi (Ed.), *Flow and the foundations of positive psychology: The collected works of Mihaly Csikszentmihalyi* (pp. 21–34). Springer Netherlands; [cited 2024 Jan 16]. Available from: https://doi.org/10.1007/978-94-017-9088-8_2

Lenntorp, B. (1976). *Paths in Space-time environments: A time-geographic study of movement possibilities of individuals* (156 p). Royal University of Lund, Department of Geography.

Matthews, S. A., & Yang, T. C. (2013). Spatial polygamy and contextual exposures (SPACEs): Promoting activity Space approaches in research on place and health. *The American Behavioral Scientist, 57*(8), 1057–1081.

Mennis, J., Mason, M., & Ambrus, A. (2018). Urban greenspace is associated with reduced psychological stress among adolescents: A Geographic Ecological Momentary Assessment (GEMA) analysis of activity space. *Landscape and Urban Planning, 174*, 1–9.

Mereish, E. H., Kuerbis, A., & Morgenstern, J. (2018). A daily diary study of stressful and positive events, alcohol use, and addiction severity among heavy drinking sexual minority men. *Drug and Alcohol Dependence, 187*, 149–154.

Mitchell, J. T., Schick, R. S., Hallyburton, M., Dennis, M. F., Kollins, S. H., Beckham, J. C., et al. (2014). Combined ecological momentary assessment and global positioning system tracking to assess smoking behavior: A proof of concept study. *Journal of Dual Diagnosis, 10*(1), 19–29.

Nathan, A., Villanueva, K., Rozek, J., Davern, M., Gunn, L., Trapp, G., et al. (2018). The role of the built environment on health across the life course: A call for CollaborACTION. *American Journal of Health Promotion, 32*(6), 1460–1468.

Perchoux, C., Chaix, B., Cummins, S., & Kestens, Y. (2013). Conceptualization and measurement of environmental exposure in epidemiology: Accounting for activity space related to daily mobility. *Health & Place, 21*, 86–93.

Peterson, A., Charles, V., Yeung, D., & Coyle, K. (2021). The health equity framework: A science- and justice-based model for public health researchers and practitioners. *Health Promotion Practice, 22*(6), 741–746.

Ramadier T, Lee-Gosselin MEH, Frenette A. Conceptual perspectives for explaining Spatio-temporal behaviour in urban areas. In: M.E.H. Lee-Gosselin Doherty S.T., Integrated land-use and transportation models (pp. 87–100) Emerald Group Publishing Limited, 2005; [cited 2024 Jan 16]. Available from: https://doi.org/10.1108/9781786359520-004

Riis, J. A. (1890). In A. Jacob (Ed.), *How the other half lives: Studies among the tenements of New York*. C. Scribner's sons. https://www.gutenberg.org/files/45502/45502-h/45502-h.htm. [cited 2024 Jan 16]. Available from: https://www.gutenberg.org/ebooks/45502/pg45502-images.html

Riva, M., Gauvin, L., & Barnett, T. A. (2007). Toward the next generation of research into small area effects on health: A synthesis of multilevel investigations published since July 1998. *Journal of Epidemiology and Community Health, 61*(10), 853–861.

Shao, S., Canner, J. E., Everett, R. A., Bekele-Maxwell, K., Kuerbis, A., Stephenson, L., et al. (2022). A comparison of mathematical and statistical modeling with longitudinal data: An application to ecological momentary assessment of behavior change in individuals with alcohol use disorder. *Bulletin of Mathematical Biology, 85*(1), 5.

Sharp, G., Denney, J. T., & Kimbro, R. T. (2015). Multiple contexts of exposure: Activity spaces, residential neighborhoods, and self-rated health. *Social Science & Medicine, 146*, 204–213.

Shiffman, S., Stone, A. A., & Hufford, M. R. (2008). Ecological momentary assessment. *Annual Review of Clinical Psychology, 4*(1), 1–32.

Stone, A. A., & Shiffman, S. (1994). Ecological momentary assessment (Ema) in behavioral medicine. *Annals of Behavioral Medicine, 16*(3), 199–202.

Stuhrmann, C. (2022). "It felt more like a revolution." How behavioral ecology succeeded ethology, 1970–1990. *Berichte Zur Wiss., 45*(1–2), 135–163.

Tao, Y., Chai, Y., Kou, L., & Kwan, M. P. (2020). Understanding noise exposure, noise annoyance, and psychological stress: Incorporating individual mobility and the temporality of the exposure-effect relationship. *Applied Geography, 125*, 102283.

Taylor, S. E., Repetti, R. L., & Seeman, T. (1997). Health psychology: What is an unhealthy environment and how does it get under the skin? *Annual Review of Psychology, 48*(1), 411–447.

Thrift, N. (2006). *Space, Theory, Culture and Society, 23*(2–3), 139–146.

Tountas, Y. (2009). The historical origins of the basic concepts of health promotion and education: The role of ancient Greek philosophy and medicine. *Health Promotion International, 24*(2), 185–192.

Tulchinsky, T. H. (2018). John Snow, Cholera, the broad street pump; waterborne diseases then and now. *Case Studies in Public Health*, 77. https://doi.org/10.1016/B978-0-12-804571-8.00017-2

Van Eetvelde, V., & Aagaard, C. A. (2023). Theories in landscape ecology. An overview of theoretical contributions merging spatial, ecological and social logics in the study of cultural landscapes. *Landscape Ecology, 38*(12), 4033–4064.

Watkins, K. L., Regan, S. D., Nguyen, N., Businelle, M. S., Kendzor, D. E., Lam, C., et al. (2014). Advancing cessation research by integrating EMA and geospatial methodologies: Associations between tobacco retail outlets and real-time smoking urges during a quit attempt. *Nicotine & Tobacco Research, 16*(Suppl_2), S93–S101.

World Health Organization. (1986). *Ottawa charter for health promotion*. WHO; [cited 2023 Dec 12]. Available from: https://www.canada.ca/content/dam/phac-aspc/documents/services/health-promotion/population-health/ottawa-charter-health-promotion-international-conference-on-health-promotion/charter.pdf

Wrzus, C., & Neubauer, A. B. (2023). Ecological momentary assessment: A meta-analysis on designs, samples, and compliance across research fields. *Assessment*; [cited 2024 Jan 16]. Available from: https://journals.sagepub.com/doi/full/10.1177/10731911211067538?rfr_dat=cr_pub++0pubmed&url_ver=Z39.88-2003&rfr_id=ori%3Arid%3Acrossref.org#table1-10731911211067538

Zimmerman, D. (2007). The privileged present : Defending an 'a-theory' of time. In T. Sider, J. Hawthorne, & D. W. Zimmerman (Eds.), *Contemporary debates in metaphysics* (pp. 211–225). Blackwell.

Mapping the Human Brain with Computational Anatomy

Bogdan Draganski, Stéphane Joost, and Ferath Kherif

Abstract For almost five decades of its existence, imaging neuroscience—the research branch that uses noninvasive brain imaging to provide inferences about brain structure and function, has focused almost exclusively on studies pertaining to well-controlled laboratory experiments or brain pathologies. There is a recent focus on the interaction between individual and the exposome, a construct encompassing not only the natural and built environment but also social and aesthetic components of it. Some of these studies use imaging neuroscience methods to investigate the rich ensemble of associations and causal relationships between exposome variables and the human brain. In this chapter, we present the well-established domain of computational anatomy that uses magnetic resonance imaging data and sophisticated algorithms to provide a flexible framework for statistical analysis in brain space. We cover the basic principles of feature extraction and statistical analysis to point toward some limitations and corresponding new in vivo brain histology

B. Draganski (✉)
LREN—Département des neurosciences cliniques, CHUV, Université de Lausanne, Lausanne, Switzerland

Max-Planck-Institute for Human Cognitive and Brain Science, Leipzig, Germany

Neurocentre, Neurology Department, Inselspital - University Hospital Bern, University Bern, Bern, Switzerland
e-mail: bogdan.draganski@unibe.ch

S. Joost
Geospatial Molecular Epidemiology Group (GEOME), Laboratory for Biological Geochemistry (LGB), School of Architecture, Civil and Environmental Engineering (ENAC), Ecole Polytechnique Fédérale de Lausanne (EPFL), Lausanne, Switzerland

Unit of Population Epidemiology (UEP), Division of Primary Care Medicine (SMPR), Geneva University Hospitals (HUG), Geneva, Switzerland

La Source School of Nursing, University of Applied Sciences and Arts Western Switzerland (HES-SO), Lausanne, Switzerland

F. Kherif
LREN—Département des neurosciences cliniques, CHUV, Université de Lausanne, Lausanne, Switzerland

S. Kühn (ed.), *Environmental Neuroscience*,
https://doi.org/10.1007/978-3-031-64699-7_20

developments helping to circumvent previous interpretational shortcomings. The aim is to provide a sufficient methodological background that will allow for understanding computational anatomy and functional imaging findings and will motivate formulating own questions in the context of environmental neuroscience.

Keywords Voxel-based morphometry · Surface-based morphometry · Voxel-based quantification · Tissue classification · Spatial registration · Computational anatomy

Introduction

Since the early days of civilization, the inherent desire to measure and map the world naturally led to the unceasing interest in charting the human brain at various levels of granularity and complexity. While the investigation of brain anatomy was confined to the domain of post-mortem investigations, the advent of in vivo brain imaging swiftly replaced the simplistic phrenological views on the relationship between head form and behavioural traits by scientifically sound and objective measures of brain volume and shape. Computed tomography (CT) and magnetic resonance imaging (MRI) became the methods of choice with the latter dominating the imaging neuroscience field for the last few decades. In parallel, the information technology progress and particularly the growing accessibility of powerful calculating machines allowed for the development of specific software solutions for feature extraction and analysis of three-dimensional brain imaging data. Together with the insatiable curiosity of scientists, this provided the fruitful grounds for establishing computational anatomy—the study of brain structure using mathematical models and rigorous statistical methods applied on brain images.

The fields of application of computational anatomy range from addressing fundamental questions in neuroscience about the relationship between brain structure and function to probing the accuracy of imaging biomarkers of brain disorders and monitoring disease progression in chronic diseases. Exemplary initiatives for open access large-scale data in clinical cohorts or in the community-dwelling population like the Alzheimer's Disease Neuroimaging Initiative (ADNI—www.adni.loni.usc.edu), the Parkinson's Progression Markers Initiative (PPMI—www.ppmi-info.org), and the UK Biobank (www.ukbiobank.ac.uk), to name but a few, or pooling together uniformly processed brain images aiming at providing sufficient statistical power for diseases with small effect size like the ENIGMA consortium (Thompson et al., 2020), brought computational anatomy research to a new level of validity and clinical utility.

Alongside the scanner hardware developments going in the two diametrically opposite directions of ultra-high and ultra-low magnetic field machines, magnetic resonance physicists continued to develop novel acquisition protocols measuring brain tissue properties. This provided a window of opportunity to validate in vivo obtained proxies of microstructure with ex vivo histology and bridge the gap in our knowledge about the neuro-biological substrate of brain anatomy differences and changes.

In this chapter, we provide a comprehensive overview of the basic principles of computational anatomy and the emerging new developments to conclude with an outlook on the scarcely mapped territory of combined georeferenced and brain space cartography. Given the vast landscape of currently existing computational anatomy methods, we acknowledge the subjective focus on specific techniques. In our presentation, we deliberately do not expand on the steadily growing number of all existing computational anatomy methods and software solutions and refer to the relevant publications (Ashburner et al., 2003; Eickhoff et al., 2018a, b) for a broader overview. We briefly touch on the topic of functional MRI and its importance for investigating the impact of environment on neural activity and brain function.

Basic Principles of Computational Anatomy

For the readership outside the field of imaging neuroscience, the domain of computational anatomy can be easily confounded with clinical neuroradiology. In both cases, the machines used for MRI and the principle of acquiring high-resolution and high contrast images of brain's structure are identical. However, the similarities end here. While neuroradiologists look for pathology using two-dimensional images of a single brain with their naked eyes, computational anatomy uses three-dimensional data and automated methods for feature extraction followed by statistical analysis at the group level.

Imaging Parameters

This explains to a certain extent the differences in image resolution prerequisites—for the case of neuroradiology, it suffices to have an exquisite spatial resolution in only two dimensions and image intensities providing good contrast between brain and non-brain tissue. Computational anatomy requires equal spatial resolution of the voxels—a unit on a three-dimensional regular grid, in analogy to the two-dimensional pixel, to avoid partial volume effects. This becomes clear when an algorithm must assign a probability value to a given voxel to belong to a certain brain tissue type, and this voxel features unequal spatial dimensions—it quickly becomes problematic to any algorithm to assign consistent values within and across individuals along an unequal voxel axis.

Similarly, the image intensity values are of no importance to the human observer of a brain MRI—the goal is to have sufficient contrast between the different tissue types that the eye can detect and a basic notion about the interpretation of a given pathology in the context of predefined imaging protocols. On the contrary, the differential intensity distribution of brain tissue types and their consistency across space and time are primordial for the accuracy of feature extraction in computational anatomy. For the sake of interpretational consistency, the field has adopted a simplistic view of the brain consisting of gray matter—the surface ribbon consisting of up to six layers neurons and the mid-brain structures of basal ganglia and thalamus, and white matter—the substance containing the myelin wrapped projections of the neurons crossing within, between the hemispheres and the spinal cord.

Voxel-Based and Surface-Based Feature Extraction

Since its inception, computational anatomy has used T1-weighted MRI data, which provides an excellent contrast between brain's gray and white matter at roughly $1 \times 1 \times 1$ mm spatial resolution (Fig. 1). Voxel-based morphometry as implemented in the freely available Statistical Parametric Mapping SPM12 package (www.fil.ion.ucl.ac.uk/spm, Wellcome Centre for Human Neuroimaging, UCL London, UK) provides an elegant solution for an accurate and automated delineation between tissue compartments using a probabilistic Bayesian framework (Ashburner & Friston, 2000; Bullmore et al., 1995; Wright et al., 1995). Individual three-dimensional brain MRIs are classified in six tissue classes—gray matter, white matter, cerebrospinal fluid, bony skull, soft tissues of the head and space outside the head (Fig. 2). The prior knowledge about the spatial distribution of tissue classes is provided in the form of tissue probability maps derived from large-scale MRIs of individuals without known or evident brain disorders. In this Bayesian framework, each voxel is assigned a probability value for belonging to a certain tissue class depending on its location in relation to the tissue priors and its intensity. The voxel-specific probabilities across the six tissue classes sum up to 1.

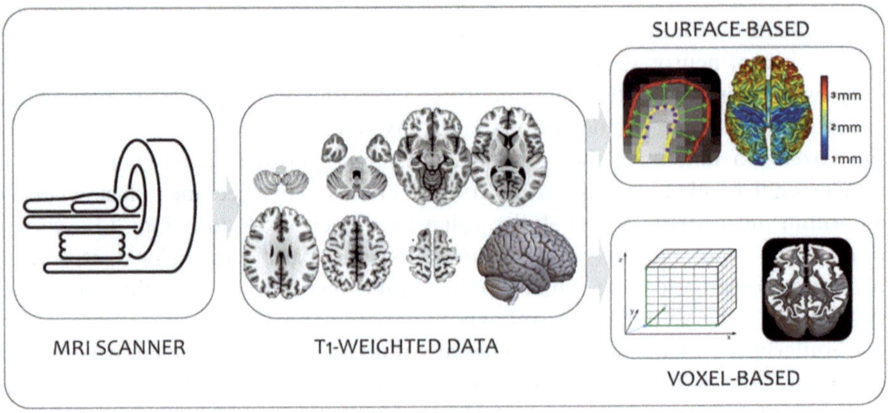

Fig. 1 Schematic representation of structural magnetic resonance imaging (MRI) T1-weighted data acquisition used in surface- and voxel-based computational anatomy techniques

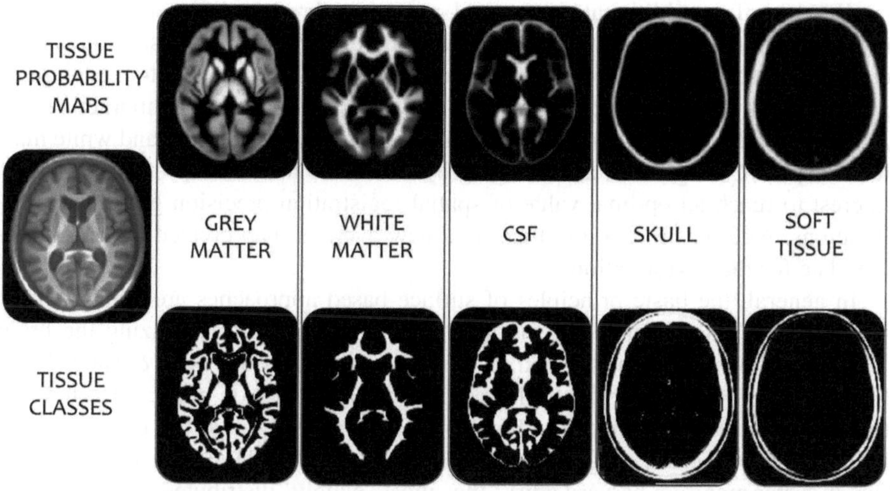

Fig. 2 Statistical parametric mapping—SPM12 (www.fil.ion.ucl.ac.uk/spm): automated Bayesian tissue classification framework with tissue probability maps (upper row) and resulting tissue classes (bottom row) from a T1-weighted magnetic resonance image

Given that the image intensities across the brain tissue are represented as partially overlapping normal distributions, we would like to emphasize here the importance of adjusting these for a phenomenon called inhomogeneity bias. In simple terms, this bias arises since we introduce a perturber in the homogenous magnetic field of the scanner—a human body, and the bias is subject to change in its extent, relative position in space and over time. The result is that parts of our organ of interest—the brain appear darker or lighter depending on the location and extent of this bias. In most computational anatomy software solutions, there are dedicated algorithms dealing with this inhomogeneity bias; however, there are also ways to acquire data quantifying this bias that we can correct the image of interest with (Weiskopf et al., 2011).

In this context, we also introduce the notion of spatial registration of individuals' data residing in their "native" space of acquisition to a standard Cartesian space that is by convention the Montreal Neurological Institute (MNI) space (Ashburner et al., 1997, 1999; Ashburner & Friston, 1999). Given SPM12s probabilistic framework, the necessity to establish a spatial correspondence between the tissue priors that are provided in MNI space, and the individual native space becomes evident. The mapping between a "template" brain in MNI space and the individual MRI is done using linear and nonlinear deformations and the calculated voxel-wise deformation field represents the accurate mapping between native and MNI space for this very individual. The deformation field maps allow the spatial registration in both directions—from native to MNI and from MNI—to native space. The mutual link between tissue classification and spatial registration motivated their combination into one expectation-maximization algorithm—SPM12s "unified segmentation" (Ashburner & Friston, 2005).

SPM12 offers additional anatomical precision through diffeomorphic spatial registration algorithms that rely on a higher number of nonlinear registration parameters to a template that is specific to a given study cohort rather than to the template provided with SPM derived from healthy young individuals (Ashburner, 2007; Ashburner & Friston, 2011). Here, the subject-specific maps of gray and white matter are spatially aligned to an average gray and white matter map of the cohort of interest to reach an optimal value of spatial registration precision going over few iterations. After each iteration, there is a new average map created that serves as template for the next iteration.

In general, the basic principles of surface-based approaches are similar to the voxel-based strategy following the general idea of compartmentalizing the brain into gray matter, white matter, and the rest (Dale et al., 1999; Fischl & Dale, 2000; Fischl et al., 1999a, b). This is followed by spatial registration of the extracted features to standard MNI space to allow for statistical analysis at the group level in brain space (Dale et al., 1999; Fischl & Dale, 2000; Fischl et al., 1999a, b). Freesurfer (www.surfer.nmr.mgh.harvard.edu), the most widely distributed surface-based computational anatomy tool relies on the accurate detection of boundaries defined by intensity differences to then derive Euclidian distances between the inner and outer boundaries of the cortex—i.e., cortical thickness and represent this on the vertices of a surface mesh. The surface-based representation of the obtained anatomical features—cortical thickness and surface area offer complementary insights into brain anatomy, such that the product of cortical thickness and surface area is supposed to equal the volume of the cortical mantle (Dale et al., 1999; Fischl & Dale, 2000; Fischl et al., 1999a).

There is also a twist in the spatial registration strategy compared with SPMs voxel-based morphometry—individual cortical surface representations are inflated to a sphere under the assumption that this will lead to a greater accuracy due to exposing the almost two-thirds of brain's surface hidden in the sulci and to the simpler geometry of a sphere. In our view, this is only valid if the space between the opposed banks of a sulcus is more than a millimetre—the nominal resolution of a typical T1-weighted image. Otherwise, we face the above-mentioned issue with partial volume effects considering that MRI data are typically reconstructed in voxel-space.

Statistical Analysis

The core difference between the diagnostic viewing of brain MRIs in neuroradiology and computational anatomy is rooted in the fact that the inferences are based on state-of-the-art statistical analysis. From a historical point of view, the benefits of a voxel- or surface-based analysis over a selective manual or semi-automated feature extraction followed by a region-of-interest approach are evident (Coffey et al., 1992, 1993). It enables the unbiased statistical analysis over the whole-brain offering insight in structure–function relationships going beyond a limited number of predefined regions. The additional gain of statistical analysis performed in standard brain space is the localization precision that comes for free—significant results can be then easily interpreted in neuro-anatomical terms.

The existing analytical frameworks offer sufficient flexibility for cross-sectional and longitudinal data, for a classical frequentist or probabilistic Bayesian analysis as well as for univariate or multivariate statistics (Kawasaki et al., 2007). The prerequisite for the statistical analysis, as outlined before, is the accurate spatial registration to MNI space, which, depending on the chosen statistical framework, will allow for inferences in brain space. For the simplest case of a mass-univariate analysis, one fits the General Linear Model for the chosen statistical design to each voxel that after correction for multiple comparisons at a threshold—e.g., $p < 0.05$ after family-wise error correction, provides results that can be further investigated using F- and T-contrasts. For a critical appraisal of the use of false-discovery rate for multiple test correction in neuroimaging analysis, we refer to previous publications on the topic (Chumbley et al., 2010; Chumbley & Friston, 2009).

Along these lines, we focus on some important considerations and prerequisite steps of computational anatomy statistical analyses. From a descriptive mathematical point of view, brain images are three-dimensional matrices. There is an additional step of spatial smoothing that renders the data within gray or white matter maps normally distributed and gives a continuous character to the otherwise discrete data (Kiebel et al., 1999; Poline et al., 1995). The recommendation when following the random field theory in SPM, is to apply a three-dimensional Gaussian smoothing kernel with a full-width-at-half-maximum at least three times the spatial resolution of the data. In a similar way, surface-based analyses allow for inferences at the vertex level after the application of a spatial smoothing kernel along the cortical surface (Fischl & Dale, 2000).

Limitations

From its inception, computational anatomy methods for feature extraction, statistical analysis, and interpretation are subject to comments spanning the full bandwidth of constructive criticism (Ashburner & Friston, 2001; Bookstein, 2001). Despite decades of cumulating empirical evidence about the validity of computational anatomy, there are still questions about the reason for "manipulating" native space MRIs when talking about spatial registration and "blurring" of data—in the case of spatial smoothing. Similarly, the existing significant gap between scales of observation at the microscopic level in the micrometre ranges and the MRI-derived mesoscopic scale in millimeters leaves room for speculations about the neuro-biological correlates of regional volume, cortical thickness, or surface area metrics.

The attentive reader will rightly question here the chance of obtaining statistically significant differences between maps of gray matter probabilities that are deformed to perfectly match a standard template with a sub-millimetric precision. In the same vein, the reader might wonder about the neuro-biological interpretation of gray or white matter probabilities. In voxel-based morphometry, there is an additional step called "modulation" before spatial smoothing that involves the voxel-wise scaling of the probability maps by the Jacobian determinants from the spatial registration to MNI space. This step preserves the consistency between gray or white matter metrics in native and MNI space, which allow for interpreting voxel-based morphometry results as volume differences or changes (Good et al., 2001; Peelle et al., 2012).

The traditionally established use of T1-weighted brain imaging offering optimal gray-white matter contrast at 1 mm spatial resolution can be also seen as a hurdle limiting the interpretability of computational anatomy findings (Lorio et al., 2016a). T1-weighted protocols are not disambiguating the differential contributions of brain tissue properties—myelin, iron, or tissue water to MR contrast. Given that tissue properties change nonuniformly across time and brain space, the features extracted from T1-weighted images will be susceptible to the effects of brain maturation, aging, or pathological processes on MR contrast. The unknown relationship between brain tissue property changes, MR contrast and derived metrics of volume or cortical thickness challenges the interpretation of computational anatomy findings as "spurious" morphometric findings (Lorio et al., 2016a; Natu et al., 2019).

Functional Brain Imaging

Even though functional MRI (fMRI) falls outside the computational anatomy domain, we will briefly describe the method given that linking brain structure and function is one of the fundamental questions in neuroscience.

Rather than acquiring a static image of brain structure that takes few minutes, in fMRI, we use rapid acquisition protocols that cover the whole brain within 2–3 s for the "price" of lower spatial resolution usually at 2–3 mm along the voxel dimensions. These protocols are sensitive to the concentration of deoxy-hemoglobin, which changes in specific brain areas following the principle of neuro-vascular coupling—increased perfusion and related oxygen consumption in regions with increased neural activity. This is known in the literature as the blood-oxygen level dependent (BOLD) effect. The underlying physiological response measured with MRI is rather sluggish—the hemo-dynamic response function reaches its peak 4–5 s after the stimulus, and it takes up to 20 s until it reaches its pre-stimulus level. The approximate estimate of signal intensity fluctuation is around 10% for the usual 3 Tesla field strength, which depending on the system of interest, requires multiple stimulus repetitions in series either within blocks—i.e., block design or scattered in time—event-related that are synchronized with the MRI scanner with millisecond precision.

The spatial alignment of the resulting time series consisting of a few hundred brain volumes for an individual is followed by the explicit attribution of time stamps corresponding to the stimulus onsets. The four-dimensional whole-brain data featuring time as the fourth dimension are convolved with a standard hemodynamic response function paralleled by specifications of additional variables and parametric modulations when deemed appropriate for the specific statistical design. After model estimation at the level of an individual, the F- or T-contrasts of interest are tested for significance at the group level provided that the individual data was warped in standard MNI space.

Functional imaging has brought novel insights in the restorative value of landscape environments and potential negative impact of built environment of neural activity and brain function (Neale et al., 2020; Sudimac et al., 2022; Sudimac & Kühn, 2022; Tang et al., 2017), for an exemplary review on the topic in Alzheimer's disease see (Besser, 2021).

New Developments—Toward In Vivo Histology

One of the major challenges to computational anatomy is closing the gap between micro- and mesoscopic scales of observation and opening toward advanced imaging techniques that quantify microstructure brain tissue properties (Amunts et al., 2014; Amunts & Zilles, 2015). For the validation with post mortem investigations of cyto- or myeloarchitectonic characteristics, we look back to an overwhelming wealth of knowledge introduced by Brodmann, Voigt, von Economo, and Koskinas at the beginning of the last century (Amunts et al., 2007; Annese et al., 2004).

The novel brain imaging developments address the critical issue of T1-weighted imaging by attempting to provide indices of myelin, iron, and MR-visible tissue water content (Draganski et al., 2011; Helms et al., 2008; Weiskopf et al., 2013; Weiskopf et al., 2021). Relaxometry (Fig. 3) is one of the well-established techniques for quantitative MRI (qMRI) that has demonstrated potential within the voxel-based and surface-based frameworks of SPM and Freesurfer (Lorio et al., 2014; Lorio et al., 2016a). Relaxometry showed improved contrast-to-noise ratio for "difficult" brain structures—e.g., iron-rich basal ganglia. The automated tissue classification using maps of magnetization transfer saturation (MTsat) and the default SPM12 parameters provided not only accurate detection of the dorso-lateral putamen but also helped delineating the pallidum, substantia nigra, subthalamic nucleus, and dentate nucleus (Helms et al., 2009). The introduction of enhanced tissue priors that allocate pallidum, substantia nigra, subthalamic nucleus, and dentate nucleus to the gray matter brought additional detection accuracy of these structures when using qMRI data (Lorio et al., 2016b).

A steadily growing number of studies showed not only the improved contrast in iron, myelin, and free tissue water brain structures when using the established relaxometry protocol, but they also brought empirical evidence for the validity of the biophysical models allowing for neuro-biological interpretation of MR contrast differences. Of note, the claim that qMRI maps are indicative for myelin, iron, and tissue unbound water is an oversimplification of the underlying biophysical models. Though, first studies correlating qMRI maps with state-of-the-art histology or established means for iron content measurements support the initial assumptions

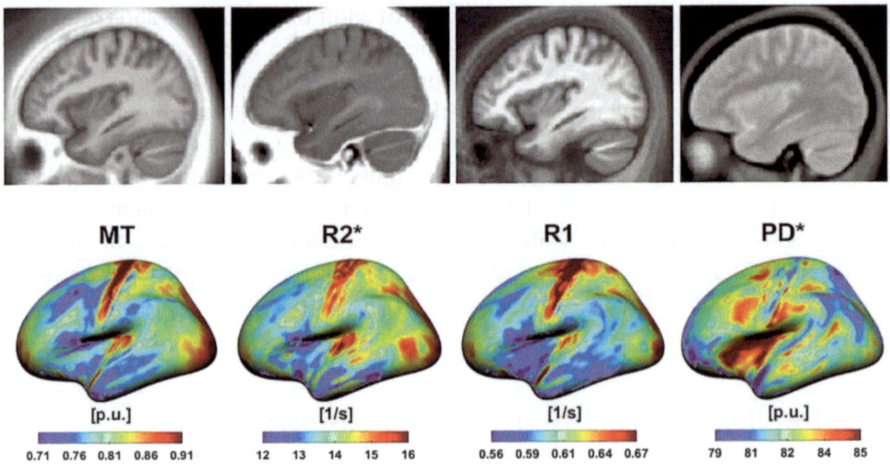

Fig. 3 Relaxometry quantitative magnetic resonance imaging (qMRI) protocol providing maps of magnetization transfer (MT) saturation, transverse relaxation rate (R2*), longitudinal relaxation rate (R1) and effective proton density (PD*). Voxel-based (upper row) and surface-based (bottom row) representation

(Brammerloh et al., 2021; Kirilina et al., 2020; Morawski et al., 2018). One of the promising developments in the new emerging field is the combination of the magnetization transfer saturation indicator of myelin content with estimates of axonal density to calculate an index close to the g-ratio of axon diameter in relation to its myelin sheath (Campbell et al., 2018; Stikov et al., 2015). The in vivo obtained g-ratio estimate opens new opportunities to link brain structure and function via the well-defined physiological parameter of signal conduction velocity (Clark et al., 2022). Another strong point of the qMRI technique is its quantitative character, which precludes cite- or vendor-related bias in the longitudinal setting or in multicentric studies. The field has already taken steps to validate this in multicentric studies; however, there are still controversies to be addressed (Gracien et al., 2020; Karakuzu et al., 2022; Seif et al., 2022; Weiskopf et al., 2013).

Current efforts aim at addressing one of the main limitations of the relaxometry qMRI protocol—the difficulty of its implementation on different vendor MRI scanners. This limits the number of centers active in the field, which hampers the progress of open science and contributions from a wider community. We have created a dedicated SPM toolbox (Tabelow et al., 2019) for data processing of qMRI data following the voxel-based quantification approach (Draganski et al., 2011) including exemplary datasets (Callaghan et al., 2019).

Outlook—Cartography of Georeferenced Data in Brain Space

We can conclude that the qMRI "in vivo histology" approach holds a great promise to bring computational anatomy to another level—in terms of interpretability of the obtained findings and in terms of bias-free data acquisition across different study sites. It also opens new opportunities for collaborations between disciplines that were distant in the past—in this context, between imaging neuroscience and data science using georeferenced data (Dadvand et al., 2018; de Keijzer et al., 2020; Kühn et al., 2017, 2020, 2022). Pioneering large-scale studies provided a proof-of-concept that should be motivating for future research (➔ chapter "Interplay Between the Physical Environment and the Human Brain: A Review of MRI Research" by Beyer et al.). The challenges to be addressed pertain to the current approach collapsing the multidimensional spatial character of either georeferenced or brain data to one dimension. Our current methodological developments attempt to resolve this issue by creating a common framework for combined analysis of georeferenced exposome and brain data without the necessity to reduce the spatial dimensions. The use of a hybrid approach with deep learning methods (e.g., dual encoder-decoder neural network) or the adaptation of multivariate techniques such as partial least squares (PLS) and canonical correlation analysis (CCA), for example, provides a comprehensive framework for the simultaneous analysis of georeferenced and neuroimaging data that can fully capture complex spatial relationships.

In summary, the current developments in computational anatomy using features-rich quantitative MRI protocols sensitive to microstructural properties, together with the increased awareness for open science, carry an enormous potential for making progress in understanding the brain. The next step is to place the individuals' brains in their natural habitat and dig deep into the complex brain—exposome relationships to make the world a better place to live.

Acknowledgments BD is supported by the Swiss National Science Foundation (project grants Nr. 32003B_135679, 32003B_159780, 324730_192755, and CRSK-3_190185), ERA_NET iSEE project, the Swiss Personalized Health Network SACR project, and the Donase and the Leenaards Foundations. LREN is very grateful to the Roger De Spoelberch and Partridge Foundations for their generous financial support.

References

Amunts, K., & Zilles, K. (2015). Architectonic Mapping of the Human Brain beyond Brodmann. *Neuron, 88*(6), 1086–1107.

Amunts, K., Schleicher, A., & Zilles, K. (2007). Cytoarchitecture of the cerebral cortex—More than localization. *NeuroImage, 37*(4), 1061–1065. discussion 1066–1068.

Amunts, K., Hawrylycz, M. J., Van Essen, D. C., Van Horn, J. D., Harel, N., Poline, J. B., et al. (2014). Interoperable atlases of the human brain. *NeuroImage, 99*, 525–532.

Annese, J., Pitiot, A., Dinov, I. D., & Toga, A. W. (2004). A myelo-architectonic method for the structural classification of cortical areas. *NeuroImage, 21*(1), 15–26.

Ashburner, J. (2007). A fast diffeomorphic image registration algorithm. *NeuroImage, 38*(1), 95–113.
Ashburner, J., & Friston, K. J. (1999). Nonlinear spatial normalization using basis functions. *Human Brain Mapping, 7*(4), 254–266.
Ashburner, J., & Friston, K. J. (2000). Voxel-based morphometry—The methods. *NeuroImage, 11*(6 Pt 1), 805–821.
Ashburner, J., & Friston, K. J. (2001). Why voxel-based morphometry should be used. *NeuroImage, 14*(6), 1238–1243.
Ashburner, J., & Friston, K. J. (2005). Unified segmentation. *NeuroImage, 26*(3), 839–851.
Ashburner, J., & Friston, K. J. (2011). Diffeomorphic registration using geodesic shooting and Gauss-Newton optimisation. *NeuroImage, 55*(3), 954–967.
Ashburner, J., Neelin, P., Collins, D. L., Evans, A., & Friston, K. (1997). Incorporating prior knowledge into image registration. *NeuroImage, 6*(4), 344–352.
Ashburner, J., Andersson, J. L., & Friston, K. J. (1999). High-dimensional image registration using symmetric priors. *NeuroImage, 9*(6 Pt 1), 619–628.
Ashburner, J., Csernansky, J. G., Davatzikos, C., Fox, N. C., Frisoni, G. B., & Thompson, P. M. (2003). Computer-assisted imaging to assess brain structure in healthy and diseased brains. *Lancet Neurology, 2*(2), 79–88.
Besser, L. (2021). Outdoor green space exposure and brain health measures related to Alzheimer's disease: a rapid review. *BMJ Open, 11*(5), e043456.
Bookstein, F. L. (2001). 'Voxel-based morphometry' should not be used with imperfectly registered images. *NeuroImage, 14*(6), 1454–1462.
Brammerloh, M., Morawski, M., Friedrich, I., Reinert, T., Lange, C., Pelicon, P., et al. (2021). Measuring the iron content of dopaminergic neurons in substantia nigra with MRI relaxometry. *NeuroImage, 239*, 118255.
Bullmore, E., Brammer, M., Rouleau, G., Everitt, B., Simmons, A., Sharma, T., et al. (1995). Computerized brain tissue classification of magnetic resonance images: a new approach to the problem of partial volume artifact. *NeuroImage, 2*(2), 133–147.
Callaghan, M. F., Lutti, A., Ashburner, J., Balteau, E., Corbin, N., Draganski, B., et al. (2019). Example dataset for the hMRI toolbox. *Data in Brief, 25*, 104132.
Campbell, J. S. W., Leppert, I. R., Narayanan, S., Boudreau, M., Duval, T., Cohen-Adad, J., et al. (2018). Promise and pitfalls of g-ratio estimation with MRI. *NeuroImage, 182*, 80–96.
Chumbley, J. R., & Friston, K. J. (2009). False discovery rate revisited: FDR and topological inference using Gaussian random fields. *NeuroImage, 44*(1), 62–70.
Chumbley, J., Worsley, K., Flandin, G., & Friston, K. (2010). Topological FDR for neuroimaging. *NeuroImage, 49*(4), 3057–3064.
Clark, I. A., Mohammadi, S., Callaghan, M. F., & Maguire, E. A. (2022). Conduction velocity along a key white matter tract is associated with autobiographical memory recall ability. *eLife, 11*, e79303.
Coffey, C. E., Wilkinson, W. E., Parashos, I. A., Soady, S. A., Sullivan, R. J., Patterson, L. J., et al. (1992). Quantitative cerebral anatomy of the aging human brain: a cross-sectional study using magnetic resonance imaging. *Neurology, 42*(3 Pt 1), 527–536.
Coffey, C. E., Wilkinson, W. E., Weiner, R. D., Parashos, I. A., Djang, W. T., Webb, M. C., et al. (1993). Quantitative cerebral anatomy in depression. A controlled magnetic resonance imaging study. *Archives of General Psychiatry, 50*(1), 7–16.
Dadvand, P., Pujol, J., Macià, D., Martínez-Vilavella, G., Blanco-Hinojo, L., Mortamais, M., et al. (2018). The Association between lifelong greenspace exposure and 3-dimensional brain magnetic resonance imaging in Barcelona schoolchildren. *Environmental Health Perspectives, 126*(2), 027012.
Dale, A. M., Fischl, B., & Sereno, M. I. (1999). Cortical surface-based analysis. I. Segmentation and surface reconstruction. *NeuroImage, 9*(2), 179–194.

de Keijzer, C., Bauwelinck, M., & Dadvand, P. (2020). Long-term exposure to residential greenspace and healthy ageing: A systematic review. *Current Environmental Health Reports, 7*(1), 65–88.

Draganski, B., Ashburner, J., Hutton, C., Kherif, F., Frackowiak, R. S. J., Helms, G., et al. (2011). Regional specificity of MRI contrast parameter changes in normal ageing revealed by voxel-based quantification (VBQ). *NeuroImage, 55*(4), 1423–1434.

Eickhoff, S. B., Constable, R. T., & Yeo, B. T. T. (2018a). Topographic organization of the cerebral cortex and brain cartography. *NeuroImage, 170*, 332–347.

Eickhoff, S. B., Yeo, B. T. T., & Genon, S. (2018b). Imaging-based parcellations of the human brain. *Nature Reviews. Neuroscience, 19*(11), 672–686.

Fischl, B., & Dale, A. M. (2000). Measuring the thickness of the human cerebral cortex from magnetic resonance images. *Proceedings of the National Academy of Sciences, 97*(20), 11050–11055.

Fischl, B., Sereno, M. I., & Dale, A. M. (1999a). Cortical surface-based analysis. II: Inflation, flattening, and a surface-based coordinate system. *NeuroImage, 9*(2), 195–207.

Fischl, B., Sereno, M. I., Tootell, R. B., & Dale, A. M. (1999b). High-resolution intersubject averaging and a coordinate system for the cortical surface. *Human Brain Mapping, 8*(4), 272–284.

Good, C. D., Johnsrude, I. S., Ashburner, J., Henson, R. N., Friston, K. J., & Frackowiak, R. S. (2001). A voxel-based morphometric study of ageing in 465 normal adult human brains. *NeuroImage, 14*(1), 21–36.

Gracien, R. M., Maiworm, M., Brüche, N., Shrestha, M., Nöth, U., Hattingen, E., et al. (2020). How stable is quantitative MRI?—Assessment of intra- and inter-scanner-model reproducibility using identical acquisition sequences and data analysis programs. *NeuroImage, 207*, 116364.

Helms, G., Dathe, H., Kallenberg, K., & Dechent, P. (2008). High-resolution maps of magnetization transfer with inherent correction for RF inhomogeneity and T1 relaxation obtained from 3D FLASH MRI. *Magnetic Resonance in Medicine, 60*(6), 1396–1407.

Helms, G., Draganski, B., Frackowiak, R., Ashburner, J., & Weiskopf, N. (2009). Improved segmentation of deep brain grey matter structures using magnetization transfer (MT) parameter maps. *NeuroImage, 47*(1), 194–198.

Karakuzu, A., Biswas, L., Cohen-Adad, J., & Stikov, N. (2022). Vendor-neutral sequences and fully transparent workflows improve inter-vendor reproducibility of quantitative MRI. *Magnetic Resonance in Medicine, 88*(3), 1212–1228.

Kawasaki, Y., Suzuki, M., Kherif, F., Takahashi, T., Zhou, S. Y., Nakamura, K., et al. (2007). Multivariate voxel-based morphometry successfully differentiates schizophrenia patients from healthy controls. *NeuroImage, 34*(1), 235–242.

Kiebel, S. J., Poline, J. B., Friston, K. J., Holmes, A. P., & Worsley, K. J. (1999). Robust smoothness estimation in statistical parametric maps using standardized residuals from the general linear model. *NeuroImage, 10*(6), 756–766.

Kirilina, E., Helbling, S., Morawski, M., Pine, K., Reimann, K., Jankuhn, S., et al. (2020). Superficial white matter imaging: Contrast mechanisms and whole-brain in vivo mapping. *Science Advances, 6*(41), eaaz9281.

Kühn, S., Düzel, S., Eibich, P., Krekel, C., Wüstemann, H., Kolbe, J., et al. (2017). In search of features that constitute an 'enriched environment' in humans: Associations between geographical properties and brain structure. *Scientific Reports, 7*(1), 11920.

Kühn, S., Banaschewski, T., Bokde, A. L. W., Büchel, C., Quinlan, E. B., Desrivières, S., et al. (2020). Brain structure and habitat: Do the brains of our children tell us where they have been brought up? *NeuroImage, 222*, 117225.

Kühn, S., Mascherek, A., Filevich, E., Lisofsky, N., Becker, M., Butler, O., et al. (2022). Spend time outdoors for your brain—An in-depth longitudinal MRI study. *World Journal of Biological Psychiatry Off Journal of World Federation of Societies of Biological Psychiatry, 23*(3), 201–207.

Lorio, S., Lutti, A., Kherif, F., Ruef, A., Dukart, J., Chowdhury, R., et al. (2014). Disentangling in vivo the effects of iron content and atrophy on the ageing human brain. *NeuroImage, 103*, 280–289.
Lorio, S., Kherif, F., Ruef, A., Melie-Garcia, L., Frackowiak, R., Ashburner, J., et al. (2016a). Neurobiological origin of spurious brain morphological changes: A quantitative MRI study. *Human Brain Mapping, 37*(5), 1801–1815.
Lorio, S., Fresard, S., Adaszewski, S., Kherif, F., Chowdhury, R., Frackowiak, R. S., et al. (2016b). New tissue priors for improved automated classification of subcortical brain structures on MRI. *NeuroImage, 130*, 157–166.
Morawski, M., Kirilina, E., Scherf, N., Jäger, C., Reimann, K., Trampel, R., et al. (2018). Developing 3D microscopy with CLARITY on human brain tissue: Towards a tool for informing and validating MRI-based histology. *NeuroImage, 182*, 417–428.
Natu, V. S., Gomez, J., Barnett, M., Jeska, B., Kirilina, E., Jaeger, C., et al. (2019). Apparent thinning of human visual cortex during childhood is associated with myelination. *Proceedings of the National Academy of Sciences of the United States of America, 116*, 20750.
Neale, C., Aspinall, P., Roe, J., Tilley, S., Mavros, P., Cinderby, S., et al. (2020). The impact of walking in different urban environments on brain activity in older people. *Cities Health., 4*(1), 94–106.
Peelle, J. E., Cusack, R., & Henson, R. N. A. (2012). Adjusting for global effects in voxel-based morphometry: Gray matter decline in normal aging. *NeuroImage, 60*(2), 1503–1516.
Poline, J. B., Worsley, K. J., Holmes, A. P., Frackowiak, R. S., & Friston, K. J. (1995). Estimating smoothness in statistical parametric maps: variability of p values. *Journal of Computer Assisted Tomography, 19*(5), 788–796.
Seif, M., Leutritz, T., Schading, S., Emmengger, T., Curt, A., Weiskopf, N., et al. (2022). Reliability of multi-parameter mapping (MPM) in the cervical cord: A multi-center multi-vendor quantitative MRI study. *NeuroImage, 264*, 119751.
Stikov, N., Campbell, J. S., Stroh, T., Lavelée, M., Frey, S., Novek, J., et al. (2015). In vivo histology of the myelin g-ratio with magnetic resonance imaging. *NeuroImage, 118*, 397–405.
Sudimac, S., & Kühn, S. (2022). A one-hour walk in nature reduces amygdala activity in women, but not in men. *Frontiers in Psychology, 13*, 931905.
Sudimac, S., Sale, V., & Kühn, S. (2022). How nature nurtures: Amygdala activity decreases as the result of a one-hour walk in nature. *Molecular Psychiatry, 27*(11), 4446–4452.
Tabelow, K., Balteau, E., Ashburner, J., Callaghan, M. F., Draganski, B., Helms, G., et al. (2019). hMRI—A toolbox for quantitative MRI in neuroscience and clinical research. *NeuroImage, 194*, 191–210.
Tang, I. C., Tsai, Y. P., Lin, Y. J., Chen, J. H., Hsieh, C. H., Hung, S. H., et al. (2017). Using functional Magnetic Resonance Imaging (fMRI) to analyze brain region activity when viewing landscapes. *Landscape and Urban Planning, 162*, 137–144.
Thompson, P. M., Jahanshad, N., Ching, C. R. K., Salminen, L. E., Thomopoulos, S. I., Bright, J., et al. (2020). ENIGMA and global neuroscience: A decade of large-scale studies of the brain in health and disease across more than 40 countries. *Translational Psychiatry, 10*(1), 100.
Weiskopf, N., Lutti, A., Helms, G., Novak, M., Ashburner, J., & Hutton, C. (2011). Unified segmentation based correction of R1 brain maps for RF transmit field inhomogeneities (UNICORT). *NeuroImage, 54*(3), 2116–2124.
Weiskopf, N., Suckling, J., Williams, G., Correia, M. M., Inkster, B., Tait, R., et al. (2013). Quantitative multi-parameter mapping of R1, PD*, MT, and R2* at 3T: a multi-center validation. *Frontiers in Neuroscience, 7*, 95.
Weiskopf, N., Edwards, L. J., Helms, G., Mohammadi, S., & Kirilina, E. (2021). Quantitative magnetic resonance imaging of brain anatomy and in vivo histology. *Nature Reviews Physics, 3*(8), 570–588.
Wright, I. C., McGuire, P. K., Poline, J. B., Travere, J. M., Murray, R. M., Frith, C. D., et al. (1995). A voxel-based method for the statistical analysis of gray and white matter density applied to schizophrenia. *NeuroImage, 2*(4), 244–252.

EEG for the Study of Environmental Neuroscience

Simone Grassini

Abstract This chapter examines the use of the electroencephalogram (EEG) as a tool in environmental neuroscience, looking at its technical attributes, data, practicality, and the exclusive advantages it possesses over other neuroscientific techniques. Traditionally, environmental psychology research has employed psychophysiological indicators such as heart rate, level of cortisol, and skin conductance, to measure the restoring potential of natural spaces. These measures have generally been preferred due to their easy use, minimal invasiveness, and adaptability to outdoor settings. Nevertheless, over the past decade, there has been a significant increase in utilizing neuroscientific methodologies within environmental neuroscience. In this context, EEG has emerged as an important tool for exploring the intricate relationship between humans and the environment. As environmental neuroscience is a scientific field where ecological validity is crucially important, the use of EEG features important attributes, as it is noninvasive, and integrates into diverse experimental designs, including those set outdoors. Recent findings from EEG studies highlight that exposure to natural settings—be they outdoor experiences or exposure to visual stimuli like images or videos in laboratory settings—is associated with an increase of the activity of some brain waves, specifically alpha and theta waves, that have been commonly linked to relaxation. As technology improves, the importance of EEG as a methodology for effectively decoding the human–nature connection may become even more important. The future of environmental neuroscience and EEG research in this context promises deeper insights into our bond with the environment and paves the way for theory development in the context of environmental psychology and knowledge transfer into practical fields such as architecture and urban planning.

S. Grassini (✉)
Department of Psychosocial Science, University of Bergen, Bergen, Norway

Cognitive and Behavioral Neuroscience Laboratory, University of Stavanger, Stavanger, Norway
e-mail: simone.grassini@uib.no

S. Kühn (ed.), *Environmental Neuroscience*,
https://doi.org/10.1007/978-3-031-64699-7_21

Keywords Electroencephalogram · Brain waves · Environmental neuroscience · Psychophysiology · Human–nature interaction

Introduction to EEG

Electroencephalography (EEG) is a method that captures the signals produced by the coordinated activity of groups of neurons in the upper layers of the cortex (Buzsáki, 2006). These electrical signals typically have the power of a few microvolts (μV) (da Silva, 2023). While the most common method of recording EEG signals involves placing electrodes on the scalp, known as scalp-EEG, it is also possible to record these signals by positioning electrodes directly on the surface of the cortex, a method called electrocorticography (ECoG), or by inserting very thin electrodes deep into the brain, referred to as intracranial EEG (iEEG). ECoG and iEEG are seldom used in humans, except during brain surgery, but are commonly employed in animal neuroscience research. On the other hand, scalp-EEG is a noninvasive and reasonably comfortable technique. In this chapter, the term "EEG" will be used to describe scalp-EEG.

EEG signals are dynamic and nonstationary. One approach to analyzing these signals is to examine their frequency and power across different bands or rhythms. These frequency bands are—in the EEG literature (Teplan, 2002)—generally defined as delta (0–4 Hz), theta (4–8 Hz), alpha (8–13 Hz), beta (13–30 Hz), and gamma (30+ Hz), and a modulation of the power (related to the amount of activity and often indicated in μV^2, or as amplitude in μV; Ng et al., 2019) in each of these bands is frequently linked to distinct brain states. For instance, alpha wave activity recorded over the occipital cortex increases when the eyes are closed, beta activity increases with alertness or cognitive demands, delta is related to deep sleep, and theta is associated with drowsiness (Abou-Khalil et al., 2022). Furthermore, activity in these frequency bands can be altered specifically by some cognitive tasks. For example, theta oscillations are also linked to navigational tasks (Watrous et al., 2013), and a modulation of alpha oscillations has often been found to be linked with meditative states (Banquet, 1973; Lo et al., 2003). The versatility and noninvasive nature of EEG make it an important tool in various applications outside clinical and research settings. For example, EEG is used in the field of neuromarketing to understand consumer behavior by analyzing brain responses to advertisements or products (Yadava et al., 2017). In addition, EEG is employed in brain-computer interfaces (BCIs) to translate brain signals into some type of data that can be used to control devices or computer applications (Cincotti et al., 2008).

Traditionally, EEG has been important in the study of neurological disorders (Vandana & Nirali, 2021; de Aguiar & Rosa, 2019). For example, EEG is regularly used to diagnose epilepsy, as it is able to detect abnormal electrical activity in the brain (Smith, 2005). It is also used to monitor the brain activity of patients under anesthesia or in a coma (Musialowicz & Lahtinen, 2014). Furthermore, it is used to assess brain function after traumatic brain injuries (Nuwer et al., 2005). In recent years, there has been an increase in interest in the use of EEG in real-time monitoring of brain functions during various common activities, such as relaxation (Jacobs & Friedman, 2004), exercise (Schneider et al., 2009), and learning (Wan et al., 2021). The use of EEG outside the traditional clinical and laboratory-based setting was supported by the relatively recent development of portable and wearable EEG devices (Casson, 2019). Advances in machine learning and signal processing techniques have supported more accurate and comprehensive analysis of EEG data (Sanei & Chambers, 2013).

EEG is part of a broader category of brain imaging methods, which also includes functional magnetic resonance imaging (fMRI) (➔ chapter "Interplay Between the Physical Environment and the Human Brain: A Review of MRI Research" by Beyer et al.), magnetoencephalography (MEG), positron emission tomography (PET), and others. Each of these methods is employed based on the brain area and neural phenomenon under investigation. While EEG has excellent temporal resolution, in the order of milliseconds, it is typically restricted to studying neural activity from areas near the skull surface. However, novel 3D source-reconstruction techniques partially address this limitation (da Silva, 2023). In contrast, fMRI can study systems deep within the brain, like the hippocampus or amygdala, noninvasively, but has the drawback of being bulky, noisy, and requiring participants to remain still and horizontal throughout the experiment. Furthermore, fMRI cannot be used outside a controlled laboratory setting. EEG is less restrictive, lightweight, potentially wireless, and increasingly tolerant of head and body movements thanks to modern analytical tools.

Compared to the equipment necessary in other neuroimaging techniques, the EEG system is generally cost effective. This characteristic makes such types of equipment more accessible for research environments with limited financial availability. Additionally, advancements in wireless technology and wearable devices have led to the development of portable EEG systems, enabling researchers to conduct studies outside of the traditional laboratory setting. This has opened new possibilities for research in various fields, including cognitive neuroscience, psychology, and human–computer interaction.

EEG Recording and Data Handling

A typical EEG system includes a range of electrodes, usually varying from 3 to 256. The density of the EEG system, which refers to the number of electrodes placed on the scalp, is usually customized based on the objectives of the specific scientific study, the data analysis requirements, or the technical capabilities of the equipment. The electrodes are placed on the scalp to cover the brain areas under examination, such as the visual or prefrontal cortex, and the temporal or parietal lobes (Binnie et al., 1982). One (in some cases more) electrode generally serves as recording reference, and the EEG is measured as the voltage difference between each electrode and the reference (or the average voltage across references). Reference electrode(s) have been traditionally positioned far from brain activity, typically on the nose tips, earlobe, or mastoid; however, it is not unusual to use other placement as close to the head vertex, as for example in Cz (Yao et al., 2019; Nunez et al., 1997). Occasionally, additional electrodes are placed near one or both eyes to measure ocular muscle activity, and then used the recorded eye activity to mitigate eye-movement-related noise from the rest of the EEG signal. To ensure consistent placement across different head sizes and shapes, the electrode placement usually follows the International 10–20 system (Cobb et al., 1958), although there can be modifications for higher electrode density (Oostenveld & Praamstra, 2001). It is vital to achieve good conductivity, or low impedance, between the scalp and the electrodes for precise measurements. This is typically achieved using conductive gels, saline solution, or—more recently—dry electrodes (Mathewson et al., 2017), making the devices quicker and less complex to use, and therefore making the methodology more suitable for commercial and user-oriented applications.

EEG systems generally operate at high sampling rates, up to several thousand Hz, for two primary reasons. First, to capture different frequency bands of the brain rhythms according to the Nyquist theorem, which states that a signal must be captured at a sampling frequency at least twice the signal under investigation (da Silva, 2023). Second, to capture rapid neural events that happen within a fraction of a second after a stimulus, typically within few hundred milliseconds from the onset of the stimulus.

The analysis of EEG signals involves distinguishing noise, or irrelevant non-brain signals, from oscillations that reflect cognitive function (Kaur & Kaur, 2015). The noise can originate from biological sources, such as electrical signals generated by muscle movements due to blinks or head movements, or non-biological sources like electrode movement on the scalp or other electrical noise sources, such as interferences generated by the power grid or electromagnetic signals from appliances or nearby machines (e.g., elevators). To analyze EEG signals, these artifacts must be filtered out, either removed or corrected. Generally, these steps of data preprocessing are nowadays done using algorithms or a combination of computer algorithms and human judgment (Islam et al., 2016). Additionally, techniques such as independent component analysis (ICA) are nowadays often employed to separate the EEG signals into independent components, which can be particularly useful in identifying and removing artifacts or isolating specific sources of neural activity (Onton et al., 2006).

Various signal processing techniques are available for evaluating EEG signals, and the chosen method often depends on the experimental paradigm in question. Two of the most commonly used techniques include the analysis of EEG brain waves and event-related potential (ERP) analysis (Teplan, 2002). The EEG brain waves analysis involves breaking down the continuous EEG signals into the spectral domain and extracting signal features, typically in the traditional frequency bands of delta, theta, alpha, beta, and gamma. This method helps in identifying and understanding the oscillatory activities in the brain during rest, sleep, or during the experience of different cognitive or physiological states. Time-frequency methods, such as wavelet transform or Fourier transform, are commonly used to analyze the dynamics of EEG signals in the frequency domain (Percival & Walden, 1993).

ERP analysis involves examining signals generated during a specific event that is time-locked, such as the presentation of an image or a sound and comparing the changes in signals between various conditions. Such approach is especially useful in studies aiming to understand the brain's response to specific stimuli and how it varies under different conditions. Psychological research in the field of perception and attentional processes (Woodman, 2010) has extensively used such methodology due to its ability of capturing fast perceptual or cognitive effects with an excellent time resolution, in the order of the milliseconds. Various stereotypical perceptual ERPs and their correlations with behavioral or cognitive states have been extensively studied (Luck & Kappenman, 2011).

EEG data are examined also in other methods. For example, analyzing the synchronization and desynchronization of brain waves after an event (event-related spectral perturbations, ERSPs; Makeig, 1993), signal localization analyses (Jatoi et al., 2014), source connectivity analysis (Schoffelen & Gross, 2009), and the analysis of EEG signal entropy (Zhang et al., 2008). Furthermore, network analysis approaches—such as graph theory—have been employed to study EEG data, focusing on the functional connectivity between various brain regions (Bullmore & Sporns, 2009). Additionally, machine-learning approaches have been increasingly employed to classify EEG signals and predict various cognitive or neurological states (Lotte et al., 2018). These approaches have shown great promise, particularly for the advancement of brain–computer interfaces (BCIs), where real-time processing and interpretation of EEG signals are essential.

A primary concern with EEG is the potential interference from non-cerebral sources, including muscle activity and eye movements. Historically, this has restricted EEG's applicability in real-world environments beyond controlled laboratory settings. Environmental artifacts and non-brain signals (e.g., from muscle movements) made the EEG data unsuitable for scientific purposes. However, advancements in computational techniques, as seen in recent years, have significantly improved data preprocessing. This has enabled researchers to interpret EEG data taken during movements (like walking) or contaminated with other types of non-brain artifacts with increased reliability (Gwin et al. 2010; Gorjan et al., 2022). As a result, there has been, during the past decade, an emerging opportunity to utilize EEG in dynamic, outdoor settings. This enhanced the possibility to perform more ecologically valid experiments and paved the way for increasing the use of EEG into environmental psychology research.

Psychophysiological Methods for the Study of Interaction Human Nature

The experience of strolling through a rushing street compared to a park can yield vastly differing experiences and, as a result, distinct psychological impacts. Environmental psychologists (see e.g., Ulrich et al., 1991; Kaplan & Kaplan, 1989) questioned whether specific common environments could facilitate recovery from stress, while others could potentially hinder or negate the recovery process This idea forms the basis of the stress reduction theory (SRT) in environmental psychology, which posits that natural environments, abundant with greeneries and/or water, positively impact mental function, reduce stress and fatigue, and build resilience. Several theories have been put forth to explain the restoration of attention or stress, involving factors such as decreased attentional demands, aesthetic preferences, or evolutionary-related inclinations (Ulrich et al., 1991; Parsons & Tassinary, 2002; Roe & Aspinall, 2011).

Several experimental methods have been employed over the years to investigate this hypothesis, including questionnaire self-evaluations, psychophysiological indicator measurements like heart rate (HR), skin conductance (SC), and cortisol levels (Berto, 2014; Haluza et al., 2014). These studies were conducted in controlled laboratory settings, using video walkthroughs or image slideshows to recreate experiences, and in general confirmed that exposure to vegetation-dominated stimuli generally accelerates recovery from stressors and builds resistance to future stressor tasks (Parsons et al., 1998; Ulrich et al., 1991).

After these early attempts, other studies have employed the use of EEG to understand how the brain reacts to natural settings. The experimental designs usually involved showing participants natural and urban stimuli (as commonly done with other physiological or behavioral methods, see Ohly et al., 2016; Stevenson et al., 2018) and recording brain activity while the participants experienced the stimuli.

EEG for Research on Urban vs. Nature in the Laboratory

In the context of scientific literature that utilizes EEG to decipher the distinctions in brain dynamics when exposed to natural and urban scenes, the use of static visual stimuli stands out notably (Bolouki, 2023). The seminal research in this domain by Ulrich (1981) examined participants' EEG responses as they observed a series of images depicting natural, urban, or aquatic environments. An important finding of this study was that the natural scenes led to a heightened power in the alpha frequency band. This modulation in the alpha range was proposed to correlate with relaxation effects induced by visuals of green surroundings. Remarkably, a replication and expansion of this important finding did not emerge until over three decades later.

Roe et al. (2013) used an EEG-based tool to analyze emotions and has reported that green spaces reduce the level negative emotions while increasing the level of positive ones. However, the data reported from the latter study is limited in terms of interpretation because the researchers did not have access to raw brain data, but instead relied on the interpretation provided by a software (which algorithms are proprietary) provided by the portable EEG manufacturer.

The study of Grassini et al. (2019) built upon the foundational EEG research of Ulrich (1981). Leveraging modern computational tools, the more recent study employed a significantly larger and more diverse image set, which included colored stimuli. Furthermore, the use of computerized stimuli presentation enabled precise timing of the stimuli presentation, and therefore made possible a more extensive analysis of the data, including the possibility to analyze time-locked ERPs. In fact, this study incorporated both the analysis of spectral decomposition analysis of brain waves and ERPSs elicited during the viewing of natural and urban images. The study results largely echoed Ulrich's findings, highlighting an augmentation in low alpha and high theta waves when participants viewed natural images. This observation corroborates the outcomes of preceding studies which proposed that alpha and theta wave activities are positively modulated during experiences of relaxation and meditative states (Lagopoulos et al., 2009; Rodriguez-Larios & Alaerts, 2021). The ERP analyses of the study showed that urban images positively modulated fast ERP components, generally linked to early attentional processes, implying that urban visuals might demand more attentional resources. The early posterior negativity (EPN) ERP component was more pronounced when viewing urban scenes. Prior research has linked this brain component to early amygdala activity (Sabatinelli et al., 2013), which suggests that urban images might incite higher levels of amygdala response than their natural counterparts.

The study of Jiang et al. (2020) showed that EEG alpha waves were positively modulated after visual immersion in natural landscapes, compared to a contrasting setting of urban traffic. Echoing this, another investigation by Elsadek et al. (2021) confirmed that alpha waves experienced a surge when participants were presented with natural images compared to when they were presented with urban ones. The recent study of Imperatori et al. (2023) indicated that viewing natural images is linked to reduced functional connectivity within the distress network. This aligns with past neurophysiological research, which suggests that interactions with natural environments correlate with brain function related to emotional restoration.

Auditory stimuli have also been used to study how natural elements modulate brain activity. The study of Li and Kang (2019) showed that natural sounds like ocean and bird sounds were associated with a higher level of brain alpha waves compared to city sounds as traffic. The study of Qi et al. (2022) showed that people experiencing birdsong resulted in higher levels of restoration compared to a control group (experiencing a white wall in silence), indexed by a variety of physiological measures, as higher-level EEG alpha and lower level of beta-waves, and that the effect become stronger when visuo-auditory stimulation was provided to the participants. Koivisto et al. (2022) showed that when participants assumed an ambiguous sound (composed of pink and white noise, inspired by the study of Haga et al., 2016) originated from a natural source, there was a significant rise in alpha and theta waves activity when EEG was registered. This study showed the possibility that source attribution of natural or urban environments (i.e., the interpretation of the stimulation and not the stimulation alone) could be of pivotal importance in mediating the positive effects of nature and suggesting the importance of the high-level cognitive aspect of the interaction with natural environments as a mediator of positive effects.

To increase the realism of stimuli while preserving the integrity of experimental control, other experimental techniques have been adopted. A recent study by Grassini et al. (2022) uses as experimental stimuli videos depicting both natural and urban settings. A distinctive characteristic of this study compared to others studying natural vs. urban stimulation was the introduction of a control condition involving videos of animated shapes. This was incorporated to discern if the purported "effect of nature" was primarily elicited by natural images or if it arose in reaction to urban ones. The findings confirmed the modulation of low-alpha and high-theta waves during the viewing of natural videos. However, no discernible difference was observed in the alpha and theta brain wave patterns between urban and the control conditions. This observation supports the hypothesis that the effect observed in the alpha-theta domain could be attributed to the positive influence of nature and not to the negative affect of urban environments, when compared to a control or neutral condition that does not display stereotypical characteristic neither of natural or urban environments.

The study of Gao et al. (2019) used virtual reality (VR) simulations while exposing the participants to different types of environments (➔ chapter "Extended Reality in Environmental Neuroscience Research" by Li et al.). However, their EEG data did not show any difference in modulation of brain waves in reaction to different environments, suggesting a possible methodological or practical difficulty to replicate studies that have found modulation of the alpha and theta waves in more controlled settings using images or videos when using a more immersive type of simulation. However, the later VR study of Lei et al. (2021) observed that individuals exposed to a biophilic design exhibited significantly higher EEG alpha power in occipital and parietal areas of their brains. This contrasted with those in environments that lacked biophilic design elements, suggesting a distinct neurophysiological response to natural-inspired environments.

Overall, the trajectory of research in this field seems to be directed toward the use of more ecologically valid stimulation. The past few years, in particular, have seen a surge in interest in leveraging the immersive capabilities of VR systems, as underscored by the recent work of Liu et al. (2023). This trend underscores the evolving nature of research methodologies and the constant pursuit of more ecologically valid and immersive ways to understand the human brain's response to the environment, while at the same time retain the ability of controlling the experimental conditions in a controlled laboratory setting.

EEG for Research on Urban vs. Nature in the Field

To further enhance the ecological validity of the findings and examine the questions in a more realistic context (Parsons et al., 1998), field studies have been conducted using various methods to assess individuals' psychological states and the effects of different environments. Historically investigating brain dynamics using EEG outdoors has not been possible due to technical limitations, and other behavioral and physiological measures have been used. Self-assessments, for example, were used to gauge mood and restoration post-walks in different settings (Roe & Aspinall, 2011). Additionally, Thompson and colleagues (2012) assessed diurnal changes in salivary cortisol concentrations, a psychological stress indicator, and found a positive association between proximity to green spaces and mental health in residents of several disadvantaged Scottish communities. However, a significant limitation of this approach is the necessity for frequent saliva samples from participants. Moreover, salivary cortisol levels change slowly throughout the day, rendering them suboptimal for detecting stress reduction in short-duration nature exposure experiments (Tyrväinen et al., 2014). This infrequent sampling (e.g., pre- and post-walk) fails to capture transient spatial experiences, such as the effects of different streets, intersections, or social and architectural experiences. Other types of wearables have over the years been used to establish a connection between psychological factors and environmental ones (Reeves et al., 2019; Chen et al., 2018; Xiang et al., 2021). Hence, while in situ studies have been offering valuable insights into the real-world impacts of different environments on psychological well-being, however, presenting several methodological challenges.

In this context, while mobile EEG is more intrusive than other tools that allow to collect physiological data as hear rate, it greatly enhances our ability to record mental and cognitive conditions in natural environments compared to most of the portable devices. Often inspired by the potential and cost-effectiveness of consumer-grade portable EEG, researchers have examined its effectiveness in providing insights into environmental preferences, a fact corroborated by naturalistic, situated studies. The literature presented several studies that have compared the impacts of in situ nature and urban exposure on human brain activities using EEG.

The study of Elsadek et al. (2020) showed that observing green areas from a building window led to a notable increase in alpha wave intensity in the frontal region, a marked boost in parasympathetic activity, and a significant reduction in skin conductivity. Their results pointed toward the significance of considering the window views in urban planning efforts. The study of Neale et al. (2020) used mobile EEG to evaluate the effects of various urban environments (namely, busy, quiet, and green spaces) on the brain activities of 95 senior participants, all aged over 65 years. Each of them participated in one of six urban neighborhood walks. The research specifically investigated changes in alpha (linked to relaxation) and beta (linked to attention) brain activity during these walks. Significant variations in brain activity were observed when participants moved between busy urban areas and green spaces. Specifically, green environments were associated with decreased beta activity, indicating shifts in attention, aligning with the attention restoration theory. Moreover, transitions between busy urban and quiet urban areas led to significant changes in alpha activity, with increased alpha levels being noted in busier surroundings. The researchers found no marked difference in EEG measurements between green and quiet urban spaces. This study underscores the idea that the contrast between busy urban environments and other settings plays a crucial role in how these spaces influence brain activity.

Chen et al. (2016) examined the effects of direct exposure to natural and urban environments on brain activity in a sample of college students. Participants were randomly assigned to spend 20 minutes in either a natural or urban setting, after which their brain activity was measured using EEG. Results suggest that short direct experiences in nature may promote more effective brain connectivity, characterized by enhanced small-world properties, than urban exposures. In a further study, Chen et al. (2020) exposed the participants to different urban and natural environments while sitting on a chair. The findings revealed enhanced and more efficient alpha-theta synchronization (functional connectivity) and increased alpha and theta wave oscillations over the occipital lobe when participants were in natural environments with restorative properties.

Aspinall et al. (2015) conducted a study using EEG to monitor the emotional responses of students during a walk in either a natural or urban environment. Utilizing an EEG-based toolbox designed for emotion categorization, the study found that walk-in natural surroundings evoked more positive emotions than a city walk. Similarly, Neale et al. (2017) reported analogous findings in a senior population aged 65 years and above using a comparable methodology. However, the interpretability of the data from these studies is somewhat limited, as the researchers did not analyze the raw brain data directly but relied on interpretations given by proprietary software.

These results underscore the necessity of integrating green spaces into urban areas, given their crucial role in mitigating negative emotions and fostering positive mental states. Furthermore, they spotlight the potential of mobile EEG technology as an instrument to gauge the real-time psychological impacts of diverse environments, yielding invaluable data to steer urban planning and public health initiatives. As technological advancements persist, it is imperative to harness these cutting-edge tools to deepen our understanding of the intricate relationship between our environment and mental well-being, ultimately facilitating the development of healthier, more sustainable urban spaces.

Though often featuring a limited small sample size, these studies' results aligned with theories suggesting the restorative effects of natural environments and showcased the potential of mobile EEG and emotion classification in capturing the emotional and cognitive states of individuals engaged in naturalistic activities, such as urban walking. In the urban design context, mobile EEG could be instrumental in understanding how people experience urban spaces, thereby informing architects, urban designers, and planners about the affective functioning of spaces and the impact of the built environment on mood (Collins et al., 2014; Mavros et al., 2016).

This paves way for an evidence-based approach to urban planning, where design decisions are not merely driven by aesthetics or functionality but are also influenced by the direct psychological responses of its inhabitants. The broader implications of this research could be monumental in tackling urban mental health crises. In an era where urbanization is accelerating, and most of the global population is projected to live in cities, understanding the cognitive and emotional nuances tied to urban spaces is not just valuable, it is essential. Moreover, the seamless integration of mobile EEG technology in everyday activities offers a nonintrusive method of gathering real-time data. This stands in stark contrast to traditional research methods that often require controlled environments, which may not always reflect genuine emotional responses. Therefore, the findings drawn from studies using mobile EEG and emotion classification tools can be considered more authentic, adding an additional layer of credibility to their results.

Future Research

Our understanding of the complex relationship between natural environments and the human brain has significant gaps. EEG technology offers a compelling approach to deepening our insights into this connection. A vital area that remains under-researched is determining the optimal duration of exposure to natural settings for achieving the best neurological benefits.

It's essential to go beyond just being in nature; we must identify the specific components or "active principles" within these natural settings that drive beneficial physiological and clinical outcomes. For example, which aspects of a forest contribute to enhancing well-being and cognitive restoration?

A promising venue is the incorporation of artificial intelligence (AI) in EEG data interpretation. This will significantly refine our ability to detect patterns and insights from the intricate neural data generated during interactions with natural settings. AI-based algorithms have the potential to sift through extensive EEG datasets, pinpointing moments when significant neurological changes occur. This capability can help us understand the ideal exposure durations. Furthermore, machine learning models can assist in identifying the environmental elements, or their combinations, that have the most profound positive neurological impact.

The adaptability and mobility of EEG technology make it possible to monitor brain activity in real time, whether outdoors or in controlled virtual reality scenarios. This positions EEG as a central neuroscientific method in the evolving field of environmental neuroscience.

Conclusion

During the past decade, EEG has emerged as an influential tool for environmental neuroscience, providing invaluable insights into the relationship between humans and the environment. The methodology showcases several possibilities for increasing our understanding of our intrinsic affinity toward natural environments.

EEG stands out in its ability to provide a high-temporal-resolution glimpse into the brain's activity, and providing reliable data even when measures are performed outside traditional laboratory settings. This allows us to observe and decode the swift cognitive and perceptual shifts triggered by varying environmental cues. Whether it is the subtle rustling of leaves in a quiet forest or the cacophonous melodies of bustling urban streets, EEG signal can be used to detect the brain's immediate reactions. EEG, with his fast time resolution, can help to understand when the various cognitive and perceptual effect of the perception of different types of environments happen, and therefore help pinpointing the psychological and physiological processes behind these effects.

By integrating state-of-the-art data analysis methods and machine learning techniques, it will become even more feasible to predict individual emotional and cognitive reactions to distinct environmental attributes. These forward-looking insights can serve architects, urban planners, and decision-makers, enabling them to shape environments that resonate with human preferences. This approach can lead to the development of environments specifically designed to increase overall well-being and support mental health.

References

Abou-Khalil, B., Misulis, K. E., Sonmezturk, H., & Ess, K. C. (2022). *Atlas of EEG, seizure semiology, and management*. Oxford University Press.

Aspinall, P., Mavros, P., Coyne, R., & Roe, J. (2015). The urban brain: Analysing outdoor physical activity with mobile EEG. *British Journal of Sports Medicine, 49*(4), 272–276.

Banquet, J. P. (1973). Spectral analysis of the EEG in meditation. *Electroencephalography and Clinical Neurophysiology, 35*(2), 143–151.

Berto, R. (2014). The role of nature in coping with psycho-physiological stress: A literature review on restorativeness. *Behavioral Sciences, 4*(4), 394–409.

Binnie, C. D., Dekker, E., Smit, A., & Van der Linden, G. (1982). Practical considerations in the positioning of EEG electrodes. *Electroencephalography and Clinical Neurophysiology, 53*(4), 453–458.

Bolouki, A. (2023). Neurobiological effects of urban built and natural environment on mental health: Systematic review. *Reviews on Environmental Health, 38*(1), 169–179.

Bullmore, E., & Sporns, O. (2009). Complex brain networks: Graph theoretical analysis of structural and functional systems. *Nature Reviews Neuroscience, 10*(3), 186–198.

Buzsaki, G. (2006). *Rhythms of the brain*. Oxford University Press.

Casson, A. J. (2019). Wearable EEG and beyond. *Biomedical Engineering Letters, 9*(1), 53–71.

Chen, Z., He, Y., & Yu, Y. (2016). Enhanced functional connectivity properties of human brains during in-situ nature experience. *PeerJ, 4*, e2210.

Chen, Z., Schulz, S., Qiu, M., Yang, W., He, X., Wang, Z., & Yang, L. (2018). Assessing affective experience of in-situ environmental walk via wearable biosensors for evidence-based design. *Cognitive Systems Research, 52*, 970–977.

Chen, Z., He, Y., & Yu, Y. (2020). Attention restoration during environmental exposure via alpha-theta oscillations and synchronization. *Journal of Environmental Psychology, 68*, 101406. https://doi.org/10.1016/j.jenvp.2020.101406. Search in Google Scholar.

Cincotti, F., Mattia, D., Aloise, F., Bufalari, S., Schalk, G., Oriolo, G., et al. (2008). Non-invasive brain–computer interface system: Towards its application as assistive technology. *Brain Research Bulletin, 75*(6), 796–803.

Cobb, W., London, G. B., Gastaut, H., Hess, R., Jr., Jung, R., Magnus, O., & Terzian, H. (1958). Report of the committee on methods of clinical examination in electroencephalography. *Electroencephalography and Clinical Neurophysiology, 10*(2), 370–375.

Collins, M., Hasegawa, T., Johnston, R., Barnes, C., & Louras, G. P. (2014). *DUMBO neural cartography*. Retrieved June 8, 2015, from http://www.thecloudlab.org/dumbo_neural_cartography.html

Da Silva, F. L. (2023). EEG: Origin and measurement. In *EEG-fMRI: Physiological basis, technique, and applications* (pp. 23–48). Springer International Publishing.

de Aguiar Neto, F. S., & Rosa, J. L. G. (2019). Depression biomarkers using non-invasive EEG: A review. *Neuroscience & Biobehavioral Reviews, 105*, 83–93.

Elsadek, M., Liu, B., & Xie, J. (2020). Window view and relaxation: Viewing green space from a high-rise estate improves urban dwellers' wellbeing. *Urban Forestry & Urban Greening, 55*, 126846.

Elsadek, M., Shao, Y., & Liu, B. (2021). Benefits of indirect contact with nature on the physio-psychological well-being of elderly people. *HERD: Health Environments Research & Design Journal, 14*, 227–241.

Gao, T., Zhang, T., Zhu, L., Gao, Y., & Qiu, L. (2019). Exploring psychophysiological restoration and individual preference in the different environments based on virtual reality. *International Journal of Environmental Research and Public Health, 16*(17), 3102.

Gorjan, D., Gramann, K., De Pauw, K., & Marusic, U. (2022). Removal of movement-induced EEG artifacts: Current state of the art and guidelines. *Journal of Neural Engineering, 19*(1), 011004.

Grassini, S., Revonsuo, A., Castellotti, S., Petrizzo, I., Benedetti, V., & Koivisto, M. (2019). Processing of natural scenery is associated with lower attentional and cognitive load compared with urban ones. *Journal of Environmental Psychology, 62*, 1–11.

Grassini, S., Segurini, G. V., & Koivisto, M. (2022). Watching nature videos promotes physiological restoration: Evidence from the modulation of alpha waves in electroencephalography. *Frontiers in Psychology, 13*, 871143.

Gwin, J. T., Gramann, K., Makeig, S., & Ferris, D. P. (2010). Removal of movement artifact from high-density EEG recorded during walking and running. *Journal of Neurophysiology, 103*(6), 3526–3534.

Haga, A., Halin, N., Holmgren, M., & Sörqvist, P. (2016). Psychological restoration can depend on stimulus-source attribution: A challenge for the evolutionary account? *Frontiers in Psychology, 7*, 1831.

Haluza, D., Schönbauer, R., & Cervinka, R. (2014). Green perspectives for public health: A narrative review on the physiological effects of experiencing outdoor nature. *International Journal of Environmental Research and Public Health, 11*(5), 5445–5461.

Imperatori, C., Massullo, C., De Rossi, E., Carbone, G. A., Theodorou, A., Scopelliti, M., ... & Panno, A. (2023). Exposure to nature is associated with decreased functional connectivity within the distress network: A resting state EEG study. *Frontiers in Psychology, 14*, 1171215.

Islam, M. K., Rastegarnia, A., & Yang, Z. (2016). Methods for artifact detection and removal from scalp EEG: A review. *Neurophysiologie Clinique/Clinical Neurophysiology, 46*(4–5), 287–305.

Jacobs, G. D., & Friedman, R. (2004). EEG spectral analysis of relaxation techniques. *Applied Psychophysiology and Biofeedback, 29*, 245–254.

Jatoi, M. A., Kamel, N., Malik, A. S., Faye, I., & Begum, T. (2014). A survey of methods used for source localization using EEG signals. *Biomedical Signal Processing and Control, 11*, 42–52.

Jiang, M., Hassan, A., Chen, Q., & Liu, Y. (2020). Effects of different landscape visual stimuli on psychophysiological responses in Chinese students. *Indoor and Built Environment, 29*, 1006–1016.

Kaplan, R., & Kaplan, S. (1989). *The experience of nature: A psychological perspective*. Cambridge University Press.

Kaur, J., & Kaur, A. (2015, March). A review on analysis of EEG signals. In *2015 international conference on advances in computer engineering and applications* (pp. 957–960). IEEE.

Koivisto, M., Jalava, E., Kuusisto, L., Railo, H., & Grassini, S. (2022). Top-down processing and nature connectedness predict psychological and physiological effects of nature. *Environment and Behavior, 54*(5), 917–945.

Lagopoulos, J., Xu, J., Rasmussen, I., Vik, A., Malhi, G. S., Eliassen, C. F., et al. (2009). Increased theta and alpha EEG activity during nondirective meditation. *The Journal of Alternative and Complementary Medicine, 15*(11), 1187–1192.

Lei, Q., Yuan, C., & Lau, S. S. Y. (2021). A quantitative study for indoor workplace biophilic design to improve health and productivity performance. *Journal of Cleaner Production, 324*, 129168.

Li, Z., & Kang, J. (2019). Sensitivity analysis of changes in human physiological indicators observed in soundscapes. *Landscape and Urban Planning, 190*, 103593.

Liu, P., Liu, J., Fernandez, J., Zou, Q., & Lin, M. (2023). Positive affect and natural landscape in virtual reality: A systematic review comparing interventions, measures, and outcomes. *Journal of Environmental Psychology, 88*, 102011.

Lo, P. C., Huang, M. L., & Chang, K. M. (2003). EEG alpha blocking correlated with perception of inner light during Zen meditation. *The American Journal of Chinese Medicine, 31*(04), 629–642.

Lotte, F., Bougrain, L., Cichocki, A., Clerc, M., Congedo, M., Rakotomamonjy, A., & Yger, F. (2018). A review of classification algorithms for EEG-based brain–computer interfaces: A 10 year update. *Journal of Neural Engineering, 15*(3), 031005.

Luck, S. J., & Kappenman, E. S. (Eds.). (2011). *The Oxford handbook of event-related potential components*. Oxford university press.

Makeig, S. (1993). Auditory event-related dynamics of the EEG spectrum and effects of exposure to tones. *Electroencephalography and Clinical Neurophysiology, 86*(4), 283–293.

Mathewson, K. E., Harrison, T. J., & Kizuk, S. A. (2017). High and dry? Comparing active dry EEG electrodes to active and passive wet electrodes. *Psychophysiology, 54*(1), 74–82.

Mavros, P., Austwick, M. Z., & Smith, A. H. (2016). Geo-EEG: Towards the use of EEG in the study of urban behaviour. *Applied Spatial Analysis and Policy, 9*, 191–212.

Musialowicz, T., & Lahtinen, P. (2014). Current status of EEG-based depth-of-consciousness monitoring during general anesthesia. *Current Anesthesiology Reports, 4*, 251–260.

Neale, C., Aspinall, P., Roe, J., Tilley, S., Mavros, P., Cinderby, S., et al. (2017). The aging urban brain: Analyzing outdoor physical activity using the emotiv affectiv suite in older people. *Journal of Urban Health, 94*, 869–880.

Neale, C., Aspinall, P., Roe, J., Tilley, S., Mavros, P., Cinderby, S., et al. (2020). The impact of walking in different urban environments on brain activity in older people. *Cities Health, 4*, 94–106. https://doi.org/10.1080/23748834.2019.1619893. Search in Google Scholar.

Ng, M. C., Jing, J., & Westover, M. B. (2019). *Atlas of intensive care quantitative EEG*. Springer Publishing Company.

Nunez, P. L., Srinivasan, R., Westdorp, A. F., Wijesinghe, R. S., Tucker, D. M., Silberstein, R. B., & Cadusch, P. J. (1997). EEG coherency: I: Statistics, reference electrode, volume conduction, Laplacians, cortical imaging, and interpretation at multiple scales. *Electroencephalography and Clinical Neurophysiology, 103*(5), 499–515.

Nuwer, M. R., Hovda, D. A., Schrader, L. M., & Vespa, P. M. (2005). Routine and quantitative EEG in mild traumatic brain injury. *Clinical Neurophysiology, 116*(9), 2001–2025.

Ohly, H., White, M. P., Wheeler, B. W., Bethel, A., Ukoumunne, O. C., Nikolaou, V., & Garside, R. (2016). Attention restoration theory: A systematic review of the attention restoration potential of exposure to natural environments. *Journal of Toxicology and Environmental Health, Part B, 19*(7), 305–343.

Onton, J., Westerfield, M., Townsend, J., & Makeig, S. (2006). Imaging human EEG dynamics using independent component analysis. *Neuroscience & Biobehavioral Reviews, 30*(6), 808–822.

Oostenveld, R., & Praamstra, P. (2001). The five percent electrode system for high-resolution EEG and ERP measurements. *Clinical Neurophysiology, 112*(4), 713–719.

Parsons, R., Tassinary, L. G., Ulrich, R. S., Hebl, M. R., & Grossman-Alexander, M. (1998). The view from the road : Implications for stress recovery and immunization. *Journal of Environmental Psychology, 18*, 113–139.

Parsons, R., & Tassinary, L. G. (2002). Environmental psychophysiology. *Handbook of Environmental Psychology*, 172.

Percival, D. B., & Walden, A. T. (1993). *Spectral analysis for physical applications*. Cambridge University Press.

Qi, Y., Chen, Q., Lin, F., Liu, Q., Zhang, X., Guo, J., et al. (2022). Comparative study on birdsong and its multi-sensory combinational effects on physio-psychological restoration. *Journal of Environmental Psychology, 83*, 101879.

Reeves, J. P., Knight, A. T., Strong, E. A., Heng, V., Neale, C., Cromie, R., & Vercammen, A. (2019). The application of wearable technology to quantify health and wellbeing co-benefits from urban wetlands. *Frontiers in Psychology, 10*, 1840.

Rodriguez-Larios, J., & Alaerts, K. (2021). EEG alpha–theta dynamics during mind wandering in the context of breath focus meditation: An experience sampling approach with novice meditation practitioners. *European Journal of Neuroscience, 53*(6), 1855–1868.

Roe, J., & Aspinall, P. (2011). The emotional affordances of forest settings: an investigation in boys with extreme behavioural problems. *Landscape Research, 36*(5), 535–552.

Roe, J., Aspinall, P., Mavros, P., & Coyne, R. (2013). Engaging the brain: The impact of natural versus urban scenes using novel EEG methods in an experimental setting. *Journal of Environmental Sciences, 1*(2), 93–104.

Sabatinelli, D., Keil, A., Frank, D. W., & Lang, P. J. (2013). Emotional perception: Correspondence of early and late event-related potentials with cortical and subcortical functional MRI. *Biological Psychology, 92*(3), 513–519.

Sanei, S., & Chambers, J. A. (2013). *EEG signal processing*. John Wiley & Sons.

Schneider, S., Askew, C. D., Diehl, J., Mierau, A., Kleinert, J., Abel, T., et al. (2009). EEG activity and mood in health orientated runners after different exercise intensities. *Physiology & Behavior, 96*(4–5), 709–716.

Schoffelen, J. M., & Gross, J. (2009). Source connectivity analysis with MEG and EEG. *Human Brain Mapping, 30*(6), 1857–1865.

Smith, S. J. (2005). EEG in the diagnosis, classification, and management of patients with epilepsy. *Journal of Neurology, Neurosurgery & Psychiatry, 76*(suppl 2), ii2-ii7.

Stevenson, M. P., Schilhab, T., & Bentsen, P. (2018). Attention restoration theory II: A systematic review to clarify attention processes affected by exposure to natural environments. *Journal of Toxicology and Environmental Health, Part B, 21*(4), 227–268.

Teplan, M. (2002). Fundamentals of EEG measurement. *Measurement Science Review, 2*(2), 1–11.

Thompson, C. W., Roe, J., Aspinall, P., Mitchell, R., Clow, A., & Miller, D. (2012). More green space is linked to less stress in deprived communities: Evidence from salivary cortisol patterns. *Landscape and Urban Planning, 105*(3), 221–229.

Tyrväinen, L., Ojala, A., Korpela, K., Lanki, T., Tsunetsugu, Y., & Kagawa, T. (2014). The influence of urban green environments on stress relief measures: A field experiment. *Journal of Environmental Psychology, 38*, 1–9.

Ulrich, R. S. (1981). Natural versus urban scenes: Some psychophysiological effects. *Environment and Behavior, 13*(5), 523–556.

Ulrich, R. S., Simons, R. F., Losito, B. D., Fiorito, E., Miles, M. A., & Zelson, M. (1991). Stress recovery during exposure to natural and urban environments. *Journal of environmental psychology, 11*(3), 201–230.

Vandana, J., & Nirali, N. (2021). A review of EEG signal analysis for diagnosis of neurological disorders using machine learning. *Journal of Biomedical Photonics & Engineering, 7*(4), 40201.

Wan, Z., Yang, R., Huang, M., Zeng, N., & Liu, X. (2021). A review on transfer learning in EEG signal analysis. *Neurocomputing, 421*, 1–14.

Watrous, A. J., Fried, I., & Ekstrom, A. D. (2011). Behavioral correlates of human hippocampal delta and theta oscillations during navigation. *Journal of Neurophysiology, 105*(4), 1747–1755.

Watrous, A. J., Lee, D. J., Izadi, A., Gurkoff, G. G., Shahlaie, K., & Ekstrom, A. D. (2013). A comparative study of human and rat hippocampal low-frequency oscillations during spatial navigation. *Hippocampus, 23*(8), 656–661.

Woodman, G. F. (2010). A brief introduction to the use of event-related potentials in studies of perception and attention. *Attention, Perception, & Psychophysics, 72*, 2031–2046.

Xiang, L., Cai, M., Ren, C., & Ng, E. (2021). Modeling pedestrian emotion in high-density cities using visual exposure and machine learning: Tracking real-time physiology and psychology in Hong Kong. *Building and Environment, 205*, 108273.

Yadava, M., Kumar, P., Saini, R., Roy, P. P., & Prosad Dogra, D. (2017). Analysis of EEG signals and its application to neuromarketing. *Multimedia Tools and Applications, 76*, 19087–19111.

Yao, D., Qin, Y., Hu, S., Dong, L., Bringas Vega, M. L., & Valdés Sosa, P. A. (2019). Which reference should we use for EEG and ERP practice? *Brain Topography, 32*, 530–549.

Zhang, A., Yang, B., & Huang, L. (2008, May). Feature extraction of EEG signals using power spectral entropy. In *2008 international conference on BioMedical engineering and informatics* (Vol. 2, pp. 435–439). IEEE.

Part VII
Sociological and Historical Perspective

The Cognitive Neuroscience of Nature?: Sociological and Historical Perspective on Environmental Neuroscience

Des Fitzgerald

Abstract This chapter concerns some of the key sociological, philosophical, and historical underpinnings of environmental neuroscience. The chapter first outlines the technological and conceptual developments that preceded environmental neuroscience and grounds the field within a wider context of the "post-genomic" sciences. It then turns to earlier attempts to relate human psychology to experiences within particular physical environments, going back to the nineteenth century, before asking whether recent concerns about urbanization and climate change might have motivated a return to such topics. Lastly, the chapter asks environmental neuroscientists to pay attention to the politics of their field, and especially to the politics of consistently advocating for more "natural" or "traditional" environments for human beings. The chapter concludes, optimistically, with an argument that environmental neuroscience is a field "in the making," with some scope yet to intervene on what will become its core facts and assumptions.

Keywords History of psychology · Politics of neuroscience · Environmental history · Traditional architecture · Urbanization

Introduction: Hiking Makes You Smarter

In May 2012, the well-known hiking and wilderness magazine, *Backpacker*, published a story about new research that showed how, beyond even the everyday pleasure and invigoration of outdoor exercise, "staying home is just plain dumb." Under the title, *Hiking Makes You Smarter,* the journalist, Elisabeth Kwak-Hefferan, described how she had recently taken part in a pilot study in a Utah canyon, to show how being exposed to nature can cause a measurable change to the brain. The

D. Fitzgerald (✉)
Radical Humanities Laboratory and Department of Sociology and Criminology, University College Cork, Cork, Ireland
e-mail: desfitzgerald@ucc.ie

S. Kühn (ed.), *Environmental Neuroscience*,
https://doi.org/10.1007/978-3-031-64699-7_22

research team, Kwak-Hefferan reported, hypothesized that being out in the wilderness would make a person think and focus more acutely; it would destress them, calm them down, make them measurably happier. The appeal for her magazine's readers was obvious. "Imagine a world," Kwak-Hefferan wrote, in which "backpacking becomes the science-recommended way to prepare for the SATs, chess tournaments, and all of life's biggest mental challenges."

Here then, in the hollow wilderness of the American west, was the frontier of a new discipline, "environmental neuroscience"—an emerging field, said the magazine, that "hones in on how one's surroundings specifically affect the way the brain works." Environmental neuroscience, the reporter went on, is grounded in the idea "that something about being in the wilderness causes physiological changes: the release of certain hormones, maybe, or the switch of activity from one brain region to another." In the experiment, participants were given a set of cognitive tests, before spending the next few days hiking in the wilderness, and taking more tests. The hypothesis was that participants' cognitive performance would improve as they became more and more immersed in the environment. This all had to do, Kwak-Hefferan told the readers of *Backpacker*, with the prefrontal cortex, an area of the brain that was becoming chronically over-burdened by the "buzzers, sirens, and ringtones of life"—that was now increasingly over-worked, tired, stressed out. "Our modern environment is something that we've invented in the last 100 years or so," David Strayer, the lead researcher, told the journalist: "Frankly, we don't know what that's going to do to us." Following some positive results, the article ends with an account of how the scientists were planning a bigger study down the line, perhaps also taking a blood measure of participants.[1] But the journalist herself needed no such reassurance: "I'm a believer," she declares, startlingly, at the end of the article: "While my weekend outings are just as much fun as they've always been, I think of these trips differently. They've become a vital investment in my own brainpower, a much-needed splash of smart sauce on an overtired frontal lobe." Sure, she concludes, she could use biological or cognitive tests to demonstrate this association—but the point is she doesn't need to. "Hell yes, I'm smarter," this journalist reports at the end of her four-day wilderness immersions: "I can feel it. Can't you?"

I begin this short essay on sociological and philosophical issues in environmental neuroscience with this example for several reasons. As an early media story on the coalescence of environmental neuroscience, the article in itself is a useful historical artifact for making sense of precisely how, and in what terms, the then emergent (and as I argue in this contribution: *still* emergent) field was being constituted by its publics. I also begin with this article because of where it appears—not the *New York Times*, not the *Guardian*, not *Die Zeit*, but the rather less-storied pages of *Backpacker* magazine, i.e., a self-conscious media accomplice of the solidification of "nature" and "the outdoors" as sources of middle-class physical and moral improvement in

[1] Indeed, Strayer, also a well-established expert on attention, has continued to publish assiduously on this topic—including, at the time of writing this essay, an EEGH-based randomized control trial of how walking in urban versus rural environments affects performance on an attention task (McDonnell & Strayer, 2024).

the middle decades of the twentieth century (indeed, a subscription to the magazine is bundled with membership of the American Hiking Society to this day. See Berg, 2015). But I begin with it especially because of that last striking, definitive claim; that whatever the quality or strength of the scientific research, however exciting the emergence of a brain measure for environmental effects may be—in some sense, all of this doesn't matter. Because the mere idea itself, that nature contact is good for our minds, that it improves our mental health and our capacity to think clearly, all of this is, somehow, something we already know intuitively—we know it already because we can *feel* it (➔ chapter "The CITY & the PSYCHE—A Chronology" by Mehran, Kluge and Adli). At stake, here, is a set of implicit philosophical and sociological commitments—about the gap between city and the countryside; about our ideas of nature; about what an environment is; about what kinds of spaces might be good for human beings; about what is good for our mental health—that come together, I argue, to make environmental neuroscience into a potent and rapidly growing field. In what follows, rather than go through a prescriptive list of "social issues," I want to make an argument that environmental neuroscience has emerged within a very specific social and historical context. Understanding the field as a social phenomenon, I suggest, means understanding how it is embedded in, and reproductive of, a set of historical trajectories and macro-social contexts—about urbanization, about the place of nature, and about the perceived catastrophe of our modern built environment.

The Cognitive Neuroscience of Nature

Of course, understanding the emergence of environmental neuroscience means, initially at least, understanding the technical and conceptual developments that have made the field thinkable in the first place. In terms of the former, large-scale funding and coordination activities over the last 10–15 years, such as the NIH's BRAIN initiative, have significantly advanced the technical capacity for what is sometimes called "real world neuroscience"—i.e., the use of either virtual environments, or mobile brain-imaging methods, to take a brain measure while a person moves around a space (see NIH, 2023). While these technologies are still at a preliminary stage, we are at least now within sight, say Krugliak and Clarke, of a "toolkit that allows for the study of neural function in uncontrolled complex visual environments, getting us closer to studying cognition in our natural habitat" (2022: 2291). In one sense, we could read environmental neuroscience as essentially technologically determined: the field emerged as it becomes possible (at least in a preliminary sense) to take a brain measure of a person moving around a changing three-dimensional environment. The "environment" is now suddenly visible and measurable at the level of the brain.

However, this technological story has to be added to a conceptual one. Because we might *also* read environmental neuroscience as simply one part of a wider realization, common to a larger set of neuro-disciplines, that the environment has a much larger role in brain function than was previously thought. As Berman and his colleagues put it in an introductory text to the field, "simply controlling for, or regressing out, environmental context is not sufficient to truly understand behavior" (2019a: 1040). More specifically, if genome-wide association studies explain only a small portion of brain phenotype variance—then environment, or environmental exposure, must explain the remainder, whether we understand this relationship as "gXe," the "exposome" (➔ chapter "Advancing the Neural Exposome" by Veasey) or some other neologism (see Liu et al., 2023; Vineis & Barouki, 2022). In this sense, environmental neuroscience might be read as another development within what sociologists of science have called the "post-genomic" turn in the life science, i.e. a general move, in the wake of the relative disappointments of both the decade of the brain and the human genome project, towards an account of human biology as no longer an isolated kingdom unto itself– but understood, now, as a system in constant transaction with, and indeed highly dependent on, the social and cultural world in which biological individuals are embedded (see Rearden, 2017).

But what, precisely, is meant by "environment" here? The word itself is a recent invention and has been used to mean a wide array of non-overlapping things (see Warde et al., 2015). Moreover, neuroscientists have been interested in different kinds of environments for some time. Indeed, one way of reading the emergence of environmental neuroscience would be as a kind of outgrowth, or expansion, of the more established field of *social* neuroscience. If the latter rested on a claim that it was now possible to integrate biological and social levels of analysis in making sense of mental behavior (see e.g., Cacioppo et al., 2000), the promise of *environmental* neuroscience, in turn, is to transform this narrowly cultural definition of the "social" into, as Berman and his colleagues put it, the "sociophysical." Environmental neuroscience is in this sense, is straightforwardly "the scientific study of brain-mediated, bidirectional relationships between organisms and their *social* and *physical* environment (Berman et al., 2019a: 1041; my emphasis). However, in another paper that same year, Berman again, now with different co-authors, draws a sharper distinction, leaving the "social" or even the "sociophysical" some distance behind: although "social psychology and social neuroscience are interested in similar questions" to environmental neuroscience, these scholars write, what distinguishes the latter is attention to "aspects of the physical environment that include but are not limited to noise, greenspace, degree or intensity of urban development, crime, low-level perceptual features (for example, colors, contours, etc.), pollutants, lighting, temperature, etc." (Berman et al., 2019b: 414). On its current [at time of writing] homepage at the University of Chicago, Marc Berman's lab describes itself as having an interest in "how the physical environment affects the brain and behavior." From a more capacious and open account of different kinds of environment, here the role of the *physical* environment is exclusively established as the conceptual grounding of the field.

Indeed, in practice, the term is generally used even more narrowly than this. While, in principle, the physical environment in environmental neuroscience means any space a person might be in, whether a modern office, a busy city center street, or an open field, in experimental and theoretical work, "environment" tends to indicate something like what a person means when they refer to the "environmental movement" or an incoming "environmental catastrophe." In other words, the "environment" usually means the strictly physical qualities of the *natural* environment, or what Kühn and her colleagues (2023a, b) call "living environments"—whether these living environments are measured by the preponderance of vegetation and greenery, by physical distance from urban centers, or by the relative absence of humans. It should be noted that this focus on nature is in contrast to the much earlier physical-environment psychology of figures such as Edward Tolman or Kurt Lewin, for whom the spatial layout of interior rooms, the arrangement of corridors etc., were key variables of interest. Certainly, contemporary researchers are still interested in built interiors (I will say more on this below) though very often these too are analyzed in terms of the role of plants or natural materials within them (see e.g., Qin et al., 2014). In one important recent intervention by a range of leading authors, environmental neuroscience is posited not only as a science of nature, but even as a science for and on behalf of nature—a science that offers a route for neuroscientists to help "tackle the existential environmental threats Earth is currently facing" (Doell et al., 2023: 1288). Such is the strength and consistency of this bias, indeed, that one might even read "environmental neuroscience" as—in the words of Madan and his colleagues (2019)—simply "the cognitive neuroscience of nature."

From Nature to Architecture

We may live in a moment of widespread scientific and popular interest in how physical environments affect minds and brains, but it is not the first such moment. Environmental neuroscience, of course, has conceptual and methodological links to environmental *psychology,* which itself crystallized at the turn of the nineteenth and twentieth centuries. Enric Pol traces the origins of this field to the work of Willy Hellpach—once a student of Wundt at Leipzig—whose *Geopsyche* (1911) was concerned with, inter alia, "the effect of the sun and the moon on people's activities; the effect of colour and form; the effect of extreme environments – such as the Tropics or the Artic; the effect of urban microclimates" (Pol, 2006: 96). As Pol points out, Hellpach was himself the product of a fruitful local ecology among a particular generation of German-language psychological researchers: interest in questions of *lebensraum* (first coined by the geographer, Friedrich Ratzel (1901), to describe how society is shaped by its physical parameters) and *umwelt* (a term invented by the aristocratic biologist, Jacob von Uexküll (1934), for whom it describes the relationship between an organism and the perceptual qualities of its physical environment) were in the air. It is not incidental to this story of how human psychology came to be seen as dependent on a particular kind of physical and social territory that both concepts would go on to be associated with National Socialism. Uexküll particularly, in the early 1930s, was an enthusiastic Nazi, and a signatory to the infamous professorial "Vow of Allegiance" to Hitler (see Schnödl & Sprenger, 2022).

However, another strand of this history—one with a very different political inflection—passes through the early behavioral work of Edward C. Tolman (1948), whose pioneering experimental experiments on spatial navigation (Tolman was initially interested in how rats would learn to complete a maze to find food) showed that rats were not simply following stimuli, but were constructing "cognitive maps" to make sense of their spatial environment. For Tolman, writing not long after the defeat of National Socialism, this capacity for learning, and for developing a contextually sensitive perception of space, was a model for cold-war optimism about the human psychological capacity for social understanding. Humans, he wrote, much like the happiest and best treated of his own rats, could be taught to develop rich, nuanced accounts of the world around them, thus abandoning the narrow, stimulus-driven simplicities of previous eras. Just as for Hellpach, Tolman was working in an environment with a range of conceptual affordances. In the contemporaneous work of figures such as John B. Calhoun, Stanley Milgram, and especially Kurt Lewin, the capacity of *space*, and of spatial complexity, to dramatically effect an organism's behaviour, became a major site of interest, culminating in O'Keefe and Dostrovsky's famous recording of "place cells" in the brains of rats in the 1970s (for a fuller account of these histories, see Rose & Fitzgerald, 2022; Fitzgerald, 2024).

But a third historical strand—and the one I really wish to emphasize here—is the relationship of environmental neuroscience to the now largely forgotten field of architectural psychology in the 1960s and 1970s. As Emina Petrović and her colleagues (2015) point out, massive postwar urban re-development, and in particular the rapid development of new housing, often constructed along daringly modernist lines, produced a widespread concern with how the designers of such spaces may "lack of a deeper understanding of the culture and mental processing of the inhabitants" (Petrović et al., 2015: 482). In one early history of environmental psychology, Canter and Kenneth (1981) argued that postwar architecture was confronted with a whole set of new design problems, but also a whole new terrain of possible investigation in the modern "totally controlled" environment. Lighting, for example, now "had to be reappraised in terms of its capacity to meet human requirements for visual performance" (1981: 3). Moreover, the rapid spread of heating and air-conditioning technology meant that, suddenly, a merely "physiological model of human thermal comfort could no longer be considered satisfactory" (ibid). In other words, with the problems of the worker's *body* now apparently resolved (at least as far as the bodies of middle- and upper-class office workers were concerned) the problems of the *mind* now moved to center stage. In his landmark 1975 monograph, *Psychology for Architects*, David Canter, unquestionably the leading figure in the field in Britain in this period, contrasts an admiring quote from the modernist architects, the Smithsons, about a new Mies van der Rohe design, with a less admiring quote "attributed to" a psychiatrist: "half the mental hospitals would be empty and psychiatrists out of a job if every home had two bathrooms and soundproof bedrooms" (1975: 1). In other words, what mental health professionals knew about how spaces affected people's mental states, and the kinds of spaces leading architect thought looked good or worked well, were often in stark contrast to one another. Into this gap, a whole new field of empirical inquiry inserted itself (➔ chapters "Aesthetics of Architecture: Emerging Insights from the Brain" by Vartanian, "The

Built Environment and the Brain: Review of Emerging Methods to Investigate the Impact of Viewing Architectural Design" by Tawil and Kühn, "Applications of Biometrics in Architectural and Environmental Design" by Lavdas and Sussman, "The Potential of Biophilic Design and Nature to Improve Health, Creativity and Well-Being" by Sjövall and Spiers).

It is noteworthy that this concern with working spaces remains a core feature of environmental neuroscience and its linked design practices. In one recent study, Mostajeran and her colleagues showed that participants in a virtual office showed improved performance on memory and creativity tasks when virtual plants were added to the workspace (Mostajeran et al., 2023). Similarly, in an experiment testing the effect of a short-term nature exposure on people's cognitive performance (participants were shown a video of an urban forest), Zhang and colleagues lamented that "limited understanding of natural environments' restorative effects from a neuro-cognitive perspective lames [sic.] the practitioners to build healthy living and working environments scientifically" (2023: 116830). In a recent special issue of *Architectural Design* on "neuroarchitecture, the architect, Ian Ritchie (2020), described his work on the Sainsbury Wellcome Centre at University College London—where John O'Keefe was the first director. This design, says Ritchie, is steeped in "the results of neuroscientific research to inform the spatial structure of the interior architecture: about how people interact with the space around them, the importance of adaptability, the nature of light and its wavelengths in the working environment, and how the spaces in which we live and work affect our mood and behavior. The nature and organization of the resulting interior spaces with long, short, and vertical views and varying volumes was described by one of the scientists as akin to the scale-free nature of the brain itself" (2020: 9). None of this is inherently objectionable. And yet as the historian of science Michelle Murphy (2006) reminds us, corporate managers have long been interested in how working environments can ensure the maximum efficiency and range of the human bodies operating within them. Murphy shows, for example, that a critical forerunner of women entering the white-collar workforce *en masse* in the mid twentieth century was the widespread use of air conditioning systems, which made it possible for middle-class women simply to *be* in an office environment, without the sociobiological disturbance of their visible sweat.

With such histories in mind, we might also see the work of environmental neuroscience in these kinds of spaces today as a way of using science to mediate the relationship between the needs of capital and the body of the worker. Indeed, bringing psychological, behavioral, and increasingly neurobiological insights to bear on the design and functioning of some large-scale enterprise, as various consulting firms now offer to do, is not a politically neutral activity; nor, indeed, is it necessarily benign from the perspective of an employee or resident within such an enterprise. After all, making working environments that are healthier, more restful, calmer, also means making spaces that are more economically efficient, that extract value from the employee with less visible friction. The employees' improved mental state is leveraged and indeed monetized by the organization for whom that same working environment is, often, little more than a tool for the generation of capital. Perhaps this goes too far. But the question of the relationship between design and intention is quite central for the field, though it is little discussed. A good deal of the energy within environmental neuroscience derives from the fact that the external environment is not, as it is for most nonhuman animals, a source of stimulus for the passive organism. Rather the human environment is something we make and create with intention. Scholars (e.g., Berman et al., 2019a: 1041) are fond of a quote from Winston Churchill: "we shape our buildings and afterwards our buildings shape us." And yet, the fact that the same Winston Churchill, in 1920, had the physical environment in which I am currently writing this paper (i.e. the city of Cork, Ireland) literally burnt down in an act of colonial reprisal, should remind us that architectural shaping is not an obviously affirmative activity. We need to think much more seriously about what, precisely, is being shaped by our science, and what economic and political purposes lie behind that shaping.

Urbanization and Climate Change

In addition to these intellectual inheritances, I want to argue that the emergence of environmental neuroscience should also be understood in terms of a distinctive macro-social context, and via a growing global attention to two phenomena in particular: (1) urbanization and (2) climate breakdown. The much-cited United Nations statistic is that the earth has recently (in fact, sometime around 2007) passed the milestone of being more than 50% urban. In other words, today more than half of the population of the planet lives in an urban area, and this proportion is expected to increase to about 70% by 2050. This constitutes, according to UN Habitat (the UN programme focused on sustainable urban development) a "global megatrend" — one that "is intertwined with the existential challenges that the world has faced in the last 50 years, including climate change, rising inequality and the rise in zoonotic viruses" (UN Habitat, 2022: xvi). Though urbanization is often thought to have positive potential for sustainable development (through, for example, the concentration of population and more efficient use of resources) the growth of cities is nonetheless usually understood, politically as well as scientifically, as a problem in need of a solution: "We have now entered the Century of the City," point out the geographer Karen C. Seto and her colleagues in a recent *Annual Review.* In this era, even if the city has positive potentials and affordances, urbanization nonetheless "creates the most human dominated landscapes and drives local and regional environmental changes by transforming land cover, hydrological systems, and biogeochemistry. Worldwide, urban expansion is one of the primary drivers of habitat loss and alteration and of plant and animal species extinction" (2010: 168). In the public health literature, similarly, scholars have moved from the analysis of urbanization as an historical, sociological, or ecological phenomenon, to the idea of "urbanicity" as a kind of atmospheric exposure. Urbanicity, say Vlahov and Galea, means "the impact of living in urban areas at a given point in time… [it] refers to the presence of conditions that are particular to urban areas or present to a much greater extent than in nonurban areas" (2006: S5).

This idea of the contemporary city as a negative psychological exposure has been critical to the development of environmental neuroscience, and it remains a central experimental topic. In a 2020 paper, for example, Mazumder and his colleagues placed a group of participants outside two kinds of buildings in central London—one a medieval church, and the other a glass and steel skyscraper—while taking a range of physiological and cognitive measures. In a now famous paper from 2011, Lederbogen and his colleagues showed not only that an urban population processes stress differently from a more rural population, but these researchers were even able to isolate the neural mechanisms implicated in this processing. What is at stake here—and this I argue is a central philosophical commitment of the field—is a hard, diagnostic line between urban and rural life, or between natural and human-made materials. More specifically, there is a growing consensus, within environmental neuroscience, that there is something about the physical parameters of *nature*, some perhaps still not fully understood visual, olfactory, or atmospheric quality of natural materials or scenes, that soothes and calms the mind—a quality that protects the modern person, and in particular, the modern *urban* person, from the stresses and strains of life in the city (see Rose & Fitzgerald, 2022, for a history). It is worth pointing out that this largely implicit philosophical distinction—between nature and non-nature— is fraught with difficulty, to say the least, and those difficulties remain a large lacuna for the field (see e.g., Vogel (2015) on weak ontological justifications for distinguishing the natural from the unnatural, or Levinovitz (2020) for a more normative critique of the inherent goodness of nature).

But it is urbanization that remains the central focus. As Kühn and colleagues again put it: urbanization is "one of the most important global health issues of the twenty-first century, in particular since mental health problems seem to be more frequent in cities… to safeguard mental and physical well-being in cities, research is needed that is targeted at identifying salutogenic factors of *living* environments" (2023a, b: 312; my emphasis). Or as Bratman and his colleagues put it in one of the field's most well-known empirical papers, "Never before has such a large percentage of humanity been so far removed from nature …more than 50% of people now live in urban areas, and by 2050, this proportion will be 70%…What are the potential mental health implications of this demographic shift?" (2015: 8567). Though they note that the picture is complex and far from unidirectional, Kühn and Gallinat (2024) similarly argue that "mental health problems typically accumulate in urban areas. This seems to be especially true for schizophrenia, affective and anxiety disorders and autism." Thus "environmental Neuroscience may inform landscape and urban planning, as well as architecture, to make living environments more salutogenic." This desire to produce the salutogenic city is, I suggest, primary to environmental neuroscience, indeed to the extent that it is difficult to see the field existing at all without a widespread cultural and political concern with the growing city and its psychological effects. This concern is hardly new. Frederick Law Olmsted, the great designer of urban parks, founded his career on a similar wave of concern in the mid nineteenth century, and indeed with much the same solution as that usually proposed today (see again Fitzgerald, 2024 for a full account), Still it remains critical to stress that environmental neuroscience, today, exists as a field whose self-given goal is in large part to heal the sickly city—to help cure the planet of at least the *psychological* consequences of its frightening march towards more and more urbanization.

The second aspect of the macro-social context that has produced the new field, environmental neuroscience' is climate breakdown, and in particular a widespread sense of climate anxiety. As Susan Clayton has pointed out, climate breakdown has quite direct psychological effects: heat increases the likelihood of anger and aggression; being caught in a major environmental disaster is likely to produce PTSD and depression; people displaced by climate events will experience stress and burnout (Clayton, 2020: 102262). But also, Clayton suggests, scientists are increasingly aware of another, less direct, and more nebulous phenomena, called climate anxiety, climate grief, ecological trauma, or sometimes, using Glenn Albrecht's and his collaborators' terms, "solastalgia" and "psychoterratic illnesses" (e.g., Albrecht et al., 2007; Clayton, 2020). The last of these describes an "earth-related mental illness where people's mental wellbeing (psyche) is threatened by the severing of healthy links between themselves and their home/territory." These kinds of terms are of course quite new, generally describe sub-clinical phenomena, and have perhaps found larger audiences in popular than in scholarly literatures. But the wider relation is nonetheless finding its way into scientific literature. Here I am on rather speculative ground, but I want to suggest, that the generalized projection of an anxious or hopeless relationship to the natural environment cannot be easily disentangled from a science that claims to now be able to fully understand—and perhaps also recuperate—humans' psychic relationship to nature. Indeed, though experimental work in environmental neuroscience rarely operationalizes climate change or climate anxiety as a variable, the phenomenon is nonetheless often in the background as the field accounts for itself. As Berman and colleagues argue, "Humans and other species are in trouble… if we do not make attempts to protect and conserve nature that currently exists. Understanding how and when humans act in more environmentally friendly ways is an important area of research for an environmental neuroscience approach" (Berman et al., 2021: 80) (➔ chapters "Leveraging the Social Neuroscience of Prosocial Behavior to Advance Our Understanding of Pro-environmental Behavior" by Lamm et al. and "Neuroscientific Approaches to Understand Barriers and Promotors of Pro-environmental Behavior" by Krebs and Sawe). For Leeuwis et al. (2022), similarly, environmental neuroscience can help to understand the neural basis of "pro-environmental behavior" in consumers. While in a review paper, Susie Wang and Barry van der Berg (2021) propose using EEG measures (➔ chapter "EEG for the Study of Environmental Neuroscience" by Grassini) to answers questions such as whether preparatory attention is linked to environmental motivation. My point here is not that the field (falsely) promotes itself by promising to resolve climate change; rather I suggest that the presence of climate change as a psychological problem gives a neuroscience of the environment critical purchase in both institutional and public contexts, and forms a major part of its wider appeal.

The Politics of the Environment

In the mid-2000s, when social scientists and philosophers began to pay sustained attention to neuroscience's expansive epistemic vision following the "Decade of the Brain," and especially as they attended to the claims made on behalf of the emerging field of social neuroscience, three consistent arguments took hold: first, that fields like social neuroscience, affective neuroscience, and their cognates, represented an unwarranted "biologisation" of phenomena that scholars elsewhere took to be determined by historical and cultural forces (for example, a person's likelihood of committing a crime; an individual's voting preferences; the emotions we feel and express about a romantic partner); second, partly but not entirely with regard to this biologisation, that the expansive neuro-disciplines were attached to a new philosophy of personhood, in which a person's desires and actions were to be represented by their brain only; and third, largely but not wholly due to this notion of personhood, scholars that neuroscience was attached to an implicit political ideology, wherein political interventions that would previously have focused on social or structural solutions now took the neurobiology of the individual person as a privileged site of political attention and intervention. Perhaps most infamous, in this last regard, was work promoted by the one-time British Conservative Party leader, Iain Duncan Smith, on the neurobiological effects of poverty, which seemed to proceed as if being poor was an unfortunate brain disease, and not the direct effect of Duncan Smith's own party's welfare policies (see Woolf, 2010; on the wider critiques, see especially Choudhury et al., 2009; Vidal & Ortega, 2017; Rose & Abi-Rached, 2013). As the historian Joelle Abi-Rached put it at the very outset of these discussions in the early 2000s, "we can already see the rise of new markets and economies based on the new findings and technologies coming from brain research. We can foresee applications in politics and in the courtroom; we have seen the continuous expansion of psychiatric disorders and diseases, and we are witnessing an intensifying debate about the ethical, moral and philosophical aspects of neuroscience and its applications. What is lacking, however, is a broad and inclusive debate about how society values and should deal with the knowledge and technology and their applications" (Abi-Rached, 2008: 1162).

These debates largely fizzled out in subsequent years, partly because the irresistible force of the "neuro age" eventually ran into the immovable object of biological complexity; and partly because philosophical and sociological energy moved away from these sometimes entrenched disciplinary disagreements, and toward attempts to find a conceptual and methodological third space (where, for example, innovative neurobiological experiments might not have to be at odds with decades of patient, qualitative work in social and political sciences). I do not wish to revive those rather stale debates here, but it is nonetheless striking how willing scholars within environmental neuroscience are to talk about policy, but without any serious reflection on how policy recommendations may be guided by their own political and ideological priors, and still further without any demonstrated awareness of the complexity and mess of policy implementation. Berman and his colleagues, for example, argue that that it should be an explicit goal of environmental neuroscience "to take a generative theoretical approach using results from (epi) genetics and neuroscience to design the physical environment to improve human psychological functioning" (2019b: 415). But once we start talking about the design of the physical environment, we are instantly in the realm of local politics, of municipal democracy, of competing interests, of ideology, and so on. It is not at all clear how well equipped the environmental neuroscience community is to intervene with skill or care in such spaces, or indeed whether skill and care are even seen as important qualities within these proposed interventions. There is sometimes a sense in the concluding sections of environmental neuroscience papers that biological data will simply overcome political interest—that if the right people simply pay attention to the science, then the right course of action will follow, and competing interests will fall away. As Kühn and her colleague put it at the end of a paper showing positive associations between the extent of tree cover around one's home, volume of gray matter in the right orbitofrontal cortex, and improved cognitive performance in verbal and spatial domains: "we anticipate that the research field of environmental neuroscience may in the long run provide a knowledge base for evidence-based urban landscape planning" (2023a, b: 6–7). This is of course laudable, but it represents, too, a perhaps unfounded optimism about the capacity of biological data to over-ride, or even to weigh significantly in, political decision making, as well as an implicit normative commitment to the idea that biological data *should* have political priority (that, for example, measures of cognitive performance or volume of gray matter should over-ride what car-drivers say about the journey times they find tolerable, or what we know about homeowners' desire to pave their front gardens, and so on).

In the conclusion of their paper cited above, which contrasted a tall modern building with a much smaller and more traditional one, Mazumder and his colleagues directly considered the position of Mayor of London, Sadiq Khan. Khan, they said, "has directed individual boroughs in London, which is seeing an upswell of high-rise buildings across the city, to decide whether how tall buildings should be [sic.] and to consider if the spatial context and character is suitable to them." They quote approvingly from London planning documents that impose visual and other neighborhood-relational constraints on the building of high rises, and they note that "it is encouraging to see this focus on the atmospheric stressors that can be caused by exposure to high-rise buildings" (2022: 1090). They thus conclude their paper with the claim that, "as cities continue to densify, it is important that a psychological lens is applied to the process of urbanization" (ibid, 1091). In another paper, Berman and his colleagues draw on the work of the idiosyncratic architect, Christopher Alexander (once described by the *New York Times* as a "fierce anti-modernist"). They contrast an image of a very modern-looking kitchen with a picture of a dining room in something like a medieval brick refectory: "the red-outlined [modern] indoor environment," they argue, "has less scaling and contrast than the blue-outlined indoor environment… these elements may affect social interactions [and] brain network behavior" (2019b: 415). They conclude that "in the future, dynamically altering the environment will also require that environmental neuroscientists work with architects, builders and planners to determine what designs are feasible" (ibid.: 416). In their work using eye-tracking methods, similarly, Justin Hollander, Ann Sussman, and their colleagues draw a sharp contrast between neighborhoods that do and do not favor "Traditional Neighborhood Design" (TND)—in other words, street design that features more old-fashioned qualities like "narrow streets, complex facades, and bilateral symmetry" (2020: 485). Non-TND, by contrast, tends to feature more modern-looking square or rectangular buildings, often with blank facades, and little detailing. Hollander and colleagues find much greater eye fixation in street-scene images that feature higher levels of TND—"suggesting features," they argue, "that draw the eye and may help people feel more connected in their built environments" (ibid.: 502).

This consistent focus on the old or the traditional as less psychologically taxing is worth paying attention to. While it is uncontroversial to note that urban nature alleviates the mental stress associated with city exposure, increasingly scholars in environmental neuroscience make a related but different claim, which is that the features associated with urban nature may also be associated with more *traditional* building styles in cities, and/or that specifically *modern* architecture and planning worsen the cognitive load of urban space. Thus, scholars in this field often find themselves either explicitly or implicitly arguing for a kind of traditionalism in urban design. There are two things to stress here. First, that this is an unavoidably political position; and, second, that a preference for traditional design is an ideological position often associated with very reactionary political actors. The conservative philosopher, Roger Scruton, for example, long railed against modern buildings in post-war Britian, which he took as a sign of both aesthetic and moral decay. The related idea that modernist or tall buildings were inherently associated with crime or social breakdown—as in the notion of "defensible space"—was part of the intellectual armature that allowed conservative governments throughout Europe to disinvest from large scale public housing, which had long been associated with often daring modern design (see Scruton, 1974; Coleman, 1985). More recently, figures like Donald Trump, drawing on the broadly fascist aesthetic and social "movement" sometimes called RETVRN, have used modern architecture as an important symbolic front for making their wider far-right political commitments clear (indeed, Trump signed an executive order before leaving office, mandating that all federal buildings must follow traditional or classical styles). Throughout Europe and North America, *The Art Newspaper* points out, "reactionary ideas hostile to the cosmopolitan, to Modernism, to modernity itself, are in the ascendant" and thus "traditional architecture has become a vehicle of choice for the Right and Far Right" (Bevan, 2022).

None of this, of course, is to argue that the scientists cited above are pursuing a conservative political agenda. But we might nonetheless reasonably ask scholars in this space to reflect upon, and indeed take responsibility for, the fact that their science has the capacity to provide encouragement and support to a deeply reactionary urban politics—indeed that it may be used to biologize or naturalize a conservative, even fascistic agenda for the city and its buildings. This politics is of course not at all concerned with mental health, or with well-being, but rather with pushing back on a whole range socio-legal rights won in the second half of the twentieth century, restoring the physical and symbolic order of the world before the welfare state, with the high rise housing block perhaps its most potent visual symbol.

Conclusion: Environments in the Making

I have argued in this short contribution that environmental neuroscience is an emergent field, and that making sense of that emergence means taking account of this social, historical, and philosophical contexts in which it is taking shape. One part of this context, I argued, is a very particular concern with *nature* as both a psychological and moral force; I suggested that this concern, in turn, should be understood in terms of a wider cultural anxiety about climate breakdown and what urban sociologists call "planetary urbanization" (Brenner & Schmid, 2015). This is how, ultimately, we should make cultural and social sense of the coming-into-prominence of a "cognitive neuroscience of nature," or the transformation of the psychology of architecture into the psychology of nature. I also suggested this at least partly accounts for the field's guiding and central problematic, which is the question of urban mental health—i.e., the question of how modern spaces and habitats, often bereft of natural materials and vistas, can negatively affect our mental states. I went on to say that there is a coincidence, here, between spaces that are thought to be *psychologically* unhealthy by researchers in environmental neuroscience, and spaces that are thought to be *morally* unhealthy by a whole range of political reactionaries. This is not to say the former bear responsibility for the latter or that they have anything to do with them. But it *is* to suggest that scholars may need to think through this problem as, at the very least, a significant ethical conundrum for the field.

The philosopher and anthropologist, Bruno Latour (1987), once argued that science always looks in two directions at the same time: in one direction, there is the world of established scientific facts, undisputed truths, agreed-upon methods and so on; in the other direction is what Latour called "science in the making," i.e., that much messier and uncertain, perhaps earlier process, in which things are not yet fully settled, in which what will become the "established facts" of the field are, still, yet to be determined. Environmental neuroscience, I suggest, is very much "in the making" in this sense, which is not at all a negative description but is rather an invitation to think reflexively, and indeed politically, about which facts are currently being made and which are not, and what researchers' responsibilities are to those facts either way. In a world that really *is* rapidly urbanizing, in which nature reality *is* under terrible threat, in which people really *do* lament the loss of ecological contact, environmental neuroscience clearly has a major role to play in making sense of the coming decades. Precisely what the role will consist of, however, is not yet clear..

References

Abi-Rached, J. M. (2008). The implications of the new brain sciences. *EMBO Reports, 9*(12), 1158–1162.

Albrecht, G., et al. (2007). Solastalgia: The distress caused by environmental change. *Australasian Psychiatry, 15*(1_suppl), S95–S98.

Berg, A. (2015). "To Conquer Myself": The new Strenuosity and the emergence of "Thru-hiking" on the Appalachian Trail in the 1970s. *Journal of Sport History, 42*(1), 1–19.

Berman, M. G., Stier, A. J., & Akcelik, G. N. (2019a). Environmental neuroscience. *The American Psychologist, 74*(9), 1039–1052.

Berman, M. G., Kardan, O., Kotabe, H. P., et al. (2019b). The promise of environmental neuroscience. *Nature Human Behaviour, 3*, 414–417.

Berman, M. G., Cardenas-Iniguez, C., & Meidenbauer, K. L. (2021). An environmental neuroscience perspective on the benefits of nature. In A. R. Schutte, J. C. Torquati, & J. R. Stevens (Eds.), *Nature and psychology. Nebraska symposium on motivation* (Vol. 67).

Bevan, R. (2022). The ugly pursuit of beauty: How traditional architecture has become a battleground for right-wing politicians. *The Art Newspaper* 07 January.

Bratman, G. N., et al. (2015). Nature experience reduces rumination and subgenual prefrontal cortex activation. *Proceedings of the National Academy of Sciences of the United States of America, 112*(28), 8567–8572.

Brenner, N., & Schmid, C. (2015). Towards a new epistemology of the urban? *City, 19*(2–3), 151–182.

Cacioppo, J. T., Berntson, G. G., Sheridan, J. F., & McClintock, M. K. (2000). Multilevel integrative analyses of human behavior: Social neuroscience and the complementing nature of social and biological approaches. *Psychological Bulletin, 126*(6), 829–843.

Canter, D. (1975). *Psychology for architects*. Applied Science.

Canter, D. V., & Kenneth, H. C. (1981). Environmental psychology. *Journal of Environmental Psychology, 1*(1), 1–11.

Choudhury, S., Nagel, S., & Slaby, J. (2009). Critical neuroscience: Linking neuroscience and society through critical practice. *BioSocieties, 4*, 61–77.

Clayton, S. (2020). Climate anxiety: Psychological responses to climate change. *Journal of Anxiety Disorders, 74*, 102263.

Coleman, A. (1985). *Utopia on trial: Vision and reality in planned housing*. Hillary Shipman.

Doell, K. C., Berman, M. G., Bratman, G. N., et al. (2023). Leveraging neuroscience for climate change research. *Nature Climate Change, 13*, 1288–1297.

Fitzgerald, D. (2024). *The city of today is a dying thing: In search of the cities of tomorrow*. Faber & Faber.

Hellpach, W. (1950 [1911]). *Geopsyche: Die Menschenseele unterm Einfluß von Wetter und Klima, Boden und Landschaft.* Enke Verlag.

Hollander, J. B., et al. (2020). Using eye-tracking to understand human responses to traditional neighborhood designs. *Planning Practice & Research, 35*(5), 485–509.

Krugliak, A., & Clarke, A. (2022). Towards real-world neuroscience using mobile EEG and augmented reality. *Scientific Reports, 12*, 2291.

Kühn, S., & Gallinat, J. (2024). Environmental neuroscience: Unraveling the pathway from the physical environment to mental health. *Nature Mental Health*. https://doi.org/10.1038/s44220-023-00137-6

Kühn, S., Banaschewski, T., Bokde, A. L. W., et al. (2023a). Trees for brains: Current residential tree cover density and its association with brain structure in young adults. *Journal of Environmental Psychology, 89*, 102047.

Kühn, S., et al. (2023b). Green is not the same as green: Differentiating between the association of trees and open green spaces with children's brain structure in The Netherlands. *Environment and Behaviour, 55*(5), 311–334.

Kwak-Hefferan, E. (2012). Hiking makes you smarter. *Backpacker* 21 May.

Latour, B. (1987). *Science in action: How to follow scientists and engineers through society*. Harvard University Press.

Lederbogen, F., Kirsch, P., Haddad, L., et al. (2011). City living and urban upbringing affect neural social stress processing in humans. *Nature, 474*, 498–501.

Leeuwis, N., et al. (2022). A framework for application of consumer neuroscience in pro-environmental behavior change interventions. *Frontiers in Human Neuroscience, 16*, 886600.

Levinovitz, A. (2020). *Natural: The seductive myth of nature's goodness*. Profile.
Liu, F., Xu, J., Guo, L., et al. (2023). Environmental neuroscience linking exposome to brain structure and function underlying cognition and behavior. *Molecular Psychiatry, 28L*, 17–27.
Madan, C. R., et al. (2019). The cognitive neuroscience of nature: From motor cognition to grounded cognition. In A. A. Donnelly & T. E. M. Intyre (Eds.), *Physical activity in natural settings: Green and blue exercise*. Routledge.
Mazumder, R., et al. (2022). Exposure to high-rise buildings negatively influences affect: Evidence from real world and 360-degree video. *Cities & Health, 6*(6), 1081–1093.
McDonnell, A. S., & Strayer, D. L. (2024). Immersion in nature enhances neural indices of executive attention. *Scientific Reports, 14*(1), 1845.
Mostajeran, F., Steinicke, F., Reinhart, S., et al. (2023). Adding virtual plants leads to higher cognitive performance and psychological well-being in virtual reality. *Scientific Reports, 13*, 8053.
Murphy, M. (2006). *Sick building syndrome and the problem of uncertainty: Environmental politics, Technoscience, and women workers*. Duke University Press.
National Institutes of Health (NIH). (2023). *BRAIN-funded researchers capture first-ever mobile intracranial EEG recordings during real-world human movements*. https://braininitiative.nih.gov/news-events/blog/brain-funded-researchers-capture-first-ever-mobile-intracranial-eeg-recordings
Petrovic, E., et al. (2015). On the rise and apparent fall of architectural psychology in the 1960s, 1970s and early 1980s. In P. Hogben & J. O'Callaghan (Eds.), *Proceedings of the society of architectural historians Australia and New Zealand* (Vol. 32).
Pol, E. (2006). Blueprints for a history of environmental psychology (I): From first birth to American transition. *Medio Ambiente y Comportamiento Humano, 7*(2), 95–113.
Qin, J., Sun, C., Zhou, X., Leng, H., & Lian, Z. (2014). The effect of indoor plants on human comfort. *Indoor and Built Environment., 23*(5), 709–723.
Ratzel, F. (1901). Der Lebensraum. Eine biogeographische Studie. In K. Bicher et al. (Eds.), *Festgaben fur Albert Schiffle zur siebensigen*. Tubingen.
Rearden, J. (2017). *The postgenomic condition: Ethics, justice, and knowledge after the genome*. University of Chicago Press.
Ritchie, I. (2020). Why do people feel more comfortable in one space than another? *Architectural Design, 90*, 6–13.
Rose, N., & Abi-Rached, J. (2013). *Neuro: The new brain scienueces and the management of the mind*. Princeton University Press.
Rose, N., & Fitzgerald, D. (2022). *The urban brain: Mental health in the vital city*. Princeton University Press.
Schnödl, G., & Sprenger, F. (2022). *Uexküll's surroundings: Umwelt theory and right-wing thought*. Meson Press.
Scruton, R. (2013 [1974]). *The aesthetics of architecture*. Princeton University Press.
Seto, K. C., et al. (2010). The new geography of contemporary urbanization and the environment. *Annual Review of Environment and Resources, 35*(1), 167–194.
Tolman, E. C. (1948). Cognitive maps in rats and men. *Psychological Bulletin, 55*(4), 189–208.
UN Habitat. (2022). *Envisaging the future of cities: World cities report 2022*. UN Habitat.
Vidal, F., & Ortega, F. (2017). *Being brains: Making the cerebral subject*. Fordham University Press.
Vineis, P., & Barouki, R. (2022). The exposome as the science of social-to-biological transitions. *Environment International, 165*, 107312.
Vlahov, D., & Galea, S. (2006). Urbanization, urbanicity, and health. *Journal of Urban Health, 79*(Suppl 1), S1–S12.
Vogel, S. (2015). *Thinking like a mall: Environmental philosophy after the end of nature*. MIT Press.
von Uexküll, J. (2010 [1934]). *A foray into the worlds of animals and humans: With a theory of meaning*. University of Minnesota Press.
Wang, S., & van den Berg, B. (2021). Neuroscience and climate change: How brain recordings can help us understand human responses to climate change. *Current Opinion in Psychology, 42*, 126–132.

Warde, P., et al. (2015). *The environment: A history of the idea*. Johns Hopkins Press.
Woolf, M. (2010). Broken homes 'damage infant brains', says Tory. *The Times* 05 March.
Zhang, G., et al. (2023). The restorative effects of short-term exposure to nature in immersive virtual environments (IVEs) as evidenced by participants' brain activities. *Journal of Environmental Management, 326*(Part B), 116830.

Correction to: The Rise and Future of Environmental Neuroscience in Environmental Psychology

Lindsay J. McCunn

Correction to: Chapter 2 in: S. Kühn (ed.), *Environmental Neuroscience*, https://doi.org/10.1007/978-3-031-64699-7_2

In Chapter 2, below correction was missed out, which has been corrected now:

- The last name should be "Berman" not "Bernman" for the 2019 reference. The correct reference is as follows: Berman, M., Steir, A. J., & Akcelik, G. N. (2019). Environmental Neuroscience. *American Psychologist, 74*, 1039–1052.

The updated version of this chapter can be found at
https://doi.org/10.1007/978-3-031-64699-7_2

S. Kühn (ed.), *Environmental Neuroscience*,
https://doi.org/10.1007/978-3-031-64699-7_23

Correction to: The Potential of Biophilic Design and Nature to Improve Health, Creativity and Well-Being

Isabelle A. K. Sjövall and Hugo J. Spiers

Correction to:
Chapter 11 in: S. Kühn (ed.), *Environmental Neuroscience*, https://doi.org/10.1007/978-3-031-64699-7_11

In Chapter 11, Figure 1 was incorrectly published; the correct figure is as follows:

The updated version of this chapter can be found at
https://doi.org/10.1007/978-3-031-64699-7_11

S. Kühn (ed.), *Environmental Neuroscience*,
https://doi.org/10.1007/978-3-031-64699-7_24

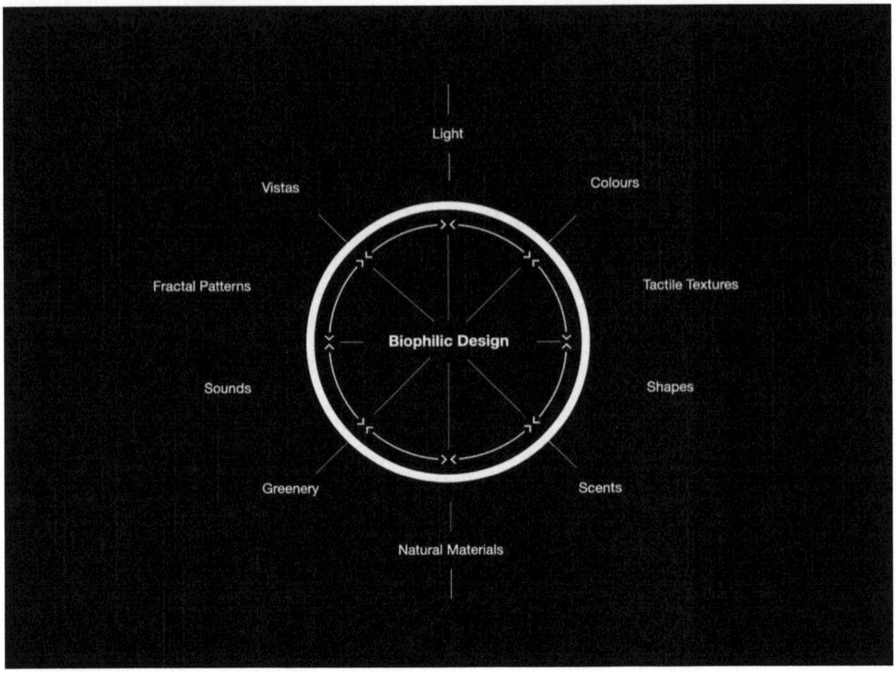

Fig. 1 Visualisation of biophilic design elements. From "The Krebs Cycle of Creativity" by N. Oxman (2016), adapted by Sjövall & Spiers, 2024

Index

S. Kühn (ed.), *Environmental Neuroscience*,
https://doi.org/10.1007/978-3-031-64699-7

W

MIX
Papier aus verantwortungsvollen Quellen
Paper from responsible sources
FSC® C105338

If you have any concerns about our products,
you can contact us on
ProductSafety@springernature.com

In case Publisher is established outside the EU,
the EU authorized representative is:
Springer Nature Customer Service Center GmbH
Europaplatz 3, 69115 Heidelberg, Germany

Printed by Libri Plureos GmbH
in Hamburg, Germany